BASIC GRAPHS

Power Functions

$f(x) = x^2$

$f(x) = x^3$

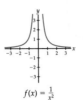

$f(x) = \frac{1}{x}$

$f(x) = \frac{1}{x^2}$

$f(x) = \sqrt{x}$

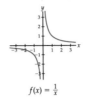

$f(x) = \left| x^{\frac{2}{3}} \right|$

Exponential and Logarithmic Functions

$f(x) = e^x$

$f(x) = e^{-x}$

$f(x) = \ln x$

Trigonometric Functions

$f(x) = \sin x$

$f(x) = \cos x$

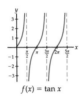

$f(x) = \tan x$

$f(x) = \sec x$

$f(x) = \csc x$

$f(x) = \cot x$

Inverse Trigonometric Functions

$f(x) = \sin^{-1} x$

$f(x) = \tan^{-1} x$

$f(x) = \sec^{-1} x$

Hyperbolic Functions

$f(x) = \sinh x$

$f(x) = \cosh x$

$f(x) = \tanh x$

GEOMETRY

Pythagorean Theorem

$a^2 + b^2 = c^2$

Law of Similar Triangles

$\dfrac{h}{b} = \dfrac{H}{B}$

$\dfrac{d}{b} = \dfrac{D}{B}$

$\dfrac{d}{h} = \dfrac{D}{H}$

Rectangular Solid

$V = xyz$

$S = 2xy + 2yz + 2xz$

Sphere

$V = \dfrac{4}{3}\pi r^3$

$S = 4\pi r^2$

Cylinder

$V = \pi r^2 h$

$S = 2\pi rh + 2\pi r^2$

Cone

$V = \dfrac{1}{3}\pi r^2 h$

$S = \pi r\sqrt{r^2 + h^2} + \pi r^2$

To test your understanding of these and other prerequisites to calculus, work through the Chapter 0 Review Exercises, in particular the Notation and Skill Certification sections. Those review exercises can serve as a Diagnostic Test for precalculus and algebra comprehension.

CALCULUS

MULTIVARIABLE

LAURA TAALMAN
James Madison University

PETER KOHN
James Madison University

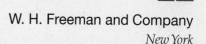

W. H. Freeman and Company
New York

Senior Publisher: Ruth Baruth
Executive Editor: Terri Ward
Marketing Manager: Steve Thomas
Market Development Manager: Steven Rigolosi
Developmental Editors: Leslie Lahr, Katrina Wilhelm
Senior Media Editor: Laura Judge
Associate Editor: Jorge Amaral
Editorial Assistant: Liam Ferguson
Photo Editor: Ted Szczepanski
Cover Photo Researcher: Elyse Rieder
Cover Designer: Vicki Tomaselli
Text Designer: Marsha Cohen
Illustrations: Network Graphics
Illustration Coordinator: Bill Page
Production Coordinator: Susan Wein
Project Management and Composition: Aptara
Printing and Binding: RR Donnelley

Library of Congress Control Number: 2012948659

Multivariable:
ISBN-13: 978-1-4641-2551-5
ISBN-10: 1-4641-2551-1

Printed in the United States of America

First printing

W. H. Freeman and Company
41 Madison Avenue
New York, NY 10010
Houndmills, Basingstoke RG21 6XS, England
www.whfreeman.com

To Leibniz and Newton

—Laura Taalman

To Newton and Leibniz

—Peter Kohn

IV VECTOR CALCULUS

V MULTIVARIABLE CALCULUS

Laura Taalman and Peter Kohn are professors of mathematics at James Madison University, where they have taught calculus for a combined total of over 30 years.

Laura Taalman received her undergraduate degree from the University of Chicago and master's and Ph.D. degrees in mathematics from Duke University. Her research includes singular algebraic geometry, knot theory, and the mathematics of games and puzzles. She is a recipient of both the Alder Award and the Trevor Evans award from the Mathematical Association of America, and the author of five books on Sudoku and the mathematics of Sudoku. In her spare time, she enjoys being a geek.

Peter Kohn received his undergraduate degree from Antioch College, a master's degree from San Francisco State University, and a Ph.D. in mathematics from the University of Texas at Austin. His main areas of research are low-dimensional topology and knot theory. He has been a national judge for MathCounts since 2001. In his spare time, he enjoys hiking and riding his bicycle in the beautiful Shenandoah Valley.

Calculus books have become full of clutter, distracting margin notes, and unneeded features. This calculus book clears out that clutter so that students can focus on the important ideas of calculus. Our goal was to create a clean, streamlined calculus book that is accessible and readable for students while still upholding the standards required in science, mathematics, and engineering programs, and that is flexible enough to accommodate different teaching and learning styles.

Linear Flow with Clean Margins

One thing that is distinctive about this calculus book is that it follows a linear writing style. Figures and equations flow with the text as part of a clear, structured exposition instead of being scattered about in the margins. We feel that this approach greatly increases the clarity of the book and encourages focused reading.

Exposition Before Calculation

Another distinctive feature of this book is that in each section we have separated the exposition and illustrative examples from the longer, more complicated calculational examples. Including these longer examples separately from the exposition increases flexibility: Students who want to read and understand the development of the material can do so without being bogged down or distracted by large examples, while students who want to use the book as a reference for looking up examples that are similar to homework problems can also do that.

Examples to Learn From

Within the exposition of each section are short examples that quickly illustrate the concepts being developed. Following the exposition is a set of detailed, in-depth examples that explore both calculations and concepts. We took great pains to provide many steps and illustrations in each example in order to aid the student, including details about how to get started on a problem and choose an appropriate solution method. One of the elements of the book that we are most proud of is the "Checking the Answer" feature, which we have included after selected examples to encourage students to learn how to check their own answers.

Building Mathematics

We were very careful in this book to approach mathematics as a discipline that is developed logically, theorem by theorem. Whenever possible, theorems are followed by proofs that are written to be understood by students. We have included these proofs because they are part of the logical development of the material, but we have clearly labeled and indented each proof to indicate that it can be covered or skipped, according to instructor preference. Each exercise set contains an optional subsection of proofs, many of which are accessible even to beginning students. In addition, we have emphasized the interconnections among topics by providing "Thinking Back" and "Thinking Forward" exercises in each section and "Capstone" problems at the end of each chapter.

Consistency and Reliability

Another improvement in this book is that it has a consistent and predictable structure. For example, instructors can rely on every section concluding with a "Test Your Understanding" feature which includes five questions that students can use to self-test and that instructors can choose to use as pre-class questions. The exercises are always consistently split into subsections of different types of problems: "Thinking Back," "Concepts," "Skills," "Applications," "Proofs," and "Thinking Forward." In addition, the "Concepts" subsection always begins with a summary exercise, eight true/false questions, and three example construction exercises. Instructors and students alike can rely on this consistent structure when assigning exercises and choosing a path of study.

Flexibility

We recognize that instructors use calculus books in many different ways and that the real direction of a calculus course comes from the instructor, not any book. The streamlined, consistent structure of this book makes it easy to use with a wide variety of courses and pedagogical styles. In particular, instructors will find it easy to include or omit sections, proofs, examples, and exercises consistently according to their preferences and course requirements. Students can focus on mathematical development or on examples and calculations as they need to throughout the course. Later, they can use the book as a reliable reference.

We think it will be immediately clear to anyone opening this book that what we have written is substantially different from the other calculus books on the market today while still following the standard topics taught in most modern science, mathematics, and engineering calculus courses. Our hope is that faculty who use the book will find it flexible for different pedagogical approaches and that students will be able to read it on different levels as they learn to understand the beauty of calculus.

A Special Taalman/Kohn Option for Underprepared Calculus Students

Do some of your calculus students struggle with algebra and precalculus material? The Taalman/Kohn *Calculus* series has a ready-made option for such students, called *Calculus I with Integrated Precalculus*. This option includes all the material in Chapters 0–6 of Taalman/Kohn *Calculus*, but in a different order and with supplementary precalculus and algebra material.

▶ Chapters 0–3 of *Calculus I with Integrated Precalculus* cover the same development of differential calculus topics as Chapters 0–3 in Taalman/Kohn *Calculus*, but the more complicated calculational examples are deferred to later chapters.

▶ Chapters 4–6 of *Calculus I with Integrated Precalculus* revisit differential calculus through the lens of studying progressively more challenging types of functions. Any exercises or examples from Taalman/Kohn *Calculus* that were left out of Chapters 0–3 of *Calculus I with Integrated Precalculus* are included in Chapters 4–6. The requisite background precalculus and algebra material is built from the ground up.

▶ Chapters 7–9 of *Calculus I with Integrated Precalculus* are identical to Chapters 4–6 of Taalman/Kohn *Calculus* and cover all topics from integral calculus.

Students who learn Calculus I from *Calculus I with Integrated Precalculus* can continue with Calculus II using Taalman/Kohn *Calculus* or any other calculus textbook. Students who have weak algebra and precalculus skills can succeed in STEM-level calculus if given the right help along the way, and *Calculus I with Integrated Precalculus* is written specifically to address the needs of those students.

For an examination copy of *Calculus I with Integrated Precalculus*, please contact your local W. H. Freeman & Company representative.

For Instructors

Instructor's Solutions Manual

Single-variable ISBN: 1-4641-5017-6

Multivariable ISBN: 1-4641-5018-4

Contains worked-out solutions to all exercises in the text.

Test Bank

Computerized (CD-ROM), ISBN: 1-4641-2547-3

Includes multiple-choice and short-answer test items.

Instructor's Resource Manual

ISBN: 1-4641-2545-7

Provides suggested class time, key points, lecture material, discussion topics, class activities, worksheets, and group projects corresponding to each section of the text.

Instructor's Resource CD-ROM

ISBN: 1-4641-2548-1

Search and export all resources by key term or chapter. Includes text images, Instructor's Solutions Manual, Instructor's Resource Manual, and Test Bank.

For Students

Student Solutions Manual

Single-variable ISBN: 1-4641-2538-4

Multivariable ISBN: 1-4641-5019-2

Contains worked-out solutions to all odd-numbered exercises in the text.

Software Manuals

Maple™ and Mathematica® software manuals are available within CalcPortal. Printed versions of these manuals are available through custom publishing. They serve as basic introductions to popular mathematical software options and guides for their use with *Calculus*.

Book Companion Web Site at www.whfreeman.com/tkcalculus

For students, this site serves as a FREE 24–7 electronic study guide, and it includes such features as self-quizzes and interactive applets.

Online Homework Options

WebAssign *Premium* **www.webassign.net/whfreeman**

WebAssign Premium integrates the book's exercises into the world's most popular and trusted online homework system, making it easy to assign algorithmically generated homework and quizzes. Algorithmic exercises offer the instructor optional algorithmic solutions. WebAssign Premium also offers access to resources, including the new Dynamic Figures, CalcClips whiteboard videos, tutorials, and "Show My Work" feature. In addition, WebAssign Premium is available with a fully customizable e-Book option that includes links to interactive applets and projects.

♪calcportal **www.yourcalcportal.com**

CalcPortal combines a fully customizable e-Book, exceptional student and instructor resources, and a comprehensive online homework assignment center. Included are algorithmically generated exercises, as well as Precalculus diagnostic quizzes, Dynamic Figures, interactive applets, CalcClips whiteboard videos, student solutions, online quizzes, Mathematica and Maple manuals, and homework management tools, all in one affordable, easy-to-use, and fully customizable learning space.

✦WeBWorK **webwork.maa.org**

W. H. Freeman offers approximately 2,500 algorithmically generated questions (with full solutions) through this free, open-source online homework system at the University of Rochester. Adopters also have access to a shared national library test bank with thousands of additional questions, including 1,500 problem sets matched to the book's table of contents.

Additional Media

SolutionMaster

This easy-to-use Web-based version of the Instructor's Solutions Manual allows instructors to generate a solution file for any set of homework exercises. Solutions can be downloaded in PDF format for convenient printing and posting.

Interactive e-Book at ebooks.bfwpub.com/tkcalculus

The Interactive e-Book integrates a complete and customizable online version of the text with its media resources. Students can quickly search the text, and they can personalize the e-Book just as they would the print version, with highlighting, bookmarking, and note-taking features. Instructors can add, hide, and reorder content, integrate their own material, and highlight key text.

Course Management Systems

W. H. Freeman and Company provides courses for Blackboard, WebCT (Campus Edition and Vista), Angel, Desire2Learn, Moodle, and Sakai course management systems. These are completely integrated solutions that you can easily customize and adapt to meet your teaching goals and course objectives. Visit www.macmillanhighered.com/catalog/other/coursepack for more information.

i·clicker

This two-way radio frequency classroom response system was developed by educators for educators. University of Illinois physicists Tim Stelzer, Gary Gladding, Mats Selen, and Benny Brown created the i-clicker system after using competing classroom responses and discovering that they were neither appropriate for the classroom nor friendly to the student. Each step of i-clicker's development has been informed by teaching and learning. i-clicker is superior to other systems from both a pedagogical and a technical standpoint. To learn more about packaging i-clicker with this textbook, contact your local sales representative or visit **www.iclicker.com.**

Each section opens with a **list of the three main section topics**. The list provides a focus and highlights key concepts.

10.4   CROSS PRODUCT

▶ Multiplying two vectors in $\mathbb{R}^3$ using the cross product
▶ The geometry of the cross product
▶ The relationship between the algebra and the geometry of the cross product

---

Definitions are clearly boxed, numbered, and labeled for easy reference. To reinforce their importance and meaning, definitions are followed by brief, often illustrated, examples.

DEFINITION 10.24   **The Determinant of a 3 × 3 Matrix**

The ***determinant*** of the 3×3 matrix $A = \begin{bmatrix} a_1 & a_2 & a_3 \\ b_1 & b_2 & b_3 \\ c_1 & c_2 & c_3 \end{bmatrix}$, denoted by $\det A$, is the sum

$$\det \begin{bmatrix} a_1 & a_2 & a_3 \\ b_1 & b_2 & b_3 \\ c_1 & c_2 & c_3 \end{bmatrix} = a_1 b_2 c_3 - a_1 b_3 c_2 + a_2 b_3 c_1 - a_2 b_1 c_3 + a_3 b_1 c_2 - a_3 b_2 c_1.$$

At first it might seem difficult to remember how to compute this sum. Fortunately, there is a nice visual trick that can be used to help. Compare the products along the six colored "diagonals" shown in the following matrix with the final sum in Definition 10.24. Note that four of the six diagonals wrap around the matrix as they descend.

Every 3 × 3 matrix has six "diagonals"

Theorems are developed intuitively before they are stated formally, and simple examples inform the discussion. **Proofs** follow most theorems, although they are optional, given instructor preference.

THEOREM 10.34   **Two Nonparallel Vectors and Their Cross Product Form a Right-Handed Triple**

Let **u** and **v** be nonparallel vectors in $\mathbb{R}^3$. Then the vectors **u**, **v**, and $\mathbf{u} \times \mathbf{v}$ form a right-handed triple.

Proof.   Let **u** and **v** be nonparallel position vectors in $\mathbb{R}^3$. We may position our coordinate axes so that **u** lies along the positive x-axis and **v** lies in the xy-plane and has a positive y-coordinate. That is, $\mathbf{u} = \langle u_1, 0, 0 \rangle$ and $\mathbf{v} = \langle v_1, v_2, 0 \rangle$ with $u_1 > 0$ and $v_2 > 0$. The sign of v_1 may be positive, zero, or negative. Two of these possibilities are illustrated here:

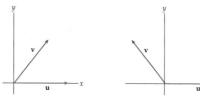

*Vectors **u** and **v** with $v_1 > 0$*   *Vectors **u** and **v** with $v_1 < 0$*

In any of the three cases, when we take the cross product we obtain

$$\mathbf{u} \times \mathbf{v} = \det \begin{bmatrix} \mathbf{i} & \mathbf{j} & \mathbf{k} \\ u_1 & 0 & 0 \\ v_1 & v_2 & 0 \end{bmatrix} = \langle 0, 0, u_1 v_2 \rangle.$$

Since $\mathbf{u} \times \mathbf{v}$ has a positive z-coordinate, we have our desired result.   ■

Color is used consistently and pedagogically in **graphs and figures** to relate like concepts. For instance, the color used for rectangles in Riemann sum approximations is also quite purposefully used for linear approximations of arc length and rectangular solid approximations of volume.

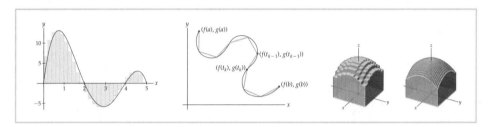

Cautions are appropriately placed at points in the exposition where students typically have questions about the nuances of mathematical thinking, processes, and notation.

> ⊗ CAUTION | We graphed the polar equation $r = 5 \csc \theta$ in Example 4 by transforming it to rectangular coordinates. Using Theorems 9.6 and 9.7, we can always transform an equation in polar coordinates to one in rectangular coordinates. However, as the next example demonstrates, we *cannot* expect that every polar equation will have a simple expression in rectangular coordinates. Therefore, we may not understand the polar equation any better by transforming it to rectangular coordinates.

Every section includes short illustrative examples as part of the discussion and development of the material. Once the groundwork has been laid, more complex **examples** and calculations are provided. Students find this approach easier to handle because the difficult calculations do not interfere with the development of why things work. Example **solutions** are explained in detail and include all the steps necessary for student comprehension.

EXAMPLE 4   Determining when two lines intersect

Do the lines $\mathcal{L}_1$ and $\mathcal{L}_2$, respectively given by

$$\mathbf{r}_1(t) = \langle -8 - 5t, 3 - t, 4 \rangle, -\infty < t < \infty, \text{ and } \mathbf{r}_2(t) = \langle 6 + t, 4 + 2t, 1 + 3t \rangle, -\infty < t < \infty,$$

intersect?

SOLUTION

We see immediately that the direction vectors for these lines are $\mathbf{d}_1 = \langle -5, -1, 0 \rangle$ and $\mathbf{d}_2 = \langle 1, 2, 3 \rangle$, which are *not* parallel. (Why?) Do the lines share a point? If so, the lines intersect. If not, they are skew.

   Finding a candidate for a point of intersection here is more delicate than in Example 3, where we could have used *any* point on either of the lines. We *cannot* just equate the corresponding components, as written, and try to solve for t. If you try this, you will see that the system

$$-8 - 5t = 6 + t, \quad 3 - t = 4 + 2t, \quad 4 = 1 + 3t$$

has no solution. However, as we will see in a moment, the lines *do* intersect! The difficulty here is with the parameters. Although we've used the letter t as the parameter for each line, there are *two* distinct parameters, one for $\mathcal{L}_1$ and a second for $\mathcal{L}_2$. To proceed, we change the name of the variable for one of the parameters. Let

$$\mathbf{r}_2(u) = \langle 6 + u, 4 + 2u, 1 + 3u \rangle, -\infty < u < \infty.$$

Now we equate the corresponding components of $\mathbf{r}_1$ and $\mathbf{r}_2$ to obtain the system

$$-8 - 5t = 6 + u, \quad 3 - t = 4 + 2u, \quad 4 = 1 + 3u.$$

This system has the unique solution $t = -3$, $u = 1$. Thus, the lines do intersect at

$$\mathbf{r}_1(-3) = \mathbf{r}_2(1). \qquad \square$$

Following many example solutions, **Checking the Answer** encourages students to learn to check their work, using technology such as a graphing calculator when appropriate.

CHECKING THE ANSWER	To verify our conclusion, we evaluate

$$\mathbf{r}_1(-3) = \langle -8 - 5(-3), 3 - (-3), 4 \rangle = \langle 7, 6, 4 \rangle$$

and

$$\mathbf{r}_2(1) = \langle 6 + 1, 4 + 2 \cdot 1, 1 + 3 \cdot 1 \rangle = \langle 7, 6, 4 \rangle.$$

These calculations confirm that the point $(7, 6, 4)$ is on both lines.

Each section closes with five **Test Your Understanding** questions that test students on the concepts and reading presented in the section. Because answers are not provided, instructors may choose to use these questions for discussion or assessment.

? TEST YOUR UNDERSTANDING

- ▶ How do you find an equation of a line determined by two points?
- ▶ How do you find an equation of a line parallel to a given vector and passing through a given point?
- ▶ Given a parametrization of a line, parametric equations of a line, or the equation of the line in symmetric form, how do you find the other forms?
- ▶ What are parallel lines, intersecting lines, and skew lines in 3-space?
- ▶ How do you find the distance from a point to a line?

Section Exercises are provided in a consistent format that offers the same types of exercises within each section. This approach allows instructors to tailor assignments to their course, goals, and student audience.

Thinking Back exercises ask students to review relevant concepts from previous sections and lessons.

Concepts exercises are consistently formatted to start with the following three problems:

- Problem 0 tests understanding.
- Problem 1 consists of eight true/false questions.
- Problem 2 asks the student to create examples based on their understanding of the reading.

Skills exercises offer ample practice, grouped into varying degrees of difficulty.

Applications exercises contain at least two in-depth real-world problems.

Proofs exercises can be completed by students in non-theoretical courses. Hints are often provided, and many exercises mimic work presented in the reading and examples. Often, these exercises are a continuation of a proof offered as a road map in the narrative.

Thinking Forward exercises plant seeds of concepts to come. In conjunction with the Thinking Back exercises, they offer a "tie together" of both past and future topics, thereby providing a seamless flow of concepts.

Chapter Review, Self-Test, and Capstones, found at the end of each chapter, present the following categories:

Definitions exercises prompt students to recall definitions and give an illustrative example.

Theorems exercises ask students to complete fill-in-the-blank theorem statements.

Formulas, Notation, and/or Rules exercises vary according to chapter content and ask students to show a working understanding of important formulas, equations, notation, and rules.

Skill Certification exercises provide practice with basic computations from the chapter.

Capstone Problems pull together the essential ideas of the chapter in more challenging mathematical and application problems.

ACKNOWLEDGMENTS

There are many people whose contributions to this project have made it immeasurably better. We are grateful to the many instructors from across the United States and Canada who have offered comments that assisted in the development of this book:

Jabir Abdulrahman, *Carleton University*

Jay Abramson, *Arizona State University*

Robert F. Allen, *University of Wisconsin–La Crosse*

Roger Alperin, *San Jose State University*

Matthew Ando, *University of Illinois*

Jorge Balbas, *California State University, Northridge*

Lynda Ballou, *New Mexico Institute of Mining and Technology*

E. N. Barron, *Loyola University Chicago*

Stavros Belbas, *University of Alabama*

Michael Berg, *Loyola Marymount University*

Geoffrey D. Birky, *Georgetown University*

Paul Blanchard, *Boston University*

Joseph E. Borzellino, *California Polytechnic State University, San Luis Obispo*

Eddie Boyd, Jr., *University of Maryland Eastern Shore*

James Brawner, *Armstrong Atlantic State University*

Jennifer Bready, *Mount Saint Mary College*

Mark Brittenham, *University of Nebraska*

Jim Brown, *Clemson University*

John Burghduff, *Lone Star College–CyFair*

Christopher Butler, *Case Western Reserve University*

Katherine S. Byler Kelm, *California State University, Fresno*

Weiming Cao, *The University of Texas at San Antonio*

Deb Carney, *Colorado School of Mines*

Lester Caudill, *University of Richmond*

Leonard Chastkofsky, *The University of Georgia*

Fengxin Chen, *University of Texas at San Antonio*

Dominic Clemence, *North Carolina A&T State University*

A. Coffman, *Indiana–Purdue Fort Wayne*

Nick Cogan, *Florida State University*

Daniel J. Curtin, *Northern Kentucky University*

Donatella Danielli-Garofalo, *Purdue University*

Shangrong Deng, *Southern Polytechnic State University*

Hamide Dogan-Dunlap, *The University of Texas at El Paso*

Alexander Engau, *University of Colorado, Denver*

Said Fariabi, *San Antonio College*

John C. Fay, *Chaffey College*

Tim Flaherty, *Carnegie Mellon University*

Stefanie Fitch, *Missouri University of Science & Technology*

Kseniya Fuhrman, *Milwaukee School of Engineering*

Robert Gardner, *East Tennessee State University*

Richard Green, *University of Colorado, Boulder*

Weiman Han, *University of Iowa*

Yuichi Handa, *California State University, Chico*

Liang (Jason) Hong, *Bradley University*

Steven Hughes, *Alabama A&M University*

Alexander Hulpke, *Colorado State University*

Colin Ingalls, *University of New Brunswick, Fredericton*

Lea Jenkins, *Clemson University*

Lenny Jones, *Shippensburg University*

Heather Jordan, *Illinois State University*

Mohammad Kazemi, *The University of North Carolina at Charlotte*

Dan Kemp, *South Dakota State University*

Boris L. Kheyfets, *Drexel University*

Alexander A. Kiselev, *University of Wisconsin–Madison*

Greg Klein, *Texas A&M University*

Evangelos Kobotis, *University of Illinois at Chicago*

Alex Kolesnik, *Ventura College*

Amy Ksir, *US Naval Academy*

Dan Kucerovsky, *University of New Brunswick*

Trent C. Kull, *Winthrop University*

Alexander Kurganov, *Tulane University*

Jacqueline LaVie, *SUNY College of Environmental Science and Forestry*

Melvin Lax, *California State University, Long Beach*

Dung Le, *The University of Texas at San Antonio*

Mary Margarita Legner, *Riverside City College*

Denise LeGrand, *University of Arkansas–Little Rock*

Mark L. Lewis, *Kent State University*

Xiezhang Li, *Georgia Southern University*

Antonio Mastroberardino, *Penn State Erie, The Behrend College*

Michael McAsey, *Bradley University*

Jamie McGill, *East Tennessee State University*

Gina Moran, *Milwaukee School of Engineering*

Abdessamad Mortabit, *Metropolitan State University*

Emilia Moore, *Wayland Baptist University*

Vivek Narayanan, *Rochester Institute of Technology*

Rick Norwood, *East Tennessee State University*

Gregor Michal Olsavsky, *Penn State Erie, The Behrend College*

Rosanna Pearlstein, *Michigan State University*

Kanishka Perera, *Florida Institute of Technology*

Cynthia Piez, *University of Idaho*

Jeffrey L. Poet, *Missouri Western State University*

Joseph P. Previte, *Penn State Erie, The Behrend College*

Jonathan Prewett, *University of Wyoming*

Elise Price, *Tarrant County College*

Stela Pudar-Hozo, *Indiana University of Northwest*

Don Redmond, *Southern Illinois University*

Dan Rinne, *California State University, San Bernardino*

Joe Rody, *Arizona State University*

John P. Roop, *North Carolina A&T State University*

Amber Rosin, *California State Polytechnic University, Pomona*

Nataliia Rossokhata, *Concordia University*

Dev K. Roy, *Florida International University*

Hassan Sedaghat, *Virginia Commonwealth University*

Asok Sen, *Indiana University–Purdue University*

Adam Sikora, *The State University of New York at Buffalo*
Mark A. Smith, *Miami University*
Shing Seung So, *University of Central Missouri*
David Stowell, *Brigham Young University–Idaho*
Jeff Stuart, *Pacific Lutheran University*
Howard Wainer, *Wharton School of the University of Pennsylvania*
Thomas P. Wakefield, *Youngstown State University*
Bingwu Wang, *Eastern Michigan University*
Lianwen Wang, *University of Central Missouri*
Antony Ware, *University of Calgary*
Talitha M. Washington, *Howard University*

Mary Wiest, *Minnesota State University, Mankato*
Mark E. Williams, *University of Maryland Eastern Shore*
G. Brock Williams, *Texas Tech University*
Dennis Wortman, *University of Massachusetts, Boston*
Hua Xu, *Southern Polytechnic State University*
Wen-Qing Xu, *California State University, Long Beach*
Yvonne Yaz, *Milwaukee School of Engineering*
Hong-Ming Yin, *Washington State University*
Mei-Qin Zhan, *University of North Florida*
Ruijun Zhao, *Minnesota State University, Mankato*
Yue Zhao, *University of Central Florida*
Jan Zijlstra, *Middle Tennessee State University*

We would also like to thank the Math Clubs at the following schools for their help in checking the accuracy of the exercises and their solutions:

CUNY Bronx Community College
Duquesne University
Fitchburg State College
Florida International University
Idaho State University

Jackson State University
Lander University
San Jose State University
Southern Connecticut State University
Texas A&M University

Texas State University–San Marcos
University of North Texas
University of South Carolina–Columbia
University of South Florida
University of Wisconsin–River Falls

Our students and colleagues at James Madison University have used preliminary versions of this text for the past two years and have helped to clarify the exposition and remove ambiguities. We would particularly like to thank our colleagues Chuck Cunningham, Rebecca Field, Bill Ingham, John Johnson, Brant Jones, Stephen Lucas, John Marafino, Kane Nashimoto, Edwin O'Shea, Ed Parker, Gary Peterson, Katie Quertermous, James Sochacki, Roger Thelwell, Leonard Van Wyk, Debra Warne, and Paul Warne for class-testing our book and for their helpful feedback. During the class-testing at JMU, hundreds of students provided feedback, made suggestions that improved the book, and, of course, showed us how they learned from the book! Thank you to all of our students, especially Lane O'Brien and Melissa Moxie for their meticulous review of an earlier draft of the text.

Chris Brazfield, now at Carroll Community College, helped with the initial development of the text and the ideas behind it. Kevin Cooper of Washington State University contributed many interesting and challenging real-world applications, and Elizabeth Brown and Dave Pruett of James Madison University contributed greatly to the development of the chapter on vector calculus. Roger Lipsett of Brandeis University wrote the excellent solution manual for the text, and at the same time eliminated any ambiguities in the exercises. We owe all of them great thanks for their expertise.

We also owe thanks to all of the people at W. H. Freeman who helped with the development of this text. Our developmental editor, Leslie Lahr, has been with this project from the beginning. Even under pressure, Leslie always maintains a positive attitude and finds a way for us to move forward. Without her support, we would not have made it through the rocky patches. Our executive editor, Terri Ward and developmental editor Katrina Wilhelm helped keep us on track while we wrote, rewrote, revised, revised, and revised some more, and we thank them for their support and patience. Brian Baker, our meticulous copy editor, made significant improvements to the text. Misplaced commas, dangling modifiers, and run-on sentences didn't stand a chance under his scrutiny. Ron Weickart and his team at Network Graphics took our graphs and sketches and turned them into the beautiful artwork in the text. Sherrill Redd and the compositing team at Aptara did a great job implementing all the design elements from our crazy LaTeX files.

Finally, we would like to thank our families and friends for putting up with us during the years of stress, turmoil, and tedium that inevitably come with any book project. Without their support, this book would not have been possible.

Learning something new can be both exciting and daunting. To gain a full understanding of the material in this text, you will have to read, you will have to think about the connections between the new topics and the topics that were previously presented, and you will have to work problems—many, many problems.

The structure of this text should help you understand the material. The material is laid out in a linear fashion that we think will facilitate your understanding. Each section is separated into two main parts: first, a presentation of new material and then second, a set of *Examples and Explorations*, where you will find problems that are carefully worked through. Working through these examples on your own, as you read the steps for guidance, will help prepare you for the exercises.

Reading a mathematics book isn't like reading a novel: You may have to read some parts more than once, and you may need to make notes or work things out on paper. Pay special attention to the "Checking Your Answer" features, so that you can learn how to check your own answers to many types of questions.

To succeed in calculus, you need to do homework exercises. The exercises in every section of this text are broken into six categories: "Thinking Back," "Concepts," "Skills," "Applications," "Proofs," and "Thinking Forward."

- As the title suggests, the *Thinking Back* problems are intended to tie the current material to material you've seen in previous sections or even previous courses.
- The *Concepts* problems are designed to help you understand the main ideas presented in the section without a lot of calculation. Every group of *Concepts* exercises begins by asking you to summarize the section, continues with eight true/false questions, and then asks for three examples illustrating ideas from the section.
- The bulk of the exercises in each section consists of *Skills* problems that may require more calculation.
- The *Applications* exercises use the concepts from the section in "real-world" problems.
- The *Proofs* exercises ask you to prove some basic theory from the section.
- Finally, the *Thinking Forward* questions use current ideas to introduce topics that you will see in subsequent sections.

We hope this structure allows you to tie together the material as you work through the book. We have supplied the answers to the odd-numbered exercises, but don't restrict yourself to those problems. You can check answers to even-numbered questions by hand or by using a calculator or an online tool such as wolframalpha.com. After all, on a quiz or test you won't have the answers, so you'll have to know how to decide for yourself whether or not your answers are reasonable.

Some students may like to work through each section "backwards," starting by attempting the exercises, then checking back to the examples as needed when they get stuck, and, finally, using the exposition as a reference when they want to see the big picture. That is fine; although we recommend that you at least try reading through the sections in order to see how things work for you. Either way, we hope that the separation of examples from exposition and the division of homework problems into subsections will help make the process of learning this beautiful subject easier. We have written this text with you, the student, in mind. We hope you enjoy using it!

CHAPTER 9

Parametric Equations, Polar Coordinates, and Conic Sections

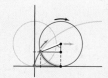

9.1   PARAMETRIC EQUATIONS

▶ Defining and graphing parametric equations

▶ Finding slopes of tangent lines on parametric curves

▶ Finding the arc length of a parametric curve

Parametric Equations

In most of this book up to now we have studied functions of the form $y = f(x)$, where y is explicitly expressed as a function of x. There are advantages to using equations of this form. In particular, they are often relatively easy to analyze and graph with the tools of calculus. However, there are disadvantages as well. One fundamental limitation is that not every curve of interest has a simple representation in this form. As we saw in Section 2.4, this limitation can be partially addressed by using implicitly defined functions of the form $F(x, y) = c$. In this section we discuss a different method for expressing relationships in the plane: parametric equations. When we use parametric equations, each variable is expressed independently as a function of a new variable, called a parameter. The knobs on an Etch A Sketch® toy show how each variable of a graph can be controlled individually. In a more complex setting, machine tools are also guided parametrically.

DEFINITION 9.1

Parametric Equations in Two Variables

Parametric equations in two variables are a pair of functions

$$x = f(t) \quad \text{and} \quad y = g(t),$$

where the *parameter* t is defined on some interval I of real numbers.

A *parametric curve* is the set of points in the coordinate plane:

$$\{(x(t), y(t)) \mid t \in I\}.$$

For example, we will soon show that the pair of equations

$$x = t + 1, \quad y = t^2 - 4 \text{ for } t \geq -2$$

has the graph shown next on the left, and that the pair of equations

$$x = t - \sin t, \quad y = 1 - \cos t \text{ for } t \in [0, 4\pi]$$

has the graph shown on the right.

The graph of $x = t + 1, y = t^2 - 4$ *The graph of $x = t - \sin t, y = 1 - \cos t$*

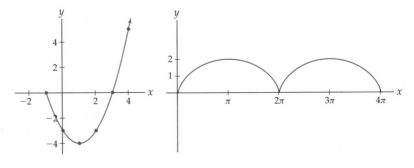

In these equations we used the letter t as the parameter, so we say that the curves are **parametrized** by t. The particular letter that we use is not important. However, we will often think of t as representing time. When this is the case, we say that the curve is **parametrized by time**. On other occasions, the parameter may represent a rotation; in that case we might use the angles θ or ϕ as our parameter.

There are two important reasons for introducing parametric equations. First, some curves are most naturally expressed with parametric equations. Second, such equations can be generalized to express curves in three-dimensional space. (See Sections 10.5 and 11.1.) In many applications, we will have a curve or phenomenon we wish to understand and parametric equations will be the easiest way to describe the motion involved.

To gain a basic understanding of parametric equations, however, we continue by ignoring these important reasons, at least for now. Instead we will start with some parametric equations and look for the curves they describe.

Graphing Parametric Equations

The most basic way to plot any curve is to plot points and then "connect the dots." This is probably the first technique you learned when you started graphing functions. It is also the way your graphing calculator and most computer algebra systems plot curves. In fact, this is the very reason that graphs of curves with discontinuities often are not plotted well by a calculator. To start we will use the "connect the dots" approach to plot the parametric curve given by

$$x = t + 1 \quad \text{and} \quad y = t^2 - 4 \quad \text{for} \quad t \in [-2, \infty).$$

Using these equations, we may generate the following table that contains the coordinates of several points on the curve:

t	-2	-1	0	1	2	3
$x = t + 1$	-1	0	1	2	3	4
$y = t^2 - 4$	0	-3	-4	-3	0	5

We then plot the points (x, y) and connect the dots with a curve, as follows:

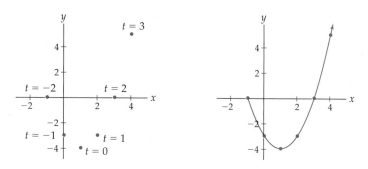

Observe that we have added an arrowhead to one end of the curve in the second figure. This is because, as the parameter t increases, it imposes a direction on the curve. For the values of t that we happened to consider, the curve begins at the smallest value, $t = -2$, and the arrow indicates the direction of motion as t increases. It is important to note that a given curve in the plane could be parametrized many different ways. Different parametric equations could traverse the same curve in the opposite direction. Other equations could traverse the same curve, but double back on portions of the curve over certain subintervals.

For example, in Exercises 14 and 15 we will ask you to show that the parametrization

$$x = 2t + 1 \quad \text{and} \quad y = 4t^2 - 4 \quad \text{for} \quad t \in [-1, \infty)$$

traces exactly the same curve as the one shown, and that the equations

$$x = \sin t + 1 \quad \text{and} \quad y = \sin^2 t - 4 \quad \text{for} \quad t \in (-\infty, \infty)$$

repeatedly trace a portion of the same parabola.

If the parametric equations are simple enough, we may also graph the curve by *eliminating the parameter*. The process of eliminating the parameter usually takes one of two forms. If the function $x = f(t)$ is invertible and we can find a simple expression for the inverse, $t = f^{-1}(x)$, then we may obtain $y = g(f^{-1}(x))$. For example, this is possible with the preceding equations by solving $x = t + 1$ for t, yielding $t = x - 1$. We then replace t with $x - 1$ in the equation for y, giving the equation

$$y = (x - 1)^2 - 4.$$

This equation defines an upwards-opening parabola whose vertex is at the point $(1, -4)$, just as we saw earlier. However, note that when we eliminated the parameter we lost information: The equation $y = (x - 1)^2 - 4$ neither contains information about the direction of motion imposed by the parameter nor tells us that the parametric equations specify only a portion of the parabola.

We may use a variation of this procedure when $y = g(t)$ is invertible and we can find a simple expression for the inverse, $t = g^{-1}(y)$. In this case, we are expressing either x or y as a function of the other variable, and that allows us to use the techniques of earlier chapters to analyze the resulting function.

Another way to eliminate the parameter results in a relationship in which y is expressed implicitly as a function of x. For example, if $x = 2 \sin t$ and $y = 2 \cos t$, then when we square both equations, add the two left-hand sides, and add the two right-hand sides, we obtain $x^2 + y^2 = 4 \sin^2 t + 4 \cos^2 t$. Using one of the Pythagorean identities, we have $x^2 + y^2 = 4$. Thus, the graph of these parametric equations lies on the circle with radius 2 and centered at the origin.

Direction of Motion and Tangent Lines

In general, the parametric equations $x = f(t)$ and $y = g(t)$ do not have to be either continuous or differentiable. For the time being we will assume that the functions are differentiable, and, to allow us to use the chain rule, we will also assume that y is locally a differentiable function of x for a point at which x is a differentiable function of t.

We may use the derivatives of the parametric equations to obtain information about the direction of motion along the parametric curve with increasing values of t. When we compute $\frac{dx}{dt}$ and $\frac{dy}{dt}$ and examine the signs of these derivatives, we easily determine the direction in which the curve is being drawn. For example, for the parametric equation $x = t + 1$ and $y = t^2 - 4$ for $t \geq -2$, we have

$$\frac{dx}{dt} = 1 \text{ and } \frac{dy}{dt} = 2t.$$

Thus, for all values of t, since $\frac{dx}{dt} > 0$, the curve is being traced from left to right. Since $\frac{dy}{dt} < 0$ for $-2 \leq t < 0$ and $\frac{dy}{dt} > 0$ for $t > 0$, the curve is moving down at first, but then starts rising.

To understand the slope of a parametric curve we may use the chain rule,

$$\frac{dy}{dt} = \frac{dy}{dx} \frac{dx}{dt}.$$

Thus, if $\frac{dx}{dt} \neq 0$, then

$$\frac{dy}{dx} = \frac{dy/dt}{dx/dt} = \frac{y'(t)}{x'(t)}.$$

For example, the slope of the tangent line to the curve given by $x = t+1$ and $y = t^2 - 4$ when $t = 4$ is

$$\left.\frac{dy/dt}{dx/dt}\right|_{t=4} = \left.\frac{2t}{1}\right|_{t=4} = 8.$$

More generally, we use this idea to define the tangent line to a parametric curve.

DEFINITION 9.2

The Slope of a Parametric Curve and the Tangent Line to a Parametric Curve

Let $x = x(t)$, $y = y(t)$ be parametric equations for t in some interval I, and let t_0 be a point in I at which $x'(t_0)$ and $y'(t_0)$ both exist.

(a) If $x'(t_0) \neq 0$, we define the **slope** of the parametric curve at the point $(x(t_0), y(t_0))$ to be

$$m = \frac{y'(t_0)}{x'(t_0)}.$$

If $x'(t_0) = 0$, we define the slope to be

$$m = \lim_{t \to t_0} \frac{y'(t)}{x'(t)},$$

provided that the limit exists.

(b) If the slope m is defined, the **tangent line** to the parametric curve at the point t_0 is given by the equation

$$y - y(t_0) = m(x - x(t_0)).$$

Note that certain tangents to parametric curves are vertical lines. As an extension to the preceding definition we may say that if $\lim_{t \to t_0} \frac{y'(t)}{x'(t)} = \pm\infty$, then the parametric curve has a **vertical tangent line** at $(x(t_0), y(t_0))$ whose equation is $x = x(t_0)$.

Even when the functions f and g are differentiable, the curve defined by the parametric equations $x = f(t)$, $y = g(t)$ may have cusps or other anomalies. For example, the function in the figure that follows at the left is not differentiable at the cusps, although we will soon see that the parametric equations we use to define the curve are differentiable everywhere. In addition, a given parametric curve may have points with two or more tangent lines; see, for example, the following figure at the right:

Cusps are points of non-differentiability

A point on a parametric curve with two tangents

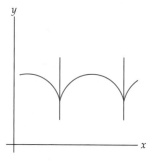

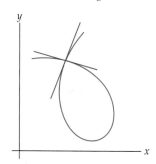

Arc Length

In Section 6.3 we saw that the arc length of a continuous function $f(x)$ on an interval $[x_1, x_2]$ is given by the formula

$$\int_{x_1}^{x_2} \sqrt{1 + (f'(x))^2}\, dx.$$

We will extend this definition to functions defined parametrically.

Suppose C is a parametric curve defined by $x = f(t)$ and $y = g(t)$ for $t \in [a, b]$. We can approximate the curve C with line segments:

Curve C can be approximated with line segments

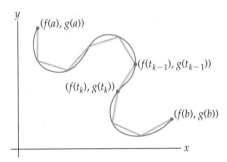

We need to use general notation to compute the sum of the lengths of the approximating line segments. Toward that end we subdivide the interval $[a, b]$ into n equal pieces and define $\Delta t = \frac{b-a}{n}$. Then, for $k = 0, 1, 2, \ldots, n$, we define $t_k = a + k\Delta t$. The length of C can be approximated by the sum of the lengths of the segments extending from the points $(f(t_{k-1}), g(t_{k-1}))$ to the points $(f(t_k), g(t_k))$. Notice that this is the same process we followed in Section 6.3, except that here we are subdividing the interval on which the parameter is defined. Using the distance formula and then adding up the lengths of each of the line segments, we arrive at the following approximation for the length of the curve C as the parameter t varies from a to b:

$$\sum_{k=1}^{n} \sqrt{(f(t_k) - f(t_{k-1}))^2 + (g(t_k) - g(t_{k-1}))^2}.$$

As in Section 6.3, the length of the curve C for $t \in [a, b]$ is defined as the limit of the approximate quantity as the number of pieces in our subdivision increases without bound.

DEFINITION 9.3   **The Arc Length of a Parametric Curve**

Let C be a curve in the plane with parametrization $x = f(t)$, $y = g(t)$ for $t \in [a, b]$, where f and g are differentiable functions of t such that the parametrization is a one-to-one function from the interval $[a, b]$ to the curve C. Then the **length of the curve** C is

$$\lim_{n \to \infty} \sum_{k=1}^{n} \sqrt{(f(t_k) - f(t_{k-1}))^2 + (g(t_k) - g(t_{k-1}))^2},$$

where $\Delta t = \frac{b-a}{n}$ and $t_k = a + k\Delta t$.

The requirement in Definition 9.3 that the parametrization be a one-to-one function from the interval $[a, b]$ to the curve C ensures that C is traversed exactly once as t increases from a to b. This means that the parametrization does not trace any part of the curve more than once. (Actually, this requirement can be relaxed somewhat; we can allow isolated points to be duplicated, and the definition will still hold.)

You may have noticed that the expression in Definition 9.3 looks almost like a limit of Riemann sums. The following theorem tells us how to interpret this quantity as a definite integral:

THEOREM 9.4

The Arc Length of a Parametric Curve

Let C be a curve in the plane with parametrization $x = f(t)$, $y = g(t)$ for $t \in [a, b]$ such that the parametrization is a one-to-one function from the interval $[a, b]$ to the curve C. If $x = f(t)$ and $y = g(t)$ are differentiable functions of t such that $f'(t)$ and $g'(t)$ are continuous on $[a, b]$, then the length of the curve C is given by

$$\int_a^b \sqrt{(f'(t))^2 + (g'(t))^2}\, dt.$$

Proof. By Definition 9.3, the length of the curve C is

$$\lim_{n \to \infty} \sum_{k=1}^n \sqrt{(f(t_k) - f(t_{k-1}))^2 + (g(t_k) - g(t_{k-1}))^2}.$$

Some simple algebra then shows that this is equivalent to

$$\lim_{n \to \infty} \sum_{k=1}^n \sqrt{\left(\frac{f(t_k) - f(t_{k-1})}{\Delta t}\right)^2 + \left(\frac{g(t_k) - g(t_{k-1})}{\Delta t}\right)^2}\, \Delta t,$$

where $\Delta t = \dfrac{b-a}{n}$ and $t_k = a + k\Delta t$. Applying the Mean Value Theorem to f and g, we find that for each k there are points t_k^* and t_k^{**} in $[t_{k-1}, t_k]$ such that

$$f'(t_k^*) = \frac{f(t_k) - f(t_{k-1})}{\Delta t} \text{ and } g'(t_k^{**}) = \frac{g(t_k) - g(t_{k-1})}{\Delta t}.$$

Thus the length of C is

$$\lim_{n \to \infty} \sum_{k=1}^n \sqrt{(f'(t_k^*))^2 + (g'(t_k^{**}))^2}\, \Delta t.$$

This last quantity is almost the limit of a Riemann sum. It is not the limit of a Riemann sum because t_k^* and t_k^{**} are, in general, different points. However, since f' and g' are both continuous, it can be shown that t_k^* and t_k^{**} get sufficiently close to each other as $n \to \infty$ for this not to matter. (The proof of this fact is outside of the scope of the text.) Applying the preceding final result, we find that the limit of our sum is the desired integral

$$\int_a^b \sqrt{(f'(t))^2 + (g'(t))^2}\, dt. \qquad \blacksquare$$

Examples and Explorations

EXAMPLE 1

Parametrizing the unit circle

Sketch the parametric curves determined by the following parametric equations:

(a) $x = \cos t$, $y = \sin t$, $t \in [0, \pi]$

(b) $x = \cos t$, $y = \sin t$, $t \in [\pi, 2\pi]$

(c) $x = \sin(3t)$, $y = \cos(3t)$, $t \in [0, 2\pi]$

(d) $x = \dfrac{1 - t^2}{1 + t^2}$, $y = \dfrac{2t}{1 + t^2}$, $t \in [0, \infty)$

SOLUTION

(a) We will start by eliminating the parameter, but also plot a few points along the way. If we square the equations for x and y, we obtain $x^2 = \cos^2 t$ and $y^2 = \sin^2 t$. Now adding these two equations and applying the Pythagorean identity, we have

$$x^2 + y^2 = \cos^2 t + \sin^2 t = 1.$$

Therefore $x^2 + y^2 = 1$, which means that the parametric curve is (at least a portion of) a circle of radius 1 and centered at the origin. We want to know where on this circle the parametric curve lies, and we also want to know the direction of motion as t increases. Consider the following three values of t: $t = 0$, $t = \frac{\pi}{2}$, and $t = \pi$. Corresponding to these values we have $(x(0), y(0)) = (\cos 0, \sin 0) = (1, 0)$,

$$\left(x\left(\frac{\pi}{2}\right), y\left(\frac{\pi}{2}\right)\right) = \left(\cos\frac{\pi}{2}, \sin\frac{\pi}{2}\right) = (0, 1), \text{ and } (x(\pi), y(\pi)) = (\cos\pi, \sin\pi) = (-1, 0).$$

In addition, as t increases (slightly) from 0, both x and y will be positive. Thus, we see that the parametric curve is the top half of the unit circle and the direction of motion is counterclockwise; see the first figure that follows at the left.

(b) We eliminated the parameter from these equations in part (a). A similar analysis of the behavior of the variables x and y as t increases from π to 2π shows that this parametrization gives us the bottom half of the unit circle, with the direction of motion still being counterclockwise; see the middle figure. The details are left to Exercise 35.

(c) We again eliminate the parameter by squaring each equation and adding. Once more, we obtain $x^2 + y^2 = 1$. As before, the curve is some portion of the unit circle centered at the origin. Now, however, when $t = 0$, we are at the point $(x(0), y(0)) = (\sin 0, \cos 0) = (0, 1)$. Furthermore, as t increases from 0, x increases and y decreases; therefore the direction of motion is clockwise. Finally, we note that for $t \in [0, 2\pi]$ both $x = \sin 3t$ and $y = \cos 3t$ go through three periods. Therefore, the unit circle is traversed three times in the clockwise direction with these equations; see the following figure at the right:

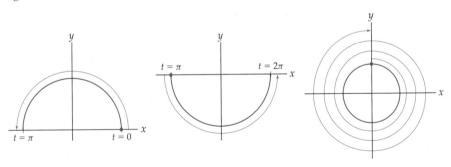

(d) We will see that this pair of equations is quite a different parametrization of the unit circle. Although it does not involve trigonometric functions, we may again eliminate the parameter by squaring each equation and adding the result:

$$x^2 + y^2 = \left(\frac{1 - t^2}{1 + t^2}\right)^2 + \left(\frac{2t}{1 + t^2}\right)^2 = \frac{1 - 2t^2 + t^4 + 4t^2}{(1 + t^2)^2}$$

$$= \frac{1 + 2t^2 + t^4}{(1 + t^2)^2} = \frac{(1 + t^2)^2}{(1 + t^2)^2} = 1.$$

Thus, the parametric curve is again at least a portion of the unit circle. To determine which portion (and in what direction and with what speed), we begin by evaluating x and y for two values of t:

t	0	1
$x = \dfrac{1-t^2}{1+t^2}$	1	0
$y = \dfrac{2t}{1+t^2}$	0	1

Since the parameter t can vary from 0 to ∞, we also evaluate the following limits:

$$\lim_{t\to\infty} x(t) = \lim_{t\to\infty} \frac{1-t^2}{1+t^2} = -1, \quad \lim_{t\to\infty} y(t) = \lim_{t\to\infty} \frac{2t}{1+t^2} = 0.$$

Putting all this information together, we arrive at the following parametric curve:

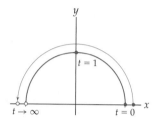

Note that every point on the top half of the unit circle is traversed exactly once, except for $(-1, 0)$, which is not part of the parametrization. In addition, unlike the motion in parts (a), (b), and (c), the motion here does *not* take place at a uniform speed. For example, as t increases from 0 to 1, the particle moves all the way from $(1, 0)$ to $(0, 1)$, but as t increases from 1, the particle moves from $(0, 1)$ toward, but never reaches $(-1, 0)$.   □

EXAMPLE 2   Graphing an astroid

Sketch the graph of the curve described by the parametric equations

$$x = \cos^3 \theta \quad \text{and} \quad y = \sin^3 \theta.$$

SOLUTION

We begin by tabulating the coordinates of several points for some convenient values of θ:

θ	0	$\dfrac{\pi}{6}$	$\dfrac{\pi}{4}$	$\dfrac{\pi}{3}$	$\dfrac{\pi}{2}$	$\dfrac{2\pi}{3}$	$\dfrac{3\pi}{4}$	$\dfrac{5\pi}{6}$
$x = \cos^3 \theta$	1	$\dfrac{3\sqrt{3}}{8}$	$\dfrac{\sqrt{2}}{4}$	$\dfrac{1}{8}$	0	$-\dfrac{1}{8}$	$-\dfrac{\sqrt{2}}{4}$	$-\dfrac{3\sqrt{3}}{8}$
$y = \sin^3 \theta$	0	$\dfrac{1}{8}$	$\dfrac{\sqrt{2}}{4}$	$\dfrac{3\sqrt{3}}{8}$	1	$\dfrac{3\sqrt{3}}{8}$	$\dfrac{\sqrt{2}}{4}$	$\dfrac{1}{8}$

θ	π	$\dfrac{7\pi}{6}$	$\dfrac{5\pi}{4}$	$\dfrac{4\pi}{3}$	$\dfrac{3\pi}{2}$	$\dfrac{5\pi}{3}$	$\dfrac{7\pi}{4}$	$\dfrac{11\pi}{6}$
$x = \cos^3 \theta$	-1	$-\dfrac{3\sqrt{3}}{8}$	$-\dfrac{\sqrt{2}}{4}$	$-\dfrac{1}{8}$	0	$\dfrac{1}{8}$	$\dfrac{\sqrt{2}}{4}$	$\dfrac{3\sqrt{3}}{8}$
$y = \sin^3 \theta$	0	$-\dfrac{1}{8}$	$-\dfrac{\sqrt{2}}{4}$	$-\dfrac{3\sqrt{3}}{8}$	-1	$-\dfrac{3\sqrt{3}}{8}$	$-\dfrac{\sqrt{2}}{4}$	$-\dfrac{1}{8}$

We plot these points in the figure that follows on the left:

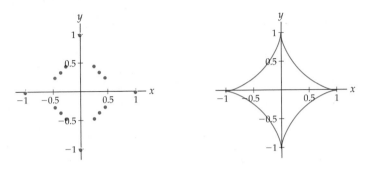

In addition, the slope of the curve at each point is a function of θ:

$$\frac{dy}{dx} = \frac{dy/d\theta}{dx/d\theta} = \frac{\sin^2\theta\cos\theta}{-\cos^2\theta\sin\theta} = -\tan\theta.$$

Thus, the slope of the tangent line to the curve when $\theta = 0$ is zero. We also see that the slope of the tangent line at every point in the first quadrant is negative, and since $\lim\limits_{\theta\to\pi/2^-}(-\tan\theta) = -\infty$, the slopes are nearly vertical close to the point $(0, 1)$. When we analyze the tangents to the curve in the other three quadrants in a similar fashion, and connect the points appropriately, we obtain the graph on the right, known as an ***astroid***.

Finally, we note that we can also eliminate the parameter. Here we have $x^{2/3} = \cos^2\theta$ and $y^{2/3} = \sin^2\theta$. When we add these two equations together, we obtain $x^{2/3} + y^{2/3} = 1$, whose graph is the same astroid. □

EXAMPLE 3 Finding parametric equations for the cycloid

Find the parametric equations that describe the curve traced by a point on the circumference of a wheel as it rolls along a straight path.

SOLUTION

We see that this is a more realistic application of parametric equations. We may visualize the given curve as the path traced by a point on a bicycle tire as it rolls along a road. In this situation we wish to understand a curve, and parametric equations provide an effective tool to describe it.

We let the radius of the wheel be r and, for convenience, we place our wheel on a track formed by the x-axis, with the point tracing the curve starting at the origin. The gray circle in the figure that follows shows the initial position of the wheel, and the black circle shows the wheel after it has rolled a short distance. The blue curve is what we wish to describe with the use of parametric equations.

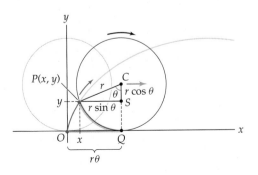

To parametrize the blue curve, we need to find a parameter that can be used to describe both the x- and y-coordinates of each point on the curve; one such parameter is the angle θ, measured in radians, through which the wheel has rotated. We need to express the coordinates of each point $P = P(x, y)$ on the curve in terms of the parameter θ. On the one hand, first note that the arc $\overset{\frown}{PQ}$ from P to Q subtended by θ has length $r\theta$ and that this length is precisely equal to the length of the horizontal segment $\overline{OQ}$. (To see this think of rolling the black circle back to its original position.) On the other hand, the length of $\overline{OQ}$ is also equal to x plus the length of segment $\overline{PS}$. Since $PS = r\sin\theta$ and $OQ = r\theta$, it follows that $x = r\theta - r\sin\theta$. Notice that we have now described the coordinate x in terms of the parameter θ (and the constant r).

Similarly, y plus the length of vertical segment $\overline{SC}$ is equal to the radius r. Since the length of SC is $r\cos\theta$, we have $y = r - r\cos\theta$. Therefore, we can express the cycloid with the parametric equations

$$x = r\theta - r\sin\theta, \quad y = r - r\cos\theta, \quad \theta \in \mathbb{R}.$$

The following graph shows the cycloid after two revolutions of the circle:

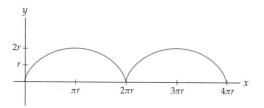

EXAMPLE 4

Finding tangent lines on the cycloid

Find the equations of the tangent lines to the cycloid $x = r\theta - r\sin\theta$, $y = r - r\cos\theta$ at $\theta = \frac{\pi}{2}$.

SOLUTION

To find the equations of the tangent lines, we must first determine the derivative $\frac{dy}{dx}$ in terms of the parameter θ. The parametric equations for the cycloid are

$$x = r\theta - r\sin\theta, \quad y = r - r\cos\theta, \quad \theta \in \mathbb{R}.$$

It is easy to find the derivatives of the coordinates x and y with respect to the parameter θ:

$$\frac{dx}{d\theta} = r - r\cos\theta, \quad \frac{dy}{d\theta} = r\sin\theta.$$

Now, using Definition 9.2, we have

$$\frac{dy}{dx} = \frac{dy/d\theta}{dx/d\theta} = \frac{r\sin\theta}{r - r\cos\theta} = \frac{\sin\theta}{1 - \cos\theta}.$$

This means that at $\theta = \frac{\pi}{2}$, the slope of the tangent line will be $\frac{\sin(\pi/2)}{1 - \cos(\pi/2)} = 1$. The point on the cycloid corresponding to $\theta = \frac{\pi}{2}$ is

$$(x, y) = \left(r\left(\frac{\pi}{2} - \sin\frac{\pi}{2}\right), r - r\cos\frac{\pi}{2}\right) = \left(r\left(\frac{\pi}{2} - 1\right), r\right).$$

Therefore, the equation of the tangent line at $\theta = \frac{\pi}{2}$ will be

$$y - r = 1\left(x - r\left(\frac{\pi}{2} - 1\right)\right), \quad \text{or equivalently,} \quad y = x + r\left(2 - \frac{\pi}{2}\right).$$

The following figure shows the tangent line from this example as well as the horizontal tangent line from Example 5:

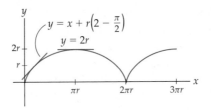

EXAMPLE 5

Horizontal and vertical tangent lines on the cycloid

Find all the points on the graph of the cycloid $x = r\theta - r\sin\theta$, $y = r - r\cos\theta$, $\theta \in \mathbb{R}$ at which the tangent line is either horizontal or vertical, and find the equations of the horizontal tangent lines.

SOLUTION

The graph of the cycloid will have a horizontal tangent line wherever the numerator of $\frac{dy}{dx} = \frac{\sin\theta}{1-\cos\theta}$ is zero and the denominator is nonzero. This occurs when θ is any odd multiple of π. Now we have to find the equations of these horizontal tangent lines. If θ is an odd multiple of π, then $\cos\theta = -1$ and therefore at such values of θ we are at the y-coordinate $y = r - r(-1) = 2r$. Thus the equation of the horizontal tangent line at this point is $y = 2r$, which should be obvious from looking at the graph of the cycloid. In addition, this horizontal line is tangent to the cycloid at every at point where the cycloid has a zero slope.

In the computations we just did, we found the locations and equations of the horizontal tangent lines on the graph of the cycloid. To find the vertical tangent lines, we must find the values of θ for which the denominator of $\frac{dy}{dx} = \frac{\sin\theta}{1-\cos\theta}$ is zero but the numerator is nonzero, and the values of α such that

$$\lim_{\theta \to \alpha} \frac{\sin\theta}{1-\cos\theta}$$

is undefined. In Exercise 73 you will show that the latter occurs at each even multiple of π.

EXAMPLE 6

Analyzing the concavity of a parametric curve

Let $x = x(t)$ and $y = y(t)$ be the parametric equations for a curve C. Show that if x and y are both twice-differentiable functions of t, then the concavity at a point on C is given by

$$\frac{d^2y}{dx^2} = \frac{\dfrac{dx}{dt}\dfrac{d^2y}{dt^2} - \dfrac{d^2x}{dt^2}\dfrac{dy}{dt}}{\left(\dfrac{dx}{dt}\right)^3}$$

when $\frac{dx}{dt} \neq 0$.

Use this result to confirm that the cycloid given by $x = r\theta - r\sin\theta$, $y = r - r\cos\theta$ is concave down everywhere except when θ is an even multiple of pi.

SOLUTION

We first note that by the chain rule we have

$$\frac{d^2x}{dt^2} = \frac{d}{dt}\left(\frac{dx}{dt}\right) = \frac{d}{dx}\left(\frac{dx}{dt}\right)\frac{dx}{dt}.$$

Therefore, when $\frac{dx}{dt} \neq 0$,

$$\frac{d}{dx}\left(\frac{dx}{dt}\right) = \frac{d^2x/dt^2}{dx/dt}.$$

Similarly, we may show that

$$\frac{d}{dx}\left(\frac{dy}{dt}\right) = \frac{d^2y/dt^2}{dx/dt}.$$

Now, since $\frac{dy}{dx} = \frac{dy/dt}{dx/dt}$, it follows that

$$\frac{d^2y}{dx^2} = \frac{d}{dx}\left(\frac{dy}{dx}\right) = \frac{d}{dx}\left(\frac{dy/dt}{dx/dt}\right)$$

$$= \frac{\frac{dx}{dt}\frac{d}{dx}\left(\frac{dy}{dt}\right) - \frac{d}{dx}\left(\frac{dx}{dt}\right)\frac{dy}{dt}}{\left(\frac{dx}{dt}\right)^2} \qquad \leftarrow \text{the quotient rule}$$

$$= \frac{\frac{dx}{dt}\frac{d^2y/dt^2}{dx/dt} - \frac{d^2x/dt^2}{dx/dt}\frac{dy}{dt}}{\left(\frac{dx}{dt}\right)^2} \qquad \leftarrow \text{the earlier equalities}$$

$$= \frac{\frac{dx}{dt}\frac{d^2y}{dt^2} - \frac{d^2x}{dt^2}\frac{dy}{dt}}{\left(\frac{dx}{dt}\right)^3}. \qquad \leftarrow \text{algebra}$$

For the cycloid, since $x = r\theta - r\sin\theta$ and $y = r - r\cos\theta$, we have

$$\frac{dx}{d\theta} = r - r\cos\theta, \quad \frac{d^2x}{d\theta^2} = r\sin\theta, \quad \frac{dy}{d\theta} = r\sin\theta, \quad \text{and} \quad \frac{d^2y}{d\theta^2} = r\cos\theta.$$

Combining these equations with the formula for $\frac{d^2y}{dx^2}$, we obtain

$$\frac{d^2y}{dx^2} = \frac{(r - r\cos\theta)(r\cos\theta) - (r\sin\theta)(r\sin\theta)}{(r - r\cos\theta)^3}$$

$$= \frac{r^2(\cos\theta - 1)}{r^3(1 - \cos\theta)^3} \qquad \leftarrow \text{algebra and } \sin^2\theta + \cos^2\theta = 1$$

$$= -\frac{1}{r(1 - \cos\theta)^2}. \qquad \leftarrow \text{algebra}$$

Since r is a positive constant, the quotient $-\frac{1}{r(1-\cos\theta)^2} < 0$ when θ is not a multiple of 2π. Therefore, the cycloid is concave down when θ is not an even multiple of π. □

EXAMPLE 7   Finding the arc length of a parametric curve

Find the arc length of one arch of the cycloid traced by a point on the circumference of a wheel with radius r units (i.e., the portion of the graph of the cycloid that represents one full rotation of the wheel).

SOLUTION

One such arch is given by the portion of the cycloid corresponding to $\theta \in [0, 2\pi]$, since one full rotation of the wheel corresponds to 2π radians. For $\theta \in [0, 2\pi]$, the desired arc length is given by

$$\int_0^{2\pi} \sqrt{(f'(\theta))^2 + (g'(\theta))^2}\, d\theta,$$

where $f(\theta) = r\theta - r\sin\theta$ and $g(\theta) = r - r\cos\theta$. Thus, the arc length is

$$\int_0^{2\pi} \sqrt{(r - r\cos\theta)^2 + (r\sin\theta)^2}\, d\theta = r\int_0^{2\pi} \sqrt{2 - 2\cos\theta}\, d\theta = 8r.$$

The details of the last step (in which the definite integral is calculated to be equal to $8r$) are left for Exercise 40. □

TEST YOUR UNDERSTANDING

▶ What is the definition of parametric equations?

▶ How do you graph parametric equations by plotting points? By eliminating the parameter?

▶ How do you find the derivative of a curve defined parametrically? How do you find the locations of any horizontal or vertical tangent lines on a parametric curve?

▶ The cycloid in Example 3 is the graph of some function of x—that is, $y = h(x)$—since the graph passes the vertical line test. Can you eliminate the parameter from the equations in Example 3 to find $h(x)$?

▶ How can a definite integral be used to calculate the arc length of a parametric curve? How does this definite integral arise from a limit of approximations arrived at with the use of line segments?

EXERCISES 9.1

Thinking Back

Eliminating a variable: Find a relationship between variables x and y by eliminating the variable t.

▶ $x = t^2,\ y = t^6$
▶ $x = \sin t,\ y = \sin t$
▶ $x = \sin t,\ y = \cos t$
▶ $x = \sinh t,\ y = \cosh t$

Arc length: Find the arc length of the following curves on the given intervals.

▶ $y = 3x - 4,\ x \in [0, 5]$
▶ $y = \sqrt{4 - x^2},\ x \in [0, 2]$
▶ $y = x^{3/2},\ x \in [0, 4]$
▶ $y = x^2,\ x \in [0, 1]$

Concepts

0. *Problem Zero:* Read the section and make your own summary of the material.

1. *True/False:* Determine whether each of the statements that follow is true or false. If a statement is true, explain why. If a statement is false, provide a counterexample.

(a) *True or False:* Every function $y = f(x)$ can be written in terms of parametric equations.

(b) *True or False:* Given parametric equations $x = x(t)$ and $y = y(t)$, the parameter can be eliminated to obtain the form $y = f(x)$.

(c) *True or False:* Every parametric curve passes the vertical line test.

(d) *True or False:* Every curve in the plane has a unique expression in terms of parametric equations.

(e) *True or False:* If the functions $x = f(t)$ and $y = g(t)$ are differentiable for every $t \in \mathbb{R}$, then the parametric curve defined by x and y is differentiable for every value of t.

(f) *True or False:* A curve parametrized by $x = x(t)$, $y = y(t)$ has a horizontal tangent line at $(x(t_0), y(t_0))$ if $y'(t) = 0$.

(g) *True or False:* A curve parametrized by $x = x(t)$, $y = y(t)$ has a horizontal tangent line at $(x(t_0), y(t_0))$ if $x'(t) \neq 0$ and $y'(t) = 0$.

(h) *True or False:* The cycloid curve associated with a circle of radius r is made up of a series of semicircles of radius r.

2. *Examples:* Construct examples of the thing(s) described in the following. Try to find examples that are different than any in the reading.

(a) Parametric equations $x = f(t)$, $y = g(t)$ on the interval $[0, 1)$ that trace the unit circle exactly once clockwise, starting at the point $(1, 0)$.

(b) Parametric equations $x = f(t)$, $y = g(t)$ on the interval $[0, 2\pi)$ that trace the circle centered at $(2, -3)$ with radius 5 exactly once counterclockwise, starting at the point $(7, -3)$.

(c) Parametric equations $x = f(t)$, $y = g(t)$ whose graph is not the graph of a function $y = f(x)$.

3. If $x = x(t)$ and $y = y(t)$ are differentiable functions of t, determine the direction of motion along the curve when

(a) $x'(t) > 0$ and $y'(t) > 0$.

(b) $x'(t) > 0$ and $y'(t) < 0$.

(c) $x'(t) < 0$ and $y'(t) > 0$.

(d) $x'(t) < 0$ and $y'(t) < 0$.

4. Complete the following definition: ***Parametric equations*** are _____ .

Use the results of Exercise 3 to analyze the direction of motion for the parametric curves given by the equations in Exercises 5–8.

5. $x = t^2$, $y = t^3$, $t \in \mathbb{R}$

6. $x = \sin t$, $y = \cos t$, $t \in \mathbb{R}$

7. $x = e^t$, $y = \ln t$, $t > 0$

8. $x = t^3 - t$, $y = t^3 + t$, $t \in \mathbb{R}$

In Exercises 9–11 parametrizations are provided for portions of the same function. For each problem do the following:

(i) Eliminate the parameter to show that the curves are portions of the same function.

(ii) Describe the portion of the graph that each parametrization describes.

(iii) Discuss the direction of motion along the graph for each parametrization.

9. (a) $x = t$, $y = t^2 - 1$, $t \geq 0$

   (b) $x = -t$, $y = t^2 - 1$, $t \geq 0$

10. (a) $x = t^2$, $y = t^3$, $t \geq 0$

    (b) $x = t^2$, $y = t^3$, $t \leq 0$

11. (a) $x = t$, $y = \sin t$, $t \geq 0$

    (b) $x = t - 1$, $y = \sin(t - 1)$, $t \geq 1$

12. Suppose a parametric curve is given by parametric equations $x = x(t), y = y(t)$ for t in some interval I. How can we find the slope of the parametric curve at some point $(x(t_0), y(t_0))$? What is the equation of the tangent line to the parametric curve at the point t_0?

13. Explain how we can find the locations at which a parametric curve determined by $x = x(t)$ and $y = y(t)$ has horizontal or vertical tangent lines.

14. Show that the parametrization $x = 2t + 1$, $y = 4t^2 - 4$ for $t \in [-1, \infty)$ has the same graph as the one we plotted point by point in the reading.

15. Explain why the parametrization $x = \sin t + 1$, $y = \sin^2 t - 4$ for $t \in (-\infty, \infty)$ repeatedly traces the same small portion of the graph of the function $y = x^2 - 2x - 3$.

Skills

In Exercises 16–23 sketch the parametric curve by plotting points.

16. $x = t$, $y = t^2$, $t \in \mathbb{R}$

17. $x = 3t + 1$, $y = t$, $t \in [-2, 5]$

18. $x = 3t + 1$, $y = 2t$, $t \in [0, 8]$

19. $x = 1 - 2t$, $y = 5 - 3t$, $t \in \mathbb{R}$

20. $x = 2t - 1$, $y = 3t + 5$, $t \in \mathbb{R}$

21. $x = t^3 - t$, $y = t^3 + t$, $t \in \mathbb{R}$

22. $x = 2\sin^3 t$, $y = 2\cos^3 t$, $t \in [0, 2\pi]$

23. $x = \cos^5 t$, $y = \sin^5 t$, $t \in [0, 2\pi]$

In Exercises 24–34 sketch the parametric curve by eliminating the parameter.

24. $x = 2t - 1$, $y = 3t + 5$, $t \in \mathbb{R}$

25. $x = 2t - 1$, $y = 3t^2 + 5$, $t \in \mathbb{R}$

26. $x = t + 2$, $y = e^t$, $t \in \mathbb{R}$

27. $x = \tan t$, $y = \tan t$, $t \in \left(-\dfrac{\pi}{2}, \dfrac{\pi}{2}\right)$

28. $x = \cos 2t$, $y = -\sin 2t$, $t \in [0, 2\pi]$

29. $x = 3\cos t$, $y = 4\sin t$, $t \in [0, 2\pi]$

30. $x = \sin t$, $y = \cos 2t$, $t \in [0, 2\pi]$

31. $x = \cosh t$, $y = \sinh t$, $t \in \mathbb{R}$

32. $x = \sec t$, $y = \tan t$, $t \in \left(-\dfrac{\pi}{2}, \dfrac{\pi}{2}\right)$

33. $x = \csc t$, $y = \cot t$, $t \in (0, \pi)$

34. $x = \log_{10} t$, $y = \ln t$, $t \in (0, \infty)$

35. Finish Example 1 (b) by showing that the graph of the parametric equations $x = \cos t$, $y = \sin t$, $t \in [\pi, 2\pi]$ is the bottom half of the unit circle centered at the origin with a counterclockwise direction of motion.

In Exercises 36–39 provide a parametrization with the given properties

36. The curve is a circle centered at the origin. It is traced once, clockwise, starting at the point $(0, 1)$ with $t \in [0, 2\pi]$.

37. The curve is a circle centered at the origin. It is traced once, counterclockwise, starting at the point $(0, 3)$ with $t \in [0, 1]$.

38. The curve is a circle centered at the point (a, b). It is traced once, counterclockwise, starting at the point $(a+r, b)$ with $t \in [0, 2\pi]$.

39. The curve is a circle centered at the origin. It is traced once, counterclockwise, and contains all points of the unit circle except for $(0, -1)$ with $t \in \mathbb{R}$.

40. Complete the calculation in Example 7 by using the trigonometric identity $\sin^2\left(\dfrac{\theta}{2}\right) = \dfrac{1}{2}(1 - \cos\theta)$ to show that $\int_0^{2\pi} \sqrt{1 - \cos\theta}\, d\theta = 4\sqrt{2}$.

In Exercises 41–44 find an equation for the line tangent to the parametric curve at the given value ot t.

41. $x = 2t - 1$, $y = 3t + 5$, $t = -1$.

42. $x = t + 2$, $y = e^t$, $t = 0$.

43. $x = t^2$, $y = (2 - t)^2$, $t = \dfrac{1}{2}$.

44. $x = \cos^3 t$, $y = \sin^3 t$, $t = \dfrac{\pi}{4}$.

In Exercises 45–48 use Example 6 to find $\dfrac{d^2y}{dx^2}$ for the parametric curve at the given value of t. Note that these are the same parametric equations as in Exercises 41–44.

45. $x = 2t - 1$, $y = 3t + 5$, $t = -1$.

46. $x = t + 2$, $y = e^t$, $t = 0$.

47. $x = t^2$, $y = (2-t)^2$, $t = \dfrac{1}{2}$.

48. $x = \cos^3 t$, $y = \sin^3 t$, $t = \dfrac{\pi}{4}$.

In Exercises 49–53 sketch the parametric curve and find its length.

49. $x = 1 + t^2$, $y = 3 + 2t^3$, $t \in [0, 1]$

50. $x = \cos\theta + \theta \sin\theta$, $y = \sin\theta - \theta \cos\theta$, $\theta \in [0, 2\pi]$

51. $x = \sin^3\theta$, $y = \cos^3\theta$, $\theta \in [0, 2\pi]$

52. $x = e^t \cos t$, $y = e^t \sin t$, $t \in [0, 1]$

53. $x = 5 + 2t$, $y = e^t + e^{-t}$, $t \in [0, 1]$

54. Show that the graph of the parametric equations

$$x = a + (c-a)t, \quad y = b + (d-b)t, \quad t \in [0, 1]$$

is a line segment from (a, b) to (c, d).

In Exercises 55–60 use the result of Exercise 54 to find parametric equations for the line segments connecting the given pairs of points in the direction indicated.

55. From $(1, -3)$ to $(6, 7)$

56. From $(6, 7)$ to $(1, -3)$

57. From $(1, 4)$ to $(-3, 5)$

58. From $(-3, 5)$ to $(1, 4)$

59. From $(\pi, 3)$ to $(\pi, 8)$

60. From $(0, e)$ to $(-6, e)$

61. Use the arc length formula for parametric equations and your answer to Exercise 55 to find the distance between the points $(1, -3)$ and $(6, 7)$. Verify your answer by using the distance formula to compute the distance.

62. A **trochoid** is a generalization of a cycloid in which the point tracing the path is on a spoke of the wheel, instead of on the circumference of the wheel. Thus, if the radius of the wheel is r, the point is k units from the center of the wheel, such that either $k < r$ or $k > r$. (When $k > r$, you can think of the point being on a flange extending the radius of the wheel. This case occurs as train wheels roll, since there is an extension of each wheel beyond the portion of the wheel rolling on the track.) Find parametric equations for the trochoid.

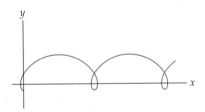

63. An **epicycloid** is another variation of a cycloid in which the point tracing the path is on the circumference of a wheel, but the wheel is rolling without slipping on the outside of another wheel, instead of along a horizontal track. If the radius of the rolling wheel is k and the radius of the fixed wheel is r, find parametric equations for the epicycloid.

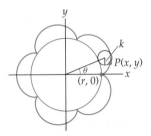

64. The **involute** of a circle is the curve described by the endpoint P of a thread as it unwinds from a fixed circular spool. For simplicity suppose that the radius of the circular spool is r and that when the spool is placed with its center at the origin, the point P starts at $(r, 0)$. Assume that the thread is unwinding counterclockwise. Find parametric equations for the point P. (*Hint: If the string is taut at all times as it unwinds, the length of segment PT is $r\theta$.*)

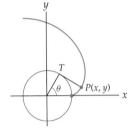

Applications

65. (a) Find an integral that represents the length of an elliptical track whose equations are given by the parametric equations $x = \sin\theta$, $y = 3\cos\theta$, $\theta \in [0, 2\pi]$, where x and y are in kilometers.

(b) Approximate the length of the track, using the midpoint method with 20 subintervals.

66. Annie needs to make a crossing in her kayak from an island to a north–south coastline 3 miles due east. It is foggy, so she cannot see any landmarks to steer by. Instead, she takes a compass heading due east and sticks to it all the way across. Tidal currents in the channel push her boat southward at a speed proportional to the distance from either shoreline. She paddles at 2 miles per hour from west to east. That is, her east–west position changes with time as $x'(t) = 2$. Her north–south position changes as $y'(t) = 1.778t^2 - 2.667t$. Her starting position was $(0, 0)$, at time $t = 0$.

(a) Solve the two given differential equations by integrating with respect to t to find a parametric description of Annie's path across the channel.

(b) How far south of her starting point does Annie make landfall?

(c) Use a numerical integration technique to determine the distance she paddles.

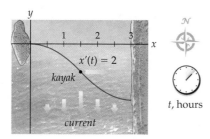

67. Annie has to make the same crossing as in Exercise 66 two weeks later, again in a fog. Remembering how far off course she was pushed previously, she takes a compass heading 15 degrees north of east this time. Thus, her eastward speed is $x'(t) = 2\cos\frac{\pi}{12}$ and her north–south velocity is $y'(t) = 0.444x(t)(x(t) - 3) + 2\sin\frac{\pi}{12}$.

   (a) Solve the two given differential equations by integrating with respect to t to find a parametrization of Annie's path across the channel.
   (b) How far south of her starting point does Annie make landfall this time?
   (c) How far does she paddle? How long does she paddle?

Proofs

68. Show that every function $y = f(x)$ can be written as parametric equations.

69. Use your result from Exercise 68 to show that the arc length formula for a function $y = f(x)$ is a special case of the arc length formula for a parametric curve.

70. Let c and d be constants, and for $t \in [a, b]$ let $f(t)$ and $g(t)$ be differentiable functions with continuous first derivatives. Prove that the arc length of the parametric curve given by $x = f(t)$, $y(t) = g(t)$ for $t \in [a, b]$ is equal to the arc length of the parametric curve defined by $x = c + f(t)$, $y(t) = d + g(t)$ for $t \in [a, b]$ for every c and d in $\mathbb{R}$.

71. Let $k > 0$ be a constant, and let $f(t)$ and $g(t)$ be differentiable functions of t with continuous first derivatives for every $t \in [a, b]$. Prove that the arc length of the curve defined by the parametric equations

$$x = kf(t), \quad y = kg(t), \quad t \in [a, b]$$

is k times as long as the arc length of the curve defined by the parametric equations

$$x = f(t), \quad y = g(t), \quad t \in [a, b].$$

What is the arc length of the curve defined by the equations

$$x = f(kt), \quad y = g(kt), \quad t \in [a/k, b/k]?$$

72. Show that the parametric curve associated with the *linear* parametric equations

$$x = m_1 t + b_1, \quad y = m_2 t + b_2, \quad t \in \mathbb{R}$$

is a line if m_1 and m_2 are not both zero. For what values of m_1 and m_2 will y be a function of x? In this case, what is the slope of the line? What do the equations describe if m_1 and m_2 are both zero? Find parametric equations $x = f(t)$, $y = g(t)$ such that neither f nor g is a linear function, but the parametric curve associated with the equations is a line.

73. In Example 5 we saw that the cycloid

$$x = r\theta - r\sin\theta, \quad y = r - r\cos\theta, \quad \theta \in \mathbb{R}$$

has a horizontal tangent line at each odd multiple of π. Show that the cycloid has a vertical tangent at each even multiple of π by showing that $\lim_{\theta \to 2k\pi} \frac{dy}{dx}$ does not exist wherever k is an integer.

Thinking Forward

Parametric equations in three dimensions: Parametric equations can also be used to describe curves in three dimensions. Try to visualize the graphs described by the following parametric equations.

▶ A line through the origin:

$$x = t, \quad y = -3t, \quad z = 2t, \quad t \in \mathbb{R}$$

▶ A helix centered on the z-axis:

$$x = \cos t, \quad y = \sin t, \quad z = t, \quad t \in \mathbb{R}$$

▶ An expanding helix:

$$x = t\cos t, \quad y = t\sin t, \quad z = t, \quad t \in [0, \infty)$$

9.2 POLAR COORDINATES

▶ Polar coordinates are defined

▶ Conversion formulas between rectangular and polar coordinates are derived

▶ Principles of graphing with polar coordinates are introduced

Plotting Points in Polar Coordinates

We begin by specifying a point called the **pole** and a ray emanating from the pole called the **polar axis**, as follows:

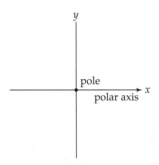

In the preceding figure you will notice that we've placed the pole at the origin of a rectangular coordinate system and the polar axis along the positive x-axis. This is the convention we always use with polar coordinates. As we will see later in the section, this convention will also allow us to derive formulas for converting between polar coordinates and rectangular coordinates.

In a polar system we use the coordinate pair (r, θ) to represent a point in the plane. The first coordinate r is the signed distance from the pole, while θ is an angular measure, in radians, from the polar axis. When $\theta > 0$, we rotate counterclockwise from the polar axis, and when $\theta < 0$, we rotate clockwise. When $r > 0$, we measure r units along the ray specified by the value of θ. When $r < 0$, we measure $|r|$ units in the opposite direction, which is equivalent to moving in the positive direction along the ray $\theta + \pi$, as shown here:

For $r > 0$ we move along the ray θ;
for $r < 0$ we move along $\theta + \pi$

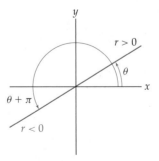

In the rectangular coordinate system, once scales on the axes are specified, there is a one-to-one correspondence between points in the plane and coordinates that name points. This relationship does not hold in the polar coordinate system. We have the following theorem:

THEOREM 9.5	**Equivalent Polar Coordinates**

(a) The polar coordinates $(r, \theta + 2\pi k)$ represent the same point for every integer k.

(b) The polar coordinates $(-r, \theta + \pi)$ represent the same point as (r, θ) for any value of θ.

(c) The polar coordinates $(0, \theta)$ represent the pole for any value of θ.

Converting Between Polar and Rectangular Coordinates

To convert from polar coordinates to rectangular coordinates, we will write the coordinates x and y as functions of the coordinates r and θ. Then, given r and θ, we will easily be able to find the corresponding x and y. The figure that follows shows a point in the first quadrant of the plane and its geometric representation in terms of both rectangular and polar coordinates. In the other three quadrants we can draw similar figures and reach the same conclusions as we do shortly; however, for simplicity we will use the first-quadrant picture as our general example.

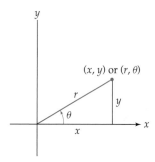

Using the triangle shown, we see that $\cos\theta = \frac{x}{r}$ and $\sin\theta = \frac{y}{r}$. This gives us the following theorem:

THEOREM 9.6	**Converting from Polar to Rectangular Coordinates**

If a point in the plane is represented by (r, θ) in polar coordinates, then the rectangular coordinates of the point are given by (x, y), where

$$x = r\cos\theta \quad \text{and} \quad y = r\sin\theta.$$

Note that given the polar coordinates (r, θ) of any point in the plane, there will be a unique pair of rectangular coordinates (x, y) representing the point.

To convert from rectangular coordinates to polar coordinates, we need to have a way of finding r and θ given any coordinates x and y. Applying the Pythagorean theorem and the definition of the tangent function to the triangle in the figure shown, we have the following relationships:

THEOREM 9.7	**Converting from Rectangular to Polar Coordinates**

If a point in the plane is represented by (x, y) in rectangular coordinates, then the polar coordinates (r, θ) of the point satisfy the following formulas:

$$r^2 = x^2 + y^2 \quad \text{and} \quad \tan\theta = \frac{y}{x}.$$

Notice that the formulas in Theorem 9.7 do not give unique values of r and θ for given values of x and y. This is to be expected, since every point in the plane has multiple polar coordinate representations. For example, the rectangular coordinates $(1, 1)$ may be expressed as $\left(\sqrt{2}, \frac{\pi}{4} + 2k\pi\right)$ or $\left(-\sqrt{2}, \frac{3\pi}{4} + 2k\pi\right)$, for any positive integer k.

The Graphs of Some Simple Polar Coordinate Equations

In rectangular coordinates, the simplest possible equations are those of the form $x = a$ and $y = b$, where a and b are constants. Appropriately enough, the graphs of these equations are very simple as well: vertical and horizontal lines, respectively. In polar coordinates we first consider the equations $r = c$ and $\theta = \alpha$, where c and α are constants. For any real number c, the polar equation $r = c$ describes the set of points $|c|$ units from the pole. Therefore, the graph of $r = c$ is a circle with radius $|c|$ centered at the pole, as shown next at the left. For any real number α, the polar equation $\theta = \alpha$ describes the set of points with polar angle α. At first you might think that this would give a graph of the ray with polar angle θ. However, remember that when $\theta = \alpha$, the value of r can be either positive or negative. Therefore, the graph of $\theta = \alpha$ is a line through the pole, as shown here at the right:

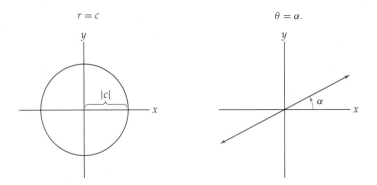

Two other simple categories of polar equations are those defined by

$$r = 2a\cos\theta \text{ and } r = 2a\sin\theta,$$

where $a \neq 0$. The following theorem tells us that these are the equations of circles passing through the pole and tangent to one of the coordinate axes.

THEOREM 9.8

Circles Tangent to the Coordinate Axes at the Pole

For $a \neq 0$, the graphs of the equations

$$r = 2a\cos\theta \text{ and } r = 2a\sin\theta$$

are the circles whose equations in rectangular coordinates are

$$(x - a)^2 + y^2 = a^2 \text{ and } x^2 + (y - a)^2 = a^2,$$

respectively.

Proof. We prove that $(x - a)^2 + y^2 = a^2$ is another equation for $r = 2a\cos\theta$ and leave the other case for Exercise 68. We multiply both sides of $r = 2a\cos\theta$ by r and use the facts that $r^2 = x^2 + y^2$ and $x = r\cos\theta$ to obtain

$$x^2 + y^2 = 2ax.$$

If we subtract $2ax$ from each side of the equation and then complete the square, we get

$$(x-a)^2 + y^2 = a^2,$$

as required. The graph of this equation is the circle with radius a and centered at the point $(a, 0)$. The circle is tangent to the y-axis and passes through the pole.   ■

Examples and Explorations

EXAMPLE 1

Plotting points with polar coordinates

Plot the points that have polar coordinates $\left(2, \frac{\pi}{3}\right)$ and $\left(-1, -\frac{\pi}{2}\right)$.

SOLUTION

To plot $\left(2, \frac{\pi}{3}\right)$, we first locate the ray $\theta = \frac{\pi}{3}$ and move 2 units (in the positive direction) along this ray, as shown next at the left. Similarly, to plot the point $\left(-1, -\frac{\pi}{2}\right)$, we first find the ray $\theta = -\frac{\pi}{2}$ and move 1 unit in the negative direction along that ray. Note that this operation is equivalent to moving one unit in the positive direction along the ray $\theta = -\frac{\pi}{2} + \pi = \frac{\pi}{2}$, as shown here at the right:

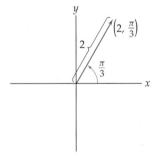 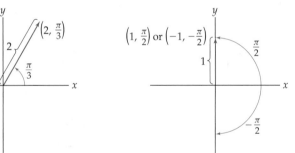

□

EXAMPLE 2

Converting from polar to rectangular coordinates

Find the rectangular coordinates of the points that have polar coordinates $\left(2, \frac{\pi}{3}\right)$ and $\left(-1, -\frac{\pi}{2}\right)$.

SOLUTION

We use the conversion formulas given in Theorem 9.6. For the first of the two pairs of coordinates we have

$$x = 2\cos\frac{\pi}{3} \text{ and } y = 2\sin\frac{\pi}{3}.$$

The rectangular coordinates are $(1, \sqrt{3})$.

The second coordinate pair converts to

$$x = -1\cos\left(-\frac{\pi}{2}\right) = 0 \text{ and } y = -1\sin\left(-\frac{\pi}{2}\right) = 1.$$

Compare these results with those you obtained in Example 1.   □

EXAMPLE 3

Finding all the polar coordinate representations for a point

Find all of the pairs of polar coordinates for the point with rectangular coordinates $(1, \sqrt{3})$.

SOLUTION

We use the conversion formulas given in Theorem 9.7 along with the fact that the point $(1, \sqrt{3})$ lies in the first quadrant. We have $r^2 = 1^2 + (\sqrt{3})^2 = 4$ and $\tan\theta = \frac{\sqrt{3}}{1} = \sqrt{3}$. Thus, either $r = 2$ and $\theta = \frac{\pi}{3} + 2\pi k$, where $k \in \mathbb{Z}$, or $r = -2$ and $\theta = \frac{4\pi}{3} + 2\pi k$, where $k \in \mathbb{Z}$. Note that although the angles $\theta = \frac{4\pi}{3} + 2\pi k$ are in the third quadrant for each k, the fact that $r < 0$ still gives us a point in the first quadrant.

This example builds on Example 2. When we converted the point $\left(2, \frac{\pi}{3}\right)$ to rectangular coordinates, we got the unique answer of that example. However, since every point in the polar plane has infinitely many names, when we convert back to polar coordinates we expect infinitely many answers.   ☐

EXAMPLE 4

Graphing polar equations by converting to rectangular coordinates

Graph the equations $r = 4\cos\theta$ and $r = 5\csc\theta$.

SOLUTION

We will begin with the polar equation $r = 4\cos\theta$. From Theorem 9.8 we know that this is the equation for the circle whose equation in rectangular coordinates is $(x - 2)^2 + y^2 = 4$, as shown next at the left.

Now consider the polar equation $r = 5\csc\theta$. We begin by multiplying both sides of the equation by $\sin\theta$ to obtain

$$r\sin\theta = 5.$$

By Theorem 9.6 we have $r\sin\theta = y$; combining this equation with the preceding one, we must have $y = 5$. Therefore, the graph of $r = 5\csc\theta$ is a horizontal line at height 5, as shown here at the right:

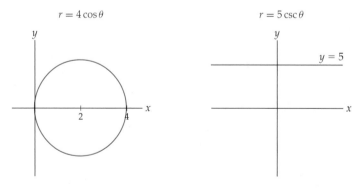

$r = 4\cos\theta$ $r = 5\csc\theta$

☐

⊗ CAUTION

We graphed the polar equation $r = 5\csc\theta$ in Example 4 by transforming it to rectangular coordinates. Using Theorems 9.6 and 9.7, we can always transform an equation in polar coordinates to one in rectangular coordinates. However, as the next example demonstrates, we *cannot* expect that every polar equation will have a simple expression in rectangular coordinates. Therefore, we may not understand the polar equation any better by transforming it to rectangular coordinates.

EXAMPLE 5

Converting a polar equation to an equation in rectangular coordinates

Convert the polar equation $r = \cos 2\theta$ to rectangular coordinates.

SOLUTION

We begin by using the double-angle identity for cosine, $\cos 2\theta = \cos^2 \theta - \sin^2 \theta$. We now have $r = \cos^2 \theta - \sin^2 \theta$. Since $\cos \theta = \frac{x}{r}$ and $\sin \theta = \frac{y}{r}$, we have

$$r = \frac{x^2}{r^2} - \frac{y^2}{r^2},$$

or equivalently,

$$r^3 = x^2 - y^2.$$

Finally, because $r = \pm(x^2 + y^2)^{1/2}$, we obtain the equation

$$\pm(x^2 + y^2)^{3/2} = x^2 - y^2.$$

We could now use the techniques of Section 2.4 to graph $\pm(x^2 + y^2)^{3/2} = x^2 - y^2$. However, that would be making the problem of graphing the relatively simple polar equation $r = \cos 2\theta$ a complicated mess involving implicit functions. In the next section we will be discussing methods for graphing polar equations like $r = \cos 2\theta$ directly.   □

TEST YOUR UNDERSTANDING

▸ What are the formulas for converting between rectangular coordinates and polar coordinates? How are these formulas derived?

▸ Why does every point in the plane have a unique pair of rectangular coordinates that represents the point, but have infinitely many pairs of polar coordinates that represent it?

▸ Why are the polar coordinates $\left(8, -\frac{\pi}{6}\right)$ *not* a representation of the point with rectangular coordinates $(-4, 4\sqrt{3})$, even though these values of r, θ, x, and y satisfy the formulas $r^2 = x^2 + y^2$ and $\tan \theta = \frac{y}{x}$?

▸ Why is the graph of the equation $r = 2a \cos \theta$ a circle in the polar plane?

▸ Why is the graph of the equation $r = b \sec \theta$ a line in the polar plane?

EXERCISES 9.2

Thinking Back

Plotting in a rectangular coordinate system:

▸ Plot the points $(1, 1)$, $(1, 2)$, $(1, 3)$, and $(1, 4)$ in a rectangular coordinate system. Fill in the blank: These points all lie on the same _____ line.

▸ Plot the points $(-1, -2)$, $(0, -2)$, $(1, -2)$, and $(2, -2)$ in a rectangular coordinate system. Fill in the blank: These points all lie on the same _____ line.

▸ Plot the equations $x = -3$ and $y = 5$ in a rectangular coordinate system. Where do the graphs of these two equations intersect?

Completing the square:

▸ Consider the equation $x^2 + y^2 = 4x$. Use the method of completing the square to show that this equation is equivalent to $(x - 2)^2 + y^2 = 4$. (This calculation was used in Example 4.)

▸ Find the center and radius of the circle with equation

$$x^2 + y^2 + 6x - 8y - 44 = 0$$

by completing the square.

Concepts

0. *Problem Zero:* Read the section and make your own summary of the material.

1. *True/False:* Determine whether each of the statements that follow is true or false. If a statement is true, explain why. If a statement is false, provide a counterexample.

(a) *True or False:* Each point in the plane has a unique representation in rectangular coordinates.

(b) *True or False:* Each point in the plane has a unique representation in polar coordinates.

(c) *True or False:* If $b \neq c$, then the graphs of the polar equations $r = b$ and $r = c$ are different.

(d) *True or False:* If $\alpha \neq \beta$, then the graphs of the polar equations $\theta = \alpha$ and $\theta = \beta$ are different.

(e) *True or False:* The graph of $r = \csc\theta$ for $-\frac{\pi}{2} < \theta < \frac{\pi}{2}$ is a horizontal line in a polar coordinate system.

(f) *True or False:* In a polar coordinate system, the coordinates (r, θ) and $(r, \theta + \pi)$ represent the same point if and only if $r = 0$.

(g) *True or False:* When A and B are nonzero constants, the graph of $r = A\sin\theta + B\cos\theta$ is a circle in a polar coordinate system.

(h) *True or False:* Every function $r = f(\theta)$ in a polar coordinate system can be expressed in terms of rectangular coordinates x and y.

2. *Examples:* Construct examples of the thing(s) described in the following. Try to find examples that are different than any in the reading.

(a) Two pairs of polar coordinates for the point $(0, 3)$ given in rectangular coordinates.

(b) Two equations for the line $y = x$ in polar coordinates.

(c) The equations of two distinct circles with radius 2 tangent to the the x-axis at the pole in polar coordinates.

3. Explain why every point in the polar coordinate plane has infinitely many different polar coordinate representations.

4. Consider the point in the plane given by polar coordinates $(r, \theta) = \left(2, \frac{\pi}{3}\right)$.

(a) Express this point in polar coordinates where $r = 2$, but $\theta \neq \frac{\pi}{3}$, if possible.

(b) Express this point in polar coordinates where $\theta = \frac{\pi}{3}$, but $r \neq 2$, if possible.

(c) Express this point in polar coordinates where $r \neq 2$ and $\theta \neq \frac{\pi}{3}$, if possible.

5. Explain why the point $(r, \theta) = \left(8, -\frac{\pi}{3}\right)$ is not a polar representation of the point with rectangular coordinates $(x, y) = (-4, 4\sqrt{3})$, even though these values of r, θ, x, and y satisfy the formulas $r^2 = x^2 + y^2$ and $\tan\theta = \frac{y}{x}$. Include a picture with your explanation.

6. Explain why the graphs of $r = 3$ and $r = -3$ are identical in a polar coordinate system.

7. Find all values of c such that the graphs of $r = c$ and $r = -c$ are the same in a polar coordinate system.

8. Explain why the graphs of $\theta = \frac{\pi}{2}$ and $\theta = -\frac{\pi}{2}$ are identical in a polar coordinate system.

9. Find all values of α such that the graphs of $\theta = \alpha$ and $\theta = -\alpha$ are the same in a polar coordinate system.

10. In Example 5 we converted the simple polar equation $r = \cos 2\theta$ into the messy rectangular equation $\pm(x^2 + y^2)^{3/2} = x^2 - y^2$. Does this rectangular equation have a simpler rectangular coordinate form? If so, what is it? If not, why not?

11. Explain why the inequalities $r > 0$ and $0 < \theta < \frac{\pi}{2}$ together describe the points in the first quadrant. Use similar inequalities to describe the points in the third quadrant.

12. Explain why the inequality $0 \leq r \leq 2$ describes the points inside or on the circle with radius 2 and centered at the origin. Use a similar inequality to describe the points in the **annulus** shown here:

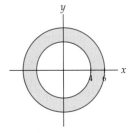

13. Find all polar coordinates that represent the point $(1, 0)$ given in rectangular coordinates.

14. Find all polar coordinates that represent the point $(0, 1)$ given in rectangular coordinates.

15. Find all polar coordinates that represent the point $(-1, 0)$ that is also given in polar coordinates.

16. Find all values of a and b such that (a, b) represents the same point whether it is given in rectangular coordinates or polar coordinates.

Skills

In Exercises 17–23 the polar coordinates for several sets of points are given. Find the rectangular coordinates for each of the points, and then plot and label the points in the same polar coordinate system.

17. $\left(3, \frac{\pi}{6}\right)$, $\left(-3, \frac{\pi}{6}\right)$, $\left(3, -\frac{\pi}{6}\right)$ and $\left(-3, -\frac{\pi}{6}\right)$

18. $(1, 0)$, $(2, 0)$, $(3, 0)$, and $(4, 0)$

19. $\left(5, \frac{\pi}{2}\right)$, $\left(-5, -\frac{3\pi}{2}\right)$, $\left(5, \frac{5\pi}{2}\right)$ and $\left(-5, -\frac{\pi}{2}\right)$

20. $\left(0, \frac{\pi}{6}\right)$, $\left(0, \frac{\pi}{3}\right)$, $(0, \pi)$ and $(0, -\pi)$

21. $\left(1, \frac{\pi}{4}\right)$, $\left(2, \frac{\pi}{4}\right)$, $\left(3, \frac{\pi}{4}\right)$ and $\left(4, \frac{\pi}{4}\right)$

22. $(2, 0)$, $\left(2, \frac{\pi}{4}\right)$, $\left(2, \frac{\pi}{2}\right)$, $(2, \pi)$ and $\left(2, \frac{3\pi}{2}\right)$

23. $\left(1, -\frac{\pi}{2}\right)$, $\left(2, -\frac{\pi}{2}\right)$, $\left(3, -\frac{\pi}{2}\right)$ and $\left(4, -\frac{\pi}{2}\right)$

In Exercises 24–31 find all polar coordinate representations for the point given in rectangular coordinates.

24. $(0, 0)$

25. $(1, 0)$

26. $(0, -3)$

27. $(0, 2)$

28. $(3, 4)$

29. $(6, 2\sqrt{3})$

30. $(-2, 0)$

31. $(3, -3)$

In Exercises 32–47 convert the equations given in polar coordinates to rectangular coordinates.

32. $\theta = \pi$

33. $\theta = \dfrac{\pi}{4}$

34. $\theta = -\dfrac{\pi}{6}$

35. $\theta = \dfrac{7\pi}{6}$

36. $r = 2\cos\theta$

37. $r = 5\sin\theta$

38. $r = -3\sec\theta$

39. $r = 6\csc\theta$

40. $r = \tan\theta$

41. $r = \sin 2\theta$

42. $r^2 = \sin\theta$

43. $r^2 = \cos\theta$

44. $r = \sin 4\theta$

45. $r = \theta$

46. $r = \cos 4\theta$

47. $r = \sin^3\theta$

In Exercises 48–55 convert the equations given in rectangular coordinates to equations in polar coordinates.

48. $x = 0$

49. $y = 0$

50. $x = 4$

51. $y = x$

52. $y = \sqrt{3}x$

53. $y = -3$

54. $y = x + 1$

55. $y = mx$

In Exercises 56–59 convert the equations given in polar coordinates to equations in rectangular coordinates.

56. $r = 1$, $r = -2$, $r = 3$.

57. $r = k$, for each positive integer k less than 10.

58. $\theta = 0$, $\theta = \dfrac{\pi}{4}$, $\theta = -\dfrac{\pi}{2}$.

59. $\theta = \dfrac{k\pi}{6}$, for each positive integer k less than 12.

Applications

In Exercises 60 and 61 Ian is a rock climber. Rock protection these days relies on a process called camming. The idea for camming devices is that they push harder against the rock when they are loaded with the force of a fall.

60. Ian has one cam that is 1 inch wide, with a point where the rope attaches on its right corner, so that the device pivots on its left corner. In other words, the left corner of the cam becomes the center of a circle, and the point where the rope attaches follows the curve $r = 1$.

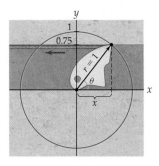

(a) If a crack is 0.75 inch wide, at what angle must Ian turn the device in order to put it into the crack? *(Hint: compute the x-coordinate of the right edge of the device.)*

(b) If Ian falls, the rope will pull on the right edge of the cam while the left edge remains wedged in a fixed position. What happens to the width of the cam in the crack?

61. Ian has another camming device, one that has two arms extending from a point where the rope attaches in the center.

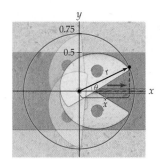

The outside edge of each arm follows the curve $r = 0.75$. The ends of the arms make an angle $\pm\theta$ with the rope, and the rope pulls along the ray $\theta = 0$. Ian can retract the arms so that θ is small in order to put the device into a crack, and then a spring pulls them back so that each arm wedges against the walls of the crack.

(a) Ian puts the device into a horizontal crack that is 1 inch tall. What angle do the arms make with the rope? *(Hint: Compute the y-coordinates of the two arm ends.)*

(b) What happens if Ian falls, causing the rope to pull outwards (rightwards) in the crack?

Proofs

62. Prove that the graph of the equation

$$r = k\sec\theta, \ -\frac{\pi}{2} < \theta < \frac{\pi}{2},$$

is a vertical line for any value of $k \neq 0$.

63. Prove that the graph of the equation

$$r = k\csc\theta, \ 0 < \theta < \pi,$$

is a horizontal line for any value of $k \neq 0$.

64. Let $a \neq 0$. Prove that the graph of the equation

$$r = \frac{a}{1 - \cos\theta}$$

is a parabola in a polar coordinate system. When $a \neq 0$, what is the graph of the equation $r = \frac{a}{1 - \sin\theta}$?

65. Let $a \neq 0$ and $0 < b < 1$. Prove that the graph of the equation

$$r = \frac{a}{1 - b\cos\theta}$$

is an ellipse in a polar coordinate system. When $a \neq 0$ and $b > 1$, what is the graph of the equation $r = \frac{a}{1 - b\cos\theta}$?

66. In this problem you will prove the three parts of Theorem 9.5:

(a) Prove that the polar coordinates $(r, \theta + 2\pi k)$ represent the same point for every integer k.

(b) Prove that the point with polar coordinates $(-r, \theta + \pi)$ represents the same point as (r, θ) for any value of θ.

(c) Prove that the polar coordinates $(0, \theta)$ represent the pole for any value of θ.

67. Show that the graph of the equation $r = k\cos\theta$ is a circle tangent to the y-axis for any $k \neq 0$. What are the center and radius of the circle?

68. Finish the proof of Theorem 9.8 by showing that the graph of the equation $r = 2k\sin\theta$ is a circle with center $(0, k)$ and radius k tangent to the x-axis for any $k \neq 0$.

69. Modify the proof of Theorem 9.8 to show that the graph of the equation $r = k\sin\theta + l\cos\theta$ is a circle. Find the center and radius in terms of k and l.

70. Find and prove a formula for the distance between the points (r_1, θ_1) and (r_2, θ_2) when they are plotted in a polar coordinate system.

Thinking Forward

▶ *Understanding symmetry:*

When a function $y = f(x)$ has the property that $f(-x) = f(x)$ for every value of x in the domain of f, the function is said to be an *even* function and its graph in a rectangular coordinate system is symmetrical with respect to the y-axis. When a function $r = f(\theta)$ has the property that $f(-\theta) = f(\theta)$ for every value of θ in the domain of f, what geometrical property would the graph of $r = f(\theta)$ have when it is plotted in a polar coordinate system?

▶ *Understanding symmetry:*

When a function $y = f(x)$ has the property that $f(-x) = -f(x)$ for every value of x in the domain of f, the function is said to be an *odd* function and its graph in a rectangular coordinate system is symmetrical with respect to the origin. When a function $r = f(\theta)$ has the property that $f(-\theta) = -f(\theta)$ for every value of θ in the domain of f, what geometrical property would the graph of $r = f(\theta)$ have when it is plotted in a polar coordinate system?

9.3   GRAPHING POLAR EQUATIONS

▶   A general method for graphing polar curves is provided

▶   Symmetry tests for polar curves are discussed

▶   Several categories of polar curves are introduced

Using the θr-plane to Get Information About a Polar Graph

In Section 9.2 we introduced polar coordinates and studied how points and some simple equations are graphed. We can always transform an equation in polar coordinates to an equation in rectangular coordinates. Some polar equations have nice rectangular coordinate representations. Since we are now experts at graphing curves in a rectangular coordinate system, the transformed equation may be easier to graph. However, as we saw in Example 5 of Section 9.2, the transformation process to rectangular coordinates does not always result in a nice equation. To aid in our understanding of graphing polar equations in the **_polar plane_** (see the figure that follows at the left), we introduce the θr-plane (figure at right), which is simply a rectangular coordinate system with the variable θ plotted on the horizontal axis and the variable r plotted on the vertical axis.

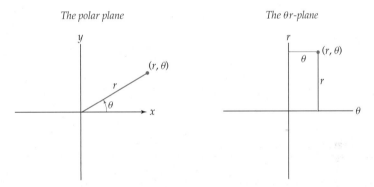

The polar plane   *The θr-plane*

In Example 4 we graphed the equation $r = 4\cos\theta$ in the polar plane. We reproduce that graph next at the left. The figure at the right shows the graph of the same equation in the θr-plane.

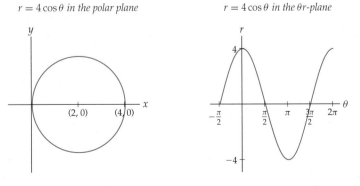

$r = 4\cos\theta$ *in the polar plane* $r = 4\cos\theta$ *in the θr-plane*

What is the connection between these two graphs? More generally, assuming that we are successful at graphing an equation $r = f(\theta)$ in the θr-plane, how can we use the properties of that graph to draw the desired curve in the polar plane? Our reason for doing this should be clear: We have a great deal of experience drawing curves in a rectangular coordinate system, and we wish to draw upon this experience to help us learn to understand curves in the polar plane.

First consider the inequalities that determine the four quadrants in the polar plane, as shown in the following table:

Quadrant	Inequality
I	$0 < \theta < \dfrac{\pi}{2}$
II	$\dfrac{\pi}{2} < \theta < \pi$
III	$\pi < \theta < \dfrac{3\pi}{2}$
IV	$\dfrac{3\pi}{2} < \theta < 2\pi$

The following diagrams show the same information in the polar plane (left) and in the θr-plane (right):

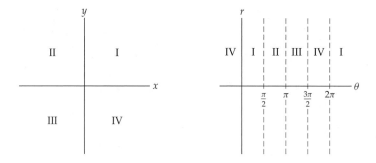

Notice that in the θr-plane the quadrant numbers start repeating; this is because, for example, quadrant I in the polar plane corresponds to angles $0 < \theta < \frac{\pi}{2}$ as well as angles $2\pi < \theta < \frac{5\pi}{2}$ and, in fact, to any angle in an interval of the form $\left(0 + 2\pi k, \frac{\pi}{2} + 2\pi k\right)$, where k is an integer.

The figure at the right illustrates the correspondence with the polar plane quadrants only when r is positive. Recall that when r is negative points are plotted in the diagonally opposite quadrant. The following figure shows this situation schematically:

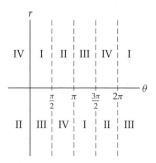

Given the graph of a polar equation in the θr-plane, what does this information tell us about the graph of the curve in the polar plane? Consider again the graph of $r = 4\cos\theta$. The following figure regraphs that equation with various pieces color-coded according to the sectors shown in the previous graph:

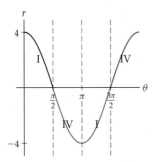

$r = 4\cos\theta$ in the θr-plane

Notice that the black and blue portions of the graph will appear in the first quadrant when graphed in polar coordinates while the red and green portions will be in the fourth quadrant of the polar plane. This analysis tells us why the graph plotted appears only in the first and fourth quadrants; however, it does not tell us why the graph is a circle.

Certain points in the θr-plane can tell us a great deal about the corresponding graph in the polar plane. For example, consider those points in the θr-plane that lie *on* the θ-axis. The r-coordinate of each of these points is zero. Since every point in the polar form $(0, \theta)$ is drawn at the pole of the polar plane, each point on the θ-axis corresponds to the pole in the polar plane. In terms of the previous color-coded graph, this means that the points at $\theta = \frac{\pi}{2}$ and $\theta = \frac{3\pi}{2}$ (and, in fact, for θ any odd multiple of $\frac{\pi}{2}$) correspond to the pole in the polar plane.

The points in the θr-plane with θ-coordinate of the form πk and $\frac{\pi}{2} + \pi k$ for some integer k are also good reference points. Points with $\theta = \pi k$ will correspond to points in the polar plane that lie on the horizontal axis. Points with $\theta = \frac{\pi}{2} + \pi k$ will correspond to points on the vertical axis in the polar plane. In terms of the color-coded graph, this means that the only vertical intercept of $r = 4\cos\theta$ in the polar plane will be at the pole (since the points where $\theta = \frac{\pi}{2} + \pi k$ have $r = 0$). Moreover, there are only two horizontal intercepts of $r = 4\cos\theta$ in the polar plane. We have already identified the horizontal intercept at the pole, and the only other horizontal intercept appears at the single point with polar coordinates $(4, 0)$ or $(-4, \pi)$ or $(4, 2\pi)$, etc. Notice that by examining certain points on the graph in the θr-plane, we have obtained all of the quadrant and intercept information for the corresponding graph in the polar plane.

Symmetry in Polar Graphs

In Section 0.2 we discussed functions whose graphs are symmetrical with respect to the y-axis and functions whose graphs are symmetrical with respect to the origin. These functions are called *even* and *odd*, respectively. Graphs that do not represent functions can also show these symmetries. For example, the graph that follows at the left is symmetrical with respect to the vertical axis, but it is not the graph of a function in rectangular coordinates because it fails the vertical line test. Similarly, the graph at the right is symmetrical with respect to the origin, but it is not the graph of a function in rectangular coordinates.

Symmetry with respect to the vertical axis

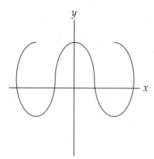

Symmetry with respect to the origin

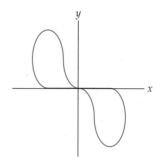

There is one more type of symmetry we'd like to consider: **symmetry with respect to the x-axis**. Using rectangular coordinates, we have the following definition (compare this with the definition of y-axis symmetry given in Definition 0.9 of Section 0.2):

DEFINITION 9.9   Symmetry with Respect to the x-axis

A graph in the xy-plane is said to be **symmetrical with respect to the x-axis** if, for every point (x, y) on the graph, the point $(x, -y)$ is also a point on the graph.

Following is an example of a graph that is symmetrical with respect to the x-axis:

Symmetry with respect to the x-axis

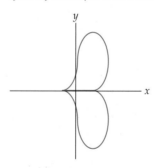

We have now considered three types of symmetry. It is an interesting fact that if a graph has any two of those types of symmetry, it also has the third! The proof is straightforward and is left for Exercise 56.

THEOREM 9.10   Relationships Among Symmetries in the Plane

Consider the following three types of symmetry:

   (a) symmetry with respect to the vertical axis;

   (b) symmetry with respect to the horizontal axis;

   (c) symmetry with respect to the origin.

If a graph in the plane has any two of these symmetries, then it has the third as well.

We are particularly concerned about symmetries in the polar plane because many of the equations we will be graphing have one or all of these symmetries. Such symmetries may be recognized in the polar plane by understanding the relationships between r and θ. By writing the definitions of these symmetries in terms of r and θ, we arrive at the properties summarized in the following theorem:

THEOREM 9.11

Symmetries in the Polar Plane

(a) A graph is symmetrical with respect to the x-axis if, for every point (r, θ) on the graph, the point $(r, -\theta)$ is also on the graph.

(b) A graph is symmetrical with respect to the y-axis if, for every point (r, θ) on the graph, the point $(-r, -\theta)$ is also on the graph.

(c) A graph is symmetrical with respect to the origin if, for every point (r, θ) on the graph, the point $(-r, \theta)$ is also on the graph.

You can visualize these symmetry properties in the figure shown next at the left.

⊗ CAUTION

Note that because every point in the polar plane has multiple representations in terms of its polar coordinates, the symmetries in Theorem 9.11 may also be expressed in terms of other relationships between coordinates on the graph. For example, if $(r, \pi - \theta)$ is on the graph whenever (r, θ) is on the graph, then the graph will be symmetrical with respect to the y-axis; see the following figure at the right:

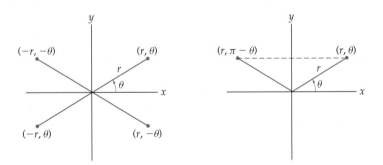

For example, we can see that the polar graph $r = 4\cos\theta$ is symmetrical with respect to the x-axis. Later we will examine a polar function that has all three types of symmetry mentioned in Theorem 9.11. Of course, polar graphs can lack all of these symmetries. For example, consider one of the simplest polar equations: $r = \theta$, $\theta \geq 0$. This simple equation has a quite beautiful graph, known as the ***spiral of Archimedes***.

The spiral of Archimedes

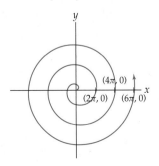

Before we proceed to our examples, we list several of the more common polar curves in the following compendium:

Equation	Name	Example
$r = a \pm b \cos\theta$ $r = a \pm b \sin\theta$, where a, $b \in \mathbb{R}$, $a \neq 0$, $b \neq 0$	Cardioid when $\left\lvert\dfrac{a}{b}\right\rvert = 1$	$r = 1 + \cos\theta$
	Limaçon when $\left\lvert\dfrac{a}{b}\right\rvert \neq 1$ ▶ with inner loop when $\left\lvert\dfrac{a}{b}\right\rvert < 1$	$r = 1 + 2\cos\theta$
	▶ dimpled when $1 < \left\lvert\dfrac{a}{b}\right\rvert < 2$	$r = 3 + 2\cos\theta$
	▶ convex when $\left\lvert\dfrac{a}{b}\right\rvert > 2$	$r = 5 + 2\cos\theta$
$r = a \sin k\theta$ $r = a \cos k\theta$, where $a \in \mathbb{R}$ and $k \in \mathbb{Z}$	Circle when $\lvert k \rvert = 1$	$r = \sin\theta$
	Rose curve with $2k$ petals when $k \neq 0$ is even	$r = \cos 2\theta$
	Rose curve with k petals when $k \neq \pm 1$ is odd	$r = \cos 3\theta$
$r^2 = \pm a \sin 2\theta$ $r^2 = \pm a \cos 2\theta$, where $a \in \mathbb{R}$ and $a \neq 0$	Lemniscate	$r^2 = \sin 2\theta$
		$r^2 = \cos 2\theta$

We will explore each of these curves shortly.

The names of several of the curves in the table are derived from the Greek and/or Latin. "Limaçon" comes from the Latin "limax," or snail. "Cardioid" is derived from the Greek word "kardia," or heart. "Lemniscate" has antecedents in both Greek and Latin; "lemni-cos" and "lemniscus" both mean "ribbon."

Examples and Explorations

EXAMPLE 1

Showing that the graphs of polar equations have symmetry

Show that $r = 4\cos\theta$ is symmetrical with respect to the x-axis and that $r = \cos 2\theta$ is symmetrical with respect to the x-axis, the y-axis, and the origin.

SOLUTION

The cosine is an even function; therefore $4\cos(-\theta) = 4\cos\theta$ for every value of θ. Thus, if (r, θ) is on the graph of $r = 4\cos\theta$, then $(r, -\theta)$ will also be on the graph. Therefore, the graph of $r = 4\cos\theta$ is symmetrical with respect to the x-axis.

We next turn to the equation $r = \cos 2\theta$. Showing that the graph of this equation is symmetrical with respect to the x-axis is similar to our analysis of $r = 4\cos\theta$ and is left for Exercise 15. To show that the graph is symmetrical with respect to the y-axis we will show that $\cos(2(\pi - \theta)) = \cos 2\theta$ for every value of θ. We have

$$\cos(2(\pi - \theta)) = \cos(2\pi - 2\theta) \qquad \leftarrow \text{distributive property}$$
$$= \cos(-2\theta) \qquad \leftarrow \text{periodicity of cosine}$$
$$= \cos 2\theta. \qquad \leftarrow \text{symmetry of cosine}$$

Therefore, $r = \cos 2\theta$ is symmetrical with respect to the y-axis. We are now done because, by Theorem 9.10, $r = \cos 2\theta$ must also be symmetrical with respect to the origin.   □

EXAMPLE 2

Graphing a cardioid

Graph the polar equation $r = 1 + \cos\theta$.

SOLUTION

We take this opportunity to show how some of the properties of the graphs of the polar functions in our earlier compendium may be constructed. First, notice that since $1 + \cos\theta$ is periodic with period 2π, we need only consider values of θ in the interval $[0, 2\pi)$. Now consider the graph of $r = 1 + \cos\theta$ in the θr-plane. This is the graph of $\cos\theta$ shifted up one unit, as shown in the next figure. From certain points on this graph, we will be able to extract quadrant and intercept information about the polar graph.

$r = 1 + \cos\theta$ *in the θr-plane*

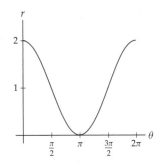

As we discussed earlier in this section, the points on the horizontal axis in the preceding graph correspond to the pole in the polar graph of the equation, so we know that the polar graph of $r = 1 + \cos\theta$ passes through the pole. Moreover, the points on the graph where θ is an integral multiple of π correspond to horizontal intercepts on the polar graph; in the given graph, the points at $\theta = 0$ and $\theta = 2\pi$ both correspond to the horizontal intercept $(r, \theta) = (2, 0)$ on the polar graph and the point at $\theta = \pi$ corresponds to the pole, as already mentioned. Finally, at odd multiples of $\theta = \frac{\pi}{2}$, the given graph indicates that the polar graph of $r = 1 + \cos\theta$ will have vertical intercepts at the polar coordinates $\left(1, \frac{\pi}{2}\right)$ and $\left(1, \frac{3\pi}{2}\right)$. This intercept information is recorded in the graph that follows at the left. We can also obtain quadrant information as shown in the following figure at the right:

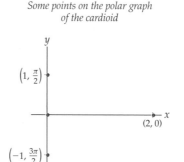

Some points on the polar graph of the cardioid

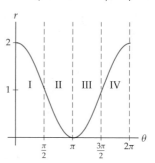

Sectors in the θr-plane indicate quadrants in the polar plane

When we test for symmetry, we see that $1 + \cos(-\theta) = 1 + \cos\theta$; therefore the graph in the polar plane will be symmetrical with respect to the horizontal axis. This means that we need only consider values of θ in the interval $[0, \pi]$. The following table shows values of r for several angles $\theta \in [0, \pi]$:

θ	0	$\dfrac{\pi}{6}$	$\dfrac{\pi}{4}$	$\dfrac{\pi}{3}$	$\dfrac{\pi}{2}$	$\dfrac{2\pi}{3}$	$\dfrac{3\pi}{4}$	$\dfrac{5\pi}{6}$	π
r	2	$1 + \dfrac{\sqrt{3}}{2}$	$1 + \dfrac{\sqrt{2}}{2}$	$\dfrac{3}{2}$	1	$\dfrac{1}{2}$	$1 - \dfrac{\sqrt{2}}{2}$	$1 - \dfrac{\sqrt{3}}{2}$	0

Plotting the intercepts and points from the table gives us the figure that follows at the left. Finally, we connect the plotted points with a smooth curve and use the symmetry of the graph to obtain the graph on the right, called a cardioid.

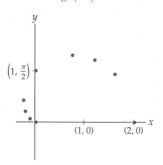

Points on the graph of $r = 1 + \cos\theta$

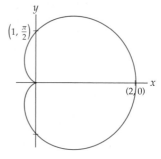

The cardioid $r = 1 + \cos\theta$

As mentioned in our chart before the examples, there are other equations whose graphs are cardioids in the polar plane. For example, the graph of the equation $r = 2(1 - \sin\theta)$ is another cardioid:

The cardioid $r = 2(1 - \sin\theta)$

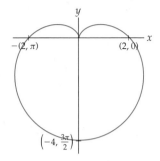

In the Exercise 39 we ask you to graph this curve.   □

EXAMPLE 3 | Graphing a limaçon

Graph the curve defined by the polar equation $r = \frac{1}{2} + \cos\theta$.

SOLUTION

The graph here is a member of the class of curves called the **limaçons**. We again start by graphing the equation $r = \frac{1}{2} + \cos\theta$ in the θr-plane:

$$r = \frac{1}{2} + \cos\theta \text{ in the } \theta r\text{-plane}$$

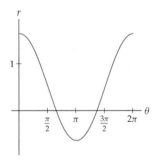

As in Example 2, $r = \frac{1}{2} + \cos\theta$ is periodic with period 2π and its graph in the polar plane will be symmetrical with respect to the x-axis. Therefore we need only consider values of θ in the interval $[0, \pi]$. Horizontal intercepts occur at integral multiples of π, resulting in the polar points $\left(\frac{3}{2}, 0\right)$ and $\left(-\frac{1}{2}, \pi\right)$. We will have a third horizontal intercept at the pole, corresponding to the value where $r = 0$. Here $0 = \frac{1}{2} + \cos\theta$ has the unique solution $\theta = \frac{2\pi}{3}$ on the interval $[0, \pi]$, and that solution corresponds to the polar point $\left(0, \frac{2\pi}{3}\right)$. The only additional vertical intercept occurs when $\theta = \frac{\pi}{2}$; here we have the polar point $\left(\frac{1}{2}, \frac{\pi}{2}\right)$.

Unlike the situation in Example 2, the sign of r varies when θ is in the interval $[0, \pi]$. We have $r > 0$ when $0 \leq \theta < \frac{2\pi}{3}$ and $r < 0$ when $\frac{2\pi}{3} < \theta \leq \pi$. Here all values of θ in the interval $\left(0, \frac{\pi}{2}\right)$ will be graphed in the first quadrant of the polar plane. Second-quadrant angles in the interval $\left(\frac{\pi}{2}, \frac{2\pi}{3}\right)$ will be graphed in the second quadrant of the polar plane, while second-quadrant angles in the interval $\left(\frac{2\pi}{3}, \pi\right)$ will be graphed in the fourth quadrant of the polar plane! The following table lists the values of r for several angles $\theta \in [0, \pi]$:

θ	0	$\dfrac{\pi}{6}$	$\dfrac{\pi}{4}$	$\dfrac{\pi}{3}$	$\dfrac{\pi}{2}$	$\dfrac{2\pi}{3}$	$\dfrac{3\pi}{4}$	$\dfrac{5\pi}{6}$	π
r	$\dfrac{3}{2}$	$\dfrac{1}{2}+\dfrac{\sqrt{3}}{2}$	$\dfrac{1}{2}+\dfrac{\sqrt{2}}{2}$	1	$\dfrac{1}{2}$	0	$\dfrac{1}{2}-\dfrac{\sqrt{2}}{2}$	$\dfrac{1}{2}-\dfrac{\sqrt{3}}{2}$	$-\dfrac{1}{2}$

The figure that follows at the left shows a plot of the intercepts and points given in the preceding table. By connecting these points with a smooth curve and using the symmetry of the graph, we obtain the limaçon shown at the right. As mentioned earlier, the term "limaçon" is derived from the Latin word for snail. The snail's "shell" is more obvious in the plotted points than in the final graph.

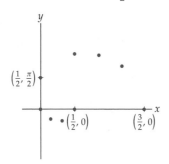

Points on the graph of $r = \dfrac{1}{2} + \cos\theta$

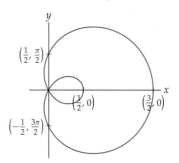

The limaçon $r = \dfrac{1}{2} + \cos\theta$

EXAMPLE 4

A rose with four petals

Graph the polar equation $r = \cos 2\theta$.

SOLUTION

In Example 1 we showed that the graph of $r = \cos 2\theta$ is symmetrical with respect to the x-axis, with respect to the y-axis, and with respect to the origin. These symmetries will save us a considerable amount of work as we graph the given curve. In particular, we need only graph the curve on the interval $\left[0, \dfrac{\pi}{2}\right]$ and then let the symmetry of the graph do the rest of the work.

We start our analysis by graphing the equation in the θr-plane:

$r = \cos 2\theta$ in the θr-plane

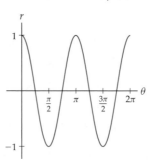

The only horizontal intercepts on the interval $\theta \in \left[0, \dfrac{\pi}{2}\right]$ will occur at the polar point $(1, 0)$ and when r is zero. We obtain the other horizontal intercept at the pole by finding the unique solution of the polar equation $0 = \cos 2\theta$ for $\theta \in \left[0, \dfrac{\pi}{2}\right]$. This occurs when $\theta = \dfrac{\pi}{4}$. On our interval there will be one vertical intercept, at the polar point $\left(-1, \dfrac{\pi}{2}\right)$.

From the graph, we can see that the sign of r varies when θ is in the interval $\left[0, \frac{\pi}{2}\right]$. We have $r > 0$ on the interval $\left[0, \frac{\pi}{4}\right)$ and $r < 0$ on $\left(\frac{\pi}{4}, \frac{\pi}{2}\right]$. Thus the values of θ in the interval $\left(0, \frac{\pi}{4}\right)$ will be graphed in the first quadrant while the values of θ in the interval $\left(\frac{\pi}{4}, \frac{\pi}{2}\right)$ will be graphed in the third quadrant. The following table lists values of r for several angles in the interval $\left[0, \frac{\pi}{2}\right]$:

θ	0	$\dfrac{\pi}{12}$	$\dfrac{\pi}{8}$	$\dfrac{\pi}{6}$	$\dfrac{\pi}{4}$	$\dfrac{\pi}{3}$	$\dfrac{3\pi}{8}$	$\dfrac{5\pi}{12}$	$\dfrac{\pi}{2}$
r	1	$\dfrac{\sqrt{3}}{2}$	$\dfrac{\sqrt{2}}{2}$	$\dfrac{1}{2}$	0	$-\dfrac{1}{2}$	$-\dfrac{\sqrt{2}}{2}$	$-\dfrac{\sqrt{3}}{2}$	-1

Plotting the intercepts and points from the table gives us the picture shown next at the left. We connect these points with a smooth curve and use the symmetry of the graph to obtain the four-petaled rose graphed at the right.

Points on the graph of $r = \cos 2\theta$ *The rose $r = \cos 2\theta$*

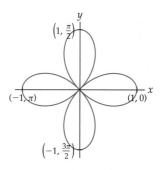

In Exercise 55 you are asked to show that when n is a positive odd integer, the polar rose $r = \cos n\theta$ or $r = \sin n\theta$ is traced twice on the interval $[0, 2\pi]$. This implies that when n is odd, we need only graph $r = \sin n\theta$ on the interval $[0, \pi]$. In addition, in Exercise 58 you are asked to show that for every integer n, the graph of $r = \sin n\theta$ is symmetrical with respect to the y-axis. These facts will prove useful in the next example.

EXAMPLE 5

A rose with three petals

Graph the polar equation $r = \sin 3\theta$.

SOLUTION

Because $r = \sin 3\theta$ is a rose defined with the sine function and the odd integer $n = 3$, we need only consider the graph of $r = \sin 3\theta$ for $\theta \in \left[0, \frac{\pi}{2}\right]$. The only horizontal intercepts of $r = \sin 3\theta$ on the interval $\left[0, \frac{\pi}{2}\right]$ will occur at the pole $(0, 0)$. However, the curve will pass through the pole twice, initially when θ is zero and again when θ is $\frac{\pi}{3}$. On our interval there will be one more vertical intercept at $\left(-1, \frac{\pi}{2}\right)$.

The sign of r varies when θ is in the interval $\left[0, \frac{\pi}{2}\right]$: $r > 0$ on $\left[0, \frac{\pi}{3}\right)$ and $r < 0$ on $\left(\frac{\pi}{3}, \frac{\pi}{2}\right]$. Thus the values of θ in the interval $\left(0, \frac{\pi}{3}\right)$ will be graphed in the first quadrant, while the values of θ in the interval $\left(\frac{\pi}{3}, \frac{\pi}{2}\right)$ will be graphed in the third quadrant. We again tabulate values of r for several angles in the interval $\left[0, \frac{\pi}{2}\right]$:

θ	0	$\dfrac{\pi}{18}$	$\dfrac{\pi}{12}$	$\dfrac{\pi}{9}$	$\dfrac{\pi}{6}$	$\dfrac{2\pi}{9}$	$\dfrac{\pi}{4}$	$\dfrac{5\pi}{18}$	$\dfrac{\pi}{3}$	$\dfrac{7\pi}{18}$	$\dfrac{5\pi}{12}$	$\dfrac{4\pi}{9}$	$\dfrac{\pi}{2}$
r	0	$\dfrac{1}{2}$	$\dfrac{\sqrt{2}}{2}$	$\dfrac{\sqrt{3}}{2}$	1	$\dfrac{\sqrt{3}}{2}$	$\dfrac{\sqrt{2}}{2}$	$\dfrac{1}{2}$	0	$-\dfrac{1}{2}$	$-\dfrac{\sqrt{2}}{2}$	$-\dfrac{\sqrt{3}}{2}$	-1

Plotting the intercepts and these points gives us the graph shown next at the left. Connecting the points with a smooth curve and using the symmetry of the graph, we obtain the three-petaled rose graphed at the right.

Points on the graph of $r = \sin 3\theta$

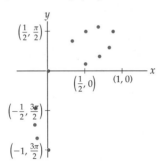

The rose $r = \sin 3\theta$

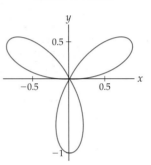

EXAMPLE 6   **A lemniscate**

Graph the polar equation $r^2 = \sin 2\theta$.

SOLUTION

First we note that θ cannot be a second- or fourth-quadrant angle, because if θ were in one of those quadrants, then $\sin 2\theta$ would be negative (since the sine is negative on the intervals $(\pi, 2\pi)$ and $(3\pi, 4\pi)$). In this case, because $r^2 \geq 0$, the equation $r^2 = \sin 2\theta$ could not hold. Therefore the graph of our equation will appear only in the first and third quadrants of the polar plane.

We next show that the graph of $r^2 = \sin 2\theta$ is symmetrical with respect to the origin. It suffices to show that for each point (r, θ) on the graph of $r^2 = \sin 2\theta$, the point $(r, \theta + \pi)$ is also on the graph. We have $\sin(2(\theta + \pi)) = \sin(2\theta + 2\pi) = \sin 2\theta = r^2$. Thus, $(r, \theta + \pi)$ satisfies the equation $r^2 = \sin 2(\theta + \pi)$, so $(r, \theta + \pi)$ is a point on the graph. Therefore the graph of $r^2 = \sin 2\theta$ is symmetrical with respect to the origin.

We again tabulate values of r for several angles in the interval $\left[0, \dfrac{\pi}{2}\right]$:

θ	0	$\dfrac{\pi}{12}$	$\dfrac{\pi}{8}$	$\dfrac{\pi}{6}$	$\dfrac{\pi}{4}$	$\dfrac{\pi}{3}$	$\dfrac{3\pi}{8}$	$\dfrac{5\pi}{12}$	$\dfrac{\pi}{2}$
r	0	$\dfrac{\sqrt{2}}{2}$	$\sqrt{\dfrac{\sqrt{2}}{2}}$	$\sqrt{\dfrac{\sqrt{3}}{2}}$	1	$\sqrt{\dfrac{\sqrt{3}}{2}}$	$\sqrt{\dfrac{\sqrt{2}}{2}}$	$\dfrac{\sqrt{2}}{2}$	0

Plotting the intercepts and points from the table gives us the graph shown next at the left. We connect these points with a smooth curve and use the symmetry of the graph to obtain the lemniscate graphed at the right.

Points on the graph of $r^2 = \sin 2\theta$

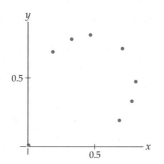

The lemniscate $r^2 = \sin 2\theta$

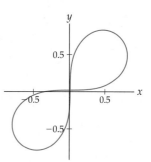

✔ CHECKING THE ANSWER

It is important to understand the basics of graphing polar curves by hand, but a graphing calculator or computer algebra system may be used to check your work. Calculators or software can also be used to graph many interesting examples that are too complicated to graph by hand. For example, the three figures that follow show some polar curves suggested by T. H. Fay (*American Mathematical Monthly*, 96 (1989), 442–443). The figures were graphed with the computer program *Mathematica*.

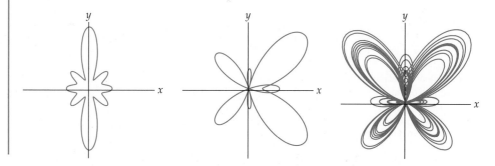

? TEST YOUR UNDERSTANDING

▶ Given a polar equation, why is it helpful to graph the equation in the θr-plane first?

▶ When θ is a first-quadrant angle, in which quadrants might the polar point corresponding to θ on the graph of an equation $F(r, \theta) = 0$ appear? What about when θ is a second-, third-, or fourth-quadrant angle? Why does this happen?

▶ How are the symmetries of a graph in polar coordinates recognized?

▶ When a polar function is symmetrical about both the x- and y-axes, why is it also symmetric about the origin?

▶ Which polar equations have polar graphs that are circles, cardioids, limaçons, roses, etc.?

EXERCISES 9.3

Thinking Back

▶ *Finding points of intersection:* Graph the functions
$$y = 1 + \sin x \text{ and } y = \cos x,$$
and find all points of intersection.

▶ *Finding points of intersection:* Graph the functions
$$y = \frac{1}{2} + \cos x \text{ and } y = \frac{1}{2} - \cos x,$$
and find all points of intersection.

▶ *Translations of graphs:* What is the relationship between the graphs of the functions
$$y = f(x) \text{ and } y = f(x - k)$$
when $k > 0$?

▶ *Translations of graphs:* What is the relationship between the graphs of the functions
$$y = f(x) \text{ and } y = f(x) + k$$
when $k > 0$?

Concepts

0. *Problem Zero:* Read the section and make your own summary of the material.

1. *True/False:* Determine whether each of the statements that follow is true or false. If a statement is true, explain why. If a statement is false, provide a counterexample.

   (a) *True or False:* If $\frac{\pi}{2} < \theta < \pi$, then the point (r, θ) is located in the second quadrant when it is plotted in a polar coordinate system.

   (b) *True or False:* The graph of $r = \sin 5\theta$ is a five-petaled rose.

   (c) *True or False:* The graph of $r = \cos 6\theta$ is a six-petaled rose.

   (d) *True or False:* If a graph in the polar plane is symmetrical with respect to the origin, then for every polar point (r, θ) on the graph, the polar point $(-r, \theta + 2\pi)$ is also on the graph.

   (e) *True or False:* The graph of a polar function $r = f(\theta)$ is symmetrical with respect to the y-axis if, for every point (r, θ) on the graph, the point $(r, -\theta)$ is also on the graph.

   (f) *True or False:* When k is a positive integer, the polar roses $r = \sin k\theta$ and $r = \cos k\theta$ are symmetrical with respect to both the x-axis and y-axis if and only if k is even.

   (g) *True or False:* In the rectangular coordinate system the graph of the equation $(x^2 + y^2)^2 = k(x^2 - y^2)$ is a lemniscate for every $k > 0$.

   (h) *True or False:* In the rectangular coordinate system the only function $y = f(x)$ that is symmetrical with respect to both the y-axis and the origin is $y = 0$.

2. *Examples:* Construct examples of the thing(s) described in the following. Try to find examples that are different than any in the reading.

   (a) An equation in polar coordinates whose graph is a cardioid.

   (b) An equation in polar coordinates whose graph is a limaçon.

   (c) An equation in polar coordinates whose graph is a lemniscate.

3. How can the graph of an equation $r = f(\theta)$ in the θr-plane be used to provide information about the graph of the same equation in the polar plane?

4. Explain the significance of the following figure:

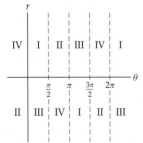

What is being shown in this figure?

5. This problem concerns the connections between the polar plane and the θr-plane.

   (a) Which points in the θr-plane correspond to the pole in the polar plane, and why?

   (b) Which points in the θr-plane correspond to the horizontal axis in the polar plane, and why?

   (c) Which points in the θr-plane correspond to the vertical axis in the polar plane, and why?

6. What is the relationship between the graphs of $r = f(\theta)$ and $r = f(\theta - k)$ in the θr-plane for $k > 0$? What is the relationship between the two graphs in polar coordinates?

7. Fill in the blanks:

   (a) On a graph that is symmetrical with respect to the x-axis in the polar plane, if the polar point $\left(2, \frac{\pi}{3}\right)$ is on the graph, then the polar point _____ is also on the graph.

   (b) On a graph that is symmetrical with respect to the y-axis in the polar plane, if the polar point $\left(2, \frac{\pi}{3}\right)$ is on the graph, then the polar point _____ is also on the graph.

   (c) On a graph that is symmetrical with respect to the y-axis in the Cartesian coordinate system, if the point $(3, 5)$ is on the graph, then the point _____ is also on the graph.

   (d) On a graph that is symmetrical with respect to the origin in the polar plane, if the polar point $\left(2, \frac{\pi}{3}\right)$ is on the graph, then the polar point _____ is also on the graph.

   (e) On a graph that is symmetrical with respect to the origin in the Cartesian coordinate system, if the point $(3, 5)$ is on the graph, then the point _____ is also on the graph.

8. Fill in the blanks:

   (a) If the point $(r, \theta + \pi)$ is on the graph of a polar curve whenever the point (r, θ) is on the graph, then the curve is symmetrical _____.

   (b) If the point _____ is on the graph of a polar curve whenever the point (r, θ) is on the graph, then the curve is symmetrical with respect to the origin. (There is more than one way to answer this question correctly!)

   (c) If the point $(-r, -\theta)$ is on the graph of a polar curve whenever the point (r, θ) is on the graph, then the curve is symmetrical _____.

9. For each of the following, give an example of a polar equation whose graph has the type(s) of symmetry listed, if possible. If such an equation doesn't exist, explain why.
   (a) symmetry about the x-axis
   (b) symmetry about the y-axis
   (c) symmetry about the origin
   (d) symmetry about the origin but not about the x-axis
   (e) symmetry about the x-axis, y-axis, and origin
   (f) symmetry about the x-axis and y-axis but not about the origin

10. For each type of polar curve, give a general form of the polar equation for that curve:
   (a) a line passing through the pole
   (b) a circle centered at the pole

(c) a cardioid

(d) a limaçon

(e) a rose with an odd number of petals

(f) a rose with an even number of petals

(g) a lemniscate

(h) the spiral of Archimedes

11. What type of polar curve is named for a heart? For a snail? For a ribbon?

12. What is the difference between a cardioid and a limaçon?

13. Which kind(s) of symmetry does the rose $r = \cos 5\theta$ have? How many petals does this curve have? Which kind(s) of symmetry does the rose $r = \sin 8\theta$ have? How many petals does this curve have?

14. Using polar coordinates, the graphs of the equations $r = \cos k\theta$ and $r = \sin k\theta$ are roses with k petals when $k \geq 3$ is an odd integer. What are the polar graphs of these equations when $k = 1$? What are the graphs of these equations when k is a negative odd integer?

15. Finish Example 1 by showing that $r = \cos 2\theta$ is symmetrical with respect to the horizontal axis.

16. In this section we graphed many polar equations, and most of our examples followed similar procedures. Describe a general procedure for sketching the graph of a polar equation.

Skills

Graph the equations in Exercises 17–24 in the θr-plane. Label each arc of your curve with the quadrant in which the corresponding polar graph will occur.

17. $r = 1 - \cos\theta$

18. $r = \dfrac{1}{2} - \sin\theta$

19. $r = 2 + \sin\theta$

20. $r = \sin 2\theta$

21. $r = \cos^2\theta$

22. $r = \theta, \ r \leq 0$

23. $r = \sec\theta$

24. $r = 1 + \sin 3\theta$

Graph the equations in Exercises 25–32 in the polar plane. Compare your graphs with the corresponding graphs in Exercises 17–24.

25. $r = 1 - \cos\theta$

26. $r = \dfrac{1}{2} - \sin\theta$

27. $r = 2 + \sin\theta$

28. $r = \sin 2\theta$

29. $r = \cos^2\theta$

30. $r = \theta, \ r \leq 0$

31. $r = \sec\theta$

32. $r = 1 + \sin 3\theta$

Find the smallest interval necessary to draw a complete graph of the functions in Exercises 33–38, and then graph each function using a graphing calculator or computer algebra system.

33. $r = \cos(2\theta/5)$

34. $r = \dfrac{4\cos 3\theta + \cos 2\theta}{\cos\theta}$

35. $r = \dfrac{4\cos\theta + \cos 9\theta}{\cos\theta}$

36. $r = e^{\sin\theta} - 2\cos 4\theta$

37. $r = e^{\sin\theta} - 2\cos 4\theta + \sin^5(\theta/12)$

38. $r = e^{\cos 2\theta} - 1.5\cos 4\theta$

39. Graph the cardioid $r = 2(1 - \sin\theta)$.

For each pair of functions in Exercises 40–45,

(a) Algebraically find all values of θ where $f_1(\theta) = f_2(\theta)$.

(b) Sketch the two curves in the same polar coordinate system.

(c) Find all points of intersection between the two curves.

40. $f_1(\theta) = \sin 2\theta$ and $f_2(\theta) = -\sin 2\theta$

41. $f_1(\theta) = \sin\theta$ and $f_2(\theta) = \cos\theta$

42. $f_1(\theta) = 1$ and $f_2(\theta) = 2\sin 2\theta$

43. $f_1(\theta) = 1 + \sin\theta$ and $f_2(\theta) = 1 - \cos\theta$

44. $f_1(\theta) = \sin\theta$ and $f_2(\theta) = \sin 2\theta$

45. $f_1(\theta) = \sin^2\theta$ and $f_2(\theta) = \cos^2\theta$

46. Graph the cardioids $r = 1 + \cos\theta$, $r = 2(1 + \cos\theta)$, and $r = 3(1 + \cos\theta)$ in the same polar coordinate system. What are the relationships among these three graphs? How does changing the constant k in the graph of the cardioid $r = k(1 + \cos\theta)$ affect the graph? What happens when $k < 0$?

47. Graph the cardioids $r = 1 + \sin\theta$ and $r = 1 - \sin\theta$. What is the relationship between these two graphs?

48. Graph the limaçons $r = \dfrac{1}{2} + \sin\theta$, $r = 1 + \sin\theta$, $r = 2 + \sin\theta$, and $r = 3 + \sin\theta$. Make a conjecture about the behavior of graphs of limaçons of the form $r = a + \sin\theta$ for various values of a. In particular, try to understand which values of a will give a limaçon with an inner loop. Which values of a will give a limaçon with a dimple? Which values of a will give a convex limaçon?

49. Graph the limaçons $r = 3 + \cos\theta$, $r = 3 + 3\cos\theta$, $r = 3 + 4\cos\theta$, and $r = 3 + 6\cos\theta$. Make a conjecture about the behavior of graphs of limaçons of the form $r = 3 + b\cos\theta$ for various values of b. In particular, try to understand which values of b will give a limaçon with an inner loop. Which values of b will give a limaçon with a dimple? Which values of b will give result in a convex limaçon?

50. Consider the polar equation $r^2 = \cos 2\theta$.

(a) Show that the polar graph of this equation is symmetrical with respect to the x-axis.

(b) Show that the polar graph of this equation is symmetrical with respect to the origin.

(c) Explain why the polar graph of the equation is symmetrical with respect to the y-axis.

(d) Use the techniques of this section to graph the equation.

(e) How could you graph the equation with a calculator or computer algebra system?

51. Find an equation for the curve obtained as the set of midpoints of every chord with one endpoint at $(1, 0)$ on the unit circle $r = 1$. Express your answer (a) with parametric equations, (b) in rectangular coordinates, and (c) in polar coordinates.

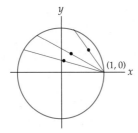

52. The graph of $r = \cos\left(\frac{5}{2}\theta\right)$ is the flower-like graph with 10 petals, shown next at the left:

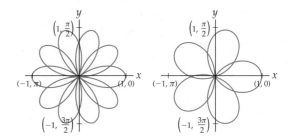

The graph of $r = \cos\left(\frac{5}{3}\theta\right)$ is also a flower-like graph, but with 5 petals, shown in the preceding figure at the right.

(a) Why does the graph of $r = \cos\left(\frac{5}{2}\theta\right)$ have 10 petals while the graph of $r = \cos\left(\frac{5}{3}\theta\right)$ has only 5?

(b) Let $\frac{p}{q}$ be a positive rational number reduced to lowest terms with $p > q$. How many petals will the curve $r = \cos\left(\frac{p}{q}\theta\right)$ have? What is the smallest interval of the form $[0, b]$ required to obtain the graphs of these curves? Explain how to calculate b in terms of p and q.

(c) What is the graph of $r = \cos k\theta$ when k is an irrational number?

Applications

53. Annie is using wood framing and a fabric shell to design a kayak. For simplicity, she considers making the cross section of the kayak follow the curve $r = A(1 - 0.5\sin\theta)$, where $A = 1$ at the middle of the kayak and becomes smaller as the cross section is taken farther from the middle of the boat. Sketch the cross section of the boat when $A = 1$.

54. Annie is concerned that her kayak will be too deep—that it will draw too much water or else have too much freeboard to be easy to paddle. She decides to investigate a cross-section function of the form

$$r = A(1 - 0.5\sin\theta)(1 + 0.2\cos^2\theta).$$

(a) How does Annie need to alter the choice of A so that the width of this boat is the same as that without the term $(1 + 0.2\cos^2\theta)$?

(b) Sketch this cross section with Annie's choice of A from part (a).

(c) Describe in words the effect that the multiplicative term $(1 + 0.2\cos^2\theta)$ has on the cross section of the kayak. How is that effect achieved?

Proofs

55. Prove that when n is a positive odd integer, the polar rose $r = \cos n\theta$ or $r = \sin n\theta$ is traced twice on the interval $[0, 2\pi]$ and thus has exactly n petals.

56. Prove Theorem 9.10. That is, show that if a graph in the plane has any two of three types of symmetry, namely, symmetry about the x-axis, symmetry about the y-axis, and symmetry about the origin, then it has the third type of symmetry as well.

57. Prove that, for every even integer n, the graph of $r = \sin n\theta$ is symmetrical with respect to the x-axis.

58. Prove that, for every integer n, the graph of $r = \sin n\theta$ is symmetrical with respect to the y-axis.

59. Prove that, for every even integer n, the graph of $r = \cos n\theta$ is symmetrical with respect to the y-axis.

60. Prove that, for every integer n, the graph of $r = \cos n\theta$ is symmetrical with respect to the x-axis.

Thinking Forward

▶ *Volume of a cylinder:* Find the volume of the right circular cylinder with height 2 units and cross-sectional circle given by the equation $r = 4\cos\theta$.

▶ *Volume of a sphere:* Find the volume of the sphere whose equator is given by the equation $r = 4\cos\theta$.

9.4   COMPUTING ARC LENGTH AND AREA WITH POLAR FUNCTIONS

▶   A formula for finding the area bounded by a polar curve is derived

▶   We explain how polar functions can be expressed with parametric equations

▶   A formula for finding the arc length of a polar curve is derived

Computing the Area Bounded by Polar Curves

We've seen that many interesting regions in the plane are bounded by polar functions of the form $r = f(\theta)$. We will start this section by discussing how to compute the area of such a region without transforming the equation into rectangular coordinates.

When working in rectangular coordinates, we used rectangles as our most "basic" shape—that is, as the building blocks for approximating areas of more complicated regions. When working in polar coordinates, it is more natural to approximate a region with sectors of circles. For example, consider a region R in the polar plane bounded by two rays $\theta = \alpha$ and $\theta = \beta$ and a polar function of the form $r = f(\theta)$, as shown next at the left. (Regions of this type will be the typical polar regions we wish to approximate; compare our work here with our work in rectangular coordinates when we considered the area under the graph of a function $y = f(x)$ from $x = a$ to $x = b$.) The following figure at the right shows the region R approximated with "wedges" (i.e., sectors of circles):

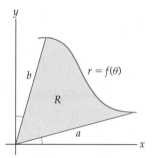
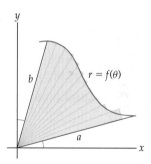

Of course, before we can use sectors of circles to approximate areas, we must determine how to find the area of a sector of a circle. Consider a sector of a circle with radius r and that has a central angle ϕ (measured in radians):

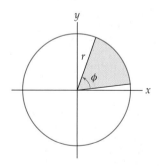

The area of the sector is a fraction of the area of the circle. Specifically, it is $\frac{\phi}{2\pi}$ times the area of the entire circle $\left(\text{i.e., } \frac{\phi}{2\pi}(\pi r^2) = \frac{1}{2}\phi r^2\right)$, as stated in the following theorem:

THEOREM 9.12   **The Area of a Sector of a Circle**

In a circle of radius r, the area of a sector with central angle ϕ, in radians, is given by

$$\text{Area} = \frac{1}{2}\phi r^2.$$

We will now return to the region R and its approximation with wedges. Although we could easily approximate the area of this region with a fixed number of wedges, we want to find the *exact* area; therefore we need to fix a general notation. The procedure is strikingly similar to our work with Riemann sums in Chapter 4, except that instead of dealing with approximating rectangles in rectangular coordinates, we will be dealing with approximating sectors in polar coordinates.

Let n be a positive integer, and define

$$\Delta\theta = \frac{\beta - \alpha}{n} \quad \text{and} \quad \theta_k = \alpha + k\Delta\theta.$$

We also choose some point θ_k^* in each interval $[\theta_{k-1}, \theta_k]$ and let S_k be the sector of the circle with radius $f(\theta_k^*)$ centered at the origin and extending from the radius with polar angle θ_{k-1} to the radius with polar angle θ_k. These are the sectors shown earlier as wedges.

By Theorem 9.12, the area of the kth sector S_k is $\frac{1}{2}(f(\theta_k^*))^2\Delta\theta$. The approximate area of the region R is the sum of the areas of the sectors:

$$\text{Area of region } R \approx \sum_{k=1}^{n} \frac{1}{2}(f(\theta_k^*))^2\Delta\theta.$$

Notice that this is a Riemann sum for the function $\frac{1}{2}(f(\theta))^2$ on the interval $[\alpha, \beta]$. Therefore, the area of R will be the limit of the Riemann sum as $n \to \infty$. That is,

$$\text{Area of region } R = \lim_{n\to\infty} \sum_{k=1}^{n} \frac{1}{2}(f(\theta_k^*))^2\Delta\theta.$$

We now have the following theorem:

THEOREM 9.13   **The Area of a Region in the Polar Plane Bounded by a Function $r = f(\theta)$**

Let α and β be real numbers such that $0 \le \beta - \alpha \le 2\pi$. Let R be the region in the polar plane bounded by the rays $\theta = \alpha$ and $\theta = \beta$ and a positive continuous function $r = f(\theta)$. Then the area of region R is

$$\frac{1}{2}\int_{\alpha}^{\beta}(f(\theta))^2\,d\theta.$$

Computing Arc Length in Polar Coordinates

Every polar function of the form $r = f(\theta)$ can be easily rewritten in terms of parametric equations of the form $x = x(\theta)$ and $y = y(\theta)$. Recall that to transform from polar coordinates to rectangular coordinates we use

$$x = r\cos\theta, \ y = r\sin\theta.$$

If $r = f(\theta)$, we immediately obtain

$$x = f(\theta)\cos\theta, \ y = f(\theta)\sin\theta.$$

These are the parametric equations that we sought.

For example, to express the cardioid with equation $r = 1 + \cos\theta$ in terms of parametric equations, we have

$$x = (1 + \cos\theta)\cos\theta, \ y = (1 + \cos\theta)\sin\theta.$$

By thinking of polar functions as parametric equations, we can use the work we just did to find the arc length of a polar curve. If we combine the parametric equations $x = f(\theta)\cos\theta$ and $y = f(\theta)\sin\theta$ with Theorem 9.4 of Section 9.1, which tells us how to compute the arc length of a curve defined by parametric equations, we arrive at the following theorem:

THEOREM 9.14

The Arc Length of a Polar Curve

Let $r = f(\theta)$ be a differentiable function of θ such that $f'(\theta)$ is continuous for all $\theta \in [\alpha, \beta]$. Furthermore, assume that $r = f(\theta)$ is a one-to-one function from $[\alpha, \beta]$ to the graph of the function. Then the length of the polar graph of $r = f(\theta)$ on the interval $[\alpha, \beta]$ is

$$\int_\alpha^\beta \sqrt{(f'(\theta))^2 + (f(\theta))^2}\, d\theta.$$

Proof. Theorem 9.4 tells us that when x and y are functions of the parameter θ, the arc length of the curve on the interval $[\alpha, \beta]$ is

$$\int_\alpha^\beta \sqrt{\left(\frac{dx}{d\theta}\right)^2 + \left(\frac{dy}{d\theta}\right)^2}\, d\theta.$$

Combining this integral with the parametric equations for the curve, $x = f(\theta)\cos\theta$ and $y = f(\theta)\sin\theta$, we have arc length

$$\int_\alpha^\beta \sqrt{\left(\frac{d}{d\theta}(f(\theta)\cos\theta)\right)^2 + \left(\frac{d}{d\theta}(f(\theta)\sin\theta)\right)^2}\, d\theta.$$

Taking the derivatives inside the radical gives

$$\int_\alpha^\beta \sqrt{(f'(\theta)\cos\theta - f(\theta)\sin\theta)^2 + (f'(\theta)\sin\theta + f(\theta)\cos\theta)^2}\, d\theta.$$

After expanding the terms under of the radical and simplifying (you will do this in Exercise 58), we obtain our result:

$$\int_\alpha^\beta \sqrt{(f'(\theta))^2 + (f(\theta))^2}\, d\theta. \qquad \blacksquare$$

Examples and Explorations

EXAMPLE 1

Finding the area bounded by a polar rose

Calculate the area bounded by the curve $r = \cos 2\theta$.

SOLUTION

This is the first of three examples in which we use Theorem 9.13 to find areas of polar regions. As with all area computations, we must understand the area we are trying to compute. When we can, we will use the symmetry of the region to simplify our work.

Recall that we graphed this polar rose in Example 4 in Section 9.3. We reproduce that graph here, with one portion of the graph shaded:

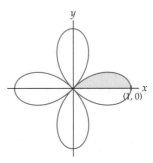

To compute the area of the curve we will use the symmetry of the rose. We will calculate only the area of the shaded region of the figure—that is one-half of one of the petals (and one-eighth of the total area). The shaded region corresponds to the values of θ in $\left[0, \frac{\pi}{4}\right]$. Therefore the area of the shaded region is

$$\frac{1}{2} \int_0^{\pi/4} \cos^2 2\theta \ d\theta.$$

The value of this integral is $\frac{\pi}{16}$. (You should be able to work out the details.) Since the area of the shaded region is one-eighth of the region bounded by the polar rose, the area bounded by the rose is $\frac{\pi}{2}$ square units.   □

⊗ CAUTION

In Example 1 we used the symmetry of the region to help us compute the area of a polar rose. Using such symmetries will often make area computations easier. We could also have computed the entire area from the integral

$$\frac{1}{2} \int_0^{2\pi} \cos^2 2\theta \ d\theta,$$

because the entire curve gets traced once on the interval $[0, 2\pi]$. However, it is not always correct to integrate over the interval $[0, 2\pi]$ to find the area of a polar rose. When $n \geq 3$ is odd, any polar rose of the form $r = \cos n\theta$ or $r = \sin n\theta$ is traced twice on the interval $[0, 2\pi]$. Therefore, to find the areas bounded by these roses, we could use integrals of the form

$$\frac{1}{2} \int_0^{\pi} \cos^2 n\theta \ d\theta \text{ or } \frac{1}{2} \int_0^{\pi} \sin^2 n\theta \ d\theta.$$

For these functions, integrating over the interval $[0, 2\pi]$ would give an answer that is twice the correct value.

EXAMPLE 2   Finding the area between the loops of a limaçon

Calculate the area between the interior and exterior loops of the limaçon $r = \frac{1}{2} + \cos\theta$.

SOLUTION

This is the limaçon we graphed in Example 3 in Section 9.3. We reproduce that graph here, again with one portion shaded:

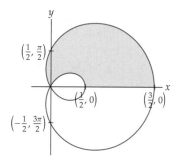

To compute the area between the loops of the limaçon, we will make use once more of the symmetry of the region. We will calculate the area of the shaded region of the figure. This area is one-half of our desired area. To do our calculation, we will first find the area between the x-axis and top half of the outer loop of the limaçon. Next, we will find the area between the x-axis and the top half of the inner loop. The difference of these areas will give us the area of the shaded region.

To compute the area between the x-axis and the top half of the outer loop of the limaçon, we use the integral

$$\frac{1}{2} \int_0^{2\pi/3} \left(\frac{1}{2} + \cos\theta \right)^2 \, d\theta.$$

The upper limit of this integral is $\frac{2\pi}{3}$, because that is the smallest positive value of θ at which $r = 0$ and the curve goes through the pole. The value of the integral is $\frac{\pi}{4} + \frac{3\sqrt{3}}{16}$. The area between the x-axis and the top half of the inner loop of the limaçon is given by the integral

$$\frac{1}{2} \int_\pi^{4\pi/3} \left(\frac{1}{2} + \cos\theta \right)^2 \, d\theta.$$

We integrate here over the interval $\left[\pi, \frac{4\pi}{3} \right]$ because this is the interval that traces the desired portion of the curve. Note that we could also have integrated over the interval $\left[\frac{2\pi}{3}, \pi \right]$. This would have given us the area between the x-axis and the bottom half of the inner loop of the limaçon. Since the two halves of the inner loop have equal area, the values of the two integrals would be equal. The value of either of these integrals is $\frac{\pi}{8} - \frac{3\sqrt{3}}{16}$. Therefore, the shaded region has area

$$\frac{\pi}{4} + \frac{3\sqrt{3}}{16} - \left(\frac{\pi}{8} - \frac{3\sqrt{3}}{16} \right) = \frac{\pi + 3\sqrt{3}}{8}.$$

The area between the two loops is twice this value, $\frac{\pi + 3\sqrt{3}}{4}$.   □

EXAMPLE 3   Finding the area between two polar curves

Calculate the area of the region in the polar plane that is inside the circle $r = 3\cos\theta$, but outside the cardioid $r = 1 + \cos\theta$.

SOLUTION

The region in question is shown as the shaded part of the following graph:

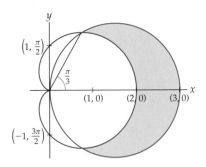

Using the symmetry of the region, we know that it is enough to compute the area of the top half of the region and then multiply by 2. Solving $3\cos\theta = 1 + \cos\theta$ for $\theta \in [0, \pi]$, we see that the two curves intersect when $\theta = \frac{\pi}{3}$. Therefore the area of the top half of the region is given by the definite integral

$$\frac{1}{2}\int_0^{\pi/3}((3\cos\theta)^2 - (1 + \cos\theta)^2)\,d\theta.$$

The value of this integral is $\frac{\pi}{2}$. Since the area of the shaded region is one-half of the original region, the desired area is π square units.   □

EXAMPLE 4   Computing the arc length of a polar curve

Compute the arc length of the cardioid defined by the polar equation $r = 1 + \cos\theta$.

SOLUTION

By Theorem 9.14, we need only evaluate the integral

$$\int_0^{2\pi}\sqrt{\left(\frac{d}{d\theta}(1 + \cos\theta)\right)^2 + (1 + \cos\theta)^2}\,d\theta.$$

After differentiating and simplifying, we find that the integral is equal to

$$\int_0^{2\pi}\sqrt{\sin^2\theta + 1 + 2\cos\theta + \cos^2\theta}\,d\theta = \int_0^{2\pi}\sqrt{2 + 2\cos\theta}\,d\theta.$$

The value of this definite integral is 8 (as you will show in Exercise 16). Therefore the arc length of the cardioid $r = 1 + \cos\theta$ is 8 units.   □

TEST YOUR UNDERSTANDING

▶ Why is the integral $\int_\alpha^\beta \frac{1}{2}(f(\theta))^2\,d\theta$ used to compute the area of a region in the polar plane bounded by the function $r = f(\theta)$? Why not just use the integral $\int_\alpha^\beta f(\theta)\,d\theta$?

▶ Given two polar equations $r = f(\theta)$ and $r = g(\theta)$, why is it insufficient to solve the equation $f(\theta) = g(\theta)$ for θ in order to find the points of intersection of the two graphs? How do you find all points of intersection on the graphs of f and g?

▶ How can the symmetries of the graphs of polar functions be used to simplify area calculations? Why is it important to know where curves intersect themselves or each other when you are calculating an area?

▶ How do you transform a polar function of the form $r = f(\theta)$ into parametric equations $x = x(\theta)$ and $y = y(\theta)$?

▶ How do you compute the arc length of a polar function $r = f(\theta)$?

EXERCISES 9.4

Thinking Back

Calculate each of the following definite integrals (these are precisely the definite integrals that were encountered in the reading for this section):

▶ $\dfrac{1}{2}\displaystyle\int_0^{\pi/4} \cos^2 2\theta \, d\theta$

▶ $\dfrac{1}{2}\displaystyle\int_0^{\pi} \cos^2 3\theta \, d\theta$

▶ $\dfrac{1}{2}\displaystyle\int_0^{2\pi/3} \left(\dfrac{1}{2} + \cos\theta\right)^2 d\theta$

▶ $\dfrac{1}{2}\displaystyle\int_\pi^{4\pi/3} \left(\dfrac{1}{2} + \cos\theta\right)^2 d\theta$

▶ $\dfrac{1}{2}\displaystyle\int_0^{\pi/3} \left((3\cos\theta)^2 - (1 + \cos\theta)^2\right) d\theta$

Concepts

0. Problem Zero: Read the section and make your own summary of the material.

1. True/False: Determine whether each of the statements that follow is true or false. If a statement is true, explain why. If a statement is false, provide a counterexample.

(a) *True or False:* To approximate the area of the region in the polar plane bounded by the function $r = f(\theta)$ and the rays $\theta = \alpha$ and $\theta = \beta$, we can use a sum of areas of sectors of circles.

(b) *True or False:* Suppose we subdivide the interval of angles $\theta \in \left[\dfrac{\pi}{4}, \dfrac{\pi}{2}\right]$ into four equal subintervals. Then $\Delta\theta = \dfrac{\pi}{16}$.

(c) *True or False:* The area of the region in the polar plane bounded by the function $r = f(\theta)$ and the rays $\theta = \alpha$ and $\theta = \beta$ is given by the definite integral $\int_\alpha^\beta f(\theta) \, d\theta$.

(d) *True or False:* The area between the continuous function $r = f(\theta)$ and the θ-axis on the interval $[\alpha, \beta]$ in the θr-plane is given by the integral $\int_\alpha^\beta |f(\theta)| \, d\theta$.

(e) *True or False:* Since the graph of $r = 2\cos\theta$ is a circle with radius 1, the value of the integral $\dfrac{1}{2}\int_0^{2\pi}(2\cos\theta)^2 \, d\theta$ is π.

(f) *True or False:* The polar equation $r = \cos 4\theta$ is traced twice as θ varies from 0 to 2π.

(g) *True or False:* Every polar function of the form $r = f(\theta)$ can be written with parametric equations.

(h) *True or False:* If $r = f(\theta)$ is a differentiable function for $0 \le \theta \le 2\pi$, then the integral $\int_0^{2\pi} \sqrt{(f(\theta))^2 + (f'(\theta))^2} \, d\theta$ represents the length of the curve $r = f(\theta)$ for $0 \le \theta \le 2\pi$.

2. Examples: Construct examples of the thing(s) described in the following. Try to find examples that are different than any in the reading.

(a) An integral that represents the area of a circle with radius a in polar coordinates.

(b) An integral that represents the circumference of a circle with radius a in polar coordinates.

(c) An integral that represents the circumference of a circle with radius a in terms of parametric equations.

3. When we use rectangular coordinates to approximate the area of a region, we subdivide the region into vertical strips and use a sum of areas of rectangles to approximate the area. Explain why we use a "wedge" (i.e., a sector of a circle) and not a rectangle when we use polar coordinates to compute an area.

4. When we investigated area in rectangular coordinates in Chapter 4, we often tried to find the areas of regions under curves $y = f(x)$ from $x = a$ to $x = b$. In the polar plane, the typical region whose area we wish to find is a region R bounded by two rays $\theta = \alpha$ and $\theta = \beta$ and a polar function of the form $r = f(\theta)$. Why is this our basic type of region in the polar plane?

5. In this section we described a method for approximating the area of a polar region bounded by two rays $\theta = \alpha$ and $\theta = \beta$ and a polar function $r = f(\theta)$. The method involved a "subdivide, approximate, and add" strategy in which we fixed some general notation. Draw a carefully labeled picture that illustrates the roles of $\Delta\theta$, θ_k^*, and $f(\theta_k^*)$ for one approximating sector S_k.

6. Explain how we arrive at the definite integral formula $\dfrac{1}{2}\int_\alpha^\beta (f(\theta))^2 \, d\theta$ in Theorem 9.13 for computing the area bounded by a polar function $r = f(\theta)$ on an interval $[\alpha, \beta]$. (Your explanation should include a limit of Riemann sums.) What would the integral $\int_\alpha^\beta f(\theta) \, d\theta$ represent?

7. Why do we require that $0 \le \beta - \alpha \le 2\pi$ in the statement of Theorem 9.13?

8. Consider the three-petaled polar rose defined by $r = \cos 3\theta$. Explain why the definite integral $\dfrac{1}{2}\int_0^{2\pi} \cos^2 3\theta \, d\theta$ calculates *twice* the area bounded by the petals of this rose.

9. Explain how the symmetries of the graphs of polar functions can be used to simplify area calculations.

10. Explain how to use parametric equations to transform a polar function $r = f(\theta)$.

11. Explain how to use parametric equations to transform the function $y = f(x)$.

12. What is the formula for computing the arc length of a polar curve $r = f(\theta)$ where $\theta \in [\alpha, \beta]$? What conditions on the polar function $f(\theta)$ are necessary for this formula to hold?

13. Rick wants to show that the circumference of a unit circle is 2π. He decides to use the arc length formula given in Theorem 9.14 with the function $r = \sin\theta$ on the interval $[0, 2\pi]$ and obtains 2π as the length. Explain the two mistakes he made and how they cancelled to give the correct answer.

14. Give a geometric explanation of why

$$\frac{n}{2}\int_0^{2\pi/n} r^2\, d\theta = \pi r^2$$

for any positive real number r and any positive integer n. Would the equation also hold for non-integer values of n?

15. The following integral expression may be used to find the area of a region in the polar coordinate plane:

$$\frac{1}{2}\int_0^{\pi/4} \sin^2\theta\, d\theta + \frac{1}{2}\int_{\pi/4}^{\pi/2} \cos^2\theta\, d\theta.$$

Sketch the region and then compute its area. (If you prefer, you may use a simpler integral to compute the same area.)

16. Complete Example 4 by evaluating the integral $\int_0^{2\pi}\sqrt{2 + 2\cos\theta}\, d\theta$.

Skills

In Exercises 17–25 find a definite integral expression that represents the area of the given region in the polar plane, and then find the exact value of the expression.

17. The region enclosed by the spiral $r = \theta$ and the x-axis on the interval $0 \le \theta \le \pi$.

18. The region inside one loop of the lemniscate $r^2 = \sin 2\theta$.

19. The region between the two loops of the limaçon $r = 1 + \sqrt{2}\cos\theta$.

20. The region between the two loops of the limaçon $r = \sqrt{3} - 2\sin\theta$.

21. The region inside the cardioid $r = 3 - 3\sin\theta$ and outside the cardioid $r = 1 + \sin\theta$.

22. The region inside both of the cardioids $r = 3 - 3\sin\theta$ and $r = 1 + \sin\theta$.

23. The region inside the circle $x^2 + y^2 = 1$ to the right of the vertical line $x = \frac{1}{2}$.

24. The region bounded by the limaçon $r = 1 + k\sin\theta$, where $0 < k < 1$. Bonus: Explain why the area approaches π as $k \to 0$.

25. The graph of the polar equation $r = \sec\theta - 2\cos\theta$ for $\theta \in \left(-\frac{\pi}{2}, \frac{\pi}{2}\right)$ is called a ***strophoid***. Graph the strophoid and find the area bounded by the loop of the graph.

In Exercises 26–30 find a definite integral that represents the length of the specified polar curve, and then find the exact value of the integral.

26. The spiral $r = \theta$ for $0 \le \theta \le 2\pi$.

27. The spiral $r = e^\theta$ for $0 \le \theta \le 2\pi$.

28. The spiral $r = e^\theta$ for $2k\pi \le \theta \le 2(k+1)\pi$.

29. The spiral $r = e^{\alpha\theta}$ for $0 \le \theta \le 2\pi$, where α is a nonzero constant.

30. The cardioid $r = 2 - 2\sin\theta$, for $0 \le \theta \le 2\pi$.

In Exercises 31–36 find a definite integral that represents the length of the specified polar curve, and then use a graphing calculator or computer algebra system to approximate the value of the integral.

31. One petal of the polar rose $r = \cos 2\theta$.

32. One petal of the polar rose $r = \cos 3\theta$.

33. One petal of the polar rose $r = \cos 4\theta$.

34. The entire limaçon $r = 1 + 2\sin\theta$.

35. The inner loop of the limaçon $r = 1 + 2\sin\theta$.

36. The limaçon $r = 2 + \cos\theta$.

37. Complete Example 2 by evaluating the integral expression

$$\int_0^{2\pi/3}\left(\frac{1}{2} + \cos\theta\right)^2 d\theta - \int_\pi^{4\pi/3}\left(\frac{1}{2} + \cos\theta\right)^2 d\theta.$$

Each of the integrals or integral expressions in Exercises 38–44 represents the area of a region in the plane. Use polar coordinates to sketch the region and evaluate the expressions.

38. $\dfrac{1}{2}\displaystyle\int_0^{2\pi}(1 + \sin\theta)^2\, d\theta$

39. $\displaystyle\int_0^{\pi}(1 + \cos\theta)^2\, d\theta$

40. $\displaystyle\int_{-\pi/2}^{\pi/2}(2 - \sin\theta)^2\, d\theta$

41. $\displaystyle\int_0^{\pi/2}\sin 2\theta\, d\theta$

42. $3\displaystyle\int_0^{\pi/2}\sin^2 3\theta\, d\theta$

43. $\dfrac{1}{2}\displaystyle\int_0^{2\pi}(2 + \sin 4\theta)^2\, d\theta$

44. $\displaystyle\int_{-\pi/4}^{\pi/2}\left(\frac{\sqrt{2}}{2} + \sin\theta\right)^2 d\theta - \int_{-\pi/2}^{-\pi/4}\left(\frac{\sqrt{2}}{2} + \sin\theta\right)^2 d\theta$

45. Use Theorem 9.13 to show that the area of the circle defined by the polar equation $r = a$ is πa^2.

46. Use Theorem 9.13 to show that the area of the circle defined by the polar equation $r = 2a\cos\theta$ is πa^2.

47. Use Theorem 9.14 to show that the circumference of the circle defined by the polar equation $r = a$ is $2\pi a$.

48. Use Theorem 9.14 to show that the area of the circle defined by the polar equation $r = 2a\sin\theta$ is $2\pi a$.

49. Find the area interior to two circles with the same radius if each circle passes through the center of the other. (*Hint: Consider the circles $r = a$ and $r = 2a\cos\theta$.*)

Applications

50. Annie is designing a kayak with central cross section given by $r = 1 - 0.5 \sin \theta$, where r is measured in feet.

Kayak and central cross section

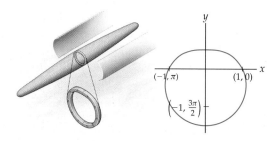

She will cover the kayak with an impermeable polyurethane fabric. The fabric comes in rolls 1 meter wide. Annie needs to use the length of the fabric to run the length of her boat, so she hopes to be able to use just two widths of the fabric to cover the boat. Will she be able to?

51. Annie's second potential design for the central cross section of her kayak is

$$r = 0.83(1 - 0.5 \sin \theta)(1 + 0.2 \cos^2 \theta).$$

New kayak and its central cross section

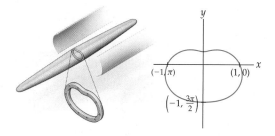

Will this design allow Annie to use just two widths of the fabric? *(Hint: Use a numerical integration technique to approximate the length of the cross section.)*

52. Ian sometimes sews his own outdoor gear. He wants to make a body-hugging climbing pack. The bottom of the pack is the area outside the circle $r = 14$, but inside $r = 14\sqrt{2} \sin \theta$, where r is measured in inches. Ian requires the pack to carry 2500 cubic inches of gear. If the pack has vertical sides, how tall does Ian need to make it?

Proofs

53. Prove that the area of a sector with central angle ϕ in a circle of radius r is given by $A = \frac{1}{2}\phi r^2$.

54. Prove that the area enclosed by one petal of the polar rose $r = \cos 3\theta$ is the same as the area enclosed by one petal of the polar rose $r = \sin 3\theta$.

55. Prove that the area enclosed by all of the petals of the polar rose $r = \cos 2n\theta$ is the same for every positive integer n.

56. Prove that the area enclosed by all of the petals of the polar rose $r = \sin(2n + 1)\theta$ is the same for every positive integer n.

57. Prove that the part of the polar curve $r = \frac{1}{\theta}$ that lies inside the circle defined by the polar equation $r = 1$ has infinite length.

58. Complete the proof of Theorem 9.14 by verifying that

$$(f'(\theta) \cos \theta - f(\theta) \sin \theta)^2 + (f'(\theta) \sin \theta + f(\theta) \cos \theta)^2$$

is equal to

$$(f'(\theta))^2 + (f(\theta))^2.$$

Thinking Forward

The cardioid defined by $r = 1 + \cos \theta$ and its interior is translated one unit perpendicular to the xy-plane to define a "cylinder."

▶ *Volume:* Find the volume of the cylinder.

▶ *Surface area:* Find the surface area of the cylinder. Remember to include the top and bottom of the solid in your calculations.

9.5   CONIC SECTIONS*

▶ Circles, ellipses, parabolas, and hyperbolas

▶ Using rectangular coordinates to express conics

▶ Using polar coordinates to express conics

Circles and Ellipses

When the line $y = x$ is revolved about the y-axis, we obtain a double cone. The point where the two cones meet is called the **vertex**. Any plane that does not pass through the vertex will intersect the cone in a curve called a **conic section**. There are four basic curves that may be formed when a plane intersects this cone:

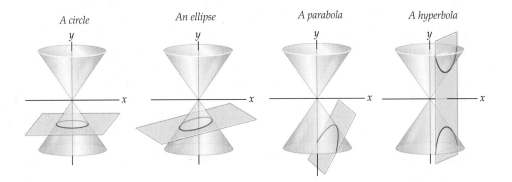

A circle   *An ellipse*   *A parabola*   *A hyperbola*

A plane through the vertex will intersect the cone in either a single point, a line, or a pair of intersecting lines. These intersections are considered to be **degenerate** conic sections. For the remainder of this section we will assume that our conic sections are not degenerate.

As shown, a plane perpendicular to the y-axis but that misses the vertex of the cone intersects the cone in a circle. If such a plane is tilted slightly, the curve of intersection will be an ellipse. As we know, the equation of the circle with center (x_0, y_0) and radius R can be written in the form $(x - x_0)^2 + (y - y_0)^2 = R^2$. When the circle is centered at the origin, its equation can be written in the form $x^2 + y^2 = R^2$. Such an equation can also be written in the form

$$\frac{x^2}{R^2} + \frac{y^2}{R^2} = 1.$$

If we allow the denominators of the two quotients in this equation to be different positive constants, we obtain the equation

$$\frac{x^2}{A^2} + \frac{y^2}{B^2} = 1.$$

You may know that the graph of such an equation is an ellipse.

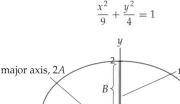

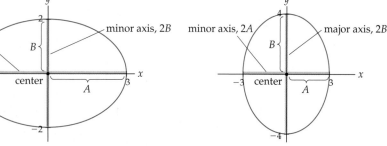

When $A > B$, the graph of the ellipse $\frac{x^2}{A^2} + \frac{y^2}{B^2} = 1$ is wider than it is tall. In this case, the segment of length $2A$ is called the **major axis** of the ellipse and the segment of length $2B$ is called the **minor axis**. When $A < B$, the ellipse is taller than it is wide. In this case, $2B$ is the length of the major axis and $2A$ is the length of the minor axis. We may consider a circle to be an ellipse in which the major and minor axes are equal.

More classically, each conic section has a geometric definition. For example, the geometric definition for an ellipse is given in Definition 9.15:

DEFINITION 9.15

Ellipse

Given two points in a plane, called **foci**, an **ellipse** is the set of points in the plane such that the sum of the distances to the two foci is a constant.

We will show that the curve with equation $\frac{x^2}{A^2} + \frac{y^2}{B^2} = 1$, where $A > B$, satisfies Definition 9.15. The argument when $A < B$ is similar.

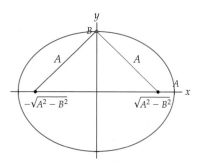

Let $\mathcal{V}$ represent the graph of the equation $\frac{x^2}{A^2} + \frac{y^2}{B^2} = 1$. From this equation, we may see that $\mathcal{V}$ is symmetric with respect to both the x- and y-axes. Furthermore, the points $(-A, 0)$, $(A, 0)$, $(0, -B)$, and $(0, B)$ are all on $\mathcal{V}$. We will show that the sum of the distances from every point on $\mathcal{V}$ to the points $(-\sqrt{A^2 - B^2}, 0)$ and $(\sqrt{A^2 - B^2}, 0)$ is the constant $2A$. This will demonstrate that $\mathcal{V}$ is an ellipse and that $(-\sqrt{A^2 - B^2}, 0)$ and $(\sqrt{A^2 - B^2}, 0)$ are the coordinates of the foci. We begin by letting $C = \sqrt{A^2 - B^2}$ and (x, y) be a point on $\mathcal{V}$. Also,

$$C^2 = A^2 - B^2 \quad \text{and} \quad y^2 = B^2 - \frac{B^2 x^2}{A^2}.$$

Let D_1 represent the distance from (x, y) to the point $(C, 0)$. Thus,

$$D_1^2 = (x - C)^2 + y^2$$

$$= x^2 - 2Cx + C^2 + B^2 - \frac{B^2 x^2}{A^2} \qquad \leftarrow y^2 = B^2 - \frac{B^2 x^2}{A^2}$$

$$= A^2 - 2Cx + x^2 - \frac{B^2 x^2}{A^2} \qquad \leftarrow C^2 = A^2 - B^2$$

$$= \frac{A^4 - 2A^2 Cx + (A^2 - B^2)x^2}{A^2} \qquad \leftarrow \text{algebra}$$

$$= \frac{A^4 - 2A^2 Cx + C^2 x^2}{A^2} \qquad \leftarrow C^2 = A^2 - B^2$$

Since both the numerator and denominator of the last expression are perfect squares, $D_1 = \frac{A^2 - Cx}{A}$. In Exercise 55 you will be asked to perform a similar computation to show that the distance D_2 from the point (x, y) to the point $(-C, 0)$ is $D_2 = \frac{A^2 + Cx}{A}$. The sum of these distances is

$$D_1 + D_2 = \frac{A^2 - Cx}{A} + \frac{A^2 + Cx}{A} = 2A.$$

This gives us the following theorem (the proof of part (b) is similar to the preceding argument and is left for Exercise 56):

THEOREM 9.16   **The Equation and Foci of an Ellipse**

(a) If $A > B > 0$, the graph of the equation $\frac{x^2}{A^2} + \frac{y^2}{B^2} = 1$ is an ellipse with foci $(\pm\sqrt{A^2 - B^2}, 0)$.

(b) If $0 < A < B$, the graph of the equation $\frac{x^2}{A^2} + \frac{y^2}{B^2} = 1$ is an ellipse with foci $(0, \pm\sqrt{B^2 - A^2})$.

For example, the foci of the ellipse $\frac{x^2}{25} + \frac{y^2}{16} = 1$ are $(\pm\sqrt{25 - 16}, 0) = (\pm 3, 0)$. Furthermore, the sum of the distances from every point on this ellipse to the foci is 10 units, the length of the major axis.

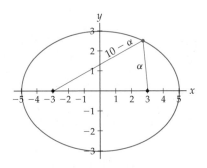

Parabolas

As a conic section, a parabola is obtained when the plane slicing a cone is parallel to a side of the cone. We studied parabolas extensively in Chapter 0. We know that the graph of the equation $y = x^2$ is an upwards-opening parabola with its vertex at the origin. We also saw that the graph of the equation $y = a(x - x_0)^2 + y_0$ is a parabola, as long as $a \neq 0$. The vertex of the new parabola is (x_0, y_0), while the constant a scales the graph of $y = x^2$ vertically. In addition, if $a < 0$, the new parabola opens downwards.

We may also define a parabola geometrically:

DEFINITION 9.17

Parabola

A ***parabola*** is the set of all points in the plane equidistant from a given fixed line in the plane, called the ***directrix*** of the parabola, and a given fixed point called the ***focus*** of the parabola.

We show that the curve $\mathcal{P}$ with equation $y = ax^2$, where $a > 0$, satisfies Definition 9.17. We will see that the focus is on the positive y-axis with coordinates $\left(0, \frac{1}{4a}\right)$ and the directrix is the horizontal line with equation $y = -\frac{1}{4a}$.

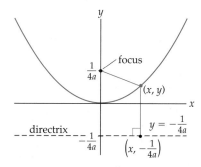

If we choose an arbitrary point (x, y) on the parabola, the distance from (x, y) to the line with equation $y = -\frac{1}{4a}$ is $y + \frac{1}{4a}$, while its distance from the point $\left(0, \frac{1}{4a}\right)$ is

$$
\begin{aligned}
\sqrt{x^2 + \left(y - \frac{1}{4a}\right)^2} &= \sqrt{x^2 + y^2 - \frac{y}{2a} + \frac{1}{16a^2}} && \leftarrow \text{algebra} \\
&= \sqrt{\frac{y}{a} + y^2 - \frac{y}{2a} + \frac{1}{16a^2}} && \leftarrow y = ax^2 \\
&= \sqrt{y^2 + \frac{y}{2a} + \frac{1}{16a^2}} && \leftarrow \text{algebra} \\
&= y + \frac{1}{4a}. && \leftarrow \text{the quantity is a perfect square}
\end{aligned}
$$

Since the distances from any point on $\mathcal{P}$ to the directrix and the focus are equal, $\mathcal{P}$ satisfies Definition 9.17. We generalize this result in the following theorem:

THEOREM 9.18

The Focus and Directrix of a Parabola

Let $y = a(x - x_0)^2 + y_0$ with $a \neq 0$ be the equation of a parabola. The coordinates of the focus of the parabola are $\left(x_0, y_0 + \frac{1}{4a}\right)$, and the equation of the directrix is $y = y_0 - \frac{1}{4a}$.

Hyperbolas

The final conic section we will discuss is the hyperbola. A hyperbola is formed when a plane intersects both halves of the double cone (in a nondegenerate way). The most familiar hyperbola is the graph of the function $y = \dfrac{1}{x}$, although the graphs of the equations $x^2 - y^2 = 1$ and $y^2 - x^2 = 1$ are also hyperbolas.

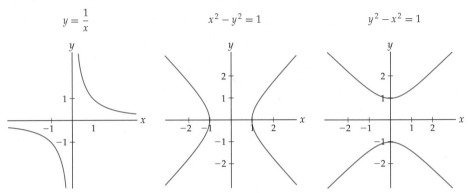

We will be discussing primarily hyperbolas with defining equations $\dfrac{x^2}{A^2} - \dfrac{y^2}{B^2} = 1$ and $\dfrac{y^2}{B^2} - \dfrac{x^2}{A^2} = 1$, where A and B are positive constants.

Hyperbolas may also be defined geometrically:

DEFINITION 9.19

Hyperbola

Given two points in a plane, called **foci**, a **hyperbola** is the set of points in the plane for which the difference of the distances to the two foci is constant. The line containing the foci is called the **focal axis**, the midpoint of the segment connecting the foci is called the **center** of the hyperbola, and the points where the focal axis intersects the hyperbola are called the **vertices** of the hyperbola.

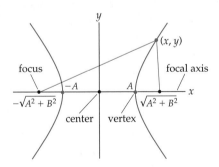

The proof of the following theorem is quite similar to the proof of Theorem 9.16 and is left for Exercises 57 and 58:

THEOREM 9.20

The Equation, Foci, and Asymptotes of a Hyperbola

Let A and B be positive.

  (a) The graph of the equation $\dfrac{x^2}{A^2} - \dfrac{y^2}{B^2} = 1$ is a hyperbola with foci $(\pm\sqrt{A^2 + B^2}, 0)$.

  (b) The graph of the equation $\dfrac{y^2}{B^2} - \dfrac{x^2}{A^2} = 1$ is a hyperbola with foci $(0, \pm\sqrt{A^2 + B^2})$.

The lines $y = \dfrac{B}{A}x$ and $y = -\dfrac{B}{A}x$ are asymptotes for these hyperbolas.

$$\frac{x^2}{A^2} - \frac{y^2}{B^2} = 1 \qquad\qquad \frac{y^2}{B^2} - \frac{x^2}{A^2} = 1$$

For example, since $A = 3$ and $B = 4$ in the equation $\frac{x^2}{9} - \frac{y^2}{16} = 1$, the foci of this hyperbola are $(\pm 5, 0)$ and the graph has asymptotes $y = \frac{4}{3}x$ and $y = -\frac{4}{3}x$.

Conic Sections in Polar Coordinates

In Definiton 9.17 we saw that a parabola is the set of all points in the plane equidistant from a given fixed line in the plane, where the line is called the directrix of the parabola and the fixed point is called the focus of the parabola. If we let F and l be the focus and directrix of the parabola, respectively, we may rephrase this definition by saying that a parabola is the set of all points P in the plane such that the ratio of the distance from F to P to the distance from l to P is 1. If we let FP represent the length of the segment from F to P, we have

$$\frac{FP}{\text{the distance from } l \text{ to } P} = 1.$$

We may generalize this idea to provide alternative geometric definitions for ellipses and hyperbolas. We have already seen that circles have very simple representations in polar coordinates. The alternative definitions will allow us to find simple polar coordinate representations for parabolas, ellipses, and hyperbolas. For consistency, we may orient any of these three conics as shown here:

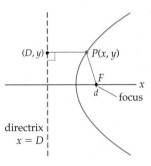

That is, we will use rightwards-opening parabolas with equations $x = c(y - y_0)^2$ for $c > 0$. For the parabola there is a single vertical directrix, l, with equation $x = D$. If we let $P = (x, y)$ be a point on the parabola, then the distance from P to l is DP and the distance from P to the focus, F, is FP. Here we have

$$\frac{FP}{DP} = 1.$$

The ellipses we consider will have equations $\frac{x^2}{A^2} + \frac{y^2}{B^2} = 1$, where $A > B$ and, thus, where the major axis of the ellipse is on the x-axis. The hyperbolas we will consider are centered at the origin, open to the left and right and, therefore, have equations $\frac{x^2}{A^2} - \frac{y^2}{B^2} = 1$. Ellipses and hyperbolas also have directrices. In fact, every ellipse and every hyperbola has two directrices.

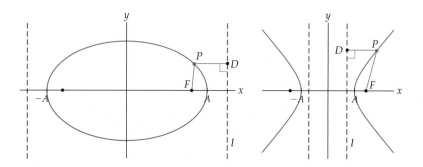

If F is the focus of an ellipse or a hyperbola $\mathcal{V}$, and D is a point on the directrix closest to that focus, then for every point P on $\mathcal{V}$, we have

$$\frac{FP}{DP} = e,$$

where e is a positive constant called the *eccentricity* of the curve. Recall that when $A > B$, the foci of the ellipse $\frac{x^2}{A^2} + \frac{y^2}{B^2} = 1$ are $(\pm\sqrt{A^2 - B^2}, 0)$ and the foci of the hyperbola $\frac{x^2}{A^2} - \frac{y^2}{B^2} = 1$ are $(\pm\sqrt{A^2 + B^2}, 0)$.

DEFINITION 9.21

The Eccentricity of Conic Sections

(a) The *eccentricity* e of a parabola is 1.

(b) When $A > B > 0$, the *eccentricity* of the ellipse $\frac{x^2}{A^2} + \frac{y^2}{B^2} = 1$ is $e = \frac{\sqrt{A^2 - B^2}}{A}$, and when $0 < A < B$, $e = \frac{\sqrt{B^2 - A^2}}{B}$.

(c) The *eccentricity* of the hyperbola $\frac{x^2}{A^2} - \frac{y^2}{B^2} = 1$ is $e = \frac{\sqrt{A^2 + B^2}}{A}$, and for $\frac{y^2}{B^2} - \frac{x^2}{A^2} = 1$, $e = \frac{\sqrt{A^2 + B^2}}{B}$.

Note that from Definition 9.21 (b), $0 < e < 1$ for an ellipse, and from Definition 9.21 (c), $e > 1$ for a hyperbola. For example, the eccentricity of the ellipse $\frac{x^2}{25} + \frac{y^2}{16} = 1$ is $e = \frac{\sqrt{25 - 16}}{5} = \frac{3}{5}$.

In Exercise 59 you will show that, for an ellipse or a hyperbola, the eccentricity is also

$$e = \frac{\text{the distance between the foci}}{\text{the distance between the vertices}}.$$

We use the eccentricity to define the directrices of ellipses and hyperbolas:

DEFINITION 9.22

Directrices of Ellipses and Hyperbolas

(a) When $A > B$, the **directrices** of the ellipse $\frac{x^2}{A^2} + \frac{y^2}{B^2} = 1$ are the vertical lines $x = \frac{A}{e}$ and $x = -\frac{A}{e}$, where e is the eccentricity of the ellipse. When $B > A$, the **directrices** of the ellipse are the horizontal lines $y = \frac{B}{e}$ and $y = -\frac{B}{e}$, where e is the eccentricity of the ellipse.

(b) The **directrices** of the hyperbola $\frac{x^2}{A^2} - \frac{y^2}{B^2} = 1$ are the vertical lines $x = \frac{A}{e}$ and $x = -\frac{A}{e}$, where e is the eccentricity of the hyperbola. The **directrices** of the hyperbola $\frac{y^2}{B^2} - \frac{x^2}{A^2} = 1$ are the horizontal lines $y = \frac{B}{e}$ and $y = -\frac{B}{e}$, where e is the eccentricity of the hyperbola.

Continuing with the brief example we started before, we see that the directrices of the ellipse $\frac{x^2}{25} + \frac{y^2}{16} = 1$ are the lines $x = \pm\frac{A}{e} = \pm\frac{5}{3/5} = \pm\frac{25}{3}$.

With these definitions for eccentricity and directrices established, we have the following theorem:

THEOREM 9.23

Eccentricity as the Ratio of Two Distances

Let F be the focus of a parabola, an ellipse, or a hyperbola, and let l be the directrix of the curve closest to F. If P is a point on the curve and D is the point on l closest to P, then

$$\frac{FP}{DP} = e.$$

The proof of Theorem 9.23 is mostly computational and is left for Exercises 60 and 61.

Note that the eccentricity of an ellipse is in the interval $(0, 1)$, the eccentricity of a parabola is 1 and the eccentricity of a hyperbola is greater than 1. We are now ready to express the equations for ellipses, parabolas, and hyperbolas in polar coordinates. We use a polar coordinate system in which the pole is at a focus, F, of the conic section. We position the parabola so that the directrix is perpendicular to the polar axis and the parabola opens to the right. For the ellipse and hyperbola, the focal axis should coincide with the polar axis. Let V represent any of these conics and P be a point on V. Let $x = u$ be the equation of the directrix closest to the pole. The following figure illustrates the situation for an ellipse:

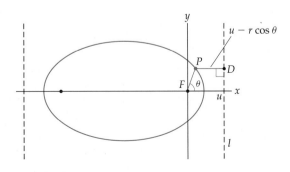

Now, the distance from the focus to the curve is $FP = r$. The distance from P to the directrix is $DP = u - r\cos\theta$. By Theorem 9.23 we have

$$\frac{FP}{DP} = \frac{r}{u - r\cos\theta} = e.$$

When we solve this equation for r, we get the following theorem:

THEOREM 9.24   **Conics in Polar Coordinates**

The graph of the polar equation

$$r = \frac{eu}{1 + e\cos\theta}$$

is a conic with eccentricity e and directrix $x = u$.

Examples and Explorations

EXAMPLE 1   **Finding the foci, eccentricity, and directrices for an ellipse**

The graph of each of the equations that follow is an ellipse. Find the foci, eccentricity, and directrices for each.

(a) $\dfrac{x^2}{16} + \dfrac{y^2}{4} = 1$ (b) $9x^2 + 25y^2 = 900$

(c) $\dfrac{x^2}{16} + \dfrac{y^2}{25} = 1$ (d) $x^2 + 4x + 9y^2 - 18y - 23 = 0$

SOLUTION

(a) Theorem 9.16 tells us that when $A > B$, the foci of the ellipse with equation $\frac{x^2}{A^2} + \frac{y^2}{B^2} = 1$ are $(\pm\sqrt{A^2 - B^2}, 0)$. Therefore, the foci of the ellipse with equation $\frac{x^2}{4^2} + \frac{y^2}{2^2} = 1$ have coordinates $(\pm\sqrt{4^2 - 2^2}, 0) = (\pm 2\sqrt{3}, 0)$.

From Definition 9.21, the eccentricity of the ellipse with equation $\frac{x^2}{A^2} + \frac{y^2}{B^2} = 1$ is $e = \frac{\sqrt{A^2 - B^2}}{A}$ when $A > B$. Here we have $e = \frac{2\sqrt{3}}{4} = \frac{\sqrt{3}}{2}$.

From Definition 9.22, the directrices of the ellipse with equation $\frac{x^2}{A^2} + \frac{y^2}{B^2} = 1$ are the vertical lines $x = \pm\frac{A}{e}$ when $A > B$. Here, the directrices have equations

$$x = \pm\frac{4}{\sqrt{3}/2} = \pm\frac{8\sqrt{3}}{3}.$$

(b) Before we use the theorem and definitions we mentioned in part (a), we need to rewrite the equation in the correct form. Here, $9x^2 + 25y^2 = 900$ is equivalent to

$$\frac{x^2}{10^2} + \frac{y^2}{6^2} = 1.$$

Therefore, the foci of this ellipse have coordinates $(\pm\sqrt{10^2 - 6^2}, 0) = (\pm 8, 0)$. The eccentricity of the ellipse is $e = \frac{8}{10} = \frac{4}{5}$, and the directrices have equations

$$x = \pm\frac{10}{4/5} = \pm\frac{25}{2}.$$

(c) Theorem 9.16 also tells us that when $A < B$, the foci of the ellipse with equation

$$\frac{x^2}{A^2} + \frac{y^2}{B^2} = 1$$

are $(0, \pm\sqrt{B^2 - A^2})$. Therefore, the foci of the ellipse with equation $\frac{x^2}{16} + \frac{y^2}{25} = 1$ have coordinates $(0, \pm\sqrt{5^2 - 4^2}) = (0, \pm 3)$.

We use Definitions 9.21 and 9.22 to find the eccentricity and directrices of this ellipse. The eccentricity will be

$$e = \frac{\sqrt{B^2 - A^2}}{B},$$

and the directrices will be horizontal lines with equations

$$y = \frac{B}{e} \quad \text{and} \quad y = -\frac{B}{e}.$$

Here, we have $e = \frac{\sqrt{5^2 - 4^2}}{5} = \frac{3}{5}$. The directrices have equations $y = \pm\frac{5}{3/5} = \pm\frac{25}{3}$.

(d) For our final equation in this group, we complete the square for each variable in the equation $x^2 + 4x + 9y^2 - 18y - 23 = 0$. After completing the squares, we have the equation

$$(x + 2)^2 + 9(y - 1)^2 = 36.$$

Dividing both sides of the equation by 36, we obtain

$$\frac{(x + 2)^2}{6^2} + \frac{(y - 1)^2}{2^2} = 1.$$

The graph of this equation is the graph of the ellipse $\frac{x^2}{6^2} + \frac{y^2}{2^2} = 1$ translated two units to the left and one unit up. The ellipse with equation $\frac{x^2}{6^2} + \frac{y^2}{2^2} = 1$ has foci $(\pm\sqrt{6^2 - 2^2}, 0) = (\pm 4\sqrt{2}, 0)$. The eccentricity of this ellipse is $e = \frac{4\sqrt{2}}{6} = \frac{2\sqrt{2}}{3}$, and the directrices have equations $x = \pm\frac{6}{2\sqrt{2}/3} = \pm\frac{9\sqrt{2}}{2}$. Therefore, the foci of the ellipse with equation $\frac{(x+2)^2}{6^2} + \frac{(y-1)^2}{2^2} = 1$ are $(-2 \pm 4\sqrt{2}, 1)$. Translating an ellipse does not change its eccentricity, but the directrices are $x = -2 \pm \frac{9\sqrt{2}}{2}$.   □

EXAMPLE 2   Finding the focus and directrix for a parabola

Find the focus and directrix for the parabola with equation $y = 4x^2 + 16x + 65$.

SOLUTION

We begin by completing the square. Here, we have $y = 4(x + 4)^2 + 1$. By Theorem 9.18, the focus of the parabola with equation $y = a(x - x_0)^2 + y_0$ is $\left(x_0, y_0 + \frac{1}{4a}\right)$ and the equation of the directrix is $y = y_0 - \frac{1}{4a}$. Here $a = 4$, $x_0 = -4$, and $y_0 = 1$. Therefore, the focus is the point $\left(-4, 1 + \frac{1}{4\cdot 4}\right) = \left(-4, \frac{17}{16}\right)$ and the directrix has equation $y = 1 - \frac{1}{4\cdot 4} = \frac{15}{16}$.   □

EXAMPLE 3   **Finding the foci, asymptotes, eccentricity, and directrices for a hyperbola**

The graph of each of the equations that follow is a hyperbola. Find the foci, asymptotes, eccentricity, and directrices for each.

(a) $\dfrac{x^2}{16} - \dfrac{y^2}{4} = 1$

(b) $9x^2 - 25y^2 = 900$

(c) $\dfrac{y^2}{25} - \dfrac{x^2}{16} = 1$

(d) $x^2 + 4x - 9y^2 - 18y - 41 = 0$

SOLUTION

(a) Theorem 9.20 tells us that the foci of the hyperbola with equation $\dfrac{x^2}{A^2} - \dfrac{y^2}{B^2} = 1$ are

$$(\sqrt{A^2 + B^2}, 0) \quad \text{and} \quad (-\sqrt{A^2 + B^2}, 0).$$

Therefore, the foci of the hyperbola with equation $\dfrac{x^2}{4^2} - \dfrac{y^2}{2^2} = 1$ have coordinates

$$(\pm\sqrt{4^2 + 2^2}, 0) = (\pm 2\sqrt{5}, 0).$$

The same theorem also tells us that the hyperbola will have asymptotes $y = \pm\dfrac{B}{A}x$. The asymptotes for this hyperbola have equations $y = \pm\dfrac{2}{4}x = \pm\dfrac{1}{2}x$.

From Definition 9.21, the eccentricity of the hyperbola with equation $\dfrac{x^2}{A^2} - \dfrac{y^2}{B^2} = 1$ is $e = \dfrac{\sqrt{A^2 + B^2}}{A}$. Here we have $e = \dfrac{2\sqrt{5}}{4} = \dfrac{\sqrt{5}}{2}$.

From Definition 9.22, the directrices of the hyperbola with equation $\dfrac{x^2}{A^2} - \dfrac{y^2}{B^2} = 1$ are the vertical lines $x = \pm\dfrac{A}{e}$. Here, the directrices have equations $x = \pm\dfrac{4}{\sqrt{5}/2} = \pm\dfrac{8\sqrt{5}}{5}$.

(b) Before we use the theorem and definitions we mentioned in part (a) we need to rewrite the equation in the correct form. Here, $9x^2 - 25y^2 = 900$ is equivalent to $\dfrac{x^2}{10^2} - \dfrac{y^2}{6^2} = 1$. Therefore, the foci of this hyperbola have coordinates $(\pm\sqrt{10^2 + 6^2}, 0) = (\pm 2\sqrt{34}, 0)$. The asymptotes have equations $y = \pm\dfrac{6}{10}x = \pm\dfrac{3}{5}x$. The eccentricity of the hyperbola is $e = \dfrac{2\sqrt{34}}{10} = \dfrac{\sqrt{34}}{5}$, and the directrices have equations $x = \pm\dfrac{10}{\sqrt{34}/5} = \pm\dfrac{25\sqrt{34}}{17}$.

(c) Theorem 9.20 also tells us that the foci of the hyperbola with equation $\dfrac{y^2}{B^2} - \dfrac{x^2}{A^2} = 1$ are $(0, \pm\sqrt{A^2 + B^2})$. Therefore, the foci of the hyperbola with equation $\dfrac{y^2}{25} - \dfrac{x^2}{16} = 1$ have coordinates $(0, \pm\sqrt{4^2 + 5^2}) = (0, \pm\sqrt{41})$. The asymptotes of a hyperbola with this type of equation are again $y = \pm\dfrac{B}{A}x$. Here, they have the equations $y = \pm\dfrac{5}{4}x$.

We use Definitions 9.21 and 9.22 to find the eccentricity and directrices of this hyperbola. The eccentricity will be

$$e = \frac{\sqrt{A^2 + B^2}}{B},$$

and the directrices will be horizontal lines with equations

$$y = \frac{B}{e} \quad \text{and} \quad y = -\frac{B}{e}.$$

Thus, $e = \dfrac{\sqrt{4^2 + 5^2}}{5} = \dfrac{\sqrt{41}}{5}$ and the directrices have equations $y = \pm\dfrac{5}{\sqrt{41}/5} = \pm\dfrac{25\sqrt{41}}{41}$.

(d) For our final equation in this group, begin by completing the square for each variable in the equation $x^2 + 4x - 9y^2 - 18y - 41 = 0$. You should confirm that this results in the equation

$$(x+2)^2 - 9(y+1)^2 = 36.$$

Dividing both sides of the equation by 36, we obtain

$$\frac{(x+2)^2}{6^2} - \frac{(y+1)^2}{2^2} = 1,$$

whose graph is the hyperbola with equation $\frac{x^2}{6^2} - \frac{y^2}{2^2} = 1$, translated two units to the left and one unit down. The hyperbola with equation $\frac{x^2}{6^2} - \frac{y^2}{2^2} = 1$ has foci $(\pm\sqrt{6^2 + 2^2}, 0) = (\pm 2\sqrt{10}, 0)$. The asymptotes have equations $y = \frac{2}{6}x = \frac{1}{3}x$ and $y = -\frac{1}{3}x$. The eccentricity of this hyperbola is $e = \frac{2\sqrt{10}}{6} = \frac{\sqrt{10}}{3}$, and the directrices have equations $x = \pm\frac{6}{\sqrt{10/3}} = \pm\frac{9\sqrt{10}}{5}$.

Therefore, the foci of our hyperbola are $(-2 \pm 2\sqrt{10}, -1)$. The asymptotes are the lines $y + 1 = \frac{1}{3}(x+2)$ and $y + 1 = -\frac{1}{3}(x+2)$, or equivalently, $y = \frac{1}{3}x - \frac{1}{3}$ and $y = -\frac{1}{3}x - \frac{5}{3}$. Translating the hyperbola did not change its eccentricity, but the directrices are $x = -2 \pm \frac{9\sqrt{10}}{5}$.   □

EXAMPLE 4

Using polar coordinates to graph conics

Use polar coordinates to graph each of the following equations:

(a) $r = \dfrac{1}{1 + \cos\theta}$ **(b)** $r = \dfrac{2}{1 + 2\cos\theta}$ **(c)** $r = \dfrac{1}{2 + \cos\theta}$

SOLUTION

By Theorem 9.24, the graph of the equation $r = \dfrac{eu}{1 + e\cos\theta}$ is a conic with a focus at the pole and with eccentricity e and directrix $x = u$.

(a) For $r = \dfrac{1}{1 + \cos\theta}$ the eccentricity is $e = 1$, so the graph is a parabola. The directrix will be the vertical line $x = 1$.

$$r = \frac{1}{1 + \cos\theta}$$

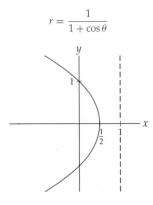

(b) For $r = \dfrac{2}{1 + 2\cos\theta}$ the eccentricity is $e = 2$, so the graph is a hyperbola. The directrix will be the vertical line $x = 1$.

$$r = \frac{2}{1 + 2\cos\theta}$$

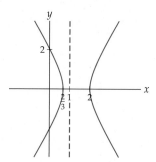

(c) Note that

$$r = \frac{1}{2 + \cos\theta} = \frac{1/2}{1 + 1/2\cos\theta}.$$

Here the eccentricity is $e = \dfrac{1}{2}$, so the graph is an ellipse. The directrix will be the vertical line $x = 1$.

$$r = \frac{1}{2 + \cos\theta}$$

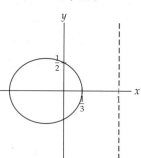

TEST YOUR UNDERSTANDING

▶ What is a conic section? How many different types of conic sections are there? What are the degenerate conic sections? What are the nondegenerate conic sections?

▶ What is the definition of a circle? What is the definition of an ellipse? What are the foci of an ellipse? What are the major and minor axes of an ellipse? What are the directrices of an ellipse? How do you recognize the equation for an ellipse in rectangular coordinates?

▶ What is the definition of a parabola? What is the focus of a parabola? What is the directrix of a parabola? How do you recognize the equation for a parabola in rectangular coordinates?

▶ What is the definition of a hyperbola? What are the foci of a hyperbola? What are the directrices of a hyperbola? How do you recognize the equation for a hyperbola in rectangular coordinates?

▶ What are the eccentricities of the conic sections? Specifically, what is the eccentricity of a parabola? In what intervals are the eccentricities for a ellipse and a hyperbola? How do you recognize the equation for a parabola, an ellipse, or a hyperbola in polar coordinates?

EXERCISES 9.5

Thinking Back

Finding the center and radius for a circle by completing the square: Each of the equations that follow is an equation for a circle. Find the center and radius for each.

▶ $x^2 + y^2 - y - \dfrac{3}{4} = 0$

▶ $x^2 + 4x + y^2 - 21 = 0$

▶ $x^2 + 3x + y^2 + y + \dfrac{3}{2} = 0$

Completing the square: Find the sets of points satisfying each of the following equations.

▶ $x^2 - 6x + y^2 + 2y + 10 = 0$

▶ $x^2 - 8x + y^2 - 10y + 40 = 0$

▶ $x^2 - 6x + y + 40 = 0$

▶ $x + y^2 - 4y - 5 = 0$

Concepts

0. *Problem Zero:* Read the section and make your own summary of the material.

1. *True/False:* Determine whether each of the statements that follow is true or false. If a statement is true, explain why. If a statement is false, provide a counterexample.

(a) *True or False:* A conic section is the intersection of a double cone and a plane.

(b) *True or False:* A point is a conic section.

(c) *True or False:* An ellipse is the set of points in the plane equidistant from two distinct points.

(d) *True or False:* Given a line $\mathcal{L}$ and a point P not on $\mathcal{L}$, a parabola is the set of points in the plane equidistant from $\mathcal{L}$ and P.

(e) *True or False:* A parabola is also a hyperbola.

(f) *True or False:* If the focus of a parabola is on the directrix of the parabola, the parabola is degenerate.

(g) *True or False:* Given two distinct points in a plane, a hyperbola is the set of points in the plane for which the quotient of the distances to the two points is constant.

(h) *True or False:* The graph of the polar equation $r = \dfrac{1}{1 + \sin\theta}$ is a parabola.

2. *Examples:* Construct examples of the thing(s) described in the following. Try to find examples that are different than any in the reading.

(a) A degenerate conic section.

(b) The equation for a hyperbola whose asymptotes are the coordinate axes.

(c) The equation for a parabola that opens to the left, in polar coordinates.

Complete the definitions in Exercises 3–5.

3. Given two points in a plane, called _____, an *ellipse* is the set of points in the plane for which _____.

4. Given a fixed line in the plane, called the _____, and a given fixed point called the _____, a *parabola* is the set of all points in the plane _____.

5. Given two points in a plane, called _____, a *hyperbola* is the set of points in the plane for which _____.

6. Explain why there are infinitely many different ellipses with the same foci.

7. Explain why there are infinitely many different hyperbolas with the same foci.

8. Three noncollinear points determine a unique circle. Do three noncollinear points determine a unique ellipse? If so, explain why. If not, provide three noncollinear points that are on two distinct ellipses.

9. The graph of the equation $\dfrac{x^2}{A^2} + \dfrac{y^2}{B^2} = 1$ is an ellipse for any nonzero constants A and B.

(a) If $A > B$, what is the eccentricity of the ellipse?

(b) If $A < B$, what is the eccentricity of the ellipse?

(c) Explain why the eccentricity, e, of an ellipse is always between 0 and 1.

(d) If $A > B$, what is $\lim_{A \to B} e$? What happens to the shape of the ellipse as $A \to B$?

(e) If $A > B$, what is $\lim_{A \to \infty} e$? What happens to the shape of the ellipse as $A \to \infty$?

10. The graph of the equation $\dfrac{x^2}{A^2} - \dfrac{y^2}{B^2} = 1$ is a hyperbola for any nonzero constants A and B.

(a) What is the eccentricity of the hyperbola?

(b) Explain why the eccentricity, e, of a hyperbola is always greater than 1.

(c) What is $\lim_{A \to 0} e$? What happens to the shape of the hyperbola as $A \to 0$?

(d) What is $\lim_{A \to \infty} e$? What happens to the shape of the hyperbola as $A \to \infty$?

11. When $A > B$, the directrices of the ellipse with equation $\dfrac{x^2}{A^2} + \dfrac{y^2}{B^2} = 1$ are the lines with equations $x = \pm\dfrac{A}{e}$, where e is the eccentricity of the ellipse. What are the equations for the directrices when $A < B$?

12. The directrices of the hyperbola with equation $\dfrac{x^2}{A^2} - \dfrac{y^2}{B^2} = 1$ are the lines with equations $x = \pm\dfrac{A}{e}$, where e is the eccentricity of the hyperbola. What are the equations for the directrices of the hyperbola with equation $\dfrac{y^2}{B^2} - \dfrac{x^2}{A^2} = 1$?

13. Sketch the graphs of the equations

$$r = \frac{1}{1 + \cos\theta} \quad \text{and} \quad r = \frac{1}{1 + \sin\theta}.$$

What is the relationship between these graphs? What is the eccentricity of each graph?

14. Sketch the graphs of the equations

$$r = \frac{2}{1 + 2\cos\theta} \quad \text{and} \quad r = \frac{2}{1 + 2\sin\theta}.$$

What is the relationship between these graphs? What is the eccentricity of each graph?

15. Sketch the graphs of the equations

$$r = \frac{2}{2 + \cos\theta} \quad \text{and} \quad r = \frac{2}{2 + \sin\theta}.$$

What is the relationship between these graphs? What is the eccentricity of each graph?

16. Let α, β, and γ be nonzero constants. Show that the graph of $r = \dfrac{\alpha}{\beta + \gamma\cos\theta}$ is a conic section with eccentricity $\left|\dfrac{\gamma}{\beta}\right|$ and directrix $x = \dfrac{\alpha}{\gamma}$.

17. Let α, β, and γ be nonzero constants. Show that the graph of $r = \dfrac{\alpha}{\beta + \gamma\sin\theta}$ is a conic section with eccentricity $\left|\dfrac{\gamma}{\beta}\right|$ and directrix $y = \dfrac{\alpha}{\gamma}$.

Skills

Complete the square to describe the conics in Exercises 18–21.

18. $2x^2 + 4x + y^2 - 6y - 3 = 0$
19. $2x^2 + 4x - y^2 - 6y - 23 = 0$
20. $4x + y^2 - 6y - 3 = 0$
21. $y^2 - 8y - 4x^2 - 8x - 13 = 0$

Use Cartesian coordinates to express the equations for the parabolas determined by the conditions specified in Exercises 22–31.

22. directrix $x = 3$, focus $(0, 1)$
23. directrix $y = 0$, focus $(0, 1)$
24. directrix $x = -1$, focus $(2, 5)$
25. directrix $y = -6$, focus $(2, -8)$
26. directrix $x = x_0$, focus (x_1, y_1), where $x_0 \neq x_1$
27. directrix $y = y_0$, focus (x_1, y_1), where $y_0 \neq y_1$

28. $r = \dfrac{3}{1 + \cos\theta}$
29. $r = \dfrac{4}{1 + \sin\theta}$
30. $r = \dfrac{\alpha}{1 + \cos\theta}$
31. $r = \dfrac{\alpha}{1 + \sin\theta}$

Use Cartesian coordinates to express the equations for the ellipses determined by the conditions specified in Exercises 32–37.

32. foci $(\pm 1, 0)$, major axis 4
33. foci $(\pm\sqrt{5}, 1)$, major axis 6

34. foci $(3, \pm 5)$, major axis 12
35. foci $\left(0, \pm\dfrac{3\sqrt{3}}{2}\right)$, minor axis 3
36. foci $(\pm\alpha, 0)$, major axis 4α
37. foci $(0, \pm\alpha)$, minor axis 2α

Use Cartesian coordinates to express the equations for the hyperbolas determined by the conditions specified in Exercises 38–43.

38. foci $(\pm 2, 0)$, directrices $x = \pm 1$
39. foci $(\pm 6, 0)$, directrices $x = \pm 1$
40. foci $(0, \pm 4)$, directrices $y = \pm 1$
41. foci $(0, \pm 4)$, directrices $y = \pm 2$
42. foci $(3, 1)$ and $(7, 1)$, directrix $x = 4$
43. foci $(3, 1)$ and $(3, 9)$, directrix $y = 4$

Use polar coordinates to graph the conics in Exercises 44–51.

44. $r = \dfrac{2}{1 + 3\cos\theta}$
45. $r = \dfrac{2}{1 - 3\cos\theta}$
46. $r = \dfrac{2}{1 + \cos\theta}$
47. $r = \dfrac{5}{1 + \cos\theta}$
48. $r = \dfrac{2}{4 + \cos\theta}$
49. $r = \dfrac{2}{4 + \sin\theta}$
50. $r = \dfrac{1}{1 - \cos\theta}$
51. $r = \dfrac{2}{2 + \sin\theta}$

Applications

In the late sixteenth century Johannes Kepler showed that the planets in our solar system revolve around the sun in elliptical orbits. Exercises 52–54 deal with the orbits of the Earth and Mars. These orbits are described by their eccentricities. If the coordinates for the orbit of a planet are oriented so that the major axis of the ellipse is aligned along the x-axis, then the Cartesian equation for the ellipse is $\dfrac{x^2}{A^2} + \dfrac{y^2}{B^2} = 1$, with $A > B$. Recall that in this case $2A$ is called the major axis of the ellipse. The **semimajor** axis has length A.

52. Show that the eccentricity satisfies the equation $B^2 = A^2(1 - e^2)$.

53. Measurements indicate that Earth's orbital eccentricity is 0.0167 and its semimajor axis is 1.00000011 astronomical units.

   (a) Write a Cartesian equation for Earth's orbit.
   (b) Give a polar coordinate equation for Earth's orbit, assuming that the sun is the focus of the elliptical orbit.

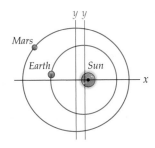

54. Measurements indicate that the orbital eccentricity of Mars is 0.0935 and its semimajor axis is 1.517323951 astronomical units.

(a) Write a Cartesian equation for the orbit of Mars.

(b) Do x and y have the same meaning as in Exercise 53?

(c) Give a polar coordinate equation for the orbit of Mars, assuming that the sun is the focus of the elliptical orbit.

Proofs

55. Let $A > B > 0$. Show that the distance from any point on the graph of the curve with equation $\frac{x^2}{A^2} + \frac{y^2}{B^2} = 1$ to the point $(-C, 0)$ is $D_2 = \frac{A^2 + Cx}{A^2}$, where $C = \sqrt{A^2 - B^2}$.

56. Prove Theorem 9.16 (b). That is, show that if $0 < A < B$, the graph of the curve with equation $\frac{x^2}{A^2} + \frac{y^2}{B^2} = 1$ is an ellipse with foci $(0, \pm\sqrt{B^2 - A^2})$.

57. Prove Theorem 9.20 (a). That is, show that the graph of the equation $\frac{x^2}{A^2} - \frac{y^2}{B^2} = 1$ satisfies Definition 9.19, where the points with coordinates $(\pm\sqrt{A^2 + B^2}, 0)$ are the foci of the hyperbola.

58. Prove Theorem 9.20 (b). That is, show that the graph of the equation $\frac{y^2}{B^2} - \frac{x^2}{A^2} = 1$ satisfies Definition 9.19, where the points with coordinates $(0, \pm\sqrt{A^2 + B^2})$ are the foci of the hyperbola.

59. Prove that for an ellipse or a hyperbola the eccentricity is given by

$$e = \frac{\text{the distance between the foci}}{\text{the distance between the vertices}}.$$

In Exercises 60 and 61 we ask you to prove Theorem 9.23 for ellipses and hyperbolas.

60. Consider the ellipse with equation $\frac{x^2}{A^2} + \frac{y^2}{B^2} = 1$, where $A > B$. Let F be the focus with coordinates $(\sqrt{A^2 - B^2}, 0)$. Let $e = \frac{\sqrt{A^2 - B^2}}{A}$ and l be the vertical line with equation $x = \frac{A}{e}$. Show that for any point P on the ellipse, $\frac{FP}{DP} = e$, where D is the point on l closest to P.

61. Consider the hyperbola with equation $\frac{x^2}{A^2} - \frac{y^2}{B^2} = 1$. Let F be the focus with coordinates $(\sqrt{A^2 + B^2}, 0)$. Let $e = \frac{\sqrt{A^2 + B^2}}{A}$ and l be the vertical line with equation $x = \frac{A}{e}$. Show that for any point P on the hyperbola, $\frac{FP}{DP} = e$, where D is the point on l closest to P.

Thinking Forward

Analogs of conic sections in three dimensions: Sketch each of the following.

▶ Given two points in three-dimensional space, a certain type of **ellipsoid** may be described as the set of points for which the sum of the distances to the two points is a constant.

▶ A certain type of **paraboloid** is the set of all points in three-dimensional space equidistant from a given plane and a given fixed point not on the plane.

▶ Given two points in three-dimensional space, a certain type of **hyperboloid** may be described as the set of points for which the difference of the distances to the two points is a constant.

CHAPTER REVIEW, SELF-TEST, AND CAPSTONES

Before you progress to the next chapter, be sure you are familiar with the definitions, concepts, and basic skills outlined here. The capstone exercises at the end bring together ideas from this chapter and look forward to future chapters.

Definitions

Give precise mathematical definitions or descriptions of each of the concepts that follow. Then illustrate the definition or description with a graph or an algebraic example.

▶ *Parametric equations* in two variables

▶ The *slope* of a parametric curve $x = x(t)$, $y = y(t)$ at the point $(x(t_0), y(t_0))$

▶ *Symmetry with respect to the x-axis*

Theorems

Fill in the blanks to complete each of the following theorem statements:

▶ Let C be a curve in the plane with parametrization $x = f(t)$, $y = g(t)$ for $t \in [a, b]$ such that the parametrization is a _____ function from the interval $[a, b]$ to the curve C. If $x = f(t)$ and $y = g(t)$ are differentiable functions of t such that $f'(t)$ and $g'(t)$ are _____ on $[a, b]$, then the length of the curve C is given by _____.

▶ For $a \neq 0$, the graph of the equation $r = 2a \cos \theta$ is the _____ whose equation in rectangular coordinates is _____.

▶ For $a \neq 0$, the graph of the equation $r = 2a \sin \theta$ is the _____ whose equation in rectangular coordinates is _____.

▶ A graph is symmetrical with respect to the x-axis if, for every point (r, θ) on the graph, the point _____ is also on the graph.

▶ A graph is symmetrical with respect to the y-axis if, for every point (r, θ) on the graph, the point _____ is also on the graph.

▶ A graph is symmetrical with respect to the origin if, for every point (r, θ) on the graph, the point _____ is also on the graph.

▶ Let α and β be real numbers such that $0 \leq \beta - \alpha \leq 2\pi$. Let R be the region in the polar plane bounded by the rays $\theta = \alpha$ and $\theta = \beta$ and a positive continuous function $r = f(\theta)$. Then the area of region R is _____.

▶ Let $r = f(\theta)$ be a differentiable function of θ such that $f'(\theta)$ is continuous for all $\theta \in [\alpha, \beta]$. Furthermore, assume that $r = f(\theta)$ is a one-to-one function from $[\alpha, \beta]$ to the graph of the function. Then the length of the polar graph of $r = f(\theta)$ on the interval $[\alpha, \beta]$ is _____.

Algebraic Rules

The calculus of parametric equations: Let $x = f(t)$ and $y = g(t)$, where f and g are differentiable functions.

▶ $\dfrac{dy}{dx} = $_____.

▶ The arc length of the curve defined by the parametric equations on the interval $[a, b]$ is_____.

Conversion formulas: Fill in the blanks to convert between rectangular coordinates and polar coordinates.

▶ $r = $_____.

▶ $\theta = $_____.

▶ $x = $_____.

▶ $y = $_____.

The calculus of polar functions: Let $r = f(\theta)$, where f is a positive differentiable function.

▶ The area enclosed by the function f and the rays $\theta = \alpha$ and $\theta = \beta$ is _____, provided that _____.

▶ The arc length of the curve defined by f on the interval $[a, b]$ is _____.

Skill Certification: Parametric Equations and Polar Coordinates

Sketching parametric equations: Sketch the curves defined by the given sets of parametric equations. Indicate the direction of motion on each curve.

1. $x = t^2$, $y = t^3$, $t \in [-2, 2]$
2. $x = 3t + 1$, $y = -2t + 3$, $t \in [0, 1]$
3. $x = \sin t$, $y = \cos t$, $t \in [0, 4\pi]$
4. $x = \sin t$, $y = \cos^2 t$, $t \in [0, 4\pi]$
5. $x = \sec t$, $y = \tan t$, $t \in \left(-\dfrac{\pi}{2}, \dfrac{\pi}{2}\right)$
6. $x = \tan t$, $y = 1 + \sec^2 t$, $t \in \left(-\dfrac{\pi}{2}, \dfrac{\pi}{2}\right)$

7. $x = \sinh t, y = \cosh t, t \in \mathbb{R}$

8. $x = \sinh t, y = \cosh^2 t, t \in \mathbb{R}$

Lengths of parametric curves: Find the arc lengths of the curves defined by the parametric equations on the specified intervals.

9. $x = t^2, y = t^3, t \in [-2, 2]$

10. $x = 3t + 1, y = -2t + 3, t \in [0, 1]$

11. $x = \sin kt, y = \cos kt, t \in [0, 2\pi]$, where k is a constant

12. $x = \cosh t, y = t, t \in [0, 1]$

Graphs of polar functions: Use polar coordinates to graph each of the following functions.

13. $r = \sin \theta$

14. $r = 3 \cos \theta$

15. $r = \sin 3\theta$

16. $r = 3 \cos 4\theta$

17. $r = 2 - 2 \sin \theta$

18. $r = 4 - 2 \cos \theta$

19. $r = 2 - \sqrt{8} \sin \theta$

20. $r = 3 - 2 \cos \theta$

21. $r^2 = -\sin 2\theta$

22. $r^2 = 3 \cos 2\theta$

The arc length of polar functions: Find the arc lengths of the following polar functions.

23. $r = a$, where a is a positive constant, for $\theta \in [0, 2\pi]$

24. $r = a \cos \theta$, where a is a positive constant, for $\theta \in [0, \pi]$

25. $r = a \sin \theta$ for $\theta \in [0, \pi]$

26. $r = \sin \theta + \cos \theta$ for $\theta \in [0, \pi]$

27. $r = 1 - \sin \theta$ for $\theta \in [0, 2\pi]$

28. $r = a \sin \theta + b \cos \theta$, where a and b are positive constants for $\theta \in [0, \pi]$

29. $r = \sec \theta$ for $\theta \in \left[-\dfrac{\pi}{6}, \dfrac{\pi}{6}\right]$

30. $r = \csc \theta$ for $\theta \in \left[\dfrac{\pi}{4}, \dfrac{3\pi}{4}\right]$

Areas of regions bounded by polar functions: Find the areas of the following regions.

31. The area bounded by the function $r = a$, where a is a positive constant.

32. The area bounded by the function $r = a \cos \theta$, where a is a positive constant.

33. The area bounded by one petal of $r = \cos 2\theta$.

34. The area bounded by one petal of $r = \sin 3\theta$.

35. The area bounded by $r = 1 - \sin \theta$

36. The area between the inner and outer loops of $r = 1 - \sqrt{2} \cos \theta$.

37. The area bounded by one loop of $r^2 = -\sin 2\theta$.

38. The area bounded by one loop of $r^2 = 4 \cos 2\theta$.

39. The area that is inside both lemniscates $r^2 = \cos 2\theta$ and $r^2 = \sin 2\theta$.

40. The area inside both polar roses $r = \sin 3\theta$ and $r = \cos 3\theta$.

Capstone Problems

A. A *hypocycloid* is another generalization of a cycloid in which the point tracing the path is on the circumference of a wheel, but the wheel is rolling without slipping on the inside of another wheel. If the radius of the rolling wheel is k and the radius of the fixed wheel is r, find parametric equations for the hypocycloid.

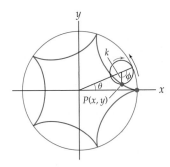

What is the path if the radius of the smaller wheel is exactly one-half the radius of the larger wheel?

B. A certain set of parametric equations has the properties that

$$x'(t) = x(t), \ y'(t) = x(t)y(t), \ x(0) = 1 \text{ and } y(0) = e.$$

(a) Solve the system of differential equations. (*Hint: Start by solving the initial-value problem $x'(t) = x(t)$ with $x(0) = 1$.*)

(b) Graph the parametric curve defined by $x = x(t)$ and $y = y(t)$ by eliminating the parameter.

C. When $k \geq 2$ is an integer, the polar graph of $r = \sin k\theta$ or $r = \cos k\theta$ is a rose.

(a) How many petals does the rose have when k is an integer?

(b) What can you say about the symmetries of either $r = \sin k\theta$ or $r = \cos k\theta$ when k is rational? Use a graphing calculator or a computer algebra system to graph several cases before answering the question.

(c) What can you say about the polar graph of $r = \sin k\theta$ or $r = \cos k\theta$ when k is irrational?

D. Let r_1 and r_2 be positive real numbers and $0 < \theta_2 - \theta_1 < \pi$. Prove that the area of the triangle with vertices $(0, 0)$, (r_1, θ_1), and (r_2, θ_2) in the polar plane is $\dfrac{r_1 r_2}{2} \sin(\theta_2 - \theta_1)$.

Vectors

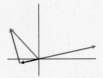

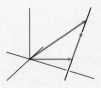

10.1   CARTESIAN COORDINATES

▶ The three-dimensional rectangular coordinate system

▶ Computing distances between two points

▶ The equation of a sphere

Three-Dimensional Space in Rectangular Coordinates

In this chapter we will study vectors and see how they are used in the descriptions of lines and planes in three dimensions. We will see that vectors have interrelated geometric and analytic properties. By studying the connections between the geometric and algebraic properties of vectors we will be able to gain a fuller understanding of vectors, lines, and planes.

In previous chapters we used a two-dimensional coordinate system almost exclusively. Until we reached Chapter 9, we focused on rectangular or Cartesian coordinates to name the points in the plane. Beginning with this section we expand this concept to describe a three-dimensional Cartesian system. In Chapter 13 we will discuss other useful coordinate systems for three-dimensional space. To construct a three-dimensional rectangular system we choose an origin O and three mutually perpendicular coordinate axes intersecting at O. The coordinate axes are ordered and usually labeled x, y, and z. We will typically draw the system as shown here:

A three-dimensional Cartesian coordinate system

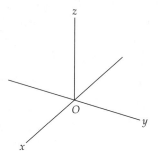

In this figure the z-axis is vertical, increasing as you go up, the y-axis increases from left to right, and the x-axis should be interpreted as pointing straight out from the page, increasing as it comes toward you. The figure shown is an example of a ***right-handed system***, because it obeys the following "right-hand rule": If the index finger of the right hand points in the positive x direction and the middle finger of the right points in the positive y direction, then the right thumb will naturally point in the positive z direction as the hand shown here illustrates:

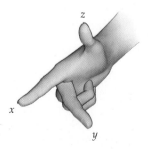

The coordinate system shown next at the left is also a right-handed system. However, there are labelings of the axes that give left-handed systems, as in the figure on the right.

Another right-handed
Cartesian coordinate system

A left-handed system

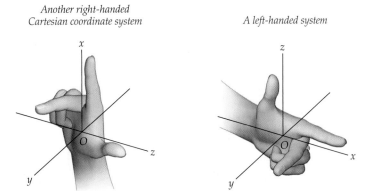

We used a different labeling of the axes when we discussed surfaces of revolution in Chapter 6. That labeling also gave us a right-handed system.

In Exercise 14 we will ask you about other possible labelings of the axes and about which labelings give right-handed systems. The notion of right-handedness will be crucial in Section 10.4, when we discuss the cross product. For now, it is just a convenient convention.

Once the axes are drawn, to complete the Cartesian system, a linear scale should be added to each axis, as is done with any two-dimensional rectangular system. Given the axes and their scales, there is a one-to-one correspondence between the points in the three-dimensional system and ordered triples of real numbers (a, b, c).

We will often use the abbreviations **2-space** or $\mathbb{R}^2$ (pronounced "R two") to denote two-dimensional space. Similarly, we will use **3-space** or $\mathbb{R}^3$ for three-dimensional space.

To locate a point (a, b, c) in a three-dimensional Cartesian coordinate system we move the appropriate number of units along each of the three coordinate axes. For example, the point $(2, 3, -5)$ is the unique point 2 units in the positive x direction from the **yz-plane**, 3 units in the positive y direction from the **xz-plane**, and 5 units in the negative z direction from the **xy-plane**. The following figure below is illustrative:

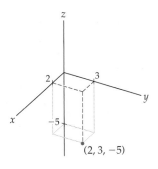

An Introduction to Planes

Two distinct intersecting lines in $\mathbb{R}^3$ determine a unique plane. In particular, when taken in pairs, the coordinate axes determine **coordinate planes**. The x- and y-axes determine the xy-plane. Similarly, the x- and z-axes determine the xz-plane, and the y- and z-axes determine the yz-plane.

As you know, in a two-dimensional rectangular system the coordinate axes divide the plane into four parts known as quadrants. Similarly, in a three-dimensional system, the three coordinate planes divide the system into eight parts, known as **octants**. The octant in which x, y, and z are all positive is called the **first octant**. The other seven octants are usually not numbered; however, they can be distinguished by the signs of the coordinates.

For example, the octant determined by the inequalities $x < 0$, $y < 0$, and $z > 0$ lies at the top (since $z > 0$), left (since $y < 0$) and back (since $x < 0$) of the following system:

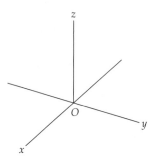

In mathematics, as in linguistics, it is important to realize that even simple statements are sensitive to the contexts in which they are made. For example, consider the mathematical sentence "$x = 3$." The three figures that follow illustrate different geometrical interpretations of this sentence. On a number line, $x = 3$ is a point, as we see on the left. In a two-dimensional system, $x = 3$ is the equation for a line parallel to the y-axis, as shown in the middle. In a three-dimensional system, the same equation $x = 3$ represents a plane parallel to the yz-plane. Every point on this plane has the property that its x-coordinate is 3.

Three interpretations of the equation $x = 3$

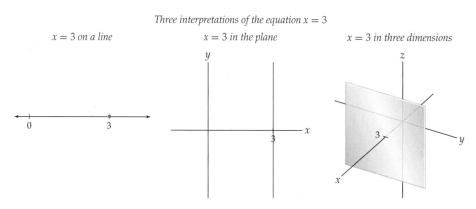

More generally, any equation of the form $x = a$, $y = b$, or $z = c$ represents a plane parallel to one of the coordinate planes. Of course there are planes in 3-space other than those parallel to the coordinate planes. We discuss planes fully in Section 10.6.

Distances, Spheres, and Cylinders

We know that in 2-space the distance between the points (x_1, y_1) and (x_2, y_2) can be computed with the Pythagorean theorem;

$$\text{distance} = \sqrt{(x_2 - x_1)^2 + (y_2 - y_1)^2}.$$

Distance in $\mathbb{R}^2$

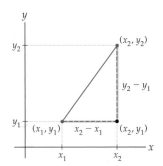

Distance in $\mathbb{R}^3$

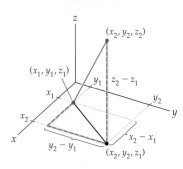

To find the distance between two points in $\mathbb{R}^3$ we need to extend this formula appropriately.

DEFINITION 10.1

The Distance Formula in $\mathbb{R}^3$

Let (x_1, y_1, z_1) and (x_2, y_2, z_2) be two points in $\mathbb{R}^3$. We define the **distance** between these points to be

$$\sqrt{(x_2 - x_1)^2 + (y_2 - y_1)^2 + (z_2 - z_1)^2}.$$

For example, to find the distance between the points $(2, 1, -4)$ and $(-3, 2, 1)$ we compute

$$\text{distance} = \sqrt{(2 - (-3))^2 + (1 - 2)^2 + (-4 - 1)^2} = \sqrt{25 + 1 + 25} = \sqrt{51}.$$

The distance between the points is $\sqrt{51} \approx 7.14$ units.

To find the **midpoint** of a line segment connecting points $P(x_1, y_1, z_1)$ and $Q(x_2, y_2, z_2)$ we may just average the corresponding coordinates. That is, the midpoint of the segment PQ is the point $\left(\frac{x_1 + x_2}{2}, \frac{y_1 + y_2}{2}, \frac{z_1 + z_2}{2}\right)$. We ask you to prove this fact in Exercise 62. We may also use the idea of the midpoint of a segment to define what it means for two points to be symmetric about a third point. We say that $P(x_1, y_1, z_1)$ and $Q(x_2, y_2, z_2)$ are symmetric about a point $R(x_3, y_3, z_3)$ if R is the midpoint of the segment PQ. For example, to find the point Q symmetric to $P(-1, 2, 4)$ about the point $R(2, 7, -1)$, we see that Q must have coordinates $(5, 12, -6)$ so that R is the midpoint of PQ.

A circle can be defined to be the set of points in a plane at a fixed distance from a given central point. Using the distance formula, we know that the equation of a circle in the xy-plane can be written as

$$(x - x_0)^2 + (y - y_0)^2 = r^2,$$

where the center of the circle is (x_0, y_0) and the radius (distance) is r. We may use a similar equation to describe the points on a sphere.

DEFINITION 10.2

Spheres

A **sphere** is the set of points in three-dimensional space at a fixed distance from a given central point. The sphere with center (x_0, y_0, z_0) and radius $r > 0$ is given by the formula

$$(x - x_0)^2 + (y - y_0)^2 + (z - z_0)^2 = r^2.$$

For example, the equation of the sphere centered at $(-2, 3, 1)$ with radius 4 is

$$(x + 2)^2 + (y - 3)^2 + (z - 1)^2 = 16.$$

In common usage a cylinder is a circle translated a finite distance in a direction out of the plane of the circle. We now define a cylinder in a more general way. The curves we translate can be any curves in the plane.

DEFINITION 10.3

Cylinders

Let C be a curve in some plane $\mathcal{P}$, and let l be a line that intersects $\mathcal{P}$, but does not lie in $\mathcal{P}$. A **cylinder** is the set of all points in $\mathbb{R}^3$ that are on lines parallel to l that intersect C. The curve C is called the **directrix** of the cylinder. The lines in the cylinder parallel to l are called **rulings** of the cylinder.

In the following examples each directrix is a curve in the xy-plane and the rulings are parallel to the z-axis:

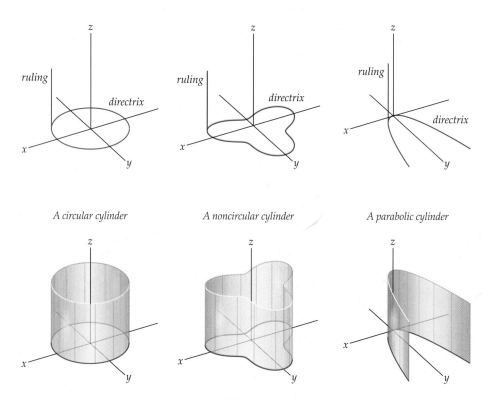

A circular cylinder   A noncircular cylinder   A parabolic cylinder

Most of the cylinders we will consider have rulings that are perpendicular to the plane containing the directrix. These cylinders are said to be **right** cylinders. The preceding examples are all right cylinders. As we mentioned earlier in this section, a given equation can have different graphs depending upon the context. When we consider the equation $y = x^2$ in $\mathbb{R}^2$, we see that the graph is the familiar parabola. But when we consider the graph of this equation in $\mathbb{R}^3$, we observe that it is the right parabolic cylinder shown in the preceding, rightmost figure. In fact, the graph of any curve given by a function $y = f(x)$ is a right cylinder when considered in $\mathbb{R}^3$. (Think of translating the graph in the direction of the z-axis.)

Quadric Surfaces

A conic section can be thought of as the graph of a second-degree polynomial in two variables,

$$Ax^2 + By^2 + Cxy + Dx + Ey + F = 0,$$

where $A, B, C, \ldots, F$ are constants. **Quadric surfaces** are the three-dimensional analogs of conic sections. A quadric surface is the graph of a second-degree polynomial in three variables; its equation is

$$Ax^2 + By^2 + Cz^2 + Dxy + Exz + Fyz + Gx + Hy + Iz + J = 0,$$

where $A, B, C, \ldots, J$ are constants. We will not need to consider this most general form for quadric surfaces because, under suitable translations and rotations, every quadric surface can be written in one of the following two forms:

$$Cz = Ax^2 + By^2 \quad \text{and} \quad Cz^2 = Ax^2 + By^2 + D.$$

If any of the constants in either of the preceding equations is zero, the corresponding quadric surface is said to be **degenerate**. The only type of degenerate quadric surface we will mention in detail is the one we get when the constant $D = 0$ in the second equation. The graphs of these surfaces are the cones we discuss next. All other degenerate quadric surfaces are right cylinders of conic sections. For example, if we let $A = C = 1$ and $B = 0$ in the equation $Cz = Ax^2 + By^2$, we have $z = x^2$. The graph of this equation in $\mathbb{R}^3$ is the right parabolic cylinder with directrix $z = x^2$ shown here:

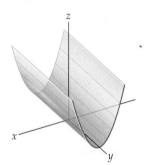

The cones and all nondegenerate quadric surfaces can be written in one of the following six forms for positive constants a, b, and c, possibly after another rotation:

1. Elliptic Cones: $z^2 = \frac{x^2}{a^2} + \frac{y^2}{b^2}$

2. Ellipsoids: $\frac{x^2}{a^2} + \frac{y^2}{b^2} + \frac{z^2}{c^2} = 1$

3. Hyperboloids of One Sheet: $\frac{x^2}{a^2} + \frac{y^2}{b^2} - \frac{z^2}{c^2} = 1$

4. Hyperboloids of Two Sheets: $\frac{x^2}{a^2} + \frac{y^2}{b^2} - \frac{z^2}{c^2} = -1$

5. Elliptic Paraboloids: $z = \frac{x^2}{a^2} + \frac{y^2}{b^2}$

6. Hyperbolic Paraboloids: $z = \frac{x^2}{a^2} - \frac{y^2}{b^2}$

To help us understand these surfaces we will look at:

▶ the **intercepts**, the places where the graphs cross the coordinate axes,

▶ the **traces**, the intersections with the coordinate planes, and

▶ the **sections**, intersections with arbitrary planes parallel to the coordinate planes.

1. **Elliptic Cones**: $z^2 = \dfrac{x^2}{a^2} + \dfrac{y^2}{b^2}$.

   The only place where the graph intersects one of the coordinate axes is the origin. The horizontal sections are ellipses with equations $c^2 = \dfrac{x^2}{a^2} + \dfrac{y^2}{b^2}$. The trace in the xz-plane is the pair of intersecting lines $z = \pm\dfrac{x}{a}$. Similarly, the trace in the yz-plane is the pair of intersecting lines $z = \pm\dfrac{y}{b}$. The following graph is one such elliptic cone:

   The elliptic cone $z^2 = \dfrac{x^2}{4} + \dfrac{y^2}{9}$

   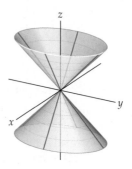

2. **Ellipsoids**: $\dfrac{x^2}{a^2} + \dfrac{y^2}{b^2} + \dfrac{z^2}{c^2} = 1$.

   An ellipsoid is a three-dimensional analog of an ellipse. The x-intercepts are $(\pm a, 0, 0)$, the y-intercepts are $(0, \pm b, 0)$, and the z-intercepts are $(0, 0, \pm c)$. To find the trace in the xy-plane we set $z = 0$ and see that we obtain the ellipse with equation $\dfrac{x^2}{a^2} + \dfrac{y^2}{b^2} = 1$. Similarly, the traces in the xz-plane and yz-plane are the ellipses $\dfrac{x^2}{a^2} + \dfrac{z^2}{c^2} = 1$ and $\dfrac{y^2}{b^2} + \dfrac{z^2}{c^2} = 1$, respectively.

   The ellipsoid $\dfrac{x^2}{9} + \dfrac{y^2}{16} + \dfrac{z^2}{4} = 1$

   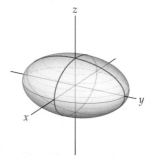

3. **Hyperboloids of One Sheet**: $\dfrac{x^2}{a^2} + \dfrac{y^2}{b^2} - \dfrac{z^2}{c^2} = 1$.

   The x-intercepts are $(\pm a, 0, 0)$ and the y-intercepts are $(0, \pm b, 0)$. There are no z-intercepts, since the equation $-\dfrac{z^2}{c^2} = 1$ has no solutions. The trace in the xy-plane is the ellipse with equation $\dfrac{x^2}{a^2} + \dfrac{y^2}{b^2} = 1$. Similarly, every section parallel to the xy-plane is an ellipse, and the size of these ellipses increases with distance from the xy-plane.

   When we set $y = 0$ to find the trace in the xz-plane, we see that we obtain a hyperbola with equation $\dfrac{x^2}{a^2} - \dfrac{z^2}{c^2} = 1$. Similarly, the trace in the yz-plane is the hyperbola with equation $\dfrac{y^2}{b^2} - \dfrac{z^2}{c^2} = 1$.

The hyperboloid $\dfrac{x^2}{9} + \dfrac{y^2}{16} - \dfrac{z^2}{4} = 1$

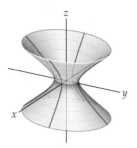

4. **Hyperboloids of Two Sheets**: $\dfrac{x^2}{a^2} + \dfrac{y^2}{b^2} - \dfrac{z^2}{c^2} = -1$.

As the name implies, the graph of a hyperboloid of two sheets consists of two pieces. There are no x- or y-intercepts because, when $z = 0$, the equation $\dfrac{x^2}{a^2} + \dfrac{y^2}{b^2} = -1$ has no real solutions. We can see that there are z-intercepts at $(0, 0, \pm c)$ and that all points on the hyperboloid have the property that $z \geq c$ or $z \leq -c$. For these values of z, the sections parallel to the xy-plane are all ellipses. The trace in the xz-plane is the hyperbola with equation $\dfrac{x^2}{a^2} - \dfrac{z^2}{c^2} = -1$. Similarly, the trace in the yz-plane is the hyperbola with equation $\dfrac{y^2}{b^2} - \dfrac{z^2}{c^2} = -1$.

The hyperboloid $\dfrac{x^2}{9} + \dfrac{y^2}{16} - \dfrac{z^2}{4} = -1$

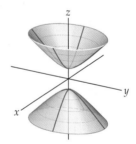

5. **Elliptic Paraboloids**: $z = \dfrac{x^2}{a^2} + \dfrac{y^2}{b^2}$.

The only place where the graph of this elliptic paraboloid intersects any of the coordinate axes is the origin. The trace in the xz-plane is the parabola with equation $z = \dfrac{x^2}{a^2}$, and the trace in the yz-plane is the parabola with equation $z = \dfrac{y^2}{b^2}$. When $a = b$, the graph is the surface of revolution formed when the graph of $z = \dfrac{x^2}{a^2}$ (in the xz-plane) is revolved around the z-axis.

The elliptic paraboloid $z = \dfrac{x^2}{4} + \dfrac{y^2}{9}$

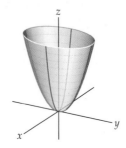

6. **Hyperbolic Paraboloids**: $z = \dfrac{x^2}{a^2} - \dfrac{y^2}{b^2}$.

Again, the only place where this paraboloid intersects any of the coordinate axes is at the origin. The trace in the xy-plane is the pair of intersecting lines $y = \pm\dfrac{b}{a}x$. The trace in the xz-plane is the (upwards-opening) parabola with equation $z = \dfrac{x^2}{a^2}$, and the trace in the yz-plane is the (downwards-opening) parabola with equation $z = -\dfrac{y^2}{b^2}$.

The hyperbolic paraboloid $z = \dfrac{x^2}{4} - \dfrac{y^2}{9}$

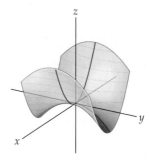

Examples and Explorations

EXAMPLE 1   Graphing a cylinder

Graph the cylinder in $\mathbb{R}^3$ defined by $z = \sin y$, $0 \le y \le 2\pi$.

SOLUTION

We begin by graphing the sine curve in the yz-plane, as shown next on the left, and then translate this graph, using the x-axis as a ruling, to obtain the graph on the right.

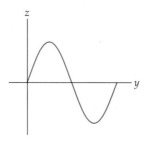
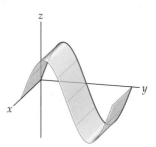

EXAMPLE 2   Inscribing a cube in a sphere

Find the side length of the largest cube that can be inscribed in a sphere of radius 1.

SOLUTION

The largest cube that can be inscribed in the sphere has a main diagonal that is a diameter of the sphere.

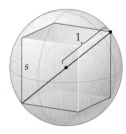

Therefore, the main diagonal of the cube will have length 2 units. If we let s be the length of each side of the cube, then, by the distance formula, we have $s^2 + s^2 + s^2 = 2^2$, or equivalently, $3s^2 = 4$. So, each side of the cube should measure $\frac{2}{3}\sqrt{3}$ units.   □

EXAMPLE 3   Inscribing a sphere between a cube and a sphere

Find the radius of the largest sphere that can be inscribed between the cube and sphere from Example 2.

SOLUTION

The following figure shows the sphere whose diameter we wish to find, along with the radius of the larger sphere:

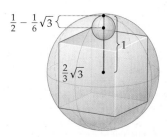

The large sphere has radius 1 and each side of the cube measures $\frac{2}{3}\sqrt{3}$ units. Therefore, the portion of this radius that extends above the cube has length $1 - \frac{1}{2} \cdot \frac{2}{3}\sqrt{3} = 1 - \frac{1}{3}\sqrt{3}$ units. This is the diameter of the smaller sphere. Therefore the radius of the smaller sphere is $\frac{1}{2} - \frac{1}{6}\sqrt{3}$ units.   □

? **TEST YOUR UNDERSTANDING**

▶ What is the difference between a right-handed coordinate system and a left-handed coordinate system?

▶ What are the coordinate planes in three-dimensional space? How are they drawn?

▶ How are the distances between two points computed in two-dimensional space and three-dimensional space? How could the distance formula be generalized to find the distance between two points in n-dimensional space?

▶ What are the graphical representations of the equations $x = a$ and $y = b$ in $\mathbb{R}^2$? What are the graphical representations of the equations $x = a$, $y = b$, and $z = c$ in 3-space?

▶ How do you write an equation of a sphere with a given center and radius?

EXERCISES 10.1

Thinking Back

Lines parallel to the coordinate axes: Provide equations for the specified lines.

▶ The line parallel to the x-axis and containing the point (π, e).

▶ The line parallel to the y-axis and containing the point (π, e).

▶ The line parallel to the y-axis and containing the point $(0, 5)$.

Circles in $\mathbb{R}^2$: Provide equations for the specified circles.

▶ The circle in $\mathbb{R}^2$ with center $(3, 6)$ and tangent to the x-axis.

▶ The circle in $\mathbb{R}^2$ with center $(3, 6)$ and tangent to the y-axis.

▶ The circle in $\mathbb{R}^2$ with center $(-4, 3)$ and containing the point $(5, -2)$.

Concepts

0. *Problem Zero:* Read the section and make your own summary of the material.

1. *True/False:* Determine whether each of the statements that follow is true or false. If a statement is true, explain why. If a statement is false, provide a counterexample.

(a) *True or False:* The distance between two distinct points can be zero.

(b) *True or False:* In the Cartesian plane the equation $x = 5$ represents a line.

(c) *True or False:* In three-dimensional space the equation $x = 5$ represents a line.

(d) *True or False:* In four-dimensional space the equation $x = 5$ represents a plane.

(e) *True or False:* The equation

$$x^2 + y^2 + z^2 + 2x - 4y - 10z + 50 = 0$$

represents a sphere with center $(-1, 2, 5)$.

(f) *True or False:* If a sphere in $\mathbb{R}^3$ has its center in the first octant and is tangent to each of the coordinate planes, then its center is at the point (c, c, c) for some constant c.

(g) *True or False:* When two distinct spheres intersect, they intersect in either a point or a circle.

(h) *True or False:* Three noncollinear points in $\mathbb{R}^3$ determine a unique plane.

2. *Examples:* Construct examples of the thing(s) described in the following. Try to find examples that are different than any in the reading.

(a) A plane parallel to the yz-plane.

(b) A sphere tangent to the xy-plane.

(c) A sphere tangent to all three coordinate planes.

3. Consider the equations $y = 5$ and $x = -3$.

(a) What do these equations represent in a two-dimensional system?

(b) What do these equations represent in a three-dimensional system?

4. Consider the equation $x + 2y = 4$.

(a) What does this equation represent in a two-dimensional system?

(b) What does this equation represent in a three-dimensional system?

5. What are the coordinates of the vertices of a cube with side length 2, whose center is at the origin, and whose faces are parallel to the coordinate planes?

6. The sides of a $2 \times 3 \times 4$ rectangular solid are parallel to the coordinate planes. The coordinates of four of its vertices are $(1, -2, 3)$, $(-1, -2, -1)$, $(-1, 1, 3)$, and $(1, -2, 3)$. What are the coordinates of the other four vertices?

7. What is the definition of a sphere?

8. What is the definition of a cylinder? What is the directrix? What is a ruling?

9. Consider the equation $x^2 + y^2 = 4$.

(a) What does this equation represent in a two-dimensional system?

(b) What does this equation represent in a three-dimensional system?

10. Consider the equation $z = y^2$.

(a) What does this equation represent in the yz-plane?

(b) What does this equation represent in a three-dimensional system?

11. What point is symmetric about the origin to the point $(5, -6, 7)$?

12. What point is symmetric to the point $(-1, 3, 6)$ with respect to the xy-plane?

13. What point is symmetric to the point $(3, -7, -4)$ with respect to the plane $z = 1$?

14. Find all of the x, y, z labelings of the axes in the diagram that follows. Determine which of your labelings are right-handed systems and which are left-handed systems.

15. Show that exchanging two of the axes labels on a right-handed system creates a left-handed system. What does exchanging two pairs of axes labels do (for example, exchanging x and y and then exchanging the "new" y and z)?

16. Sketch the point $(2, 3, 4)$ in a three-dimensional Cartesian coordinate system, and then answer the following questions:

    (a) Do the coordinates $(2, 3, 4)$ represent a unique point in a three-dimensional Cartesian coordinate system?

    (b) Are there any other coordinates (a, b, c) that would have the same location as $(2, 3, 4)$ in your graph?

    (c) How far from the xy-plane is the point $(2, 3, 4)$?

    (d) How far from the xz-plane is the point $(2, 3, 4)$?

    (e) How far from the yz-plane is the point $(2, 3, 4)$?

    (f) How far from the x-axis is the point $(2, 3, 4)$?

    (g) How far from the origin is the point $(2, 3, 4)$?

    (h) How far from the point $(1, -2, 3)$ is the point $(2, 3, 4)$?

    (i) What is the equation of the sphere with center $(2, 3, 4)$ and passing through the point $(1, -2, 3)$?

    (j) What is the equation of the plane parallel to the xy-plane and that contains the point $(2, 3, 4)$?

    (k) What is the equation of the plane parallel to the xz-plane and that contains the point $(2, 3, 4)$?

    (l) What is the equation of the plane parallel to the yz-plane and that contains the point $(2, 3, 4)$?

17. How does the Pythagorean theorem generalize to higher dimensions? In particular, how would you compute the distance between two points in four-dimensional space? Five-dimensional space? n-dimensional space?

18. The points in the first octant satisfy the inequalities $x > 0$, $y > 0$, $z > 0$. For each of the other seven octants, find a set of inequalities that describes the points in the octant. Use the inequalities to label the octants of the following right-handed coordinate system:

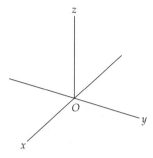

19. Use inequalities to describe each of the following sets, and sketch the regions:

    (a) The region in the first octant above the plane $z = 3$.

    (b) The region inside the sphere with center $(1, 2, 4)$ and radius 5.

    (c) The region inside the right circular cylinder with directrix $x^2 + y^2 - 4x + 6y - 12 = 0$.

20. Sketch the right circular cylinders:

$$x^2 + y^2 = 1, \ x^2 + z^2 = 4, \ \text{and} \ y^2 + z^2 = 9.$$

Skills

Find the distance between the given pair of points in Exercises 21–28.

21. $(-2, 3)$ and $(5, -6)$

22. $(4, 0)$ and $(-5, 12)$

23. $(1, 4, 7)$ and $(-2, 3, 5)$

24. $(3, 0, -1)$ and $(2, -8, 0)$

25. $(-1, 4, -3)$ and $(-4, 3, 1)$

26. $(4, 5, 8, -2)$ and $(-1, 3, -3, 6)$

27. $(-1, 3, 5, 2, 0)$ and $(0, 6, 1, -2, 3)$

28. $(0, 0, \ldots, 0)$ and $(1, 1, \ldots, 1)$ in n-space

In Exercises 29–35 find an equation of a sphere with the specified characteristics.

29. center $(3, -2, 5)$ and radius 5

30. center $(4, -2, -3)$ and radius 3

31. center $(2, 5, -7)$ and tangent to the xy-plane

32. center $(2, 5, -7)$ and tangent to the yz-plane

33. center $(2, 5, -7)$ and containing the origin

34. containing the point $(1, 4, 7)$ and whose center is $(-2, 3, 5)$

35. containing the point $(3, 0, -1)$ and whose center is $(2, -8, 0)$

Use the midpoint formula to find the equations of the spheres in Exercises 36 and 37.

36. the sphere in which the segment with endpoints $(3, -2, 6)$ and $(5, 6, 4)$ is a diameter

37. the sphere in which the segment with endpoints $(6, -1, 4)$ and $(-3, 3, 1)$ is a diameter

In Exercises 38 and 39 find the center and radius of the sphere with the given equation.

38. $x^2 + y^2 + z^2 - 2x + 6y - 8z + 17 = 0$

39. $x^2 + y^2 + z^2 + 3y - 5z + 3 = 0$

40. Find the equations of the intersections of the sphere

$$x^2 + y^2 + z^2 - 2x + 6y - 8z + 17 = 0$$

with each of the coordinate planes.

Graph the quadric surfaces given by the equations in Exercises 41–48.

41. $x^2 = \dfrac{y^2}{9} + \dfrac{z^2}{9}$

42. $z^2 = \dfrac{x^2}{9} + \dfrac{y^2}{25}$

43. $x^2 + y^2 + 1 = z^2$

44. $x^2 + y^2 - z^2 = 1$

45. $z = \dfrac{x^2}{9} + \dfrac{y^2}{25}$

46. $9x^2 + 16y^2 + 16z^2 = 144$

47. $z = x^2 - y^2$

48. $z = x^2 + y^2$

49. A circle is inscribed in a square so that each side of the square is tangent to the circle. A smaller circle is inscribed in the square so that this circle is tangent to two sides of the square and is tangent to the larger circle, as shown in the following figure:

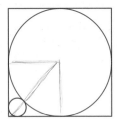

What is the ratio of the radius of the smaller circle to the radius of the larger circle?

50. A sphere is inscribed in a cube so that each face of the cube is tangent to the sphere. A smaller sphere is inscribed in the cube so that this sphere is tangent to three sides of the cube and is tangent to the larger sphere, as shown in the following figure:

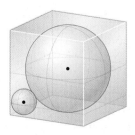

What is the ratio of the radius of the smaller sphere to the radius of the larger sphere? (*Hint: Try Exercise 49 first.*)

51. Show that the triangle with vertices $(5, 4, -1)$, $(3, 6, -1)$, and $(3, 4, 1)$ is equilateral.

52. Show that the triangle with vertices $(1, 2, -2)$, $(-3, 2, -6)$, and $(-3, 6, -2)$ is equilateral.

53. Show that the points $(1, 5, 0)$, $(3, 8, 6)$, and $(7, -7, 4)$ are the vertices of a right triangle and find its area.

A *regular tetrahedron* is a solid with four faces in which each face is an equilateral triangle of the same size. In Exercises 54 and 55 you are asked some basic questions about regular tetrahedra.

54. (a) Show that the three points $(1, 0, 0)$, $(0, 1, 0)$, and $(0, 0, 1)$ are the vertices of an equilateral triangle.

(b) Determine the two values of a so that the four points $(1, 0, 0)$, $(0, 1, 0)$, $(0, 0, 1)$, and (a, a, a) are the vertices of a regular tetrahedron.

55. Find the equations of the spheres that circumscribe the two tetrahedra you determined in Exercise 54 (b). (*Hint: The center of the sphere is the point (x_0, y_0, z_0), where x_0 is the mean of the x-coordinates of the four vertices of the tetrahedron, etc.*)

56. Show that the six points $(1, 0, 0)$, $(-1, 0, 0)$, $(0, 1, 0)$, $(0, -1, 0)$, $(0, 0, 1)$, and $(0, 0, -1)$ form the vertices of a regular octahedron. (A *regular octahedron* is an eight-sided solid in which each face is an equilateral triangle of the same size, and in which four triangles come together at each vertex.)

57. Find the volume of the octahedron given in Exercise 56. (*Hint: Recall that the volume of a pyramid is $\dfrac{1}{3} \cdot$ height $\cdot$ area of base.*)

Applications

58. Ian is doing a high traverse. One morning he looks at the map and notes that if he considers his camp to be at the origin, then his objective is at $(5.9, 3.3, -0.37)$. All distances are in miles.

(a) How far away is his objective, as the crow flies?

(b) In order to reach his objective, Ian has to go over a high pass that lies at $(4.2, 4.4, 0.15)$ relative to his camp. Find a more realistic estimate of how far he has to go to his objective than that from part (a).

59. Annie is in a kayak in the middle of a channel between Orcas Island and Blakely Island, two of the San Juan Islands in Washington State. She knows from readings she has taken in the past that if she considers the town of Deer Harbor to be at the origin, then the summit of Constitution Peak is at $(7.9, 4.0)$, while the summit of Blakely Peak is at $(9.3, -3.4)$. All distances are given in miles. She pulls out her compass and finds that the summit of Constitution Peak is 15 degrees east of north from her, while the summit of Blakely Peak is at 100 degrees from north. How far is Annie from Deer Harbor?

60. The Subaru reflecting telescope on Mauna Kea, Hawaii, has a mirror in the shape of a circular paraboloid with a diameter of 8.3 meters and a focal length of 15 meters. While there are some tricks to how that focal length plays out in practice, if we put the center of the telescope at the origin, pointed straight up, then the effective focus would be at $(0, 0, p)$, where p satisfies $4pz = x^2 + y^2$, with all distances given in meters. How high above the xy-plane is the edge of the telescope?

A reflecting telescope

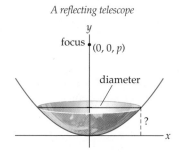

61. The Hyper Potato Chip Company makes potato chips in the shape of hyperbolic paraboloids. Each chip satisfies the equation $z = \dfrac{x^2}{a^2} - \dfrac{y^2}{a^2}$ when the center of the chip is placed at the origin. The height of each Hyper chip is 0.16 inch, the length is 2.6 inches, and the width is 1.6 inches. Find the parameter a and write the equation of a Hyper chip.

A Hyper chip

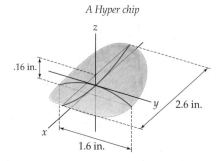

Proofs

62. Prove that the midpoint of the line segment connecting the point (x_1, y_1, z_1) to the point (x_2, y_2, z_2) is $\left(\dfrac{x_1+x_2}{2}, \dfrac{y_1+y_2}{2}, \dfrac{z_1+z_2}{2}\right)$.

63. Given any edge $\mathcal{E}$ of a tetrahedron, there is exactly one edge $\mathcal{E}'$ that does not share a face with $\mathcal{E}$. We will call $\mathcal{E}$ and $\mathcal{E}'$ **opposite** edges of the tetrahedron. Prove that the line segments connecting the midpoints of the opposite edges of a regular tetrahedron bisect each other.

64. Prove that the midpoint of the line segment connecting the points $(x_1, x_2, \ldots, x_n)$ and $(y_1, y_2, \ldots, y_n)$ in $\mathbb{R}^n$ is $\left(\dfrac{x_1+y_1}{2}, \dfrac{x_2+y_2}{2}, \ldots, \dfrac{x_n+y_n}{2}\right)$.

Thinking Forward

Hyperspheres: A four-dimensional **hypersphere** is the set of all points in $\mathbb{R}^4$ that are the same distance from a given central point.

▶ What is the equation of the hypersphere with center $(1, 2, 3, 4)$ and passing through the point $(2, 5, 3, -4)$?

▶ What is the equation of the hypersphere in which the segment with endpoints $(1, 2, 3, 4)$ and $(2, 5, 3, -4)$ is a diameter?

▶ Hyperspheres can be defined in $\mathbb{R}^n$ for any integer $n > 3$. How would you define a hypersphere in $\mathbb{R}^n$?

10.2   VECTORS

▶ Vectors in $\mathbb{R}^2$ as ordered pairs and in $\mathbb{R}^3$ as ordered triples

▶ The geometry of vectors

▶ Using vectors to analyze forces

The Algebra and Geometry of Vectors

The mass and temperature of an object are examples of physical quantities that can be represented adequately with a single numeric value, a magnitude known as a ***scalar***. Other quantities, such as displacements, velocities, and forces, need two measures—a magnitude *and* a direction—to represent them. Such quantities are represented by ***vectors***.

DEFINITION 10.4
> **Vectors in $\mathbb{R}^2$ and $\mathbb{R}^3$**
>
> A ***vector*** in $\mathbb{R}^2$ is an ordered pair $\langle x, y \rangle$ subject to the following operations of addition and scalar multiplication:
>
> **(a)** The ***sum*** of vectors $\langle x_1, y_1 \rangle$ and $\langle x_2, y_2 \rangle$ is given by
>
> $$\langle x_1, y_1 \rangle + \langle x_2, y_2 \rangle = \langle x_1 + x_2, y_1 + y_2 \rangle.$$
>
> **(b)** The ***scalar multiple*** of vector $\langle x, y \rangle$ by a real number c, a ***scalar***, is given by
>
> $$c\langle x, y \rangle = \langle cx, cy \rangle.$$
>
> Similarly, a ***vector*** in $\mathbb{R}^3$ is an ordered triple $\langle x, y, z \rangle$ subject to the following operations of addition and scalar multiplication:
>
> **(c)** The ***sum*** of vectors $\langle x_1, y_1, z_1 \rangle$ and $\langle x_2, y_2, z_2 \rangle$ is given by
>
> $$\langle x_1, y_1, z_1 \rangle + \langle x_2, y_2, z_2 \rangle = \langle x_1 + x_2, y_1 + y_2, z_1 + z_2 \rangle.$$
>
> **(d)** The ***scalar multiple*** of vector $\langle x, y, z \rangle$ by scalar c is given by
>
> $$c\langle x, y, z \rangle = \langle cx, cy, cz \rangle.$$

We will often use boldface type to denote vectors; for example $\mathbf{u} = \langle 1, -5 \rangle$ is a vector in $\mathbb{R}^2$ and $\mathbf{v} = \langle \pi, e, -0.3 \rangle$ is a vector in $\mathbb{R}^3$. The entries of a vector are referred to as ***components***. Thus, for the preceding vector $\mathbf{v}$, the x-component is π, the y-component is e, and the z-component is -0.3.

We will (almost) exclusively deal with vectors in $\mathbb{R}^2$ and $\mathbb{R}^3$; however, it should be clear that vectors with n components may be defined in an analogous fashion, as a set of ordered n-tuples with an addition and a scalar multiplication.

There are two geometric models that we may use for vectors in $\mathbb{R}^2$ and $\mathbb{R}^3$. First, we may think of the vectors as ***position vectors.*** In this model we interpret the vector $\langle a, b \rangle$ from $\mathbb{R}^2$ as an arrow in the plane whose ***initial point*** is at the origin and whose ***terminal point*** is (a, b). Similarly, we interpret the vector $\langle a, b, c \rangle$ from $\mathbb{R}^3$ as an arrow in 3-space whose initial point is at the origin and whose terminal point is (a, b, c).

The position vector $\langle a, b \rangle$ in $\mathbb{R}^2$ *The position vector $\langle a, b, c \rangle$ in $\mathbb{R}^3$*

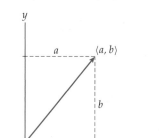 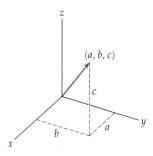

Our second geometric model is a variation of the preceding one. We may think of a vector $\langle a, b \rangle$ as any translation of the the position vector from the previous model. Thus, any parallel arrow with the same length pointing in the same direction is another model for the vector $\langle a, b \rangle$.

Any parallel translation of a vector is the same vector

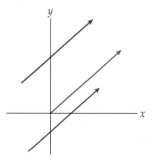

A similar model may be applied to vectors in $\mathbb{R}^3$.

Using the second model, we may geometrically add two vectors **u** and **v** as follows: As we show next, we place vector **u** in the diagram and then place the initial point of **v** at the terminal point of **u**. The vector **u** + **v** is the vector that extends from the initial point of **u** to the terminal point of **v**.

*Constructing **u** + **v** geometrically*

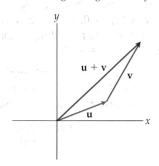

Another important concept is the magnitude of a vector, along with its synonyms "norm" and "length." We use the Pythagorean theorem in our definition.

DEFINITION 10.5   The Magnitude, Norm, or Length of a Vector

The **magnitude, *norm*,** or **length** of a vector **v** is denoted by $\|\mathbf{v}\|$.
In $\mathbb{R}^2$, when $\mathbf{v} = \langle a, b \rangle$,

$$\|\mathbf{v}\| = \sqrt{a^2 + b^2}.$$

In $\mathbb{R}^3$, when $\mathbf{v} = \langle a, b, c \rangle$,

$$\|\mathbf{v}\| = \sqrt{a^2 + b^2 + c^2}.$$

The magnitude of $\langle a, b \rangle$ is $\sqrt{a^2 + b^2}$

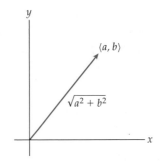

The magnitude of $\langle a, b, c \rangle$ is $\sqrt{a^2 + b^2 + c^2}$

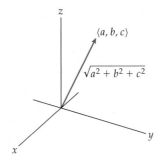

For example, the norm of the vector $\mathbf{v} = \langle 1, -2 \rangle$ is $\|\mathbf{v}\| = \sqrt{1^2 + (-2)^2} = \sqrt{5}$.

$\|\mathbf{v}\| = \sqrt{5}$

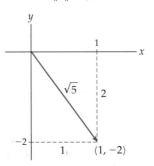

When we multiply a vector by a scalar $k > 0$, we obtain the vector $k\mathbf{v}$ that has the same direction as **v** but whose magnitude is equal to the magnitude of **v** multiplied by k. When $k < 0$, $k\mathbf{v}$ is the vector with direction opposite to the direction of the vector **v** that has magnitude equal to the magnitude of **v** multiplied by $|k|$. The following figure shows a vector **v** along with two scalar multiples of **v**:

*A vector **v** and two scalar multiples of **v***

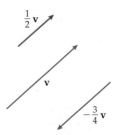

DEFINITION 10.6   Equal Vectors

Two vectors are said to be **equal** if and only if their corresponding components are equal.

In $\mathbb{R}^2$, $\langle x_1, y_1 \rangle = \langle x_2, y_2 \rangle$ if and only if $x_1 = x_2$ and $y_1 = y_2$.

In $\mathbb{R}^3$, $\langle x_1, y_1, z_1 \rangle = \langle x_2, y_2, z_2 \rangle$ if and only if $x_1 = x_2$, $y_1 = y_2$, and $z_1 = z_2$.

DEFINITION 10.7   The Zero Vector in $\mathbb{R}^2$ and $\mathbb{R}^3$

The **zero vector** in $\mathbb{R}^2$ is $\mathbf{0} = \langle 0, 0 \rangle$ and the **zero vector** in $\mathbb{R}^3$ is $\mathbf{0} = \langle 0, 0, 0 \rangle$.

Vectors in $\mathbb{R}^2$ and $\mathbb{R}^3$ obey many of the familiar algebraic properties that real numbers do.

THEOREM 10.8   Algebraic Properties of Vectors

(a) *Vector addition is commutative:*

For any two vectors **u** and **v** with the same number of components,

$$\mathbf{u} + \mathbf{v} = \mathbf{v} + \mathbf{u}.$$

(b) *Vector addition is associative:*

For any three vectors **u**, **v**, and **w**, each with the same number of components,

$$(\mathbf{u} + \mathbf{v}) + \mathbf{w} = \mathbf{u} + (\mathbf{v} + \mathbf{w}).$$

(c) *Scalar multiplication distributes over vector addition:*

For any scalar c and any two vectors **u** and **v** with the same number of components,

$$c(\mathbf{u} + \mathbf{v}) = c\mathbf{u} + c\mathbf{v}.$$

In Exercises 57, 58, and 59 we ask you to use Definition 10.4 to prove the three parts of Theorem 10.8. The following figure illustrates geometrically why vector addition is commutative. Since we may get from point A to point B via either of the paths $\mathbf{u} + \mathbf{v}$ or $\mathbf{v} + \mathbf{u}$, and the figure formed is a parallelogram, addition is commutative.

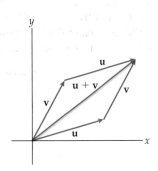

By simply relabeling two of the vectors from the preceding illustration of addition, we see how we may subtract vectors geometrically. If we let $\mathbf{w} = \mathbf{u} + \mathbf{v}$, then $\mathbf{v} = \mathbf{w} - \mathbf{u}$, as the following figure illustrates:

Constructing **w** − **u** *geometrically*

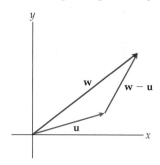

We see that when **u** and **w** are placed at the same initial point, **w** − **u** is the vector that extends from the terminal point of **u** to the terminal point of **w**.

We may use scalar multiplication to define what it means for two vectors to be parallel.

DEFINITION 10.9   **Parallel Vectors**

Two vectors are said to be *parallel* if and only if one of the vectors is a scalar multiple of the other.

As a consequence of this definition, the zero vector is parallel to every other vector, since, given any vector **v**, $0\mathbf{v} = \mathbf{0}$.

We also mention that there is a vector extending from a point $P(x_0, y_0, z_0)$ to a point $Q(x_1, y_1, z_1)$. In this case we use the notation

$$\overrightarrow{PQ} = \langle x_1 - x_0, y_1 - y_0, z_1 - z_0 \rangle.$$

That is, $\overrightarrow{PQ}$ is the vector whose initial point is P and whose terminal point is Q.

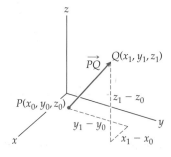

For example, if $P(8, 5, -3)$ and $Q(13, 5, -8)$, then $\overrightarrow{PQ} = \langle 13 - 8, 5 - 5, -8 - (-3) \rangle = \langle 5, 0, -5 \rangle$.

Unit Vectors

Often our primary interest is the direction of a nonzero vector **v**. At such times we may use a vector with length 1 and that is parallel to **v**. Vectors with length equal to 1 unit are called *unit* vectors.

DEFINITION 10.10   Unit Vectors

A vector **u** is said to be a ***unit vector*** if $\|\mathbf{u}\| = 1$.

To find a unit vector **u** with the same direction as a given vector $\mathbf{v} \neq \mathbf{0}$, we may multiply **v** by the reciprocal of its norm, as the following theorem states:

THEOREM 10.11   Unit Vectors

Given any nonzero vector **v**, the vector $\frac{1}{\|\mathbf{v}\|}\mathbf{v}$ is a unit vector in the direction of **v**.

We ask you to prove Theorem 10.11 in Exercise 60.

It is convenient to have a special notation for the unit vectors pointing in the positive direction of each of the coordinate axes. Hence, we make the following definition:

DEFINITION 10.12   Standard Basis Vectors in $\mathbb{R}^2$ and $\mathbb{R}^3$

The ***standard basis vectors*** in $\mathbb{R}^2$ are

$$\mathbf{i} = \langle 1, 0 \rangle \text{ and } \mathbf{j} = \langle 0, 1 \rangle.$$

The ***standard basis vectors*** in $\mathbb{R}^3$ are

$$\mathbf{i} = \langle 1, 0, 0 \rangle, \ \mathbf{j} = \langle 0, 1, 0 \rangle, \text{ and } \mathbf{k} = \langle 0, 0, 1 \rangle.$$

The standard basis vectors in $\mathbb{R}^2$

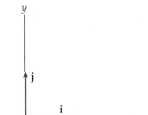

The standard basis vectors in $\mathbb{R}^3$

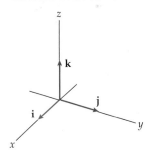

Any vector in $\mathbb{R}^2$ or $\mathbb{R}^3$ can be written in terms of the appropriate standard basis vectors. The vector $\mathbf{v} = \langle a, b \rangle$ can be written as $\mathbf{v} = a\mathbf{i} + b\mathbf{j}$. Thus, $\langle 8, -9 \rangle = 8\mathbf{i} - 9\mathbf{j}$. Similarly, the vector $\mathbf{v} = \langle a, b, c \rangle$ can be written as $\mathbf{v} = a\mathbf{i} + b\mathbf{j} + c\mathbf{k}$. Thus, $\langle -7, 2, -3 \rangle = -7\mathbf{i} + 2\mathbf{j} - 3\mathbf{k}$.

Using Vectors to Analyze Forces

There are many applications involving vectors. Any quantity with both a magnitude and a direction may be analyzed with the use of vectors. Two common vector applications are ***displacements*** and ***forces***. We discuss the analysis of forces here.

According to Newton's second law of motion, the vector sum of all of the forces acting on a body equals the mass m of the body times its acceleration: $\mathbf{F} = m\mathbf{a}$. Thus, if a body is not in motion, the sum of the forces acting upon it must be zero. As a simple example, the figure that follows shows a weight of 50 pounds suspended by a single rope. The rope exerts a force equal to 50 pounds, acting upwards, as the force of gravity acts downwards.

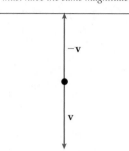

The vector forces in each direction must have the same magnitude

This same principle applies in more complicated contexts, as we will see shortly. A mass m is suspended from a ceiling by two chains that form angles α and β. The sum of the three vectors **u**, **v**, and **w** must be zero as long as the object is not in motion.

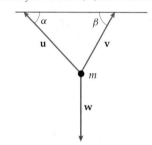

*The sum of the vectors **u**, **v**, and **w** must be zero*

We explore a specific example of this type in Example 3.

Examples and Explorations

EXAMPLE 1   **Adding and subtracting vectors**

Let $\mathbf{u} = \langle 3, 8 \rangle$ and $\mathbf{v} = \langle 2, -4 \rangle$. Compute $\mathbf{u} + \mathbf{v}$ and $\mathbf{u} - \mathbf{v}$, and then sketch all four of these vectors.

SOLUTION

Using Definition 10.4, we have

$$\mathbf{u} + \mathbf{v} = \langle 3 + 2, 8 + (-4) \rangle = \langle 5, 4 \rangle \text{ and } \mathbf{u} - \mathbf{v} = \langle 3 - 2, 8 - (-4) \rangle = \langle 1, 12 \rangle.$$

To add the two vectors geometrically we place the initial point of one of the vectors at the terminal point of the other. In the figure that follows at the left we have placed **v** at the tip of **u**. The sum is the vector extending from the initial point of **u** to the terminal point of **v**.

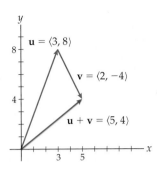

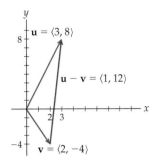

To subtract two vectors geometrically we position their initial points together, as we see in the right-hand figure. The vector $\mathbf{u} - \mathbf{v}$ extends from the terminal point of $\mathbf{v}$ to the terminal point of $\mathbf{u}$.   □

EXAMPLE 2   **Finding a unit vector in the same direction as a given nonzero vector**

Find a unit vector with the same direction as the vector $\mathbf{v} = \langle 4, -3, 5 \rangle$.

SOLUTION

The vector $\mathbf{v}$ has norm $\|\mathbf{v}\| = \sqrt{4^2 + (-3)^2 + 5^2} = \sqrt{50} = 5\sqrt{2}$. The unit vector in the direction of $\mathbf{v}$ is the vector $\dfrac{1}{\|\mathbf{v}\|} \mathbf{v} = \dfrac{1}{5\sqrt{2}} \langle 4, -3, 5 \rangle$. If we wish, we may distribute the scalar and rationalize the denominators to obtain $\left\langle \dfrac{2}{5}\sqrt{2}, -\dfrac{3}{10}\sqrt{2}, \dfrac{1}{2}\sqrt{2} \right\rangle$.   □

EXAMPLE 3   **Finding the forces in cables suspending an object**

The figure that follows shows a 50-pound weight suspended by two cables. Find the force in each of the cables attached to the ceiling.

The sum of the vector forces equals zero

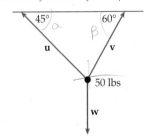

SOLUTION

Since the weight is stationary, the sum of the three vectors is $\mathbf{u} + \mathbf{v} + \mathbf{w} = \mathbf{0}$. We decompose each vector into its horizontal and vertical components, using the standard convention that motion to the right is positive on the x-axis and motion upward is positive on the y-axis. Therefore,

$$\mathbf{u} = -\cos 45° \|\mathbf{u}\| \mathbf{i} + \sin 45° \|\mathbf{u}\| \mathbf{j} = -\frac{\sqrt{2}}{2} \|\mathbf{u}\| \mathbf{i} + \frac{\sqrt{2}}{2} \|\mathbf{u}\| \mathbf{j},$$

$$\mathbf{v} = \cos 60° \|\mathbf{v}\| \mathbf{i} + \sin 60° \|\mathbf{v}\| \mathbf{j} = \frac{1}{2} \|\mathbf{v}\| \mathbf{i} + \frac{\sqrt{3}}{2} \|\mathbf{v}\| \mathbf{j}, \text{ and}$$

$$\mathbf{w} = -50\mathbf{j}.$$

In order to have $\mathbf{u} + \mathbf{v} + \mathbf{w} = \mathbf{0}$, the sum of the three x-components and the sum of the three y-components must both be zero. So,

$$\frac{\sqrt{2}}{2}\|\mathbf{u}\| = \frac{1}{2}\|\mathbf{v}\| \text{ and } \frac{\sqrt{2}}{2}\|\mathbf{u}\| + \frac{\sqrt{3}}{2}\|\mathbf{v}\| = 50 \text{ pounds}.$$

From the first of these equations, we have $\sqrt{2}\|\mathbf{u}\| = \|\mathbf{v}\|$, which we can substitute into the second equation to obtain $\|\mathbf{u}\| = \frac{100}{\sqrt{2}+\sqrt{6}} \approx 25.9$ pounds. Therefore $\|\mathbf{v}\| \approx 36.6$ pounds. □

? TEST YOUR UNDERSTANDING

▶ What are the geometric and analytic interpretations of a vector in $\mathbb{R}^2$? In $\mathbb{R}^3$?

▶ How are vectors added and subtracted geometrically? Algebraically?

▶ What does it mean for one vector to be a scalar multiple of another vector? How can you tell whether two vectors are parallel?

▶ What is a unit vector? How do you find a unit vector with the same direction as a given vector?

▶ How can you use vectors to analyze the forces acting on an object?

EXERCISES 10.2

Thinking Back

Properties of addition: Provide definitions for the following properties.

▶ the commutative property of addition for real numbers

▶ the associative property of addition for real numbers

▶ the distributive property of multiplication over addition for real numbers

Computing distances: Find the distances between the specified pairs of points.

▶ $(-3, 4)$ and $(6, -5)$

▶ $(11, 12, 13)$ and $(11, 12, -2)$

▶ $(5, -2, -1)$ and $(3, 0, -4)$

Concepts

0. *Problem Zero:* Read the section and make your own summary of the material.

1. *True/False:* Determine whether each of the statements that follow is true or false. If a statement is true, explain why. If a statement is false, provide a counterexample.

   (a) *True or False:* Every vector has a norm.

   (b) *True or False:* Given any vector $\mathbf{v}$, there is a unit vector in the direction of $\mathbf{v}$.

   (c) *True or False:* The unit vector in the direction of a given nonzero vector $\mathbf{v}$ is always shorter than $\mathbf{v}$.

   (d) *True or False:* Given two points A and B in 3-space, the vector from A to B is denoted by $\overrightarrow{BA}$.

   (e) *True or False:* The vector $\langle a, b \rangle$ in $\mathbb{R}^2$ can be interpreted as the vector $\langle a, b, 0 \rangle$ in $\mathbb{R}^3$.

   (f) *True or False:* Standard basis vectors are unit vectors.

   (g) *True or False:* For any nonzero vector $\mathbf{v}$, the vector $-\mathbf{v}$ has the same length as $\mathbf{v}$ but points in the opposite direction.

   (h) *True or False:* If $\mathbf{v} = \overrightarrow{AB}$, then $-\mathbf{v} = \overrightarrow{BA}$.

2. *Examples:* Construct examples of the thing(s) described in the following. Try to find examples that are different than any in the reading.

   (a) A unit vector in $\mathbb{R}^2$ that is not parallel to either $\mathbf{i}$ or $\mathbf{j}$.

   (b) A vector in $\mathbb{R}^3$ with magnitude 5.

   (c) Two distinct vectors parallel to $\langle 1, 2, -3 \rangle$, each with norm 3.

In Exercises 3–8, let A, B, C, D, ..., Z be points in $\mathbb{R}^3$. Simplify the given quantity.

3. $\overrightarrow{AB} + \overrightarrow{BC}$

4. $\overrightarrow{AB} + \overrightarrow{BC} + \overrightarrow{CD}$

5. $\overrightarrow{AB} + \overrightarrow{BA}$

6. $\overrightarrow{BD} - \overrightarrow{CD}$

7. $\overrightarrow{AB} + \overrightarrow{BC} + \overrightarrow{CA}$

8. $\overrightarrow{AB} + \overrightarrow{BC} + \cdots + \overrightarrow{YZ}$

9. How do you add two vectors algebraically? Geometrically?

10. How do you subtract two vectors algebraically? Geometrically?

11. Find a vector parallel to $\langle a, b, c \rangle$ but twice as long.

12. Find a vector parallel to $\langle a, b, c \rangle$ but half as long and pointing in the opposite direction.

13. If the initial point of the vector $\langle 2, 3, -5 \rangle$ is the point $(-3, 2, 4)$, what is the terminal point of the vector?

14. If the terminal point of the vector $\langle 2, 3, -5 \rangle$ is the point $(-3, 2, 4)$, what is the initial point of the vector?

15. Find the terminal point of a vector of magnitude 5 that is parallel to the vector $\langle 1, 2, 3 \rangle$ and whose initial point is $(0, 3, -2)$.

16. Let $\mathbf{v}_0 = \langle a, b \rangle$ and let $\mathbf{v} = \langle x, y \rangle$. Describe the sets of points in $\mathbb{R}^2$ satisfying the following properties:
  (a) $\|\mathbf{v}\| = 4$
  (b) $\|\mathbf{v}\| \leq 4$
  (c) $\|\mathbf{v} - \mathbf{v}_0\| = 4$

17. Let $\mathbf{v}_0 = \langle a, b, c \rangle$ and let $\mathbf{v} = \langle x, y, z \rangle$. Describe the sets of points in $\mathbb{R}^3$ satisfying the following properties:
  (a) $\|\mathbf{v}\| = 4$
  (b) $\|\mathbf{v}\| \leq 4$
  (c) $\|\mathbf{v} - \mathbf{v}_0\| = 4$

18. Let $\mathbf{v} = \langle w, x, y, z \rangle$. Describe the sets of points in $\mathbb{R}^4$ satisfying $\|\mathbf{v}\| = 4$.

19. How do you generalize the ideas of this section to vectors with four components? To vectors with n components?

20. What is the set of all position vectors in $\mathbb{R}^2$ of magnitude 5?

21. What is the set of all position vectors in $\mathbb{R}^3$ of magnitude 5?

22. Consider the vector $\mathbf{v} = \langle 2, 3, 7 \rangle$.
  (a) Graph $\mathbf{v}$.
  (b) What vector is symmetric about the origin to $\mathbf{v}$? Graph that vector.
  (c) What vector is symmetric about the xy-plane to $\mathbf{v}$? Graph that vector.
  (d) What vector is symmetric about the point $(0, 3, 0)$ to $\mathbf{v}$? Graph that vector.
  (e) What vector is symmetric about the plane $z = 2$ to $\mathbf{v}$? Graph that vector.

Skills

In each of Exercises 23–28, find $\mathbf{u} + \mathbf{v}$ and $\mathbf{u} - \mathbf{v}$. Also, sketch $\mathbf{u}, \mathbf{v}, \mathbf{u} + \mathbf{v}$, and $\mathbf{u} - \mathbf{v}$.

23. $\mathbf{u} = \langle 2, -6 \rangle$, $\mathbf{v} = \langle 6, 2 \rangle$

24. $\mathbf{u} = \langle 3, -4 \rangle$, $\mathbf{v} = \langle -1, 5 \rangle$

25. $\mathbf{u} = \langle 1, -2 \rangle$, $\mathbf{v} = \langle -3, 6 \rangle$

26. $\mathbf{u} = \langle 4, 0 \rangle$, $\mathbf{v} = \langle 0, 3 \rangle$

27. $\mathbf{u} = \langle 1, -4, 6 \rangle$, $\mathbf{v} = \langle 2, -4, 7 \rangle$

28. $\mathbf{u} = \langle 3, 6, 11 \rangle$, $\mathbf{v} = \langle 1, -2, 3 \rangle$

In Exercises 29–32 find $\overrightarrow{PQ}$.

29. $P = (3, 6)$, $Q = (-3, -2)$

30. $P = (5, -3)$, $Q = (3, -6)$

31. $P = (1, -2, 5)$, $Q = (1, -7, -2)$

32. $P = (-1, 5, 4)$, $Q = (-1, 6, 4)$

In Exercises 33–36 find the norm of the vector.

33. $\mathbf{v} = \langle 3, -4 \rangle$

34. $\mathbf{v} = \left\langle \frac{1}{2}, \frac{1}{3}, \frac{1}{4} \right\rangle$

35. $\mathbf{v} = \left\langle \frac{1}{3}, \frac{1}{3}, -\frac{1}{3} \right\rangle$

36. $\mathbf{v} = \langle 0, 0, 6 \rangle$

In Exercises 37–42, find $\|\mathbf{v}\|$ and find the unit vector in the direction of $\mathbf{v}$.

37. $\mathbf{v} = \langle 3, -4 \rangle$

38. $\mathbf{v} = \left\langle \sqrt{\frac{1}{3}}, -\sqrt{\frac{2}{3}} \right\rangle$

39. $\mathbf{v} = \left\langle \frac{1}{5}, \frac{1}{3} \right\rangle$

40. $\mathbf{v} = \langle 2, 1, -5 \rangle$

41. $\mathbf{v} = \langle 1, 1, 1 \rangle$

42. $\mathbf{v} = \left\langle \frac{\sqrt{2}}{2}, \frac{1}{2}, \frac{1}{2} \right\rangle$

In Exercises 43–51 find a vector with the given properties.

43. Find a vector in the direction of $\langle 3, 1, 2 \rangle$ and with magnitude 5.

44. Find a vector in the direction of $\langle -1, 2, 3 \rangle$ and with magnitude 3.

45. Find a vector in the direction of $\langle 8, -7, 2 \rangle$ and with magnitude 2.

46. Find a vector in the direction of $\langle 9, 0, -6 \rangle$ and with magnitude 7.

47. Find a vector in the direction opposite to $\langle -1, -4, -6 \rangle$ and with magnitude 7.

48. Find a vector in the direction opposite to $\langle -4, 5, -1 \rangle$ and with magnitude 3.

49. Find a vector in the direction opposite to $\langle 0, -3, 4 \rangle$ and with magnitude 10.

50. Find a unit vector in the direction opposite to $\langle 3, 4, 5 \rangle$.

51. Find a vector of length 3 that points in the direction opposite to $\langle 1, -2, 3 \rangle$.

52. Let $P = (2, 3, 0, -1)$ and $Q = (3, -2, 1, 6)$ be points in four-dimensional space.
  (a) Find $\overrightarrow{PQ}$.
  (b) Find $\|\overrightarrow{PQ}\|$.
  (c) Find the unit vector in the direction of $\overrightarrow{PQ}$.

Applications

53. A weight of 100 pounds is suspended by two ropes as shown in the accompanying figure. What are the magnitudes of the forces in each of the ropes?

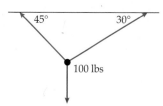

54. Each bolt attached in a certain ceiling is known to be able to withstand a force of 200 pounds before it pulls out. What is the maximum weight of an object that can be suspended in the following system?

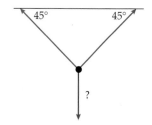

55. Annie has to make a crossing by kayak from Stuart Island to Speiden Island, two islands in the San Juan archipelago that are part of Washington State.

    (a) If she crosses due south during a slack tide at a leisurely 2 mph, what is a vector describing her velocity?

    (b) If she crosses in a perfectly southeastern direction at slack tide at 2 mph, what is a vector describing her velocity?

    (c) The tides can pump through the Stuart–Speiden channel at a good clip. The outgoing tide can move due west at 2 mph. When faced with this situation, Annie turns on the jets and paddles southeast at 3 mph. What is a vector describing her velocity?

56. Ian is descending Middle Cascade Glacier in North Cascades National Park in Washington State.

    (a) Near the top of the glacier, he is descending a 30-degree slope due northwards at 3 mph. Give a vector describing his velocity.

    (b) Toward the middle of the glacier the slope steepens. Now Ian is descending directly northeast at a 45-degree angle. His speed is only 2 mph. What is a vector describing his velocity?

    (c) Now Ian needs to exit the glacier before he runs into the icefall. He contours due east, so that he is neither ascending nor descending. He is moving only at 1.5 mph now. What is a vector describing his velocity?

Proofs

57. Prove part (a) of Theorem 10.8 for vectors in $\mathbb{R}^3$; that is, show that for $\mathbf{u} = \langle u_1, u_2, u_3 \rangle$ and $\mathbf{v} = \langle v_1, v_2, v_3 \rangle$,

$$\mathbf{u} + \mathbf{v} = \mathbf{v} + \mathbf{u}.$$

58. Prove part (b) of Theorem 10.8 for vectors in $\mathbb{R}^3$; that is, show that for $\mathbf{u} = \langle u_1, u_2, u_3 \rangle$, $\mathbf{v} = \langle v_1, v_2, v_3 \rangle$ and $\mathbf{w} = \langle w_1, w_2, w_3 \rangle$,

$$(\mathbf{u} + \mathbf{v}) + \mathbf{w} = \mathbf{u} + (\mathbf{v} + \mathbf{w}).$$

59. Prove part (c) of Theorem 10.8 for vectors in $\mathbb{R}^3$; that is, show that for $\mathbf{u} = \langle u_1, u_2, u_3 \rangle$, $\mathbf{v} = \langle v_1, v_2, v_3 \rangle$ and scalar c,

$$c(\mathbf{u} + \mathbf{v}) = c\mathbf{u} + c\mathbf{v}.$$

60. Prove Theorem 10.11; that is, show that when $\mathbf{v} \neq \mathbf{0}$, the scaled vector $\dfrac{1}{\|\mathbf{v}\|}\mathbf{v}$ is a unit vector with the same direction as $\mathbf{v}$.

61. Let c and d be a scalars and let $\mathbf{v}$ be a vector in $\mathbb{R}^3$. Show that the following distributive property holds:

$$(c + d)\mathbf{v} = c\mathbf{v} + d\mathbf{v}.$$

62. Use a vector argument to prove that the segment connecting the midpoints of two sides of a triangle is parallel to the third side of the triangle and half of its length.

63. Use vector methods to show that the diagonals of a parallelogram bisect each other.

64. Let Quad($PQRS$) denote the quadrilateral in the xy-plane with vertices P, Q, R, and S. If P' is the midpoint of side PQ, Q' is the midpoint of side QR, R' is the midpoint of side RS, and S' is the midpoint of side SP, prove that Quad($P'Q'R'S'$) is a parallelogram.

Thinking Forward

Perpendicular vectors in $\mathbb{R}^2$

▶ How many vectors are there in $\mathbb{R}^2$ that are perpendicular to a given nonzero vector $\langle a, b \rangle$?

▶ How many unit vectors are there in $\mathbb{R}^2$ that are perpendicular to a given nonzero vector $\langle a, b \rangle$?

▶ Find a vector in $\mathbb{R}^2$ that is perpendicular to the vector $\langle 1, 3 \rangle$.

▶ Find a unit vector in $\mathbb{R}^2$ that is perpendicular to the vector $\langle 1, 3 \rangle$.

Perpendicular vectors in $\mathbb{R}^3$

▶ How many vectors are there in $\mathbb{R}^3$ that are perpendicular to a given nonzero vector $\langle a, b, c \rangle$?

▶ How many unit vectors are there in $\mathbb{R}^3$ that are perpendicular to a given nonzero vector $\langle a, b, c \rangle$?

▶ Find a unit vector in $\mathbb{R}^3$ that is perpendicular to both vectors $\mathbf{i}$ and $\mathbf{k}$.

10.3   DOT PRODUCT

▶   The dot product

▶   The geometry of the dot product

▶   Projecting one vector onto another

The Dot Product

In Section 10.2 we discussed how to multiply a vector $\mathbf{v}$ by a scalar k. Recall that $k\mathbf{v}$ is a vector parallel to $\mathbf{v}$ with magnitude $|k|$ times the magnitude of $\mathbf{v}$ and with the same direction if $k > 0$ and the opposite direction if $k < 0$. Here we will define the dot product, which allows us to multiply two vectors with the same number of components. In Section 10.4 we will discuss the cross product, a different method for multiplying two vectors in $\mathbb{R}^3$.

DEFINITION 10.13

Dot Product

Let $\mathbf{u} = \langle u_1, u_2 \rangle$ and $\mathbf{v} = \langle v_1, v_2 \rangle$ be vectors in $\mathbb{R}^2$. We define the **dot product, $\mathbf{u} \cdot \mathbf{v}$,** to be

$$\mathbf{u} \cdot \mathbf{v} = u_1 v_1 + u_2 v_2.$$

Let $\mathbf{u} = \langle u_1, u_2, u_3 \rangle$ and $\mathbf{v} = \langle v_1, v_2, v_3 \rangle$ be vectors in $\mathbb{R}^3$. We define the **dot product, $\mathbf{u} \cdot \mathbf{v}$,** to be

$$\mathbf{u} \cdot \mathbf{v} = u_1 v_1 + u_2 v_2 + u_3 v_3.$$

For example, the dot product of the vectors $\langle 1, 4 \rangle$ and $\langle 7, -3 \rangle$ from $\mathbb{R}^2$ is

$$\langle 1, 4 \rangle \cdot \langle 7, -3 \rangle = 1 \cdot 7 + 4 \cdot (-3) = 7 - 12 = -5,$$

and the dot product of the vectors $\langle 2, 7, 6 \rangle$ and $\langle -1, 2, -2 \rangle$ from $\mathbb{R}^3$ is

$$\langle 2, 7, 6 \rangle \cdot \langle -1, 2, -2 \rangle = 2 \cdot (-1) + 7 \cdot 2 + 6 \cdot (-2) = -2 + 14 - 12 = 0.$$

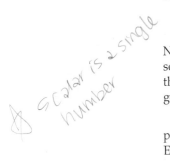 *scalar is a single number*

Note that in both cases the <u>dot product is a scalar</u>, not a vector. For most of the rest of this section we will discuss the significance of the dot product and what this scalar tells us about the relationship between the two vectors. Note also that Definition 10.13 could easily be generalized to allow us to find the dot product of two vector in $\mathbb{R}^n$.

Theorem 10.14 collects several basic algebraic properties of the dot product. We prove part (b) for vectors in $\mathbb{R}^3$ and leave the proofs of the other parts of the theorem for Exercise 62.

THEOREM 10.14

Algebraic Properties of the Dot Product

For any vectors $\mathbf{u}$, $\mathbf{v}$, and $\mathbf{w}$, and any scalar k,

(a) $\mathbf{u} \cdot \mathbf{v} = \mathbf{v} \cdot \mathbf{u}$

(b) $\mathbf{u} \cdot (\mathbf{v} + \mathbf{w}) = \mathbf{u} \cdot \mathbf{v} + \mathbf{u} \cdot \mathbf{w}$

(c) $k(\mathbf{u} \cdot \mathbf{v}) = (k\mathbf{u}) \cdot \mathbf{v} = \mathbf{u} \cdot (k\mathbf{v})$

(d) $\mathbf{v} \cdot \mathbf{v} = \|\mathbf{v}\|^2$

Proof. Let $\mathbf{u} = \langle u_1, u_2, u_3 \rangle$, $\mathbf{v} = \langle v_1, v_2, v_3 \rangle$, and $\mathbf{w} = \langle w_1, w_2, w_3 \rangle$. Using the definitions of vector addition and the dot product, we have

$$
\begin{aligned}
\mathbf{u} \cdot (\mathbf{v} + \mathbf{w}) &= \langle u_1, u_2, u_3 \rangle \cdot (\langle v_1, v_2, v_3 \rangle + \langle w_1, w_2, w_3 \rangle) \\
&= \langle u_1, u_2, u_3 \rangle \cdot \langle v_1 + w_1, v_2 + w_2, v_3 + w_3 \rangle \\
&= u_1(v_1 + w_1) + u_2(v_2 + w_2) + u_3(v_3 + w_3) \\
&= u_1 v_1 + u_1 w_1 + u_2 v_2 + u_2 w_2 + u_3 v_3 + u_3 w_3 \\
&= (u_1 v_1 + u_2 v_2 + u_3 v_3) + (u_1 w_1 + u_2 w_2 + u_3 w_3) \\
&= \langle u_1, u_2, u_3 \rangle \cdot \langle v_1, v_2, v_3 \rangle + \langle u_1, u_2, u_3 \rangle \cdot \langle w_1, w_2, w_3 \rangle \\
&= \mathbf{u} \cdot \mathbf{v} + \mathbf{u} \cdot \mathbf{w}.
\end{aligned}
$$

Note that this proof may be generalized to vectors in $\mathbb{R}^n$. Similarly, the other parts of Theorem 10.14 may be generalized to vectors in $\mathbb{R}^n$.

We obtain a scalar as the dot product of the two vectors $\mathbf{u}$ and $\mathbf{v}$. What does this number tell us about the geometric relationship between the two vectors? Before we answer that question, we need to mention that the angle between two nonzero vectors $\mathbf{u}$ and $\mathbf{v}$ is the angle $0 \le \theta \le \pi$ created when $\mathbf{u}$ and $\mathbf{v}$ are considered as position vectors. (If either $\mathbf{u} = \mathbf{0}$ or $\mathbf{v} = \mathbf{0}$, then the angle between $\mathbf{u}$ and $\mathbf{v}$ is undefined.)

The angle between two nonzero vectors is $0 \le \theta \le \pi$

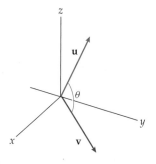

THEOREM 10.15

The Geometry of the Dot Product

Let $\mathbf{u}$ and $\mathbf{v}$ be vectors. Then the dot product

$$
\mathbf{u} \cdot \mathbf{v} = \begin{cases} 0, & \text{if either } \mathbf{u} = \mathbf{0} \text{ or } \mathbf{v} = \mathbf{0} \\ \|\mathbf{u}\| \|\mathbf{v}\| \cos\theta, & \text{if } \mathbf{u} \ne \mathbf{0} \text{ and } \mathbf{v} \ne \mathbf{0}, \text{ where } \theta \text{ is the angle between } \mathbf{u} \text{ and } \mathbf{v}. \end{cases}
$$

This theorem provides a relationship between the dot product of two vectors, their lengths, and the angle between them.

In Exercise 56 we ask you to prove Theorem 10.15 for vectors in $\mathbb{R}^2$. We will prove it for vectors in $\mathbb{R}^3$ in a moment. Before we do, let us remind you of the ***Law of Cosines***, which we will use in our proof.

THEOREM 10.16

Law of Cosines

In a triangle with side lengths a, b, and c, where θ is the angle between the sides of length a and b,

$$
a^2 + b^2 - 2ab \cos\theta = c^2.
$$

We are now ready to prove Theorem 10.15 for vectors in $\mathbb{R}^3$.

Proof. Let $\mathbf{u} = \langle u_1, u_2, u_3 \rangle$ and $\mathbf{v} = \langle v_1, v_2, v_3 \rangle$. If either $\mathbf{u}$ or $\mathbf{v}$ is $\mathbf{0}$, then

$$u_1 v_1 + u_2 v_2 + u_3 v_3 = 0 + 0 + 0 = 0,$$

and we are done.

If neither $\mathbf{u}$ nor $\mathbf{v}$ is $\mathbf{0}$ but they are parallel, then $\mathbf{u} = k\mathbf{v}$ for some scalar k, and θ, the angle between $\mathbf{u}$ and $\mathbf{v}$, is either 0 or π. If $\theta = 0$, then $\cos\theta = 1$ and $\mathbf{u} = k\mathbf{v}$ for some positive scalar k. Using the appropriate algebraic properties of Theorem 10.14, we have

$$\mathbf{u} \cdot \mathbf{v} = k\mathbf{v} \cdot \mathbf{v} = k\|\mathbf{v}\|^2 = \|\mathbf{u}\|\|\mathbf{v}\|.$$

In Exercise 58 we ask you to analyze the case where $\mathbf{u} = k\mathbf{v}$ and $k < 0$.

If neither $\mathbf{u}$ nor $\mathbf{v}$ is $\mathbf{0}$, and if $0 < \theta < \pi$, consider the following triangle formed from the vectors $\mathbf{u}$ and $\mathbf{v}$:

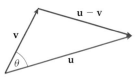

From the Law of Cosines we know that

$$\|\mathbf{u}\|^2 + \|\mathbf{v}\|^2 - 2\|\mathbf{u}\|\|\mathbf{v}\|\cos\theta = \|\mathbf{u} - \mathbf{v}\|^2.$$

Rearranging, we get

$$\|\mathbf{u}\|\|\mathbf{v}\|\cos\theta = \frac{1}{2}(\|\mathbf{u}\|^2 + \|\mathbf{v}\|^2 - \|\mathbf{u} - \mathbf{v}\|^2).$$

To finish our proof we will show that the right-hand side of this last equality is the dot product $\mathbf{u} \cdot \mathbf{v}$.

Since $\mathbf{u} = \langle u_1, u_2, u_3 \rangle$ and $\mathbf{v} = \langle v_1, v_2, v_3 \rangle$, we have $\mathbf{u} - \mathbf{v} = \langle u_1 - v_1, u_2 - v_2, u_3 - v_3 \rangle$. We also have the following norms:

$$\|\mathbf{u}\|^2 = u_1^2 + u_2^2 + u_3^2,$$

$$\|\mathbf{v}\|^2 = v_1^2 + v_2^2 + v_3^2, \quad \text{and}$$

$$\|\mathbf{u} - \mathbf{v}\|^2 = (u_1 - v_1)^2 + (u_2 - v_2)^2 + (u_3 - v_3)^2.$$

Therefore, $\frac{1}{2}(\|\mathbf{u}\|^2 + \|\mathbf{v}\|^2 - \|\mathbf{u} - \mathbf{v}\|^2)$ equals

$$\frac{1}{2}((u_1^2 + u_2^2 + u_3^2) + (v_1^2 + v_2^2 + v_3^2) - (u_1^2 - 2u_1 v_1 + v_1^2 + u_2^2 - 2u_2 v_2 + v_2^2 + u_3^2 - 2u_3 v_3 + v_3^2))$$

$$= u_1 v_1 + u_2 v_2 + u_3 v_3 = \mathbf{u} \cdot \mathbf{v}.$$

We now have our result.   ∎

As a corollary to Theorem 10.15 we may immediately determine when the angle between two vectors is acute, right, or obtuse:

THEOREM 10.17

The Angle Between Two Vectors and the Dot Product

Let θ be the angle between nonzero vectors $\mathbf{u}$ and $\mathbf{v}$. Then

$$\theta \text{ is } \begin{cases} \text{acute,} & \text{if and only if } \mathbf{u} \cdot \mathbf{v} > 0 \\ \text{right,} & \text{if and only if } \mathbf{u} \cdot \mathbf{v} = 0 \\ \text{obtuse,} & \text{if and only if } \mathbf{u} \cdot \mathbf{v} < 0. \end{cases}$$

The details of the proof are left for Exercise 60.

DEFINITION 10.18

Orthogonal Curves and Vectors

(a) Two curves are said to be **orthogonal** at a point of intersection if the tangent lines to the curves at the point of intersection are perpendicular.

(b) Two nonzero vectors are **orthogonal** if the angle between them is a right angle.

(c) The zero vector is **orthogonal** to every vector.

Orthogonal curves   *Orthogonal vectors*

We have the following theorem, whose proof is left for Exercise 61:

THEOREM 10.19

The Dot Product Test for Orthogonality

Vectors $\mathbf{u}$ and $\mathbf{v}$ are orthogonal if and only if $\mathbf{u} \cdot \mathbf{v} = 0$.

Projections

Although our coordinate systems consist of mutually perpendicular axes, in certain applications one nonzero vector $\mathbf{u}$ is of particular significance. When this is the case, we may wish to write another vector $\mathbf{v}$ as a sum of two vectors, one of which is parallel to $\mathbf{u}$, the other orthogonal to $\mathbf{u}$.

DEFINITION 10.20

Vector Projections and Vector Components

Let $\mathbf{u}$ be a nonzero vector and let $\mathbf{v}$ be any vector. We define the *vector projection of $\mathbf{v}$ onto $\mathbf{u}$*, denoted $\mathbf{v}_{\parallel}$, and the *vector component of $\mathbf{v}$ orthogonal to $\mathbf{u}$*, denoted $\mathbf{v}_{\perp}$, as the pair of vectors having the following three properties:

(a) $\mathbf{v}_{\parallel}$ is parallel to $\mathbf{u}$.

(b) $\mathbf{v}_{\perp}$ is orthogonal to $\mathbf{u}$.

(c) $\mathbf{v}_{\parallel} + \mathbf{v}_{\perp} = \mathbf{v}$.

(Note that we will be using the notations $\mathbf{v}_{\parallel}$ and $\mathbf{v}_{\perp}$ only temporarily. The formal notations are given shortly.)

In Exercise 63 we ask you to prove that such a decomposition of $\mathbf{v}$ is unique.

Geometrically we have one of the situations shown in the following two figures:

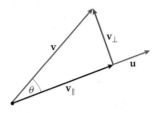 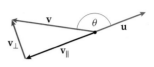

Since $\mathbf{u} \cdot \mathbf{v} = \|\mathbf{u}\|\|\mathbf{v}\| \cos\theta$, we have

$$\|\mathbf{v}\| \cos\theta = \frac{1}{\|\mathbf{u}\|} \mathbf{u} \cdot \mathbf{v} = \frac{\mathbf{u} \cdot \mathbf{v}}{\|\mathbf{u}\|}.$$

Observe that the signed quantity $\|\mathbf{v}\| \cos\theta$ is positive when $0 \le \theta < 90°$ and negative when $90° < \theta \le 180°$. In either case, $\|\mathbf{v}_{\parallel}\| = \|\mathbf{v}\| |\cos\theta|$.

We use $\|\mathbf{v}\| \cos\theta = \frac{\mathbf{u} \cdot \mathbf{v}}{\|\mathbf{u}\|}$ in the following definition:

DEFINITION 10.21

> ### The Component of a Projection
>
> Let $\mathbf{u}$ be any nonzero vector. Then the ***component of the projection of v onto u*** is the scalar
>
> $$\text{comp}_{\mathbf{u}}\mathbf{v} = \frac{\mathbf{u} \cdot \mathbf{v}}{\|\mathbf{u}\|}.$$

It bears repeating that $\text{comp}_{\mathbf{u}}\mathbf{v}$ is a signed quantity and its sign depends upon the size of the angle between $\mathbf{u}$ and $\mathbf{v}$.

From our earlier discussion, $\mathbf{v}_{\parallel}$ equals $\text{comp}_{\mathbf{u}}\mathbf{v}$ times the unit vector in the direction of $\mathbf{u}$, and we have

$$\mathbf{v}_{\parallel} = \frac{\mathbf{u} \cdot \mathbf{v}}{\|\mathbf{u}\|} \frac{\mathbf{u}}{\|\mathbf{u}\|} = \frac{\mathbf{u} \cdot \mathbf{v}}{\|\mathbf{u}\|^2} \mathbf{u}.$$

We now introduce the more formal notation for $\mathbf{v}_{\parallel}$.

DEFINITION 10.22

> ### The Vector Projection
>
> Let $\mathbf{u}$ be any nonzero vector. Then the ***vector projection of v onto u*** is
>
> $$\text{proj}_{\mathbf{u}}\mathbf{v} := \frac{\mathbf{u} \cdot \mathbf{v}}{\|\mathbf{u}\|^2} \mathbf{u}.$$

Finally, since $\mathbf{v}_{\parallel} + \mathbf{v}_{\perp} = \mathbf{v}$, the vector component of $\mathbf{v}$ orthogonal to $\mathbf{u}$ is

$$\mathbf{v}_{\perp} = \mathbf{v} - \text{proj}_{\mathbf{u}}\mathbf{v}.$$

The Triangle Inequality

You are probably familiar with the statement: "The shortest distance between two points is a straight line." This is the famous *triangle inequality*. In the figure shown next, we have two paths from point A to point B. One is $\mathbf{u} + \mathbf{v}$, while the other is $\mathbf{u}$ followed by $\mathbf{v}$. Since

$\mathbf{u} + \mathbf{v}$ is the straight path between the two points, its length must be shorter than the length of the path given by $\mathbf{u}$ followed by $\mathbf{v}$. This inequality may be phrased in terms of vectors, as in the next theorem.

The shortest distance between points A and B is the straight line

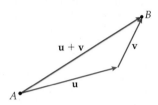

This figure illustrates why the statement about the shortest distance is a statement about triangles!

THEOREM 10.23

The Triangle Inequality

Given vectors $\mathbf{u}$ and $\mathbf{v}$, $\|\mathbf{u} + \mathbf{v}\| \leq \|\mathbf{u}\| + \|\mathbf{v}\|$, and $\|\mathbf{u} + \mathbf{v}\| = \|\mathbf{u}\| + \|\mathbf{v}\|$ if and only if $\mathbf{u} = \mathbf{0}$ or $\mathbf{v} = \mathbf{0}$ or $\mathbf{u} = k\mathbf{v}$ for some positive scalar k.

Proof. First note that if $\mathbf{u} = \mathbf{0}$ or $\mathbf{v} = \mathbf{0}$, then $\|\mathbf{u} + \mathbf{v}\| = \|\mathbf{u}\| + \|\mathbf{v}\|$. If neither $\mathbf{u}$ nor $\mathbf{v}$ is $\mathbf{0}$, but $\mathbf{u} = k\mathbf{v}$ for some positive scalar k, then

$$\|\mathbf{u} + \mathbf{v}\| = \|k\mathbf{v} + \mathbf{v}\| = (k+1)\|\mathbf{v}\| = k\|\mathbf{v}\| + \|\mathbf{v}\| = \|\mathbf{u}\| + \|\mathbf{v}\|.$$

We now use the dot product to prove the inequality. We have

$$\|\mathbf{u} + \mathbf{v}\|^2 = (\mathbf{u} + \mathbf{v}) \cdot (\mathbf{u} + \mathbf{v})$$
$$= \mathbf{u} \cdot \mathbf{u} + \mathbf{u} \cdot \mathbf{v} + \mathbf{v} \cdot \mathbf{u} + \mathbf{v} \cdot \mathbf{v}$$
$$= \|\mathbf{u}\|^2 + 2\mathbf{u} \cdot \mathbf{v} + \|\mathbf{v}\|^2$$
$$= \|\mathbf{u}\|^2 + 2\|\mathbf{u}\|\|\mathbf{v}\| \cos\theta + \|\mathbf{v}\|^2$$
$$\leq \|\mathbf{u}\|^2 + 2\|\mathbf{u}\|\|\mathbf{v}\| + \|\mathbf{v}\|^2$$
$$= (\|\mathbf{u}\| + \|\mathbf{v}\|)^2.$$

So, $\|\mathbf{u} + \mathbf{v}\|^2 \leq (\|\mathbf{u}\| + \|\mathbf{v}\|)^2$. Since magnitudes are nonnegative, we have our result,

$$\|\mathbf{u} + \mathbf{v}\| \leq \|\mathbf{u}\| + \|\mathbf{v}\|. \qquad \blacksquare$$

Examples and Explorations

EXAMPLE 1

Finding the angle between two vectors

Find the angle between $\mathbf{u} = \langle 1, 2, 3 \rangle$ and $\mathbf{v} = \langle 2, 1, -4 \rangle$.

SOLUTION

Here,

$$\mathbf{u} \cdot \mathbf{v} = 2 + 2 - 12 = -8,$$
$$\|\mathbf{u}\| = \sqrt{1^2 + 2^2 + 3^2} = \sqrt{14}, \quad \text{and}$$
$$\|\mathbf{v}\| = \sqrt{2^2 + 1^2 + (-4)^2} = \sqrt{21}.$$

Therefore, if θ is the angle between $\mathbf{u}$ and $\mathbf{v}$, then $\cos\theta = \dfrac{\mathbf{u} \cdot \mathbf{v}}{\|\mathbf{u}\|\|\mathbf{v}\|} = \dfrac{-8}{\sqrt{14}\sqrt{21}} \approx -0.4666$. Thus, $\theta \approx 117.8°$. $\qquad \square$

EXAMPLE 2

Using the dot product to show orthogonality

Show that the triangle with vertices $P = (6, 2, 2)$, $Q = (2, 0, -1)$, and $R = (5, 1, -2)$ is a right triangle.

SOLUTION

We can form the vectors

$$\overrightarrow{PQ} = \langle -4, -2, -3 \rangle, \ \overrightarrow{PR} = \langle -1, -1, -4 \rangle, \ \text{and} \ \overrightarrow{QR} = \langle 3, 1, -1 \rangle.$$

For $\triangle PQR$ to be a right triangle, we need one right angle. Using the dot product, we see that although $\overrightarrow{PQ}$ is orthogonal to neither $\overrightarrow{PR}$ nor $\overrightarrow{QR}$ (check this), we do have

$$\overrightarrow{PR} \cdot \overrightarrow{QR} = (-1)(3) + (-1)(1) + (-4)(-1) = 0.$$

Therefore the angle at vertex R is a right angle. Thus $\triangle PQR$ is a right triangle.   □

EXAMPLE 3

Finding the projection of one vector onto another

Given vectors $\mathbf{u} = \langle 8, 2 \rangle$ and $\mathbf{v} = \langle -4, 5 \rangle$, find $\text{comp}_{\mathbf{u}}\mathbf{v}$, $\text{proj}_{\mathbf{u}}\mathbf{v}$ the vector projection of $\mathbf{v}$ onto $\mathbf{u}$, and the component of $\mathbf{v}$ orthogonal to $\mathbf{u}$.

SOLUTION

The component of the projection of $\mathbf{v}$ onto $\mathbf{u}$ is

$$\text{comp}_{\mathbf{u}}\mathbf{v} = \frac{\langle 8, 2 \rangle \cdot \langle -4, 5 \rangle}{\|\langle 8, 2 \rangle\|} = \frac{-22}{\sqrt{68}}.$$

The vector projection of $\mathbf{v}$ onto $\mathbf{u}$ is $\text{comp}_{\mathbf{u}}\mathbf{v}$ times the unit vector in the direction of $\mathbf{u}$. Thus,

$$\text{proj}_{\mathbf{u}}\mathbf{v} = \frac{\mathbf{u} \cdot \mathbf{v}}{\|\mathbf{u}\|^2}\mathbf{u} = \frac{\langle 8, 2 \rangle \cdot \langle -4, 5 \rangle}{\|\langle 8, 2 \rangle\|^2}\langle 8, 2 \rangle = -\frac{11}{34}\langle 8, 2 \rangle.$$

The vector component of $\mathbf{v}$ orthogonal to $\mathbf{u}$ is given by

$$\mathbf{v} - \text{proj}_{\mathbf{u}}\mathbf{v} = \langle -4, 5 \rangle - \left(-\frac{11}{34}\langle 8, 2 \rangle\right) = \left\langle -\frac{24}{17}, \frac{96}{17} \right\rangle.$$

The following figure shows these vectors:

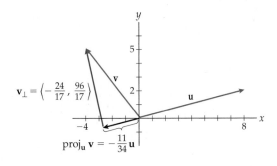

□

EXAMPLE 4

Finding the distance from a point to a line

Find the distance from the point $P = (1, 4, -2)$ to the line determined by the points $Q = (3, -2, 5)$ and $R = (0, 4, -3)$.

SOLUTION

The diagram that follows is a schematic illustrating the given situation. The distance we seek is merely the magnitude of the vector component of

$$\overrightarrow{QP} = \langle 1 - 3, 4 - (-2), -2 - 5 \rangle = \langle -2, 6, -7 \rangle$$

orthogonal to

$$\overrightarrow{QR} = \langle 0 - 3, 4 - (-2), -3 - 5 \rangle = \langle -3, 6, -8 \rangle.$$

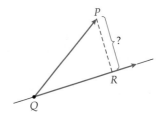

We may first find

$$\text{proj}_{\overrightarrow{QR}} \overrightarrow{QP} = \frac{\overrightarrow{QR} \cdot \overrightarrow{QP}}{\|\overrightarrow{QR}\|^2} \overrightarrow{QR} = \frac{\langle -3, 6, -8 \rangle \cdot \langle -2, 6, -7 \rangle}{\|\langle -3, 6, -8 \rangle\|^2} \langle -3, 6, -8 \rangle = \frac{98}{109} \langle -3, 6, -8 \rangle.$$

Thus the vector component of $\overrightarrow{QP}$ orthogonal to $\overrightarrow{QR}$ is

$$\overrightarrow{QP} - \text{proj}_{\overrightarrow{QR}} \overrightarrow{QP} = \langle -2, 6, -7 \rangle - \frac{98}{109} \langle -3, 6, -8 \rangle = \left\langle \frac{76}{109}, \frac{66}{109}, \frac{21}{109} \right\rangle.$$

The distance from P to the line determined by the points Q and R is the magnitude of this vector, or

$$\left\| \left\langle \frac{76}{109}, \frac{66}{109}, \frac{21}{109} \right\rangle \right\| = \sqrt{\left(\frac{76}{109}\right)^2 + \left(\frac{66}{109}\right)^2 + \left(\frac{21}{109}\right)^2} = \frac{\sqrt{10573}}{109} \approx 0.94. \qquad \square$$

? TEST YOUR UNDERSTANDING

▶ What is the definition of the dot product $\mathbf{u} \cdot \mathbf{v}$, and what is the geometric relationship among $\mathbf{u}$, $\mathbf{v}$, and $\mathbf{u} \cdot \mathbf{v}$?

▶ How is the angle between two vectors defined and how is it computed?

▶ What does it mean to project one vector onto another? When would you want to compute such a projection? How is the projection computed?

▶ How do you find the vector component of one vector orthogonal to another vector?

▶ What is the triangle inequality? How is it proved? Why is the inequality $|a+b| \leq |a|+|b|$, where a and b are real numbers, also called the triangle inequality?

EXERCISES 10.3

Thinking Back

▶ *Law of Cosines:* Use the Law of Cosines to find the measures of the angles of a triangle with side lengths 2, 3, and 4.

▶ *The distance between a point and a sphere:* Find the shortest distance from the point $(11, -3, 5)$ to the sphere $(x + 2)^2 + (y - 3)^2 + z^2 = 16$.

▶ *The Triangle Inequality:* Given $a = 5$ and $b = -3$, verify that the triangle inequality $|a + b| \leq |a| + |b|$ holds.

Concepts

0. *Problem Zero:* Read the section and make your own summary of the material.

1. *True/False:* Determine whether each of the statements that follow is true or false. If a statement is true, explain why. If a statement is false, provide a counterexample.

   (a) *True or False:* For all vectors $\mathbf{u}$ and $\mathbf{v}$ in $\mathbb{R}^3$, $\mathbf{u} \cdot \mathbf{v} = \mathbf{v} \cdot \mathbf{u}$.

   (b) *True or False:* The component of the projection of $\mathbf{v}$ onto $\mathbf{u}$, $\text{comp}_{\mathbf{u}}\mathbf{v}$, is a vector.

   (c) *True or False:* The projection of $\mathbf{v}$ onto $\mathbf{u}$, $\text{proj}_{\mathbf{u}}\mathbf{v}$, is a vector in the direction of $\mathbf{v}$.

   (d) *True or False:* Two curves are said to be *orthogonal* at a point if they intersect at the point.

   (e) *True or False:* If θ is the angle between two nonzero vectors $\mathbf{u}$ and $\mathbf{v}$, then $\cos\theta = \dfrac{\mathbf{u} \cdot \mathbf{v}}{\|\mathbf{u}\|\|\mathbf{v}\|}$.

   (f) *True or False:* If $\mathbf{u}$ and $\mathbf{v}$ are nonzero vectors such that $\text{proj}_{\mathbf{u}}\mathbf{v} = \text{proj}_{\mathbf{v}}\mathbf{u}$, then $\mathbf{u}$ and $\mathbf{v}$ are either equal or orthogonal.

   (g) *True or False:* If $\mathbf{u} \cdot \mathbf{v} = \|\mathbf{u}\|\|\mathbf{v}\|$, where $\mathbf{u}$ and $\mathbf{v}$ are nonzero, then $\mathbf{u} = k\mathbf{v}$, where $k > 0$.

   (h) *True or False:* The product $(\mathbf{u} \cdot \mathbf{v}) \cdot \mathbf{w}$ is defined for vectors $\mathbf{u}$, $\mathbf{v}$, and $\mathbf{w}$.

2. *Examples:* Construct examples of the thing(s) described in the following. Try to find examples that are different than any in the reading.

   (a) Two nonzero vectors in $\mathbb{R}^3$ whose dot product is zero.

   (b) A function $y = f(x)$ that is orthogonal to the function $y = \sin x$ at $x = 0$.

   (c) A vector $\mathbf{v} \neq \mathbf{i}$ such that $\text{proj}_{\mathbf{i}}\mathbf{v} = \mathbf{i}$.

3. State the definition of the dot product.

4. What does it mean geometrically for two vectors to be orthogonal at a point? What does it mean algebraically? What do we mean when we say that two curves are orthogonal at a point of intersection?

5. Why is the Law of Cosines a generalization of the Pythagorean theorem?

6. Let $\mathbf{v} = a\mathbf{i} + b\mathbf{j}$ and $\mathbf{w} = c\mathbf{i} + d\mathbf{j}$. Give conditions on the constants a, b, c, and d that guarantee that

   (a) $\mathbf{v}$ is parallel to $\mathbf{w}$.

   (b) $\mathbf{v}$ is perpendicular to $\mathbf{w}$.

7. Let $\mathbf{v} = a\mathbf{i} + b\mathbf{j} + c\mathbf{k}$ and $\mathbf{w} = \alpha\mathbf{i} + \beta\mathbf{j} + \gamma\mathbf{j}$. Give conditions on the constants a, b, c, α, β, and γ that guarantee that

   (a) $\mathbf{v}$ is parallel to $\mathbf{w}$.

   (b) $\mathbf{v}$ is perpendicular to $\mathbf{w}$.

8. What is the relationship between $\|\mathbf{v}\|$ and $\mathbf{v} \cdot \mathbf{v}$?

9. Let $\mathbf{u}$ be a nonzero vector.

   (a) Show that $\mathbf{u} \cdot \mathbf{v} = \mathbf{u} \cdot \mathbf{w}$ does *not* necessarily imply that $\mathbf{v} = \mathbf{w}$.

   (b) What geometric relationship must $\mathbf{u}$, $\mathbf{v}$, and $\mathbf{w}$ satisfy if $\mathbf{u} \cdot \mathbf{v} = \mathbf{u} \cdot \mathbf{w}$?

10. Let $\mathbf{u}$ and $\mathbf{v}$, be nonzero vectors.

   (a) When does $\text{comp}_{\mathbf{u}}\mathbf{v} = \text{comp}_{\mathbf{v}}\mathbf{u}$?

   (b) When does $\text{proj}_{\mathbf{u}}\mathbf{v} = \text{proj}_{\mathbf{v}}\mathbf{u}$?

11. What geometric relationship must two vectors have in order for $\|\mathbf{u} + \mathbf{v}\| = \|\mathbf{u}\| + \|\mathbf{v}\|$?

12. Consider the position vector $\mathbf{i} = \langle 1, 0 \rangle$ in $\mathbb{R}^2$. Describe the set of position vectors $\mathbf{v}$ in $\mathbb{R}^2$ with the property that $\mathbf{v} \cdot \mathbf{i} = 0$.

13. Consider the position vector $\mathbf{i} = \langle 1, 0, 0 \rangle$ in $\mathbb{R}^3$. Describe the set of position vectors $\mathbf{v}$ in $\mathbb{R}^3$ with the property that $\mathbf{v} \cdot \mathbf{i} = 0$.

14. Consider the position vector $\mathbf{i} = \langle 1, 0 \rangle$ in $\mathbb{R}^2$. Describe the set of position vectors $\mathbf{v}$ in $\mathbb{R}^2$ with the property that $\text{proj}_{\mathbf{i}}\mathbf{v} = \mathbf{i}$.

15. Consider the position vector $\mathbf{i} = \langle 1, 0, 0 \rangle$ in $\mathbb{R}^3$. Describe the set of position vectors $\mathbf{v}$ in $\mathbb{R}^3$ with the property that $\text{proj}_{\mathbf{i}}\mathbf{v} = \mathbf{i}$.

16. Let $\mathbf{v}_0 = \langle a, b \rangle$. Describe the set of points (x, y) such that, for $\mathbf{v} = \langle x, y \rangle$,

   (a) $\mathbf{v} \cdot \mathbf{v}_0 = 0$.

   (b) $(\mathbf{v} - \mathbf{v}_0) \cdot \mathbf{v}_0 = 0$.

   (c) $(\mathbf{v} - \mathbf{v}_0) \cdot \mathbf{v} = 0$.

17. Recall that when two lines are perpendicular, their slopes are negative reciprocals.

   (a) Find a vector parallel to the line $y = mx + b$.

   (b) If $m \neq 0$, find the slope of any line perpendicular to $y = mx + b$ and find a vector parallel to that perpendicular line.

   (c) Show that the dot product of the vectors in parts (a) and (b) is zero.

18. Let $\mathbf{v}_1$ and $\mathbf{v}_2$ be two nonzero position vectors in $\mathbb{R}^2$ that are not scalar multiples of each other. Explain why, given any vector $\mathbf{w}$ in $\mathbb{R}^2$, there are scalars c_1 and c_2 such that $\mathbf{w} = c_1\mathbf{v}_1 + c_2\mathbf{v}_2$.

19. To illustrate the concept in Exercise 18:
    (a) Explain why the vectors $\mathbf{v}_1 = \langle 1, -2 \rangle$ and $\mathbf{v}_2 = \langle 3, 5 \rangle$ are not scalar multiples of each other.
    (b) Find scalars c_1 and c_2 such that $\langle 5, 1 \rangle = c_1\mathbf{v}_1 + c_2\mathbf{v}_2$.

Skills

In Exercises 20-23, find the dot product of the given pairs of vectors and the angle between the two vectors.

20. $\mathbf{u} = \langle 1, 2 \rangle$, $\mathbf{v} = \langle 3, 5 \rangle$
21. $\mathbf{u} = \langle 2, 0, -5 \rangle$, $\mathbf{v} = \langle -3, 7, -1 \rangle$
22. $\mathbf{u} = \langle 3, -1, 2 \rangle$, $\mathbf{v} = \langle -4, -6, 3 \rangle$
23. $\mathbf{u} = \langle -5, 1, 3 \rangle$, $\mathbf{v} = \langle -3, 2, 7 \rangle$

In Exercises 24-27, find $\text{comp}_{\mathbf{u}}\mathbf{v}$, $\text{proj}_{\mathbf{u}}\mathbf{v}$, and the component of $\mathbf{v}$ orthogonal to $\mathbf{u}$.

24. $\mathbf{u} = \langle 1, 2 \rangle$, $\mathbf{v} = \langle 3, 5 \rangle$
25. $\mathbf{u} = \langle 3, -1, 2 \rangle$, $\mathbf{v} = \langle -4, -6, 3 \rangle$
26. $\mathbf{u} = \langle 2, 0, -5 \rangle$, $\mathbf{v} = \langle -3, 7, -1 \rangle$
27. $\mathbf{u} = \langle 3, 1, -2 \rangle$, $\mathbf{v} = \langle -6, -2, 4 \rangle$

In Exercises 28-31, find $\text{proj}_{\mathbf{u}}\mathbf{v}$ and $\text{proj}_{\mathbf{v}}\mathbf{u}$.

28. $\mathbf{u} = \langle 1, 4 \rangle$, $\mathbf{v} = \langle 2, -3 \rangle$
29. $\mathbf{u} = \langle 3, -4 \rangle$, $\mathbf{v} = \langle 16, 12 \rangle$
30. $\mathbf{u} = \langle 3, 0, 1 \rangle$, $\mathbf{v} = \langle 2, 2, -5 \rangle$
31. $\mathbf{u} = \langle 1, -5, -1 \rangle$, $\mathbf{v} = \langle 0, 1, 0 \rangle$

In Exercises 32–36, (a) compute $\mathbf{u} \cdot \mathbf{v}$, (b) find the angle between $\mathbf{u}$ and $\mathbf{v}$, and (c) find $\text{proj}_{\mathbf{u}}\mathbf{v}$.

32. Let $\mathbf{u} = \langle 1, 5 \rangle$ and $\mathbf{v} = \langle 2, 7 \rangle$.
33. Let $\mathbf{u} = \langle -2, 3, 5 \rangle$ and $\mathbf{v} = \langle 13, -5, 8 \rangle$.
34. Let $\mathbf{u} = \langle 0, 3, -4 \rangle$ and $\mathbf{v} = \langle -5, 6, 0 \rangle$.
35. Let $\mathbf{u} = \langle 2, 4, -1, 2 \rangle$ and $\mathbf{v} = \langle -1, 3, -2, 6 \rangle$.
36. Let $\mathbf{u} = \langle 5, -2, 3, 4 \rangle$ and $\mathbf{v} = \langle 0, -1, 1, 7 \rangle$.

For Exercises 37 and 38, let $P = (2, 5, 7)$, $Q = (-2, 1, -5)$, and $R = (-3, 0, 4)$.

37. Find the distance between the point P and the line determined by the points Q and R.

38. Find the altitude of triangle $\triangle PQR$ from vertex R to side $\overline{PQ}$.

39. Find the angle between the diagonal of a face of a cube and the adjoining edge of the cube that is *not* an edge of that face.

40. Find the angle between the diagonal of a cube and an adjoining edge of the cube.

41. Find the angle between the diagonal of a cube and an adjoining diagonal of one of the faces of the cube.

42. Find the angle between two distinct diagonals of a cube.

Exercises 43–53 deal with **direction angles** and **direction cosines**. Let $\mathbf{v}$ be a nonzero vector in $\mathbb{R}^3$. The **direction angles** α, β, and γ of $\mathbf{v}$ are the angles that $\mathbf{v}$ makes with the positive x-, y- and z-axes, respectively. The **direction cosines** of $\mathbf{v}$ are $\cos\alpha$, $\cos\beta$, and $\cos\gamma$.

Find the direction angles and direction cosines for the vectors given in Exercises 43–46.

43. $\langle 1, 2, 3 \rangle$
44. $\langle -2, 0, 3 \rangle$
45. $\langle -1, 1, -4 \rangle$
46. $\langle -3, 4, 2 \rangle$

47. Show that for any vector $\mathbf{v}$ in $\mathbb{R}^3$,
$$\mathbf{v} = \|\mathbf{v}\| \left((\cos\alpha)\mathbf{i} + (\cos\beta)\mathbf{j} + (\cos\gamma)\mathbf{k} \right),$$
where α, β, and γ are the direction angles of $\mathbf{v}$.

48. Use Exercise 47 to show that if $\cos\alpha$, $\cos\beta$, and $\cos\gamma$ are the direction cosines of a vector $\mathbf{v}$, then
$$\cos^2\alpha + \cos^2\beta + \cos^2\gamma = 1.$$

In Exercises 49-51, two direction cosines are given. Use Exercise 48 to find the third direction cosine.

49. $\cos\alpha = \dfrac{1}{2}$, $\cos\beta = \dfrac{1}{2}$.
50. $\cos\beta = \dfrac{1}{4}$, $\cos\gamma = \dfrac{1}{3}$.
51. $\cos\alpha = \dfrac{1}{2}$, $\cos\gamma = \dfrac{\sqrt{3}}{4}$.

52. Let $\mathbf{v}$ be a vector in $\mathbb{R}^n$.
    (a) How would you compute the direction angles and direction cosines for $\mathbf{v}$?
    (b) Show that if α_1, α_2, ..., α_n are the direction angles for $\mathbf{v}$, then $\cos^2\alpha_1 + \cos^2\alpha_2 + \cdots + \cos^2\alpha_n = 1$.

53. Let $\mathbf{v} = \langle 1, 1, 1, \ldots, 1 \rangle$ be a vector in $\mathbb{R}^n$.
    (a) Use Exercise 52 to find the direction angles α_1, α_2, ..., α_n.
    (b) Show that $\alpha_i \to \dfrac{\pi}{2}$ as $n \to \infty$.

Applications

54. Ian is climbing a glacier to a col, a gap in a ridge. The snow is steep, so he is zigzagging up, first northeast, then southeast, always 45 degrees away from due east, and always at 30 degrees from the horizontal. He travels at 0.5 mile per hour in whichever direction he heads.

    (a) What is the component of his velocity horizontally, due east?

    (b) The map shows that the col is $\frac{1}{4}$ mile east of him. How long will it take Ian to get there?

55. Annie is making a north-to-south crossing from one island to another, with a tidal current in the channel. The current in the channel is moving at 1 mph due west. Annie has pointed her kayak in the direction $\langle 1, -4 \rangle$ and is paddling at 2 mph.

    (a) What angle does Annie's boat make with a direct southerly heading?

    (b) What is the component of Annie's velocity toward the south?

    (c) The crossing is 2 miles due south. How long will it take Annie?

Proofs

56. Let $\mathbf{u}$ and $\mathbf{v}$ be two nonzero vectors in $\mathbb{R}^2$. Prove that $\mathbf{u} \cdot \mathbf{v} = \|\mathbf{u}\|\|\mathbf{v}\|\cos\theta$, where θ is the angle between $\mathbf{u}$ and $\mathbf{v}$.

57. Show that for any vector $\mathbf{v}$ in $\mathbb{R}^3$,

$$\mathbf{v} = (\mathbf{v} \cdot \mathbf{i})\mathbf{i} + (\mathbf{v} \cdot \mathbf{j})\mathbf{j} + (\mathbf{v} \cdot \mathbf{k})\mathbf{k}.$$

58. Show that $\mathbf{u} \cdot \mathbf{v} = \|\mathbf{u}\|\|\mathbf{v}\|\cos\theta$ when $\mathbf{u}$ and $\mathbf{v}$ are nonzero vectors such that $\mathbf{u} = k\mathbf{v}$ with $k < 0$.

59. Use the fact that $\mathbf{u} \cdot \mathbf{v} = \|\mathbf{u}\|\|\mathbf{v}\|\cos\theta$ to prove the Cauchy–Schwarz inequality $|\mathbf{u} \cdot \mathbf{v}| \leq \|\mathbf{u}\|\|\mathbf{v}\|$. What relationship must $\mathbf{u}$ and $\mathbf{v}$ have in order for $|\mathbf{u} \cdot \mathbf{v}| = \|\mathbf{u}\|\|\mathbf{v}\|$?

60. Let θ be the angle between nonzero vectors $\mathbf{u}$ and $\mathbf{v}$. Prove each of the following:

    (a) θ is acute if and only if $\mathbf{u} \cdot \mathbf{v} > 0$

    (b) θ is right if and only if $\mathbf{u} \cdot \mathbf{v} = 0$

    (c) θ is obtuse if and only if $\mathbf{u} \cdot \mathbf{v} < 0$

61. Prove that vectors $\mathbf{u}$ and $\mathbf{v}$ are orthogonal if and only if $\mathbf{u} \cdot \mathbf{v} = 0$. (This is Theorem 10.19.)

62. Prove the following statements from Theorem 10.14 for vectors in $\mathbb{R}^3$:

    (a) $\mathbf{u} \cdot \mathbf{v} = \mathbf{v} \cdot \mathbf{u}$

    (b) $k(\mathbf{u} \cdot \mathbf{v}) = (k\mathbf{u}) \cdot \mathbf{v} = \mathbf{u} \cdot (k\mathbf{v})$

    (c) $\mathbf{v} \cdot \mathbf{v} = \|\mathbf{v}\|^2$

63. Let $\mathbf{u}$ be a nonzero vector and let $\mathbf{v}$ be any vector. Show that the decomposition $\mathbf{v} = \mathbf{v}_\parallel + \mathbf{v}_\perp$, where $\mathbf{v}_\parallel$ is parallel to $\mathbf{u}$ and $\mathbf{v}_\perp$ is orthogonal to $\mathbf{u}$, is unique. (*Hint: Assume that there is another decomposition with these properties, and show that the two decompositions must be identical.*)

64. Use a vector argument to prove that a parallelogram is a rectangle if and only if the diagonals have the same length.

65. Use a vector argument to prove that a parallelogram is a rhombus if and only if the diagonals are perpendicular.

Thinking Forward

Lines and vectors: Find the specified unit vectors.

    ▶ Two unit vectors parallel to the line $y = \frac{3}{5}x - 7$.

    ▶ Two unit vectors parallel to the line $y = mx + b$.

    ▶ Two unit vectors perpendicular to the line $y = \frac{3}{5}x - 7$.

    ▶ Two unit vectors perpendicular to the line $y = mx + b$.

10.4   CROSS PRODUCT

▶ Multiplying two vectors in $\mathbb{R}^3$ using the cross product

▶ The geometry of the cross product

▶ The relationship between the algebra and the geometry of the cross product

Determinants of 3×3 Matrices

Before we get to the definition of the cross product, it will be convenient to discuss how to compute the determinant of a 3-by-3 matrix, or, more briefly, a 3×3 matrix. First, a **matrix** is a rectangular array of entries. Here we are interested primarily in 3×3 matrices (i.e., arrays with 3 rows and 3 columns):

$$\begin{bmatrix} a_1 & a_2 & a_3 \\ b_1 & b_2 & b_3 \\ c_1 & c_2 & c_3 \end{bmatrix}.$$

The **determinant** of a square matrix is a value derived from the entries of the matrix. In particular, for a 3×3 matrix we may use the following definition:

DEFINITION 10.24

The Determinant of a 3 × 3 Matrix

The **determinant** of the 3×3 matrix $A = \begin{bmatrix} a_1 & a_2 & a_3 \\ b_1 & b_2 & b_3 \\ c_1 & c_2 & c_3 \end{bmatrix}$, denoted by $\det A$, is the sum

$$\det \begin{bmatrix} a_1 & a_2 & a_3 \\ b_1 & b_2 & b_3 \\ c_1 & c_2 & c_3 \end{bmatrix} = a_1 b_2 c_3 - a_1 b_3 c_2 + a_2 b_3 c_1 - a_2 b_1 c_3 + a_3 b_1 c_2 - a_3 b_2 c_1.$$

At first it might seem difficult to remember how to compute this sum. Fortunately, there is a nice visual trick that can be used to help. Compare the products along the six colored "diagonals" shown in the following matrix with the final sum in Definition 10.24. Note that four of the six diagonals wrap around the matrix as they descend.

Every 3×3 matrix has six "diagonals"

If the diagonal goes down from left to right, add the product. If the diagonal goes down from right to left, subtract the product. The resulting sum is the determinant of the 3×3 matrix.

For example, to compute the determinant of the matrix

$$M = \begin{bmatrix} 1 & -2 & 5 \\ 0 & 3 & -1 \\ -3 & 2 & 4 \end{bmatrix},$$

we add the products along the six diagonals, making sure to use the appropriate signs:

$$\det M = (1)(3)(4) + (-2)(-1)(-3) + (5)(0)(2) - (1)(-1)(2) - (-2)(0)(4) - (5)(3)(-3)$$
$$= 12 - 6 + 0 + 2 + 0 + 45 = 53.$$

The determinant is defined for all square matrices, but for larger square matrices a recursive definition is used. For example, the determinant of a 4×4 matrix is a modified sum of the determinants of four 3×3 matrices, the determinant of a 5×5 matrix is a modified sum of the determinants of five 4×4 matrices, etc. As a result, these determinants have more summands than those that lie along diagonals. In general, the determinant of an $n \times n$ matrix involves the sum of $n!$ products, with each product containing n factors. For example, computing the determinant of a 5×5 matrix involves the sum of $5! = 120$ products, each with 5 factors. In this text we will be using only the determinants of 3×3 matrices to compute cross products. The study of matrices and determinants belongs to a branch of mathematics called *linear algebra*. Here we are introducing the minimum necessary to help us define the cross product.

The Cross Product

Our final vector product, the cross product, will be defined in terms of the determinant of a matrix containing the components of two vectors. Later in this section we will examine geometric properties of this product. We will also see that the cross product can be used to determine areas of parallelograms and, along with the dot product, to compute the volumes of solids known as parallelepipeds. In Section 10.6 we will see how the cross product is used to find equations of planes in $\mathbb{R}^3$.

The cross product differs from our two previous vector products in that it is defined only for vectors in $\mathbb{R}^3$. (Recall that when we multiply a vector by a scalar, the vector can have any number of components and that we may take the dot product on any two vectors with the same number of components.)

DEFINITION 10.25

The Cross Product

Let $\mathbf{u} = \langle u_1, u_2, u_3 \rangle = u_1\mathbf{i} + u_2\mathbf{j} + u_3\mathbf{k}$ and $\mathbf{v} = \langle v_1, v_2, v_3 \rangle = v_1\mathbf{i} + v_2\mathbf{j} + v_3\mathbf{k}$. Then the *cross product* of $\mathbf{u}$ and $\mathbf{v}$, denoted by $\mathbf{u} \times \mathbf{v}$, is

$$\mathbf{u} \times \mathbf{v} = \det \begin{bmatrix} \mathbf{i} & \mathbf{j} & \mathbf{k} \\ u_1 & u_2 & u_3 \\ v_1 & v_2 & v_3 \end{bmatrix} = (u_2v_3 - u_3v_2)\mathbf{i} + (u_3v_1 - u_1v_3)\mathbf{j} + (u_1v_2 - u_2v_1)\mathbf{k}.$$

For example, we'll compute the cross product of the vectors $\mathbf{u} = \langle -1, 3, 4 \rangle$ and $\mathbf{v} = \langle 2, 0, -5 \rangle$:

$$\mathbf{u} \times \mathbf{v} = \det \begin{bmatrix} \mathbf{i} & \mathbf{j} & \mathbf{k} \\ -1 & 3 & 4 \\ 2 & 0 & -5 \end{bmatrix}$$
$$= (3 \cdot (-5) - 4 \cdot 0)\mathbf{i} + (4 \cdot 2 - (-1) \cdot (-5))\mathbf{j} + (-1 \cdot 0 - 3 \cdot 2)\mathbf{k}$$
$$= -15\mathbf{i} + 3\mathbf{j} - 6\mathbf{k} = \langle -15, 3, -6 \rangle.$$

Shortly we will discuss the geometric relationship between $\mathbf{u}$, $\mathbf{v}$, and $\mathbf{u} \times \mathbf{v}$, but right now at least we know how to compute a cross product.

One simple geometric consequence of the definition is that the cross product of two parallel vectors is $\mathbf{0}$.

THEOREM 10.26

> ### The Cross Product of Parallel Vectors
> The cross product of two parallel vectors $\mathbf{u}$ and $\mathbf{v}$ in $\mathbb{R}^3$ is $\mathbf{u} \times \mathbf{v} = \mathbf{0}$.

The proof of Theorem 10.26 follows from the definition of the cross product and is left for Exercise 65.

Algebraic Properties of the Cross Product

The cross product may be the first product you've encountered that is *not* commutative. However, it is ***anticommutative***. That is, for every two vectors $\mathbf{u}$ and $\mathbf{v}$ in $\mathbb{R}^3$, we have the following:

THEOREM 10.27

> ### The Cross Product Is Anticommutative
> For any vectors $\mathbf{u}$ and $\mathbf{v}$ in $\mathbb{R}^3$,
>
> $$\mathbf{v} \times \mathbf{u} = -(\mathbf{u} \times \mathbf{v}).$$

The proof of Theorem 10.27 follows from the definition of the cross product and is left for Exercise 66.

THEOREM 10.28

> ### Multiplication by a Scalar and the Cross Product
> For any vectors $\mathbf{u}$ and $\mathbf{v}$ in $\mathbb{R}^3$ and any scalar c,
>
> $$c(\mathbf{u} \times \mathbf{v}) = (c\,\mathbf{u}) \times \mathbf{v} = \mathbf{u} \times (c\,\mathbf{v}).$$

Again the proof follows from the definition of the cross product. This proof is left for Exercise 67.

As with multiplication and addition of scalars, the cross product is distributive over addition. Because the cross product is not commutative, we need to state *two* distributive properties. Again, their proofs follow from the definition of the cross product; they are left for Exercise 68.

THEOREM 10.29

> ### Distributive Properties of the Cross Product
> Let $\mathbf{u}$, $\mathbf{v}$, and $\mathbf{w}$ be vectors in $\mathbb{R}^3$. Then
>
> **(a)** $\mathbf{u} \times (\mathbf{v} + \mathbf{w}) = \mathbf{u} \times \mathbf{v} + \mathbf{u} \times \mathbf{w}$.
>
> **(b)** $(\mathbf{u} + \mathbf{v}) \times \mathbf{w} = \mathbf{u} \times \mathbf{w} + \mathbf{v} \times \mathbf{w}$.

The final algebraic property we discuss here provides a relationship between the dot product and cross product and will be used shortly to help us understand the geometry of the cross product. It is known as Lagrange's identity.

THEOREM 10.30

> ### Lagrange's Identity
> Let $\mathbf{u}$ and $\mathbf{v}$ be vectors in $\mathbb{R}^3$. Then
>
> $$\|\mathbf{u} \times \mathbf{v}\|^2 = \|\mathbf{u}\|^2\|\mathbf{v}\|^2 - (\mathbf{u} \cdot \mathbf{v})^2.$$

The proof of Theorem 10.30 entails expanding the quantities on both sides of Lagrange's identity. This task is left for Exercise 70.

The Geometry of the Cross Product

We begin by showing that the cross product $\mathbf{u} \times \mathbf{v}$ is orthogonal to both $\mathbf{u}$ and $\mathbf{v}$. As an immediate consequence the cross product is orthogonal to any plane containing both $\mathbf{u}$ and $\mathbf{v}$.

THEOREM 10.31

The Cross Product $\mathbf{u} \times \mathbf{v}$ is Orthogonal to Both $\mathbf{u}$ and $\mathbf{v}$

Let $\mathbf{u}$ and $\mathbf{v}$ be vectors in $\mathbb{R}^3$. Then

(a) $\mathbf{u} \cdot (\mathbf{u} \times \mathbf{v}) = 0$.

(b) $\mathbf{v} \cdot (\mathbf{u} \times \mathbf{v}) = 0$.

Proof. We prove part (a) and leave part (b) for Exercise 69. Let $\mathbf{u} = \langle u_1, u_2, u_3 \rangle$ and $\mathbf{v} = \langle v_1, v_2, v_3 \rangle$. Then, by the definition of the cross product,

$$\mathbf{u} \times \mathbf{v} = \langle u_2 v_3 - u_3 v_2, u_3 v_1 - u_1 v_3, u_1 v_2 - u_2 v_1 \rangle$$

and

$$\begin{aligned} \mathbf{u} \cdot (\mathbf{u} \times \mathbf{v}) &= \langle u_1, u_2, u_3 \rangle \cdot \langle u_2 v_3 - u_3 v_2, u_3 v_1 - u_1 v_3, u_1 v_2 - u_2 v_1 \rangle \\ &= u_1(u_2 v_3 - u_3 v_2) + u_2(u_3 v_1 - u_1 v_3) + u_3(u_1 v_2 - u_2 v_1) \\ &= 0. \end{aligned}$$

Recall that for vectors $\mathbf{u}$ and $\mathbf{v}$, we have $\mathbf{u} \cdot \mathbf{v} = \|\mathbf{u}\|\|\mathbf{v}\| \cos\theta$, where θ is the angle between the vectors. The magnitude of the cross product is related to the magnitudes of the vectors and the angle between them.

THEOREM 10.32

The Cross Product $\mathbf{u} \times \mathbf{v}$ and the Angle Between $\mathbf{u}$ and $\mathbf{v}$

Let $\mathbf{u}$ and $\mathbf{v}$ be nonzero vectors in $\mathbb{R}^3$ with the same initial point. Then

$$\|\mathbf{u} \times \mathbf{v}\| = \|\mathbf{u}\|\|\mathbf{v}\| \sin\theta,$$

where θ is the angle between $\mathbf{u}$ and $\mathbf{v}$.

Proof. By Lagrange's identity

$$\|\mathbf{u} \times \mathbf{v}\|^2 = \|\mathbf{u}\|^2\|\mathbf{v}\|^2 - (\mathbf{u} \cdot \mathbf{v})^2.$$

But since $\mathbf{u} \cdot \mathbf{v} = \|\mathbf{u}\|\|\mathbf{v}\| \cos\theta$, we have

$$\|\mathbf{u} \times \mathbf{v}\|^2 = \|\mathbf{u}\|^2\|\mathbf{v}\|^2 - (\|\mathbf{u}\|\|\mathbf{v}\| \cos\theta)^2 = \|\mathbf{u}\|^2\|\mathbf{v}\|^2(1 - \cos^2\theta) = \|\mathbf{u}\|^2\|\mathbf{v}\|^2 \sin^2\theta.$$

Taking the square root of the leftmost and rightmost quantities produces our desired result.

Two nonparallel vectors determine a parallelogram. The cross product can help us find the area of that parallelogram.

The area of the parallelogram is $\|\mathbf{u}\|\|\mathbf{v}\| \sin\theta$

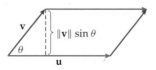

THEOREM 10.33

> ### The Area of a Parallelogram
>
> Let $\mathbf{u}$ and $\mathbf{v}$ be vectors in $\mathbb{R}^3$. Then the area of the parallelogram determined by $\mathbf{u}$ and $\mathbf{v}$ is given by $\|\mathbf{u} \times \mathbf{v}\|$.

Proof. We first consider the case where $\mathbf{u}$ and $\mathbf{v}$ are parallel vectors. Here the parallelogram determined by $\mathbf{u}$ and $\mathbf{v}$ is degenerate; that is, it is collapsed into a line segment and has zero area. By Theorem 10.26 $\mathbf{u} \times \mathbf{v} = \mathbf{0}$, so $\|\mathbf{u} \times \mathbf{v}\| = \|\mathbf{0}\| = 0$.

If $\mathbf{u}$ and $\mathbf{v}$ are not parallel, then the area of the parallelogram determined by $\mathbf{u}$ and $\mathbf{v}$ is the product of the length of one of the sides and the distance between that side and the opposite side.

In our parallelogram, $\|\mathbf{u}\|$ is the length of one side. In the parallelogram shown in the previous figure, we see that the distance from that side to the opposite side is $\|\mathbf{v}\| \sin \theta$. Therefore the area of the parallelogram is $\|\mathbf{u}\| \|\mathbf{v}\| \sin \theta = \|\mathbf{u} \times \mathbf{v}\|$. ∎

We have already discussed what it means for a three-dimensional coordinate system to be left- or right-handed. Similarly, three vectors $\mathbf{u}$, $\mathbf{v}$, and $\mathbf{w}$ that cannot be translated into the same plane are said to form a ***right-handed triple*** if, when the index finger of the right hand points in the direction of $\mathbf{u}$ and the middle finger of the right hand points in the direction of $\mathbf{v}$, then the right thumb will naturally point in the direction of $\mathbf{w}$. In particular, for any two nonparallel vectors $\mathbf{u}$ and $\mathbf{v}$ in $\mathbb{R}^3$, the vectors $\mathbf{u}$, $\mathbf{v}$, and $\mathbf{u} \times \mathbf{v}$ always form a right-handed triple.

The vectors $\mathbf{u}$, $\mathbf{v}$, and $\mathbf{u} \times \mathbf{v}$ form a right-handed triple

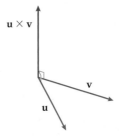

THEOREM 10.34

> ### Two Nonparallel Vectors and Their Cross Product Form a Right-Handed Triple
>
> Let $\mathbf{u}$ and $\mathbf{v}$ be nonparallel vectors in $\mathbb{R}^3$. Then the vectors $\mathbf{u}$, $\mathbf{v}$, and $\mathbf{u} \times \mathbf{v}$ form a right-handed triple.

Proof. Let $\mathbf{u}$ and $\mathbf{v}$ be nonparallel position vectors in $\mathbb{R}^3$. We may position our coordinate axes so that $\mathbf{u}$ lies along the positive x-axis and $\mathbf{v}$ lies in the xy-plane and has a positive y-coordinate. That is, $\mathbf{u} = \langle u_1, 0, 0 \rangle$ and $\mathbf{v} = \langle v_1, v_2, 0 \rangle$ with $u_1 > 0$ and $v_2 > 0$. The sign of v_1 may be positive, zero, or negative. Two of these possibilities are illustrated here:

Vectors $\mathbf{u}$ and $\mathbf{v}$ with $v_1 > 0$

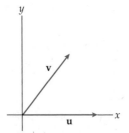

Vectors $\mathbf{u}$ and $\mathbf{v}$ with $v_1 < 0$

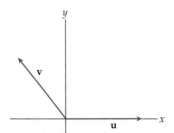

In any of the three cases, when we take the cross product we obtain

$$\mathbf{u} \times \mathbf{v} = \det \begin{bmatrix} \mathbf{i} & \mathbf{j} & \mathbf{k} \\ u_1 & 0 & 0 \\ v_1 & v_2 & 0 \end{bmatrix} = \langle 0, 0, u_1 v_2 \rangle.$$

Since $\mathbf{u} \times \mathbf{v}$ has a positive z-coordinate, we have our desired result. ∎

The next theorem reiterates several of the basic geometric properties of the cross product that we have already proved. We summarize these properties in Theorem 10.35 because they may be used as an alternative (geometric) definition of the cross product. When they are used as such, the algebraic properties follow as a consequence.

THEOREM 10.35

The Geometry of the Cross Product

Let $\mathbf{u}$ and $\mathbf{v}$ be vectors in $\mathbb{R}^3$.

If $\mathbf{u}$ and $\mathbf{v}$ are parallel then $\mathbf{u} \times \mathbf{v} = \mathbf{0}$.

If $\mathbf{u}$ and $\mathbf{v}$ are not parallel, then $\mathbf{u} \times \mathbf{v}$ has the following properties:

(a) $\|\mathbf{u} \times \mathbf{v}\| = \|\mathbf{u}\| \|\mathbf{v}\| \sin\theta$.

(b) $\mathbf{u} \times \mathbf{v}$ is perpendicular to any plane containing both $\mathbf{u}$ and $\mathbf{v}$.

(c) $\mathbf{u}$, $\mathbf{v}$, and $\mathbf{u} \times \mathbf{v}$ form a right-handed triple.

Triple Scalar Product

We cannot randomly combine three vectors $\mathbf{u}$, $\mathbf{v}$, and $\mathbf{w}$ from $\mathbb{R}^3$ with dot and cross products. For example, the combination $(\mathbf{u} \cdot \mathbf{v}) \cdot \mathbf{w}$ is *not* defined, since $\mathbf{u} \cdot \mathbf{v}$ is a scalar and we need two vectors to form a dot product. Similarly, $(\mathbf{u} \cdot \mathbf{v}) \times \mathbf{w}$ is not defined. (Why?) However, the product $\mathbf{u} \cdot (\mathbf{v} \times \mathbf{w})$ is defined and has an important geometrical interpretation. In addition, we will see that $\mathbf{u} \cdot (\mathbf{v} \times \mathbf{w}) = (\mathbf{u} \times \mathbf{v}) \cdot \mathbf{w}$. Both of these are examples of *triple scalar products*. In Exercise 80 we discuss the *vector triple product* $\mathbf{u} \times (\mathbf{v} \times \mathbf{w})$.

A *parallelepiped* is a three-dimensional analog of a parallelogram, in much the same way that a cube is a three-dimensional analog of a square. Specifically, a parallelepiped is a six-sided solid whose surface consists of three pairs of parallel faces, each of which is a parallelogram. Any three vectors $\mathbf{u}$, $\mathbf{v}$, and $\mathbf{w}$ in $\mathbb{R}^3$ that do not lie in the same plane will determine a parallelepiped.

The parallelepiped determined by $\mathbf{u}$, $\mathbf{v}$, and $\mathbf{w}$

In Theorem 10.33 we saw that the area of a parallelogram involves a cross product. Similarly, the volume of the parallelepiped involves a triple scalar product.

THEOREM 10.36

The Triple Scalar Product and the Volume of Parallelepipeds

Let $\mathbf{u}$, $\mathbf{v}$, and $\mathbf{w}$ be vectors in $\mathbb{R}^3$. Then $|\mathbf{u} \cdot (\mathbf{v} \times \mathbf{w})|$ is the volume of the parallelepiped determined by $\mathbf{u}$, $\mathbf{v}$, and $\mathbf{w}$. Furthermore, the volume of the parallelepiped is $\mathbf{u} \cdot (\mathbf{v} \times \mathbf{w})$ if and only if $\mathbf{u}$, $\mathbf{v}$, and $\mathbf{w}$ form a right-handed triple.

Proof. Let $\mathcal{F}$ denote one of the sides of the parallelepiped determined by vectors $\mathbf{v}$ and $\mathbf{w}$. From Theorem 10.33 we know that the area of $\mathcal{F}$ is $\|\mathbf{v} \times \mathbf{w}\|$. The distance between $\mathcal{F}$ and the opposite face is the distance from the terminal end of $\mathbf{u}$ to $\mathcal{F}$. There are two cases to consider: $\mathbf{u}$ and $\mathbf{v} \times \mathbf{w}$ are either on the same side of $\mathcal{F}$ or on opposite sides of $\mathcal{F}$.

Suppose $\mathbf{u}$ and $\mathbf{v} \times \mathbf{w}$ are on the same side of $\mathcal{F}$. Then $\mathbf{u}$, $\mathbf{v}$, and $\mathbf{w}$ must form a right-handed triple. Since $\mathbf{v} \times \mathbf{w}$ is perpendicular to $\mathcal{F}$, the distance from the terminal end of $\mathbf{u}$ to $\mathcal{F}$ is the component, $\text{comp}_{\mathbf{v} \times \mathbf{w}} \mathbf{u} = \|\mathbf{u}\| \cos \phi$, of the projection of $\mathbf{u}$ onto $\mathbf{v} \times \mathbf{w}$, where ϕ is the angle between $\mathbf{u}$ and $\mathbf{v} \times \mathbf{w}$. The following figure shows an example of this case:

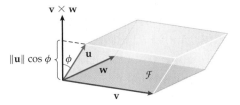

A parallelepiped determined by the right-handed triple $\mathbf{u}$, $\mathbf{v}$, and $\mathbf{w}$

Here, since $\mathbf{u}$ and $\mathbf{v} \times \mathbf{w}$ are on the same side of $\mathcal{F}$, $\phi < 90°$ and $\cos \phi > 0$. Thus the volume of the parallelepiped is $\|\mathbf{u}\| \|\mathbf{v} \times \mathbf{w}\| \cos \phi = \mathbf{u} \cdot (\mathbf{v} \times \mathbf{w})$.

In the case where $\mathbf{u}$ and $\mathbf{v} \times \mathbf{w}$ are on opposite sides of $\mathcal{F}$, $\phi > 90°$ and $\cos \phi < 0$. Thus the volume of the parallelepiped is $-\|\mathbf{u}\| \|\mathbf{v} \times \mathbf{w}\| \cos \phi = -\mathbf{u} \cdot (\mathbf{v} \times \mathbf{w})$.

We now have our desired result: The volume of the parallelepiped is $\mathbf{u} \cdot (\mathbf{v} \times \mathbf{w})$ if and only if $\mathbf{u}$, $\mathbf{v}$, and $\mathbf{w}$ form a right-handed triple, and in any case the volume is $|\mathbf{u} \cdot (\mathbf{v} \times \mathbf{w})|$. ∎

Theorem 10.36 also gives us a simple computational criterion for determining when three vectors lie in the same plane: The triple scalar product $\mathbf{u} \cdot (\mathbf{v} \times \mathbf{w})$ is equal to zero if and only if $\mathbf{u}$, $\mathbf{v}$, and $\mathbf{w}$ are coplanar.

The following theorem describes relationships between dot and cross products:

THEOREM 10.37

Properties of the Triple Scalar Product

Let $\mathbf{u}$, $\mathbf{v}$, and $\mathbf{w}$ be vectors in $\mathbb{R}^3$. Then

(a) $\mathbf{u} \cdot (\mathbf{v} \times \mathbf{w}) = \mathbf{v} \cdot (\mathbf{w} \times \mathbf{u}) = \mathbf{w} \cdot (\mathbf{u} \times \mathbf{v})$

(b) $\mathbf{u} \cdot (\mathbf{v} \times \mathbf{w}) = (\mathbf{u} \times \mathbf{v}) \cdot \mathbf{w}$

Part (b) of Theorem 10.37 says, roughly, that in a triple scalar product the dot and cross products can be "exchanged." We prove part (a) next and leave part (b) for Exercise 76.

Proof. First suppose that $\mathbf{u}$, $\mathbf{v}$, $\mathbf{w}$ form a right-handed triple. Then $\mathbf{v}$, $\mathbf{w}$, $\mathbf{u}$, and $\mathbf{w}$, $\mathbf{u}$, $\mathbf{v}$ do also. (Why?) Thus, from Theorem 10.36, the three triple scalar products $\mathbf{u} \cdot (\mathbf{v} \times \mathbf{w})$, $\mathbf{v} \cdot (\mathbf{w} \times \mathbf{u})$, and $\mathbf{w} \cdot (\mathbf{u} \times \mathbf{v})$ all give the volume of the parallelepiped determined by $\mathbf{u}$, $\mathbf{v}$ and $\mathbf{w}$. Therefore they are all equal.

For the converse, if $\mathbf{u}, \mathbf{v}, \mathbf{w}$ do not form a right-handed triple, then neither $\mathbf{v}, \mathbf{w}, \mathbf{u}$ nor $\mathbf{w}, \mathbf{u}, \mathbf{v}$ do. Here the volume of the parallelepiped determined by $\mathbf{u}, \mathbf{v}$, and $\mathbf{w}$ is

$$-\mathbf{u} \cdot (\mathbf{v} \times \mathbf{w}) = -\mathbf{v} \cdot (\mathbf{w} \times \mathbf{u}) = -\mathbf{w} \cdot (\mathbf{u} \times \mathbf{v}).$$

Therefore in this case we also have $\mathbf{u} \cdot (\mathbf{v} \times \mathbf{w}) = \mathbf{v} \cdot (\mathbf{w} \times \mathbf{u}) = \mathbf{w} \cdot (\mathbf{u} \times \mathbf{v}).$ ∎

Examples and Explorations

EXAMPLE 1   Using a cross product to find the area of a parallelogram

(a) Use Definition 10.25 to compute $\mathbf{u} \times \mathbf{v}$ for vectors $\mathbf{u} = \langle 2, 0, -3 \rangle$ and $\mathbf{v} = \langle -1, 4, 2 \rangle$.

(b) Find the area of the parallelogram determined by $\mathbf{u}$ and $\mathbf{v}$.

SOLUTION

(a) We use Definition 10.25 to compute the cross product:

$$\mathbf{u} \times \mathbf{v} = \det \begin{bmatrix} \mathbf{i} & \mathbf{j} & \mathbf{k} \\ 2 & 0 & -3 \\ -1 & 4 & 2 \end{bmatrix}$$

$$= ((0)(2) - (-3)(4))\mathbf{i} + ((-3)(-1) - (2)(2))\mathbf{j} + ((2)(4) - (0)(-1))\mathbf{k}$$

$$= 12\mathbf{i} - \mathbf{j} + 8\mathbf{k}.$$

(b) The area of the parallelogram determined by $\mathbf{u}$ and $\mathbf{v}$ is

$$\|\mathbf{u} \times \mathbf{v}\| = \|12\mathbf{i} - \mathbf{j} + 8\mathbf{k}\| = \sqrt{12^2 + (-1)^2 + 8^2} = \sqrt{209} \approx 14.5 \text{ square units.} \quad \square$$

 CHECKING THE ANSWER

We may check the plausibility of our cross product $\mathbf{u} \times \mathbf{v}$ with the dot products:

$$\mathbf{u} \cdot (\mathbf{u} \times \mathbf{v}) = \langle 2, 0, -3 \rangle \cdot \langle 12, -1, 8 \rangle = 2 \cdot 12 + 0(-1) - 3 \cdot 8 = 0 \quad \text{and}$$

$$\mathbf{v} \cdot (\mathbf{u} \times \mathbf{v}) = \langle -1, 4, 2 \rangle \cdot \langle 12, -1, 8 \rangle = -1 \cdot 12 + 4(-1) + 2 \cdot 8 = 0.$$

These calculations show that the cross product we obtained is orthogonal to both of the vectors $\mathbf{u}$ and $\mathbf{v}$.

EXAMPLE 2   Cross products involving the standard basis vectors

Find the cross products of the standard basis vectors:

$$\begin{array}{ccc} \mathbf{i} \times \mathbf{i} & \mathbf{i} \times \mathbf{j} & \mathbf{i} \times \mathbf{k} \\ \mathbf{j} \times \mathbf{i} & \mathbf{j} \times \mathbf{j} & \mathbf{j} \times \mathbf{k} \\ \mathbf{k} \times \mathbf{i} & \mathbf{k} \times \mathbf{j} & \mathbf{k} \times \mathbf{k} \end{array}$$

SOLUTION

The cross product of two parallel vectors is **0**. Since every vector is parallel to itself, we immediately have

$$\mathbf{i} \times \mathbf{i} = \mathbf{j} \times \mathbf{j} = \mathbf{k} \times \mathbf{k} = \mathbf{0}.$$

Each of the basis vectors is a unit vector, and the angle between two *distinct* basis vectors is $90°$. Therefore the norm of the cross product of two distinct basis vectors is 1 by property (a) of Theorem 10.35.

By property (b) of the same theorem, we know that the cross product of any two distinct basis vectors must be (plus or minus) the remaining basis vector. For example, $\mathbf{i} \times \mathbf{j} = \pm\mathbf{k}$. We need only determine which of $\mathbf{i}$, $\mathbf{j}$, and $\pm\mathbf{k}$ forms a right-handed triple. Here, $\mathbf{i}$, $\mathbf{j}$, and $\mathbf{k}$ form such a triple (check this), so $\mathbf{i} \times \mathbf{j} = \mathbf{k}$. Also, because the cross product is anticommutative, if we reverse the order within $\mathbf{i} \times \mathbf{j} = \mathbf{k}$, we will have $\mathbf{j} \times \mathbf{i} = -\mathbf{k}$.

The complete list of the cross products using the basis vectors is

$$
\begin{array}{lll}
\mathbf{i} \times \mathbf{i} = 0 & \mathbf{i} \times \mathbf{j} = \mathbf{k} & \mathbf{i} \times \mathbf{k} = -\mathbf{j} \\
\mathbf{j} \times \mathbf{i} = -\mathbf{k} & \mathbf{j} \times \mathbf{j} = 0 & \mathbf{j} \times \mathbf{k} = \mathbf{i} \\
\mathbf{k} \times \mathbf{i} = \mathbf{j} & \mathbf{k} \times \mathbf{j} = -\mathbf{i} & \mathbf{k} \times \mathbf{k} = 0.
\end{array}
$$

☐

EXAMPLE 3   **Finding the area of a triangle determined by three points**

Find the area of the triangle with vertices $A = (4, -2)$, $B = (7, 3)$, and $C = (-1, 3)$.

SOLUTION

We first find the vectors

$$
\begin{aligned}
\overrightarrow{AB} &= \langle 7 - 4, 3 - (-2) \rangle = \langle 3, 5 \rangle \text{ and} \\
\overrightarrow{AC} &= \langle -1 - 4, 3 - (-2) \rangle = \langle -5, 5 \rangle.
\end{aligned}
$$

The area of the triangle with vertices at A, B, and C is one-half of the area of the parallelogram determined by $\overrightarrow{AB}$ and $\overrightarrow{AC}$. We would like to use the technique of Example 1 to find the area of the parallelogram, but in order to take the cross product of two vectors, they must have three components. Here we can treat the xy-plane as part of 3-space by thinking of the xy-plane as the plane in $\mathbb{R}^3$ with $z = 0$. That is, we let $\overrightarrow{AB} = \langle 3, 5, 0 \rangle$ and $\overrightarrow{AC} = \langle -5, 5, 0 \rangle$.

Now,

$$
\text{Area } \triangle ABC = \frac{1}{2} \| \overrightarrow{AB} \times \overrightarrow{AC} \| = \frac{1}{2} \left\| \det \begin{bmatrix} \mathbf{i} & \mathbf{j} & \mathbf{k} \\ 3 & 5 & 0 \\ -5 & 5 & 0 \end{bmatrix} \right\|
$$

$$
= \frac{1}{2} \|(0 - 0)\mathbf{i} + (0 - 0)\mathbf{j} + (15 - (-25))\mathbf{k}\| = \frac{1}{2} \|40\mathbf{k}\|
$$

$$
= 20 \text{ square units.}
$$

☐

EXAMPLE 4   **Finding the volume of a parallelepiped**

(a) Find the volume of the parallelepiped determined by $\mathbf{u} = \langle 0, 3, -2 \rangle$, $\mathbf{v} = \langle 5, 3, -1 \rangle$, and $\mathbf{w} = \langle -3, 2, 7 \rangle$.

(b) Do the vectors $\mathbf{u}$, $\mathbf{v}$, and $\mathbf{w}$ form a right-handed triple or a left-handed triple?

SOLUTION

(a) The volume of the parallelepiped determined by $\mathbf{u}$, $\mathbf{v}$, and $\mathbf{w}$ is the absolute value of the triple scalar product $\mathbf{u} \cdot (\mathbf{v} \times \mathbf{w})$. Although we could first evaluate the cross product $\mathbf{v} \times \mathbf{w}$ and then take the dot product of the resulting vector with $\mathbf{u}$, it is slightly more efficient to just take the absolute value of the determinant of the 3×3 matrix formed from the components of $\mathbf{u}$, $\mathbf{v}$, and $\mathbf{w}$ as the rows. (In Exercise 78, we ask you to explain why this always works.)

Thus, the required volume is

$$|\mathbf{u} \cdot (\mathbf{v} \times \mathbf{w})| = \left| \det \begin{bmatrix} 0 & 3 & -2 \\ 5 & 3 & -1 \\ -3 & 2 & 7 \end{bmatrix} \right| = |0 + 9 - 20 + 0 - 105 - 18| = |-134|$$

$$= 134 \text{ cubic units.}$$

(b) By Theorem 10.36, the vectors $\mathbf{u}$, $\mathbf{v}$, and $\mathbf{w}$ form a left-handed triple, since the triple scalar product $\mathbf{u} \cdot (\mathbf{v} \times \mathbf{w}) = -134 < 0$. □

? **TEST YOUR UNDERSTANDING**

▶ How do you find the determinant of a 3×3 matrix?

▶ What is the definition of the cross product?

▶ What are the geometric properties of the cross product?

▶ How do you find the area of a parallelogram determined by two vectors? How do you find the volume of a parallelepiped determined by three vectors?

▶ How do you find the area of a triangle determined by three points?

EXERCISES 10.4

Thinking Back

▶ *Coordinate system:* Explain what it means for a coordinate system to be right-handed and what it means for a coordinate system to be left-handed.

▶ *Orthogonal vectors:* Let $\mathbf{u}$ and $\mathbf{v}$ be two nonparallel position vectors in $\mathbb{R}^3$ lying in the xy-plane. Find two unit vectors orthogonal to both $\mathbf{u}$ and $\mathbf{v}$ *without* using the cross product.

Concepts

0. *Problem Zero:* Read the section and make your own summary of the material.

1. *True/False:* Determine whether each of the statements that follow is true or false. If a statement is true, explain why. If a statement is false, provide a counterexample.

(a) *True or False:* For any two vectors $\mathbf{u}$ and $\mathbf{v}$ in $\mathbb{R}^3$, $\mathbf{u} \times \mathbf{v} = \mathbf{v} \times \mathbf{u}$.

(b) *True or False:* If $\mathbf{u}$ and $\mathbf{v}$ are two vectors in $\mathbb{R}^3$, then $\mathbf{u} \times \mathbf{v} = \mathbf{u} \cdot \mathbf{v}$.

(c) *True or False:* If $\mathbf{u} \times \mathbf{v} = \mathbf{v} \times \mathbf{u}$, then $\mathbf{u}$ and $\mathbf{v}$ are parallel.

(d) *True or False:* If $\mathbf{u}$, $\mathbf{v}$, and $\mathbf{w}$ are vectors in $\mathbb{R}^3$, then $(\mathbf{u} \times \mathbf{v}) \times \mathbf{w} = \mathbf{u} \times (\mathbf{v} \times \mathbf{w})$.

(e) *True or False:* The triple scalar product can be used to find the volume of a parallelepiped.

(f) *True or False:* If $\mathbf{u}$, $\mathbf{v}$, and $\mathbf{w}$ are vectors in $\mathbb{R}^3$, then $\mathbf{u} \cdot (\mathbf{v} \times \mathbf{w}) = -\mathbf{v} \cdot (\mathbf{u} \times \mathbf{w})$.

(g) *True or False:* If $\mathbf{u}$ and $\mathbf{v}$ are nonparallel vectors in $\mathbb{R}^3$, then $\dfrac{\mathbf{u} \cdot \mathbf{v}}{\|\mathbf{u} \times \mathbf{v}\|} = \cot\theta$, where θ is the angle between $\mathbf{u}$ and $\mathbf{v}$.

(h) *True or False:* If $\mathbf{u}$ and $\mathbf{v}$ are unit vectors in $\mathbb{R}^3$, then $\mathbf{u} \times \mathbf{v}$ is also a unit vector.

2. *Examples:* Construct examples of the thing(s) described in the following. Try to find examples that are different than any in the reading.

(a) Two nonzero vectors $\mathbf{u}$ and $\mathbf{v}$ in $\mathbb{R}^3$ such that $\mathbf{u} \times \mathbf{v} = \mathbf{v} \times \mathbf{u}$.

(b) Three vectors $\mathbf{u}$, $\mathbf{v}$, and $\mathbf{w}$ in $\mathbb{R}^3$ such that $(\mathbf{u} \times \mathbf{v}) \times \mathbf{w} \neq \mathbf{u} \times (\mathbf{v} \times \mathbf{w})$.

(c) Three vectors $\mathbf{u}$, $\mathbf{v}$, and $\mathbf{w}$ in $\mathbb{R}^3$ such that $(\mathbf{u} \times \mathbf{v}) \times \mathbf{w} = \mathbf{u} \times (\mathbf{v} \times \mathbf{w})$.

3. What is the definition of the cross product?

4. How is the determinant of a 3×3 matrix used in the computation of the determinant of two vectors $\mathbf{u} = \langle u_1, u_2, u_3 \rangle$ and $\mathbf{v} = \langle v_1, v_2, v_3 \rangle$?

5. If $\mathbf{u}$ and $\mathbf{v}$ are nonzero vectors in $\mathbb{R}^3$, what is the geometric relationship between $\mathbf{u}$, $\mathbf{v}$, and $\mathbf{u} \times \mathbf{v}$?

6. What is Lagrange's identity? How is it used to understand the geometry of the cross product?

7. If $\mathbf{u}$ and $\mathbf{v}$ are nonzero vectors in $\mathbb{R}^3$, why do the equations $\mathbf{u} \cdot (\mathbf{u} \times \mathbf{v}) = 0$ and $\mathbf{v} \cdot (\mathbf{u} \times \mathbf{v}) = 0$ tell us that the cross product is orthogonal to both $\mathbf{u}$ and $\mathbf{v}$?

8. What is meant by the parallelogram determined by vectors $\mathbf{u}$ and $\mathbf{v}$ in $\mathbb{R}^3$? How do you find the area of this parallelogram?

9. Sketch the parallelogram determined by the two vectors $\langle 1, 2 \rangle$ and $\langle 3, -1 \rangle$. How can you use the cross product to find the area of this parallelogram?

10. What is meant by the triangle determined by vectors $\mathbf{u}$ and $\mathbf{v}$ in $\mathbb{R}^3$? How do you find the area of this triangle?

11. If $\|\mathbf{u} \times \mathbf{v}\| = \|\mathbf{u}\| \|\mathbf{v}\|$, what is the geometric relationship between $\mathbf{u}$ and $\mathbf{v}$?

12. Give an example of three vectors in $\mathbb{R}^3$ that form a right-handed triple. Explain how you can use the same three vectors to form a left-handed triple.

13. Give an example of three nonzero vectors $\mathbf{u}$, $\mathbf{v}$, and $\mathbf{w}$ in $\mathbb{R}^3$ such that $\mathbf{u} \times \mathbf{v} = \mathbf{u} \times \mathbf{w}$ but $\mathbf{v} \neq \mathbf{w}$. What geometric relationship must the three vectors have for this to happen?

14. What is the definition of the triple scalar product for vectors $\mathbf{u}$, $\mathbf{v}$, and $\mathbf{w}$ in $\mathbb{R}^3$?

15. If the triple scalar product $\mathbf{u} \cdot (\mathbf{v} \times \mathbf{w})$ is equal to zero, what geometric relationship do the vectors $\mathbf{u}$, $\mathbf{v}$, and $\mathbf{w}$ have?

16. What is a parallelepiped? What is meant by the parallelepiped determined by the vectors $\mathbf{u}$, $\mathbf{v}$, and $\mathbf{w}$? How do you find the volume of the parallelepiped determined by $\mathbf{u}$, $\mathbf{v}$, and $\mathbf{w}$?

17. If $\mathbf{u}$, $\mathbf{v}$, and $\mathbf{w}$ are three vectors in $\mathbb{R}^3$, what is wrong with the expression $\mathbf{u} \times \mathbf{v} \cdot \mathbf{w}$?

18. If $\mathbf{u}$, $\mathbf{v}$, and $\mathbf{w}$ are three vectors in $\mathbb{R}^3$, which of the following products make sense and which do not?
(a) $\mathbf{u} \cdot (\mathbf{v} \cdot \mathbf{w})$
(b) $\mathbf{u} \cdot (\mathbf{v} \times \mathbf{w})$
(c) $\mathbf{u} \times (\mathbf{v} \cdot \mathbf{w})$
(d) $\mathbf{u} \times (\mathbf{v} \times \mathbf{w})$

19. If $\mathbf{u}$ and $\mathbf{v}$ are vectors in $\mathbb{R}^3$ such that $\mathbf{u} \cdot \mathbf{v} = 0$ and $\mathbf{u} \times \mathbf{v} = \mathbf{0}$, what can we conclude about $\mathbf{u}$ and $\mathbf{v}$?

20. If $\mathbf{u}$, $\mathbf{v}$, and $\mathbf{w}$ are three mutually orthogonal vectors in $\mathbb{R}^3$, explain why $\mathbf{u} \times (\mathbf{v} \times \mathbf{w}) = \mathbf{0}$.

21. If $\mathbf{u}$ and $\mathbf{v}$ are vectors in $\mathbb{R}^3$ such that $\mathbf{u} \times \mathbf{v} = \mathbf{v} \times \mathbf{u}$, what can we conclude about $\mathbf{u}$ and $\mathbf{v}$?

Skills

In Exercises 22–29 compute the indicated quantities when $\mathbf{u} = \langle 2, 1, -3 \rangle$, $\mathbf{v} = \langle 4, 0, 1 \rangle$, and $\mathbf{w} = \langle -2, 6, 5 \rangle$.

22. $\mathbf{u} \times \mathbf{v}$ and $\mathbf{v} \times \mathbf{u}$

23. $\mathbf{u} \times \mathbf{w}$ and $\mathbf{w} \times \mathbf{u}$

24. $\mathbf{v} \times \mathbf{w}$ and $\mathbf{w} \times \mathbf{v}$

25. $(\mathbf{u} \times \mathbf{v}) \times \mathbf{w}$ and $\mathbf{u} \times (\mathbf{v} \times \mathbf{w})$

26. $(\mathbf{u} \times \mathbf{v}) \cdot \mathbf{w}$ and $\mathbf{w} \cdot (\mathbf{v} \times \mathbf{u})$

27. Find the area of the parallelogram determined by the vectors $\mathbf{u}$ and $\mathbf{v}$.

28. Find the area of the parallelogram determined by the vectors $\mathbf{v}$ and $\mathbf{w}$.

29. Find the volume of the parallelepiped determined by vectors $\mathbf{u}$, $\mathbf{v}$, and $\mathbf{w}$. Do $\mathbf{u}$, $\mathbf{v}$, and $\mathbf{w}$ form a right-handed triple?

In Exercises 30–35 compute the indicated quantities when $\mathbf{u} = \langle -3, 1, -4 \rangle$, $\mathbf{v} = \langle 2, 0, 5 \rangle$, and $\mathbf{w} = \langle 1, 3, 13 \rangle$.

30. $\mathbf{u} \times \mathbf{v}$ and $\mathbf{v} \times \mathbf{u}$

31. $\mathbf{u} \times \mathbf{w}$ and $\mathbf{w} \times \mathbf{u}$

32. $\mathbf{v} \times \mathbf{w}$ and $\mathbf{w} \times \mathbf{v}$

33. $(\mathbf{u} \times \mathbf{v}) \cdot \mathbf{w}$ and $\mathbf{u} \cdot (\mathbf{v} \times \mathbf{w})$

34. Find the area of the parallelogram determined by the vectors $\mathbf{u}$ and $\mathbf{v}$.

35. Find the volume of the parallelepiped determined by the vectors $\mathbf{u}$, $\mathbf{v}$, and $\mathbf{w}$.

In Exercises 36–41 use the given sets of points to find:

(a) A nonzero vector $\mathbf{N}$ perpendicular to the plane determined by the points.

(b) Two unit vectors perpendicular to the plane determined by the points.

(c) The area of the triangle determined by the points.

36. $P(1, 4, 6)$, $Q(-3, 5, 0)$, $R(3, 2, -1)$

37. $P(4, 3, -2)$, $Q(0, 0, 3)$, $R(-1, 3, 6)$

38. $P(3, 1, 8)$, $Q(0, 6, -1)$, $R(-3, 5, -3)$

39. $P(2, -5, 1)$, $Q(-4, 5, 8)$, $R(-1, -5, 3)$

40. $P(4, -2)$, $Q(-2, 0)$, $R(1, -5)$
(Hint: Think of the xy-plane as part of $\mathbb{R}^3$.)

41. $P(1, 6)$, $Q(0, -3)$, $R(-5, 4)$
(Hint: Think of the xy-plane as part of $\mathbb{R}^3$.)

Suppose a triangle has side lengths a, b, and c. The **semiperimeter** of the triangle is defined to be $s = \frac{1}{2}(a+b+c)$. We can use the side lengths of the triangle to calculate its area by applying **Heron's formula**:

$$\text{Area} = \sqrt{s(s - a)(s - b)(s - c)}.$$

Use Heron's formula to compute the areas of the triangles determined by the points P, Q, and R in Exercises 42–44:

42. P, Q, and R from Exercise 36

43. P, Q, and R from Exercise 37

44. P, Q, and R from Exercise 41

45. Heron's formula allows us to compute the area of a triangle from its side lengths.
(a) Explain why knowing the side lengths of a quadrilateral is *not* sufficient to compute the area of the quadrilateral.
(b) Explain why knowing the side lengths of a quadrilateral and the length of one diagonal is sufficient to compute the area of the quadrilateral.
(c) Explain why knowing the side lengths of a quadrilateral and the measure of the angle at one vertex is also sufficient to compute the area of the quadrilateral.

Use the results of Exercise 45 to find the areas of the quadrilaterals $PQRS$ specified in Exercises 46–49.

46. $PQ = 6$, $QR = 7$, $RS = 8$, $SP = 9$, $PR = 10$

47. $PQ = 6$, $QR = 7$, $RS = 8$, $SP = 9$, $QS = 10$

48. $PQ = 6$, $QR = 7$, $RS = 8$, $SP = 6$, $\angle P = 60°$

49. $PQ = 7$, $QR = 8$, $RS = 8$, $SP = 9$, $\angle R = 60°$

In Exercises 50–53 the coordinates of points P, Q, R, and S are given. (a) Determine whether quadrilateral $PQRS$ is a parallelogram. (b) Find the area of quadrilateral $PQRS$.

50. $P(0,0)$, $Q(-1,3)$, $R(-4,-1)$, $S(-3,-2)$
51. $P(-1,3)$, $Q(2,5)$, $R(4,1)$, $S(1,-1)$
52. $P(2,7)$, $Q(3,-1)$, $R(1,-10)$, $S(-1,-2)$
53. $P(-1,3)$, $Q(2,5)$, $R(6,3)$, $S(4,-2)$

In Exercises 54–57 the coordinates of points P, Q, R, and S are given. (a) Show that the four points are coplanar. (b) Determine whether quadrilateral $PQRS$ is a parallelogram. (c) Find the area of quadrilateral $PQRS$.

54. $P(0,0,0)$, $Q(1,-2,5)$, $R(-1,2,11)$, $S(-2,4,6)$
55. $P(2,-3,8)$, $Q(-2,4,6)$, $R(7,18,-7)$, $S(15,4,-3)$
56. $P(1,2,6)$, $Q(4,1,-5)$, $R(3,6,8)$, $S(0,4,13)$
57. $P(3,4,-2)$, $Q(7,0,6)$, $R(2,1,7)$, $S(5,-2,13)$

Applications

Some crystals have rhombohedral structures. A rhombohedron is a parallelepiped in which all of the edge lengths are equal and each of the six faces is a congruent rhombus. Find the volumes of the rhombohedral crystals described in Exercises 58 and 59.

A rhombohedral crystal

58. Each side length is 1 cm, and the acute angles in each face measure $60°$. (*Hint: Let vector* $\mathbf{i}$ *form one of the edges of the rhombohedron, and let a second nonparallel edge be in the xy-plane.*)
59. Each side length is 2 cm, and the acute angles in each face measure $45°$. (*See the hint in the previous exercise.*)

Turning a bolt with a wrench produces a **torque** vector that drives the bolt forward. The magnitude of the torque vector is $\|\mathbf{r}\|\|\mathbf{F}\|\sin\theta$, where $\mathbf{r}$ is the vector along the handle of the wrench, $\mathbf{F}$ is the force vector applied to the handle of the wrench, and θ is the angle between these two vectors. Therefore, the magnitude of the torque is $\|\mathbf{r} \times \mathbf{F}\|$. In Exercises 60 and 61, find the magnitude of the torque. Express each answer in foot-pounds.

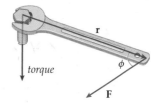

The torque vector

60. A force of 20 lb is applied to a wrench with a 6-inch handle at an angle of $60°$.
61. A force of 40 lb is applied to a wrench with a 9-inch handle at an angle of $90°$.

Proofs

62. Prove that the determinant of a 3×3 matrix with integer entries is an integer.
63. Let A be a 3×3 matrix with determinant D, and let A' be a 3×3 matrix obtained from A by exchanging two rows. Prove that $\det A' = -D$.
64. Let B be a 3×3 matrix with determinant d, and let B' be a 3×3 matrix obtained from B by exchanging two columns. Prove that $\det B' = -d$.
65. Use the definition of the cross product to prove that the cross product of two parallel vectors is $\mathbf{0}$. (This is Theorem 10.26.)
66. Use the definition of the cross product to prove that the cross product of two vectors $\mathbf{u}$ and $\mathbf{v}$ is anticommutative; that is, prove that $\mathbf{u} \times \mathbf{v} = -\mathbf{v} \times \mathbf{u}$. (This is Theorem 10.27.)

67. Let $\mathbf{u}$ and $\mathbf{v}$ be vectors in $\mathbb{R}^3$ and let c be a scalar. Prove that $c(\mathbf{u} \times \mathbf{v}) = (c\mathbf{u}) \times \mathbf{v} = \mathbf{u} \times (c\mathbf{v})$. (This is Theorem 10.28).
68. Let $\mathbf{u}$, $\mathbf{v}$, and $\mathbf{w}$ be vectors in $\mathbb{R}^3$. Prove:

$$\mathbf{u} \times (\mathbf{v} + \mathbf{w}) = \mathbf{u} \times \mathbf{v} + \mathbf{u} \times \mathbf{w} \text{ and}$$

$$(\mathbf{u} + \mathbf{v}) \times \mathbf{w} = \mathbf{u} \times \mathbf{w} + \mathbf{v} \times \mathbf{w}.$$

(This is Theorem 10.29.)

69. Let $\mathbf{u}$ and $\mathbf{v}$ be vectors in $\mathbb{R}^3$. Prove that $\mathbf{v} \cdot (\mathbf{u} \times \mathbf{v}) = 0$. (This is Theorem 10.31(b).)
70. Let $\mathbf{u}$ and $\mathbf{v}$ be vectors in $\mathbb{R}^3$. Prove Lagrange's identity, Theorem 10.30:

$$\|\mathbf{u} \times \mathbf{v}\|^2 = \|\mathbf{u}\|^2\|\mathbf{v}\|^2 - (\mathbf{u} \cdot \mathbf{v})^2.$$

71. Let $\mathbf{u}$, $\mathbf{v}$, and $\mathbf{w}$ be three vectors in $\mathbb{R}^3$ in which the components of each vector are integers.

    (a) Prove that the volume of the parallelepiped determined by $\mathbf{u}$, $\mathbf{v}$ and $\mathbf{w}$ is an integer.

    (b) Find examples of vectors $\mathbf{u}$ and $\mathbf{v}$ with integer components that show that the area of the parallelogram determined by $\mathbf{u}$ and $\mathbf{v}$ can be either an integer or an irrational number.

72. Let $\mathbf{u}$, $\mathbf{v}$, and $\mathbf{w}$ be three mutually perpendicular vectors in $\mathbb{R}^3$.

    (a) Prove that $\mathbf{u} \times (\mathbf{v} \times \mathbf{w}) = \mathbf{0}$.

    (b) Show that $|\mathbf{u} \cdot (\mathbf{v} \times \mathbf{w})| = \|\mathbf{u}\|\|\mathbf{v}\|\|\mathbf{w}\|$.

73. Let $\mathbf{u}$ and $\mathbf{v}$ be vectors in $\mathbb{R}^3$ such that $\mathbf{u} \cdot \mathbf{v} \neq 0$. Prove that if θ is the angle between $\mathbf{u}$ and $\mathbf{v}$, then $\tan\theta = \dfrac{\|\mathbf{u} \times \mathbf{v}\|}{\mathbf{u} \cdot \mathbf{v}}$.

74. Let $\mathbf{u}$, $\mathbf{v}$, and $\mathbf{w}$ be vectors in $\mathbb{R}^3$. Prove that $\mathbf{u} \times \mathbf{v} = \mathbf{u} \times \mathbf{w}$ if and only if $\mathbf{u}$ is parallel to $\mathbf{v} - \mathbf{w}$.

75. Let $\mathbf{u}$, $\mathbf{v}$, and $\mathbf{w}$ be vectors in $\mathbb{R}^3$ with $\mathbf{u} \neq \mathbf{0}$. Show that if $\mathbf{u} \times \mathbf{v} = \mathbf{u} \times \mathbf{w}$ and $\mathbf{u} \cdot \mathbf{v} = \mathbf{u} \cdot \mathbf{w}$, then $\mathbf{v} = \mathbf{w}$.

76. Let $\mathbf{u}$, $\mathbf{v}$, and $\mathbf{w}$ be vectors in $\mathbb{R}^3$. Prove that

$$\mathbf{u} \cdot (\mathbf{v} \times \mathbf{w}) = (\mathbf{u} \times \mathbf{v}) \cdot \mathbf{w}.$$

(This is part (b) of Theorem 10.37.)

77. Prove that, for vectors $\mathbf{r}$, $\mathbf{s}$, $\mathbf{u}$, and $\mathbf{v}$ in $\mathbb{R}^3$,

$$(\mathbf{r} \times \mathbf{s}) \cdot (\mathbf{u} \times \mathbf{v}) = (\mathbf{r} \cdot \mathbf{u})(\mathbf{s} \cdot \mathbf{v}) - (\mathbf{r} \cdot \mathbf{v})(\mathbf{s} \cdot \mathbf{u}).$$

78. Let $\mathbf{u} = \langle u_1, u_2, u_3 \rangle$, $\mathbf{v} = \langle v_1, v_2, v_3 \rangle$, and $\mathbf{w} = \langle w_1, w_2, w_3 \rangle$. Show that

$$\mathbf{u} \cdot (\mathbf{v} \times \mathbf{w}) = \det \begin{bmatrix} u_1 & u_2 & u_3 \\ v_1 & v_2 & v_3 \\ w_1 & w_2 & w_3 \end{bmatrix}.$$

79. Prove that if vectors $\mathbf{r}$, $\mathbf{s}$, $\mathbf{u}$, and $\mathbf{v}$ in $\mathbb{R}^3$ can all be translated to the same plane, then

$$(\mathbf{r} \times \mathbf{s}) \times (\mathbf{u} \times \mathbf{v}) = \mathbf{0}.$$

80. The product $\mathbf{u} \times (\mathbf{v} \times \mathbf{w})$ is an example of a vector called a ***vector triple product.***

    (a) Show that if $\mathbf{v} = \langle v_1, v_2, v_3 \rangle$ and $\mathbf{w} = \langle w_1, w_2, w_3 \rangle$, then $\mathbf{i} \times (\mathbf{v} \times \mathbf{w}) = w_1 \mathbf{v} - v_1 \mathbf{w}$.

    (b) Derive similar expressions from $\mathbf{j} \times (\mathbf{v} \times \mathbf{w})$ and $\mathbf{k} \times (\mathbf{v} \times \mathbf{w})$.

    (c) Use your results from parts (a) and (b) to show that $\mathbf{u} \times (\mathbf{v} \times \mathbf{w}) = (\mathbf{u} \cdot \mathbf{w})\mathbf{v} - (\mathbf{u} \cdot \mathbf{v})\mathbf{w}$.

    (d) Use your results from part (c) and the anticommutativity of the cross product to derive a similar expression for the vector triple product $(\mathbf{u} \times \mathbf{v}) \times \mathbf{w}$.

    (e) Use your results from parts (c) and (d) to show that the cross product is *not* associative.

    (f) Under what conditions is

$$\mathbf{u} \times (\mathbf{v} \times \mathbf{w}) = (\mathbf{u} \times \mathbf{v}) \times \mathbf{w}?$$

Thinking Forward

Planes in $\mathbb{R}^3$: Different geometric conditions can be used to specify a plane. The following questions ask you to explain why the specified conditions uniquely determine a plane.

▶   Explain why two nonparallel vectors and a point uniquely determine a plane containing both vectors and the point.

▶   Explain why a single nonzero vector and a point uniquely determine a plane containing the point. (*Hint: Think of the collection of vectors orthogonal to the given vector with the given point as the initial point of all of the vectors.*)

▶   Give two other sets of geometric conditions that would uniquely determine a plane in $\mathbb{R}^3$.

10.5   LINES IN THREE-DIMENSIONAL SPACE

▶ Using vectors to construct equations for lines in $\mathbb{R}^3$

▶ Parallel, intersecting, and skew lines

▶ Computing the distance from a point to a line

Equations for Lines

Recall that the general form for the equation of a line in the xy-plane is $ax+by = d$, where a, b, and d are scalars. An equation of this form is called *a linear equation in two variables.* In this section we will discuss how to write equations of lines in three-dimensional space.

It seems natural to think that the graph of a *linear equation in three variables*,

$$ax + by + cz = d, \text{ where } a, \ b, \ c, \text{ and } d \text{ are scalars,}$$

would be a line in 3-space. However, as we will see in Section 10.6, the graphs of such equations are planes, not lines! (Recall our discussion of these equations in Section 10.2. There we examined the linear equation $x = 3$ and saw that in 3-space the graph of the equation was a plane parallel to the yz-plane. The equation $x = 3$ is a simple example of an equation of the form $ax + by + cz = d$, with $a = 1$, $b = c = 0$, and $d = 3$.) So, if the graph of a linear equation in three variables is not a line, what types of equations *do* give lines?

As you know, a line in the plane can be determined by two distinct points or by a point and a direction, specified by a slope. In 3-space the situation is analogous: A line can be determined by two distinct points or by a single point and a direction, specified by a vector. To determine an equation for a line, $\mathcal{L}$, in 3-space we will use a point, P_0, on $\mathcal{L}$ and a *direction vector*, $\mathbf{d}$, parallel to $\mathcal{L}$. If we are given two points, we can immediately compute a direction vector and start by finding a parametrization for $\mathcal{L}$. If $P_0 = (x_0, y_0, z_0)$ and $P_1 = (x_1, y_1, z_1)$ are two distinct points on $\mathcal{L}$, then a direction vector for $\mathcal{L}$ is

$$\mathbf{d} = \overrightarrow{P_0P_1} = \langle x_1 - x_0, y_1 - y_0, z_1 - z_0 \rangle.$$

We now let P_0 be the point (x_0, y_0, z_0) and $\mathbf{d}$ be the vector $\langle a, b, c \rangle$. Point P_0 corresponds to the position vector $\mathbf{P_0} = \langle x_0, y_0, z_0 \rangle$. Every point on $\mathcal{L}$ can be obtained by adding a suitable multiple of $\mathbf{d}$ to the position vector $\mathbf{P_0}$, as in the following figure:

A line in a three-dimensional Cartesian coordinate system

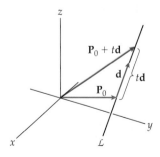

That is, every point on $\mathcal{L}$ is of the form $\mathbf{P_0} + t\mathbf{d}$, for some value of our parameter t. This result immediately gives us a *vector function,* $\mathbf{r}(t)$, for the line $\mathcal{L}$. That is,

$$\mathbf{r}(t) = \mathbf{P_0} + t\mathbf{d}, \quad \text{or}$$
$$\mathbf{r}(t) = \langle x_0, y_0, z_0 \rangle + t \langle a, b, c \rangle, \quad \text{or}$$
$$\mathbf{r}(t) = \langle x_0 + at, y_0 + bt, z_0 + ct \rangle,$$

where $-\infty < t < \infty$. Any of these three equations is a parametrization for $\mathcal{L}$. From the last of the equations, we may immediately find parametric equations for the line $\mathcal{L}$:

$$x(t) = x_0 + at, \quad y(t) = y_0 + bt, \quad z(t) = z_0 + ct, \quad \text{where } -\infty < t < \infty.$$

Assuming that none of the components of the vector $\mathbf{d}$ is zero, we can solve each of the preceding equations for the parameter t to obtain

$$t = \frac{x - x_0}{a}, \; t = \frac{y - y_0}{b}, \; t = \frac{z - z_0}{c}.$$

We may use these equations to eliminate the parameter t to find a **symmetric form** for the line $\mathcal{L}$:

$$\frac{x - x_0}{a} = \frac{y - y_0}{b} = \frac{z - z_0}{c}.$$

Thus, we may use a vector equation, parametric equations, or the symmetric form to specify a line, and given any one of these, we may quickly obtain the others.

For example, to find an equation of the line $\mathcal{L}$ containing the points $P = (3, 2, -3)$ and $Q = (-1, 4, 2)$, we immediately have a direction vector

$$\mathbf{d} = \overrightarrow{PQ} = \langle -1 - 3, 4 - 2, 2 - (-3) \rangle = \langle -4, 2, 5 \rangle.$$

Using P for our point on $\mathcal{L}$, we obtain

$$\mathbf{r}(t) = \langle 3, 2, -3 \rangle + t \langle -4, 2, 5 \rangle, -\infty < t < \infty, \quad \text{and}$$
$$\mathbf{r}(t) = \langle 3 - 4t, 2 + 2t, -3 + 5t \rangle, -\infty < t < \infty.$$

Either of these two equations is a vector function whose graph is $\mathcal{L}$. If we prefer to express $\mathcal{L}$ by means of parametric equations, we have

$$x(t) = 3 - 4t, \; y(t) = 2 + 2t, \; z(t) = -3 + 5t, -\infty < t < \infty.$$

Finally, if we want to use the symmetric form, we have

$$\frac{x - 3}{-4} = \frac{y - 2}{2} = \frac{z + 3}{5}.$$

Two Lines in $\mathbb{R}^3$

Given two lines in the plane, the lines either intersect, are parallel, or are identical. In three-dimensional space there is a fourth possibility: The lines may be **skew**; that is, they neither are parallel nor intersect. For example, on a cube, some edges intersect and some are parallel, but the lines containing the highlighted edges in the following cube are skew:

Skew lines

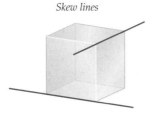

Given the equations of two lines, we can use the direction vectors of the lines to start our analysis of the relationship between the lines. If the direction vectors of the two lines are scalar multiples of each other, then the lines are either parallel or identical. If the direction vectors of the two lines are not multiples of each other, the lines either intersect or are skew. To complete the analysis, we determine whether or not the lines share a point. Parallel lines that share a point are identical. Nonparallel lines that share a point (obviously) intersect.

The Distance from a Point to a Line (Revisited)

The distance from a point, P, to a line, $\mathcal{L}$, is the distance from P to the closest point on $\mathcal{L}$. This is the distance along a line through P and perpendicular to $\mathcal{L}$. That distance can be computed with the technique of Example 4 from Section 10.3. Here, we provide an alternative method.

Let $\mathbf{r}(t) = \mathbf{P}_0 + t\mathbf{d}$, $-\infty < t < \infty$, be a parametrization for $\mathcal{L}$. Thus, P_0 is on $\mathcal{L}$ and $\mathbf{d}$ is parallel to $\mathcal{L}$, as in the following figure:

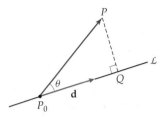

Now, suppose Q is the point on the line $\mathcal{L}$ that is closest to the point P, as in the figure shown. If we knew the location of the point Q, then we could immediately use the distance formula to calculate the length of $\overrightarrow{PQ}$. Typically, though, we do not know the location of Q. However, we do see that $\|\overrightarrow{PQ}\| = \|\overrightarrow{P_0P}\| \sin\theta$, where θ is the angle between $\overrightarrow{P_0P}$ and $\mathbf{d}$. Now, by Theorem 10.32,

$$\|\mathbf{d} \times \overrightarrow{P_0P}\| = \|\mathbf{d}\|\|\overrightarrow{P_0P}\| \sin\theta.$$

Thus, we have

$$\|\overrightarrow{PQ}\| = \|\overrightarrow{P_0P}\| \sin\theta = \frac{\|\mathbf{d} \times \overrightarrow{P_0P}\|}{\|\mathbf{d}\|}.$$

To summarize, we have the following theorem:

THEOREM 10.38

The Distance from a Point to a Line

Given a point P and a line $\mathcal{L}$ parameterized by $\mathbf{r}(t) = \mathbf{P}_0 + t\mathbf{d}$, the distance from P to $\mathcal{L}$ is

$$\frac{\|\mathbf{d} \times \overrightarrow{P_0P}\|}{\|\mathbf{d}\|}.$$

Examples and Explorations

EXAMPLE 1   **Expressing the equation of a line in symmetric form when one component of the direction vector is zero**

Find an equation of the line $\mathcal{L}$ containing the points $P = (-5, 6, -1)$ and $Q = (4, 6, -7)$. Write the answer as a parametrization, in terms of parametric equations, and in symmetric form.

SOLUTION

A direction vector for the line is

$$\overrightarrow{PQ} = \langle 4 - (-5), 6 - 6, -7 - (-1) \rangle = \langle 9, 0, -6 \rangle.$$

However, any nonzero multiple of $\overrightarrow{PQ}$ may be used instead. Here we will take $\mathbf{d} = \frac{1}{3}\overrightarrow{PQ} = \langle 3, 0, -2 \rangle$ as our direction vector. Now, using P for our point on $\mathcal{L}$, we have

$$\mathbf{r}(t) = \langle -5, 6, -1 \rangle + t \langle 3, 0, -2 \rangle, \quad -\infty < t < \infty,$$
$$\mathbf{r}(t) = \langle -5 + 3t, 6, -1 - 2t \rangle, \quad -\infty < t < \infty.$$

Either of these two equations is a parametrization.

Using the second of the two forms, we have the parametric equations

$$x(t) = -5 + 3t, \ y(t) = 6, \ z(t) = -1 - 2t, \quad -\infty < t < \infty.$$

There is also a symmetric form for a line $\mathcal{L}$ if one of the three components of the direction vector $\mathbf{d}$ is zero. When $\mathbf{P_0} = \langle x_0, y_0, z_0 \rangle$ and $\mathbf{d} = \langle a, 0, c \rangle$, then, as before, we have the parametrization

$$\mathbf{r}(t) = \mathbf{P_0} + t\mathbf{d}, -\infty < t < \infty$$
$$= \langle x_0 + at, y_0, z_0 + ct \rangle, -\infty < t < \infty.$$

We have the following parametric equations for $\mathcal{L}$:

$$x(t) = x_0 + at, \ y(t) = y_0, \ z(t) = z_0 + ct, -\infty < t < \infty.$$

The symmetric form for $\mathcal{L}$ is

$$\frac{x - x_0}{a} = \frac{z - z_0}{c}, \ y = y_0.$$

Thus, for our line $\mathcal{L}$, we have

$$\frac{x + 5}{3} = \frac{z + 1}{-2}, \ y = 6.$$

There are analogous expressions when one of the other components of the direction vector is zero. □

EXAMPLE 2   **Converting from symmetric form to parametric equations**

Given the symmetric equations

$$\frac{x - 4}{5} = \frac{y + 2}{-3} = \frac{z - 1}{7}$$

for a line $\mathcal{L}$, find parametric equations for $\mathcal{L}$.

SOLUTION

We introduce the parameter, t, as follows:

$$t = \frac{x - 4}{5} = \frac{y + 2}{-3} = \frac{z - 1}{7}.$$

So,

$$x - 4 = 5t$$
$$y + 2 = -3t$$
$$z - 1 = 7t,$$

where $-\infty < t < \infty$. These equations are parametric equations for $\mathcal{L}$. We may also write the system in the form

$$x(t) = 4 + 5t, \ y(t) = -2 - 3t, \ z(t) = 1 + 7t, -\infty < t < \infty. \quad □$$

EXAMPLE 3

Determining when two lines are parallel

Consider the lines $\mathcal{L}_1$ and $\mathcal{L}_2$ respectively given by the parametrizations

$$\mathbf{r}_1(t) = \langle 3 + 5t, -1 - 6t, 4t \rangle, -\infty < t < \infty, \text{ and}$$

$$\mathbf{r}_2(t) = \langle 4 - 10t, 7 + 12t, -1 - 8t \rangle, -\infty < t < \infty.$$

Are lines $\mathcal{L}_1$ and $\mathcal{L}_2$ parallel? If they are, find the distance between them.

SOLUTION

Two distinct lines $\mathcal{L}_1$ and $\mathcal{L}_2$ are parallel if they have (nonzero) parallel direction vectors, $\mathbf{d}_1$ and $\mathbf{d}_2$, respectively. That is, the lines are parallel if there is a scalar, k, such that $\mathbf{d}_2$ is equal to $k\mathbf{d}_1$.

Lines $\mathcal{L}_1$ and $\mathcal{L}_2$ have direction vectors $\mathbf{d}_1 = \langle 5, -6, 4 \rangle$ and $\mathbf{d}_2 = \langle -10, 12, -8 \rangle$, respectively. Since $\mathbf{d}_2 = -2\mathbf{d}_1$, the lines are either parallel or identical. Two parallel lines are identical if they share a point. We may use any point on either of the lines and check whether that point is on the other line. One convenient point on $\mathcal{L}_1$ is $\mathbf{r}_1(0) = \langle 3, -1, 0 \rangle$. Is $(3, -1, 0)$ also on $\mathcal{L}_2$? It is if there is a solution of the system of equations

$$4 - 10t = 3$$
$$7 + 12t = -1$$
$$-1 - 8t = 0.$$

Note that the unique solution of the last of these equations is $t = -\frac{1}{8}$, but this value does not satisfy either of the other equations. Therefore, there is no value of t that simultaneously satisfies all three of the preceding equations. This fact tells us that the lines $\mathcal{L}_1$ and $\mathcal{L}_2$ are parallel, but not identical.

To find the distance between the two lines, we choose any point on one of the lines and find the distance from that point to the other line. We already know that $(3, -1, 0)$ is on $\mathcal{L}_1$. We will therefore use this point as P. We need a point on $\mathcal{L}_2$, so we will use $\mathbf{r}_2(0) = \langle 4, 7, -1 \rangle$. That is, we will let $P_0 = (4, 7, -1)$. Thus $\overrightarrow{P_0 P} = \langle -1, -8, 1 \rangle$. Finally, the direction vector $\mathbf{d}_2 = \langle -10, 12, -8 \rangle$. We are now ready to use the distance formula given in Theorem 10.38. The distance from the point P to the line $\mathcal{L}_2$, and therefore the distance between $\mathcal{L}_1$ and $\mathcal{L}_2$, is

$$\frac{\|\langle -10, 12, -8 \rangle \times \langle -1, -8, 1 \rangle\|}{\|\langle -10, 12, -8 \rangle\|} = \frac{\|\langle -52, 18, 92 \rangle\|}{\sqrt{308}} = \sqrt{\frac{2873}{77}} \approx 6.1 \text{ units.}$$

As an extension of this example, consider the parametrization

$$\mathbf{r}_3(t) = \langle -12 + 15t, 17 - 18t, -12 + 12t \rangle, -\infty < t < \infty.$$

A direction vector for this parametrization is $\mathbf{d}_3 = \langle 15, -18, 12 \rangle$, which is a scalar multiple of $\mathbf{d}_1$. Furthermore, the point $(3, -1, 0)$ is on the line determined by $\mathbf{r}_3(t)$, since the system of equations

$$-12 + 15t = 3$$
$$17 - 18t = -1$$
$$-12 + 12t = 0$$

has the solution $t = 1$. Thus, $\mathbf{r}_3$ is a different parametrization for $\mathcal{L}_1$.

Note that the two parameters are *not* equal at the point $(3, -1, 0)$; that is,

$$\mathbf{r}_1(0) = \langle 3, -1, 0 \rangle = \mathbf{r}_3(1).$$

This is irrelevant! The significant concepts here are that the lines are parallel and they share a point; therefore they must represent the same line.   □

EXAMPLE 4

Determining when two lines intersect

Do the lines $\mathcal{L}_1$ and $\mathcal{L}_2$, respectively given by

$$\mathbf{r}_1(t) = \langle -8 - 5t, 3 - t, 4 \rangle, -\infty < t < \infty, \text{ and } \mathbf{r}_2(t) = \langle 6 + t, 4 + 2t, 1 + 3t \rangle, -\infty < t < \infty,$$

intersect?

SOLUTION

We see immediately that the direction vectors for these lines are $\mathbf{d}_1 = \langle -5, -1, 0 \rangle$ and $\mathbf{d}_2 = \langle 1, 2, 3 \rangle$, which are *not* parallel. (Why?) Do the lines share a point? If so, the lines intersect. If not, they are skew.

Finding a candidate for a point of intersection here is more delicate than in Example 3, where we could have used *any* point on either of the lines. We *cannot* just equate the corresponding components, as written, and try to solve for t. If you try this, you will see that the system

$$-8 - 5t = 6 + t, \quad 3 - t = 4 + 2t, \quad 4 = 1 + 3t$$

has no solution. However, as we will see in a moment, the lines *do* intersect! The difficulty here is with the parameters. Although we've used the letter t as the parameter for each line, there are *two* distinct parameters, one for $\mathcal{L}_1$ and a second for $\mathcal{L}_2$. To proceed, we change the name of the variable for one of the parameters. Let

$$\mathbf{r}_2(u) = \langle 6 + u, 4 + 2u, 1 + 3u \rangle, -\infty < u < \infty.$$

Now we equate the corresponding components of $\mathbf{r}_1$ and $\mathbf{r}_2$ to obtain the system

$$-8 - 5t = 6 + u, \quad 3 - t = 4 + 2u, \quad 4 = 1 + 3u.$$

This system has the unique solution $t = -3$, $u = 1$. Thus, the lines do intersect at

$$\mathbf{r}_1(-3) = \mathbf{r}_2(1).$$

 **CHECKING THE ANSWER**

To verify our conclusion, we evaluate

$$\mathbf{r}_1(-3) = \langle -8 - 5(-3), 3 - (-3), 4 \rangle = \langle 7, 6, 4 \rangle$$

and

$$\mathbf{r}_2(1) = \langle 6 + 1, 4 + 2 \cdot 1, 1 + 3 \cdot 1 \rangle = \langle 7, 6, 4 \rangle.$$

These calculations confirm that the point $(7, 6, 4)$ is on both lines.

EXAMPLE 5

Finding the distance from a point to a line

Find the distance from the point $P = (2, 0, 1)$ to the line given by

$$\mathbf{r}(t) = \langle 3 - t, -5 + 4t, 6 \rangle, -\infty < t < \infty.$$

SOLUTION

Here, $P_0 = (3, -5, 6)$ corresponds to $\mathbf{r}(0)$ and $\mathbf{d} = \langle -1, 4, 0 \rangle$ is the direction vector for the line. Thus, $\|\mathbf{d}\| = \sqrt{17}$ and $\overrightarrow{P_0P} = \langle -1, 5, -5 \rangle$. The norm of $\mathbf{d} \times \overrightarrow{P_0P}$ is computed as follows:

$$\|\mathbf{d} \times \overrightarrow{P_0P}\| = \left\| \det \begin{bmatrix} \mathbf{i} & \mathbf{j} & \mathbf{k} \\ -1 & 4 & 0 \\ -1 & 5 & -5 \end{bmatrix} \right\| = \|-20\mathbf{i} - 5\mathbf{j} - \mathbf{k}\| = \sqrt{426}.$$

By Theorem 10.38, the distance from P to the line is $\dfrac{\|\mathbf{d} \times \overrightarrow{P_0P}\|}{\|\mathbf{d}\|} = \sqrt{\dfrac{426}{17}} \approx 5.0$ units.

? TEST YOUR UNDERSTANDING

▶ How do you find an equation of a line determined by two points?

▶ How do you find an equation of a line parallel to a given vector and passing through a given point?

▶ Given a parametrization of a line, parametric equations of a line, or the equation of the line in symmetric form, how do you find the other forms?

▶ What are parallel lines, intersecting lines, and skew lines in 3-space?

▶ How do you find the distance from a point to a line?

EXERCISES 10.5

Thinking Back

Lines in the plane: Write the equation of the specified lines in $\mathbb{R}^2$ as vector parametrizations and in symmetric form.

▶ $y = 2x + 5$

▶ The line perpendicular to the line $y = 2x + 5$ and containing the point $(2, -1)$.

▶ $y = mx + b$

▶ The line perpendicular to the line $y = mx + b$ and containing the point $(0, b)$.

▶ *The distance between two points in the plane:* What is the formula for computing the distance between points (x_1, y_1) and (x_2, y_2)?

▶ *The distance between a point and a line in the plane:* Describe a method for computing the distance between the point (x_0, y_0) and the line $y = mx + b$.

Concepts

0. *Problem Zero:* Read the section and make your own summary of the material.

1. *True/False:* Determine whether each of the statements that follow is true or false. If a statement is true, explain why. If a statement is false, provide a counterexample.

(a) *True or False:* Two parallel lines that share a point are identical.

(b) *True or False:* If the direction vector for a line is not a multiple of one of the standard basis vectors, then the line will intersect all three coordinate planes.

(c) *True or False:* If a line $\mathbf{r}(t) = \mathbf{P}_1 + t\mathbf{d}, -\infty < t < \infty$, is parallel to the xz-plane then $\mathbf{d}$ must have the form $\langle a, 0, c \rangle$ for some real numbers a and c.

(d) *True or False:* If $\mathbf{r}_1(t) = \mathbf{P}_1 + t\mathbf{d}_1, -\infty < t < \infty$, and $\mathbf{r}_2(t) = \mathbf{P}_2 + t\mathbf{d}_2, -\infty < t < \infty$, are vector parametrizations for lines in 3-space with $\mathbf{d}_1 \neq \mathbf{d}_2$, then the lines do not intersect.

(e) *True or False:* If $\mathbf{r}_1(t) = \mathbf{P}_1 + t\mathbf{d}_1, -\infty < t < \infty$, and $\mathbf{r}_2(t) = \mathbf{P}_2 + t\mathbf{d}_2, -\infty < t < \infty$, are vector parametrizations for lines in 3-space with $\mathbf{P}_1 = \mathbf{P}_2$, then the lines intersect.

(f) *True or False:* Let $\mathbf{r}_1(t) = \mathbf{P}_1 + t\mathbf{d}_1, -\infty < t < \infty$, and $\mathbf{r}_2(t) = \mathbf{P}_2 + t\mathbf{d}_2, -\infty < t < \infty$, be vector parametrizations for lines in 3-space. If $\mathbf{P}_1 \neq \mathbf{P}_2$ and $\mathbf{d}_1 \neq \mathbf{d}_2$, then the lines cannot be identical.

(g) *True or False:* Let $\mathbf{r}_1(t) = \mathbf{P}_1 + t\mathbf{d}_1, -\infty < t < \infty$, and $\mathbf{r}_2(t) = \mathbf{P}_2 + t\mathbf{d}_2, -\infty < t < \infty$, be vector parametrizations for lines in 3-space. If $\mathbf{P}_1 \neq \mathbf{P}_2$ and $\mathbf{d}_1 \neq \mathbf{d}_2$, then the lines cannot be parallel.

(h) *True or False:* If $\mathbf{r}_1(t) = \mathbf{P}_1 + t\mathbf{d}_1, -\infty < t < \infty$, and $\mathbf{r}_2(t) = \mathbf{P}_2 + t\mathbf{d}_2, -\infty < t < \infty$, are vector parametrizations for lines in 3-space with $\mathbf{d}_1 \neq \mathbf{d}_2$, then the lines are skew.

2. *Examples:* Construct examples of the thing(s) described in the following. Try to find examples that are different than any in the reading.

(a) A line in $\mathbb{R}^3$ that is parallel to the xz-plane.

(b) A line in $\mathbb{R}^3$ through the origin.

(c) A line parallel to the z-axis.

3. What is a linear equation in three variables? Give an example. What is the graph of a linear equation in three variables?

4. Provide two sets of geometric conditions that can be used to determine a line.

5. Let P and Q be distinct points in $\mathbb{R}^3$. Provide a step-by-step procedure for finding the equation of the line containing P and Q.

6. Let $P = (a, b, c)$ and $Q = (\alpha, \beta, \gamma)$ be distinct points in $\mathbb{R}^3$. Explain why the parametrization

$$x = a + (\alpha - a)t, \quad y = b + (\beta - b)t, \quad z = c + (\gamma - c)t,$$

for $0 \leq t \leq 1$, describes the line *segment* connecting P and Q. What parametrization would describe the segment from Q to P?

7. Explain how the slopes of two lines in $\mathbb{R}^2$ can be used to determine whether the lines are parallel or identical. If the two lines are not parallel or identical, what must be true about the lines?

8. Explain how the direction vectors of two lines in $\mathbb{R}^3$ can be used to determine whether the two lines are parallel or identical. If the two lines are not parallel or identical, what are the possibilities for them?

9. Find an equation of the line containing the point $(-1, 3, 7)$ with direction vector $\langle 2, -4, 9 \rangle$.
   (a) Use a different direction vector to find another equation for the same line.
   (b) Find an equation for the same line with the form
   $$x = at + 5, \ y = bt + y_0, \ z = ct + z_0.$$

10. Find an equation of the line containing the point (x_0, y_0, z_0) with direction vector $\langle a, b, c \rangle$.
    (a) Use a different direction vector to find another equation for the same line.
    (b) Assuming that $a \neq 0$, find an equation for the same line with the form
    $$x = at + 5, \ y = bt + y_0, \ z = ct + z_0.$$

11. Find the points where the line
    $$\mathbf{r}(t) = \langle 4 - 3t, 8 + 7t, 5 + t \rangle, -\infty < t < \infty,$$
    intersects each of the coordinate planes.

12. If a, b, and c are nonzero, find the points where the line
    $$\mathbf{r}(t) = \langle x_0 + at, y_0 + bt, z_0 + ct \rangle$$
    intersects each of the coordinate planes. If $a = 0$, explain why the line is parallel to the yz-plane.

13. Find the points where the line
    $$x = t + 2, \ y = 2t - 5, \ z = -4t - 7$$
    intersects each of the coordinate planes.

14. Find the points where the line
    $$x = -3t + 5, \ y = 4, \ z = 2t + 11$$
    intersects the xy-plane and yz-plane. Explain why the line does *not* intersect the xz-plane.

15. Find the point where the line
    $$x = -7, \ y = t, \ z = 5$$
    intersects the xz-plane. Explain why the line does *not* intersect the other two coordinate planes.

16. Let
    $$x = at + x_0, \ y = bt + y_0, \ z = ct + z_0$$

be parametric equations for a line $\mathcal{L}$ in $\mathbb{R}^3$. If x_0, y_0, and z_0 are all nonzero, give conditions on a, b, and c so that
   (a) $\mathcal{L}$ intersects all three coordinate planes.
   (b) $\mathcal{L}$ intersects the xy- and yz-planes, but not the xz-plane.
   (c) $\mathcal{L}$ intersects exactly one of three coordinate planes.

17. Let $\mathcal{L}$ be the line determined by the equation
    $$\mathbf{r}(t) = \langle 2 + 7t, 3 - 5t, 2t \rangle, -\infty < t < \infty.$$
    (a) Give parametric equations for $\mathcal{L}$.
    (b) Write an equation for $\mathcal{L}$ in symmetric form.

18. Let $\mathcal{L}$ be the line determined by the equation
    $$\frac{x}{4} = \frac{y+2}{5} = -\frac{z-8}{3}.$$
    (a) Provide a vector parametrization for $\mathcal{L}$.
    (b) Give parametric equations for $\mathcal{L}$.

19. Let $\mathcal{L}$ be the line determined by the system of equations
    $$x(t) = 4, \ y(t) = 3 - 5t, \ z(t) = t, -\infty < t < \infty.$$
    (a) Provide a vector parametrization for $\mathcal{L}$.
    (b) Write an equation for $\mathcal{L}$ in symmetric form.

20. We wish to find the distance from the point P to the line $\mathcal{L}$ as shown in the figure that follows. We know the coordinates of points P and P_0, but we do not know the coordinates of point Q.
    (a) If you knew the measure of angle θ, explain how you would find the distance from point P to line $\mathcal{L}$.

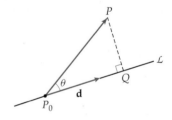

   (b) Using a cross product, explain how you can find the distance from point P to line $\mathcal{L}$ even if you do not know the measure of angle θ.

Skills

In Exercises 21 and 22, find the equation of the line containing the given points in slope–intercept form. Then, use the technique of this section to find a vector parametrization for the same line. Finally, show that your equations are equivalent.

21. $P(0, 5), \ Q(2, -1)$

22. $P(3, -2), \ Q(6, 4)$

In Exercises 23–28, find an equation of the line containing the given point and parallel to the given vector. Express your answer (a) as a vector parametrization, (b) in terms of parametric equations, and (c) in symmetric form.

23. $P(0, 0, 0), \ \mathbf{d} = \langle 1, 2, -4 \rangle$

24. $P(2, 3, 5), \ \mathbf{d} = \langle 2, 3, 5 \rangle$

25. $P(-1, 3, 7), \ \mathbf{d} = \langle 2, 0, 4 \rangle$

26. $P(x_0, y_0, z_0), \ \mathbf{d} = \langle a, b, c \rangle$

27. $P(3, 1), \ \mathbf{d} = \langle 2, 5 \rangle$

28. $P(1, 3, -2, 4), \ \mathbf{d} = \langle 4, -1, 5, 8 \rangle$

In Exercises 29–34, find an equation of the line containing the given pair of points. Express your answer (a) as a vector parametrization, (b) in terms of parametric equations, and (c) in symmetric form.

29. $P(0, 0, 0), \ Q(4, -1, 6)$

30. $P(3, -1, 7), \ Q(5, 8, -2)$

31. $P(-4, 11, 0)$, $Q(4, 11, 2)$

32. $P(x_0, y_0, z_0)$, $Q(x_1, y_1, z_1)$

33. $P(1, 6)$, $Q(4, 5)$

34. $P(3, -1, 2, 6)$, $Q(1, 4, 5, -2)$

35. (a) Find a vector parametrization for the line containing the points $P(x_0, y_0, z_0)$ and $Q(x_1, y_1, z_1)$.

   (b) Apply a restriction to your parameter from part (a) so that the result parametrizes the *segment* from P to Q.

In Exercises 36–39, use the result of Exercise 35 to find parametric equations for the line segment connecting point P to point Q.

36. $P(0, 2, 3)$, $Q(4, 5, -1)$

37. $P(1, 7, 3)$, $Q(-1, -2, 5)$

38. $P(0, 0, 0)$, $Q(1, 2, 3)$

39. $P(3, -1, 4)$, $Q(-1, 5, 9)$

In Exercises 40–45, determine whether the given pairs of lines are parallel, identical, intersecting, or skew. If the lines are parallel, compute the distance between them. If the lines intersect, find the point of intersection and the angle at which the lines intersect.

40. $\mathbf{r}_1(t) = \langle 2 + t, 5 - 3t, 6 + 7t \rangle$,
   $\mathbf{r}_2(u) = \langle 3 - u, 4 + 3u, 5 - 2u \rangle$

41. $\mathbf{r}_1(t) = \langle 2t + 6, -t + 1, 3t \rangle$,
   $\mathbf{r}_2(t) = \langle -4t + 3, 2t - 1, -6t + 2 \rangle$

42. $\mathbf{r}_1(t) = \langle 3 - t, 7 + t, 4 + 5t \rangle$,
   $\mathbf{r}_2(t) = \langle 3 - t, 4 + 3t, 5 - 2t \rangle$

43. $\mathbf{r}_1(t) = \langle 5t + 2, -4t, t - 7 \rangle$,
   $\mathbf{r}_2(t) = \langle -3t + 4, -t + 12, -2t - 1 \rangle$

44. $\mathbf{r}_1(t) = \langle 4 + 5t, 6, 7 - 2t \rangle$,
   $\mathbf{r}_2(t) = \langle 6 - 4t, -3 + 3t, -1 + 4t \rangle$

45. $\mathbf{r}_1(t) = \langle 2 + 5t, 6 - 4t, 8 + t \rangle$,
   $\mathbf{r}_2(u) = \langle 2 - 10u, 3 + 8u, -2u \rangle$

In Exercises 46–49, calculate the distance from the given point to the given line in the following two ways: (a) using the method of Example 4 from Section 10.3 and (b) using Theorem 10.38 from this section.

46. Point $P(-6, 3, 0)$ to the line determined by the points $Q(3, -1, 5)$ and $R(4, 5, -2)$.

47. Point $P(0, 1, -2)$ to the line given by

$$\mathbf{r}(t) = \langle 1 + 5t, -6 + t, -4t \rangle.$$

48. Point $P(-6, 3, 0)$ to the line given by

$$\frac{x-3}{4} = \frac{y+2}{6} = z.$$

49. Point $P(2, 5)$ to the line $y = 2x - 3$.

50. Find the distance from the point $P(2, 3, -1, 4)$ to the line determined by the points $Q(1, 0, 4, 8)$ and $R(3, 4, -1, 6)$.

51. Find values for α such that the lines determined by

$$\mathbf{r}_1(t) = \langle 4 - t, -6 + 2t, 6 + 5t \rangle \quad \text{and}$$

$$\mathbf{r}_2(t) = \langle \alpha t, 2 - 2\alpha t, 3 + 15t \rangle$$

are (a) parallel and (b) orthogonal.

Applications

52. Emmy is a civil engineer at the Hanford Nuclear Reservation in Washington State. She has discovered a leak of toxic wastes in one of the tank farms of the facility. The tank farm is huge, and she does not know which tank is leaking. Worse yet, the tanks are all underground. Inspecting the tanks would require digging up the entire tank farm, an operation that is considered too expensive. Instead, Emmy has wells dug in several locations around the tank farm, to try to trace the leak back to its source. If the earth's surface is considered to be the xy-plane, then the bottom of the tank farm is the plane $z = -40$, where the distance is given in feet. Emmy's wells find contaminated groundwater at the points $(758, 60, -49)$ and $(1033, 247, -55)$. If the waste is leaking from the bottom of one of the tanks and is moving along a straight path, what is the location of the tank she should check first for the leak?

53. Ian is climbing Mount Logan in Canada's Icefield Range. He has made a second camp high on a ridge of the mountain. He does not know how high his camp is,

but he notices that a summit far away lines up perfectly with a nearer minor point whose height he can read on his map. When Ian considers the summit of Mount Logan to be $(0, 0, 5.96)$, he finds that the far summit is at $(11.2, -5.6, 4.2)$ and the nearer point is at $(5.4, -2.5, 4.5)$. All the coordinates are given in kilometers. Ian's camp is due east of the summit of Mount Logan. How high is Ian on the mountain? How far is he from the summit?

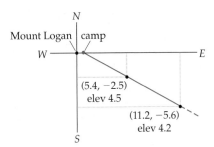

Proofs

54. Let $\mathcal{L}_1$ and $\mathcal{L}_2$ be lines in $\mathbb{R}^3$, with P_1 and Q_1 points on $\mathcal{L}_1$ and P_2 and Q_2 points on $\mathcal{L}_2$. Show that $\mathcal{L}_1$ is parallel to $\mathcal{L}_2$ if and only if $\overrightarrow{P_1Q_1}$ is parallel to $\overrightarrow{P_2Q_2}$.

55. Prove that the distance from the point P to the line given by the equation $\mathbf{r}(t) = \mathbf{P}_0 + t\mathbf{d}$ is given by $\dfrac{\|\mathbf{d} \times \overrightarrow{P_0P}\|}{\|\mathbf{d}\|}$.

Thinking Forward

▶ *Lines tangent to spheres*: Let $\mathcal{S}$ denote the unit sphere centered at the origin. How many lines are there that are tangent to $\mathcal{S}$? How many lines are there that are tangent to $\mathcal{S}$ at each point on the sphere? How many lines are there that are tangent to $\mathcal{S}$ and parallel to one of the coordinate planes? How many lines are there that are tangent to $\mathcal{S}$ and parallel to one of the coordinate axes?

▶ *Planes tangent to spheres*: Let $\mathcal{S}$ denote the unit sphere centered at the origin. How many planes are there that are tangent to $\mathcal{S}$? How many planes are there that are tangent to $\mathcal{S}$ at each point on the sphere? How many planes are there that are tangent to $\mathcal{S}$ and parallel to one of the coordinate planes? How many planes are there that are tangent to $\mathcal{S}$ and parallel to one of the coordinate axes?

10.6   PLANES

▶ Using a normal vector and a point to construct the equation for a plane

▶ Computing the distance from a point to a plane

▶ Determining whether two planes are parallel or intersect

The Equation of a Plane

In our final section of the chapter we discuss planes in three-dimensional space. Just as curves in two dimensions have tangent lines, surfaces have tangent planes, and those planes can be used as approximations for the surface. We will visit these topics in Chapter 12.

There are several geometric conditions that determine a unique plane. For example, each of the following sets of information uniquely determines one plane in $\mathbb{R}^3$:

▶ three noncollinear points;

▶ a line and a point not on the line;

▶ two distinct intersecting lines;

▶ two (distinct) parallel lines.

Our path to the equation of a plane, however, involves a point on the plane and a single vector, $\mathbf{N}$, called a **normal vector**, orthogonal to the plane. We will also see, in examples and exercises, how we can find an equation for the plane given any of the conditions just listed.

Let $P_0 = (x_0, y_0, z_0)$ be a given point on a plane $\mathcal{P}$ and let $\mathbf{N} = \langle a, b, c \rangle$ be a vector orthogonal to $\mathcal{P}$. Place $\mathbf{N}$ so that its initial point is at P_0, and consider the vector $\overrightarrow{P_0P}$, where P is an arbitrary point on $\mathcal{P}$.

*A plane is determined by a point on the plane
and a vector orthogonal to the plane*

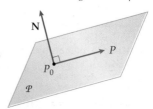

Note that since **N** is orthogonal to $\mathcal{P}$, it is orthogonal to every vector lying in $\mathcal{P}$. In particular, **N** is orthogonal to the vector $\overrightarrow{P_0P}$, and thus $\mathbf{N} \cdot \overrightarrow{P_0P} = 0$. If $P = (x, y, z)$, then

$$\mathbf{N} \cdot \overrightarrow{P_0P} = \langle a, b, c \rangle \cdot \langle x - x_0, y - y_0, z - z_0 \rangle = 0.$$

Carrying out the dot product, we obtain the following equation for a plane:

$$a(x - x_0) + b(y - y_0) + c(z - z_0) = 0.$$

This equation is analogous to the point–slope form for the equation of a line that we studied in Chapter 0.

We may rearrange the last equation to obtain

$$ax + by + cz = ax_0 + by_0 + cz_0.$$

Since a, b, c, x_0, y_0, and z_0 are all constants, if we let $d = ax_0 + by_0 + cz_0$, then

$$ax + by + cz = d,$$

the **general form for the equation of a plane**. Recall from the last section that equations of this form are called linear equations and the graph of a linear equation in three variables is a plane.

For example, to find the equation of the plane that has the normal vector $\mathbf{N} = \langle 5, -3, 8 \rangle$ and contains the point $(2, -7, 1)$, we set up the equation

$$\langle 5, -3, 8 \rangle \cdot \langle x - 2, y - (-7), z - 1 \rangle = 0.$$

We can then write the equation in either of the forms

$$5(x - 2) - 3(y + 7) + 8(z - 1) = 0 \quad \text{and} \quad 5x - 3y + 8z = 39.$$

Geometric Conditions That Determine Planes

We now know how to find the equation of a plane given a normal vector to the plane and a point on the plane. When we are given different geometric conditions that determine a plane, we will follow this same procedure to find the equation of the plane.

Suppose we know three noncollinear points on a plane and we wish to find the equation for the plane. We can use the three points to find a normal vector for the plane and then use this normal vector and any one of the three points to find the equation of the plane, as we did at the beginning of this section. One normal vector to the plane containing P, Q, and R is the cross product $\mathbf{N} = \overrightarrow{PQ} \times \overrightarrow{PR}$, as shown in the following figure:

A plane is determined by three noncollinear points

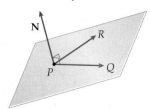

We will discuss how to find the equation of a plane by using one more of the sets of geometric conditions that determine the plane. You should then consider how to find equations for planes by using the other sets of geometric conditions.

A given line and a point not on the line define a plane uniquely. When you have the equation for a line, $\mathbf{r}(t) = \mathbf{P}_0 + t\mathbf{d}$, together with a point, P, not on the line, you immediately know that P and P_0 are two distinct points on some plane $\mathcal{P}$. In addition, the direction vector **d** for the line gives you one vector parallel to the plane. We already know that we can find a vector normal to $\mathcal{P}$ if we have two nonparallel vectors that are parallel to the plane (since we can then use their cross product). Vectors **d** and $\overrightarrow{P_0P}$ meet our requirements. Thus, we will use $\mathbf{N} = \mathbf{d} \times \overrightarrow{P_0P}$ and point P_0 to find the equation of the plane.

A plane is determined by a line and a point not on the line

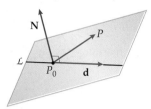

Distances

The distance from a point P, to a plane, $\mathcal{P}$, is the distance from P to the closest point on $\mathcal{P}$. This is the distance along a line through P and orthogonal to $\mathcal{P}$.

When we have the equation of a plane, we can immediately find a vector, **N**, normal to the plane. One way to find the distance from point P to plane $\mathcal{P}$ would be to find the equation of the line $\mathcal{L}$ containing P with direction vector **N**. We could then locate the point Q of intersection of $\mathcal{L}$ and $\mathcal{P}$. Finally we could use the distance formula to calculate the distance from P to Q.

Finding the distance from a point to a plane

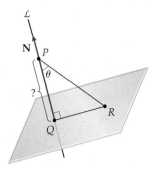

The method just outlined would work, but is more than we need to just find the distance from P to the plane. The extra work comes when we perform several steps to find the point of intersection, Q. It turns out that we do not need to find Q. We do need to find $\|\overrightarrow{QP}\|$, and this magnitude is $|\text{comp}_{\mathbf{N}}\,\overrightarrow{RP}|$, where R can be any point on the plane $\mathcal{P}$.

To summarize, we have the following theorem:

THEOREM 10.39

The Distance from a Point to a Plane

Given a point P and a plane $\mathcal{P}$ containing a point R with normal vector **N**, the distance from P to $\mathcal{P}$ is

$$|\text{comp}_{\mathbf{N}}\,\overrightarrow{RP}| = \frac{|\mathbf{N} \cdot \overrightarrow{RP}|}{\|\mathbf{N}\|}.$$

The following theorem is a direct consequence of Theorem 10.39:

THEOREM 10.40

The Distance Between a Plane and the Origin

Let $P = (x_0, y_0, z_0)$ be any point on the plane $\mathcal{P}$. The distance between the origin and $\mathcal{P}$ is given by $|\langle x_0, y_0, z_0 \rangle \cdot \mathbf{n}|$ where $\mathbf{n}$ is a unit normal vector for $\mathcal{P}$.

The proof of Theorem 10.40 is left for Exercise 62. We may also adapt Theorem 10.39 to find the distance between parallel planes $\mathcal{P}_1$ and $\mathcal{P}_2$.

Finding the distance between two parallel planes

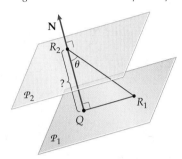

Since $\mathcal{P}_1$ and $\mathcal{P}_2$ are parallel, they have a common normal vector $\mathbf{N}$. We need a single point on each of the planes. Suppose that $R_1 = (x_1, y_1, z_1)$ is on $\mathcal{P}_1$ and $R_2 = (x_2, y_2, z_2)$ is on $\mathcal{P}_2$. We may now apply Theorem 10.39 to compute the distance to obtain the following:

THEOREM 10.41

The Distance Between Parallel Planes

Given a point R_1 on a plane $\mathcal{P}_1$ and a point R_2 on a parallel plane $\mathcal{P}_2$ with common normal vector $\mathbf{N}$, the distance from $\mathcal{P}_1$ to $\mathcal{P}_2$ is

$$\frac{|\mathbf{N} \cdot \overrightarrow{R_1 R_2}|}{\|\mathbf{N}\|}.$$

For our final distance computation we compute the distance between two skew lines. When two lines $\mathcal{L}_1$ and $\mathcal{L}_2$ are skew, we can find a unique pair of parallel planes such that each of the parallel planes contains one of the skew lines. To visualize this situation, think of a translation that takes one of the lines onto the x-axis. The same translation moves the other line to another line somewhere in the coordinate system that is skew in relation to the x-axis. Now, rotate this new system around the x-axis until the skew line lies in a plane parallel to the xy-plane, say, $z = c$. This rotation leaves the x-axis fixed. The planes $z = 0$ and $z = c$ are the unique parallel planes that contain these transformed lines. Since our translation and rotation did not change the relative orientations of the lines, if you rotate the lines back and then translate the lines back to their original positions, these two operations together will take the planes $z = 0$ and $z = c$ to the unique pair of parallel planes containing the original skew lines. Furthermore, since the distance between the planes $z = 0$ and $z = c$ is $|c|$, that is also the distance between the original skew lines. This *will not* be our method for computing the distance between skew lines, but we hope that it makes the geometry easier to understand.

Our method computes the distance between the unique parallel planes that contain the skew lines. If we had the normal vector to these parallel planes, we could use Theorem 10.41

to compute the distance. The normal vector $\mathbf{N}$ may be computed from the direction vectors $\mathbf{d}_1$ and $\mathbf{d}_2$ for $\mathcal{L}_1$ and $\mathcal{L}_2$. As the following figure shows, the normal vector $\mathbf{N} = \mathbf{d}_1 \times \mathbf{d}_2$:

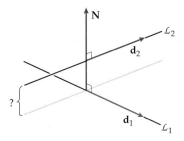

Combining this reasoning with Theorem 10.41, we have

THEOREM 10.42

The Distance Between Skew Lines

Let $\mathcal{L}_1$ and $\mathcal{L}_2$ be nonparallel lines with equations $\mathbf{r}_1(t) = P_1 + t\mathbf{d}_1$ and $\mathbf{r}_2(t) = P_2 + t\mathbf{d}_2$, respectively. Then the distance between $\mathcal{L}_1$ and $\mathcal{L}_2$ is given by

$$\frac{|(\mathbf{d}_1 \times \mathbf{d}_2) \cdot \overrightarrow{P_1P_2}|}{\|\mathbf{d}_1 \times \mathbf{d}_2\|}.$$

Intersecting Planes

We now turn our attention to intersections; two planes may intersect, and a line may intersect a plane. Two distinct planes in 3-space either intersect or are parallel. Given the equations of two planes, we can immediately tell whether or not they intersect by examining the normal vectors of the planes. If the normal vectors are parallel, then the planes are parallel and possibly even identical. The planes are identical if and only if the original equations are equivalent. As the following figure illustrates, when the normal vectors are not parallel, the planes will intersect in a line. To find the equation of the line of intersection, we will need a direction vector for the line and a point on the line. The direction vector can be found with the use of the cross product.

The line of intersection of two planes is orthogonal to each of the normal vectors

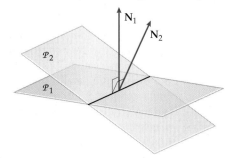

Thus, to find the equation of the line of intersection of two intersecting planes $\mathcal{P}_1$ and $\mathcal{P}_2$, we find their respective normal vectors $\mathbf{N}_1$ and $\mathbf{N}_2$. We also determine a single point on the line of intersection. We use the cross product $\mathbf{N}_1 \times \mathbf{N}_2$ as the direction vector for the line of intersection in order to determine the equation of the line.

Given a line and a plane in 3-space, one of three things must occur: The line and plane can intersect in a unique point, the line and plane can be parallel, or the line may lie *in* the plane. We can determine which of these situations occurs by looking at the dot product of the direction vector for the line and the normal vector for the plane. If $\mathbf{d}$ is the direction vector for the line and $\mathbf{N}$ is the normal vector for the plane, we consider $\mathbf{d} \cdot \mathbf{N}$. If $\mathbf{d} \cdot \mathbf{N} = 0$, the two vectors are orthogonal, so either the line and plane are parallel or the line is in the plane. If $\mathbf{d} \cdot \mathbf{N} \neq 0$, the line and plane intersect in a unique point.

Examples and Explorations

EXAMPLE 1 **Finding the equation of a plane determined by three points**

Show that the three points $P = (2, -1, 5)$, $Q = (9, 3, 7)$, and $R = (4, 7, -5)$ are noncollinear, and find the equation of the plane they determine.

SOLUTION

We form the vectors

$$\vec{PQ} = \langle 9 - 2, 3 - (-1), 7 - 5 \rangle = \langle 7, 4, 2 \rangle \text{ and } \vec{PR} = \langle 4 - 2, 7 - (-1), -5 - 5 \rangle = \langle 2, 8, -10 \rangle$$

and note that they are not parallel, since neither of them is a scalar multiple of the other. This implies that the three points are noncollinear. Next, we compute our normal vector. To find a normal vector we may take the cross product of the two vectors we just found. We let $\mathbf{N} = \vec{PQ} \times \vec{PR} = \langle 7, 4, 2 \rangle \times \langle 2, 8, -10 \rangle = \langle -56, 74, 48 \rangle$. Finally, since P, Q, and R are all in the plane, we are free to choose any of them as our designated point. We will use P. We now have an equation for the plane:

$$-56(x - 2) + 74(y + 1) + 48(z - 5) = 0.$$

Note that the components of $\mathbf{N}$ are all multiples of 2. This is because the components of both vectors are integers and all components of $\vec{PR}$ are multiples of 2. To make our calculation (slightly) simpler, we could have scaled $\vec{PR}$ by $\frac{1}{2}$ before we took the cross product. Scaling either $\vec{PQ}$ or $\vec{PR}$ by a nonzero constant leaves the relative geometry of the vectors unchanged. Certainly, if $\vec{PQ}$ and $\vec{PR}$ are not parallel, then $\vec{PQ}$ and $k(\vec{PR})$ are not parallel for any nonzero scalar k. Had we scaled $\vec{PR}$ by $\frac{1}{2}$, we would have obtained the normal vector

$$\tilde{\mathbf{N}} = \vec{PQ} \times \left(\frac{1}{2} \vec{PR} \right) = \langle -28, 37, 24 \rangle.$$

Using this as our normal vector, we would have gotten

$$-28(x - 2) + 37(y + 1) + 24(z - 5) = 0$$

as the equation for our plane. This equation is equivalent to the one we found before. □

EXAMPLE 2 **Finding the equation of a plane determined by a line and a point not on the line**

Show that the point $Q(3, -4, -1)$ is not on the line, $\mathcal{L}$, determined by

$$\mathbf{r}(t) = \langle 8 - 3t, -2 + 5t, 6 + 2t \rangle.$$

Give a general form for the equation of the plane $\mathcal{P}$ determined by Q and $\mathcal{L}$.

SOLUTION

If P were on $\mathcal{L}$, there would be a value of t that simultaneously satisfied the system of equations

$$8 - 3t = 3, \quad -2 + 5t = -4, \quad 6 + 2t = -1.$$

Since the solutions of each of the three equations are different, Q is not on $\mathcal{L}$.

We quickly note that the vector $\mathbf{d} = \langle -3, 5, 2 \rangle$ is a direction vector for $\mathcal{L}$ and point $Q_0(8, -2, 6)$ is on $\mathcal{L}$. Another vector parallel to the plane of interest is $\overrightarrow{Q_0Q}$. Here,

$$\overrightarrow{Q_0Q} = \langle 3 - 8, -4 - (-2), -1 - 6 \rangle = \langle -5, -2, -7 \rangle.$$

To find a normal vector to the plane, we use the cross product of $\mathbf{d}$ and $\overrightarrow{Q_0Q}$:

$$\mathbf{d} \times \overrightarrow{Q_0Q} = \langle -3, 5, 2 \rangle \times \langle -5, -2, -7 \rangle = \langle -31, -31, 31 \rangle.$$

Since we are interested primarily in the direction of this vector, we may scale it by any nonzero value to obtain simpler coefficients. We will scale the cross product by $-\dfrac{1}{31}$ and let $\mathbf{N} = \langle 1, 1, -1 \rangle$.

We can use either Q or Q_0 as our point on the plane. If we use Q to obtain the equation for the plane, we arrive at the equation

$$1(x - 3) + 1(y + 4) - 1(z + 1) = 0.$$

A general form for the equation of the plane is $x + y - z = 0$. □

EXAMPLE 3 | **Finding the distance from a point to a plane**

Find the distance from the point $P = (6, 3, -2)$ to the plane given by $3x - y + 2z = 4$.

SOLUTION

We need a point on the plane. Verify that $R = (0, 0, 2)$ is on the plane. We can immediately find the vector $\overrightarrow{RP}$ and a normal vector $\mathbf{N}$ to the plane. In this example we have $\overrightarrow{RP} = \langle 6, 3, -4 \rangle$ and $\mathbf{N} = \langle 3, -1, 2 \rangle$. Thus, the distance from P to the plane can be computed as follows:

$$|\text{comp}_{\mathbf{N}} \overrightarrow{RP}| = \frac{|\mathbf{N} \cdot \overrightarrow{RP}|}{\|\mathbf{N}\|} = \frac{|\langle 3, -1, 2 \rangle \cdot \langle 6, 3, -4 \rangle|}{\|\langle 3, -1, 2 \rangle\|} = \frac{|18 - 3 - 8|}{\sqrt{3^2 + 1^2 + 2^2}} = \frac{7}{\sqrt{14}} = \frac{\sqrt{14}}{2}. \quad □$$

EXAMPLE 4 | **Finding the distance between two parallel planes**

Show that the plane $\mathcal{P}_1$ with equation $-4x + 6y + 2z = 1$ and the plane $\mathcal{P}_2$ given by $2x - 3y - z = 5$ are parallel. Then compute the distance between $\mathcal{P}_1$ and $\mathcal{P}_2$.

SOLUTION

The plane $\mathcal{P}_1$ has the normal vector $\mathbf{N}_1 = \langle -4, 6, 2 \rangle$ and the plane $\mathcal{P}_2$ has the normal vector $\mathbf{N}_2 = \langle 2, -3, -1 \rangle$. Notice that the vectors $\mathbf{N}_1$ and $\mathbf{N}_2$ are scalar multiples of each other. This means that the planes $\mathcal{P}_1$ and $\mathcal{P}_2$ are parallel.

Now, the point $R_1 = (0, 0, 1/2)$ is on plane $\mathcal{P}_1$ and the point $R_2 = (0, 0, -5)$ is on plane $\mathcal{P}_2$. (Check these.) We use Theorem 10.41 to compute the distance between the planes:

$$\frac{|\langle 2, -3, -1 \rangle \cdot \overrightarrow{R_1R_2}|}{\|\langle 2, -3, -1 \rangle\|} = \frac{|\langle 2, -3, -1 \rangle \cdot \langle 0, 0, -11/2 \rangle|}{\|\langle 2, -3, -1 \rangle\|} = \frac{11\sqrt{14}}{28}. \quad □$$

EXAMPLE 5

Finding the distance between two skew lines

Find the distance between the skew lines $\mathcal{L}_1$ and $\mathcal{L}_2$, respectively given by the vector functions

$$\mathbf{r}_1(t) = \langle 5 + 3t, -7 + t, 2 - 5t \rangle \text{ and } \mathbf{r}_2(t) = \langle -2t, 7, 1 - 3t \rangle.$$

SOLUTION

Before we begin to compute the distance, we should really think about the relationship of the lines $\mathcal{L}_1$ and $\mathcal{L}_2$. Are they really skew? Might they intersect, be parallel, or even be identical?

Direction vectors for $\mathcal{L}_1$ and $\mathcal{L}_2$ are $\mathbf{d}_1 = \langle 3, 1, -5 \rangle$ and $\mathbf{d}_2 = \langle -2, 0, -3 \rangle$, respectively. Since these direction vectors are not scalar multiples of each other, the lines either intersect or are skew. Now, at this point we could look for a point of intersection. However, let us think about the process we are about to start. We are about to compute the distance between $\mathcal{L}_1$ and $\mathcal{L}_2$. If this distance is zero, the lines intersect. If the distance is positive, the lines are skew.

We first find the normal vector $\mathbf{N}$:

$$\mathbf{N} = \mathbf{d}_1 \times \mathbf{d}_2 = \langle 3, 1, -5 \rangle \times \langle -2, 0, -3 \rangle = \langle -3, 19, 2 \rangle.$$

(Check this.) We will use $P_1 = \mathbf{r}_1(0) = (5, -7, 2)$ and $P_2 = \mathbf{r}_2(0) = (0, 7, 1)$ as our points on $\mathcal{L}_1$ and $\mathcal{L}_2$, respectively. Thus, $\overrightarrow{P_1 P_2} = \langle -5, 14, -1 \rangle$. Finally, we use Theorem 10.42 to find the distance between $\mathcal{L}_1$ and $\mathcal{L}_2$. This distance is

$$\frac{|\mathbf{N} \cdot \overrightarrow{P_1 P_2}|}{\|\mathbf{N}\|} = \frac{|\langle -3, 19, 2 \rangle \cdot \langle -5, 14, -1 \rangle|}{\|\langle -3, 19, 2 \rangle\|} = \frac{279}{374}\sqrt{374}.$$

Since the distance is nonzero, the lines are skew. $\square$

EXAMPLE 6

Finding the line of intersection of two planes

Show that the planes whose equations are $-x + 2y - 4z = 6$ and $3x + 5y - z = 4$ intersect, and find parametric equations for the line of intersection.

SOLUTION

The normal vector to $-x + 2y - 4z = 6$ is $\mathbf{N}_1 = \langle -1, 2, -4 \rangle$, while the normal vector to $3x + 5y - z = 4$ is $\mathbf{N}_2 = \langle 3, 5, -1 \rangle$. These normal vectors are not parallel; therefore the planes must intersect. The line of intersection must be orthogonal to each of the normal vectors. To find the direction vector, $\mathbf{d}$, for the line of intersection, we use the cross product. Here,

$$\mathbf{d} = \mathbf{N}_1 \times \mathbf{N}_2 = \langle -1, 2, -4 \rangle \times \langle 3, 5, -1 \rangle = \langle 18, -13, -11 \rangle.$$

To find the equation for the line of intersection, we now need only a point on the line, which will, of course, lie on each of the planes. Unless the line is parallel to one of the coordinate planes, it will intersect all three coordinate planes. In this example, our line will intersect all three coordinate planes, because all three components of $\mathbf{d}$ are nonzero. To simplify our search for a point of intersection, we can decide to find such a point in one of the coordinate planes. Here, we will look for the point of intersection in the xy-plane. Thus, we will set $z = 0$ in each of the equations $-x + 2y - 4z = 6$ and $3x + 5y - z = 4$, to obtain the system

$$-x + 2y = 6 \text{ and } 3x + 5y = 4,$$

which has the solution $x = -2$ and $y = 2$. Thus, the point $(-2, 2, 0)$ is in each of the planes and therefore on our line of intersection. Parametric equations for the line of intersection are

$$x(t) = -2 + 18t, \ y(t) = 2 - 13t, \ z(t) = -11t, \ -\infty < t < \infty.$$

EXAMPLE 7   **Understanding the geometry of a line and a plane**

Let $\mathcal{L}$ be the line given by $\mathbf{r}(t) = \langle 4 + t, -3 + 5t, 2 - 3t \rangle$ and $\mathcal{P}$ be the plane with equation $-x + 5y + 6z = 5$. Show that $\mathcal{L}$ and $\mathcal{P}$ intersect and find the point of intersection.

SOLUTION

The vector $\mathbf{d} = \langle 1, 5, -3 \rangle$ is a direction vector for $\mathcal{L}$, while the vector $\mathbf{N} = \langle -1, 5, 6 \rangle$ is normal to $\mathcal{P}$. We compute the dot product:

$$\mathbf{d} \cdot \mathbf{N} = \langle 1, 5, -3 \rangle \cdot \langle -1, 5, 6 \rangle = -1 + 25 - 18 = 6 \neq 0.$$

Since the dot product is *not* zero, the line and plane intersect. Because $\mathcal{L}$ and $\mathcal{P}$ intersect, there will be a value of the parameter $t = t_0$ such that $\mathbf{r}(t_0)$ is on the plane. That is, the coordinates $(x(t_0), y(t_0), z(t_0))$ given by

$$x(t_0) = 4 + t_0, \ y(t_0) = -3 + 5t_0, \ z(t_0) = 2 - 3t_0$$

satisfy the equation of the plane. Since $-x + 5y + 6z = 5$, we have

$$-x(t_0) + 5y(t_0) + 6z(t_0) = 5.$$

Equivalently,

$$-(4 + t_0) + 5(-3 + 5t_0) + 6(2 - 3t_0) = 5.$$

Solving this equation for t_0, we obtain $t_0 = 2$. We now have the point of intersection,

$$(x(2), y(2), z(2)) = (6, 7, -4).$$

TEST YOUR UNDERSTANDING

▶ How do you use a point on a plane and a vector normal to the plane to find the equation of the plane?

▶ How do you use other geometric conditions to find the equation of a plane? For example, how do you find the equation of a plane containing three noncollinear points or a plane containing two intersecting lines?

▶ How do you find the distance from a point to a plane? How do you find the distance between two parallel planes? How do you find the distance between two skew lines?

▶ How do you determine when two planes intersect? How do you find the line of intersection when they do?

▶ How do you determine when a line intersects a plane? How do you find the point of intersection when it does?

EXERCISES 10.6

Thinking Back

▶ *Linear equations:* Explain why the equation

$$2x - 3y = 5$$

represents a line in the xy-plane, but represents a plane in a three-dimensional coordinate system.

▶ *Orthogonal vectors:* Show that $\mathbf{u} = \langle 1, 2, -3 \rangle$ is orthogonal to $a\mathbf{v} + b\mathbf{w}$, where $\mathbf{v} = \langle 1, 1, 1 \rangle$, $\mathbf{w} = \langle -1, 2, 1 \rangle$, and a and b are any real numbers. Interpret this statement geometrically.

Concepts

0. *Problem Zero:* Read the section and make your own summary of the material.

1. *True/False:* Determine whether each of the statements that follow is true or false. If a statement is true, explain why. If a statement is false, provide a counterexample.

 (a) *True or False:* The graph of every linear equation in $\mathbb{R}^3$ is a line.

 (b) *True or False:* Three distinct points determine a plane.

 (c) *True or False:* Two distinct lines in $\mathbb{R}^3$ always determine a plane.

 (d) *True or False:* Three concurrent lines determine a plane. (**Concurrent** means that the three lines intersect in a common point.)

 (e) *True or False:* If two distinct lines in $\mathbb{R}^3$ are not skew, they determine a unique plane.

 (f) *True or False:* If two distinct planes in $\mathbb{R}^3$ intersect, they intersect in a unique line.

 (g) *True or False:* If the direction vector of a line is parallel to the normal vector to a plane, the line and plane are parallel.

 (h) *True or False:* Let $\mathcal{P}_1$ and $\mathcal{P}_2$ be distinct planes with normal vectors $\mathbf{N}_1$ and $\mathbf{N}_2$, respectively. If $\mathbf{N}_1 \times \mathbf{N}_2 = \mathbf{0}$, the planes are parallel.

2. *Examples:* Construct examples of the thing(s) described in the following. Try to find examples that are different than any in the reading.

 (a) A plane parallel to $x + 3y - 4z = 7$.

 (b) A line orthogonal to the plane $x + 3y - 4z = 7$.

 (c) A plane orthogonal to the line

$$x = 3t - 5, \ y = -2t + 7, \ z = -4.$$

3. Let $ax + by + cz = d$ be the equation of a plane with a, b, c, and d all nonzero. What are the coordinates of the intersection of the plane and the x-, y-, and z-axes? Explain how to use these points to sketch the plane.

4. Explain why two planes orthogonal to the same vector are either parallel or identical.

5. Let $\mathbf{v} = \langle a, b, c \rangle$ and $\mathbf{w} = \langle \alpha, \beta, \gamma \rangle$, where $\mathbf{v}$ and $\mathbf{w}$ are *not* parallel vectors. Explain why the planes $ax + by + cz = d$ and $\alpha x + \beta y + \gamma z = \delta$ intersect.

6. Explain why there are infinitely many different planes containing any given line in $\mathbb{R}^3$. What form does the equation of a plane containing the x-axis have?

7. What does it mean for three points to be collinear? How do you determine that three given points are collinear? What does it mean for three points to be noncollinear?

8. Explain why three noncollinear points determine a unique plane. Explain how you would use the coordinates of the points to find the equation of the plane. Explain why three collinear points do not determine a unique plane.

9. Explain why a line $\mathcal{L}$ and a point P not on $\mathcal{L}$ determine a unique plane. Explain how you would use the equation of $\mathcal{L}$ and the coordinates P to find the equation of the plane. Explain why P and $\mathcal{L}$ do *not* determine a unique plane if P is on $\mathcal{L}$.

10. Explain why two intersecting lines determine a unique plane. Explain how you would use the equations of the lines to find the equation of the plane.

11. Explain why two distinct parallel lines determine a unique plane. Explain how you would use the equations of the lines to find the equation of the plane.

12. Explain why two skew lines do *not* determine a plane.

13. Explain why any two skew lines lie on a unique pair of parallel planes.

14. The angle θ between two intersecting planes, called the **dihedral angle**, is defined to be the angle between the two normal vectors to the planes, where

$$\theta = \cos^{-1} \frac{\mathbf{N}_1 \cdot \mathbf{N}_2}{\|\mathbf{N}_1\| \|\mathbf{N}_2\|}.$$

Draw a figure that illustrates the dihedral angle and explain why the definition given is a reasonable definition.

15. Given the equations

$$x = at + x_0, \ y = bt + y_0, \ z = ct + z_0$$

for a line $\mathcal{L}$ and

$$\alpha x + \beta y + \gamma z = \delta$$

for a plane $\mathcal{P}$, explain how to determine whether $\mathcal{L}$ is orthogonal to $\mathcal{P}$.

16. Explain how to tell when two planes are perpendicular.

17. When a line $\mathcal{L}$ intersects a plane $\mathcal{P}$ the angle between them is defined to be the complement of the acute angle between the direction vector for the line and the normal vector to the plane. Draw a figure that illustrates this angle, and explain why the definition given is a reasonable definition.

18. Explain the similarities in the derivations of the formulas for the distances from a point to a plane and from a point to a line.

19. Explain the derivation of the formula for finding the distance between two skew lines $\mathcal{L}_1$ and $\mathcal{L}_2$. Why does this formula work?

20. Two distinct nonparallel planes intersect in a line. Outline a procedure for finding the equation of the line of intersection.

Skills

In Exercises 21–30, find the equations of the planes determined by the given conditions.

21. The plane contains the origin and is normal to the vector $\langle 4, -1, 5 \rangle$.

22. The plane contains the point $(-2, 3, -1)$ and is normal to the vector $3\mathbf{i} - 2\mathbf{j}$.

23. The plane contains the point $(2, -1, 6)$ and is normal to the vector $\langle 2, -1, 6 \rangle$.

24. The plane contains the points $(1, 0, 0)$, $(0, 1, 0)$, and $(0, 0, 1)$.

25. The plane contains the points $(2, 4, 3)$, $(3, -5, 0)$, and $(-4, 1, 6)$.

26. The plane contains the points $(-4, 0, 0)$, $(0, 3, 0)$, and $(0, 0, 5)$.

27. The plane contains the points $(x_0, 0, 0)$, $(0, y_0, 0)$, and $(0, 0, z_0)$.

28. The plane contains the point $(1, 2, -5)$ and the line determined by $\mathbf{r}(t) = \langle -4 + t, 3 + 5t, 2 - 3t \rangle$.

29. The plane contains the point $(-4, 1, 3)$ and the line determined by $\mathbf{r}(t) = \langle 3 + 7t, -2 + t, 3 + 4t \rangle$.

30. The plane contains the point $(-4, 1, 3)$ and is normal to the line determined by $\mathbf{r}(t) = \langle -4 + t, 3 + 5t, 2 - 3t \rangle$.

31. Show that the lines determined by

$$\mathbf{r}_1(t) = \langle -2 - 5t, 3 + 2t, 4t \rangle \text{ and}$$

$$\mathbf{r}_2(t) = \langle 8 + 15t, 1 - 6t, 3 - 12t \rangle$$

are parallel, and then find an equation of the plane containing both lines.

32. Show that the lines determined by

$$\mathbf{r}_1(t) = \langle 3 - 5t, -2 + t, 6 \rangle \text{ and}$$

$$\mathbf{r}_2(t) = \langle 4 + 15t, 5 - 3t, 4 \rangle$$

are parallel, and then find an equation of the plane containing both lines.

33. Show that the lines determined by

$$\mathbf{r}_1(t) = \langle 7, 3 - 4t, 2 + 6t \rangle \text{ and}$$

$$\mathbf{r}_2(t) = \langle 6 - t, 3 + 8t, 9 - 5t \rangle$$

intersect, and then find an equation of the plane containing the two lines.

34. Show that the lines determined by

$$\mathbf{r}_1(t) = \langle 3 - t, 4 - 4t, -3 + 4t \rangle \text{ and}$$

$$\mathbf{r}_2(t) = \langle 5 - t, -6 + 2t, -2 + t \rangle$$

intersect, and then find an equation of the plane containing both lines.

In Exercises 35–38, find an equation of the line of intersection of the two given planes.

35. $x + 2y + 3z = 4$ and $-2x + y - 4z = 6$

36. $y - 5z = 3$ and $6x - 7y = 5$

37. $x = 4$ and $3x - 5y + 2z = -3$

38. $x - 2z = 7$ and $x - 3y - 4z = 0$

39. Find the distance from the point $(2, 0, -3)$ to the plane $3x - 4y + 5z = 1$.

40. Show that the planes given by $2x - 4y - 3z = 5$ and $-4x + 8y + 6z = 1$ are parallel, and find the distance between the planes.

41. Show that the planes given by $2x - 3y + 5z = 7$ and $-6x + 9y - 15z = 8$ are parallel, and find the distance between the planes.

42. Show that the planes given by $y - 7z = 16$ and $2y - 14z = 5$ are parallel, and find the distance between the planes.

43. Show that the lines with the equations

$$\frac{x+1}{2} = \frac{y-3}{-4} = \frac{z-2}{5} \text{ and } \frac{x-4}{3} = \frac{y+1}{2} = \frac{z}{3}$$

are skew, find the equations of the parallel planes containing the lines, and find the distance between the lines.

Use your answers from Exercise 14 to find the angle between the indicated planes in Exercises 44 and 45.

44. $7x - 3y + 5z = 6$ and $2x + 3y - z = 1$

45. $-x + 7y - 2z = 5$ and $3x + 5y - 4z = 2$

Use Exercise 17 to find the angle between the indicated lines and planes in Exercises 46 and 47.

46. $\mathbf{r}(t) = \langle 5 + 2t, 6 - t, 4 + 5t \rangle$ and $3y + 5z = -4$

47. $\mathbf{r}(t) = \langle 3 + 5t, 2 - t, 4 - 3t \rangle$ and $-10x + 2y + 6z = 7$

In Exercises 48–51, determine whether the given line is parallel to, intersects, or lies in the given plane. If the line is parallel to the plane, calculate its distance from the plane. If the line intersects the plane, find the point and angle at which they intersect.

48. $\mathbf{r}(t) = \langle t, -5 - 2t, -1 + 3t \rangle$ and $4x - 5y + 2z = 5$

49. $\mathbf{r}(t) = \langle 3 - 2t, -4, 5 - 4t \rangle$ and $2x + 5y - z = 7$

50. $\mathbf{r}(t) = \langle 4 + t, 6 - 5t, -3t \rangle$ and $3x + 3y - 4z = 30$

51. $\mathbf{r}(t) = \langle 5 + 6t, 3 - t, 0 \rangle$ and $7y - 5z = 3$

52. At every point on a sphere $(x-a)^2 + (y-b)^2 + (z-c)^2 = r^2$, there is some plane tangent to the sphere. Explain how to find the equation of the tangent plane at any given point.

53. Use your answer in Exercise 52 to find the equations of the planes tangent to the given spheres at the specified points.

(a) $x^2 + y^2 + z^2 = 1$ at $\left(\frac{1}{2}, \frac{1}{4}, \frac{\sqrt{11}}{4} \right)$.

(b) $(x - 1)^2 + (y + 2)^2 + z^2 = 9$ at $(3, 0, 1)$.

Applications

54. Emmy is trying to get information about the water table below the Hanford reservation. She has drilled wells that show that the water table can be found at $(0, 0, -35)$ and $(300, 0, -38)$. She drills one more well and finds the water table at $(0, 300, -37)$.

(a) Find a plane that approximates the water table.

(b) If she drills another hole at $x = 300$, $y = 300$, how deep does she expect to find the water table?

55. Annie is sitting on a beach in the evening, looking out at a mooring buoy. She wonders how deep the water is at the buoy. She assumes that the beach slopes out as a plane. She is sitting at a point $(150, 30, 5)$ relative to the buoy, where a z-coordinate of zero represents sea level and the coordinates are given in feet. The point on the shore that is closest to the buoy looks as if it is around $(120, 40, 0)$ relative to the buoy. There is a piece of driftwood down the beach that seems to be at about $(140, 60, 5)$. How deep is the water at the buoy?

Proofs

56. Let $\mathcal{L}_1$ and $\mathcal{L}_2$ be two skew lines. Prove that no plane contains both $\mathcal{L}_1$ and $\mathcal{L}_2$.

57. Prove that the planes determined by the equations $ax + by + cz = d$ and $\alpha x + \beta y + \gamma z = \delta$ are perpendicular if and only if $a\alpha + b\beta + c\gamma = 0$.

58. Let $\mathbf{a}$, $\mathbf{b}$, and $\mathbf{c}$ be position vectors terminating in some plane $\mathcal{P}$. Show that $(\mathbf{a} \times \mathbf{b}) + (\mathbf{b} \times \mathbf{c}) + (\mathbf{c} \times \mathbf{a})$ is normal to $\mathcal{P}$.

59. (a) Show that the distance from the point $P(x_0, y_0, z_0)$ to the plane with equation $ax + by + cz + d = 0$ is

$$\frac{|ax_0 + by_0 + cz_0 + d|}{\sqrt{a^2 + b^2 + c^2}}.$$

(b) Use the result of part (a) to recalculate the distance in Exercise 39.

60. Let $\mathbf{r}_1(t) = \mathbf{P}_1 + t\mathbf{d}_1$ and $\mathbf{r}_2(t) = \mathbf{P}_2 + t\mathbf{d}_2$ be parametrizations for two nonparallel lines. Prove that the lines intersect if and only if the three vectors $\mathbf{P}_1 - \mathbf{P}_2$, $\mathbf{d}_1$, and $\mathbf{d}_2$ are coplanar.

61. Let $\mathbf{r}_1(t) = \mathbf{P}_0 + t\mathbf{d}_1$ and $\mathbf{r}_2(u) = \mathbf{Q}_0 + u\mathbf{d}_2$ respectively be the equations of lines $\mathcal{L}_1$ and $\mathcal{L}_2$. Show that $(\mathbf{P}_0 - \mathbf{Q}_0) \cdot (\mathbf{d}_1 \times \mathbf{d}_2) = 0$ if and only if $\mathcal{L}_1$ and $\mathcal{L}_2$ lie in the same plane.

62. Prove Theorem 10.40. That is, show that if $P = (x_0, y_0, z_0)$ is a point on plane $\mathcal{P}$, then the distance between the origin and $\mathcal{P}$ is given by $|\langle x_0, y_0, z_0 \rangle \cdot \mathbf{n}|$, where $\mathbf{n}$ is a unit normal vector to $\mathcal{P}$.

Thinking Forward

▶ *A plane tangent to a surface:* A particular smooth surface has tangent vectors $\mathbf{v}_x = \mathbf{i} - 3\mathbf{j}$ and $\mathbf{v}_y = \mathbf{i} + 4\mathbf{k}$ at the point $P(2, -3, 4)$. Find the equation of the tangent plane to the surface by finding the normal vector to the plane $\mathbf{N} = \mathbf{v}_x \times \mathbf{v}_y$ containing the point P.

▶ *A plane tangent to a surface:* A particular smooth surface has tangent vectors $\mathbf{v}_x = \mathbf{i} + \alpha\mathbf{j}$ and $\mathbf{v}_y = \mathbf{i} + \beta\mathbf{k}$ at the point $P(x_0, y_0, z_0)$. Find the equation of the tangent plane to the surface by finding the normal vector to the plane $\mathbf{N} = \mathbf{v}_x \times \mathbf{v}_y$ containing the point P.

CHAPTER REVIEW, SELF-TEST, AND CAPSTONES

Before you progress to the next chapter, be sure you are familiar with the definitions, concepts, and basic skills outlined here. The capstone exercises at the end bring together ideas from this chapter and look forward to future chapters.

Definitions

Give precise mathematical definitions or descriptions of each of the concepts that follow. Then illustrate the definition with a sketch or an algebraic example.

▶ a *sphere* in $\mathbb{R}^3$

▶ a *cylinder* in $\mathbb{R}^3$, along with the *directrix* and the *rulings* of the cylinder

▶ *vectors* in $\mathbb{R}^2$ and $\mathbb{R}^3$

▶ the *scalar* multiple of a vector

▶ the *magnitude, norm,* or *length* of a vector

▶ the *dot product* of two vectors

▶ the *standard basis vectors* in $\mathbb{R}^2$ and $\mathbb{R}^3$

▶ *orthogonal* curves and vectors

▶ the *projection* of one vector onto another

▶ the *determinant* of a 3×3 matrix

▶ the *cross product* of two vectors from $\mathbb{R}^3$

Theorems

Fill in the blanks to complete each of the following theorem statements:

▶ Given any nonzero vector $\mathbf{v}$, the vector _____ is a unit vector in the direction of $\mathbf{v}$.

▶ For any vectors $\mathbf{u} = \langle u_1, u_2, u_3 \rangle$ and $\mathbf{v} = \langle v_1, v_2, v_3 \rangle$, $\mathbf{u} \cdot \mathbf{v} = $ _____.

▶ If $\mathbf{u}$ and $\mathbf{v}$ are two nonzero vectors, then $\mathbf{u} \cdot \mathbf{v} = $ _____ $\cos \theta$, where θ is _____.

▶ *Law of Cosines*: In a triangle with side lengths a, b, and c, where θ is the angle between the sides of length a and b, $a^2 + b^2 - $ _____ $= c^2$.

▶ Let θ be the angle between nonzero vectors $\mathbf{u}$ and $\mathbf{v}$. Then

$$\theta \text{ is } \begin{cases} \underline{\hspace{1.5cm}} & \text{if and only if } \mathbf{u} \cdot \mathbf{v} > 0 \\ \underline{\hspace{1.5cm}} & \text{if and only if } \mathbf{u} \cdot \mathbf{v} = 0 \\ \underline{\hspace{1.5cm}} & \text{if and only if } \mathbf{u} \cdot \mathbf{v} < 0. \end{cases}$$

▶ *The Triangle Inequality*: Given any vectors **u** and **v**, $\|\mathbf{u} + \mathbf{v}\|$ _____ $\|\mathbf{u}\| + \|\mathbf{v}\|$.

▶ The cross product of two parallel vectors **u** and **v** in $\mathbb{R}^3$ is _____.

▶ For any vectors **u** and **v** in $\mathbb{R}^3$, $\mathbf{v} \times \mathbf{u} =$ ____ $(\mathbf{u} \times \mathbf{v})$.

▶ For any vectors **u** and **v** in $\mathbb{R}^3$ and any scalar c, $c(\mathbf{u} \times \mathbf{v}) =$ _____ $=$ _____.

▶ For vectors **u** and **v** in $\mathbb{R}^3$, $\mathbf{u} \cdot (\mathbf{u} \times \mathbf{v}) =$ _____ and $\mathbf{v} \cdot (\mathbf{u} \times \mathbf{v}) =$ _____.

▶ Let **u** and **v** be nonzero vectors in $\mathbb{R}^3$ with the same initial point. Then $\|\mathbf{u} \times \mathbf{v}\| =$ _____ $\sin\theta$, where θ is _____.

▶ If **u** and **v** are nonparallel vectors in $\mathbb{R}^3$, then **u**, **v**, and $\mathbf{u} \times \mathbf{v}$ form a _____ triple.

▶ If **u**, **v**, and **w** are vectors in $\mathbb{R}^3$, then _____ is the volume of the parallelepiped determined by **u**, **v**, and **w**.

▶ The distance from a point P to a line $\mathcal{L}$ parametrized by $\mathbf{r}(t) = \mathbf{P}_0 + t\mathbf{d}$ is _____.

▶ The distance from a point P to a plane $\mathcal{P}$ containing a point R with normal vector **N** is _____.

Notation and Algebraic Rules for Vectors

Notation: Describe the meanings of each of the following mathematical expressions:

▶ $\mathbb{R}^2$ ▶ $\mathbb{R}^3$ ▶ $\langle a, b \rangle$

▶ $\langle a, b, c \rangle$ ▶ $\overrightarrow{PQ}$ ▶ $\|\mathbf{v}\|$

▶ **i**   ▶ **j**   ▶ **k**

▶ $\mathbf{u} \cdot \mathbf{v}$ ▶ $\mathbf{u} \times \mathbf{v}$ ▶ $\mathrm{comp}_{\mathbf{u}}\mathbf{v}$

▶ $\mathrm{proj}_{\mathbf{u}}\mathbf{v}$

Algebraic Properties of Vector Arithmetic: Each of the statements that follow demonstrates a commutative rule, an associative rule, or a distributive rule of vector arithmetic. Fill in the blanks and give the name of the relevant property.

▶ For any two vectors **u** and **v** with the same number of components, $\mathbf{u} + \mathbf{v} =$ _____.

▶ For any three vectors **u**, **v** and **w**, each with the same number of components, $(\mathbf{u} + \mathbf{v}) + \mathbf{w} =$ _____.

▶ For any scalar c and any two vectors **u** and **v** with the same number of components, $c(\mathbf{u} + \mathbf{v}) =$ _____.

▶ For any vectors **u**, **v**, and **w**, $\mathbf{u} \cdot (\mathbf{v} + \mathbf{w}) =$ _____.

▶ For any vectors **u** and **v**, and any scalar k, $k(\mathbf{u} \cdot \mathbf{v}) =$ _____ $=$ _____.

▶ Let **u**, **v**, and **w** be vectors in $\mathbb{R}^3$. Then $\mathbf{u} \times (\mathbf{v} + \mathbf{w}) =$ _____.

▶ **u**, **v**, and **w** be vectors in $\mathbb{R}^3$. Then $(\mathbf{u}+\mathbf{v}) \times \mathbf{w} =$ _____.

Skill Certification: Working in $\mathbb{R}^3$

Distances between points: Find the distance between each pair of points.

1. $(1, 2, -3)$ and $(4, 7, -3)$ 2. $(-5, 7, 0)$ and $(0, 6, -3)$

3. $(1, 5, -2)$ and $(3, 9, -1)$ 4. $(4, 6, 2)$ and $(1, 3, -5)$

Equations of Spheres: Find the equation of the specified sphere.

5. center $(2, -3, 4)$, radius 6
6. center $(2, -3, 4)$, tangent to the xz-plane
7. the segment with endpoints $(1, 5, -2)$ and $(3, 9, -1)$ is a diameter
8. center $(4, 6, 2)$, $(1, 3, -5)$ is a point on the sphere

Products and Norms: In Exercises 9–24 let $\mathbf{u} = \mathbf{i}$, $\mathbf{v} = 2\mathbf{j}$, and $\mathbf{w} = \mathbf{i} + 2\mathbf{j} + \frac{1}{2}\mathbf{k}$.

9. Sketch the position vectors **u**, **v**, and **w**.
10. Sketch the parallelogram determined by **u** and **v**.
11. Sketch the parallelepiped determined by **u**, **v**, and **w**.
12. Compute $\|\mathbf{u}\|$, and use the result to label your sketches in Exercises 9–11.

13. Compute $\|\mathbf{v}\|$, and use the result to label your sketches in Exercises 9–11.
14. Compute $\|\mathbf{w}\|$, and use the result to label your sketches in Exercises 9–11.
15. Compute $\mathbf{u} \cdot \mathbf{v}$.
16. Compute $\mathbf{u} \times \mathbf{v}$.
17. Compute the angle between **u** and **v**.
18. Compute the area of the parallelogram determined by **u** and **v**.
19. Compute the lengths of the diagonals of the parallelogram determined by **u** and **v**.
20. Find a vector orthogonal to both **u** and **v**.
21. Find a unit vector orthogonal to both **u** and **v**.
22. Compute the volume of the parallelepiped determined by **u**, **v**, and **w**.
23. Compute the areas of the six faces of the parallelepiped determined by **u**, **v**, and **w**.
24. Compute the lengths of the four diagonals of the parallelepiped determined by **u**, **v**, and **w**. (*Hint:* $\mathbf{u} + \mathbf{v} + \mathbf{w}$ is one of the diagonals.)

Products and Norms: Let $\mathbf{u} = \langle 2, 4, -1 \rangle$, $\mathbf{v} = \langle 0, -3, 2 \rangle$, and $\mathbf{w} = \langle -1, 1, 5 \rangle$. Use these vectors to find the specified quantities in Exercises 25–36.

25. $\|\mathbf{u}\|$ **26.** $\|\mathbf{v}\|$

27. $\mathbf{u} \cdot \mathbf{v}$ **28.** $\mathbf{u} \times \mathbf{v}$

29. the angle between $\mathbf{u}$ and $\mathbf{v}$

30. the area of the parallelogram determined by $\mathbf{u}$ and $\mathbf{v}$

31. the lengths of the diagonals of the parallelogram determined by $\mathbf{u}$ and $\mathbf{v}$

32. a vector orthogonal to both $\mathbf{u}$ and $\mathbf{v}$

33. a unit vector orthogonal to both $\mathbf{u}$ and $\mathbf{v}$

34. the volume of the parallelepiped determined by $\mathbf{u}$, $\mathbf{v}$, and $\mathbf{w}$

35. the areas of the six faces of the parallelepiped determined by $\mathbf{u}$, $\mathbf{v}$, and $\mathbf{w}$

36. the lengths of the four diagonals of the parallelepiped determined by $\mathbf{u}$, $\mathbf{v}$, and $\mathbf{w}$

Lines and Planes Determined by Points: Consider the three points $P(2, -3, 5)$, $Q(3, 1, 0)$, and $R(-1, 0, 7)$. Find the following.

37. an equation for the line containing P and Q

38. an equation for the line containing P and R

39. an equation for the plane containing P, Q, and R

40. the area of the triangle determined by P, Q, and R

Intersections and Distances: Let P be the point with coordinates $(1, 2, -3)$, $\mathcal{L}$ be the line with equation $\mathbf{r}(t) = \langle 2 + 5t, -3 + t, -4t \rangle$, and $\mathcal{P}$ be the plane with equation $x - 3y + 4z = 5$. Find the following.

41. the distance from P to $\mathcal{L}$

42. the distance from P to $\mathcal{P}$

43. the point at which $\mathcal{L}$ intersects $\mathcal{P}$

44. the angle at which $\mathcal{L}$ intersects $\mathcal{P}$

Capstone Problems

A. A weight of p pounds is suspended by two ropes as shown in the figure that follows. What are the magnitudes of the forces in each of the ropes?

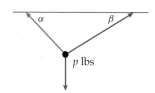

B. Explain why the solid with vertices $(1, 0, 0)$, $(-1, 0, 0)$, $(0, 1, 0)$, $(0, -1, 0)$, $(0, 0, 1)$, and $(0, 0, -1)$ is a regular octahedron.

  (a) Find the area of each face.

  (b) Find the equations of the three lines that form the edges of the face of the octahedron with vertices $(1, 0, 0)$, $(0, 1, 0)$, and $(0, 0, 1)$.

  (c) Find the equation of each face of the octahedron.

  (d) Find the volume of the octahedron.

C. Find the equation of the sphere with center $(1, -3, 5)$ and tangent to the plane with equation $3x - 5y - 2z = 6$. Find the equation of the sphere with center (α, β, γ) and tangent to the plane with equation $ax + by + cz = d$.

D. One molecule of methane is composed of one carbon atom (chemical symbol C) and four hydrogen atoms (chemical symbol H). Therefore the chemical formula for methane is CH_4. Geometrically, if we picture the carbon atom at the center of the molecule, the four hydrogen atoms would lie at the four vertices of a regular tetrahedron. (Recall that a regular tetrahedron is a solid with four congruent faces, each of which is an equilateral triangle.) Each of the hydrogen atoms is bonded to the carbon atom, but the hydrogens are not bonded to each other. Following is a schematic of methane inside a regular tetrahedron:

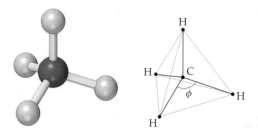

To aid you in visualizing the molecule, it might help to place the carbon atom at the origin of a three-dimensional coordinate system with two of the hydrogen atoms in the xz-plane at $(a, 0, c)$ and $(-a, 0, c)$. Then the other two hydrogen atoms must lie in the yz-plane at $(0, a, -c)$ and $(0, -a, -c)$, as shown here:

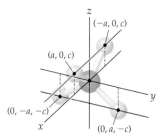

Since the carbon is at the origin, the coordinates of these four points also give the components of the vectors from the carbon atom to the hydrogen atoms.

  (a) In methane the H–C–H bond angles are all equal. Explain why this means that

$$\langle a, 0, c \rangle \cdot \langle -a, 0, c \rangle = \langle a, 0, c \rangle \cdot \langle 0, a, -c \rangle.$$

  (b) What are the H–C–H bond angles in methane?

Vector Functions

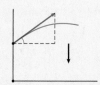

11.1   VECTOR-VALUED FUNCTIONS

▶ Defining curves in $\mathbb{R}^3$ with parametric equations in three variables

▶ Vector functions in $\mathbb{R}^2$ and $\mathbb{R}^3$

▶ Curves defined with parametric equations in $\mathbb{R}^2$ and $\mathbb{R}^3$

Parametric Equations in $\mathbb{R}^3$

For most of this book we have studied functions $f : X \to Y$ whose domains and codomains are subsets of the real numbers. In this chapter we broaden our study to include functions whose codomains are subsets of $\mathbb{R}^2$ or $\mathbb{R}^3$. Such functions are called vector-valued functions or vector functions. The graph of such a function is a curve in the plane, with codomain $\mathbb{R}^2$, or a curve in 3-space, with codomain $\mathbb{R}^3$. We will still use the basic concepts of calculus to analyze these functions, their graphs, and their applications.

Recall that in Chapter 9 we defined parametric equations in $\mathbb{R}^2$ to be a pair of functions

$$x = x(t) \quad \text{and} \quad y = y(t),$$

where t is a *parameter* defined on some interval I of real numbers. At the time we also remarked that the definition could be extended to three-dimensional space, which we do here:

DEFINITION 11.1   Parametric Equations in $\mathbb{R}^3$

Parametric equations in $\mathbb{R}^3$ are triples of functions

$$x = x(t), y = y(t), \text{ and } z = z(t),$$

where the *parameter* t is defined on some interval I of real numbers.

The *parametric curve*, or *space curve*, associated with the equations is the set of points

$$\{(x(t), y(t), z(t)) \in \mathbb{R}^3 \mid t \in I\}.$$

For example, the equations

$$x = \cos t, \ y = \sin t, \ z = t, \ t \in [0, 4\pi]$$

are parametric equations in three variables. (See Example 1.)

Vector-Valued Functions

Another way to express a curve defined by parametric equations, using slightly different notation, is with a vector-valued function. Our definition follows:

DEFINITION 11.2   Vector Functions with Three Components
Let
$$x = x(t), y = y(t), \text{ and } z = z(t)$$
be three real-valued functions, each of which is defined on some interval $I \subseteq \mathbb{R}$. A ***vector function***, or ***vector-valued function***, $\mathbf{r}(t)$, with three components is a function of the form
$$\mathbf{r}(t) = \langle x(t), y(t), z(t) \rangle = x(t)\mathbf{i} + y(t)\mathbf{j} + z(t)\mathbf{k}.$$
The variable t in the vector function is called the ***parameter***. The functions $x = x(t)$, $y = y(t)$, and $z = z(t)$ are called the ***components*** of $\mathbf{r}(t)$. The ***vector curve***, or ***space curve***, associated with the vector function is the set of points
$$\{(x(t), y(t), z(t)) \mid t \in I\}$$
in $\mathbb{R}^3$.

Note that for every value of $t \in I$, the value of the function is a vector. When we consider a vector-valued function that represents the position of a particle, we always interpret the vector as a position vector and the vector curve as the set of terminal points of all the position vectors of the particle. That is, the collection of points
$$(x(t), y(t), z(t)) \text{ for } t \in I,$$
is the vector curve for $\mathbf{r}(t) = \langle x(t), y(t), z(t) \rangle$ since these are the terminal points of the vectors
$$\langle x(t), y(t), z(t) \rangle \text{ for } t \in I.$$
A vector curve is illustrated in the following figure:

The collection of terminal points of the position vectors forms the graph of a vector function

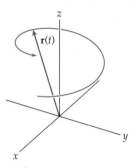

We could similarly define a vector function with two components to be a function of the form $\mathbf{r}(t) = \langle x(t), y(t) \rangle = x(t)\mathbf{i} + y(t)\mathbf{j}$, where $x(t)$ and $y(t)$ are functions of t on some interval $I \subseteq \mathbb{R}$. This is an alternative way of expressing a function defined by two parametric equations.

Note that we will sometimes consider a vector function and a real-valued function of a single variable, $f(t)$, simultaneously. When we do, we will refer to $f(t)$ as a ***scalar function***.

Parametrized Curves

The space curve associated with the vector function $\mathbf{r}(t) = \langle x(t), y(t), z(t) \rangle$ is the set of points $\{(x(t), y(t), z(t)) \mid t \in I \subseteq \mathbb{R}\}$. Along with the curve itself, the parameter, t, imposes directionality on the curve, pointing in the direction in which t is increasing. This directionality is indicated with an arrow.

To understand a space curve, it can be helpful to analyze its projection onto one of the coordinate planes. For example, rather than immediately drawing $\mathbf{r}(t) = \langle x(t), y(t), z(t) \rangle$ in 3-space, we may consider the image of the parametric curve defined by $x = x(t)$ and $y = y(t)$ in the xy-plane. We could similarly consider the projections onto the yz- and xz-planes. If we judiciously select which two components to consider, we will obtain the best information.

For example, consider the function $\mathbf{r}(t) = \langle \cos t, \sin t, t \rangle$ for $t \in [0, 4\pi]$. Temporarily ignoring the z-component of this function, we have the parametric equations $x = \cos t$ and $y = \sin t$. From our work in Chapter 9, we recognize that these equations describe the unit circle centered at the origin. In addition, the curve is traced twice, counterclockwise, starting at the point $(1, 0)$ as t increases on the interval $[0, 4\pi]$, as in the following figure:

The projection of the vector function onto the xy-plane

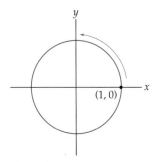

To incorporate the z-coordinate, we note that z increases as t increases. The complete vector curve will spiral around the z-axis as t increases, and we obtain the following circular helix:

To visualize a vector curve, it may help to tabulate the points in 3-space that correspond to a few values of the parameter t. We do this in the examples that follow.

Scalars, scalar functions, and vector-valued functions may be combined algebraically in various ways. For example, the product of the scalar k and the vector function $\mathbf{r}(t) = \langle x(t), y(t), z(t) \rangle$ is

$$k\,\mathbf{r}(t) = k\,\langle x(t), y(t), z(t) \rangle = \langle k\,x(t), k\,y(t), k\,z(t) \rangle.$$

The product of a scalar function $f(t)$ and the vector function $\mathbf{r}(t) = \langle x(t), y(t), z(t) \rangle$ is

$$f(t)\,\mathbf{r}(t) = f(t)\,\langle x(t), y(t), z(t) \rangle = \langle f(t)\,x(t), f(t)\,y(t), f(t)\,z(t) \rangle.$$

The dot product and cross product of the vector functions $\mathbf{r}_1(t) = \langle x_1(t), y_1(t), z_1(t) \rangle$ and $\mathbf{r}_2(t) = \langle x_2(t), y_2(t), z_2(t) \rangle$ are, respectively,

$$\mathbf{r}_1(t) \cdot \mathbf{r}_2(t) = \langle x_1(t), y_1(t), z_1(t) \rangle \cdot \langle x_2(t), y_2(t), z_2(t) \rangle$$
$$= x_1(t) x_2(t) + y_1(t) y_2(t) + z_1(t) z_2(t) \quad \text{and}$$

$$\mathbf{r}_1(t) \times \mathbf{r}_2(t) = \langle x_1(t), y_1(t), z_1(t) \rangle \times \langle x_2(t), y_2(t), z_2(t) \rangle$$
$$= \langle y_1(t) z_2(t) - y_2(t) z_1(t), x_2(t) z_1(t) - x_1(t) z_2(t), x_1(t) y_2(t) - x_2(t) y_1(t) \rangle.$$

Note that the dot product is a scalar function and the cross product is a vector function. The structures of the various products formed with two component vector functions are similar.

Limits and Continuity of Vector Functions

The limit of a vector-valued function may be defined in terms of the limits of the components.

DEFINITION 11.3   The Limit of a Vector Function

Let $\mathbf{r}(t) = \langle x(t), y(t), z(t) \rangle$ be a vector-valued function defined on a punctured interval around t_0. The *limit* of $\mathbf{r}(t)$ as t approaches t_0, denoted by $\lim_{t \to t_0} \mathbf{r}(t)$, is defined by

$$\lim_{t \to t_0} \mathbf{r}(t) = \lim_{t \to t_0} \langle x(t), y(t), z(t) \rangle = \lim_{t \to t_0} x(t)\, \mathbf{i} + \lim_{t \to t_0} y(t)\, \mathbf{j} + \lim_{t \to t_0} z(t)\, \mathbf{k},$$

provided that each of the limits, $\lim_{t \to t_0} x(t)$, $\lim_{t \to t_0} y(t)$, and $\lim_{t \to t_0} z(t)$, exists.

Similarly, the *limit* of the vector function $\mathbf{r}(t) = \langle x(t), y(t) \rangle$ is

$$\lim_{t \to t_0} \mathbf{r}(t) = \lim_{t \to t_0} \langle x(t), y(t) \rangle = \lim_{t \to t_0} x(t)\mathbf{i} + \lim_{t \to t_0} y(t)\mathbf{j},$$

provided that each of the limits, $\lim_{t \to t_0} x(t)$ and $\lim_{t \to t_0} y(t)$, exists.

As in Definition 11.3, our definition of the continuity of a vector-valued function relies on the continuity of the component functions.

DEFINITION 11.4   The Continuity of a Vector Function

Let $\mathbf{r}(t)$ be a vector-valued function defined on an open interval $I \subseteq \mathbb{R}$, and let $t_0 \in I$. The function $\mathbf{r}(t)$ is said to be *continuous* at t_0 if

$$\lim_{t \to t_0} \mathbf{r}(t) = \mathbf{r}(t_0).$$

For a vector function $\mathbf{r}(t) = \langle x(t), y(t), z(t) \rangle$ this definition means that $\mathbf{r}$ is continuous at t_0 if and only if $\lim_{t \to t_0} \mathbf{r}(t) = \langle x(t_0), y(t_0), z(t_0) \rangle$. The analogous statement holds for a vector function with two components. We also define continuity on an interval, using the ideas of Chapter 1.

DEFINITION 11.5   The Continuity of a Vector Function on an Interval

A vector function $\mathbf{r}(t)$ is ***continuous on an interval*** I if it is continuous at every interior point of I, right continuous at any closed left endpoint, and left continuous at any closed right endpoint.

Examples and Explorations

EXAMPLE 1   Graphing a circular helix by plotting points

Graph the circular helix defined by $\mathbf{r}(t) = \langle \cos t, \sin t, t \rangle$ for $t \in [0, 4\pi]$ by plotting points on the curve.

SOLUTION

We've tabulated the coordinates of several points on the curve for values of t in the interval $[0, 4\pi]$:

t	0	$\pi/2$	π	$3\pi/2$	2π
(x, y, z)	$(1, 0, 0)$	$(0, 1, \pi/2)$	$(-1, 0, \pi)$	$(0, -1, 3\pi/2)$	$(1, 0, 2\pi)$

t		$5\pi/2$	3π	$7\pi/2$	4π
(x, y, z)		$(0, 1, 5\pi/2)$	$(-1, 0, 3\pi)$	$(0, -1, 7\pi/2)$	$(1, 0, 4\pi)$

These points and the helix are shown in the following figure:

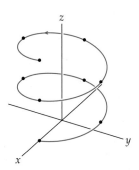

EXAMPLE 2   Graphing a spiral

Graph the vector curve defined by $\mathbf{r}(t) = \langle t \sin t, t \cos t \rangle$ for $t \in [0, 4\pi]$.

SOLUTION

We have already seen that the graph of the parametric equations $x = \sin t$, $y = \cos t$ for $t \in [0, 4\pi]$ is a circle with radius 1 and centered at the origin. The graph of this parametrization starts at $(0, 1)$ when $t = 0$, and the circle is traced twice, in a clockwise direction for

$t \in [0, 4\pi]$. The graph of $\mathbf{r}(t)$ is related. If we square $x = t \sin t$ and $y = t \cos t$ and add, we obtain

$$x^2 + y^2 = t^2 \sin^2 t + t^2 \cos^2 t = t^2.$$

This is not the equation of a circle, but we know that $\sqrt{x^2 + y^2} = t$ is the distance from the origin. That is, the distance from the origin increases with t as the particle revolves around the origin. When $t = 0$, we have $(x, y) = (0, 0)$, and the graph will spiral out clockwise as t increases. To obtain a more precise graph, we will plot a few values of (x, y) corresponding to values of the parameter in the table.

We also evaluate a few reference points for the space curve in the following table:

t	0	$\pi/2$	π	$3\pi/2$	2π
(x, y)	$(0, 0)$	$(\pi/2, 0)$	$(0, -\pi)$	$(-3\pi/2, 0)$	$(0, 2\pi)$

t		$5\pi/2$	3π	$7\pi/2$	4π
(x, y)		$(5\pi/2, 0)$	$(0, -3\pi)$	$(-7\pi/2, 0)$	$(0, 4\pi)$

We plot these points along with the spiral in the following figure:

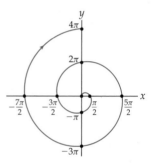

EXAMPLE 3   Graphing a conical helix

Graph the vector curve defined by $\mathbf{r}(t) = \langle t \sin t, t, t \cos t \rangle$ for $t \in [0, 4\pi]$.

SOLUTION

We begin by considering a projection of this curve onto the xz-plane. The graph in the xz-plane will be defined by the parametric equations $x = t \sin t$, $z = t \cos t$. The graph of these equations is the spiral we graphed in Example 2, except in the xz-plane rather than in the xy-plane.

Note that when we include the y-coordinate, we obtain a helix that winds around the y-axis. We evaluate a few reference points for the space curve in the following table:

t	0	π	2π	3π	4π
(x, y, z)	$(0, 0, 0)$	$(0, \pi, -\pi)$	$(0, 2\pi, 2\pi)$	$(0, 3\pi, -3\pi)$	$(0, 4\pi, 4\pi)$

These points and the curve are shown next. This curve is an example of a *conical helix* because the graph can be drawn on the surface of a cone.

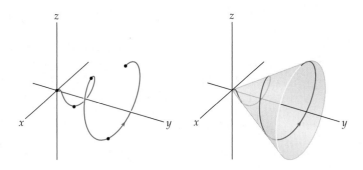

CHECKING THE ANSWER | Elegant graphs such as the ones we have shown may be produced with a computer algebra system like Maple, Mathematica, or Matlab. However, many simple curves may be drawn by hand using the techniques introduced in this section.

EXAMPLE 4

Evaluating the limit of a vector function

Evaluate the limits of the following vector functions if they exist:

$$\lim_{t \to 0} \left\langle t \sin \frac{1}{t}, \frac{t}{e^t - 1} \right\rangle \quad \text{and} \quad \lim_{t \to 2} \left\langle \frac{t - 2}{t^2 - 4}, \frac{2^t - 4}{t - 2}, \frac{t}{t - 2} \right\rangle$$

SOLUTION

The limit of the first vector function exists if and only if the two limits

$$\lim_{t \to 0} t \sin \frac{1}{t} \quad \text{and} \quad \lim_{t \to 0} \frac{t}{e^t - 1}$$

exist. We may use the Squeeze Theorem to show that $\lim_{t \to 0} t \sin \frac{1}{t} = 0$ and L'Hôpital's Rule to show that $\lim_{t \to 0} \frac{t}{e^t - 1} = 1$. Therefore, $\lim_{t \to 0} \left\langle t \sin \frac{1}{t}, \frac{t}{e^t - 1} \right\rangle = \langle 0, 1 \rangle$.

For the other vector function, the limits of the first two components exist, since

$$\lim_{t \to 2} \frac{t - 2}{t^2 - 4} = \lim_{t \to 2} \frac{1}{t + 2} = \frac{1}{4} \quad \text{and} \quad \lim_{t \to 2} \frac{2^t - 4}{t - 2} = \lim_{t \to 2} (\ln 2) 2^t = 4 \ln 2.$$

However, the limit of the third component, $\lim_{t \to 2} \frac{t}{t - 2}$, does not exist. Therefore,

$$\lim_{t \to 2} \left\langle \frac{t - 2}{t^2 - 4}, \frac{2^t - 4}{t - 2}, \frac{t}{t - 2} \right\rangle$$

does not exist.

TEST YOUR UNDERSTANDING

▶ What are parametric equations in two variables? How does the definition generalize to parametric equations in three variables?

▶ What techniques can you use to graph parametric equations in three variables?

▶ What is a vector-valued function? How is the definition related to the definition of parametric equations?

▶ What techniques can you use to graph a vector function in three variables?

▶ If one or more of the components of a vector function is discontinuous, how does that affect the graph of the function?

EXERCISES 11.1

Thinking Back

Parametric equations for the unit circle: Find parametric equations for the unit circle centered at the origin of the xy-plane that satisfy the given conditions.

▶ The graph is traced counterclockwise once on the interval $[0, 2\pi]$ starting at the point $(1, 0)$.

▶ The graph is traced clockwise once on the interval $[0, 2\pi]$ starting at the point $(1, 0)$.

▶ The graph is traced counterclockwise twice on the interval $[0, 2\pi]$ starting at the point $(1, 0)$.

▶ *Parametric equations for a circle:* Find parametric equations whose graph is the circle with radius ρ centered at the point (a, b) in the xy-plane such that the graph is traced counterclockwise $k > 0$ times on the interval $[0, 2\pi]$ starting at the point $(a + \rho, b)$.

Concepts

0. *Problem Zero:* Read the section and make your own summary of the material.

1. *True/False:* Determine whether each of the statements that follow is true or false. If a statement is true, explain why. If a statement is false, provide a counterexample.

   (a) *True or False:* Every curve in the plane has a unique expression in terms of a vector-valued function with two components.

   (b) *True or False:* Every space curve has a unique expression in terms of a vector-valued function with three components.

   (c) *True or False:* The graph of two parametric equations in two variables is also the graph of a vector-valued function with two components.

   (d) *True or False:* Every vector-valued function in three variables can be expressed in terms of three parametric equations.

   (e) *True or False:* The graph of three parametric equations is also the graph of a vector-valued function with three components.

   (f) *True or False:* If $\mathbf{r}(t)$ is a vector-valued function with domain $\mathbb{R}$, $\lim\limits_{t \to c} \mathbf{r}(t)$ exists for every $c \in \mathbb{R}$.

   (g) *True or False:* If a point t_0 is in the domain of a vector function $\mathbf{r}(t)$, then $\lim\limits_{t \to t_0} \mathbf{r}(t) = \mathbf{r}(t_0)$.

   (h) *True or False:* Every vector function is continuous at every point in its domain.

2. *Examples:* Construct examples of the thing(s) described in the following. Try to find examples that are different than any in the reading.

   (a) Parametric equations with three components for a circular helix winding around the x-axis

   (b) A vector-valued function with two components whose graph is a spiral starting at the origin

   (c) A vector-valued function with three components whose graph is a circle contained in the plane 5 units from the yz-plane.

3. Let $y = f(x)$. What is the definition of $\lim\limits_{x \to c} f(x) = L$?

4. Let $\mathbf{r}(t) = \langle x(t), y(t), z(t) \rangle$. What is the definition of $\lim\limits_{t \to c} \mathbf{r}(t)$?

5. Explain why we do *not* need an "epsilon–delta" definition for the limit of a vector-valued function.

6. Let $y = f(x)$. State the definition for the continuity of the function f at a point c in the domain of f.

7. Let $\mathbf{r}(t) = \langle x(t), y(t), z(t) \rangle$. Provide a definition for the continuity of the vector function $\mathbf{r}$ at a point c in the domain of $\mathbf{r}$.

8. Most of the parametric equations and vector-valued functions we have studied have component functions that are continuous. What happens when one of the component functions is discontinuous at a point? For example, the "floor" function $z(t) = \lfloor t \rfloor$ has a jump discontinuity for every integer t. What is the graph of the equations

$$x = \cos 2\pi t, \ y = \sin 2\pi t, z = \lfloor t \rfloor, \ t \in \mathbb{R}?$$

9. Let $\mathbf{r}(t) = \langle x(t), y(t) \rangle$, $t \in [a, b]$, be a vector-valued function, where $a < b$ are real numbers and the functions $x(t)$ and $y(t)$ are continuous. Explain why the graph of $\mathbf{r}$ is contained in some circle centered at the origin. (*Hint: Think about the Extreme Value Theorem.*)

10. Let $\mathbf{r}(t) = \langle x(t), y(t) \rangle$, $t \in [a, \infty)$, be a vector-valued function, where a is a real number. Explain why the graph of $\mathbf{r}$ may or may not be contained in a circle centered at the origin. (*Hint: Graph the functions* $\mathbf{r}_1(t) = \left\langle \frac{1}{t}, \frac{1}{t} \right\rangle$ *and* $\mathbf{r}_2(t) = \langle t, t \rangle$, *both with domain* $[1, \infty)$.)

11. Let $\mathbf{r}(t) = \langle x(t), y(t), z(t) \rangle$, $t \in [a, b]$, be a vector-valued function, where $a < b$ are real numbers and the functions $x(t)$, $y(t)$, and $z(t)$ are continuous. Explain why the graph of $\mathbf{r}$ is contained in some sphere centered at the origin.

12. Let $\mathbf{r}(t) = \langle x(t), y(t), z(t) \rangle$, $t \in [a, \infty)$, be a vector-valued function, where a is a real number. Explain why the graph of $\mathbf{r}$ may or may not be contained in some sphere centered at the origin. (*Hint: Consider the functions* $\mathbf{r}_1(t) = \langle \cos t, \sin t, 1/t \rangle$ *and* $\mathbf{r}_2(t) = \langle \cos t, \sin t, t \rangle$, *both with domain* $[1, \infty)$.)

13. As we saw in Example 1, the graph of the vector-valued function $\mathbf{r}(t) = \langle \cos t, \sin t, t \rangle$, for $t \in [0, 2\pi]$ is a circular helix that spirals counterclockwise around the z-axis and climbs as t increases. Find another parametrization for this helix so that the motion along the helix is faster for a given change in the parameter.

14. As we saw in Example 1, the graph of the vector-valued function $\mathbf{r}(t) = \langle \cos t, \sin t, t \rangle$, for $t \in [0, 2\pi]$ is a circular helix that spirals counterclockwise around the z-axis and climbs as t increases. Find another parametrization for this helix so that the motion is downwards.

15. Let $\mathbf{r}(t) = \langle x(t), y(t) \rangle$, $t \in [a, \infty)$, be a vector-valued function, where a is a real number. Under what conditions would the graph of $\mathbf{r}$ have a horizontal asymptote as $t \to \infty$? Provide an example illustrating your answer.

16. Let $\mathbf{r}(t) = \langle x(t), y(t) \rangle$, $t \in [a, \infty)$, be a vector-valued function, where a is a real number. Under what conditions would the graph of $\mathbf{r}$ have a vertical asymptote as $t \to \infty$? Provide an example illustrating your answer.

17. What is the dot product of the vector functions $\mathbf{r}_1(t) = \langle x_1(t), y_1(t) \rangle$ and $\mathbf{r}_2(t) = \langle x_2(t), y_2(t) \rangle$?

18. Compute the cross product of the vector functions $\mathbf{r}_1(t) = \langle x_1(t), y_1(t) \rangle$ and $\mathbf{r}_2(t) = \langle x_2(t), y_2(t) \rangle$ by thinking of $\mathbb{R}^2$ as the xy-plane in $\mathbb{R}^3$. That is, let $\mathbf{r}_1(t) = \langle x_1(t), y_1(t), 0 \rangle$ and $\mathbf{r}_2(t) = \langle x_2(t), y_2(t), 0 \rangle$, and take the cross product of these vector functions.

In Exercises 19–21 sketch the graph of a vector-valued function $\mathbf{r}(t) = \langle x(t), y(t) \rangle$ with the specified properties. Be sure to indicate the direction of increasing values of t.

19. Domain $t \geq 0$, $\mathbf{r}(0) = \langle 0, 3 \rangle$, $\mathbf{r}(1) = \langle -2, 1 \rangle$, $\lim\limits_{t \to \infty} x(t) = -5$, and $\lim\limits_{t \to \infty} y(t) = -\infty$.

20. Domain $t \geq 0$, $\mathbf{r}(0) = \langle 1, 0 \rangle$, $\mathbf{r}(2) = \langle 0, -1 \rangle$, $\lim\limits_{t \to \infty} x(t) = -2$, and $\lim\limits_{t \to \infty} y(t) = -3$.

21. Domain $t \in \mathbb{R}$, $\mathbf{r}(2) = \langle 1, 1 \rangle$, $\mathbf{r}(0) = \langle -3, 3 \rangle$, $\mathbf{r}(-2) = \langle -5, -5 \rangle$, $\lim\limits_{t \to -\infty} x(t) = \infty$, $\lim\limits_{t \to -\infty} y(t) = -\infty$, and $\lim\limits_{t \to \infty} \mathbf{r}(t) = \langle 0, 0 \rangle$.

22. Given a vector-valued function $\mathbf{r}(t)$ with domain $\mathbb{R}$, what is the relationship between the graph of $\mathbf{r}(t)$ and the graph of $k\mathbf{r}(t)$, where $k > 1$ is a scalar?

23. Given a vector-valued function $\mathbf{r}(t)$ with domain $\mathbb{R}$, what is the relationship between the graph of $\mathbf{r}(t)$ and the graph of $\mathbf{r}(kt)$, where $k > 1$ is a scalar?

24. Explain why the graph of every vector-valued function $\mathbf{r}(t) = \langle \cos t, \sin t, f(t) \rangle$ lies on the surface of the cylinder $x^2 + y^2 = 1$ for every continuous function f.

25. Explain why the graph of every vector-valued function $\mathbf{r}(t) = \langle \cos t, \sin t, \cos t \rangle$ lies on the intersection of the two cylinders $x^2 + y^2 = 1$ and $y^2 + z^2 = 1$.

Skills

Find parametric equations for each of the vector-valued functions in Exercises 26–34, and sketch the graphs of the functions, indicating the direction for increasing values of t.

26. $\mathbf{r}(t) = \langle \sin 3t, \cos 3t \rangle$ for $t \in [0, 2\pi]$

27. $\mathbf{r}(t) = \langle 2 - \sin t, 4 + \cos t \rangle$ for $t \in [0, 2\pi]$

28. $\mathbf{r}(t) = \langle \sin t, \cos 2t \rangle$ for $t \in [0, 2\pi]$

29. $\mathbf{r}(t) = \langle 1 + \sin t, 3 - \cos 2t \rangle$ for $t \in [0, 2\pi]$

30. $\mathbf{r}(t) = (3 + t)\mathbf{i} + \left(3 - \dfrac{1}{t}\right)\mathbf{j}$ for $t > 0$

31. $\mathbf{r}(t) = \langle t, t^2, t^3 \rangle$ for $t \in [0, 2]$

32. $\mathbf{r}(t) = \langle \cos^2 t, 4 \sin t, t \rangle$ for $t \in [0, 2\pi]$

33. $\mathbf{r}(t) = \langle \cos^2 t, \sin 2t \rangle$ for $t \in [0, 2\pi]$

34. $\mathbf{r}(t) = \langle t \sin t, t \cos t, t \rangle$ for $t \in [0, 4\pi]$

Evaluate and simplify the indicated quantities in Exercises 35–41.

35. $\langle 1, 3t, t^3 \rangle + \langle t, t^2, t^3 \rangle$

36. $\langle 1, t, t^2 \rangle - \langle t, t^2, t^3 \rangle$

37. $5 \langle \cos t, \sin t \rangle$

38. $t \langle \sin t, \cos t \rangle$

39. $\langle \sin t, \cos t \rangle \cdot \langle \cos t, -\sin t \rangle$

40. $((\sin t)\mathbf{i} + (\cos t)\mathbf{j} + t\mathbf{k}) \times ((\cos t)\mathbf{i} + (\sin t)\mathbf{j})$

41. $\langle 1, t, t^2 \rangle \times \langle t, t^2, t^3 \rangle$

Evaluate the limits in Exercises 42–45.

42. $\lim\limits_{t \to 0} \langle \sin 3t, \cos 3t \rangle$

43. $\lim\limits_{t \to \pi} \langle \sin t, \cos t, \sec t \rangle$

44. $\lim\limits_{t \to 1^-} \left\langle \ln t, \dfrac{e^t - 1}{t - 1}, e^t \right\rangle$

45. $\lim\limits_{t \to 0^+} \left\langle \dfrac{\sin t}{t}, \dfrac{1 - \cos t}{t}, \left(1 + \dfrac{1}{t}\right)^t \right\rangle$

Find and graph the vector function $\mathbf{r}(t) = \langle x(t), y(t) \rangle$ determined by the differential equations in Exercises 46–48.

46. $x'(t) = x$, $y'(t) = x^2$, $x(0) = 1$, $y(0) = 2$. (*Hint: Start by solving the initial-value problem $x'(t) = x$, $x(0) = 1$.*)

47. $x'(t) = 1 + x^2$, $y'(t) = x^2$, $x(0) = 0$, $y(0) = 1$ (*Hint: Start by solving the initial-value problem $x'(t) = 1 + x^2$, $x(0) = 0$.*)

48. $x'(t) = -y$, $y'(t) = x$, $x(0) = 1$, $y(0) = 0$ (*Hint: What familiar pair of functions have the given properties?*)

49. Show that the graph of the vector function $\mathbf{r}(t) = \langle 3 \sin t, 5 \cos t, 4 \sin t \rangle$ is a circle. (*Hint: Show that the graph lies on a sphere and in a plane.*)

Applications

50. Annie is conscious of tidal currents when she is sea kayaking. This activity can be tricky in an area south-southwest of Cattle Point on San Juan Island in Washington State. Annie is planning a trip through that area and finds that the velocity of the current changes with time and can be expressed by the vector function

$$\left\langle 0.4 \cos\left(\dfrac{\pi(t - 8)}{6}\right), -1.1 \cos\left(\dfrac{\pi(t - 11)}{6}\right) \right\rangle,$$

where t is measured in hours after midnight, speeds are given in knots, and $\langle 0, 1 \rangle$ points due north.

Cattle Point on San Juan Island

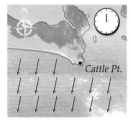 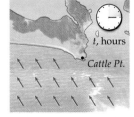

(a) What is the velocity of the current at 8:00 A.M.?

(b) What is the velocity of the current at 11:00 A.M.?

(c) Annie needs to paddle through here heading southeast, 135 degrees from north. She wants the current to push her. What is the best time for her to pass this point? (*Hint: Find the dot product of the given vector function with a vector in the direction of Annie's travel, and determine when the result is maximized.*)

51. Arne is a wingsuit base jumper in Norway. He is working out a jump from a cliff high above a fjord. After a couple of seconds, he will be 1500 meters above the bottom of the fjord, will reach his terminal velocity of 14 meters per second towards the ground, and will travel 30 meters per second horizontally. He calls the time when this happens $t = 0$. Below the cliff from which he jumped, the ground slopes to a second cliff, 300 meters above the water of the fjord. Arne must clear that second cliff, 2000 meters due south from his point at $t = 0$. For $t \geq 0$, his position function is given by

$$\left\langle A \sin\left(\frac{\pi t}{21}\right), \frac{A}{2}t, 1200 - 14t \right\rangle.$$

Approximately how large can A be while still allowing Arne to clear the second cliff?

The cliffs above the fjord

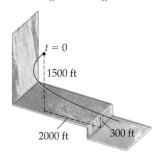

52. The DNA molecule takes the shape of a double helix—two helices that stay a roughly uniform distance apart.

(a) Neglecting actual dimensions, we can model one strand of DNA using the vector function

$$h_1(t) = \langle \cos(t), \sin(t), \alpha t \rangle.$$

Sketch the graph of h_1. What is the effect of the parameter α?

(b) A second strand of DNA can be constructed by shifting the first. Does the graph of

$$h_2(t) = \langle \cos(t), \sin(t), \alpha t + \beta \rangle$$

ever intersect that of h_1?

(c) The distance between two curves is the minimum distance between any two points on the curves. What is the distance between h_1 and h_2 if $\alpha = 1$ and $\beta = \pi$? (*Hint: Write two points on the curves using parameters t_1 and t_2, expand the formula for the distance between them, and then use a trigonometric identity for addition. Then let $s = t_1 - t_2$ and minimize.*)

A DNA molecule

53. Every description of the DNA molecule says that the strands of the helices run in opposite directions. This is meant as a statement about chemistry, not about the geometric shape of the double helix. Consider two helices

$$h_1(t) = \langle \cos t, \sin t, \alpha t \rangle, \text{ and}$$

$$h_2(t) = \langle \sin t, \cos t, \alpha t \rangle.$$

(a) Sketch these two helices in the same coordinate system, and show that they run geometrically in different directions.

(b) Explain why it is impossible for these two helices to fail to intersect, and hence why they could not form a configuration for DNA.

Proofs

54. Let $\mathbf{r}(t) = \langle x(t), y(t) \rangle$ be a vector-valued function defined on an open interval containing the point t_0. Prove that $\mathbf{r}(t)$ is continuous at t_0 if and only if $x(t)$ and $y(t)$ are both continuous at t_0.

55. Prove that the graph of the vector function $\mathbf{r}(t) = \langle t \sin t, t \cos t, t \rangle$, where $t \geq 0$, is a conical helix by showing that it lies on the graph of the cone described by $z = \sqrt{x^2 + y^2}$.

56. Let c_1 and c_2 be scalars, $\mathbf{r}_1(t)$ and $\mathbf{r}_2(t)$ be continuous vector functions with two components, and t_0 be a point in the domains of both $\mathbf{r}_1$ and $\mathbf{r}_2$. Prove that
$$\lim_{t \to t_0} (c_1 \mathbf{r}_1(t) + c_2 \mathbf{r}_2(t)) = c_1 \mathbf{r}_1(t_0) + c_2 \mathbf{r}_2(t_0).$$

57. Let $\mathbf{r}_1(t)$ and $\mathbf{r}_2(t)$ be continuous vector functions with two components, and let t_0 be a point in the domains of both $\mathbf{r}_1$ and $\mathbf{r}_2$. Prove that
$$\lim_{t \to t_0} (\mathbf{r}_1(t) \cdot \mathbf{r}_2(t)) = \mathbf{r}_1(t_0) \cdot \mathbf{r}_2(t_0).$$

58. Let $\mathbf{r}_1(t)$ and $\mathbf{r}_2(t)$ be continuous vector functions with three components, and let t_0 be a point in the domains of both $\mathbf{r}_1$ and $\mathbf{r}_2$. Prove that
$$\lim_{t \to t_0} (\mathbf{r}_1(t) \times \mathbf{r}_2(t)) = \mathbf{r}_1(t_0) \times \mathbf{r}_2(t_0).$$

59. Prove that the dot product of the continuous vector-valued functions $\mathbf{r}_1(t) = \langle x_1(t), y_1(t) \rangle$ and $\mathbf{r}_2(t) = \langle x_2(t), y_2(t) \rangle$ is a continuous scalar function.

60. If α, β, and γ are nonzero constants, the graph of a vector function of the form $\mathbf{r}(t) = \langle \alpha t, \beta t^2, \gamma t^3 \rangle$ is called a *twisted cubic*. Prove that a twisted cubic intersects any plane in at most three points.

61. Let $x_1(t)$, $x_2(t)$, $y_1(t)$, and $y_2(t)$ be differentiable scalar functions. Prove that the dot product of the vector-valued functions $\mathbf{r}_1(t) = \langle x_1(t), y_1(t) \rangle$ and $\mathbf{r}_2(t) = \langle x_2(t), y_2(t) \rangle$ is a differentiable scalar function.

Thinking Forward

▶ *The derivative of a vector function:* Give a definition for the differentiability of a vector-valued function.

▶ *The integral of a vector function:* Give a definition for the definite integral of a vector-valued function.

11.2   THE CALCULUS OF VECTOR FUNCTIONS

▶ Differentiation and integration of vector functions

▶ Velocity and acceleration for vector-valued functions

▶ The geometry of the derivative of a vector-valued function

The Derivative of a Vector Function

In order for the vector function $\mathbf{r}(t) = \langle x(t), y(t), z(t) \rangle$ to be differentiable, we need each of the component functions x, y, and z to be differentiable.

DEFINITION 11.6   **The Derivative of a Vector Function**

Let
$$x = x(t), y = y(t), \text{ and } z = z(t)$$

be three real-valued functions, each of which is differentiable at every point in some interval $I \subseteq \mathbb{R}$. The ***derivative*** of the vector function $\mathbf{r}(t) = \langle x(t), y(t), z(t) \rangle$ is
$$\mathbf{r}'(t) = \langle x'(t), y'(t), z'(t) \rangle = x'(t)\mathbf{i} + y'(t)\mathbf{j} + z'(t)\mathbf{k}.$$

Similarly, the ***derivative*** of the vector function $\mathbf{r}(t) = \langle x(t), y(t) \rangle$ is
$$\mathbf{r}'(t) = \langle x'(t), y'(t) \rangle = x'(t)\mathbf{i} + y'(t)\mathbf{j}.$$

In these cases we say that the function $\mathbf{r}$ is ***differentiable***.

For example, if $\mathbf{r}(t) = \langle \sin t, \cos t, t^2 \rangle$ for $t \in \mathbb{R}$, then $\mathbf{r}'(t) = \langle \cos t, -\sin t, 2t \rangle$.

Recall that in Chapter 2 the derivative of a (scalar) function was defined in terms of a limit. Thus, the derivatives of the component functions, $x'(t)$, $y'(t)$, and $z'(t)$, are defined in terms of a limit, and we have the following theorem, the proof of which is left for Exercise 62:

THEOREM 11.7

The Derivative of a Vector-Valued Function

Let $\mathbf{r}(t)$ be a differentiable vector function with either two or three components. The derivative of $\mathbf{r}(t)$ is given by

$$\mathbf{r}'(t) = \lim_{h \to 0} \frac{\mathbf{r}(t+h) - \mathbf{r}(t)}{h}.$$

When you prove Theorem 11.7, be sure to interpret $\mathbf{r}(t+h)$ correctly. For a vector function in $\mathbb{R}^3$,

$$\mathbf{r}(t+h) = \langle x(t+h), y(t+h), z(t+h) \rangle.$$

In $\mathbb{R}^2$, $\mathbf{r}(t+h)$ has the analogous meaning.

THEOREM 11.8

The Chain Rule for Vector-Valued Functions

Let $t = f(\tau)$ be a differentiable real-valued function of τ, and let $\mathbf{r}(t)$ be a differentiable vector function with either two or three components such that $f(\tau)$ is in the domain of $\mathbf{r}$ for every value of τ on some interval I. Then

$$\frac{d\mathbf{r}}{d\tau} = \frac{d\mathbf{r}}{dt}\frac{dt}{d\tau}.$$

Proof. We will prove the Chain Rule for the case where $\mathbf{r}(t) = \langle x(t), y(t) \rangle$ and leave the case where $\mathbf{r}(t)$ has three components to Exercise 67. In either case, the hard work was actually done in Chapter 2 when we proved the Chain Rule for the composition of two scalar functions. By the definition of the derivative of a vector-valued function,

$$\frac{d}{d\tau}(\mathbf{r}(t)) = \frac{d}{d\tau}(\langle x(t), y(t) \rangle) \qquad \leftarrow \text{the definition of } \mathbf{r}$$

$$= \left\langle \frac{d}{d\tau}(x(t)), \frac{d}{d\tau}(y(t)) \right\rangle \qquad \leftarrow \text{the definition of the derivative of a vector function}$$

$$= \left\langle \frac{dx}{dt}\frac{dt}{d\tau}, \frac{dy}{dt}\frac{dt}{d\tau} \right\rangle \qquad \leftarrow \text{the Chain Rule for scalar functions}$$

$$= \left\langle \frac{dx}{dt}, \frac{dy}{dt} \right\rangle \frac{dt}{d\tau} \qquad \leftarrow \text{multiplication of a vector function by a scalar function}$$

$$= \frac{d\mathbf{r}}{dt}\frac{dt}{d\tau} \qquad \leftarrow \text{the definition of the derivative of a vector function}$$

Returning to the example we used earlier, let $\mathbf{r}(t) = \langle \sin t, \cos t, t^2 \rangle$ for $t \in \mathbb{R}$, and let $t = e^{3\tau}$. Then

$$\frac{d\mathbf{r}}{d\tau} = \langle \cos t, -\sin t, 2t \rangle (3e^{3\tau}) = \langle \cos e^{3\tau}, -\sin e^{3\tau}, 2e^{3\tau} \rangle (3e^{3\tau})$$

$$= \langle 3e^{3\tau}\cos e^{3\tau}, -3e^{3\tau}\sin e^{3\tau}, 6e^{6\tau} \rangle.$$

The Geometry of the Derivative

We use Theorem 11.7 to interpret the geometry of the derivative. Recall from Section 10.2 that the difference $\mathbf{v} - \mathbf{w}$ of two vectors with the same initial point may be interpreted as the vector that extends from the terminal point of $\mathbf{w}$ to the terminal point of $\mathbf{v}$. Thus, for position vectors $\mathbf{r}(t+h)$ and $\mathbf{r}(t)$, the numerator of the quotient $\frac{\mathbf{r}(t+h) - \mathbf{r}(t)}{h}$ is the vector that points from the terminal point of $\mathbf{r}(t)$ to the terminal point of $\mathbf{r}(t+h)$, as shown here at the left:

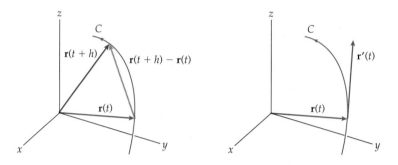

If the scalar, h, in the denominator of the quotient is positive, the direction of $\frac{\mathbf{r}(t+h) - \mathbf{r}(t)}{h}$ is the same as the direction of $\mathbf{r}(t+h) - \mathbf{r}(t)$, although their magnitudes may differ. (A similar argument may be made when $h < 0$.) As with the derivative of a function of a single variable, the derivative is tangent to the curve at $\mathbf{r}(t)$. As shown in the graph on the right, we will always assume that the initial point of the derivative $\mathbf{r}'(t)$ is positioned at the terminal point of the vector $\mathbf{r}(t)$. This ensures that the derivative is tangent to the curve C at $\mathbf{r}(t)$.

We may also use the derivative of the vector function to construct a tangent line to the vector curve.

DEFINITION 11.9   **The Tangent Line to a Vector Curve**

Let $\mathbf{r}(t) = \langle x(t), y(t), z(t) \rangle$ be a differentiable vector function on some interval $I \subseteq \mathbb{R}$, and let t_0 be a point in I such that $\mathbf{r}'(t_0) \neq \mathbf{0}$. The **tangent line** to the vector curve defined by $\mathbf{r}(t)$ at $\mathbf{r}(t_0)$ is given by

$$\mathcal{L}(t) = \mathbf{r}(t_0) + t\,\mathbf{r}'(t_0).$$

From Definition 11.9 we see that the tangent line is the line containing the point $\mathbf{r}(t_0)$ whose direction vector is $\mathbf{r}'(t_0)$.

Velocity and Acceleration

Recall that when t represents time and $y = s(t)$ represents the position function of a particle moving along a straight path, $s'(t)$ and $s''(t)$ represent the velocity and acceleration of the particle, respectively. We define the velocity and acceleration of a particle moving along a space curve determined by a vector function in an analogous fashion.

DEFINITION 11.10   Velocity, Speed, and Acceleration along a Space Curve

Let $\mathbf{r}(t) = \langle x(t), y(t), z(t) \rangle$ be a differentiable vector-valued function defined at every point in some time interval $I \subseteq \mathbb{R}$, and let C be the space curve defined by $\mathbf{r}(t)$.

(a) The *velocity*, **v**, of the particle as it moves along C is given by

$$\mathbf{v}(t) = \mathbf{r}'(t) = \langle x'(t), y'(t), z'(t) \rangle.$$

(b) The *speed* of the particle as it moves along C is given by

$$\|\mathbf{v}(t)\| = \|\langle x'(t), y'(t), z'(t) \rangle\|.$$

(c) In addition, if $\mathbf{r}(t)$ is twice differentiable, the *acceleration*, **a**, of the particle as it moves along C is given by

$$\mathbf{a}(t) = \mathbf{v}'(t) = \mathbf{r}''(t) = \langle x''(t), y''(t), z''(t) \rangle.$$

Note that velocity and acceleration are vectors, but speed is a scalar, since it is the norm of the velocity.

Derivatives of Vector Products

Recall that there are three products that apply to vectors: Vectors may be multiplied by scalars, two vectors with the same number of components may be multiplied by means of the dot product, and two three-component vectors may be multiplied via the cross product. The same products apply to vector functions. Fortunately, the derivatives of these product functions follow the rules we would predict.

THEOREM 11.11   Derivatives of Products of Vector Functions

Let k be a scalar, $f(t)$ be a differentiable scalar function, and $\mathbf{r}(t)$ be a differentiable vector function. Then

(a) $\dfrac{d}{dt}(k\mathbf{r}(t)) = k\mathbf{r}'(t)$.

(b) $\dfrac{d}{dt}(f(t)\mathbf{r}(t)) = f'(t)\mathbf{r}(t) + f(t)\mathbf{r}'(t)$.

Furthermore, if $\mathbf{r}_1(t)$ and $\mathbf{r}_2(t)$ are differentiable vector functions with both having either two or three components, then

(c) $\dfrac{d}{dt}(\mathbf{r}_1(t) \cdot \mathbf{r}_2(t)) = \mathbf{r}_1'(t) \cdot \mathbf{r}_2(t) + \mathbf{r}_1(t) \cdot \mathbf{r}_2'(t)$.

Finally, if $\mathbf{r}_1(t)$ and $\mathbf{r}_2(t)$ are both differentiable three-component vector functions, then

(d) $\dfrac{d}{dt}(\mathbf{r}_1(t) \times \mathbf{r}_2(t)) = \mathbf{r}_1'(t) \times \mathbf{r}_2(t) + \mathbf{r}_1(t) \times \mathbf{r}_2'(t)$.

Recall that the dot product is commutative, so the order of the products in Theorem 11.11 (c) is not significant, but since the cross product is not commutative, the order of the products in Theorem 11.11 (d) is significant.

The proof of each part of Theorem 11.11 follows directly from the definitions of the derivative and the relevant product. We will prove Theorem 11.11(c) when $\mathbf{r}_1(t)$ and $\mathbf{r}_2(t)$

are differentiable vector functions with two components. The proofs of all other parts of Theorem 11.11 are left for Exercises 63–66.

Proof. Let $\mathbf{r}_1(t) = \langle x_1(t), y_1(t) \rangle$ and $\mathbf{r}_2(t) = \langle x_2(t), y_2(t) \rangle$. Then

$$\mathbf{r}_1(t) \cdot \mathbf{r}_2(t) = \langle x_1(t), y_1(t) \rangle \cdot \langle x_2(t), y_2(t) \rangle = x_1(t) x_2(t) + y_1(t) y_2(t).$$

Thus, when we take the derivative, we have

$$\begin{aligned}
\frac{d}{dt}(\mathbf{r}_1(t) \cdot \mathbf{r}_2(t)) &= \frac{d}{dt}(x_1(t) x_2(t) + y_1(t) y_2(t)) \\
&= x_1'(t) x_2(t) + x_1(t) x_2'(t) + y_1'(t) y_2(t) + y_1(t) y_2'(t) \\
&= (x_1'(t) x_2(t) + y_1'(t) y_2(t)) + (x_1(t) x_2'(t) + y_1(t) y_2'(t)) \\
&= \langle x_1'(t), y_1'(t) \rangle \cdot \langle x_2(t), y_2(t) \rangle + \langle x_1(t), y_1(t) \rangle \cdot \langle x_2'(t), y_2'(t) \rangle \\
&= \mathbf{r}_1'(t) \cdot \mathbf{r}_2(t) + \mathbf{r}_1(t) \cdot \mathbf{r}_2'(t).
\end{aligned}$$
∎

When the magnitude of a vector function is constant, the function and its derivative are orthogonal, as stated in the next theorem. This result is not terribly surprising: In $\mathbb{R}^2$, any vector function with constant magnitude k has a graph that is at least a portion of a circle with radius k centered at the origin, and every tangent to a circle is orthogonal to the radius at the point of tangency. The three-dimensional case is similar: Every vector function with constant magnitude k has a graph that lies on the boundary of a sphere of radius k and that is centered at the origin. The following figures illustrate the two- and three-dimensional cases (left and right, respectively).

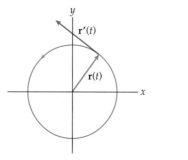 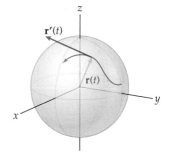

THEOREM 11.12   **A Vector Function with a Constant Magnitude Is Orthogonal to Its Derivative**

Let $\mathbf{r}(t)$ be a differentiable vector function such that $\|\mathbf{r}(t)\| = k$ for some constant k. Then

$$\mathbf{r}(t) \cdot \mathbf{r}'(t) = 0.$$

Proof. Since the magnitude of $\mathbf{r}(t)$ is constant, so is its square. That is,

$$\|\mathbf{r}(t)\|^2 = k^2,$$

where k^2 is a constant. However, recall that

$$\|\mathbf{r}(t)\|^2 = \mathbf{r}(t) \cdot \mathbf{r}(t).$$

So we have

$$\mathbf{r}(t) \cdot \mathbf{r}(t) = k^2.$$

We now take the derivative of each side of the latter equation, using Theorem 11.11 (c) to take the derivative of the left side, and obtain

$$\mathbf{r}(t) \cdot \mathbf{r}'(t) + \mathbf{r}'(t) \cdot \mathbf{r}(t) = 2\mathbf{r}(t) \cdot \mathbf{r}'(t) = 0.$$

Since their dot product is zero, the vectors $\mathbf{r}(t)$ and $\mathbf{r}'(t)$ are orthogonal. ∎

Integration of a Vector Function

Recall from Chapter 4 that every function that is continuous on an interval $[a, b]$ is integrable on $[a, b]$. We use this property in the following definition:

DEFINITION 11.13

The Integral of a Vector Function

Let $x = x(t)$, $y = y(t)$, and $z = z(t)$ be three real-valued functions with antiderivatives

$$\int x(t)\, dt, \quad \int y(t)\, dt, \quad \text{and} \quad \int z(t)\, dt,$$

respectively, on some interval $I \subseteq \mathbb{R}$. Then the vector function

$$\int \mathbf{r}(t)\, dt = \int (x(t)\mathbf{i} + y(t)\mathbf{j} + z(t)\mathbf{k})\, dt = \mathbf{i} \int x(t)\, dt + \mathbf{j} \int y(t)\, dt + \mathbf{k} \int z(t)\, dt$$

is an ***antiderivative*** of the vector function $\mathbf{r}(t) = \langle x(t), y(t), z(t) \rangle$.

Similarly, if $x(t)$, $y(t)$, and $z(t)$ are all integrable on the interval $[a, b]$, then the ***definite integral*** of the vector function $\mathbf{r}(t)$ from a to b is the vector

$$\int_a^b \mathbf{r}(t)\, dt = \int_a^b (x(t)\mathbf{i} + y(t)\mathbf{j} + z(t)\mathbf{k})\, dt = \mathbf{i} \int_a^b x(t)\, dt + \mathbf{j} \int_a^b y(t)\, dt + \mathbf{k} \int_a^b z(t)\, dt.$$

Note that every antiderivative of a vector function is another vector function and any two antiderivatives of the same function differ by a constant (vector). However, the definite integral of a vector function is unique and is a constant (vector). In order for a vector function $\mathbf{r}(t) = \langle x(t), y(t), z(t) \rangle$ to be integrable, each of its component functions must be integrable. This condition tells us that we could also find $\int_a^b \mathbf{r}(t)\, dt$ as the limit of a Riemann sum if necessary. We explore antiderivatives and definite integrals in the examples and exercises.

Examples and Explorations

EXAMPLE 1

The derivative of a vector function

Find the derivative of the vector function $\mathbf{r}(t) = \langle \cos t, \sin t, t \rangle$, and find the tangent lines to the curve defined by $\mathbf{r}$ at $t = \frac{\pi}{4}$ and $t = \pi$.

SOLUTION

To find the derivative of $\mathbf{r}(t)$, we take the derivative of each component function to obtain $\mathbf{r}'(t) = \langle -\sin t, \cos t, 1 \rangle$.

We now find the tangent line when $t = \frac{\pi}{4}$. Since

$$\mathbf{r}\left(\frac{\pi}{4}\right) = \left\langle \frac{\sqrt{2}}{2}, \frac{\sqrt{2}}{2}, \frac{\pi}{4} \right\rangle \quad \text{and} \quad \mathbf{r}'\left(\frac{\pi}{4}\right) = \left\langle -\frac{\sqrt{2}}{2}, \frac{\sqrt{2}}{2}, 1 \right\rangle,$$

the equation of the tangent line at $t = \frac{\pi}{4}$ is

$$\mathcal{L}_1(\tau) = \left\langle \frac{\sqrt{2}}{2}, \frac{\sqrt{2}}{2}, \frac{\pi}{4} \right\rangle + \tau \left\langle -\frac{\sqrt{2}}{2}, \frac{\sqrt{2}}{2}, 1 \right\rangle.$$

Because we may use any nonzero multiple of the direction vector as the direction vector for our line, we may also express the line as

$$\mathcal{L}_1(\tau) = \left\langle \frac{\sqrt{2}}{2}, \frac{\sqrt{2}}{2}, \frac{\pi}{4} \right\rangle + \tau \langle -1, 1, \sqrt{2} \rangle.$$

Similarly, when $t = \pi$, the tangent line will pass through the terminal point of the position vector $\mathbf{r}(\pi) = \langle -1, 0, \pi \rangle$ and will have the direction vector $\mathbf{r}'(\pi) = \langle 0, -1, 1 \rangle$. The equation of this line is

$$\mathcal{L}_2(\tau) = \langle -1, 0, \pi \rangle + \tau \langle 0, -1, 1 \rangle. \qquad \square$$

EXAMPLE 2

Using derivatives to understand the graph of a vector function

Graph the vector function $\mathbf{r}(t) = t^2 \mathbf{i} + (3t - 2t^3) \mathbf{j}$.

SOLUTION

Let $\mathcal{C}$ represent the graph of the function. We begin by noting that $\mathcal{C}$ will lie entirely in the first and fourth quadrants, since $x = t^2$. We also note that because $x = t^2$ is an even function and $y = 3t - 2t^3$ is an odd function, $\mathcal{C}$ will be symmetric with respect to the x-axis. To understand the complete graph we need only analyze the vector function for positive values of t. We first note that the only time $\mathcal{C}$ will intersect the y-axis is when $x = 0$, but $\mathcal{C}$ will intersect the x-axis for the three values of t when $3t - 2t^3 = 0$, namely, $t = 0$ and $t = \pm \sqrt{\frac{3}{2}}$.

We take the derivative and obtain

$$\mathbf{r}'(t) = 2t \mathbf{i} + (3 - 6t^2) \mathbf{j}.$$

We can find where the curve $\mathcal{C}$ has horizontal tangents by finding the values where $\frac{dy}{dx} = 0$. By the chain rule we know that $\frac{dy}{dx} = \frac{dy/dt}{dx/dt}$. Here we have

$$\frac{dy}{dx} = \frac{dy/dt}{dx/dt} = \frac{3 - 6t^2}{2t}.$$

We see that the slope of the tangent lines to $\mathcal{C}$ will be zero when $t = \pm \frac{1}{\sqrt{2}}$ and will be undefined when $t = 0$. For $t > 0$, we note that $\frac{dx}{dt}$ is positive, $\frac{dy}{dt} > 0$ when $t \in (0, 1/\sqrt{2})$, and $\frac{dy}{dt} < 0$ when $t > \frac{1}{\sqrt{2}}$. Therefore, the graph will have a relative maximum when $t = \frac{1}{\sqrt{2}}$. We also note that $y > 0$ for $t \in (0, \sqrt{3/2})$ and $y < 0$ for $t > \sqrt{3/2}$. Incorporating this information into a graph, we obtain the figure that follows on the left. We then use symmetry to obtain the complete graph on the right.

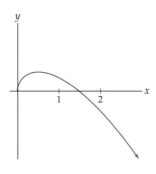
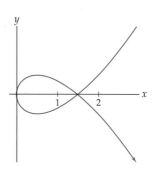

EXAMPLE 3   The integral of a vector function

Find the antiderivative of the vector function $\mathbf{r}(t) = \langle \cos t, \sin t, t \rangle$, and compute the definite integral of $\mathbf{r}$ from 0 to π.

SOLUTION

We integrate

$$\int \mathbf{r}(t)\, dt = \int \langle \cos t, \sin t, t \rangle\, dt = \mathbf{i} \int \cos t\, dt + \mathbf{j} \int \sin t\, dt + \mathbf{k} \int t\, dt = \left\langle \sin t, -\cos t, \frac{t^2}{2} \right\rangle + \mathbf{C},$$

where $\mathbf{C}$ is a vector constant and has the form $\mathbf{C} = \langle c_1, c_2, c_3 \rangle$ for scalar constants c_1, c_2, and c_3.

We use the Fundamental Theorem of Calculus to evaluate the definite integral

$$\int_0^\pi \mathbf{r}(t)\, dt.$$

We already have an antiderivative $\left\langle \sin t, -\cos t, \frac{t^2}{2} \right\rangle$ for $\mathbf{r}(t)$. Thus, to compute the definite integral, we evaluate this antiderivative at 0 and π and take the difference. That is,

$$\int_0^\pi \mathbf{r}(t)\, dt = \left\langle \sin t, -\cos t, \frac{t^2}{2} \right\rangle \Bigg|_0^\pi = \left\langle 0, 1, \frac{\pi^2}{2} \right\rangle - \langle 0, -1, 0 \rangle = \left\langle 0, 2, \frac{\pi^2}{2} \right\rangle.$$

Example 3 was a rather mechanical and artificial use of the integral. In a slightly more satisfying example, we are given a formula for a tangent vector and are asked to find a vector function that contains a particular point.

EXAMPLE 4   Finding a position vector given its velocity vector

A particle is moving along a space curve. The velocity vector for the curve is given by $\mathbf{v}(t) = \langle \sin t, t^2, t^3 \rangle$. When $t = 0$, the position of the particle is $(3, -2, 5)$. Find the position function for the curve.

SOLUTION

We begin by integrating $\mathbf{v}(t) = \mathbf{r}'(t) = \langle \sin t, t^2, t^3 \rangle$:

$$\mathbf{r}(t) = \int \mathbf{v}(t)\, dt = \int \langle \sin t, t^2, t^3 \rangle\, dt = (-\cos t + c_1)\, \mathbf{i} + \left(\frac{t^3}{3} + c_2 \right) \mathbf{j} + \left(\frac{t^4}{4} + c_3 \right) \mathbf{k}.$$

We now use the initial position $\mathbf{r}(0) = 3\mathbf{i} - 2\mathbf{j} + 5\mathbf{k} = (-\cos 0 + c_1)\, \mathbf{i} + \left(\frac{0^3}{3} + c_2 \right) \mathbf{j} + \left(\frac{0^4}{4} + c_3 \right) \mathbf{k}$. Solving the system

$$3 = -1 + c_1, \quad -2 = 0 + c_2, \quad 5 = 0 + c_3,$$

we obtain $c_1 = 4$, $c_2 = -2$, and $c_3 = 5$. Thus,

$$\mathbf{r}(t) = (-\cos t + 4)\,\mathbf{i} + \left(\frac{t^3}{3} - 2\right)\mathbf{j} + \left(\frac{t^4}{4} + 5\right)\mathbf{k}.$$

□

CHECKING
THE ANSWER

We may check our answer to Exercise 4 in two steps. First we take the derivative of $\mathbf{r}(t) = (-\cos t + 4)\,\mathbf{i} + \left(\frac{t^3}{3} - 2\right)\mathbf{j} + \left(\frac{t^4}{4} + 5\right)\mathbf{k}$ and obtain

$$\mathbf{r}'(t) = (\sin t)\mathbf{i} + t^2\mathbf{j} + t^3\mathbf{k},$$

giving us the correct tangent vector function.

Next, we evaluate

$$\mathbf{r}(0) = (-\cos 0 + 4)\,\mathbf{i} + \left(\frac{0^3}{3} - 2\right)\mathbf{j} + \left(\frac{0^4}{4} + 5\right)\mathbf{k} = \langle 3, -2, 5\rangle.$$

Thus, we have the correct solution.

EXAMPLE 5   Finding the angle between two intersecting vector functions

Show that the conical helices defined by the vector functions $\mathbf{r}_1(t) = \langle t\cos t, t\sin t, t\rangle$ and $\mathbf{r}_2(t) = \langle t\sin t, t, t\cos t\rangle$ intersect at the origin, and then find the angle of intersection between the curves.

SOLUTION

Since both $\mathbf{r}_1(0) = (0, 0, 0)$ and $\mathbf{r}_2(0) = (0, 0, 0)$, the two helices intersect at the origin. We now find the tangent vectors to both graphs. The derivatives are

$$\mathbf{r}_1'(t) = \langle \cos t - t\sin t, \sin t + t\cos t, 1\rangle \quad \text{and} \quad \mathbf{r}_2'(t) = \langle \sin t + t\cos t, 1, \cos t - t\sin t\rangle.$$

The tangent vectors when $t = 0$ are $\mathbf{r}_1'(0) = \langle 1, 0, 1\rangle$ and $\mathbf{r}_2'(0) = \langle 0, 1, 1\rangle$. Recall that the cosine of the angle θ between two vectors $\mathbf{u}$ and $\mathbf{v}$ is given by $\cos\theta = \dfrac{\mathbf{u}\cdot\mathbf{v}}{\|\mathbf{u}\|\|\mathbf{v}\|}$. Thus, the cosine of the angle between the helices at the origin is

$$\cos\theta = \frac{\langle 1, 0, 1\rangle \cdot \langle 0, 1, 1\rangle}{\|\langle 1, 0, 1\rangle\|\|\langle 0, 1, 1\rangle\|} = \frac{1}{\sqrt{2}\sqrt{2}} = \frac{1}{2}.$$

Therefore, the (acute) angle of intersection between the two curves is $60°$.

□

TEST YOUR
UNDERSTANDING

▸ We defined the derivative of a vector-valued function in terms of the derivatives of its component functions. Why do we bother stating and proving Theorem 11.7, which says that we can find a derivative by using the limit of a difference quotient?

▸ Given a vector-valued function $\mathbf{r}$ and its graph C, what geometric relationship does the derivative function have to C?

▸ Under what conditions are a vector-valued function and its derivative orthogonal?

▸ For a vector-valued function $\mathbf{r}(t)$, when would the velocity vector $\mathbf{v}(t)$ and acceleration vector $\mathbf{a}(t)$ be orthogonal?

▸ Given an acceleration vector $\mathbf{a}(t)$, how would you find the position vector $\mathbf{r}(t)$?

EXERCISES 11.2

Thinking Back

Differentiability: Use the definition of the derivative to find the derivatives of the following functions.

▶ $f(x) = x^2$

▶ $f(x) = x^3 - 1$

▶ $f(x) = \sqrt{x+1}$

Integrability: Use the definition of the definite integral to evaluate the following integral.

▶ $\displaystyle\int_1^4 (x+3)\,dx$

▶ $\displaystyle\int_0^2 x^2\,dx$

▶ $\displaystyle\int_1^3 (x^2 - 3)\,dx$

Concepts

0. *Problem Zero:* Read the section and make your own summary of the material.

1. *True/False:* Determine whether each of the statements that follow is true or false. If a statement is true, explain why. If a statement is false, provide a counterexample.

   (a) *True or False:* If $\mathbf{r}(t)$ is a vector-valued function with domain $\mathbb{R}$, then $\lim_{h \to 0} \mathbf{r}(t+h)$ exists.

   (b) *True or False:* If $\mathbf{r}(t)$ is a vector-valued function with domain $\mathbb{R}$, then $\lim_{h \to 0} \dfrac{\mathbf{r}(t+h) - \mathbf{r}(t)}{h}$ exists.

   (c) *True or False:* Given a differentiable vector-valued function $\mathbf{r}(t)$, if $\|\mathbf{r}(t)\|$ is constant, then $\mathbf{r}(t) \cdot \mathbf{r}'(t) = 0$.

   (d) *True or False:* Given a differentiable vector-valued function $\mathbf{r}(t)$, if $\mathbf{r}(t) \cdot \mathbf{r}'(t) = 0$, then $\|\mathbf{r}(t)\|$ is constant.

   (e) *True or False:* Given a twice-differentiable vector-valued function $\mathbf{r}(t)$, if $\|\mathbf{r}(t)\|$ is constant, then $\mathbf{r}'(t) \cdot \mathbf{r}''(t) = 0$.

   (f) *True or False:* Given a differentiable vector-valued function $\mathbf{r}(t)$, if $\|\mathbf{r}(t)\|$ is constant, then $\mathbf{r}(t) \times \mathbf{r}'(t) \neq 0$.

   (g) *True or False:* Given a twice-differentiable vector-valued function, the velocity and acceleration vectors are never equal.

   (h) *True or False:* If $\mathbf{r}(t)$ is a vector-valued function defined on the interval $[a, b]$, then $\int_a^b \mathbf{r}(t)\,dt$ exists.

2. *Examples:* Construct examples of the thing(s) described in the following. Try to find examples that are different than any in the reading.

   (a) A differentiable vector function.

   (b) A differentiable vector function $\mathbf{r}(t) = \langle x(t), y(t), z(t) \rangle$ in which none of the component functions is a constant but $\|\mathbf{r}(t)\|$ is constant.

   (c) A twice-differentiable vector function $\mathbf{r}(t) = \langle x(t), y(t), z(t) \rangle$ in which none of the component functions is a constant but $\mathbf{r}(t) = \mathbf{r}'(t)$ for every t.

3. State what it means for a scalar function $y = f(x)$ to be differentiable at a point c.

4. State what it means for a vector function $\mathbf{r}(t) = \langle x(t), y(t), z(t) \rangle$ to be differentiable.

5. Explain why we do not need to explicitly use a limit in the definition for the derivative of a vector function $\mathbf{r}(t) = \langle x(t), y(t), z(t) \rangle$.

6. Given a differentiable vector-valued function $\mathbf{r}(t)$, explain why $\mathbf{r}'(t_0)$ is tangent to the curve defined by $\mathbf{r}(t)$ when the initial point of $\mathbf{r}'(t_0)$ is placed at the terminal point of $\mathbf{r}(t_0)$.

7. Give an example of a vector-valued function $\mathbf{r}(t)$ that is *not* differentiable at at least one point in its domain. Explain why your example is not differentiable at that point. Graph the function, and discuss the problem there is with constructing a tangent vector at the point of nondifferentiability.

8. In our discussion of the geometry of the derivative we graphed a differentiable vector function $\mathbf{r}(t)$ and the difference quotient $\dfrac{\mathbf{r}(t+h) - \mathbf{r}(t)}{h}$ for positive values of h. Draw the corresponding picture for negative values of h, and explain why $\lim_{h \to 0^-} \dfrac{\mathbf{r}(t+h) - \mathbf{r}(t)}{h}$ will result in the same tangent vector.

9. State what it means for a scalar function $y = f(x)$ to be integrable on an interval $[a, b]$.

10. State what it means for a vector function $\mathbf{r}(t) = \langle x(t), y(t), z(t) \rangle$ to be integrable.

11. Explain why we do not need to explicitly use a limit in the definition for the definite integral of a vector function $\mathbf{r}(t) = \langle x(t), y(t), z(t) \rangle$.

12. This exercise has to do with the integrability of vector functions.

   (a) Explain why every vector function $\mathbf{r}(t) = \langle x(t)y(t), z(t) \rangle$ defined on $[a, b]$ is integrable if each of the component functions $x(t)$, $y(t)$, and $z(t)$ is continuous.

   (b) When a vector function $\mathbf{r}(t)$ defined on $[a, b]$ is integrable, $\mathbf{r}(t) = \int_a^b \mathbf{r}(t)\,dt$ may be difficult to evaluate. Explain why.

13. Complete the following definition: If $\mathbf{r}(t) = \langle x(t), y(t), z(t) \rangle$ is a differentiable position function, then the **velocity** vector $\mathbf{v}(t)$ is . . . .

14. Complete the following definition: If $\mathbf{r}(t) = \langle x(t), y(t), z(t) \rangle$ is a twice-differentiable position function, then the **acceleration** vector $\mathbf{a}(t)$ is . . . .

15. Let $t = f(\tau)$ be a differentiable real-valued function of τ, and let $\mathbf{r}(t)$ be a differentiable vector function with either two or three components such that $f(\tau)$ is in the domain of **r** for every value of τ on some interval I. Find $\frac{d}{d\tau}(\mathbf{r}(f(\tau)))$.

16. Find the derivative of the vector-valued function $\mathbf{r}(t) = t^2\mathbf{i} - 6t^2\mathbf{j}$, $t > 0$, and show that the derivative at any point t_0 is a scalar multiple of the derivative at any other point. What does this property say about the graph of **r**? Use that information to sketch the graph of **r**.

17. Find the derivative of the vector-valued function $\mathbf{r}(t) = \langle t^3, 5t^3, -2t^3 \rangle$, $t > 0$, and show that the derivative at any point t_0 is a scalar multiple of the derivative at any other

point. What does this property say about the graph of **r**? Sketch the graph of **r**.

18. The graph of every vector-valued function $\mathbf{r}(t)$ is a curve in the xy-plane. If **r** is twice differentiable, explain how information about $\mathbf{r}'(t)$ and $\mathbf{r}''(t)$ can be used to graph **r**.

19. What is the relationship between the graph of a differentiable vector function $\mathbf{r}(t)$ and the graph of $\int \mathbf{r}'(t)\, dt$, one of the antiderivatives of $\mathbf{r}'(t)$?

20. Given a differentiable vector function $\mathbf{r}(t)$ defined on $[a, b]$, explain why the integral $\int_a^b \|\mathbf{r}'(t)\|\, dt$ would be a scalar, not a vector.

Skills

In Exercises 21–23 you are given a vector function **r** and a scalar function $t = f(\tau)$. Compute $\frac{d\mathbf{r}}{d\tau}$ in the following two ways:

(a) By using the chain rule $\frac{d\mathbf{r}}{d\tau} = \frac{d\mathbf{r}}{dt}\frac{dt}{d\tau}$.

(b) By substituting $t = f(\tau)$ into the formula for **r**.

Ensure that your two answers are consistent.

21. $\mathbf{r}(t) = \langle t, t^2, t^3 \rangle$, $t = \sin \tau$

22. $\mathbf{r}(t) = \langle \cos t, \sin t, t \rangle$, $t = \tau^3$

23. $\mathbf{r}(t) = \langle t^2 \sin t^2, t, t^2 \cos t^2 \rangle$, $t = \sqrt{\tau}$

In Exercises 24–29 a vector function and a point on the graph of the function are given. Find an equation for the line tangent to the curve at the specified point, and then find an equation for the plane orthogonal to the tangent line containing the given point.

24. $\mathbf{r}(t) = \langle t, t^2, t^3 \rangle$, $(2, 4, 8)$

25. $\mathbf{r}(t) = te^t\mathbf{i} + t \ln t\,\mathbf{j}$, $(e, 0, 0)$

26. $\mathbf{r}(t) = \langle \sec t, \frac{1}{t+1}, e^t \ln(t+1) \rangle$, $(1, 1, 0)$

27. $\mathbf{r}(t) = \langle \sin t, \cos t, 2 \sin 2t \rangle$, $(1, 0, 0)$

28. $\mathbf{r}(t) = \langle t, t, t^{3/2} \rangle$, $(4, 4, 8)$

29. $\mathbf{r}(t) = \cos 3t\,\mathbf{i} + \sin 4t\,\mathbf{j} + t\,\mathbf{k}$, $\left(0, 0, \frac{\pi}{2}\right)$

Find the velocity and acceleration vectors for the position vectors given in Exercises 30–34.

30. $\mathbf{r}(t) = \langle t, t^2, t^3 \rangle$

31. $\mathbf{r}(t) = te^t\mathbf{i} + t \ln t\,\mathbf{j}$

32. $\mathbf{r}(t) = \langle \sec t, \frac{1}{t}, e^t \ln t \rangle$

33. $\mathbf{r}(t) = \langle \sin t, \cos t, 2 \sin 2t \rangle$

34. $\mathbf{r}(t) = \cos 3t\,\mathbf{i} + \sin 4t\,\mathbf{j} + t\,\mathbf{k}$

In Exercises 35–39 a vector function $\mathbf{r}(t)$ and scalar function $t = f(\tau)$ are given. Find $\frac{d\mathbf{r}}{d\tau}$.

35. $\mathbf{r}(t) = \langle t, t^2, t^3 \rangle$, $t = \tau^3 + 1$

36. $\mathbf{r}(t) = te^t\mathbf{i} + t \ln t\,\mathbf{j}$, $t = e^\tau$

37. $\mathbf{r}(t) = \langle \sec t, 1/t, e^t \ln t \rangle$, $t = \tau^{-1}$

38. $\mathbf{r}(t) = \langle \sin t, \cos t, 2 \sin 2t \rangle$, $t = \sqrt{\tau^2 + 1}$

39. $\mathbf{r}(t) = \cos 3t\,\mathbf{i} + \sin 4t\,\mathbf{j} + t\,\mathbf{k}$, $t = 5\tau - 2$

Evaluate the integrals in Exercises 40–44.

40. $\int \langle t, t^2, t^3 \rangle\, dt$

41. $\int \langle \sin t, \cos t, \tan t \rangle\, dt$

42. $\int_1^2 \langle te^t, \ln t, \tan^{-1} t \rangle\, dt$

43. $\int_0^{2\pi} \langle \sin t, \cos t, t \rangle\, dt$

44. $\int_0^{2\pi} \|\langle \sin t, \cos t, t \rangle\|\, dt$

Use the given velocity vectors $\mathbf{v}(t) = \mathbf{r}'(t)$ and initial positions in Exercises 45–48 to find the position function $\mathbf{r}(t)$.

45. $\mathbf{v}(t) = \langle t, t^2 \rangle$, $\mathbf{r}(0) = \langle 3, -4 \rangle$

46. $\mathbf{v}(t) = \langle 0, \sec t \tan t, \sec^2 t \rangle$, $\mathbf{r}(\pi) = \langle -2, 4, 3 \rangle$

47. $\mathbf{v}(t) = e^t\mathbf{i} + \ln t\,\mathbf{j}$, $\mathbf{r}(1) = \mathbf{i} - 6\mathbf{j}$

48. $\mathbf{v}(t) = \cos t\,\mathbf{i} + \sin t\,\mathbf{j} + t\,\mathbf{k}$, $\mathbf{r}(0) = -\mathbf{i} + 5\mathbf{j}$

Use the given acceleration vectors $\mathbf{a}(t) = \mathbf{v}'(t) = \mathbf{r}''(t)$ and initial conditions in Exercises 49–52 to find the position function $\mathbf{r}(t)$.

49. $\mathbf{a}(t) = \langle 2t, 3t^2 \rangle$, $\mathbf{v}(0) = \langle 1, -2 \rangle$, $\mathbf{r}(0) = \langle 2, -3 \rangle$

50. $\mathbf{a}(t) = \langle 1, \sec t \tan t, \sec^2 t \rangle$, $\mathbf{v}(\pi) = \langle 0, 3, -1 \rangle$, $\mathbf{r}(\pi) = \langle 4, 5, -2 \rangle$

51. $\mathbf{a}(t) = -32\mathbf{j}$, $\mathbf{v}(0) = 5\mathbf{i} + 5\mathbf{j}$, $\mathbf{r}(0) = 26\mathbf{j}$

52. $\mathbf{a}(t) = \langle \sin 3t, t^2, \cos 3t \rangle$, $\mathbf{v}(0) = \langle 0, 0, 0 \rangle$, $\mathbf{r}(0) = \langle 2, -5, 3 \rangle$

53. Find all points of intersection between the vector function $\mathbf{r}(t) = \langle t, t^2, t^3 \rangle$ and the plane defined by $3x - 3y + z = 1$.

Find all points of intersection between the graphs of the vector functions in Exercises 54–56, and find the acute angle of intersection of the curves at those points.

54. $\mathbf{r}_1(t) = \langle t, t^2 \rangle$ and $\mathbf{r}_2(t) = \langle t^2, t \rangle$

55. $\mathbf{r}_1(t) = \langle 3\cos t, 3\sin t \rangle$ and $\mathbf{r}_2(t) = \langle 2 + 2\sin t, 2\cos t \rangle$

56. $\mathbf{r}_1(t) = \langle \cos t, \sin t, t \rangle$ and $\mathbf{r}_2(t) = \langle \sin t, \cos t, t \rangle$

57. A certain vector function has the properties that its graph is a space curve passing through the point $(0, 1, 2)$ and that its tangent vector at every point on the curve is equal to the position vector. Find the position function and sketch its graph.

Applications

58. Annie is making a crossing from south to north between two islands. The distance between the islands is 2 miles. A current in the channel pushes her off her line and slows her progress towards the opposite island. Her position t hours after leaving the southern island is

$$(0.255t^3 - 0.479t^2, 2t - 0.958t^2 + 0.511t^3).$$

Space coordinates are measured in miles from her starting position; time is measured in hours.

(a) What is Annie's actual velocity vector at any time t?

(b) Where is Annie's speed a minimum? What is her speed at that point?

59. Arne is a wingsuit base jumper in Norway. He wants to jump off a cliff above a fjord, 900 meters atop the valley floor. He knows that when he jumps, his downward acceleration due to gravity will be constant while there will be an upward acceleration depending on the air catching his wingsuit. He will convert about half of that upward acceleration into forward motion, which will cause drag. We ignore that, since he is still basically falling. Thus, his acceleration will be approximately

$$\langle -0.5d y'(t), -9.8 - d y'(t) \rangle,$$

where $y(t)$ denotes his vertical distance (in meters) from the jump point. The constant $d \approx 0.36$ is a parameter describing air resistance. Note that after he jumps, y will always be negative.

(a) What is Arne's velocity at time t in terms of his position relative to the start?

(b) Observe that the second coordinate of his velocity is his vertical speed, which is actually $y'(t)$. Use this fact to write an expression for $y(t)$ in terms of $y'(t)$.

(c) Arne's horizontal speed away from his jump point is the first coordinate of his velocity; call it $x'(t)$. Write an expression for $x'(t)$ in terms of $y'(t)$.

(d) Use parts (b) and (c) and integrate one more time to find a vector in terms of $y(t)$ describing Arne's position at any time t.

(e) The cliff Arne jumps from is vertical for the first 100 meters down, but it quickly slopes out below that. To survive the jump, Arne must clear a point 200 meters below him and 80 meters horizontally away. He quickly makes a calculation showing that it will take him more than 7.4 seconds to fall 200 meters. Can he make it?

Proofs

60. Prove that the tangent vector is always orthogonal to the position vector for the vector-valued function $\mathbf{r}(t) = \langle \sin t, \sin t \cos t, \cos^2 t \rangle$.

61. Let $\mathbf{r}(t)$ be a vector-valued function whose graph is a curve C, and let $\mathbf{a}(t)$ be the acceleration vector. Prove that if $\mathbf{a}(t)$ is always zero, then C is a straight line.

62. Prove Theorem 11.7 for vectors in $\mathbb{R}^2$. That is, let $x(t)$ and $y(t)$ be two scalar functions, each of which is differentiable on an interval $I \subseteq \mathbb{R}$, and let $\mathbf{r}(t) = \langle x(t), y(t) \rangle$ be a vector function. Prove that

$$\mathbf{r}'(t) = \lim_{h \to 0} \frac{\mathbf{r}(t+h) - \mathbf{r}(t)}{h}.$$

63. Let k be a scalar and $\mathbf{r}(t)$ be a differentiable vector function. Prove that $\frac{d}{dt}(k\mathbf{r}(t)) = k\mathbf{r}'(t)$. (This is Theorem 11.11 (a).)

64. Let $f(t)$ be a differentiable scalar function and $\mathbf{r}(t)$ be a differentiable vector function. Prove that

$$\frac{d}{dt}(f(t)\mathbf{r}(t)) = f'(t)\mathbf{r}(t) + f(t)\mathbf{r}'(t).$$

(This is Theorem 11.11 (b).)

65. Let $\mathbf{r}_1(t)$ and $\mathbf{r}_2(t)$ be differentiable vector functions with three components each. Prove that

$$\frac{d}{dt}(\mathbf{r}_1(t) \cdot \mathbf{r}_2(t)) = \mathbf{r}_1'(t) \cdot \mathbf{r}_2(t) + \mathbf{r}_1(t) \cdot \mathbf{r}_2'(t).$$

(This is Theorem 11.11 (c).)

66. Let $\mathbf{r}_1(t)$ and $\mathbf{r}_2(t)$ both be differentiable three-component vector functions. Prove that

$$\frac{d}{dt}(\mathbf{r}_1(t) \times \mathbf{r}_2(t)) = \mathbf{r}_1'(t) \times \mathbf{r}_2(t) + \mathbf{r}_1(t) \times \mathbf{r}_2'(t).$$

(This is Theorem 11.11 (d).)

67. Let $t = f(\tau)$ be a differentiable real-valued function of τ, and let $\mathbf{r}(t)$ be a differentiable vector function with three components such that $f(\tau)$ is in the domain of $\mathbf{r}$ for every value of τ on some interval I. Prove that $\frac{d\mathbf{r}}{d\tau} = \frac{d\mathbf{r}}{dt}\frac{dt}{d\tau}$. (This is Theorem 11.8.)

68. Let $\mathbf{r}(t)$ be a differentiable vector function. Prove that $\frac{d}{dt}(\|\mathbf{r}\|) = \frac{1}{\|\mathbf{r}\|}\mathbf{r} \cdot \mathbf{r}'$. (*Hint:* $\|\mathbf{r}\|^2 = \mathbf{r} \cdot \mathbf{r}$.)

69. Let $\mathbf{r}(t)$ be a differentiable vector function such that $\mathbf{r}(t) \cdot \mathbf{r}'(t) = 0$ for every value of t. Prove that $\|\mathbf{r}(t)\|$ is a constant.

70. For constants α, β, and γ, the graph of a vector-valued function of the form

$$\mathbf{r}(t) = \alpha e^{\gamma t}(\cos t)\mathbf{i} + \alpha e^{\gamma t}(\sin t)\mathbf{j} + \beta e^{\gamma t}\mathbf{k}, \ t \geq 0$$

is called a **_concho-spiral_**. Prove that the angle between the tangent vector to a concho-spiral and the vector $\mathbf{k}$ is constant.

71. Prove that if a particle moves along a curve at a constant speed, then the velocity and acceleration vectors are orthogonal.

Thinking Forward

▶ *Unit Tangent Vectors in $\mathbb{R}^2$:* Given the vector function $\mathbf{r}(t) = \langle t^2, t^3 \rangle$, find a vector of length 1 tangent to the curve at the point $\mathbf{r}(1)$. How many vectors in $\mathbb{R}^2$ are there of length 1 that are orthogonal to this tangent vector?

▶ *Unit Tangent Vectors in $\mathbb{R}^3$:* Given the vector function $\mathbf{r}(t) = \langle t, t^2, t^3 \rangle$, find a vector of length 1 tangent to the curve at the point $\mathbf{r}(1)$. How many vectors in $\mathbb{R}^3$ are there of length 1 that are orthogonal to this tangent vector?

▶ *Orthogonal Vectors:* For any constants a, b, and c, show that the velocity and acceleration vectors for

$$\mathbf{r}(t) = \langle a \sin bt, a \cos bt, ct \rangle$$

are orthogonal.

▶ *Horizontal Vectors:* For any constants a, b, and c, show that the acceleration vector for

$$\mathbf{r}(t) = \langle a \sin bt, a \cos bt, ct \rangle$$

is parallel to the xy-plane.

11.3   UNIT TANGENT AND UNIT NORMAL VECTORS

▶ The unit tangent and principal unit normal vectors to a space curve

▶ The osculating plane, the plane of "best fit" to a space curve

▶ The binormal vector to a space curve

Unit Tangent and Unit Normal Vectors

Our goal for the remainder of this chapter is to understand the graphs of vector functions and the motion of particles whose trajectories can be specified by a vector function. No matter what parametrization, $\mathbf{r}(t)$, we use for a curve C, the derivative vector $\mathbf{r}'(t)$ will be tangent to C at the terminal point of $\mathbf{r}(t)$, assuming that $\mathbf{r}(t)$ is differentiable and $\mathbf{r}'(t) \neq \mathbf{0}$. We will be interested primarily in the direction of the tangent vector, not its magnitude, so we make the following definition:

DEFINITION 11.14   The Unit Tangent Vector

Let $\mathbf{r}(t)$ be a differentiable vector function on some interval $I \subseteq \mathbb{R}$ such that $\mathbf{r}'(t) \neq \mathbf{0}$ on I. The **unit tangent function** is defined to be

$$\mathbf{T}(t) = \frac{\mathbf{r}'(t)}{\|\mathbf{r}'(t)\|}.$$

As we do with the tangent vector $\mathbf{r}'(t)$, we will always position the initial point of the unit tangent vector $\mathbf{T}(t)$ at the terminal point of $\mathbf{r}(t)$ to ensure that $\mathbf{T}(t)$ is tangent to the curve C, as shown in the following figure:

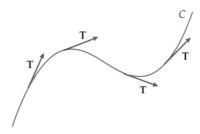

Since the magnitude of the unit tangent vector is constant, Theorem 11.12 tells us that $\mathbf{T}(t)$ and its derivative are always orthogonal. We use this property in the following definition:

DEFINITION 11.15   **The Principal Unit Normal Vector**

Let $\mathbf{r}(t)$ be a differentiable vector function on some interval $I \subseteq \mathbb{R}$ such that the derivative of the unit tangent vector $\mathbf{T}'(t) \neq \mathbf{0}$ on I. The *principal unit normal vector* at $\mathbf{r}(t)$ is defined to be

$$\mathbf{N}(t) = \frac{\mathbf{T}'(t)}{\|\mathbf{T}'(t)\|}.$$

Definition 11.15 ensures that the principal unit normal vector is a unit vector, that it is orthogonal to the tangent vectors, and that it points in the direction in which C is bending. The following figure illustrates these properties:

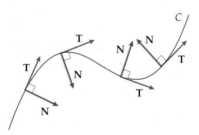

The Osculating Plane and the Binormal Vector

We do not expect curves, even curves in the plane, to contain straight segments. For example, the parabola defined by the equation $y = x^2$ does not contain any linear pieces. However, we have mentioned that the tangent line is the best linear approximation to a differentiable function at the point of tangency. Analogously, we do not expect portions of space curves to be planar. That is, there may not be a plane that contains even a small segment of a space curve, but we may use the unit tangent and principal unit normal vectors to define a plane, the *osculating plane*, in which the curve fits "best."

DEFINITION 11.16   **The Osculating Plane and the Binormal Vector**

Let $\mathbf{r}(t) = \langle x(t), y(t), z(t) \rangle$ be a differentiable vector function on some interval $I \subseteq \mathbb{R}$ such that the derivative of the unit tangent vector $\mathbf{T}'(t_0) \neq 0$ where $t_0 \in I$. The *binormal vector* $\mathbf{B}$ at $\mathbf{r}(t_0)$ is defined to be

$$\mathbf{B}(t_0) = \mathbf{T}(t_0) \times \mathbf{N}(t_0),$$

and the *osculating plane* at $\mathbf{r}(t_0)$ is defined by

$$\mathbf{B}(t_0) \cdot \langle x - x(t_0), y - y(t_0), z - z(t_0) \rangle = 0.$$

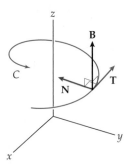

The preceding figure illustrates Definition 11.16. At a point on the curve C the vectors $\mathbf{T}$ and $\mathbf{N}$ determine a unique plane, the osculating plane. (*Osculating* comes from the Latin "osculum" meaning "kiss.") From its definition, the binormal vector $\mathbf{B}$ is orthogonal to both $\mathbf{T}$ and $\mathbf{N}$, and therefore to the osculating plane. Vectors $\mathbf{T}$, $\mathbf{N}$, and $\mathbf{B}$ are three mutually perpendicular unit vectors. (We ask you to prove this in Exercise 47.) In the order $\mathbf{T}, \mathbf{N}, \mathbf{B}$ they also form a right-hand coordinate system, called the **TNB**-frame, *moving frame*, or *Frenet frame*. This frame allows for a more in-depth understanding of the motion of a particle moving along a space curve.

When we are working with a curve in the xy-plane, since both $\mathbf{T}$ and $\mathbf{N}$ lie in the plane, the xy-plane *is* the osculating plane. In this case, we may think of the xy-plane as part of a three-dimensional coordinate system, with every binormal vector parallel to the z-axis. Thus, for a planar curve, at every point where the binormal vector exists, $\mathbf{B}(t_0) = \pm\mathbf{k}$.

Examples and Explorations

EXAMPLE 1

Finding a unit tangent vector and a unit normal vector in the plane

Find the unit tangent and principal unit normal vectors to the graph of the sine function $y = \sin x$ at $x = \dfrac{\pi}{2}$ and at $x = \dfrac{5\pi}{4}$.

SOLUTION

Any function of the form $y = f(x)$ has the parametrization $x = t$, $y = f(t)$. Thus, we may use the vector function $\mathbf{r}(t) = \langle t, \sin t \rangle$ to compute the desired vectors. The unit tangent vector

$$\mathbf{T}(t) = \frac{\mathbf{r}'(t)}{\|\mathbf{r}'(t)\|} = \frac{\langle 1, \cos t \rangle}{\|\langle 1, \cos t \rangle\|} = \frac{\langle 1, \cos t \rangle}{\sqrt{1 + \cos^2 t}}.$$

To compute the principal unit normal vector $\mathbf{N}(t) = \dfrac{\mathbf{T}'(t)}{\|\mathbf{T}'(t)\|}$, we first use the quotient rule to find $\mathbf{T}(t)$. You may check that:

$$\mathbf{T}'(t) = \frac{\sin t \, \langle \cos t, -1 \rangle}{(1 + \cos^2 t)^{3/2}}.$$

Dividing $\mathbf{T}'(t)$ by its magnitude and simplifying, we obtain

$$\mathbf{N}(t) = \frac{\sin t \, \langle \cos t, -1 \rangle}{|\sin t| \sqrt{1 + \cos^2 t}}.$$

Evaluating the unit tangent vector and principal unit normal vector at $t = \dfrac{\pi}{2}$, we get

$$\mathbf{T}\!\left(\frac{\pi}{2}\right) = \langle 1, 0 \rangle \ \text{ and } \ \mathbf{N}\!\left(\frac{\pi}{2}\right) = \langle 0, -1 \rangle.$$

This result should not be a surprise. At $\dfrac{\pi}{2}$ we expect there to be a horizontal tangent and a vertical normal vector pointing into the curve. In addition, since we chose a parametrization that traced the curve from left to right, the tangent vector should (and does) point to the right.

At $t = \frac{5\pi}{4}$ we obtain

$$\mathbf{T}\left(\frac{5\pi}{4}\right) = \frac{\langle 1, \cos(5\pi/4)\rangle}{\sqrt{1 + \cos^2(5\pi/4)}} = \frac{\langle 1, -\sqrt{2}/2\rangle}{\sqrt{1 + 1/2}} = \frac{1}{\sqrt{3}}\langle \sqrt{2}, -1\rangle \text{ and } \mathbf{N}\left(\frac{5\pi}{4}\right) = \frac{1}{\sqrt{3}}\langle 1, \sqrt{2}\rangle. \ \square$$

✔ **CHECKING THE ANSWER**

Now, given that we are dealing with the function $y = \sin x$, perhaps it seemed that the preceding computations were (unnecessarily) complicated. If they did, you were right! For a differentiable function of the form $y = f(x)$, we have other techniques for finding the desired unit vectors. Since $f'(x_0)$ is the slope of f at x_0, we can immediately obtain the tangent vector, $\langle f'(x), 1\rangle$. If we divide this vector by its magnitude, we will obtain the equation of the unit tangent vector at x_0. Recall that two nonvertical lines in the plane are perpendicular if and only if their slopes are negative reciprocals. Therefore, a normal vector at x_0 will be a multiple of $\pm\langle 1, -f'(x_0)\rangle$. To obtain the principal unit normal vector at x_0, we again divide by the magnitude and choose the correct sign, the one that points *into* the curve. Example 2 repeats these computations with a space curve, for which the additional complexity *is* necessary. Before continuing to that example, repeat the calculations in Example 1, using the analysis just outlined to find the unit tangent and principal unit normal vectors to $y = \sin x$ at $x = \frac{\pi}{2}$ and at $x = \frac{5\pi}{4}$. Make sure your results agree with the answers we found in Example 1.

EXAMPLE 2

Finding the equation of an osculating plane

Find the unit tangent and principal unit normal vectors to the graph of the helix defined by the vector function $\mathbf{r}(t) = \langle \cos t, \sin t, t\rangle$ at $t = \frac{\pi}{2}$ and at $t = \frac{5\pi}{4}$. Then, use these vectors to construct the equations of the osculating planes at those points on the curve.

SOLUTION

We start by computing

$$\mathbf{r}'(t) = \langle -\sin t, \cos t, 1\rangle.$$

Thus,

$$\mathbf{T}(t) = \frac{\mathbf{r}'(t)}{\|\mathbf{r}'(t)\|} = \frac{\langle -\sin t, \cos t, 1\rangle}{\|\langle -\sin t, \cos t, 1\rangle\|} = \frac{\langle -\sin t, \cos t, 1\rangle}{\sqrt{2}}.$$

To compute $\mathbf{N}(t)$ we will divide $\mathbf{T}'(t)$ by its magnitude. In this example, since the radical in the denominator of the final quotient in the preceding equation is a constant, it will cancel when we divide. Thus, to make our computation slightly simpler, we will divide the derivative of $\langle -\sin t, \cos t, 1\rangle$ by its magnitude. We obtain

$$\mathbf{N}(t) = \frac{\langle -\cos t, -\sin t, 0\rangle}{\|\langle -\cos t, -\sin t, 0\rangle\|} = \frac{\langle -\cos t, -\sin t, 0\rangle}{\sqrt{1}} = \langle -\cos t, -\sin t, 0\rangle.$$

We now evaluate these functions to obtain the required vectors at $t = \frac{\pi}{2}$ and at $t = \frac{5\pi}{4}$. At $t = \frac{\pi}{2}$ we have

$$\mathbf{T}\left(\frac{\pi}{2}\right) = \frac{\langle -\sin(\pi/2), \cos(\pi/2), 1\rangle}{\sqrt{2}} = \frac{\langle -1, 0, 1\rangle}{\sqrt{2}} \text{ and}$$

$$\mathbf{N}\left(\frac{\pi}{2}\right) = \langle -\cos(\pi/2), -\sin(\pi/2), 0\rangle = \langle 0, -1, 0\rangle.$$

At $t = \frac{5\pi}{4}$ we have

$$\mathbf{T}\left(\frac{5\pi}{4}\right) = \frac{\langle -\sin(5\pi/4), \cos(5\pi/4), 1 \rangle}{\sqrt{2}} = \frac{\langle \sqrt{2}/2, -\sqrt{2}/2, 1 \rangle}{\sqrt{2}} = \left\langle \frac{1}{2}, -\frac{1}{2}, \frac{\sqrt{2}}{2} \right\rangle \quad \text{and}$$

$$\mathbf{N}\left(\frac{5\pi}{4}\right) = \left\langle -\cos\left(\frac{5\pi}{4}\right), -\sin\left(\frac{5\pi}{4}\right), 0 \right\rangle = \left\langle \frac{\sqrt{2}}{2}, \frac{\sqrt{2}}{2}, 0 \right\rangle.$$

Finally, we construct the equations of the osculating planes. For the equation of a plane, we need a point on the plane and a vector orthogonal to the plane. For an osculating plane at a point t_0, we will use the terminal point of the vector $\mathbf{r}(t_0)$ and the binormal vector $\mathbf{B}(t_0) = \mathbf{T}(t_0) \times \mathbf{N}(t_0)$, respectively.

At $t = \frac{\pi}{2}$ we have $\mathbf{r}\left(\frac{\pi}{2}\right) = \left\langle 0, 1, \frac{\pi}{2} \right\rangle$ and

$$\mathbf{B}\left(\frac{\pi}{2}\right) = \mathbf{T}\left(\frac{\pi}{2}\right) \times \mathbf{N}\left(\frac{\pi}{2}\right) = \frac{\langle -1, 0, 1 \rangle}{\sqrt{2}} \times \langle 0, -1, 0 \rangle = \frac{\langle 1, 0, 1 \rangle}{\sqrt{2}}.$$

For simplicity we may use $\langle 1, 0, 1 \rangle$ as the vector orthogonal to the osculating plane and obtain the equation $x + z = \frac{\pi}{2}$ for the osculating plane.

Similarly, at $t = \frac{5\pi}{4}$ we have $\mathbf{r}\left(\frac{5\pi}{4}\right) = \left\langle -\frac{\sqrt{2}}{2}, -\frac{\sqrt{2}}{2}, \frac{5\pi}{4} \right\rangle$ and

$$\mathbf{B}\left(\frac{5\pi}{4}\right) = \left\langle \frac{1}{2}, -\frac{1}{2}, \frac{\sqrt{2}}{2} \right\rangle \times \left\langle \frac{\sqrt{2}}{2}, \frac{\sqrt{2}}{2}, 0 \right\rangle = \left\langle -\frac{1}{2}, \frac{1}{2}, \frac{\sqrt{2}}{2} \right\rangle.$$

Again, for simplicity we may use $\langle -1, 1, \sqrt{2} \rangle$ as the vector orthogonal to the osculating plane. Therefore, when $t = \frac{5\pi}{4}$, we obtain the equation $-x + y + \sqrt{2}z = \frac{5\pi\sqrt{2}}{4}$ for the osculating plane. □

CHECKING THE ANSWER

There are several things we can check in our answers to Example 2.

▶ As the parameter t increases, the helix is spiraling upwards. Therefore, each tangent vector should have a positive z-component. Note that $\mathbf{T}(t) = \frac{\langle -\sin t, \cos t, 1 \rangle}{\sqrt{2}}$. The z-component is always $\frac{1}{\sqrt{2}}$.

▶ Each of the answers should be a unit vector. (They are, as you can check.)

▶ At each of the points, the tangent and normal vectors should be orthogonal. Here we can use the dot product. For example, at $t = \frac{5\pi}{4}$,

$$\mathbf{T}\left(\frac{5\pi}{4}\right) \cdot \mathbf{N}\left(\frac{5\pi}{4}\right) = \left\langle \frac{1}{2}, -\frac{1}{2}, \frac{\sqrt{2}}{2} \right\rangle \cdot \left\langle \frac{\sqrt{2}}{2}, \frac{\sqrt{2}}{2}, 0 \right\rangle = 0.$$

? TEST YOUR UNDERSTANDING

▶ How is the unit tangent vector defined? Why is the unit tangent vector useful?

▶ How is the principal unit normal vector defined? Why are the unit tangent vector at a point P and the principal unit normal vector at P always orthogonal?

▶ Why does the principal unit normal vector always point "into" the curve?

▶ How is the osculating plane defined? What is its geometric relationship to a space curve at a point?

▶ What is the binormal vector?

EXERCISES 11.3

Thinking Back

▶ *Unit vectors:* If $\mathbf{v}$ is a nonzero vector, explain why $\dfrac{\mathbf{v}}{\|\mathbf{v}\|}$ is a unit vector.

▶ *Equation of a plane:* Find the equation of the plane determined by the vectors $\langle 1, -3, 5 \rangle$ and $\langle 2, 4, -1 \rangle$ and containing the point $(0, 3, -2)$.

Concepts

0. *Problem Zero:* Read the section and make your own summary of the material.

1. *True/False:* Determine whether each of the statements that follow is true or false. If a statement is true, explain why. If a statement is false, provide a counterexample.

(a) *True or False:* For every position vector, $\mathbf{r}(t)$, $\mathbf{r}'(t) \cdot \mathbf{r}''(t) = 0$.

(b) *True or False:* If $\|\mathbf{T}'(t)\| = 0$, then $\mathbf{N}(t)$ is not defined.

(c) *True or False:* The cross product of two unit vectors is another unit vector.

(d) *True or False:* If C is a curve in the xy-plane and the principal unit normal vector, $\mathbf{N}$, is defined at the point P_0 on C, then the xy-plane is the osculating plane for C at P_0.

(e) *True or False:* If a vector function has a unit tangent vector and a principal unit normal vector at a point, then it has a binormal vector at that point also.

(f) *True or False:* If the unit tangent vector, $\mathbf{T}(t)$, the principal unit normal vector, $\mathbf{N}(t)$, and the binormal vector $\mathbf{B}(t)$, are all defined for some value of t, then $\mathbf{B}(t) \times \mathbf{N}(t) = -\mathbf{T}(t)$.

(g) *True or False:* If the unit tangent vector, $\mathbf{T}(t)$, the principal unit normal vector, $\mathbf{N}(t)$, and the binormal vector $\mathbf{B}(t)$, are all defined at some point t, then $\mathbf{T}(t) \cdot (\mathbf{N}(t) \times \mathbf{B}(t)) = 1$.

(h) *True or False:* A space curve can have infinitely many different osculating planes.

2. *Examples:* Construct examples of the thing(s) described in the following. Try to find examples that are different than any in the reading.

(a) A differentiable vector-valued function $\mathbf{r}(t)$ that does not have a unit tangent vector at any point.

(b) A twice-differentiable vector-valued function with a unit tangent vector at every point but that does not have a unit normal vector at any point.

(c) A nonconstant differentiable vector function $\mathbf{r}(t) = \langle x(t), y(t) \rangle$ for which the xy-plane is *not* the osculating plane.

3. Imagine that you are driving on a twisting mountain road. Describe the unit tangent vector, principal unit normal vector, and binormal vector as you ascend, descend, twist right, and twist left.

4. Given a differentiable vector-valued function $\mathbf{r}(t)$, what is the definition of the unit tangent vector $\mathbf{T}(t)$?

5. Under what conditions does a differentiable vector-valued function $\mathbf{r}(t)$ *not* have a unit tangent vector at a point in the domain of $\mathbf{r}(t)$?

6. Let $\mathcal{L}$ be a straight line in $\mathbb{R}^3$. Find vector functions with the following properties:

(a) The graph of $\mathbf{r}_1(t)$ is $\mathcal{L}$, and $\mathbf{r}_1(t)$ has a unit tangent vector for every value of t.

(b) The graph of $\mathbf{r}_2(t)$ is $\mathcal{L}$, and there is a least one value of t at which $\mathbf{r}_2(t)$ does not have a unit tangent vector.

7. Given a differentiable vector-valued function $\mathbf{r}(t)$, what is the relationship between $\mathbf{r}'(t_0)$ and $\mathbf{T}(t_0)$ at a point t_0 in the domain of $\mathbf{r}(t)$?

8. Given a differentiable vector-valued function $\mathbf{r}(t)$, what are the advantages and disadvantages in using $\mathbf{r}'$ or $\mathbf{T}$ to analyze the behavior of $\mathbf{r}(t)$?

9. Given a twice-differentiable vector-valued function $\mathbf{r}(t)$, what is the definition of the principal unit normal vector $\mathbf{N}(t)$?

10. Given a twice-differentiable vector-valued function $\mathbf{r}(t)$, why does the principal unit normal vector $\mathbf{N}(t)$ point into the curve? (*Hint: Use the definition!*)

11. Given a twice-differentiable vector-valued function $\mathbf{r}(t)$ and a point t_0 in its domain, what are the geometric relationships between the unit tangent vector $\mathbf{T}(t_0)$, the principal unit normal vector $\mathbf{N}(t_0)$, and $\dfrac{d\mathbf{N}}{dt}\Big|_{t_0}$?

12. Under what conditions does a twice-differentiable vector-valued function $\mathbf{r}(t)$ *not* have a principal unit normal vector at a point in the domain of $\mathbf{r}(t)$?

13. Given a twice-differentiable vector-valued function $\mathbf{r}(t)$, what is the definition of the binormal vector $\mathbf{B}(t)$?

14. We've seen that the graph of a continuous function $y = f(x)$ is the same as the graph of the vector-valued function $\mathbf{r}(t) = \langle t, f(t) \rangle$. What is the direction of the binormal vector $\mathbf{B}(t)$ when the graph of f is concave up? When the graph is concave down? What happens to $\mathbf{B}(t)$ at a point of inflection?

15. Under what conditions does a twice-differentiable vector-valued function $\mathbf{r}(t)$ not have a binormal vector at a point in the domain of $\mathbf{r}(t)$?

16. Given a twice-differentiable vector-valued function $\mathbf{r}(t)$ and a point t_0 in its domain, what is the osculating plane at $\mathbf{r}(t_0)$?

17. Carefully outline the steps you would use to find the equation of the osculating plane at a point $\mathbf{r}(t_0)$ on the graph of a vector function.

18. If the osculating plane is the same at every point on a space curve, what must be true about the curve?

19. Given the three mutually perpendicular vectors $\mathbf{T}(t_0)$, $\mathbf{N}(t_0)$, and $\mathbf{B}(t_0)$ at a point $\mathbf{r}(t_0)$, how many distinct planes contain the point and two of the vectors?

20. Let C be the graph of a vector-valued function $\mathbf{r}$. The plane determined by the vectors $\mathbf{N}(t_0)$ and $\mathbf{B}(t_0)$ and containing the point $\mathbf{r}(t_0)$ is called the *normal* plane for C at $\mathbf{r}(t_0)$. Find the equation of the normal plane to the helix determined by $\mathbf{r}(t) = \langle \cos t, \sin t, t \rangle$ for $t = \pi$.

21. Let C be the graph of a vector-valued function $\mathbf{r}$. The plane determined by the vectors $\mathbf{T}(t_0)$ and $\mathbf{B}(t_0)$ and containing the point $\mathbf{r}(t_0)$ is called the *rectifying* plane for C at $\mathbf{r}(t_0)$. Find the equation of the rectifying plane to the helix determined by $\mathbf{r}(t) = \langle \cos t, \sin t, t \rangle$ when $t = \pi$.

Skills

For each of the vector-valued functions in Exercises 22–28, find the unit tangent vector.

22. $\mathbf{r}(t) = \langle t, t^2 \rangle$

23. $\mathbf{r}(t) = \langle t^2 + 5, 5t, 4t^3 \rangle$

24. $\mathbf{r}(t) = \langle \cos \alpha t, \sin \alpha t \rangle$

25. $\mathbf{r}(t) = \langle \cos^3 t, \sin^3 t \rangle$

26. $\mathbf{r}(t) = \langle t, t^2, t^3 \rangle$

27. $\mathbf{r}(t) = \langle 3 \sin t, 5 \cos t, 4 \sin t \rangle$

28. $\mathbf{r}(t) = \langle \sin 2t, \cos 2t, t \rangle$

For each of the vector-valued functions in Exercises 29–34, find the unit tangent vector and the principal unit normal vector at the specified value of t.

29. $\mathbf{r}(t) = \langle t, t^2 \rangle$, $t = 1$

30. $\mathbf{r}(t) = \langle a \sinh t, b \cosh t \rangle$, where a and b are positive, $t = 0$

31. $\mathbf{r}(t) = \langle \cos \alpha t, \sin \alpha t \rangle$, where $\alpha > 0$, $t = \pi$

32. $\mathbf{r}(t) = \langle \cos^3 t, \sin^3 t \rangle$, $t = \dfrac{\pi}{4}$

33. $\mathbf{r}(t) = \langle 3 \sin t, 5 \cos t, 4 \sin t \rangle$, $t = \pi$

34. $\mathbf{r}(t) = \langle \sin 2t, \cos 2t, t \rangle$, $t = \dfrac{\pi}{4}$

For each of the vector-valued functions in Exercises 35–39, find the unit tangent vector, the principal unit normal vector, the binormal vector, and the equation of the osculating plane at the specified value of t.

35. $\mathbf{r}(t) = \left\langle t, t^2, \dfrac{2}{3}t^3 \right\rangle$ at $t = 1$

36. $\mathbf{r}(t) = \left\langle \dfrac{t}{\sqrt{2}}, \sqrt{t}, \dfrac{1}{3}t^{3/2} \right\rangle$ at $t = 1$

37. $\mathbf{r}(t) = \langle e^t, e^{-t}, \sqrt{2}t \rangle$ at $t = 0$

38. $\mathbf{r}(t) = \langle 3 \sin t, 5 \cos t, 4 \sin t \rangle$ at $t = \dfrac{\pi}{4}$

39. $\mathbf{r}(t) = \langle \sin 2t, \cos 2t, t \rangle$ at $t = \dfrac{\pi}{2}$

Using the definitions of the normal plane and rectifying plane in Exercises 20 and 21, respectively, find the equations of these planes at the specified points for the vector functions in Exercises 40–42. Note: These are the same functions as in Exercises 35, 37, and 39.

40. $\mathbf{r}(t) = \left\langle t, t^2, \dfrac{2}{3}t^3 \right\rangle$ at $t = 1$

41. $\mathbf{r}(t) = \langle e^t, e^{-t}, \sqrt{2}t \rangle$ at $t = 0$

42. $\mathbf{r}(t) = \langle \sin 2t, \cos 2t, t \rangle$ at $t = \dfrac{\pi}{2}$

Applications

43. Annie is kayaking with her friend Jim. Jim is inexperienced, and soon his boat is upside down. Being inexperienced, he also fails to execute a roll, so he has to swim. Since they are both in single kayaks, they cannot get him back in the boat, so Jim grabs his kayak and Annie's kayak, and Annie tows the whole lot towards shore. There is a current, so if they did not paddle, they would follow a path defined by the vector function $\mathbf{r}(t) = \langle t^{1.5}, t(1.7 - t) \rangle$, where t represents time in hours.

(a) For $0 \le t \le 1.7$, where would they move fastest if they did not paddle at all?

(b) Their best strategy to get to shore is to paddle exactly orthogonally to the current, towards the land that is on the inside of the path of the current. If they get things in order and start paddling at $t = 0.1$, in what direction are they paddling?

44. Ian is planning a return from a mountain in the Cascades. He will descend to around 6000 feet and then follow a contour, staying high and out of the trees to make travel easy, until he is at a point directly above a trail. He will then descend directly to the trail. The contour at 6000 feet follows the curve $\mathbf{r}(x) = \langle x, 0.45x^2 + x \rangle$. Ian uses his map to determine that he will be at the point where he needs to descend to the trail when he can first see a mountain peak that is on a compass heading of 30 degrees; that is, when the tangent to the level curve is 30 degrees east of north.

(a) Where does that happen? What is the unit tangent vector at that point?

(b) When he reaches that point, he will descend directly down the slope, orthogonally to the contour. Downhill is generally north. What compass heading will that be? What is the unit normal vector there?

Proofs

45. Prove that the cross product of two orthogonal unit vectors is a unit vector.

46. Let $y = f(x)$ be a function with domain $\mathbb{R}$ and that is differentiable at every point in its domain, and let $\mathbf{r}(t) = \langle t, f(t) \rangle$.

(a) Explain why the graph of $\mathbf{r}$ is the same as the graph of $y = f(x)$.

(b) Prove that the unit tangent vector at $\mathbf{r}(\alpha)$ is parallel to the tangent line to f at α.

(c) Prove that the principal unit normal vector for $\mathbf{r}$ is undefined at every inflection point of f.

47. Let $\mathbf{r}(t) = \langle x(t), y(t), z(t) \rangle$ be a differentiable vector function on some interval $I \subseteq \mathbb{R}$ such that the derivative of the unit tangent vector $\mathbf{T}'(t_0) \neq 0$, where $t_0 \in I$. Prove that the binormal vector

$$\mathbf{B}(t_0) = \mathbf{T}(t_0) \times \mathbf{N}(t_0)$$

(a) is a unit vector;

(b) is orthogonal to both $\mathbf{T}(t_0)$ and $\mathbf{N}(t_0)$.

Also, prove that $\mathbf{T}(t_0)$, $\mathbf{N}(t_0)$, and $\mathbf{B}(t_0)$ form a right-handed coordinate system.

Thinking Forward

▶ *Equation of a circle:* Find an equation for the circle of radius ρ whose center is ρ units from the point (a, b) in the direction of the vector $\langle \alpha, \beta \rangle$.

▶ *Equation of a sphere:* Find an equation for the sphere of radius ρ whose center is ρ units from the point (a, b, c) in the direction of the vector $\langle \alpha, \beta, \gamma \rangle$.

11.4 ARC LENGTH PARAMETRIZATIONS AND CURVATURE

▶ The arc length of a curve in $\mathbb{R}^3$

▶ Arc length parametrizations

▶ The curvature of planar curves and space curves

The Arc Length of Space Curves

We begin this section with a reminder of some of the work we did in Section 9.1. Let $x = x(t)$ and $y = y(t)$ be differentiable functions of t. In Definition 9.3 we said that the arc length of the parametric curve determined by x and y on an interval $[a, b]$ was

$$\lim_{n \to \infty} \sum_{k=1}^{n} \sqrt{(x(t_k) - x(t_{k-1}))^2 + (y(t_k) - y(t_{k-1}))^2},$$

where $\Delta t = \dfrac{b-a}{n}$ and $t_k = a + k\Delta t$. Then, in Theorem 9.4 we argued that the arc length could be evaluated with the use of $\int_a^b \sqrt{(x'(t))^2 + (y'(t))^2}\, dt$.

We now express Definition 9.3 and Theorem 9.4 in the language of vector functions. The vector function $\mathbf{r}(t) = x(t)\mathbf{i} + y(t)\mathbf{j}$ has the same graph as the parametric equations. Definition 9.3 would say that the arc length of the vector curve on the interval $[a, b]$ would be $\lim_{n \to \infty} \sum_{k=1}^{n} \|\mathbf{r}(t_k) - \mathbf{r}(t_{k-1})\|$, where $\Delta t = \dfrac{b-a}{n}$ and $t_k = a + k\Delta t$. Similarly, $\int_a^b \|\mathbf{r}'(t)\|\, dt$ represents the length of the vector curve for $t \in [a, b]$. These formulas are valid, since

$$\|\mathbf{r}(t_k) - \mathbf{r}(t_{k-1})\| = \sqrt{(x(t_k) - x(t_{k-1}))^2 + (y(t_k) - y(t_{k-1}))^2} \text{ and}$$

$$\|\mathbf{r}'(t)\| = \sqrt{(x'(t))^2 + (y'(t))^2},$$

by the Pythagorean theorem and the definition of the norm of a vector, respectively.

The great thing about these forms is that they carry over to space curves (or even to curves in higher dimensions)! Thus, we have the following definition:

DEFINITION 11.17   **The Arc Length of a Vector Curve**

Let C be a curve in $\mathbb{R}^3$ given by the vector function $\mathbf{r}(t) = \langle x(t), y(t), z(t) \rangle$ for $t \in [a, b]$, where x, y, and z are differentiable functions of t and such that the function is one-to-one from the interval $[a, b]$ to the curve C. The **length of the curve** C from a to b, denoted by $l(a, b)$, is

$$l(a, b) = \lim_{n \to \infty} \sum_{k=1}^{n} \| \mathbf{r}(t_k) - \mathbf{r}(t_{k-1}) \|,$$

where $\Delta t = \frac{b-a}{n}$ and $t_k = a + k\Delta t$.

The following figure illustrates the idea behind the definition of arc length:

The arc length of a vector curve may be approximated by using segments

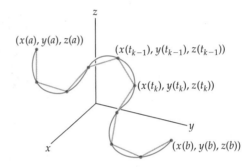

We also have the following theorem:

THEOREM 11.18   **The Arc Length of a Vector Curve**

Let C be a curve in $\mathbb{R}^3$ given by the vector function $\mathbf{r}(t) = \langle x(t), y(t), z(t) \rangle$ such that the function is one-to-one from the interval $[a, b]$ to the curve C. If x, y, and z are differentiable functions of t such that x', y', and z' are continuous on $[a, b]$, then the length of the curve C from $\mathbf{r}(a)$ to $\mathbf{r}(b)$ is given by

$$l(a, b) = \int_a^b \| \mathbf{r}'(t) \| \, dt.$$

A proof of Theorem 11.18 may be adapted from the proof of Theorem 9.4 in Section 9.1 and is left to you.

Arc Length Parametrizations

We have already discussed the geometry of the relationship between a differentiable vector function and its derivative. The derivative gives us information about the rate and direction of motion of a particle moving along a vector curve. Recall that for a real-valued function of the form $y = f(x)$, the second derivative gives us more subtle information about the shape of the curve and the sign of f'' tells us whether a curve is concave up or concave down. Space curves are even more complicated than planar curves. A single number cannot tell us everything about how a space curve is bending. However, we do want to know about the "curvature" of a space curve. We will get to that definition later in this section.

To help us define curvature in a relatively easy way, we need to settle on a particular type of parametrization for a given vector curve. You should have already realized that a given vector curve has many different parametrizations. For example, in Chapter 9 we examined several parametrizations for the unit circle. In fact, every vector curve has infinitely many different parametrizations. For example, for every nonzero value of k, the vector function $\mathbf{r}(t) = \langle \cos kt, \sin kt \rangle$ for $t \geq 0$ provides a different parametrization of the unit circle. The differences between two of these parametrizations are the rate at which the particle traces the circle and the direction of motion. For every value of k, the particle starts at the point $(1, 0)$ when $t = 0$. The motion is clockwise when $k < 0$ and counterclockwise when $k > 0$. The rate at which the particle moves increases linearly with k. For example, when $k = 1$, the particle traces the unit circle once on the interval $[0, 2\pi)$, but if $k = 5$, the particles traces the circle 5 times on the same interval.

Similarly, every vector curve in $\mathbb{R}^3$ has infinitely many parametrizations. The particular type of parametrization that we need for our upcoming definition of curvature is called an "arc length parametrization."

DEFINITION 11.19

Arc Length Parametrization

Let C be the graph of a differentiable vector function $\mathbf{r}(t)$ defined on an interval I. The function $\mathbf{r}(t)$ is said to be an ***arc length parametrization*** for C if

$$\int_{c}^{d} \|\mathbf{r}'(t)\| \, dt = d - c$$

for every c and $d \in I$. Also, C is said to be ***parametrized by arc length.***

When a curve is parametrized by arc length, the length of every segment of the curve is equal to the change in the value of the parameter. For example, $\mathbf{r}_1(t) = \langle \cos t, \sin t \rangle$ and $\mathbf{r}_2(t) = \langle \cos 2t, \sin 2t \rangle$ are both parametrizations of the unit circle. The function $\mathbf{r}_1(t)$ is an arc length parametrization of the unit circle, since

$$\int_{c}^{d} \|\mathbf{r}_1'(t)\| \, dt = \int_{c}^{d} \|\langle -\sin t, \cos t \rangle\| \, dt = \int_{c}^{d} \sqrt{\sin^2 t + \cos^2 t} \, dt = \int_{c}^{d} dt = d - c.$$

However, the function $\mathbf{r}_2(t)$ is not an arc length parametrization of the unit circle, since

$$\int_{c}^{d} \|\mathbf{r}_2'(t)\| \, dt = \int_{c}^{d} \|\langle -2 \sin 2t, 2 \cos 2t \rangle\| \, dt$$

$$= \int_{c}^{d} \sqrt{4 \sin^2 2t + 4 \cos^2 2t} \, dt = \int_{c}^{d} 2 \, dt = 2(d - c).$$

In the next theorem we will see that when we use an arc length parametrization, the derivative of $\mathbf{r}$ with respect to arc length is particularly simple.

THEOREM 11.20

The Derivative of a Vector Function with Respect to Arc Length

Let C be a curve in $\mathbb{R}^3$ given by the differentiable vector function $\mathbf{r}(t) = \langle x(t), y(t), z(t) \rangle$ such that $\mathbf{r}(t)$ is one-to-one from the interval $[a, b]$ to the curve C and has a nonzero derivative. If we define the real-valued function $s(t)$ by

$$s(t) = \int_{a}^{t} \|\mathbf{r}'(\tau)\| \, d\tau,$$

then

$$\frac{d\mathbf{r}}{ds} = \mathbf{T}(t)$$

is the unit tangent vector function for C.

Proof. We first note that

$$\int_a^t \|\mathbf{r}'(\tau)\| d\tau = l(a, t) = s(t).$$

That is, $s(t)$ represents the arc length of C between the terminal points of $\mathbf{r}(a)$ and $\mathbf{r}(t)$.

Now, by the Fundamental Theorem of Calculus

$$s'(t) = \frac{ds}{dt} = \|\mathbf{r}'(t)\|,$$

and by the chain rule,

$$\frac{d\mathbf{r}}{dt} = \frac{d\mathbf{r}}{ds}\frac{ds}{dt}.$$

Solving for $\frac{d\mathbf{r}}{ds}$ and using the fact that $\frac{ds}{dt} = \|\mathbf{r}'(t)\|$, we have

$$\frac{d\mathbf{r}}{ds} = \frac{\mathbf{r}'(t)}{\|\mathbf{r}'(t)\|}.$$

We see that $\frac{d\mathbf{r}}{ds}$ has the same direction as $\mathbf{r}'(t)$ and remind you that any nonzero vector divided by its magnitude is a unit vector. Thus, $\frac{d\mathbf{r}}{ds} = \mathbf{T}(t)$, the unit tangent vector to C. ∎

With most parametrizations of a curve, both the direction *and* magnitude of the tangent vectors change as a particle traverses the curve. However, Theorem 11.20 tells us that when a curve is parametrized by arc length, since the derivative is a unit vector, only the directions of the tangent vectors are changing. This change in the unit tangent vectors is the basis for the following definition:

DEFINITION 11.21

Curvature

Let C be the graph of a vector function $\mathbf{r}(s)$ defined on an interval I, parametrized by arc length s, and with unit tangent vector $\mathbf{T}$. The ***curvature*** κ of C at a point on the curve is the scalar given by

$$\kappa = \left\| \frac{d\mathbf{T}}{ds} \right\|.$$

The magnitude of the rate of change of the unit tangent vector defines curvature

C

The following theorem tells us the relationship between the derivative of the unit tangent vector, curvature, and the principal unit normal vector:

THEOREM 11.22

The Relationship Between $\frac{d\mathbf{T}}{ds}$, κ, and $\mathbf{N}$

Let C be the graph of a vector function $\mathbf{r}(s)$ defined on an interval I and parametrized by arc length s. If the unit tangent vector $\mathbf{T}$ has a nonzero derivative, then

$$\frac{d\mathbf{T}}{ds} = \kappa \mathbf{N}.$$

The proof of Theorem 11.22 follows from the facts that the vector functions $\frac{d\mathbf{T}}{ds}$, $\frac{d\mathbf{T}}{dt}$, and $\mathbf{N}$ are scalar multiples of each other and that $\|\mathbf{N}\| = 1$. The details are left for Exercise 64.

Although Definition 11.21 provides an elegant way to express and understand curvature, it is usually impractical in computations, because, unfortunately, arc length parametrizations are frequently difficult to construct.

The next two theorems provide more computationally efficient alternatives for computing the curvature.

THEOREM 11.23

Formulas for the Curvature of a Space Curve

Let C be the graph of a twice-differentiable vector function $\mathbf{r}(t)$ defined on an interval I with unit tangent vector $\mathbf{T}(t)$. Then the curvature κ of C at a point on the curve is given by

(a) $\kappa = \dfrac{\|\mathbf{T}'(t)\|}{\|\mathbf{r}'(t)\|}.$

(b) $\kappa = \dfrac{\|\mathbf{r}'(t) \times \mathbf{r}''(t)\|}{\|\mathbf{r}'(t)\|^3}.$

Proof. Let s be an arc length parameter for C.

(a) By the chain rule,

$$\kappa = \left\|\frac{d\mathbf{T}}{ds}\right\| = \left\|\frac{d\mathbf{T}/dt}{ds/dt}\right\| = \frac{\|\mathbf{T}'(t)\|}{\|\mathbf{r}'(t)\|}.$$

(b) By the chain rule and Theorem 11.20,

$$\mathbf{r}'(t) = \frac{d\mathbf{r}}{dt} = \frac{d\mathbf{r}}{ds}\frac{ds}{dt} = \mathbf{T}\frac{ds}{dt}.$$

Differentiating with respect to t by the product rule, we obtain

$$\mathbf{r}''(t) = \frac{d\mathbf{T}}{dt}\frac{ds}{dt} + \mathbf{T}\frac{d^2s}{dt^2}.$$

Now, first by the chain rule and then by Theorem 11.22, we have

$$\mathbf{r}''(t) = \frac{d\mathbf{T}}{ds}\frac{ds}{dt}\frac{ds}{dt} + \mathbf{T}\frac{d^2s}{dt^2} = \kappa\mathbf{N}\left(\frac{ds}{dt}\right)^2 + \mathbf{T}\frac{d^2s}{dt^2}.$$

Using the preceding equations to form the cross product $\mathbf{r}'(t) \times \mathbf{r}''(t)$, we obtain

$$\mathbf{r}'(t) \times \mathbf{r}''(t) = \mathbf{T}\frac{ds}{dt} \times \left(\kappa\mathbf{N}\left(\frac{ds}{dt}\right)^2 + \mathbf{T}\frac{d^2s}{dt^2}\right)$$

$$= \kappa(\mathbf{T} \times \mathbf{N})\left(\frac{ds}{dt}\right)^3 + (\mathbf{T} \times \mathbf{T})\frac{ds}{dt}\frac{d^2s}{dt^2}$$

$$= \kappa\mathbf{B}\left(\frac{ds}{dt}\right)^3 = \kappa\mathbf{B}\|\mathbf{r}'(t)\|^3,$$

since $\mathbf{T} \times \mathbf{N} = \mathbf{B}$ and the cross product of any three-component vector with itself is $\mathbf{0}$. Finally, if we take the norms of the first and last parts in the preceding chain of equalities and divide by $\|\mathbf{r}'(t)\|^3$, we have our result.   ∎

Even with Theorem 11.23, the computation of the curvature can be quite laborious. We provide a complete calculation in Example 3.

It is also convenient to have formulas for the curvature of planar curves. The following theorem gives us a simpler method for computing the curvature of a function of the form $y = f(x)$ and for finding the curvature of a curve defined by a vector function with two components:

THEOREM 11.24

> **Formulas for Curvature in the Plane**
>
> **(a)** Let $y = f(x)$ be a twice-differentiable function. Then the curvature of the graph of f is given by
>
> $$\kappa = \frac{|f''(x)|}{(1 + (f'(x))^2)^{3/2}}.$$
>
> **(b)** Let C be the graph of a vector function $\mathbf{r}(t) = \langle x(t), y(t) \rangle$ in the xy-plane, where x and y are twice-differentiable functions of t such that $x'(t)$ and $y'(t)$ are not simultaneously zero. Then the curvature κ of C at a point on the curve is given by
>
> $$\kappa = \frac{|x'(t)y''(t) - x''(t)y'(t)|}{((x'(t))^2 + (y'(t))^2)^{3/2}}.$$

We ask you to prove Theorem 11.24 in Exercises 65 and 66.

We make two final definitions to aid our understanding of curves. For a vector curve C with a positive curvature at a point, we define the radius of curvature and osculating circle.

DEFINITION 11.25

> **Radius of Curvature and Osculating Circle**
>
> Let C be the graph of a vector function $\mathbf{r}$, and let $\mathbf{r}(t_0)$ be the position vector for a point on C at which the curvature $\kappa > 0$.
>
> **(a)** The **radius of curvature** ρ of C at $\mathbf{r}(t_0)$ is given by
>
> $$\rho = \frac{1}{\kappa}.$$
>
> **(b)** The **osculating circle** to the curve C at $\mathbf{r}(t_0)$ is the circle in the osculating plane, with radius ρ, and whose center is the terminal point of the position vector $\mathbf{r}(t_0) + \rho \mathbf{N}(t_0)$.

As we mentioned in Section 11.3, when a curve C is planar, the osculating plane *is* the plane. In this case, the osculating circle also lies in the plane. The figure appearing next at the left shows a portion of the sine curve and the osculating circles when $x = \frac{\pi}{2}$ and when $x = \frac{5\pi}{4}$. In Example 2 we discuss how to find the equations of these circles. The figure at the right is a space curve, also with two osculating circles.

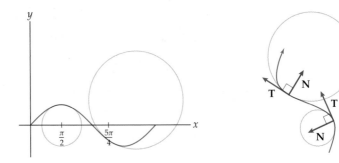

In two or three dimensions, an osculating circle is the circle that fits the shape of the curve "the best," in that the curvatures of the curve and the osculating circle are the same at the point of tangency.

Examples and Explorations

EXAMPLE 1   Finding the arc length of a space curve

Find the arc length of the helix defined by $\mathbf{r}(t) = \langle \cos \alpha t, \sin \alpha t, \beta t \rangle$ on the interval $[a, b]$, where a, b, α, and β are constants.

SOLUTION
The integral

$$\int_a^b \|\mathbf{r}'(t)\| \, dt = \int_a^b \|\langle -\alpha \sin \alpha t, \alpha \cos \alpha t, \beta \rangle \| \, dt$$

represents the specified arc length and equals

$$\int_a^b \sqrt{\alpha^2 \sin^2 \alpha t + \alpha^2 \cos^2 \alpha t + \beta^2} \, dt = \int_a^b \sqrt{\alpha^2 + \beta^2} \, dt = (b - a)\sqrt{\alpha^2 + \beta^2}.$$ $\square$

EXAMPLE 2   Finding osculating circles in the plane

Find the equations of the osculating circles to the graph of the sine function $y = \sin x$ at $x = \frac{\pi}{2}$ and at $x = \frac{5\pi}{4}$.

SOLUTION
This is a continuation of Example 1 in Section 11.3. In that example, we found the unit tangent and principal unit normal vectors to the curve at $x = \frac{\pi}{2}$ and at $x = \frac{5\pi}{4}$. Next, we need the curvature. Using Theorem 11.24(a), we have

$$\kappa = \frac{|f''(x)|}{(1 + (f'(x))^2)^{3/2}} = \frac{|-\sin x|}{(1 + \cos^2 x)^{3/2}}.$$

At $x = \frac{\pi}{2}$, we have $\kappa = 1$. The radius of curvature $\rho = \frac{1}{\kappa} = 1$. Therefore, the center of the osculating circle will be one unit from the point $\left(\frac{\pi}{2}, \sin\left(\frac{\pi}{2}\right)\right) = \left(\frac{\pi}{2}, 1\right)$ in the direction of the principal unit normal vector $\mathbf{N}\left(\frac{\pi}{2}\right) = \langle 0, -1 \rangle$. Thus, the osculating circle will have its center at $\left(\frac{\pi}{2}, 0\right)$ and have radius 1. The equation of this circle is $\left(x - \frac{\pi}{2}\right)^2 + y^2 = 1$.

At $x = \frac{5\pi}{4}$,

$$\kappa = \frac{|-\sin(5\pi/4)|}{\left(1 + \cos^2(5\pi/4)\right)^{3/2}} = \frac{\sqrt{2}/2}{(3/2)^{3/2}} = \frac{2}{3\sqrt{3}}.$$

The radius of curvature is the reciprocal of κ, so $\rho = \frac{3\sqrt{3}}{2}$. The principal unit normal vector $\mathbf{N}\left(\frac{5\pi}{4}\right) = \frac{1}{\sqrt{3}}\langle 1, \sqrt{2} \rangle$. The osculating circle will have its center at the terminal point of the position vector $\left\langle \frac{5\pi}{4}, \sin\left(\frac{5\pi}{4}\right)\right\rangle + \rho \mathbf{N}\left(\frac{5\pi}{4}\right) = \left\langle \frac{5\pi}{4} + \frac{3}{2}, \sqrt{2} \right\rangle$ and have radius $\frac{3\sqrt{3}}{2}$. The equation of the circle is $\left(x - \frac{5\pi}{4} - \frac{3}{2}\right)^2 + (y - \sqrt{2})^2 = \frac{27}{4}$. The graph of the function, along with these two osculating circles, appears on the left preceding Example 1. $\square$

EXAMPLE 3   Finding the center and radius of an osculating circle to a space curve

Find the center and radius of the osculating circle to the graph of the helix defined by the vector function $\mathbf{r}(t) = \langle \cos t, \sin t, t \rangle$ at $t = \frac{\pi}{2}$ and at $t = \frac{5\pi}{4}$.

SOLUTION

This is also a continuation of an example from Section 11.3. In Example 2 of that section, we found the unit tangent and principal unit normal vectors to the helix at $t = \frac{\pi}{2}$ and at $t = \frac{5\pi}{4}$. Now we need the curvature. Using Theorem 11.23(b), we have

$$\kappa = \frac{\|\mathbf{r}'(t) \times \mathbf{r}''(t)\|}{\|\mathbf{r}'(t)\|^3} = \frac{\|\langle -\sin t, \cos t, 1 \rangle \times \langle -\cos t, -\sin t, 0 \rangle\|}{\|\langle -\sin t, \cos t, 1 \rangle\|^3} = \frac{\|\langle \sin t, -\cos t, 1 \rangle\|}{\sqrt{2}^3} = \frac{1}{2}.$$

The curvature at every point on the helix is $\frac{1}{2}$. (Why does it make sense that the curvature on this helix is constant?) The radius of curvature is constantly 2. Finally, we only need the center points for the two circles. At $t = \frac{\pi}{2}$, the center of the osculating circle will be

$$\mathbf{r}\left(\frac{\pi}{2}\right) + 2\mathbf{N}\left(\frac{\pi}{2}\right) = \left\langle 0, 1, \frac{\pi}{2} \right\rangle + 2\langle 0, -1, 0 \rangle = \left\langle 0, -1, \frac{\pi}{2} \right\rangle.$$

The osculating circle will therefore have its center at $\left(0, -1, \frac{\pi}{2}\right)$ and have radius 2.

At $t = \frac{5\pi}{4}$, the center of the osculating circle will be

$$\mathbf{r}\left(\frac{5\pi}{4}\right) + 2\mathbf{N}\left(\frac{5\pi}{4}\right) = \left\langle -\frac{\sqrt{2}}{2}, -\frac{\sqrt{2}}{2}, \frac{5\pi}{4} \right\rangle + 2\left\langle \frac{\sqrt{2}}{2}, \frac{\sqrt{2}}{2}, 0 \right\rangle = \left\langle \frac{\sqrt{2}}{2}, \frac{\sqrt{2}}{2}, \frac{5\pi}{4} \right\rangle.$$

The osculating circle will therefore have its center at $\left(\frac{\sqrt{2}}{2}, \frac{\sqrt{2}}{2}, \frac{5\pi}{4}\right)$ and have radius 2.   □

? TEST YOUR UNDERSTANDING

▶ How is the arc length of a vector-valued function defined? Why is this definition consistent with the definitions of arc length you learned for differentiable functions of the form $y = f(x)$ and for parametric equations $x = x(t)$ and $y = y(t)$?

▶ What does it mean for a vector function to be parametrized by arc length?

▶ How is curvature defined? Why is the definition of curvature difficult to use?

▶ For a twice-differentiable function $y = f(x)$, how is the curvature at a point x_0 related to the concavity of the graph at x_0?

▶ What is the radius of curvature at a point P on a curve? How is the osculating circle at P defined?

EXERCISES 11.4

Thinking Back

▶ *The arc length of the sine function:* Find the arc length of the sine function $y = e^{x/2} + e^{-x/2}$ for $0 \leq x \leq 1$.

▶ *The arc length of the cycloid:* Find the arc length of the cycloid $x = \theta - \sin\theta$, $y = 1 - \cos\theta$, for $0 \leq \theta \leq 2\pi$.

Concepts

0. *Problem Zero:* Read the section and make your own summary of the material.

1. *True/False:* Determine whether each of the statements that follow is true or false. If a statement is true, explain why. If a statement is false, provide a counterexample.

(a) *True or False:* For every differentiable vector-valued function $\mathbf{r}$, $\int_a^b \|\mathbf{r}'(t)\| \, dt \geq (b - a)$.

(b) *True or False:* Every vector-valued function has an arc length parametrization.

(c) *True or False:* Every straight line has constant curvature.

(d) *True or False:* The radius of curvature is the reciprocal of the curvature.

(e) *True or False:* The radius of curvature is zero at every point on a straight line.

(f) *True or False:* If a vector-valued function $\mathbf{r}$ is differentiable for some value of t, its curvature at that point is positive.

(g) *True or False:* If a vector-valued function $\mathbf{r}$ is twice differentiable at some value t_0, then the osculating circle is defined at $\mathbf{r}(t_0)$.

(h) *True or False:* If a vector-valued function **r** has a principal unit normal vector for some value t_0, then the osculating circle at $\mathbf{r}(t_0)$ is defined.

2. *Examples:* Construct examples of the thing(s) described in the following. Try to find examples that are different than any in the reading.

   (a) An arc length parametrization for a straight line in the plane.

   (b) A space curve with constant curvature.

   (c) A space curve with varying curvature.

3. Define what it means for a curve to be parametrized by arc length.

4. Explain how a ruler made of string could be used to understand an arc length parametrization.

5. Why are the concepts of "concave up" and "concave down" insufficient for quantifying how a space curve bends?

6. Let C be the graph of the vector-valued function $\mathbf{r}(t)$. Define the curvature at a point on C.

7. What makes Definition 11.21 for curvature easy to understand? What makes it difficult to use?

8. Let P_0 be a point on a curve C with positive curvature κ. Define the radius of curvature at P_0.

9. Compare the definition of curvature with the four formulas given in Theorems 11.23 and 11.24. Discuss when each formula would be the easiest to use.

10. Sam and Ben are arguing about the curvature of a function $y = f(x)$. Sam claims that the maximum curvature of every such function occurs at a relative extremum. Ben disagrees. Who is right? If Sam is right, prove the result. If Ben is right, provide a counterexample that disproves Sam's claim.

11. Recall that the graph of the vector function $\mathbf{r}(t) = \langle a + \alpha t, b + \beta t, c + \gamma t \rangle$ is a straight line in $\mathbb{R}^3$ if constants α, β, and γ are not all zero.

    (a) Show that the curvature at every point on this line is zero and the radius of curvature at every point is undefined.

    (b) Explain why what you showed in part (a) makes sense.

12. Let $\alpha > 0$.

    (a) Explain why the graph of the vector function $\mathbf{r}(t) = \langle a + \alpha \cos t, b + \alpha \sin t \rangle$ is a circle C with center (a, b) and radius α.

    (b) Show that the curvature at every point on C is $\dfrac{1}{\alpha}$ and the radius of curvature at every point is α.

    (c) Explain why the osculating circle at every point on C *is* the circle C.

    (d) Explain why the answers to parts (a), (b), and (c) make sense.

13. Give a step-by-step procedure for finding the equation of the osculating circle at a point on a planar curve where the curvature is nonzero.

14. Give a step-by-step procedure for finding the center and radius of the osculating circle at a point on a space curve where the curvature is nonzero.

15. Show that the second derivative of the function of $y = x^2$ is constant, but its curvature varies with x.

16. Show that the curvature of the function of $y = \sqrt{1 - x^2}$, $x \in (-1, 1)$, is constant, but its second derivative varies with x.

In Exercises 17–21 sketch the graph of a vector-valued function $\mathbf{r}(t) = x(t)\mathbf{i} + y(t)\mathbf{j}$ with the specified properties.

17. $\mathbf{r}(1) = \langle 1, 2 \rangle$, $\kappa(1) = 3$, $\mathbf{N}(1) = \left\langle \dfrac{3}{5}, \dfrac{4}{5} \right\rangle$

    $\mathbf{r}(3) = \langle -3, 5 \rangle$, $\kappa(3) = 3$, $\mathbf{N}(3) = \left\langle \dfrac{1}{\sqrt{2}}, \dfrac{1}{\sqrt{2}} \right\rangle$

18. $\mathbf{r}(0) = \langle 0, 0 \rangle$, $\kappa(0) = 0$, $\mathbf{N}(0)$ is undefined

    $\mathbf{r}(2) = \langle 3, 7 \rangle$, $\kappa(2) = 1$, $\mathbf{N}(2) = \langle 1, 0 \rangle$

19. $\mathbf{r}(1) = \langle 3, 5 \rangle$, $\rho(1) = 3$, $\mathbf{N}(1) = \langle 1, 0 \rangle$

    $\mathbf{r}(4) = \langle -2, 4 \rangle$, $\rho(4) = 2$, $\mathbf{N}(4) = \left\langle \dfrac{\sqrt{3}}{3}, \dfrac{\sqrt{6}}{3} \right\rangle$

20. κ is always 0, $\mathbf{r}(0) = \langle 2, 0 \rangle$, $\mathbf{r}(1) = \langle 0, 2 \rangle$

21. κ is always $\dfrac{1}{2}$, $\mathbf{r}(0) = \langle 2, 0 \rangle$, $\mathbf{r}(1) = \langle 0, 2 \rangle$

Skills

Find the arc length of the curves defined by the vector-valued functions on the specified intervals in Exercises 22–27.

22. $\mathbf{r}(t) = \langle 3t - 4, -2t + 5, t + 3 \rangle$, $[1, 5]$

23. $\mathbf{r}(t) = \langle 3 \cos 4t, 3 \sin 4t \rangle$, $\left[0, \dfrac{\pi}{2} \right]$

24. $\mathbf{r}(t) = \langle t - \sin t, 1 - \cos t \rangle$, $[0, 2\pi]$

25. $\mathbf{r}(t) = \left\langle 4 \sin t, t^{3/2}, -4 \cos t \right\rangle$, $[0, 4]$

26. $\mathbf{r}(t) = \left\langle e^t, \sqrt{2}t, e^{-t} \right\rangle$, $[0, 1]$

27. $\mathbf{r}(t) = \left\langle e^t \sin t, e^t \cos t, e^t \right\rangle$, $[0, \pi]$

In Exercises 28–30 show that the given vector-valued functions are not arc length parametrizations. Then find an arc length parametrization for the curves defined by those functions.

28. $\mathbf{r}(t) = \langle 3 \cos t, 3 \sin t \rangle$

29. $\mathbf{r}(t) = \langle 3 + 2t, 4 - t, -1 + 5t \rangle$

30. $\mathbf{r}(t) = \langle \cos t, \sin t, t \rangle$

In Exercises 31–35 find the curvature of the given function at the indicated value of x. Then sketch the curve and the osculating circle at the indicated point.

31. $y = e^x$, $x = 0$

32. $y = \sin x$, $x = \dfrac{\pi}{2}$

33. $y = \csc x$, $x = \dfrac{\pi}{2}$

34. $y = \sqrt{x}$, $x = 2$

35. $y = \dfrac{1}{3}(x^2 + 2)^{3/2}$, $x = 1$

Find the curvature of each of the functions defined by the parametric equations in Exercises 36–38.

36. $x = k\cos t$, $y = k\sin t$

37. $x = a\cos t$, $y = b\sin t$

38. $x = \cos t + t\sin t$, $y = \sin t - t\cos t$

Find the curvature of each of the vector-valued functions defined in Exercises 39–44.

39. $\mathbf{r}(t) = \langle t - \sin t, 1 - \cos t\rangle$

40. $\mathbf{r}(t) = \langle t, t^2, t^3\rangle$

41. $\mathbf{r}(t) = \langle t\sin t, t\cos t, t\rangle$

42. $\mathbf{r}(t) = \langle t\sin t, t\cos t, 2t\rangle$

43. $\mathbf{r}(t) = \langle t\sin t, t\cos t, \alpha t\rangle$, where α is a constant

44. $\mathbf{r}(t) = \langle \sin t, \cos t, \cosh t\rangle$

45. Find the value(s) of x where the curvature is greatest on the graph of $y = x^3$.

46. Show that the curvature on the parabola defined by $y = x^2$ is greatest at the origin.

47. Show that the curvature is constant at every point on the circular helix defined by $\mathbf{r}(t) = \langle a\cos t, a\sin t, bt\rangle$, where a and b are positive constants.

48. Find the points on the graphs of $y = e^x$ and $y = \ln x$ where the curvature is maximal. Explain how the two answers are related.

49. Let $\mathbf{r}(t) = \left\langle \cos t, \sin t, \dfrac{1}{t}\right\rangle$, $t > 0$. Show that $\lim\limits_{t\to\infty} \kappa = 1$ and $\lim\limits_{t\to 0^+} \kappa = 0$.

50. Find the curvature on the graph of the elliptical helix defined by $\mathbf{r}(t) = \langle a\cos t, b\sin t, ct\rangle$, where a, b, and c are positive constants.

Exercises 51 and 52 derive the equations necessary to define the **torsion** τ of a space curve. (Torsion measures the rate at which a space curve twists away from the osculating plane.) In each of these exercises, let C be a space curve and let $\mathbf{r}(s)$ be an arc length parametrization for C such that $\dfrac{d\mathbf{T}}{ds}$ and $\dfrac{d\mathbf{N}}{ds}$ exist at every point on C.

51. Explain why $\dfrac{d\mathbf{B}}{ds}$ exists at every point on C. (*Hint: Differentiate* $\mathbf{B} = \mathbf{T} \times \mathbf{N}$ *with respect to arc length.*)

52. Use Exercise 51 and Theorem 11.12 to show that $\dfrac{d\mathbf{B}}{ds}$ is parallel to $\mathbf{N}$. (*Hint: Recall that* $\dfrac{d\mathbf{T}}{ds} = \kappa\mathbf{N}$.) Since $\dfrac{d\mathbf{B}}{ds}$ and $\mathbf{N}$ are parallel, we define the torsion τ to be the scalar such that $\dfrac{d\mathbf{B}}{ds} = -\tau\mathbf{N}$.

53. Use Exercise 52 to show that $\dfrac{d\mathbf{N}}{ds} = -\kappa\mathbf{T} + \tau\mathbf{B}$, where κ is the curvature. (*Hint: Differentiate* $\mathbf{N} = \mathbf{B} \times \mathbf{T}$ *with respect to arc length.*)

Use the definition of torsion in Exercise 52 to compute the torsion of the vector functions in Exercises 54–56.

54. $\mathbf{r}(t) = \langle \cos t, \sin t, t\rangle$

55. $\mathbf{r}(t) = \langle \cosh t, \sinh t, t\rangle$

56. $\mathbf{r}(t) = \langle 3\sin t, 5\cos t, 4\sin t\rangle$

57. Use the results of Exercise 52 to show that the torsion of a planar curve C is zero at every point on C.

Applications

58. Ian will walk along a contour on the side of a mountain to get from a point he calls the origin to the point $(2, 0)$, with distances measured in miles. Unfortunately, the mountainside is riven by gullies and ridges. Thus, his route will follow the curve

$$\mathbf{r}(t) = \langle 1.2t - 0.5t^2 + 0.1t^4, 0.15\sin(6\pi t)\rangle.$$

How far does Ian actually have to walk?

59. Travelling along a contour on a mountainside is usually most difficult at the points of greatest curvature of the

route, since these indicate deep gullies or sharp, rocky promontories that must be passed.

(a) Sketch the graph of Ian's route from the previous problem.

(b) Sketch each component of the curve describing his route. What information does that give you about the curvature?

(c) Roughly where does Ian expect the greatest difficulties?

Proofs

60. Use Theorem 11.24 to prove that the curvature of a linear function $y = mx + b$ is zero for every value of x.

61. Use Theorem 11.24 to prove that the curvature is zero at a point of inflection of a twice-differentiable function $y = f(x)$.

62. Prove that a planar curve with constant curvature is either a line or a circle. Give an example of a space curve with constant curvature that is neither a line nor a circle.

63. Prove that the magnitude of the acceleration of a particle moving along a curve, C, with a constant speed is proportional to the curvature of C.

64. Let C be the graph of a vector function $\mathbf{r}(s)$ defined on an interval I and parametrized by arc length s.

(a) If the unit tangent vector $\mathbf{T}$ has a nonzero derivative, use the chain rule and the definitions of the unit tangent vector $\mathbf{T}$ and principal unit normal vector $\mathbf{N}$ to show that $\dfrac{d\mathbf{T}}{ds}$, $\dfrac{d\mathbf{T}}{dt}$, and $\mathbf{N}$ are scalar multiples of each other. In particular, argue that $\dfrac{d\mathbf{T}}{ds} = \alpha\mathbf{N}$ for some scalar α.

(b) Use the fact that $\mathbf{N}$ is a unit vector to show that the constant α in part (a) must equal κ.

65. Let $y = f(x)$ be a twice-differentiable function. Show that the curvature of the graph of f is given by

$$\kappa = \frac{|f''(x)|}{(1 + (f'(x))^2)^{3/2}}.$$

This is part (a) of Theorem 11.24. (*Hint: Use the parametrization $x = t$, $y = f(t)$.*)

66. Let C be the graph of a vector function $\mathbf{r}(t) = \langle x(t), y(t)\rangle$ in the xy-plane, where $x(t)$ and $y(t)$ are twice-differentiable functions of t. Show that the curvature κ of C at a point on the curve is given by

$$\kappa = \frac{|x'(t)y''(t) - x''(t)y'(t)|}{((x'(t))^2 + (y'(t))^2)^{3/2}}.$$

This is part (b) of Theorem 11.24.

Thinking Forward

A decomposition of the acceleration vector: Find $\text{comp}_{\mathbf{v}(t)}\mathbf{a}(t)$, where $\mathbf{v}$ and $\mathbf{a}$ are the velocity and acceleration vectors, respectively, of the following functions.

▶ $\mathbf{r}(t) = \langle t, t^2, t^3\rangle$

▶ $\mathbf{r}(t) = \langle \cos t, \sin t, t\rangle$

▶ $\mathbf{r}(t) = \langle e^t \sin t, e^t \cos t, e^t\rangle$

11.5 MOTION

▶ The behavior of objects subject to the force of gravity

▶ The displacement and distance of projectiles

▶ The tangential and normal components of the acceleration vector

Motion Under the Force of Gravity

In Chapter 2 we discussed motion along a straight path. The motions we examined were of two types: straight up and down, and straight back and forth. Now that we have studied vector functions, we are ready to widen the discussion to motion along curves in a plane and motion along a space curve. We begin by discussing motion under the force of gravity.

Imagine a catapult that can propel a rock with an initial velocity of 30 miles per hour. A number of questions arise:

▶ If the catapult always releases the rock at the same angle to the horizontal, how far will the rock travel?

▶ Along what path would the rock travel?

▶ What is the maximum height reached by the rock?

▶ How long will it take until the rock hits the ground?

▶ How fast will the rock be travelling upon impact?

▶ If the catapult can be moved to the top of a hill, would the rock travel farther?

▶ If the angle at which the catapult releases the rock can be adjusted, what angle would maximize the distance the rock travels?

▶ Will the motion be planar?

We will answer some of these questions in this section and leave others to the exercises.

To begin our analysis we will assume that wind resistance and friction are negligible and that our object is a point mass. Recall that (on Earth) the gravitational constant g is 32 feet per second per second in the English system, and 9.8 meters per second per second in the metric system. We know from experience that a thrown object travels in a curved path. When we ignore wind resistance, we may assume that the motion is planar and use a two-dimensional coordinate system that places the initial position of the object at $(0, h)$, where h is the starting height of the projectile (as it is released). The following diagram illustrates the starting point of our analysis:

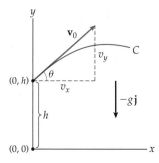

We assume that the projectile moves along a planar curve C given by the vector function $\mathbf{r}(t) = \langle x(t), y(t) \rangle$. The initial velocity of the vector function is

$$\mathbf{v}_0 = \mathbf{v}(0) = \mathbf{r}'(0) = v_x \mathbf{i} + v_y \mathbf{j} = \|\mathbf{v}_0\| \cos\theta \, \mathbf{i} + \|\mathbf{v}_0\| \sin\theta \, \mathbf{j}.$$

We also assume that the acceleration of the projectile after it is released is due solely to gravity, which is constant and acts downwards. Therefore, $\mathbf{r}''(t) = \mathbf{a}(t) = -g\mathbf{j}$. We will now be able to find the vector function $\mathbf{v}$ by integrating $\mathbf{a}$. That is,

$$\mathbf{v}(t) = \int \mathbf{a}(t) \, dt = \int \langle 0, -g \rangle \, dt = \langle c_1, -gt + c_2 \rangle,$$

where c_1 and c_2 are constants of integration. We may evaluate these constants because we know the initial velocity $\mathbf{v}(0) = \langle \|\mathbf{v}_0\| \cos\theta, \|\mathbf{v}_0\| \sin\theta \rangle = \langle c_1, c_2 \rangle$. Consequently,

$$c_1 = \|\mathbf{v}_0\| \cos\theta \quad \text{and} \quad c_2 = \|\mathbf{v}_0\| \sin\theta.$$

We have the velocity function

$$\mathbf{v}(t) = \langle \|\mathbf{v}_0\| \cos\theta, -gt + \|\mathbf{v}_0\| \sin\theta \rangle.$$

Now, we integrate $\mathbf{v}$ to find $\mathbf{r}$:

$$\mathbf{r}(t) = \int \mathbf{v}(t) \, dt = \int \langle \|\mathbf{v}_0\| \cos\theta, -gt + \|\mathbf{v}_0\| \sin\theta \rangle \, dt$$
$$= \left\langle (\|\mathbf{v}_0\| \cos\theta) \, t + C_1, -\frac{1}{2}gt^2 + (\|\mathbf{v}_0\| \sin\theta) \, t + C_2 \right\rangle.$$

Here C_1 and C_2 are new constants of integration that may be evaluated by using the initial position of the projectile. We positioned our coordinate system so that $\mathbf{r}(0) = \langle 0, h \rangle$. So, we obtain $C_1 = 0$ and $C_2 = h$; thus, the parametrization for C is

$$\mathbf{r}(t) = \left\langle (\|\mathbf{v}_0\| \cos\theta) \, t, -\frac{1}{2}gt^2 + (\|\mathbf{v}_0\| \sin\theta) \, t + h \right\rangle.$$

It follows that the path of the projectile is a parabola, because when we eliminate the parameter t from the equations

$$x = (\|\mathbf{v}_0\| \cos\theta) \, t, \quad y = -\frac{1}{2}gt^2 + (\|\mathbf{v}_0\| \sin\theta) \, t + h,$$

we obtain the quadratic function

$$y = -\frac{g}{2\|\mathbf{v}_0\|^2 \cos^2\theta} x^2 + (\tan\theta) x + h.$$

(Recall that $\mathbf{v}_0$, θ and h are all constants.)

We will consider extended examples next.

Displacement and Distance Travelled

We are all familiar with the sentence:

The shortest distance between two points is a straight line.

This shortest distance is the displacement, while the length of the curve traversed by an object is the distance travelled. In terms of vector functions we have the following definition:

DEFINITION 11.26   Displacement Versus Distance Travelled

Let the vector-valued function $\mathbf{r}(t)$ represent the position function for a particle travelling on a curve C. If the domain of $\mathbf{r}$ is an interval I containing points a and b:

(a) The ***displacement*** of the the particle from a to b is given by the vector $\mathbf{r}(b) - \mathbf{r}(a)$.

(b) The ***distance travelled*** by the particle from a to b is given by the scalar $\int_a^b \|\mathbf{r}'(t)\|dt$.

The following figure illustrates these definitions:

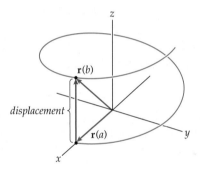

The displacement vector is the vector with initial point $\mathbf{r}(a)$ and terminal point $\mathbf{r}(b)$. The distance travelled is this length of the curve connecting $\mathbf{r}(a)$ and $\mathbf{r}(b)$. As a quick example, for the helix defined by the vector function $\mathbf{r}(t) = \langle \cos t, \sin t, t \rangle$, the displacement vector from 0 to 2π is

$$\mathbf{r}(2\pi) - \mathbf{r}(0) = \langle \cos 2\pi, \sin 2\pi, 2\pi \rangle - \langle \cos 0, \sin 0, 0 \rangle = \langle 1, 0, 2\pi \rangle - \langle 1, 0, 0 \rangle = \langle 0, 0, 2\pi \rangle,$$

while the distance travelled by a particle on the helix from 0 to 2π is

$$\int_0^{2\pi} \|\mathbf{r}'(t)\|dt = \int_0^{2\pi} \|\langle -\sin t, \cos t, 1 \rangle\|\, dt = \int_0^{2\pi} \sqrt{\sin^2 t + \cos^2 t + 1}\ dt$$

$$= \int_0^{2\pi} \sqrt{2}\ dt = 2\sqrt{2}\pi.$$

Tangential and Normal Components of Acceleration

We have already seen that every vector in $\mathbb{R}^3$ can be decomposed into a sum of three vectors that are directed along the x-, y-, and z-axes. Doing this requires nothing more than writing a vector $\mathbf{v}$ in the form $\mathbf{v} = a\mathbf{i} + b\mathbf{j} + c\mathbf{k}$. Rather than using this type of decomposition, however, when we are working with curves resulting from the motion of objects in $\mathbb{R}^3$, it is often convenient to decompose the acceleration vector into a sum of two vectors, one of which is in the direction of the unit tangent vector $\mathbf{T}$ and the other is in the direction of the principal unit normal vector **N**. These two components are known as the ***tangential component of acceleration*** and the ***normal component of acceleration***, respectively, and are illustrated as follows:

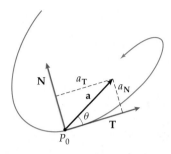

Even the graph of a very complicated vector function in $\mathbb{R}^3$ is locally planar, as long as the function is twice differentiable. As we mentioned in Section 11.3, the osculating plane fits a space curve "best" at a point P_0 on the curve, and the osculating plane is determined by the point P_0 and the unit tangent and principal unit normal vectors at P_0.

THEOREM 11.27

The Tangential and Normal Components of Acceleration

Let $\mathbf{r}(t)$ be a twice-differentiable vector function with either two or three components and with derivatives $\mathbf{r}'(t) = \mathbf{v}(t)$ and $\mathbf{r}''(t) = \mathbf{a}(t)$. Then the tangential component of acceleration a_T is given by

$$a_{\mathbf{T}} = \frac{\mathbf{v} \cdot \mathbf{a}}{\|\mathbf{v}\|}$$

and the normal component of acceleration a_N is given by

$$a_{\mathbf{N}} = \frac{\|\mathbf{v} \times \mathbf{a}\|}{\|\mathbf{v}\|}.$$

Proof. Since vectors **v** and **T** are scalar multiples, $\text{comp}_{\mathbf{T}}\mathbf{a} = \text{comp}_{\mathbf{v}}\mathbf{a}$. From the figure shown, if θ is the angle between vectors **a** and **v**, then

$$a_{\mathbf{T}} = \|\mathbf{a}\| \cos\theta = \frac{\|\mathbf{v}\|\|\mathbf{a}\| \cos\theta}{\|\mathbf{v}\|} = \frac{\mathbf{v} \cdot \mathbf{a}}{\|\mathbf{v}\|}.$$

Similarly,

$$a_{\mathbf{N}} = \|\mathbf{a}\| \sin\theta = \frac{\|\mathbf{v}\|\|\mathbf{a}\| \sin\theta}{\|\mathbf{v}\|} = \frac{\|\mathbf{v} \times \mathbf{a}\|}{\|\mathbf{v}\|}. \qquad \blacksquare$$

Note that if $\mathbf{r}(t) = \langle x(t), y(t) \rangle$, then to take the cross product in Theorem 11.27, we would let $\mathbf{v}(t) = \langle x'(t), y'(t), 0 \rangle$, and let $\mathbf{a}(t) = \langle x''(t), y''(t), 0 \rangle$. The normal component of acceleration is also called ***centripetal acceleration***.

Examples and Explorations

EXAMPLE 1

The Human Cannonball

Chuckles the Human Cannonball is to be shot from the circus cannon. The muzzle of the cannon is 10 feet above ground level and is initially at an angle of 30° with the horizontal. The muzzle velocity of the cannon is 30 mph. The square net to catch Chuckles measures 20 feet on each side and is placed at a height of 6 feet, 30 feet directly in front of the muzzle of the cannon. Should Chuckles call an ambulance?

SOLUTION

We place our coordinate system so that the muzzle of the cannon is at $(0, 10)$. The leading edge of the net will then be at $(30, 6)$. The muzzle speed of 30 mph is equivalent to 44 feet/sec. At the angle of $30°$, Chuckles' initial velocity function will be

$$\mathbf{v}_0 = 44 \cos 30°\mathbf{i} + 44 \sin 30°\mathbf{j} = 22\sqrt{3}\mathbf{i} + 22\mathbf{j}.$$

Since, in general, $\mathbf{r}(t) = \left\langle (\|\mathbf{v}_0\| \cos\theta)\, t, -\frac{1}{2}gt^2 + (\|\mathbf{v}_0\| \sin\theta)\, t + h \right\rangle$, here we have

$$\mathbf{r}(t) = \left\langle (\|\mathbf{v}_0\| \cos\theta)\, t, -\frac{1}{2}gt^2 + (\|\mathbf{v}_0\| \sin\theta)\, t + h \right\rangle = \left\langle 22\sqrt{3}t, -16t^2 + 22t + 10 \right\rangle,$$

where we have used the gravitational constant $g = 32$ feet/sec^2. We next find the time it takes until Chuckles would reach the net's height of 6 feet. To do this we set the y-component of $\mathbf{r}(t)$ equal to 6 and solve for t.

That is, we solve $-16t^2 + 22t + 10 = 6$ and obtain the roots $t = \frac{11 \pm \sqrt{185}}{16}$. In this context, only the positive root $t = \frac{11 + \sqrt{185}}{16}$ is relevant. It would take Chuckles approximately 1.53 seconds before he reached a height of 6 feet. We now use the x-component of $\mathbf{r}(t)$ to find the horizontal distance that Chuckles would travel in 1.53 seconds and obtain the distance $22\sqrt{3}(1.53) \approx 58.3$ feet. Since the net extends only from 30 feet to 50 feet from the cannon muzzle, Chuckles should prepare for a bad landing.   □

EXAMPLE 2

The Human Cannonball, part 2

Chuckles the Human Cannonball is unhappy with the planned scenario of Example 1. If the angle of the cannon can be adjusted, at what angle should Chuckles ask to be shot so that he hits the middle of the net? You may assume that all other constants given in Example 1 remain the same.

SOLUTION

We now have the vector function

$$\mathbf{r}(t) = \left\langle (44 \cos\theta)\, t, -16t^2 + (44 \sin\theta)\, t + 10 \right\rangle.$$

We need to find the values of t and θ such that the x-component is 40 and the y-component is 6. That is, we need to solve the system of equations

$$(44 \cos\theta)\, t = 40 \quad \text{and} \quad -16t^2 + (44 \sin\theta)\, t + 10 = 6$$

simultaneously for t and θ. There are four pairs of values for t and θ that satisfy this system of equations, but only the two pairs in which t is positive make sense contextually. One pair is $t \approx 0.94$ second and $\theta \approx 14°$. The other pair is $t \approx 2.68$ seconds and $\theta \approx 70°$. (We ask you to solve the same system of equations in Exercise 25.)   □

CHECKING THE ANSWER

In Example 1 Chuckles would have gone over 55 feet if he had been shot at an angle of 30°. It makes sense that if he is shot at a smaller angle of elevation, he wouldn't travel as far. In Exercise 33 we will ask you to prove that in a ground-to-ground trajectory, a projectile always travels the greatest horizontal distance when it is shot at an angle of 45° with the horizontal. Therefore, it also makes sense that there would be a second solution in Example 2 in which the angle of elevation would be greater than 45°. This doesn't, of course, tell us that the numbers we found are correct, only that they are reasonable.

EXAMPLE 3

The Human Cannonball, part 3

Chuckles the Human Cannonball is happier with the scenario of Example 2, but he needs to decide which of the two angles would be preferable. The big top tent he is under is 50 feet tall.

SOLUTION

If Chuckles is shot at the smaller angle, he would hit the net at a more shallow angle, which might lead to his skidding off the net. For this reason the steeper angle might be preferable. However, we should also consider his speed upon impact and the maximum height he would reach if he were shot at the two angles. We consider his speed upon impact and the maximum height he would reach if he were shot at an angle of 70°. We leave the analysis at the initial angle of 14° for Exercise 26.

Since the velocity function under the force of gravity is

$$\mathbf{v}(t) = \langle \|\mathbf{v}_0\| \cos\theta, -gt + \|\mathbf{v}_0\| \sin\theta \rangle,$$

we have

$$\mathbf{v}(t) = \langle \|\mathbf{v}_0\| \cos\theta, -gt + \|\mathbf{v}_0\| \sin\theta \rangle = \langle 44\cos 70°, -32t + 44\sin 70° \rangle$$
$$= \langle 15.0, -32t + 41.3 \rangle.$$

To find Chuckles' speed upon impact, recall that speed is the magnitude of velocity and that, at an angle of 70°, he will hit the net after about 2.7 seconds. Therefore, his speed upon impact will be

$$\|\mathbf{v}(2.7)\| = \|\langle 15.0, -32(2.7) + 41.3 \rangle\| = \|\langle 15.0, -45 \rangle\| \approx 47 \text{ feet per second.}$$

To find the maximum height we consider the y-component of the position function $\mathbf{r}(t)$. This is

$$y(t) = -16t^2 + (44\sin 70°)t + 10 = -16t^2 + 41.3t + 10.$$

The maximum height will be attained when $y'(t) = -32t + 41.3 = 0$. (Why?) Thus, the maximum height occurs at about 1.3 seconds. The corresponding height is

$$y(1.3) = -16(1.3)^2 + 41.3(1.3) + 10 \approx 37 \text{ feet.}$$

He has plenty of room to spare!   □

EXAMPLE 4

Finding centripetal acceleration

A girl is swinging a small plane attached to a string 2 feet long in a circle over her head at a rate of 20 revolutions per minute. If the circle is horizontal at a height of 6 feet above the ground, find a vector function that models the path of the plane and find the centripetal acceleration of the plane.

SOLUTION

The vector function

$$\mathbf{r}(t) = \left\langle 2\cos\left(\frac{2\pi}{3}t\right), 2\sin\left(\frac{2\pi}{3}t\right), 6 \right\rangle$$

has the motion described, where t is measured in seconds. Note that every 3 seconds the plane would complete exactly one revolution, as specified. The velocity and acceleration vectors are, respectively,

$$\mathbf{v}(t) = \left\langle -\frac{4\pi}{3}\sin\left(\frac{2\pi}{3}t\right), \frac{4\pi}{3}\cos\left(\frac{2\pi}{3}t\right), 0 \right\rangle \text{ and}$$

$$\mathbf{a}(t) = \left\langle -\frac{8\pi^2}{9}\cos\left(\frac{2\pi}{3}t\right), -\frac{8\pi^2}{9}\sin\left(\frac{2\pi}{3}t\right), 0 \right\rangle.$$

The centripetal acceleration, or equivalently, the normal component of acceleration, $a_{\mathbf{N}}$, is given by

$$a_{\mathbf{N}} = \frac{\|\mathbf{v}\times\mathbf{a}\|}{\|\mathbf{v}\|} = \frac{\left\|\left\langle -\frac{4\pi}{3}\sin\left(\frac{2\pi}{3}t\right), \frac{4\pi}{3}\cos\left(\frac{2\pi}{3}t\right), 0 \right\rangle \times \left\langle -\frac{8\pi^2}{9}\cos\left(\frac{2\pi}{3}t\right), -\frac{8\pi^2}{9}\sin\left(\frac{2\pi}{3}t\right), 0 \right\rangle\right\|}{\left\|\left\langle -\frac{4\pi}{3}\sin\left(\frac{2\pi}{3}t\right), \frac{4\pi}{3}\cos\left(\frac{2\pi}{3}t\right), 0 \right\rangle\right\|}$$

$$= \frac{\|\langle 0, 0, 32\pi^3/27\rangle\|}{4\pi/3} = \frac{32\pi^3/27}{4\pi/3} = \frac{8\pi^2}{9}.$$

Note that in this example the velocity and acceleration vectors are orthogonal. When this occurs, the normal component of acceleration will be the magnitude of the acceleration vector, as we find here.   □

? TEST YOUR UNDERSTANDING |

▶ How are the vector functions giving the velocity and position of an object moving under the acceleration due to gravity derived from the gravitational constant?

▶ Why does a projectile subject to gravity move in a parabolic arch?

▶ How are the component functions for the position and velocity of a projectile used to determine the height reached by the object, the distance travelled by the object, and the travel time of the projectile?

▶ What are the normal and tangential components of acceleration? How is the osculating plane related to the components of acceleration?

▶ What is centripetal acceleration?

EXERCISES 11.5

Thinking Back

▶ *Projecting one vector onto another:* Show that the formula for the projection of a vector $\mathbf{v}$ onto a nonzero vector $\mathbf{u}$ is given by $\text{proj}_{\mathbf{u}}\mathbf{v} = \frac{\mathbf{u}\cdot\mathbf{v}}{\|\mathbf{u}\|}$, where $\mathbf{u}\neq\mathbf{0}$.

▶ *The difference between a vector and its projection onto another vector:* If $\mathbf{u}\neq\mathbf{0}$ and $\mathbf{v}$ is an arbitrary vector, what is the geometric relationship between $\mathbf{v}$, $\text{proj}_{\mathbf{u}}\mathbf{v}$, and $\mathbf{v} - \text{proj}_{\mathbf{u}}\mathbf{v}$?

Concepts

0. *Problem Zero:* Read the section and make your own summary of the material.

1. *True/False:* Determine whether each of the statements that follow is true or false. If a statement is true, explain why. If a statement is false, provide a counterexample.

   (a) *True or False:* When the wind resistance is negligible, a projectile travels on a parabolic path if its initial velocity has a nonzero horizontal component.

   (b) *True or False:* The velocity of a projectile is zero when it has reached its maximum height.

   (c) *True or False:* If the wind resistance is negligible, the horizontal component of velocity is constant.

   (d) *True or False:* The acceleration of a projectile is $-g\mathbf{k}$, where g is the gravitational constant.

   (e) *True or False:* The velocity of a moving particle is defined to be the absolute value of its speed.

(f) *True or False:* Centripetal acceleration and the normal component of acceleration are identical.

(g) *True or False:* The sum of the tangential component of acceleration and the normal component of acceleration is the acceleration.

(h) *True or False:* If $\|\mathbf{r}(b) - \mathbf{r}(a)\| = \int_a^b \|\mathbf{r}'(t)\|\, dt$ for every a and b in the domain of $\mathbf{r}$, then the curve defined by $\mathbf{r}$ is parametrized by arc length.

2. *Examples:* Construct examples of the thing(s) described in the following. Try to find examples that are different than any in the reading.

(a) A nonconstant vector function $\mathbf{r}(t)$ defined on an interval $[a, b]$ for which
$$\|\mathbf{r}(b) - \mathbf{r}(a)\| = \int_a^b \|\mathbf{r}'(t)\| dt.$$

(b) A vector function with a nonzero acceleration vector such that the normal component of acceleration is always zero.

(c) A vector function with a nonzero acceleration vector such that the tangential component of acceleration is always zero.

3. Explain why an object thrown horizontally on a planet with a smaller gravitational force than Earth's would travel farther.

In our derivation of the position function $\mathbf{r}(t) = \left\langle (\|\mathbf{v}_0\| \cos\theta)\, t, -\frac{1}{2} g t^2 + (\|\mathbf{v}_0\| \sin\theta)\, t + h \right\rangle$ for the motion of a projectile, we ignored air resistance and wind effects. In Exercises 4 and 5 we ask you to think about how these effects would change the model.

4. Air resistance is a type of friction. It is a vector that is proportional to, and acting in the opposite direction from, the velocity. How would incorporating air resistance change the development of our model of the motion of a projectile?

5. We may model the wind as a vector with a direction. How would including the wind change the development of our model of the motion of a projectile?

6. A projectile reaches its maximum height H at time $t_0 > 0$. What fraction of H does it reach at time $\frac{1}{2} t_0$?

7. Given that an object is moving along the graph of a vector function $\mathbf{r}(t)$ defined on an interval $[a, b]$, what is meant by the displacement of the object from $t = a$ to $t = b$?

8. If an object is moving along the graph of a vector function $\mathbf{r}(t)$ defined on an interval $[a, b]$, how can the distance the object travels from $t = a$ to $t = b$ be calculated?

9. Define each of the following: the tangential component of acceleration, the normal component of acceleration, and centripetal acceleration. How are each of these quantities computed?

10. In the graph that follows, a portion of a curve is drawn, along with the unit tangent vector, $\mathbf{T}$, principal unit normal vector, $\mathbf{N}$, and acceleration vector, $\mathbf{a}$, at point P. Add the vectors $\text{proj}_\mathbf{T}\mathbf{a}$ and $\text{proj}_\mathbf{N}\mathbf{a}$ to the graph, and label the magnitudes of these vectors appropriately.

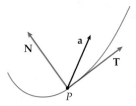

11. Imagine that you are driving on a twisting mountain road. Describe the tangential and normal components of acceleration as you ascend, descend, twist right, and twist left.

12. Two marbles roll off a table at the same moment. One of the marbles has a horizontal speed of 1 inch per second, and the other has a horizontal speed of 4 inches per second.

(a) Which marble hits the ground first?

(b) If the surface of the table is 3 feet from the ground, what are the speeds of the marbles as they hit?

13. Let $f(t)$ be a nonconstant continuous function.

(a) Explain why a particle moving along the vector function $\mathbf{r}(t) = \langle f(t), (f(t))^2 \rangle$ is on a parabolic path.

(b) What condition must $f(t)$ satisfy so that $\mathbf{r}(t) \cdot \mathbf{a}(t) = 0$?

Skills

In Exercises 14–17, find (a) the displacement vectors from $\mathbf{r}(a)$ to $\mathbf{r}(b)$, (b) the magnitude of the displacement vector, and (c) the distance travelled by a particle on the curve from a to b.

14. $\mathbf{r}(t) = \langle t, t^2 \rangle,\ a = 2,\ b = 3$

15. $\mathbf{r}(t) = \langle t \sin t, t \cos t \rangle,\ a = 0,\ b = \pi$

16. $\mathbf{r}(t) = \langle e^t \sin t, e^t \cos t, e^t \rangle,\ a = 0,\ b = 1$

17. $\mathbf{r}(t) = \langle \alpha \sin \beta t, \alpha \cos \beta t, \gamma t \rangle,\ a = 0,\ b = 1$

Find the tangential and normal components of acceleration for the position functions in Exercises 18–22

18. $\mathbf{r}(t) = \langle t, t^2 \rangle$

19. $\mathbf{r}(t) = \langle \sin 3t, \cos 4t \rangle$

20. $\mathbf{r}(t) = \langle t \sin t, t \cos t \rangle$

21. $\mathbf{r}(t) = \langle e^t \sin t, e^t \cos t, e^t \rangle$

22. $\mathbf{r}(t) = \langle 2 \sin t, 3 \cos t, \cos 2t \rangle$

23. Find the tangential and normal components of acceleration for a particle moving along the circular helix defined by $\mathbf{r}(t) = \langle \cos t, \sin t, t \rangle$.

24. Find the tangential and normal components of acceleration for a particle moving along the conical helix defined by $\mathbf{r}(t) = \langle t \cos t, t \sin t, t \rangle$.

Applications

25. Finish Example 2 by solving the system of equations

    $$(44 \cos \theta) \, t = 40 \quad \text{and} \quad -16t^2 + (44 \sin \theta) \, t + 10 = 6$$

    for θ and t. (Hint: Solve the first equation for $\cos \theta$ and the second equation for $\sin \theta$, and use a Pythagorean identity.)

26. Finish Example 3 by calculating the maximum height Chuckles the Human Cannonball will reach and the speed at which he will hit the net if he is shot from the cannon at an elevation of $14°$.

Complete Exercises 27–31 using the appropriate gravitational constants from the table that follows. In each exercise you may ignore wind resistance.

	g in m/sec^2	g in ft/sec^2
Earth	9.8	32
Moon	1.6	5.2
Mars	3.7	12.1
Jupiter	24.8	81

27. In 1968 Bob Beamon set an Olympic record in the long jump in the Mexico City Olympic Games with a jump of 8.90 meters.

    (a) Assuming that his angle of elevation was $30°$, how fast was he running as he jumped?

    (b) Assuming that he could have performed the long jump on the moon with the same initial velocity and same angle of elevation, how far would Beamon have gone?

28. There is considerable controversy surrounding who hit the longest home run in professional baseball. However, any ball that travels 500 feet in the air is considered "monumental." Assuming that a hitter contacts a ball 4 feet from the ground and that the ball leaves the bat with a $45°$ angle of elevation and subsequently travels a horizontal distance of 500 feet, answer the following questions:

    (a) How fast is the speed of the ball right after contact?

    (b) What is the maximum height reached by the ball?

    (c) How long does the ball spend in the air after being hit?

    (d) On May 22, 1963, Mickey Mantle, playing for the New York Yankees, hit a home run that struck a point on the façade of the right-field roof of the old Yankee Stadium approximately 115 feet high and 370 feet from home plate. Explain why this is not enough information to determine how far the ball would have travelled if it had not hit the façade.

29. In 1974 Nolan Ryan, then playing for the California Angels, was credited with throwing one of the fastest pitches in professional baseball history. The pitch reached a speed of 100.9 mph.

    (a) Assume that the horizontal distance from the pitcher's mound to the catcher's glove was 60 feet, that the ball left Ryan's hand at a height of 6 feet, and that it was caught by the catcher with his hand at a height of 3 feet. What was the angle of elevation of the pitch as it left Ryan's hand?

    (b) If Ryan had thrown the ball at the same speed and from the same height at an angle of elevation of $45°$, what horizontal distance would the ball have travelled before it hit the ground?

30. A certain type of tank gun can be shot with a muzzle velocity of 1500 meters per second.

    (a) At what angle should the cannon be shot to hit a target 1 kilometer away? You may assume that the target and the muzzle of the cannon are at the same height.

    (b) What is the range of the cannon? Assume that the muzzle of the cannon is 5 meters high.

    (c) If the tank is repositioned atop a small hill so that the muzzle of the cannon is now 20 meters above the surrounding area, what is the new range of the cannon?

    (d) If the cannon were on Jupiter with its muzzle 5 meters high, what would its range be?

31. An aircraft is flying toward the tank gun discussed in Exercise 30. Assume that the muzzle of the gun is 5 meters high and the gun can fire with a maximum angle of elevation of $60°$. At what altitude should the pilot of the plane fly to keep out of the gun's range?

32. In 1971, during NASA's Apollo 14 mission to the moon, astronaut Alan Shepard hit a golf ball that is estimated to have travelled 2400 feet on the lunar surface, almost a half mile. Assuming a similar terrain, and ignoring wind resistance, how far would his shot have travelled on Earth?

Proofs

33. Show that a ball thrown with an angle of elevation of $45°$ will travel farther than if it is thrown at any other angle. (Hint: Assume that the initial and final heights of the ball are equal, that the initial velocity is fixed, and that friction and wind resistance are negligible.)

34. Prove that an object thrown at an acute angle of elevation travels along a parabolic path, if wind resistance is ignored.

35. Let $\mathbf{r}(t)$ be a vector function whose graph is a space curve containing distinct points P and Q. Prove that if the acceleration is always $\mathbf{0}$, then the graph of $\mathbf{r}$ is a straight line.

36. Assuming that the initial height of a projectile is zero, prove that doubling the initial velocity of the projectile has the effect of multiplying the maximum height of the projectile and the horizontal distance travelled by the projectile by a factor of 4.

37. Prove that the normal component of acceleration is 0 at a point of inflection on the graph of a twice-differentiable function $y = f(x)$. (Hint: The graph of $y = f(x)$ is also the graph of the vector function $\mathbf{r}(t) = \langle t, f(t) \rangle$.)

38. Prove that if a particle moves along a curve at a constant speed, then $a_{\mathbf{N}} = \|\mathbf{a}\|$ and $a_{\mathbf{T}} = 0$. Is the converse true? That is, if $a_{\mathbf{N}} = \|\mathbf{a}\|$ and $a_{\mathbf{T}} = 0$, must the particle be moving at a constant speed?

Thinking Forward

▶ *The graph of a function $f : \mathbb{R}^2 \to \mathbb{R}$*

A vector-valued function with two components can be thought of as a function with a domain that is a subset of $\mathbb{R}$ and a codomain $\mathbb{R}^2$. We've seen that the graph of such a function is a curve in the plane. We may also define functions that have a subset of $\mathbb{R}^2$ as the domain and the codomain $\mathbb{R}$. In general terms, what is the graph of such a function? (One such function is $f(x, y) = x^2 - 3x \sin y^2$.)

▶ *The graph of a function $f : \mathbb{R}^3 \to \mathbb{R}$*

A vector-valued function with three components can be thought of as a function with a domain that is a subset of $\mathbb{R}$ and a codomain $\mathbb{R}^3$. The graph of such a function is a space curve. We may also define functions that have a subset of $\mathbb{R}^3$ as the domain and the codomain $\mathbb{R}$. In general terms, what is the graph of such a function? (One such function is $f(x, y, z) = x + 2y - 5xyz$.)

CHAPTER REVIEW, SELF-TEST, AND CAPSTONES

Before you progress to the next chapter, be sure you are familiar with the definitions, concepts, and basic skills outlined here. The capstone exercises at the end bring together ideas from this chapter and look forward to future chapters.

Definitions

Give precise mathematical definitions or descriptions of each of the concepts that follow. Then illustrate the definition with a graph or an algebraic example.

▶ *parametric equations* in $\mathbb{R}^2$ and $\mathbb{R}^3$

▶ a *vector function* or *vector-valued function* in $\mathbb{R}^2$ and $\mathbb{R}^3$

▶ a *space curve*

▶ the *limit* of a vector function $\mathbf{r}(t)$ as t approaches t_0

▶ the *derivative* of a vector function $\mathbf{r}(t)$

▶ an *antiderivative* of a vector function $\mathbf{r}(t)$

▶ the *definite integral* of a vector function $\mathbf{r}(t)$ on an interval $[a, b]$

▶ the *unit tangent function* for a differentiable vector function $\mathbf{r}(t)$

▶ the *principal unit normal vector* at $\mathbf{r}(t)$

▶ the *binormal vector* at $\mathbf{r}(t_0)$

▶ the *osculating plane* at $\mathbf{r}(t_0)$

▶ an *arc length parametrization* for the graph of a differentiable function $\mathbf{r}(t)$

▶ the *curvature* at a point on the graph of a vector function $\mathbf{r}(s)$

▶ the *radius of curvature* at a point on the graph of a vector function $\mathbf{r}(t)$

▶ the *osculating circle* at a point on the graph of a vector function $\mathbf{r}(t)$

Theorems

Fill in the blanks to complete each of the following theorem statements:

▶ The derivative of a vector function $\mathbf{r}(t)$ is given by $\mathbf{r}'(t) = \lim_{h \to 0}$ _____.

▶ *The Chain Rule for Vector-Valued Functions:* Let $t = f(\tau)$ be a differentiable real-valued function of τ, and let $\mathbf{r}(t)$ be a differentiable vector function with either two or three components such that $f(\tau)$ is in the domain of $\mathbf{r}$ for every value of τ on some interval I. Then $\dfrac{d\mathbf{r}}{d\tau} = $ _____.

▶ Let C be the graph of a twice-differentiable vector function $\mathbf{r}(t)$ defined on an interval I and with unit tangent vector $\mathbf{T}(t)$. Then the curvature κ of C at a point on the curve is given by $\kappa = \dfrac{\|\quad\|}{\|\quad\|}$ and $\kappa = \dfrac{\|\quad \times \quad\|}{\|\quad\|^3}$.

▶ Let $y = f(x)$ be a twice-differentiable function. Then the curvature of the graph of f is given by $\kappa = \dfrac{|\quad|}{(\qquad)^{3/2}}$.

▶ Let C be the graph of a vector function $\mathbf{r}(t) = \langle x(t), y(t) \rangle$ in the xy-plane, where $x(t)$ and $y(t)$ are twice-differentiable functions of t such that $x'(t)$ and $y'(t)$ are not simultaneously zero. Then the curvature κ of C at a point on the curve is given by $\kappa = \dfrac{|\quad|}{(\qquad)^{3/2}}$.

Notation and Rules

Notation: Describe the meanings of each of the following mathematical expressions.

▶ $\mathbf{r}(t)$ ▶ $\mathbf{v}(t)$ ▶ $\mathbf{a}(t)$

▶ $\mathbf{T}(t)$ ▶ $\mathbf{N}(t)$ ▶ $\mathbf{B}(t)$

▶ κ ▶ ρ ▶ $a_{\mathbf{T}}$

▶ $a_{\mathbf{N}}$

Derivatives of Products: Fill in the following blanks to complete a differentiation rule for vectors.

▶ Let k be a scalar and $\mathbf{r}(t)$ be a differentiable vector function. Then $\frac{d}{dt}(k\mathbf{r}(t)) =$ _____ .

▶ Let $f(t)$ be a differentiable scalar function and $\mathbf{r}(t)$ be a differentiable vector function. Then $\frac{d}{dt}(f(t)\mathbf{r}(t)) =$ _____ .

▶ If $\mathbf{r}_1(t)$, $\mathbf{r}_2(t)$ are differentiable vector functions, both having either two or three components, then $\frac{d}{dt}(\mathbf{r}_1(t) \cdot \mathbf{r}_2(t)) =$ _____ .

▶ If $\mathbf{r}_1(t)$ and $\mathbf{r}_2(t)$ are both differentiable three-component vector functions, then $\frac{d}{dt}(\mathbf{r}_1(t) \times \mathbf{r}_2(t)) =$ _____ .

▶ Let $\mathbf{r}(t)$ be a differentiable vector function such that $\|\mathbf{r}(t)\| = k$ for some constant k. Then $\mathbf{r}(t) \cdot \mathbf{r}'(t) =$ _____ .

Skill Certification: The Calculus of Vector Functions

Sketching vector functions: Sketch the following vector functions.

1. $\mathbf{r}(t) = \langle t, t^3 \rangle$, $t \in \mathbb{R}$
2. $\mathbf{r}(t) = \langle \cos t, t \rangle$, $t \in \mathbb{R}$
3. $\mathbf{r}(t) = \langle t, \sin 2t, \cos 2t \rangle$, $t \in [0, 4\pi]$
4. $\mathbf{r}(t) = \langle t, t \cos t, t \sin t \rangle$, $t \in [0, 4\pi]$

Finding limits: Find the given limits if they exist. If a limit does not exist, explain why.

5. $\lim\limits_{t \to 0} \left\langle \dfrac{1/(3+t) - 1/3}{t}, \dfrac{(3+t)^2 - 9}{t} \right\rangle$

6. $\lim\limits_{t \to -3} \left\langle \dfrac{|t+3|}{t+3}, \dfrac{|t+3|}{t-3} \right\rangle$

7. $\lim\limits_{t \to 0} \left\langle \dfrac{\sin t}{t}, \dfrac{1 - \cos t}{t}, (1+t)^{1/t} \right\rangle$

8. $\lim\limits_{t \to 1} \left\langle (\ln t)^{t-1}, t^{(1/t)-1}, \dfrac{\sin(\ln t)}{t-1} \right\rangle$

Velocity and acceleration vectors: Find the velocity and acceleration vectors for the given vector functions.

9. $\mathbf{r}(t) = \langle t, 2t - 3, 3t + 5 \rangle$ 10. $\mathbf{r}(t) = \langle t, 2t^2, 3t^3 \rangle$

11. $\mathbf{r}(t) = \langle \cos 2t, \sin 3t \rangle$ 12. $\mathbf{r}(t) = \langle e^t, t, e^{-t} \rangle$

Unit tangent vectors: Find the unit tangent vector for the given function at the specified value of t.

13. $\mathbf{r}(t) = \langle t, t^3 \rangle$, $t = 2$
14. $\mathbf{r}(t) = \langle 3 \sin 2t, 3 \cos 2t \rangle$, $t = \dfrac{\pi}{3}$
15. $\mathbf{r}(t) = \langle \alpha \sin \beta t, \alpha \cos \beta t \rangle$, $t = 0$
16. $\mathbf{r}(t) = \langle t, 5 \sin 3t, 5 \cos 3t \rangle$, $t = \dfrac{\pi}{6}$
17. $\mathbf{r}(t) = \langle t, 2t \sin t, 2t \cos t \rangle$, $t = \pi$
18. $\mathbf{r}(t) = \langle t, \alpha \sin \beta t, \alpha \cos \beta t \rangle$, $t = 0$

Principal unit normal vectors: Find the principal unit normal vector for the given function at the specified value of t.

19. $\mathbf{r}(t) = \langle t, t^3 \rangle$, $t = 2$
20. $\mathbf{r}(t) = \langle 3 \sin 2t, 3 \cos 2t \rangle$, $t = \dfrac{\pi}{3}$
21. $\mathbf{r}(t) = \langle \alpha \sin \beta t, \alpha \cos \beta t \rangle$, where α and β are positive, $t = 0$
22. $\mathbf{r}(t) = \langle t, 5 \sin 3t, 5 \cos 3t \rangle$, $t = \dfrac{\pi}{6}$
23. $\mathbf{r}(t) = \langle t, 2t \sin t, 2t \cos t \rangle$, $t = \pi$
24. $\mathbf{r}(t) = \langle t, \alpha \sin \beta t, \alpha \cos \beta t \rangle$, $t = 0$

Binormal vectors and osculating planes: Find the binormal vector and equation of the osculating plane for the given function at the specified value of t.

25. $\mathbf{r}(t) = \langle t, t^3 \rangle$, $t = 2$
26. $\mathbf{r}(t) = \langle 3 \sin 2t, 3 \cos 2t \rangle$, $t = \dfrac{\pi}{3}$
27. $\mathbf{r}(t) = \langle \alpha \sin \beta t, \alpha \cos \beta t \rangle$, $t = 0$
28. $\mathbf{r}(t) = \langle t, 5 \sin 3t, 5 \cos 3t \rangle$, $t = \dfrac{\pi}{6}$
29. $\mathbf{r}(t) = \langle t, 2t \sin t, 2t \cos t \rangle$, $t = \pi$
30. $\mathbf{r}(t) = \langle t, \alpha \sin \beta t, \alpha \cos \beta t \rangle$, $t = 0$

Osculating circles: Find the equation of the osculating circle to the given function at the specified value of t.

31. $\mathbf{r}(t) = \langle t, t^3 \rangle$, $t = 2$
32. $\mathbf{r}(t) = \langle 3 \sin 2t, 3 \cos 2t \rangle$, $t = \dfrac{\pi}{3}$
33. $\mathbf{r}(t) = \langle \alpha \sin \beta t, \alpha \cos \beta t \rangle$, $t = 0$

Osculating circles: Find the center and radius of the osculating circle to the given vector function at the specified value of t.

34. $\mathbf{r}(t) = \langle t, 5 \sin 3t, 5 \cos 3t \rangle$, $t = \dfrac{\pi}{6}$
35. $\mathbf{r}(t) = \langle t, 2t \sin t, 2t \cos t \rangle$, $t = \pi$
36. $\mathbf{r}(t) = \langle t, \alpha \sin \beta t, \alpha \cos \beta t \rangle$, $t = 0$

Osculating circles: Find the equation of the osculating circle to the given scalar function at the specified point.

37. $f(x) = x^2$, $(0, 0)$

38. $f(x) = \cos x$, $(0, 1)$

39. $f(x) = e^x$, $(0, 1)$

40. $f(x) = \ln x$, $(1, 0)$

Capstone Problems

A. A certain vector function $\mathbf{r}(t)$ has the properties that its graph is a space curve passing through the point $(4, -3, 6)$ and that its derivative is $\mathbf{r}'(t) = -2\mathbf{r}(t)$. Find $\mathbf{r}(t)$.

B. Let $y = f(x)$ be a twice-differentiable function.

(a) Carefully outline the steps required to find the equation of the osculating circle at an arbitrary point x_0 in the domain of f.

(b) Write a program, using a computer algebra system such as Maple, Mathematica, or Matlab, to animate the osculating circle moving along the graph of the function $y = \sin x$.

C. Let $\mathbf{r}(t) = \langle x(t), y(t), z(t) \rangle$ be a twice-differentiable function.

(a) Carefully outline the steps required to find the unit tangent vector, principal unit normal vector, and binormal vector at an arbitrary point t_0 in the domain of $\mathbf{r}$.

(b) Write a program, using a computer algebra system such as Maple, Mathematica, or Matlab, to animate the Frenet frame moving along the graph of the function $\mathbf{r}(t) = \langle \cos t, \sin t, t \rangle$.

D. Suppose a small planet is orbiting the sun. Its position relative to the center of the sun is given by some vector function $\mathbf{r} = \mathbf{r}(t)$. When we put Newton's second law of motion together with his law of gravitation, with the sun at the origin, we find that the position of the planet at time t satisfies $\dfrac{d^2\mathbf{r}}{dt^2} = -\dfrac{GM\mathbf{r}}{r^3}$, where G is the gravitational constant, M is the mass of the sun, and $r = \|\mathbf{r}\|$. Define $\mathbf{u} = \mathbf{r} \times \dfrac{d\mathbf{r}}{dt}$. It turns out that $\mathbf{u}$ is a constant vector.

(a) Use the relation $\mathbf{v}_1 \times (\mathbf{v}_1 \times \mathbf{v}_2) = -\|\mathbf{v}_1\|^2 \mathbf{v}_2$ for orthogonal vectors $\mathbf{v}_1$ and $\mathbf{v}_2$ to show that

$$\frac{d^2\mathbf{r}}{dt^2} \times \mathbf{u} = \frac{GM}{r}\frac{d\mathbf{r}}{dt}.$$

(b) Antidifferentiate the result of part (a), using the fact that $\mathbf{u}$ is constant to show that

$$\frac{d\mathbf{r}}{dt} \times \mathbf{u} = GM\frac{\mathbf{r}}{r} + \mathbf{w}$$

for some constant vector $\mathbf{w}$.

(c) Since $\dfrac{d\mathbf{r}}{dt} \times \mathbf{u}$ is orthogonal to $\mathbf{u}$ and $\mathbf{r}$ is orthogonal to $\mathbf{u}$, it follows that $\mathbf{w}$ is orthogonal to $\mathbf{u}$. If we let θ denote the angle from $\mathbf{w}$ to $\mathbf{r}$, then (r, θ) represents polar coordinates for the position of the planet in the $\mathbf{rw}$-plane. Use this fact and the relation

$$(\mathbf{v}_1 \times \mathbf{v}_2) \cdot \mathbf{v}_3 = \mathbf{v}_1 \cdot (\mathbf{v}_2 \times \mathbf{v}_3)$$

to show that

$$\|\mathbf{u}\|^2 = GMr + \|\mathbf{w}\|r\cos\theta.$$

(d) Convert the last expression in part (c) to Cartesian coordinates to conclude that the satellite's orbit around the planet is an ellipse (or circle) whenever $\|\mathbf{w}\| < GM$. This is Kepler's first law of planetary motion: Planets follow an elliptical path around the sun. The law can be generalized to describe the orbit of any relatively small body around a much larger body.

CHAPTER 12

Multivariable Functions

12.1   FUNCTIONS OF TWO AND THREE VARIABLES

▶   Functions of two and three variables

▶   The graphs of linear functions of two variables and surfaces of revolution

▶   Level curves and level surfaces

Functions of Two Variables

Recall that a function $f : A \to B$ is an assignment that associates to each element x of the domain set A exactly one element $f(x)$ of the codomain set B. Calculus is a tool that allows us to analyze and understand functions. Thus far we have focused primarily on the following types of functions:

▶   For the first third of this text, we studied primarily the calculus of functions for which the sets A and B are subsets of the real numbers.

▶   When we studied sequences in Chapter 7, we insisted that the domain be the set of natural numbers while the target set was usually a subset of the real numbers.

▶   The vector-valued functions we studied in Chapter 11 had domains that were subsets of the real numbers and target sets that were subsets of $\mathbb{R}^n$ for a fixed value of $n \geq 2$.

We now turn our attention to functions whose domains are subsets of either $\mathbb{R}^2$ or $\mathbb{R}^3$ but that have a target set that is a subset of the real numbers. We will first consider functions whose domains are subsets of $\mathbb{R}^2$ and soon discuss functions whose domains are subsets of $\mathbb{R}^3$. We refer to these as functions of two variables and functions of three variables, respectively.

For example, consider the addition of two real numbers x and y. We may take addition to be a function $f : \mathbb{R}^2 \to \mathbb{R}$ that uses the rule

$$f(x, y) = x + y.$$

Note that we already have a slight, but common, abuse of notation. Since we denote elements of $\mathbb{R}^2$ in one of the notations $\langle x, y \rangle$ or (x, y), perhaps we should write $f(\langle x, y \rangle)$ or $f((x, y))$; however, we use the simpler and more common notation $f(x, y)$.

The order of the input variables x and y is significant in a function of two variables. Consider the function $g : R^2 \to \mathbb{R}$ defined by the rule $g(x, y) = 5x - y$. For this function we do not generally have $g(a, b) = g(b, a)$. For example, $g(1, 0) = 5 \cdot 1 - 0 = 5$ and $g(0, 1) = 5 \cdot 0 - 1 = -1$. (When would we have $g(a, b) = g(b, a)$?)

As with functions whose domains are subsets of $\mathbb{R}$, we often do not specify the domain of a function when it is a subset of $\mathbb{R}^2$. If the function is defined by an equation, we assume that the domain is the largest subset of $\mathbb{R}^2$ for which the equation is defined. The range of a function of two variables is the set of possible outputs.

DEFINITION 12.1   **The Domain and Range of a Function of Two Variables**

If f is a function that takes an element from an unspecified subset of $\mathbb{R}^2$ into $\mathbb{R}$, then we will take the **domain** of f to be the largest subset of $\mathbb{R}^2$ for which f is defined:

$$\text{Domain}(f) = \{(x, y) \in \mathbb{R}^2 | f(x, y) \text{ is defined}\}.$$

The **range** of f is the set of all possible outputs of f. That is,

$$\text{Range}(f) = \{z \in \mathbb{R} \mid \text{ there exists } (x, y) \in \text{ Domain}(f) \text{ with } f(x, y) = z\}.$$

For example, the domain of the function $h(x, y) = \sqrt{xy}$ consists of all points $(x, y) \in \mathbb{R}^2$ such that the product $xy \geq 0$. The latter is the case on the coordinate axes and when the signs of x and y are the same (i.e., in the first and third quadrants). The range of h consists of all nonnegative real numbers.

The domain of $h(x, y) = \sqrt{xy}$

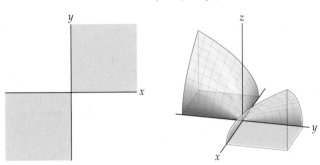

Graphing a Function of Two Variables

When we wish to analyze the behavior of a function, it is often helpful to understand the graph of the function.

DEFINITION 12.2   The Graph of a Function of Two Variables

The **graph** of a function of two variables, f, is the collection of ordered triples whose first two coordinates are in the domain of f. That is,

$$\text{Graph}(f) = \{(x, y, f(x, y)) \in \mathbb{R}^3 \mid (x, y) \in \text{Domain}(f)\}.$$

In other words the graph of a function of two variables is a particular subset of a three-dimensional Cartesian coordinate system. Consider the first function we mentioned in this chapter: $f(x, y) = x + y$. The graph of f is the subset of $\mathbb{R}^3$ given by

$$\text{Graph}(f) = \{(x, y, x + y) \mid (x, y) \in \mathbb{R}^2\}.$$

If we let $z = f(x, y) = x + y$, we see that the graph of f consists of exactly those points in the plane $z = x + y$.

The graph of $f(x, y) = x + y$

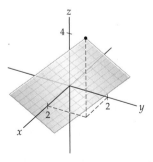

In Exercise 69 we ask you to prove that when a, b, and c are constants, the graph of the function $f(x, y) = ax + by + c$ is a plane.

In Chapter 6 we discussed another type of surface that may be readily expressed by a function of two variables. Recall that when we revolve the graph of a continuous function f that exists on an interval $[a, b]$, where $0 \le a < b$, around the y-axis, we obtain a surface of revolution. As a reminder, consider the surface we obtain when we revolve the graph of the function $f(x) = x^2$, on the interval $[0, 2]$, around the z-axis. Following are the graphs of the function and the surface:

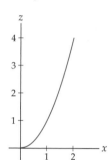

The function $z = x^2$

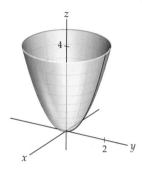

The surface of revolution $z = x^2 + y^2$

It is quite easy to express this surface as a function of two variables. When the interval we are using is $[a, b]$ with $0 \le a < b$, we replace every occurrence of the variable x by the quantity $\sqrt{x^2 + y^2}$ in the formula for the function and evaluate $z = f(\sqrt{x^2 + y^2})$. In our simple example here we have $z = (\sqrt{x^2 + y^2})^2 = x^2 + y^2$. We provide other examples in the subsections that follow.

We will see that graphs of functions of two variables can be considerably more complicated than graphs of functions of a single variable. As we progress through this chapter, we will discuss limits, continuity, and differentiability for functions of two and three variables. Understanding these concepts will enable us to analyze the functions more thoroughly. In general, if a function of two variables is sufficiently well behaved, its graph will be a surface in three-dimensional space $\mathbb{R}^3$. The graph of the plane $z = x + y$ is one such example. But graphs of functions of two variables can be much more complicated. Here we see a graph of a portion of the surface defined by the function $f(x, y) = \sin(xy)$:

The graph of $f(x, y) = \sin(xy)$

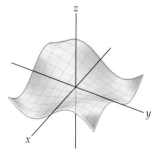

Functions of Three Variables

We now turn our attention to functions of three variables (i.e., those functions whose domains are subsets of $\mathbb{R}^3$ and whose codomains are subsets of $\mathbb{R}$). An example of a function of three variables is given by the formula $f(x, y, z) = xy + z - 7$. As before, we usually do not specify the domain of a function when it is a subset of $\mathbb{R}^3$. Rather, we assume that the domain is the largest subset of $\mathbb{R}^3$ for which the rule is defined. The range of a function of three variables is the set of possible outputs of the function.

DEFINITION 12.3 **The Domain and Range of a Function of Three Variables**

If f is a function that takes an element from an unspecified subset of $\mathbb{R}^3$ into $\mathbb{R}$, then we will take the ***domain*** of f to be the largest subset of $\mathbb{R}^3$ for which f is defined:

$$\text{Domain}(f) = \{(x, y, z) \in \mathbb{R}^3 \mid f(x, y, z) \text{ is defined}\}.$$

The ***range*** of f is the set of all possible outputs of f. That is,

$$\text{Range}(f) = \{w \in \mathbb{R} \mid \text{ there exists } (x, y, z) \in \text{ Domain}(f) \text{ with } f(x, y, z) = w\}.$$

For example, the domain of the function $f(x, y, z) = \sqrt{xyz}$ consists of all points $(x, y, z) \in \mathbb{R}^3$ such that the product $xyz \geq 0$. The latter is the case on the coordinate planes, in the first octant (when x, y, and z are all positive), and in those octants where exactly one of the signs of x, y, and z is the positive. The range of this function consists of all nonnegative real numbers.

DEFINITION 12.4 **The Graph of a Function of Three Variables**

The ***graph*** of a function of three variables, f, is the collection of ordered quadruples whose first three coordinates are in the domain of f. That is,

$$\text{Graph}(f) = \{(x, y, z, f(x, y, z)) \in \mathbb{R}^4 \mid (x, y, z) \in \text{Domain}(f)\}.$$

In other words the graph of a function of three variables is a particular subset of a four-dimensional Cartesian coordinate system! Note that we are not providing even a single example of the graph of a function of three variables here. We have mentioned that the graph of a sufficiently well-defined function of two variables is a surface (basically a two-dimensional object) in three-dimensional space $\mathbb{R}^3$. The graph of even the simplest function of three variables is a so-called hypersurface (three-dimensional object) existing in four-dimensional space, $\mathbb{R}^4$. Later in this section we will discuss how to use level surfaces to help visualize such objects, but it is often more efficient to try to understand them by using the techniques of calculus and *not* try to sketch them.

Level Curves and Level Surfaces

A topographic map renders a portion of the surface of the Earth in a two-dimensional form:

The map shows the Fuego and Acatenango volcanic complexes in Guatemala. Each curve on the map displays those points on the surface of these complexes that are at the same elevation. For example, the curve marked 3100 displays the points on the surface at 3100 meters. Thus, if you walked along that curve, you would neither ascend nor descend as you circumnavigated the volcanos. This topographic map gives us a way to understand the surface of the indicated region in Guatemala, although the map itself is flat. We will use this idea to help us understand functions of two variables. We make the following definition:

DEFINITION 12.5   **Level Curves**

Let f be a function of two variables and let c be a point in the range of f. The **level curve** for f at height c is the curve in the plane with equation $f(x, y) = c$.

Thus, each point on the level curve is at the same height on the surface determined by the function. The collection of level curves for a function of two variables, f, in effect, gives us a topographic map for the surface of the graph of f. For example, consider $f(x, y) = \sqrt{9 - (x^2 + y^2)}$, a function of two variables. The range of f is the interval $[0, 3]$. If we choose the values 0, 1, 2, and 3 for c, we have the equations

$$\sqrt{9 - (x^2 + y^2)} = 0, \ \sqrt{9 - (x^2 + y^2)} = 1, \ \sqrt{9 - (x^2 + y^2)} = 2 \text{ and } \sqrt{9 - (x^2 + y^2)} = 3.$$

The graphs of these four equations are the following level curves, although note that the only point that satisfies the equation $\sqrt{9 - (x^2 + y^2)} = 3$ is the origin $(0, 0)$:

Level curves for $f(x, y) = \sqrt{9 - (x^2 + y^2)}$

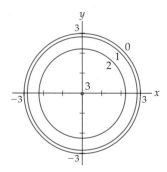

Now, if we stack these circles at the appropriate heights and interpolate similar circles between them, we can envision the hemisphere that is the graph of f:

The graph of $f(x, y) = \sqrt{9 - (x^2 + y^2)}$

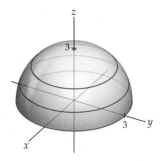

As we mentioned earlier, the graph of a function of three variables is a four-dimensional object. To aid our understanding of these functions, we can sometimes use level surfaces analogous to the level curves we have defined.

DEFINITION 12.6   Level Surfaces

Let f be a function of three variables and let c be a point in the range of f. The **level surface** for f at height c is the surface in three-dimensional space with equation $f(x, y, z) = c$.

We now consider an example quite similar to the one we just discussed. Let the function $g(x, y, z) = \sqrt{9 - (x^2 + y^2 + z^2)}$. The range of g is the interval $[0, 3]$. We will find the level surfaces for $c = 0$, 1, 2, and 3. When $c = 0$, we have $0 = \sqrt{9 - (x^2 + y^2 + z^2)}$, or equivalently, $x^2 + y^2 + z^2 = 9$. Thus, the level surface for $c = 0$ is a sphere of radius 3 and centered at the origin. You may check that the level surface for $c = 1$ is the sphere of radius $\sqrt{8}$ and centered at the origin; the level surface for $c = 2$ is the sphere of radius $\sqrt{5}$, also centered at the origin; and the level "surface" for $c = 3$ is the single point $(0, 0, 0)$. We see that each level surface of the graph of g is a sphere and that the radii of the level spheres get smaller as c increases from 0 to 3. The level surfaces are concentric spheres centered at the origin.

Three level surfaces for $g(x, y, z) = \sqrt{9 - (x^2 + y^2 + z^2)}$

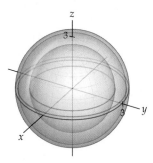

In fact, the graph of g is the "top" half of the hypersphere defined by the equation $w^2 + x^2 + y^2 + z^2 = 9$. The spheres that are the level surfaces for g are analogous to the circles that are the level curves for the function f we discussed before. Understanding the geometry of a function with one fewer variable can help us understand the graph of a figure that exists in a higher dimension.

Examples and Explorations

EXAMPLE 1   Understanding the basic properties of functions of two variables

For each of the following functions of two variables, evaluate the function at the given values and find the domain and range of the function:

(a) $f(x, y) = 3x - 4y + 5$; $(3, 4)$, $(4, 3)$ **(b)** $g(x, y) = \dfrac{x^2 - y^2}{x}$; $(2, 5)$, $(3, 0)$

(c) $h(r, s) = \dfrac{\ln(2r + s)}{2r + s}$; $(1, 0)$, $(0, e)$

SOLUTION

(a) We first evaluate the function for the specified pairs of coordinates:

$$f(3, 4) = 3 \cdot 3 - 4 \cdot 4 + 5 = 9 - 16 + 5 = -2 \quad \text{and}$$
$$f(4, 3) = 3 \cdot 4 - 4 \cdot 3 + 5 = 12 - 12 + 5 = 5.$$

In particular, we note that the order of the coordinates is significant when we evaluate this function. The domain of the function $f(x, y) = 3x - 4y + 5$ is $\mathbb{R}^2$, since the function is defined for every ordered pair in $\mathbb{R}^2$. Finally, the range of f is $\mathbb{R}$, because every real number can be obtained (in infinitely many different ways). For example, if we let r be an arbitrary nonzero real number, the ordered pair $\left(\frac{r-5}{3}, 0\right)$ has the property that $f\left(\frac{r-5}{3}, 0\right) = r$. Thus, the function f maps $\mathbb{R}^2$ *onto* $\mathbb{R}$.

(b) We evaluate the function g for the specified pairs of coordinates:

$$g(2, 5) = \frac{2^2 - 5^2}{2} = \frac{4 - 25}{2} = -\frac{21}{2} \quad \text{and} \quad g(3, 0) = \frac{3^2 - 0^2}{3} = 3.$$

The domain of the function g is the set of all ordered pairs $\{(x, y) \in \mathbb{R}^2 \mid x \neq 0\}$, since g is defined for all ordered pairs except when $x = 0$. As with f, the range of g is $\mathbb{R}$, because every real number can be obtained as an output from g. For example, $g(r, 0) = r$ and $g(r, r) = 0$ for every nonzero real number r.

(c) Before we evaluate the function h for the specified pairs of coordinates, note that we are using different letters to represent our variables, but h is just another function of two variables. When we evaluate h for a particular coordinate pair, we assume that the first value in the pair is the r-value and the second is the s-value, since we wrote the function as $h(r, s)$. Thus,

$$h(1, 0) = \frac{\ln(2 \cdot 1 + 0)}{2 \cdot 1 + 0} = \frac{\ln 2}{2} \quad \text{and} \quad h(0, e) = \frac{\ln(2 \cdot 0 + e)}{2 \cdot 0 + e} = \frac{1}{e}.$$

Because the domain of the natural logarithm is all positive real numbers, the domain of the function h is the set of all ordered pairs $\{(r, s) \in \mathbb{R}^2 \mid 2r + s > 0\}$, shown here:

The domain of $h(r, s) = \dfrac{\ln(2r + s)}{2r + s}$

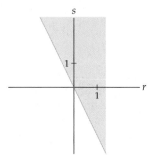

Finding the range of h is slightly more difficult. As an aid, we note that if we let $w = 2r + s$, we may represent the function in the form $h(w) = \frac{\ln w}{w}$. Using the techniques of single-variable calculus, you may show that this function has the following graph:

The graph of $h(w) = \dfrac{\ln w}{w}$

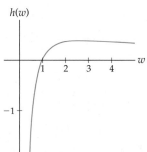

It can be shown that the maximum of this function, $1/e$, occurs when $w = e$ and that $\lim\limits_{w \to 0^+} \dfrac{\ln w}{w} = -\infty$. Therefore the range of the function $\dfrac{\ln w}{w}$ is $(-\infty, 1/e]$. Since the quantity $2r + s$ can take on any real value, this interval will also be the range of $h(r, s) = \dfrac{\ln(2r+s)}{2r+s}$.   □

EXAMPLE 2   **Composing functions of two variables**

Consider the functions

$$g(t) = t^3, \; h(t) = t^2 - 7, \; f(x, y) = x^2 - y, \; w(x, y) = \frac{x}{y}, \; \text{and} \; \mathbf{r}(t) = \langle \sin t, \; \cos t \rangle$$

For each of the following, simplify the expression or explain why it is not possible to evaluate that particular expression:

(a) $f(g(t))$ **(b)** $g(f(x, y))$ **(c)** $w(g(t), h(t))$ **(d)** $w(\mathbf{r}(t))$ **(e)** $f(w(x, y))$

SOLUTION

(a) This composition is not defined. The range of the function g is $\mathbb{R}$, but the domain of f is $\mathbb{R}^2$.

(b) This composition is defined, and
$$g(f(x, y)) = g(x^2 - y) = (x^2 - y)^3.$$

(c) This composition is also defined, and
$$w(g(t), h(t)) = \frac{g(t)}{h(t)} = \frac{t^3}{t^2 - 7}.$$

(d) The output of the vector function $\mathbf{r}$ is the ordered pair $\langle \sin t, \cos t \rangle$, so this composition is defined and
$$w(\mathbf{r}(t)) = \frac{\sin t}{\cos t} = \tan t.$$

(e) This composition is not defined. The range of the function w is a subset of $\mathbb{R}$, but the domain of f is $\mathbb{R}^2$.   □

EXAMPLE 3   **Graphing surfaces of revolution**

Sketch the surfaces formed when the graphs of the following functions on the specified intervals are revolved around the z-axis:

(a) $f(x) = x - 1$ on the interval $[0, 2]$ **(b)** $g(x) = x^3 - x$ on the interval $[-1, 0]$

Also, find the equation for each surface as a function of two variables.

SOLUTION

We begin by noting that, for each function, we are considering the dependent variable to be z. Thus the graph of the function f on the interval $[0, 2]$ is a line segment, and the graph of g on the interval $[-1, 0]$ is a portion of the cubic function, both in the xz-plane.

(a) For our first function, since our interval is $[0, 2]$, all we need to do to obtain the equation of the first surface of revolution as a function of two variables is to replace x with $\sqrt{x^2 + y^2}$ in the formula for the function f and obtain

$$z = f\left(\sqrt{x^2 + y^2}\right) = \sqrt{x^2 + y^2} - 1.$$

The graphs of the function and the surface are as follows:

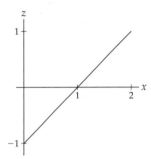

The function $f(x) = x - 1$

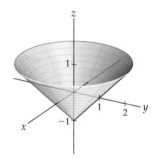

The surface of revolution $z = \sqrt{x^2 + y^2} - 1$

(b) For our second function, our interval is $[-1, 0]$. When we have an interval $[a, b]$ with $a < b \leq 0$, we consider the function $g(-x)$ on the interval $[-b, -a]$ instead. Here we have

$$g(-x) = ((-x)^3 - (-x)) = x - x^3 \text{ on } [0, 1],$$

and we proceed as we did in part (a). To obtain the equation of the second surface of revolution as a function of two variables, we replace the occurrences of x with $\sqrt{x^2 + y^2}$ in the formula for $g(-x)$ and obtain

$$z = g\left(-\sqrt{x^2 + y^2}\right) = (x^2 + y^2)^{1/2} - (x^2 + y^2)^{3/2}.$$

Here are the graphs of the functions and the surface:

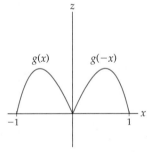

The functions
$g(x)$ on $[-1, 0]$ and $g(-x)$ on $[0, 1]$

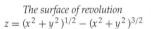
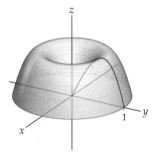

The surface of revolution
$z = (x^2 + y^2)^{1/2} - (x^2 + y^2)^{3/2}$

$\square$

EXAMPLE 4

Drawing level curves

Let $f(x, y) = -\dfrac{8x}{x^2 + y^2 + 1}$. Sketch level curves $f(x, y) = c$ for $c = -3, -2, -1, 0, 1, 2,$ and 3.

SOLUTION

For each value of c, we will graph the set of points that are solutions of the equation $-\dfrac{8x}{x^2 + y^2 + 1} = c$. We begin with $c = 0$. The equation $-\dfrac{8x}{x^2 + y^2 + 1} = 0$ holds exactly when $x = 0$. Thus, the level curve here is the y-axis.

Next, when $c = 1$, we have the equation

$$-\frac{8x}{x^2 + y^2 + 1} = 1.$$

The denominator of the function is never zero. Therefore, we obtain the equivalent equation $-8x = x^2 + y^2 + 1$ when we multiply both sides of the given equation by $x^2 + y^2 + 1$. Then,

$$-8x = x^2 + y^2 + 1$$
$$x^2 + 8x + y^2 = -1 \qquad \leftarrow \text{arithmetic}$$
$$(x + 4)^2 + y^2 = 15 \qquad \leftarrow \text{completing the square}$$

At this point we recognize that the level curve when $c = 1$ is the circle with radius $\sqrt{15}$ and centered at $(-4, 0)$. Nearly identical calculations reveal that the level curve when $c = -1$ is given by the equation $(x - 4)^2 + y^2 = 15$, whose graph is the circle with radius $\sqrt{15}$ and centered at $(4, 0)$.

The other level curves we wish to find are also circles. Similar calculations provide the information in the following table:

c	equation	radius	center
2	$(x + 2)^2 + y^2 = 3$	$\sqrt{3}$	$(-2, 0)$
-2	$(x - 2)^2 + y^2 = 3$	$\sqrt{3}$	$(2, 0)$
3	$\left(x + \frac{4}{3}\right)^2 + y^2 = \frac{7}{9}$	$\frac{\sqrt{7}}{3}$	$\left(-\frac{4}{3}, 0\right)$
-3	$\left(x - \frac{4}{3}\right)^2 + y^2 = \frac{7}{9}$	$\frac{\sqrt{7}}{3}$	$\left(\frac{4}{3}, 0\right)$

Therefore, we have the following graph of the level curves:

Level curves for $f(x, y) = -\dfrac{8x}{x^2 + y^2 + 1}$

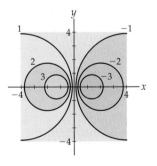

We may extend the analysis we used to obtain these level curves to find the range of the function f. (See Exercise 8.)

Finally, the graph of f shown next was generated by a computer algebra system. The breaks between the colors correspond to the level curves we just found. If you have access to such a program, it can greatly aid your understanding of the graphs of functions of two variables.

$$f(x, y) = -\frac{8x}{x^2 + y^2 + 1}$$

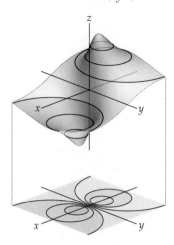

EXAMPLE 5 | **Understanding the basic properties of functions of three variables**

For each of the following functions of three variables, evaluate the function at the given values and find the domain and range of the function:

(a) $f(x, y, z) = xyz + xy + z$; $(1, 2, 3)$ **(b)** $g(x, y, z) = \dfrac{x^2 - y^2}{x + z}$; $(-3, 2, 4)$

(c) $h(r, s, t) = \dfrac{\ln(r + s + t)}{r + s + t}$; $(e, e^2, -e)$

SOLUTION

(a) We first evaluate the function for the specified triple of coordinates:

$$f(1, 2, 3) = 1 \cdot 2 \cdot 3 + 1 \cdot 2 + 3 = 6 + 2 + 3 = 11.$$

The domain of the function $f(x, y, z) = xyz + xy + z$ is $\mathbb{R}^3$, since the function is defined for every ordered triple in $\mathbb{R}^3$. Finally, the range of f is $\mathbb{R}$, because every real number can be obtained (in infinitely many different ways). For example, let r be an arbitrary real number, then the ordered triple $(0, 0, r)$ has the property that $f(0, 0, r) = r$. Thus, the function f maps $\mathbb{R}^3$ *onto* $\mathbb{R}$.

(b) We first evaluate the function g for the specified triple of coordinates:

$$g(-3, 2, 4) = \frac{(-3)^2 - 2^2}{-3 + 4} = \frac{9 - 4}{1} = 5.$$

The domain of the function g is the set of all ordered triples $\{(x, y, z) \in \mathbb{R}^3 \mid x + z \neq 0\}$, since g is defined for all triples except when $x + z = 0$. That is, the domain of g consists of all points in $\mathbb{R}^3$ except those points on the plane defined by the equation $x + z = 0$. The range of g is $\mathbb{R}$, because every real number can be obtained as an output from g. For example, if we let r be an arbitrary nonzero real number, the ordered pair $(r, 0, 0)$ has the property that $g(r, 0, 0) = r$.

(c) This function is quite similar to the function found in part (c) of Example 1. Thus, our analysis will be quite similar. We have

$$h(e, e^2, -e) = \frac{\ln(e + e^2 - e)}{e + e^2 - e} = \frac{2}{e^2}.$$

Since the domain of the natural logarithm is all positive real numbers, the domain of the function h is the set of all ordered triples $\{(r, s, t) \in \mathbb{R}^3 \mid r + s + t > 0\}$. That is, the domain consists of all points in $\mathbb{R}^3$ above the plane defined by the equation $r + s + t = 0$.

As in part (c) of Example 1, the range of $h(r, s, t) = \frac{\ln(r+s+t)}{r+s+t}$ is the interval $(-\infty, 1/e]$.

$\square$

? TEST YOUR UNDERSTANDING

▶ What is a function of two variables? What is a function of three variables? What is a function of n variables?

▶ How do you determine the domain of a function of two or three variables? What is the range of a function?

▶ Why is the graph of a linear function of two variables a plane?

▶ How many dimensions are required to sketch the graph of a function of two variables? How many dimensions are required to sketch the graph of a function of three variables?

▶ What is a level curve? What is a level surface? How can level curves be used to understand a function of two variables? How can level surfaces be used to understand a function of three variables?

EXERCISES 12.1

Thinking Back

▶ *The definition of a function:* Explain why Definition 0.1 is general enough to include functions of two and three variables.

▶ *The domain and range of a function of two variables:* Explain why Definition 0.2 is not general enough to define the domain or range of a function of two or three variables.

Concepts

0. *Problem Zero:* Read the section and make your own summary of the material.

1. *True/False:* Determine whether each of the statements that follow is true or false. If a statement is true, explain why. If a statement is false, provide a counterexample.

(a) *True or False:* The domain of a function of two variables is a subset of $\mathbb{R}^2$.

(b) *True or False:* The range of a function of two variables is a subset of $\mathbb{R}^2$.

(c) *True or False:* The graph of a function of two variables is a subset of $\mathbb{R}^3$.

(d) *True or False:* The domain of a function of three variables is a subset of $\mathbb{R}$.

(e) *True or False:* The range of a function of three variables is a subset of $\mathbb{R}$.

(f) *True or False:* The graph of a function of three variables is a subset of $\mathbb{R}^4$.

(g) *True or False:* The graph of a linear function of two variables is a plane.

(h) *True or False:* If a function $f : \mathbb{R} \to \mathbb{R}$ is continuous on an interval $[0, p]$ then the surface formed when the graph of f is rotated about the y-axis may be expressed as a function of two variables.

2. *Examples:* Construct examples of the thing(s) described in the following. Try to find examples that are different than any in the reading.

(a) A function of two variables whose graph is a plane.

(b) A function of two variables whose graph is the surface of revolution formed when the graph of a function of a single variable is revolved around the y-axis.

(c) A function of two variables for which each level curve is a parabola.

3. Let $f : \mathbb{R} \to \mathbb{R}$ be a function of a single variable. Explain why the graph of f is a subset of $\mathbb{R}^2$.

4. Let $f : \mathbb{R}^2 \to \mathbb{R}$ be a function of two variables. Explain why the graph of f is a subset of $\mathbb{R}^3$.

5. Let $f : \mathbb{R}^3 \to \mathbb{R}$ be a function of three variables. Explain why the graph of f is a subset of $\mathbb{R}^4$.

6. (a) Graph $f(x) = x$.

(b) Graph the function $f(x, y) = x + y$. Explain why the graph of $f(x, y) = x + y$ contains the origin, $(0, 0, 0)$.

(c) The graph of the function

$$f(x, y, z) = x + y + z$$

is a *hyperplane* of dimension 3 in $\mathbb{R}^4$. Explain why "hyperplane" is a good name for the graph. ("Hyper" is from a Greek word meaning *over* or *beyond*. The Latin equivalent is super, meaning essentially the same thing.) Explain why the graph of f contains the origin, $(0, 0, 0, 0)$.

(d) Fill in the blanks: The graph of the function $f(x_1, x_2, \ldots, x_n) = x_1 + x_2 + \cdots + x_n$ is a hyperplane of dimension _____ in _____.

7. (a) Graph the function $f(x) = \sqrt{1 - x^2}$.

(b) Graph the function $f(x, y) = \sqrt{1 - x^2 - y^2}$. (*Hint: Let* $z = \sqrt{1 - x^2 - y^2}$, *and then square both sides of the equation.*)

(c) The graph of the function

$$f(x, y, z) = \sqrt{1 - x^2 - y^2 - z^2}$$

is half of a *hypersphere* of dimension 3. Explain why "hypersphere" is a good name for the graph and why the graph is only half of the hypersphere. What equation defines the entire hypersphere?

(d) Fill in the blanks: The graph of the function $f(x_1, x_2, \ldots, x_n) = \sqrt{1 - x_1^2 - x_2^2 - \cdots - x_n^2}$ is half of a _____ of dimension _____ in _____.

8. Find the range of the function $f(x, y) = -\dfrac{8x}{x^2 + y^2 + 1}$ from Example 4 by finding the largest value, M, and smallest value, m, of c for which the equation

$$-\frac{8x}{x^2 + y^2 + 1} = c$$

has a solution. (*Hint: Find the two values for c such that the level "curves" are a single point. Then show that the given equation has a solution for every value of c between those two values.*)

9. Let $z = f(x, y)$ be a function of two variables. Explain why the two sets $\{(x, y) \mid (x, y, z) \in \text{Graph}(f)\}$ and $\text{Domain}(f)$ are identical.

10. Let $z = f(x, y)$ be a function of two variables. Explain why the two sets $\{z \mid (x, y, z) \in \text{Graph}(f)\}$ and $\text{Range}(f)$ are identical.

11. Let $w = f(x, y, z)$ be a function of three variables. Explain why the two sets $\{(x, y, z) \mid (x, y, z, w) \in \text{Graph}(f)\}$ and $\text{Domain}(f)$ are identical.

12. Let $w = f(x, y, z)$ be a function of three variables. Explain why the two sets $\{w \mid (x, y, z, w) \in \text{Graph}(f)\}$ and $\text{Range}(f)$ are identical.

In Exercises 13–21, provide a rough sketch of the graph of a function of two variables with the specified level "curve(s)." (There are many possible correct answers to each question.)

13. One level curve consists of a single point.

14. One level curve consists of exactly two points.

15. One level curve is a circle together with the point that is the center of the circle.

16. One level curve consists of all the points (m, n) where m and n are both integers.

17. All of the level curves are circles except one, which is a point.

18. All of the level curves are circles.

19. Some level curves consist of two concentric circles.

20. Some level curves consist of infinitely many concentric circles.

21. All of the level curves are squares.

Skills

In Exercises 22–28, evaluate the given function at the specified points in the domain, and then find the domain and range of the function.

22. $f(x, y) = x^2 y^3$, $\left(\dfrac{1}{2}, \dfrac{4}{3}\right)$, $(0, \pi)$

23. $f(x, y) = x^2 - y^2$, $(1, 5)$, $(-3, -2)$

24. $g(x, y) = \dfrac{x^2 + y^2}{x + y}$, $(\pi, 1)$, $(4, 5)$

25. $g(x, y) = \dfrac{\ln(xy)}{\sqrt{1 - x}}$, $(-e, -1)$, $\left(\dfrac{1}{2}, 2\right)$

26. $f(x, y) = \dfrac{\sin(x - y)}{x - y}$, $(0, \pi)$, $\left(\dfrac{\pi}{2}, \dfrac{\pi}{3}\right)$

27. $f(x, y, z) = x^2 + y^2 + z^2$, $(1, 0, -5)$, $\left(\dfrac{1}{2}, -1, \dfrac{1}{3}\right)$

28. $f(x, y, z) = \dfrac{x + y + z}{xyz}$, $(1, -2, 6)$, $(-3, -2, 4)$

In Exercises 29–36, let

$$g_1(t) = \sin t, \ g_2(t) = \cos t, \ g_3(t) = 1 - t,$$

$$f_1(x, y) = x^2 + y^2, \ f_2(x, y) = \frac{x^2}{y^2}, \ f_3(x, y, z) = \frac{x + y}{y + z},$$

$$\mathbf{r}_1(t) = \langle 1 + t, t - 1 \rangle, \mathbf{r}_2(t) = \langle t, t^2, t^3 \rangle.$$

Either simplify the specified composition or explain why the composition cannot be formed.

29. $f_1(g_1(t), g_2(t))$

30. $g_1(f_1(x, y), f_2(x, y))$

31. $g_1(g_2(t))$

32. $f_1(\mathbf{r}_1(t))$

33. $f_3(g_2(t), g_1(t), g_3(t))$

34. $f_1(f_2(x, y), f_3(x, y, z))$

35. $r_2(g_3(t))$ **36.** $r_2(f_3(x, y, z))$

In Exercises 37–42, sketch the surface of revolution formed when the given function on the specified interval is revolved around the z-axis and find a function of two variables with the surface as its graph.

37. $f(x) = x$, $[0, 3]$ **38.** $f(x) = \sqrt{x}$, $[0, 4]$

39. $f(x) = x^2$, $[0, 2]$ **40.** $f(x) = x^{3/2}$, $[0, 1]$

41. $f(x) = \sin x$, $\left[0, \dfrac{\pi}{2}\right]$ **42.** $f(x) = \cos x$, $\left[0, \dfrac{\pi}{2}\right]$

In Exercises 43–52, sketch the level curves $c = -3, -2, -1, 0, 1, 2, 3$ if they exist for the specified function.

43. $f(x, y) = x + y$ **44.** $f(x, y) = \dfrac{3x}{y}$

45. $f(x, y) = \dfrac{3x}{y^2}$ **46.** $f(x, y) = \sqrt{xy}$

47. $f(x, y) = 4 - (x^2 + y^2)$ **48.** $f(x, y) = x^2 - y^2$

49. $f(x, y) = \sin(x + y)$ **50.** $f(x, y) = \cos(xy)$

51. $f(x, y) = y \csc x$ **52.** $f(x, y) = y \sec x$

In Exercises 53–58, determine the level surfaces $c = -3, -2, -1, 0, 1, 2, 3$ if they exist for the specified function.

53. $f(x, y, z) = x + 2y + 3z$ **54.** $f(x, y, z) = -2x - y + 4z$

55. $f(x, y, z) = \dfrac{x}{y - z}$ **56.** $f(x, y, z) = \dfrac{x^2 + y^2}{z}$

57. $f(x, y, z) = x^2 + y^2 + z^2$

58. $f(x, y, z) = \sqrt{1 - (x^2 + y^2 + z^2)}$

In Exercises 59–62, match the given function of two variables with the surfaces in Figures I–IV and with the level curves in Figures A–D. Make sure you explain your reasoning.

59. $f(x, y) = x^2 - y^2$

60. $f(x, y) = e^{-(x+1)^2 - (y+1)^2} + e^{-(x-1)^2 - (y-1)^2}$

61. $f(x, y) = e^{-x^2} + e^{-y^2}$

62. $f(x, y) = \sin x + \sin y$

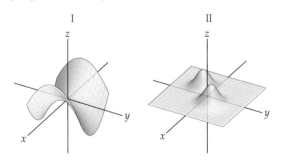

I   II

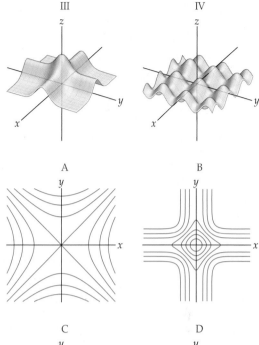

III   IV

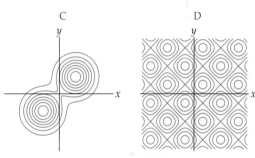

A   B

C   D

63. We may express the volume of a right circular cylinder by using the function of two variables, $V(r, h) = \pi r^2 h$, where r is the radius of either end of the cylinder and h is the height of the cylinder. What is the domain of the function $V(r, h)$? Express the surface area, S, of a cylinder as a function of variables r and h.

64. Express the volume, V, and surface area, S, of a right circular cone with radius r and height h as functions of two variables. What is the domain of each function?

65. Express the volume, V, and surface area, S, of a rectangular parallelepiped (i.e., a box) with side lengths x, y, and z as functions of three variables. What is the domain of each function?

66. Express the formulas for converting from polar coordinates to rectangular coordinates found in Section 9.2 as functions of two variables. What is the domain of each function?

Applications

67. Leila has been gathering data on the population density of caribou in a valley of the Selkirk Range in British Columbia, Canada. In winter, the caribou stay close to the bottom of the valley. Leila models the population for February with the function

$$p_F(x, y) = 13e^{-(y+0.5x-2)^2}e^{-0.2(x-1)^2},$$

where x and y are measured in miles from the center of the valley.

(a) Where are the caribou most likely to be found?

(b) During the summer, there is a stream in the valley. What is a vector in the direction of the stream?

68. Emmy is still trying to track down a leak in her Hanford tank farm. She has modeled the depth of a layer of basalt under the farm as a plane, but she has come to realize that there is a joint in the basalt that drops the level by a foot north of the curve with equation $x = 50 + 0.3y^{1.25}$. Thus, her model for the basalt layer is now

$$b(x, y) = -40 - \frac{1}{350}x + \frac{1}{725}y - u(x - 50 + 0.3y^{1.25}),$$

where u is the unit step function. The toxic solution must flow along the impenetrable layer of basalt. What is the route of its flow, and what is a vector giving the direction of its flow at any point?

Proofs

69. For constants a, b, and c, a function of two variables of the form $f(x, y) = ax + by + c$ is called a **linear function of two variables**. Show that the graph of the linear function $f(x, y) = ax + by + c$ is a plane with normal vector $\langle a, b, -1 \rangle$ containing the point $(0, 0, c)$.

70. Let $z = f(x, y)$ be a function of two variables. Prove that when $c_1 \neq c_2$, the level curves defined by the equations $f(x, y) = c_1$ and $f(x, y) = c_2$ do not intersect.

71. Let $z = f(x, y)$ be a function of two variables. Prove that if the level curves defined by the equations $f(x, y) = c_1$ and

$f(x, y) = c_2$ intersect, then the curves are identical. (*Hint: See Exercise 70.*)

72. Let $w = f(x, y, z)$ be a function of three variables. Prove that when $c_1 \neq c_2$, the level surfaces defined by the equations $f(x, y, z) = c_1$ and $f(x, y, z) = c_2$ do not intersect.

73. Let $w = f(x, y, z)$ be a function of three variables. Prove that if the level surfaces defined by the equations $f(x, y, z) = c_1$ and $f(x, y, z) = c_2$ intersect, then the surfaces are identical. (*Hint: See Exercise 72.*)

Thinking Forward

▶ *A kind of derivative for a function of two variables:* Explain why the derivative of the function $\frac{3x}{y^4}$ is $\frac{3}{y^4}$ if x is the variable and y is a constant and is $-\frac{12x}{y^5}$ if y is the variable and x is a constant. What is the derivative if both x and y are constants?

▶ *A kind of derivative for a function of three variables:* Explain why the derivative of the function $xe^{-4z}\sin y$ is $e^{-4z}\sin y$ if x is the variable and y and z are constants, and the derivative is $xe^{-4z}\cos y$ if y is the variable and x and z are constants, and the derivative is $-4xe^{-4z}\sin y$ if z is the variable and x and y are constants. What is the derivative if x, y, and z are all constants?

12.2   OPEN SETS, CLOSED SETS, LIMITS, AND CONTINUITY

▶ Open sets and closed sets

▶ The limit for a function of two or more variables

▶ The continuity of functions of two or more variables

Open Sets and Closed Sets

The tools of calculus introduced in Chapters 1 and 2 allow us to analyze functions of a single variable. The definitions of the limit, continuity, and differentiability require that our functions be defined on an open interval. Some theorems, such as the Extreme Value Theorem, require that our functions be defined on a closed interval. To understand these concepts for functions of two and three variables, we will spend the next few pages defining terms like *open sets, closed sets,* and *bounded sets.* It may be helpful to review the basic calculus concepts and theorems you studied in those earlier chapters before proceeding through this section.

We first define the following:

DEFINITION 12.7

> **Open Disks and Open Balls**
>
> Let $\epsilon > 0$.
>
> **(a)** Let $(x_0, y_0) \in \mathbb{R}^2$. A subset of $\mathbb{R}^2$ of the form
>
> $$\{(x, y) \mid (x - x_0)^2 + (y - y_0)^2 < \epsilon\}$$
>
> is said to be an **open disk** in $\mathbb{R}^2$.
>
> **(b)** Let $(x_0, y_0, z_0) \in \mathbb{R}^3$. A subset of $\mathbb{R}^3$ of the form
>
> $$\{(x, y, z) \mid (x - x_0)^2 + (y - y_0)^2 + (z - z_0)^2 < \epsilon\}$$
>
> is said to be an **open ball** in $\mathbb{R}^3$.

Thus, an open disk is a subset of $\mathbb{R}^2$ contained within a circle, not including the points on the circumference of the circle. Similarly, an open ball is a subset of $\mathbb{R}^3$ contained within a sphere, not including the points on the surface of the sphere.

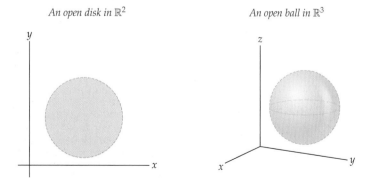

An open disk in $\mathbb{R}^2$ *An open ball in $\mathbb{R}^3$*

These basic objects allow us to define open sets in $\mathbb{R}^2$ and $\mathbb{R}^3$.

DEFINITION 12.8   Open Sets in $\mathbb{R}^2$ and $\mathbb{R}^3$

(a) A subset S of $\mathbb{R}^2$ is said to be ***open*** if, for every point $(x, y) \in S$, there is an open disk D such that $(x, y) \in D \subseteq S$.

(b) A subset S of $\mathbb{R}^3$ is said to be ***open*** if, for every point $(x, y, z) \in S$, there is an open ball B such that $(x, y, z) \in B \subseteq S$.

A set S in the plane is open if, for every point in S, there is an open disk that both contains the point and is a subset of the set S

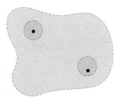

The following four examples illustrate the definition of an open set in $\mathbb{R}^2$ (consider the analogous examples in $\mathbb{R}^3$):

1. Every open disk is itself an open set. For example, the unit open disk
$$U = \{(x, y) \mid x^2 + y^2 < 1\}$$
is an open set because, for every point (x, y) in U, there is another smaller, open disk D such that $(x, y) \in D \subseteq U$.

The unit disk $U = \{(x, y) \mid x^2 + y^2 < 1\}$ is an open set

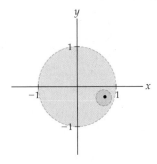

2. If we let W be the unit disk together with any point on the circumference of the circle, say, $(1, 0)$, then the set W will not be an open set. That is,
$$W = \{(x, y) \mid x^2 + y^2 < 1\} \cup \{(1, 0)\}.$$
The point $(1, 0)$ is in W, so this set is not open, because there is no open disk containing $(1, 0)$ that is a subset of W.

The set $W = \{(x, y) \mid x^2 + y^2 < 1\} \cup \{(1, 0)\}$ is not an open set

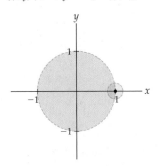

3. Similarly, if we let $\overline{U} = \{(x, y) \mid x^2 + y^2 \leq 1\}$, our set includes all points inside and on the circumference of the unit circle. Like the set W, this set is also not open.

4. The **empty set**, denoted by $\emptyset$, contains no elements and is a subset of every set. As a subset of $\mathbb{R}^2$, the empty set is open, since it vacuously satisfies Definition 12.8 (a).

DEFINITION 12.9

The Complement of a Set in $\mathbb{R}^2$ and $\mathbb{R}^3$

(a) Let A be a subset of $\mathbb{R}^2$. The **complement** of A, denoted A^c, is the set
$$A^c = \{(x, y) \in \mathbb{R}^2 \mid (x, y) \notin A\}.$$

(b) Let A be a subset of $\mathbb{R}^3$. The **complement** of A, denoted A^c, is the set
$$A^c = \{(x, y, z) \in \mathbb{R}^3 \mid (x, y, z) \notin A\}.$$

That is, the complement of a set A is the set of all points *not in* A. For example, the complement of the unit disk $U = \{(x, y) \mid x^2 + y^2 < 1\}$ is the set $U^c = \{(x, y) \mid x^2 + y^2 \geq 1\}$.

The complement of the unit disk in $\mathbb{R}^2$

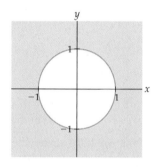

We ask you to prove the following theorem in Exercise 63:

THEOREM 12.10

The Complement of the Complement of a Set Is the Original Set

If S is a subset of $\mathbb{R}^2$ or $\mathbb{R}^3$, then $(S^c)^c = S$.

We use the idea of the complement of a set to define a closed set.

DEFINITION 12.11

Closed Sets in $\mathbb{R}^2$ and $\mathbb{R}^3$

A subset S of $\mathbb{R}^2$ or $\mathbb{R}^3$ is said to be **closed** if its complement, S^c, is open.

For example, the set $U^c = \{(x, y) \mid x^2 + y^2 \geq 1\}$ is a closed set because it is the complement of the open set $U = \{(x, y) \mid x^2 + y^2 < 1\}$. Note that a set is not closed just because it is not open. Later, in Example 1, we will show that the set $\{(x, y) \mid x \geq 0 \text{ and } y > 0\}$ is neither open nor closed. In that respect, this set is similar to a half-open, half-closed interval like $[0, 1)$.

We ask you to prove the following theorem in Exercise 64:

THEOREM 12.12

The Complement of a Closed Set Is an Open Set

If S is a closed subset of $\mathbb{R}^2$ or $\mathbb{R}^3$, then S^c is an open set.

We now define the boundary of a set.

DEFINITION 12.13

The Boundary of a Subset of $\mathbb{R}^2$ or $\mathbb{R}^3$

(a) Let A be a subset of $\mathbb{R}^2$. A point (x, y) is said to be a **boundary point** of A if every open disk containing (x, y) in $\mathbb{R}^2$ intersects both A and A^c.

(b) Let A be a subset of $\mathbb{R}^3$. A point (x, y, z) is said to be a **boundary point** of A if every open ball containing (x, y, z) in $\mathbb{R}^3$ intersects both A and A^c.

The set of all boundary points of a set A is called the **boundary** of A and is denoted by ∂A.

The boundary of the unit disk $U = \{(x, y) \mid x^2 + y^2 < 1\}$ consists exactly of the points on the unit circle. That is, $\partial U = \{(x, y) \mid x^2 + y^2 = 1\}$.

Every open disk containing a point on the unit circle intersects both U and U^c

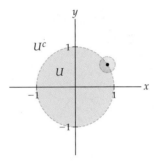

There are many interesting theorems about open and closed sets in $\mathbb{R}^2$ and $\mathbb{R}^3$. We ask you to prove several of these in the exercises.

Recall that we have already defined what it means for a function to be bounded. We now define what it means for a set to be bounded.

DEFINITION 12.14

Bounded Subsets of $\mathbb{R}^2$ and $\mathbb{R}^3$

(a) A subset A of $\mathbb{R}^2$ is said to be **bounded** if A is a subset of some open disk D in $\mathbb{R}^2$.

(b) A subset A of $\mathbb{R}^3$ is said to be **bounded** if A is a subset of some open ball B in $\mathbb{R}^3$.

A subset of $\mathbb{R}^2$ or $\mathbb{R}^3$ that is not bounded is said to be **unbounded**.

For example, the set A in the following figure is bounded, while the set consisting of the coordinate axes in $\mathbb{R}^2$ or $\mathbb{R}^3$ is unbounded.

A is a bounded set   *The points on the coordinate axes form an unbounded set*

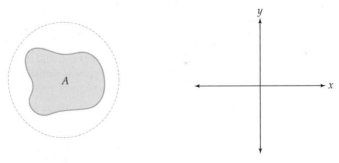

Now that we have set forth these preliminary definitions, after a brief review of the definition of the limit of a function of a single variable, we will be ready to define the limit of a function of two or three variables.

The Limit of a Function of Two or More Variables

Recall that for a function, f, of a single variable, we defined $\lim_{x \to a} f(x) = L$ if, for every $\epsilon > 0$, there is a $\delta > 0$ such that $|f(x) - L| < \epsilon$ whenever $0 < |x - a| < \delta$. The definition makes precise the informal idea that a function f has a limit L at a point a if the output values of f are "close to" L for values of the input variable x that are "close to" a. We will use this definition as a model to define the limit for functions of two and three variables. Before we get to the new definition, however, note that the definition of the limit implies that the function f be defined on a punctured open interval. In Exercise 7 we ask how this is accomplished.

There is very little that needs to be changed in the definition of a limit for a function of a single variable to form a definition for the limit of a function of two or three variables. Certainly the types of domains are different: The domain of a function of two variables consists of a set of ordered pairs, and the domain of a function of three variables consists of a set of ordered triples.

DEFINITION 12.15

The Limit of a Function of Two or More Variables

Let f be a function of two or more variables. The limit of f at $\mathbf{a}$ is L if, for every $\epsilon > 0$, there is a $\delta > 0$ such that $|f(\mathbf{x}) - L| < \epsilon$ whenever $0 < \|\mathbf{x} - \mathbf{a}\| < \delta$. In this case we write $\lim_{\mathbf{x} \to \mathbf{a}} f(\mathbf{x}) = L$.

In the case where f is a function of two variables, we may let $\mathbf{x} = \langle x, y \rangle$ and $\mathbf{a} = \langle a, b \rangle$, therefore $\lim_{\mathbf{x} \to \mathbf{a}} f(\mathbf{x}) = L$ means that for every $\epsilon > 0$, there is a $\delta > 0$ such that

$$|f(x, y) - L| < \epsilon \text{ whenever } 0 < \sqrt{(x - a)^2 + (y - b)^2} < \delta.$$

Similarly, if f is a function of three variables, we may let $\mathbf{x} = \langle x, y, z \rangle$ and $\mathbf{a} = \langle a, b, c \rangle$. Here, $\lim_{\mathbf{x} \to \mathbf{a}} f(\mathbf{x}) = L$ means that for every $\epsilon > 0$, there is a $\delta > 0$ such that

$$|f(x, y, z) - L| < \epsilon \text{ whenever } 0 < \sqrt{(x - a)^2 + (y - b)^2 + (z - c)^2} < \delta.$$

The vector notation of Definition 12.15 can be used in place of the preceding two expanded statements. We will use the more succinct notation when appropriate.

Limits of combinations of functions of two or three variables obey the same types of algebraic properties we saw for limits of functions of a single variable in Chapter 1.

THEOREM 12.16

Rules for Calculating Limits of Combinations

If $\lim_{\mathbf{x} \to \mathbf{a}} f(\mathbf{x})$ and $\lim_{\mathbf{x} \to \mathbf{a}} g(\mathbf{x})$ exist, then the following rules hold for their combinations:

Constant Multiple Rule: $\lim_{\mathbf{x} \to \mathbf{a}} kf(\mathbf{x}) = k \lim_{\mathbf{x} \to \mathbf{a}} f(\mathbf{x})$ for any real number k.

Sum Rule: $\lim_{\mathbf{x} \to \mathbf{a}} (f(\mathbf{x}) + g(\mathbf{x})) = \lim_{\mathbf{x} \to \mathbf{a}} f(\mathbf{x}) + \lim_{\mathbf{x} \to \mathbf{a}} g(\mathbf{x})$.

Difference Rule: $\lim_{\mathbf{x} \to \mathbf{a}} (f(\mathbf{x}) - g(\mathbf{x})) = \lim_{\mathbf{x} \to \mathbf{a}} f(\mathbf{x}) - \lim_{\mathbf{x} \to \mathbf{a}} g(\mathbf{x})$.

Product Rule: $\lim_{\mathbf{x} \to \mathbf{a}} (f(\mathbf{x})g(\mathbf{x})) = \left(\lim_{\mathbf{x} \to \mathbf{a}} f(\mathbf{x}) \right) \left(\lim_{\mathbf{x} \to \mathbf{a}} g(\mathbf{x}) \right)$.

Quotient Rule: $\lim_{\mathbf{x} \to \mathbf{a}} \dfrac{f(\mathbf{x})}{g(\mathbf{x})} = \dfrac{\lim_{\mathbf{x} \to \mathbf{a}} f(\mathbf{x})}{\lim_{\mathbf{x} \to \mathbf{a}} g(\mathbf{x})}$, if $\lim_{\mathbf{x} \to \mathbf{a}} g(\mathbf{x}) \neq 0$.

Another important theorem concerning limits of functions of a single variable has an analog for functions of two or more variables. Theorem 1.8 tells us that the function $f : \mathbb{R} \to \mathbb{R}$ has the two-sided limit $\lim_{x \to a} f(x) = L$ if and only if the two one-sided limits $\lim_{x \to a^+} f(x)$ and $\lim_{x \to a^-} f(x)$ are both equal to L. Before we state the analogous theorem for functions of two or more variables, we define what we mean by the limit of a function of two or more variables *along a path*.

DEFINITION 12.17   The Limit of a Function of Two or Three Variables Along a Path

(a) Let f be a function of two variables defined on an open set $S \subseteq \mathbb{R}^2$. If the graph of the continuous vector function $\langle x(t), y(t) \rangle$ is a curve $C \subset S$ such that $\lim_{t \to t_0} \langle x(t), y(t) \rangle = \langle a, b \rangle$, then we define the **limit of f as (x, y) approaches (a, b) along** C, denoted by $\lim_{\substack{(x,y) \to (a,b) \\ C}} f(x, y)$, to be $\lim_{t \to t_0} f(x(t), y(t))$.

(b) Let f be a function of three variables defined on an open set $S \subseteq \mathbb{R}^3$. If the graph of the continuous vector function $\langle x(t), y(t), z(t) \rangle$ is a curve $C \subset S$ such that $\lim_{t \to t_0} \langle x(t), y(t), z(t) \rangle = \langle a, b, c \rangle$, then we define the **limit of f as (x, y, z) approaches (a, b, c) along** C, denoted by $\lim_{\substack{(x,y,z) \to (a,b,c) \\ C}} f(x, y, z)$, to be $\lim_{t \to t_0} f(x(t), y(t), z(t))$.

The following schematic illustrates three such curves containing a point (a, b) in an open subset of $\mathbb{R}^2$.

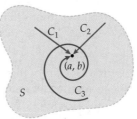

THEOREM 12.18   The Limit of a Function of Two or More Variables Along Distinct Paths

Let f be a function of two or more variables defined on an open set S. Then the limit $\lim_{\mathbf{x} \to \mathbf{a}} f(\mathbf{x}) = L$ exists if and only if $\lim_{\substack{\mathbf{x} \to \mathbf{a} \\ C}} f(\mathbf{x}) = L$ for every path C containing the point (a, b).

Given an open set S containing a point $\mathbf{a}$, there are infinitely many paths containing $\mathbf{a}$ that are also contained in S. Thus, it is impractical to prove that a limit of two or more variables exists by showing that the limit of the function along each path is the same. Theorem 12.18 is usually applied when the limit does not exist and we can find two distinct paths C_1 and C_2 such that

$$\lim_{\substack{\mathbf{x} \to \mathbf{a} \\ C_1}} f(\mathbf{x}) \neq \lim_{\substack{\mathbf{x} \to \mathbf{a} \\ C_2}} f(\mathbf{x}).$$

For example, consider the function $f(x, y) = \dfrac{xy}{x^2 + y^2}$. We will show that $\lim_{(x,y) \to (0,0)} f(x, y)$ does not exist by evaluating the limit along two paths containing the origin and getting different values. First consider the limit along the x-axis (i.e., when $y = 0$):

$$\lim_{\substack{(x,y) \to (0,0) \\ x\text{-axis}}} f(x, y) = \lim_{\substack{(x,y) \to (0,0) \\ x\text{-axis}}} \frac{xy}{x^2 + y^2} = \lim_{x \to 0} \frac{x \cdot 0}{x^2 + 0} = \lim_{x \to 0} 0 = 0.$$

However, along the line $y = x$,

$$\lim_{\substack{(x,y)\to(0,0)\\y=x}} f(x,y) = \lim_{\substack{(x,y)\to(0,0)\\y=x}} \frac{xy}{x^2+y^2} = \lim_{x\to 0} \frac{x \cdot x}{x^2+x^2} = \lim_{x\to 0} \frac{x^2}{2x^2} = \lim_{x\to 0} \frac{1}{2} = \frac{1}{2}.$$

Since the two limits do not agree, it follows from Theorem 12.18 that $\lim\limits_{(x,y)\to(0,0)} f(x,y)$ does not exist.

The Continuity of a Function of Two or More Variables

Recall that if f is a function of a single variable, f is said to be continuous at a point a in the domain of f if $\lim\limits_{x\to a} f(x) = f(a)$. We will define what it means for a function of two or more variables to be continuous at a point in its domain with an analogous definition.

DEFINITION 12.19

The Continuity of Functions of Two or Three Variables

Let f be a function of two or three variables defined on an open set S, and let $\mathbf{a}$ be a point in S. Then f is ***continuous*** at $\mathbf{a}$ if $\lim\limits_{\mathbf{x}\to\mathbf{a}} f(\mathbf{x}) = f(\mathbf{a})$.

Also, f is ***continuous on*** S if f is continuous at every point $\mathbf{a} \in S$.

If the domain of f is $\mathbb{R}^2$ or $\mathbb{R}^3$ and f is continuous at every point in its domain, then f is ***continuous everywhere***.

Since continuity is defined in terms of a limit, it follows that sums, differences, and products of continuous functions will also be continuous. In addition, quotients will be continuous as long as we do not divide by zero. Most of the functions we work with on a regular basis are continuous at every point in their domains. For example, the function $f(x,y) = \frac{xy}{x^2+y^2}$ we discussed just after Theorem 12.18 is continuous at every point in its domain. This domain is the set of all points in $\mathbb{R}^2$ except the origin, so if a and b are not both zero, we will have

$$\lim_{(x,y)\to(a,b)} \frac{xy}{x^2+y^2} = \frac{ab}{a^2+b^2}.$$

This function is discontinuous at the origin for two reasons: The function is not defined at the origin, and $\lim\limits_{(x,y)\to(0,0)} \frac{xy}{x^2+y^2}$ does not exist.

Examples and Explorations

EXAMPLE 1

Understanding open sets, closed sets, and boundaries

For each of the following subsets of $\mathbb{R}^2$, determine whether the set is open, closed, or neither open nor closed. Find the boundary of each set as well.

(a) $S_1 = \{(x,y) \in \mathbb{R}^2 \mid x \geq 0\}$ **(b)** $S_2 = \{(x,y) \in \mathbb{R}^2 \mid y \neq 0\}$

(c) $S_3 = \{(x,y) \in \mathbb{R}^2 \mid x \geq 0 \text{ and } y > 0\}$

SOLUTION

(a) The set S_1 is the right half-plane. The complement of S_1 is $S_1^c = \{(x,y) \in \mathbb{R}^2 \mid x < 0\}$. Any point (a,b) with $a < 0$ is in S_1^c. The open ball $B_1 = \{(x,y) \mid (x-a)^2 + (y-b)^2 < a^2\}$ contains (a,b) and is a subset of S_1^c. Therefore, S_1^c is an open set. Its complement $(S_1^c)^c = S_1$ is a closed set. The boundary of S_1 is the y-axis, since every open ball containing a point on the y-axis contains points of both S_1 and S_1^c but points off the y-axis do not share this property. The following figure is illustrative:

The set S_1 is closed because S_1^c is open

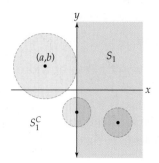

(b) The set S_2 consists of all points in the plane, except for those on the x-axis. We will show that this set is open. If we choose the point $(a, b) \in S_2$, then $b \neq 0$. The open ball $B_2 = \{(x, y) \mid (x - a)^2 + (y - b)^2 < b^2\}$ contains (a, b) and is a subset of S_2 whether $b > 0$ or $b < 0$. (In the illustration that follows, $b < 0$.) Therefore, S_2 is an open set. The boundary of S_2 is the x-axis, since every open ball containing a point on the x-axis contains points of both S_2 and S_2^c but points not on the x-axis do not share this property.

The set S_2 is open

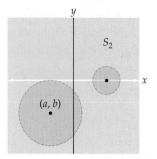

(c) The set S_3 consists of all points in the first quadrant, along with those points on the positive y-axis. We will show that S_3 is neither open nor closed. S_3 is not open because, although $(0, 1) \in S_3$, we cannot find an open ball containing $(0, 1)$ that is also a subset of S_3, since every open ball containing $(0, 1)$ contains points of both S_3 and S_3^c. To show that S_3 is not closed, we will show that S_3^c is not open. Although $(1, 0) \in S_3^c$, every open ball that contains $(1, 0)$ intersects both S_3 and S_3^c. Therefore, S_3 is neither open nor closed. The boundary of S_3 is $\partial S_3 = \{(x, 0) \mid x \geq 0\} \cup \{(0, y) \mid y \geq 0\}$.

The set S_3 is neither open nor closed

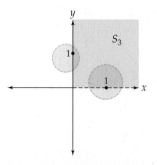

EXAMPLE 2

Evaluating limits of functions of two and three variables

Evaluate the limits:

(a) $\lim\limits_{(x,y,z)\to(2,-2,3)}(xy^2+4xyz)$ **(b)** $\lim\limits_{(x,y)\to(0,0)}\dfrac{x^2y^3+5xy-3}{x+y-1}$ **(c)** $\lim\limits_{(x,y)\to(0,3)}\dfrac{(\sin x)(\cos y)}{e^{xy}-2}$

SOLUTION

(a) The function in the first limit is a polynomial function of three variables. Just as polynomials of a single variable are continuous for every real number, polynomials of two variables are continuous for every point in $\mathbb{R}^2$ and polynomials of three variables are continuous for every point in $\mathbb{R}^3$. Therefore, to determine the limit, we may evaluate the function $f(x,y,z)=xy^2+4xyz$ at $(2,-2,3)$. We obtain

$$\lim\limits_{(x,y,z)\to(2,-2,3)}(xy^2+4xyz)=2(-2)^2+4\cdot 2(-2)3=-40.$$

(b) Similarly, rational functions of two or three variables are also continuous where they are defined. The values where the second function $g(x,y)=\dfrac{x^2y^3+5xy-3}{x+y-1}$ is discontinuous are those points where the denominator is zero, namely, all points along the line defined by the equation $x+y=1$. Since the origin does not satisfy this equation, the function g is continuous at $(0,0)$. Therefore,

$$\lim\limits_{(x,y)\to(0,0)}\dfrac{x^2y^3+5xy-3}{x+y-1}=\dfrac{0^20^3+5\cdot 0\cdot 0-3}{0+0-1}=3.$$

(c) Since the function in our final limit is an algebraic combination of the sine, cosine, and exponential functions, all of which are transcendental functions, it will also be continuous where it is defined. The points where the function $h(x,y)=\dfrac{(\sin x)(\cos y)}{e^{xy}-2}$ will be discontinuous will be the solutions of the equation $e^{xy}=2$, or equivalently, points on the hyperbola defined by the equation $y=\dfrac{\ln 2}{x}$. The point $(0,3)$ does not satisfy this equation, so the function h is continuous at $(0,3)$. Therefore,

$$\lim\limits_{(x,y)\to(0,3)}\dfrac{(\sin x)(\cos y)}{e^{xy}-2}=\dfrac{(\sin 0)(\cos 3)}{e^{0.3}-2}=0.$$ ☐

EXAMPLE 3

Showing that a limit does not exist

Explain why $\lim\limits_{(x,y)\to(6,3)}\dfrac{x^2+xy-6y^2}{x^2-4y^2}$ does not exist.

SOLUTION

The function $f(x,y)=\dfrac{x^2+xy-6y^2}{x^2-4y^2}$ is a rational function of two variables, but the point $(6,3)$ is not in the domain of f because the polynomial that forms the denominator of f evaluates to 0 at $(6,3)$. However, the numerator also evaluates to 0 at that point. Thus, this limit is an indeterminate form of the type $\dfrac{0}{0}$. As we did with rational functions of a single variable, we may try to simplify the quotient, so that we may use a multivariable version of the Cancellation Theorem to evaluate the limit. Since $\dfrac{x^2+xy-6y^2}{x^2-4y^2}=\dfrac{(x-2y)(x+3y)}{(x-2y)(x+2y)}$, we have

$$\lim\limits_{(x,y)\to(6,3)}\dfrac{x^2+xy-6y^2}{x^2-4y^2}=\lim\limits_{(x,y)\to(6,3)}\dfrac{(x-2y)(x+3y)}{(x-2y)(x+2y)}.$$

Now, we are tempted to cancel the common factor that occurs in the numerator and denominator and say that

$$\lim\limits_{(x,y)\to(6,3)}\dfrac{(x-2y)(x+3y)}{(x-2y)(x+2y)}=\lim\limits_{(x,y)\to(6,3)}\dfrac{x+3y}{x+2y}.$$

However, the equality between the two limits is not valid. There is an analog to the Cancellation Theorem for limits of functions of more than one variable, but the analog would apply only when the cancellation of factors introduces a single point into the domain of the reduced function. In our example, although the functions $f(x, y) = \frac{x^2 + xy - 6y^2}{x^2 - 4y^2}$ and $g(x, y) = \frac{x + 3y}{x + 2y}$ are equivalent when $x \neq 2y$, they are not equivalent for points on the line $x = 2y$. However, Definition 12.15 requires that, for $\lim\limits_{(x,y) \to (a,b)} f(x, y)$ to exist, the function f must be defined on an open set containing the point (a, b), with the possible exception that f does not have to be defined at the point (a, b) itself. Thus, although $\lim\limits_{(x,y) \to (6,3)} \frac{x + 3y}{x + 2y} = \frac{5}{4}$, we cannot conclude that our original limit has the same value. The function $f(x, y) = \frac{x^2 + xy - 6y^2}{x^2 - 4y^2}$ is undefined along the line defined by $x = 2y$, and thus f is undefined for infinitely many points in every open set containing the point $(6, 3)$. Therefore $\lim\limits_{(x,y) \to (6,3)} \frac{x^2 + xy - 6y^2}{x^2 - 4y^2}$ does not exist.

However, if C is any curve in the plane that intersects the line $x = 2y$ only at the point $(6, 3)$, then

$$\lim_{\substack{(x,y) \to (6,3) \\ C}} \frac{x^2 + xy - 6y^2}{x^2 - 4y^2} = \lim_{(x,y) \to (6,3)} \frac{x + 3y}{x + 2y} = \frac{5}{4}.$$ □

EXAMPLE 4   **Evaluating limits along paths**

Show that $\lim\limits_{(x,y) \to (0,0)} \frac{x^2 y}{x^4 + y^2}$ does not exist by evaluating the limit along the curves defined by the parabolas $y = mx^2$.

SOLUTION

In Exercise 13 you will show that

$$\lim_{\substack{(x,y) \to (0,0) \\ C}} \frac{x^2 y}{x^4 + y^2} = 0$$

when C is either the x- or y-axis. However, the fact that these limits exist and agree is not enough to conclude that the limit is zero. When we evaluate the limit of the function along a parabola of the form $y = mx^2$, we will see that the result depends upon the value of m. From this observation, we conclude that the limit does not exist. We have

$$\lim_{\substack{(x,y) \to (0,0) \\ y = mx^2}} \frac{x^2 y}{x^4 + y^2} = \lim_{x \to 0} \frac{x^2 \cdot mx^2}{x^4 + (mx^2)^2} = \lim_{x \to 0} \frac{mx^4}{x^4(1 + m^2)} = \lim_{x \to 0} \frac{m}{1 + m^2} = \frac{m}{1 + m^2}.$$

Since this limit is a nonconstant function of m, $\lim\limits_{(x,y) \to (0,0)} \frac{x^2 y}{x^4 + y^2}$ does not exist. □

EXAMPLE 5   **Using polar coordinates to evaluate limits**

Evaluate the limits

(a) $\lim\limits_{(x,y) \to (0,0)} \frac{x^2 y}{x^2 + y^2}$ **(b)** $\lim\limits_{(x,y) \to (0,0)} \frac{xy}{x^2 + y^2}$

if they exist.

SOLUTION

In Chapter 9 we saw that we could translate between rectangular coordinates and polar coordinates by using the equations $r^2 = x^2 + y^2$, $\tan \theta = \frac{y}{x}$, $x = r \cos \theta$, and $y = r \sin \theta$.

Changing to polar coordinates can be useful when a function contains the factor $x^2 + y^2$, which we find in both of the given limits.

(a) First,

$$\lim_{(x,y)\to(0,0)} \frac{x^2 y}{x^2 + y^2} = \lim_{r\to 0} \frac{(r\cos\theta)^2 (r\sin\theta)}{r^2} = \lim_{r\to 0} \frac{r^3 \sin\theta \cos^2\theta}{r^2} = \lim_{r\to 0} r\sin\theta \cos^2\theta.$$

Since we are taking the limit as $r \to 0$, and the limit contains a factor of r, its value is zero for every value of θ.

(b) We also convert the second limit to polar coordinates and obtain

$$\lim_{(x,y)\to(0,0)} \frac{xy}{x^2 + y^2} = \lim_{r\to 0} \frac{(r\cos\theta)(r\sin\theta)}{r^2} = \lim_{r\to 0} \frac{r^2 \sin\theta \cos\theta}{r^2} = \lim_{r\to 0} \sin\theta \cos\theta.$$

Here, the value of the limit depends upon the angle θ. Therefore, the limit can be different along different paths approaching the origin. For example,

$$\lim_{\substack{(x,y)\to(0,0)\\ \theta=0}} \frac{xy}{x^2 + y^2} = \lim_{r\to 0} 0 = 0 \quad \text{and} \quad \lim_{\substack{(x,y)\to(0,0)\\ \theta=\pi/4}} \frac{xy}{x^2 + y^2} = \lim_{r\to 0} \frac{1}{2} = \frac{1}{2}.$$

Thus $\lim\limits_{(x,y)\to(0,0)} \frac{xy}{x^2+y^2}$ does not exist. □

EXAMPLE 6   **Examining the continuity of a piecewise-defined function**

Determine where the functions

(a) $f(x, y) = \begin{cases} \dfrac{x^2 y}{x^2 + y^2}, & \text{if } (x, y) \neq (0, 0) \\ 0, & \text{if } (x, y) = (0, 0) \end{cases}$ **(b)** $g(x, y) = \begin{cases} \dfrac{xy}{x^2 + y^2}, & \text{if } (x, y) \neq (0, 0) \\ 0, & \text{if } (x, y) = (0, 0) \end{cases}$

are continuous.

SOLUTION

These are the same functions we discussed in Example 5, except that their values have been defined at the origin. Since rational functions are continuous everywhere in their domains, both of the given functions are continuous everywhere, with the possible exception of the origin.

(a) Because we already showed that

$$\lim_{(x,y)\to(0,0)} \frac{x^2 y}{x^2 + y^2} = \lim_{(x,y)\to(0,0)} f(x, y) = 0 = f(0, 0),$$

the function f is also continuous at the origin. Therefore, f is continuous everywhere.

(b) Although g is continuous everywhere except the origin, there is no way to define the function at $(0, 0)$ to make g continuous, since we have already seen that $\lim\limits_{(x,y)\to(0,0)} g(x, y)$ does not exist. □

? TEST YOUR UNDERSTANDING

▶ What are open and closed sets in $\mathbb{R}$? In $\mathbb{R}^2$? In $\mathbb{R}^3$?

▶ What is a bounded set in $\mathbb{R}$? In $\mathbb{R}^2$? In $\mathbb{R}^3$?

▶ What is the definition of the limit of a function of one variable? Of two variables? Of three variables? What does it mean intuitively for these limits to exist?

▶ What do we mean by the limit of a function of two or three variables along a path? What is the analogous concept for a function of a single variable?

▶ What is the definition of the continuity of a function of one variable at a point? Of two variables? Of three variables?

EXERCISES 12.2

Thinking Back

▶ *Intervals:* What is meant by an open interval in $\mathbb{R}$? What is meant by a closed interval in $\mathbb{R}$? Is every interval either open or closed?

▶ *Limits:* If f is a function of a single variable, what is the intuitive interpretation of $\lim_{x \to a} f(x)$? What is the formal definition?

Concepts

0. *Problem Zero:* Read the section and make your own summary of the material.

1. *True/False:* Determine whether each of the statements that follow is true or false. If a statement is true, explain why. If a statement is false, provide a counterexample.

  (a) *True or False:* The complement of a closed subset of $\mathbb{R}^2$ is an open subset of $\mathbb{R}^2$.

  (b) *True or False:* Every subset of $\mathbb{R}^2$ is either open or closed.

  (c) *True or False:* There are subsets of $\mathbb{R}^2$ that are both open and closed.

  (d) *True or False:* If a subset S of $\mathbb{R}^3$ is neither open nor closed, then S^c is neither open nor closed.

  (e) *True or False:* If $\lim_{(x,y) \to (0,0)} f(x,y) = L$, then $f(0,0) = L$.

  (f) *True or False:* If

$$\lim_{(x,y) \to (a,b)} f(x,y) = L, \quad \text{then} \quad \lim_{\substack{(x,y) \to (a,b) \\ C}} f(x,y) = L$$

  for every path C containing (a,b).

  (g) *True or False:* If

$$\lim_{\substack{(x,y) \to (a,b) \\ C}} f(x,y) = L$$

  for every path C containing (a,b), then $f(x,y)$ is continuous at (a,b).

  (h) *True or False:* If $f(x,y)$ is continuous everywhere and $\lim_{(x,y) \to (0,0)} f(x,y) = L$, then $f(0,0) = L$.

2. *Examples:* Construct examples of the thing(s) described in the following. Try to find examples that are different than any in the reading.

  (a) An open subset of $\mathbb{R}^3$.

  (b) A closed subset of $\mathbb{R}^3$.

  (c) A subset of $\mathbb{R}^3$ that is both open and closed.

3. If

$$\lim_{\substack{(x,y) \to (1,-3) \\ C_1}} f(x,y) = 5 \quad \text{and} \quad \lim_{\substack{(x,y) \to (1,-3) \\ C_2}} f(x,y) = 5,$$

where C_1 and C_2 are two distinct curves in $\mathbb{R}^2$ containing the point $(1,-3)$, what can you say about $\lim_{(x,y) \to (1,-3)} f(x,y)$?

4. If

$$\lim_{\substack{(x,y) \to (-2,0) \\ C_1}} g(x,y) = 5 \quad \text{and} \quad \lim_{\substack{(x,y) \to (-2,0) \\ C_2}} g(x,y) = 8,$$

where C_1 and C_2 are two distinct curves in $\mathbb{R}^2$ containing the point $(-2,0)$, what can you say about $\lim_{(x,y) \to (-2,0)} g(x,y)$?

5. If

$$\lim_{\substack{(x,y) \to (3,-7) \\ C}} f(x,y) = 5$$

for every curve C in $\mathbb{R}^2$ containing the point $(3,-7)$, what can you say about $\lim_{(x,y) \to (3,-7)} f(x,y)$? What can you say about $f(3,-7)$?

6. If $g : \mathbb{R}^3 \to \mathbb{R}$ is continuous at the point (a,b,c) and C is a curve in $\mathbb{R}^3$ containing the point (a,b,c), what can we say about

$$\lim_{\substack{(x,y,z) \to (a,b,c) \\ C}} g(x,y,z)?$$

7. How does the definition of the limit of a function of a single variable, f, imply that f is defined on the union of two open intervals?

8. How does the definition of the limit of a function of two variables, f, imply that f is defined on an open subset of $\mathbb{R}^2$?

9. How does the definition of the limit of a function of three variables, f, imply that f is defined on an open subset of $\mathbb{R}^3$?

10. Explain how the definition of the limit of a function of two or three variables along a path simplifies to the limit of a function of a single variable.

11. Review the definition of continuity of a function of a single variable, f, at a point. Why is it necessary for f to be defined on the union of two open intervals?

12. Review the definition of continuity of a function of two or three variables, f, at a point. Why is it necessary for f to be defined on an open subset of $\mathbb{R}^2$ or $\mathbb{R}^3$?

13. Show that when C is either the x- or y-axis, we have

$$\lim_{\substack{(x,y) \to (0,0) \\ C}} \frac{x^2 y}{x^4 + y^2} = 0.$$

14. Copy the figure that follows onto a sheet of paper. Now cut a slit along the dashed line, leave the left side of the paper on the table, and gently raise the right side of the paper along the slit.

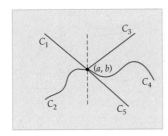

(a) Explain how the paper may be interpreted as the graph of a function of two variables, $f(x, y)$.

(b) If the tabletop is the xy-plane, explain why

$$\lim_{\substack{(x,y)\to(a,b)\\ C_1}} f(x, y) = \lim_{\substack{(x,y)\to(a,b)\\ C_2}} f(x, y) = 0.$$

(c) Explain why

$$\lim_{\substack{(x,y)\to(a,b)\\ C_3}} f(x, y) = \lim_{\substack{(x,y)\to(a,b)\\ C_4}} f(x, y) = \lim_{\substack{(x,y)\to(a,b)\\ C_5}} f(x, y) > 0.$$

(d) Explain why $\lim_{(x,y)\to(a,b)} f(x, y)$ does not exist.

15. Provide a definition for $\lim_{(x,y)\to(a,b)} f(x, y) = \infty$. Model your definition on Definitions 1.9 and 12.15.

16. Provide a definition for $\lim_{(x,y,z)\to(a,b,c)} f(x, y, z) = \infty$. Model your definition on Definitions 1.9 and 12.15.

17. Find functions $f(x, y)$ and $g(x, y)$ and a point $(a, b) \in \mathbb{R}^2$ such that

$$\lim_{(x,y)\to(a,b)} f(x, y) + \lim_{(x,y)\to(a,b)} g(x, y) \neq \lim_{(x,y)\to(a,b)} (f(x, y) + g(x, y)).$$

Does this example contradict the sum rule for limits of a function of two variables?

18. Find functions $f(x, y)$ and $g(x, y)$ and a point $(a, b) \in \mathbb{R}^2$ such that

$$\left(\lim_{(x,y)\to(a,b)} f(x, y)\right)\left(\lim_{(x,y)\to(a,b)} g(x, y)\right) \neq \lim_{(x,y)\to(a,b)} (f(x, y) g(x)).$$

Does this example contradict the product rule for limits of a function of two variables?

19. Let f be a function of two variables that is continuous everywhere.

(a) Explain why the function $\dfrac{f(x,y)}{x - y}$ is continuous if and only if $x \neq y$.

(b) Use Definition 12.15 to explain why $\lim_{(x,y)\to(a,a)} \dfrac{f(x,y)}{x - y}$ does not exist for any real number a.

20. Let f be a function of three variables that is continuous everywhere.

(a) Explain why the function $\dfrac{f(x,y,z)}{x + y + z}$ is continuous if and only if $x + y + z \neq 0$.

(b) Use Definition 12.15 to explain why

$$\lim_{(x,y,z)\to(a,b,-(a+b))} \dfrac{f(x,y,z)}{x + y + z}$$

does not exist for any pair (a, b) of real numbers.

Skills

In Exercises 21–26, (a) determine whether the given subset of $\mathbb{R}^2$ is open, closed, both open and closed, or neither open nor closed, (b) find the complement of the set, and (c) find the boundary of the given set.

21. all points (x, y) such that $x > 0$ and $y > 0$

22. all points on the coordinate axes

23. all points satisfying the inequality $|x| + |y| \leq 1$

24. all points (x, y) such that $|x| < 1$ and $|y| \leq 2$

25. the empty set

26. $\mathbb{R}^2$

In Exercises 27–32, (a) determine whether the given subset of $\mathbb{R}^3$ is open, closed, both open and closed, or neither open nor closed, (b) find the complement of the set, and (c) find the boundary of the given set.

27. all points (x, y, z) such that $x > 0$, $y < 0$ and $z < 0$

28. all points on the coordinate axes

29. all points on the xy-plane

30. all points (x, y, z) such that $|x| < 1$, $|y| < 2$, and $|z| > 3$

31. the empty set

32. $\mathbb{R}^3$

Evaluate the limits in Exercises 33–40 if they exist.

33. $\lim\limits_{(x,y)\to(-2,\pi)} x^2 y^3 \sin y$

34. $\lim\limits_{(x,y,z)\to(3,-4,\pi/4)} \dfrac{x^2 y}{\tan z}$

35. $\lim\limits_{(x,y)\to(1,2)} \dfrac{x^2 + y^2}{x^2 - y^2}$

36. $\lim\limits_{(x,y)\to(-2,1)} \dfrac{x^3 - y^3}{x^2 - y^2}$

37. $\lim\limits_{\substack{(x,y)\to(3,3)\\ x=3}} \dfrac{x^3 - y^3}{x^2 - y^2}$

38. $\lim\limits_{\substack{(x,y)\to(3,3)\\ y=3}} \dfrac{x^3 - y^3}{x^2 - y^2}$

39. $\lim\limits_{\substack{(x,y)\to(3,3)\\ y=x}} \dfrac{x^3 - y^3}{x^2 - y^2}$

40. $\lim\limits_{(x,y)\to(3,3)} \dfrac{x^3 - y^3}{x^2 - y^2}$

In Exercises 41–46, use polar coordinates to analyze the given limits.

41. $\lim\limits_{(x,y)\to(0,0)} \dfrac{x^2}{x^2 + y^2}$

42. $\lim\limits_{(x,y)\to(0,0)} \dfrac{y^2}{x^2 + y^2}$

43. $\lim\limits_{(x,y)\to(0,0)} \dfrac{x^2 y^2}{x^2 + y^2}$

44. $\lim\limits_{(x,y)\to(0,0)} \dfrac{x^2 y^3}{x^4 + 2x^2 y^2 + y^4}$

45. $\lim\limits_{(x,y)\to(0,0)} \dfrac{xy}{\sqrt{x^2 + y^2}}$

46. $\lim\limits_{(x,y)\to(0,0)} \dfrac{\sin(x^2 + y^2)}{x^2 + y^2}$

Determine the domains of the functions in Exercises 47–56, and find where the functions are continuous.

47. $f(x, y) = \dfrac{x^2}{x^2 - y^2}$

48. $f(x, y, z) = \dfrac{xy^2}{x + y - z}$

49. $f(x, y) = \sqrt{x^2 + y}$

50. $f(x, y, z) = \ln(x^2 + y^2 + z^2)$

51. $f(x, y, z) = \dfrac{\sin(x + y + z)}{\sqrt{x + y + z}}$

52. $f(x, y, z) = e^{-xyz}$

53. $f(x, y) = \begin{cases} \dfrac{xy}{\sqrt{x^2 + y^2}}, & \text{if } (x, y) \neq (0, 0) \\ 0, & \text{if } (x, y) = (0, 0) \end{cases}$

54. $f(x, y) = \begin{cases} \dfrac{xy}{x^2 + y^2}, & \text{if } (x, y) \neq (0, 0) \\ 0, & \text{if } (x, y) = (0, 0) \end{cases}$

55. $f(x, y) = \begin{cases} \dfrac{\sin(x^2 + y^2)}{x^2 + y^2}, & \text{if } (x, y) \neq (0, 0) \\ 1, & \text{if } (x, y) = (0, 0) \end{cases}$

56. $f(x, y) = \begin{cases} e^{-1/(x^2 + y^2)}, & \text{if } (x, y) \neq (0, 0) \\ 0, & \text{if } (x, y) = (0, 0) \end{cases}$

Applications

The *ideal gas law* states that the pressure, P, volume, V, temperature in degrees Kelvin, T, and amount, n, of a gas are related by the equation $P = \dfrac{nRT}{V}$, where R is the *universal gas constant*. In Exercises 57–60, assume that the amount of the gas is constant and evaluate the specified limits.

57. If the volume is held fixed, what is $\lim_{T \to 0} P$? Explain why this makes sense.

58. If the volume is held fixed, what is $\lim_{T \to \infty} P$? Explain why this makes sense.

59. If the temperature is held fixed, what is $\lim_{V \to 0^+} P$? Explain why this makes sense.

60. If the temperature is held fixed, what is $\lim_{V \to \infty} P$? Explain why this makes sense.

61. Emmy is charting a layer of basalt beneath a Hanford tank farm. She has determined that on the south end of the tank farm the basalt lies at

$$b_1(x, y) = -40 - \frac{1}{350}x + \frac{1}{725}y,$$

while on the north the plane of the top of the basalt seems to be at

$$b_2(x, y) = -39.5 - \frac{1}{150}x + \frac{1}{300}y.$$

(a) If the surface of the basalt layer is continuous, what is the line of intersection of these planes on the Earth's surface?

(b) If the surface of the basalt layer is continuous, where should it be when $x = 150$ at the intersection of the planes?

(c) Emmy drills a test hole at that point and finds that the basalt lies at -41.2 feet. What conclusions might she draw from this result?

62. Annie is intrigued by the currents off the western point of Patos Island, one of the San Juan Islands that are part of Washington State. The currents coming off the north side of the island can be faster than those coming off the south side. She models the speed of the current as a function of x and y, using $s(x, y) = 1 + \dfrac{0.25y}{\sqrt{x^2 + y^2}}$, where distances are given in miles, the origin is a point in the water just west of the point of the island, and Patos Island can be considered to lie to the right of the origin. For simplicity Annie just lets the current speed function go right over the island, although obviously there is no water there.

(a) Show that the speed of the current far north of the island is approximately 1.25 mph.

(b) Show that the speed of the current far south of the island is approximately 0.75 mph.

(c) Show that if Annie paddles through the origin along a line $x = my$ with m constant, then her speed is constant.

(d) The speed of the current is evidently not defined at the origin, but does $\lim_{(x,y) \to (0,0)} s(x, y)$ exist? Explain. Can you imagine what the water looks like at the origin?

Proofs

63. Prove Theorem 12.10. That is, show that $(S^c)^c = S$ when S is a subset of $\mathbb{R}^2$ or $\mathbb{R}^3$.

64. Prove that if S is a closed subset of $\mathbb{R}^2$ or $\mathbb{R}^3$, then S^c is an open set. This is Theorem 12.12.

65. Let S be a subset of $\mathbb{R}^2$ or $\mathbb{R}^3$. Prove that a set S is open if and only if $\partial S \cap S = \emptyset$.

66. Let S be a subset of $\mathbb{R}^2$ or $\mathbb{R}^3$. Prove that a set S is closed if and only if $\partial S \subseteq S$.

67. Let S be a subset of $\mathbb{R}^2$ or $\mathbb{R}^3$. Prove that $\partial S = \partial (S^c)$.

68. Let S be a subset of $\mathbb{R}^2$ or $\mathbb{R}^3$. Prove that ∂S is a closed set.

69. Let S be a subset of $\mathbb{R}^2$ or $\mathbb{R}^3$. Prove that $\partial (\partial S) \subseteq \partial S$.

70. Prove that the empty set is both an open subset and a closed subset of $\mathbb{R}^2$.

71. Prove that $\mathbb{R}^2$ is both an open subset and a closed subset of $\mathbb{R}^2$.

Thinking Forward

▶ *The derivative along a cut edge:* Let $f(x, y) = x^2y^3$. Find the rate of change of f in the (positive) y direction when the surface is cut by the plane with equation $x = 2$. Find the rate of change of f in the (positive) x direction when the surface is cut by the plane with equation $y = 1$.

▶ *The derivative when two variables are held fixed:* Let $f(x, y, z) = x^2y^3\sqrt{z}$. Find the rate of change of f in the (positive) z-direction when the values of x and y are constant. Find the rate of change of f in the (positive) y direction when the values of x and z are constant. Find the rate of change of f in the (positive) x direction when the values of y and z are constant.

12.3   PARTIAL DERIVATIVES

▶ Partial derivatives of functions of two and three variables

▶ The geometry of partial derivatives

▶ Higher order partial derivatives

Partial Derivatives of Functions of Two and Three Variables

Recall that a function f of a single variable is differentiable at a point c in its domain if $\lim_{h \to 0} \frac{f(c+h)-f(c)}{h}$ exists. When this limit exists, the derivative $f'(c)$ gives us the slope of the tangent line to the graph of the function at the point $(c, f(c))$. In the next section we will discuss what it means for a function of two variables, $f(x, y)$, to be differentiable at a point in its domain. However, right now we remark that the graph of a function of two variables is a surface and the analogous tangent "object" to a function of two variables is a plane, not a line. In fact, a function of two variables whose graph is sufficiently smooth at a point (x_0, y_0) in the domain of f has infinitely many tangent lines at the point $(x_0, y_0, f(x_0, y_0))$. Each of these tangent lines lies in the plane tangent to the surface at $(x_0, y_0, f(x_0, y_0))$. In order to understand the plane tangent to a surface, we begin by defining the partial derivatives of a function of two variables.

DEFINITION 12.20   **Partial Derivatives for a Function of Two Variables**

Let f be a function of two variables defined on an open set S containing the point (x_0, y_0).

(a) The ***partial derivative with respect to x at*** (x_0, y_0), denoted by $f_x(x_0, y_0)$, is the limit

$$\lim_{h \to 0} \frac{f(x_0 + h, y_0) - f(x_0, y_0)}{h},$$

provided that this limit exists.

(b) Similarly, the ***partial derivative with respect to y at*** (x_0, y_0), denoted by $f_y(x_0, y_0)$, is the limit

$$\lim_{h \to 0} \frac{f(x_0, y_0 + h) - f(x_0, y_0)}{h},$$

provided that this limit exists.

We may also define the partial-derivative functions $f_x(x, y)$ and $f_y(x, y)$. The domains of these functions are the sets of all points where the respective limits mentioned in Definition 12.22 exist. The partial derivatives measure the rates of change of the function f in the x and y directions. For example, when $f(x, y) = 9 - x^2 - y^2$,

$$f_x(2, 1) = \lim_{h \to 0} \frac{f(2 + h, 1) - f(2, 1)}{h} = \lim_{h \to 0} \frac{(9 - (2 + h)^2 - 1^2) - 4}{h}$$

$$= \lim_{h \to 0} \frac{-4h - h^2}{h} = \lim_{h \to 0}(-4 - h) = -4.$$

The following figure shows the graph of f:

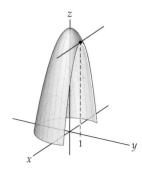

The cut edge toward the right is the intersection of the surface with the plane $y = 1$. When we compute $f_x(2, 1)$, we are finding the rate of change of this curve in the (positive) x direction.

In Exercise 23 we ask you to use the definition to show that $f_y(2, 1) = -2$. This partial derivative represents the rate of change of the surface in the positive y direction at the point $(2, 1)$. We may visualize this partial derivative as the rate of change of the curve created when the surface is cut by the plane $x = 2$.

More generally, a partial derivative with respect to x provides the rate of change of the function in the x direction. That is, if we fix the y-variable at y_0, we may think of the function $f(x, y_0)$ as a function of a single variable x, and the partial derivative tells us the rate of change of that function. To visualize this, we may think of the curve created when the surface $f(x, y)$ is cut by the vertical plane $y = y_0$. The partial derivative $f_x(x_0, y_0)$ is the slope of the tangent line of this function at (x_0, y_0). Similarly, when we take the partial derivative of the function with respect to y, we fix x_0 and the result is a function of the single variable y.

We may also define partial derivatives for functions of three variables.

DEFINITION 12.21

Partial Derivatives for a Function of Three Variables

Let f be a function of three variables defined on an open set S containing the point (x_0, y_0, z_0).

(a) The **partial derivative with respect to x at** (x_0, y_0, z_0), denoted by $f_x(x_0, y_0, z_0)$, is the limit

$$\lim_{h \to 0} \frac{f(x_0 + h, y_0, z_0) - f(x_0, y_0, z_0)}{h},$$

provided that this limit exists.

(b) The **partial derivative with respect to y at** (x_0, y_0, z_0), denoted by $f_y(x_0, y_0, z_0)$, is the limit

$$\lim_{h \to 0} \frac{f(x_0, y_0 + h, z_0) - f(x_0, y_0, z_0)}{h},$$

provided that this limit exists.

(c) The **partial derivative with respect to z at** (x_0, y_0, z_0), denoted by $f_z(x_0, y_0, z_0)$, is the limit

$$\lim_{h \to 0} \frac{f(x_0, y_0, z_0 + h) - f(x_0, y_0, z_0)}{h},$$

provided that this limit exists.

From the preceding definitions, we see that to compute the partial derivative f_x for a function of two variables, we treat the variable y like a constant and take the derivative of the function as if it were a function of the single variable x. Similarly, to compute f_y, we treat the variable x like a constant and take the derivative of the function as if it were a function of the single variable y. When f is a function of three variables, to find f_x, we treat *both* y and z like constants and take the derivative of the function as though it were a function of the single variable x. Analogous statements hold for finding partial derivatives with respect to y and z. This discussion leads to the following shortcuts:

▶ Let $f(x, y)$ be a function of two variables. The partial derivative $f_x(x, y)$ may be computed by treating y like a constant and taking the derivative of the resulting function of the single variable x, using any appropriate rules for finding derivatives. Similarly, the partial derivative $f_y(x, y)$ may be computed by treating x like a constant and taking the derivative of the resulting function of the single variable y, using any appropriate rules for finding derivatives.

▶ Let $f(x, y, z)$ be a function of three variables. The partial derivative $f_x(x, y, z)$ may be computed by treating y and z like constants and taking the derivative of the resulting function of the single variable x, using any appropriate rules for finding derivatives. The partial derivative $f_y(x, y, z)$ may be computed by treating x and z like constants and taking the derivative of the resulting function of the single variable y, using any appropriate rules for finding derivatives. Finally, the partial derivative $f_z(x, y, z)$ may be computed by treating x and y like constants and taking the derivative of the resulting function of the single variable z, using any appropriate rules for finding derivatives.

For example, if $f(x, y, z) = x\sin(yz) + y^2$, we have the following partial derivatives:

$$f_x(x, y, z) = \sin(yz), \quad f_y(x, y, z) = xz\cos(yz) + 2y, \quad \text{and } f_z(x, y, z) = xy\cos(yz).$$

There are alternative notations for partial derivatives. For example, when $w = f(x, y, z)$, we may use the following notation:

$$\frac{\partial w}{\partial x} = \frac{\partial}{\partial x}(f) = \frac{\partial f}{\partial x} = f_x(x, y, z),$$

$$\frac{\partial w}{\partial y} = \frac{\partial}{\partial y}(f) = \frac{\partial f}{\partial y} = f_y(x, y, z),$$

$$\frac{\partial w}{\partial z} = \frac{\partial}{\partial z}(f) = \frac{\partial f}{\partial z} = f_z(x, y, z).$$

Note that we are using the notation $\frac{\partial}{\partial x}$ as an operator meaning "take the partial derivative with respect to x" of the function, and $\frac{\partial w}{\partial x}$ is the partial derivative.

Higher Order Partial Derivatives

In Section 12.6 we will see how partial derivatives are used to locate points in the domain where the relative extremes of a function of two variables occur. We will also see how higher order partial derivatives may be used to classify those points as maxima, minima, or neither maxima nor minima.

DEFINITION 12.22   Second-Order Partial Derivatives

Let $f(x, y)$ be a function of two variables. We define the following four second-order partial derivatives for f:

(a) $\dfrac{\partial^2 f}{\partial x^2} = \dfrac{\partial}{\partial x}\left(\dfrac{\partial f}{\partial x}\right) = f_{xx}(x, y).$
  (b) $\dfrac{\partial^2 f}{\partial x \partial y} = \dfrac{\partial}{\partial x}\left(\dfrac{\partial f}{\partial y}\right) = f_{yx}(x, y).$

(c) $\dfrac{\partial^2 f}{\partial y \partial x} = \dfrac{\partial}{\partial y}\left(\dfrac{\partial f}{\partial x}\right) = f_{xy}(x, y).$
  (d) $\dfrac{\partial^2 f}{\partial y^2} = \dfrac{\partial}{\partial y}\left(\dfrac{\partial f}{\partial y}\right) = f_{yy}(x, y).$

Similarly, there are nine different second-order partial derivatives for a function of three variables. Note that in parts (b) and (c) of the definition above we have $\dfrac{\partial^2 f}{\partial x \partial y} = f_{yx}(x, y)$ and $\dfrac{\partial^2 f}{\partial y \partial x} = f_{xy}(x, y)$, respectively. These are *not* typographical errors. When we write $\dfrac{\partial^2 f}{\partial x \partial y}$, we mean take the partial derivative of the function f first with respect to variable y and then take the partial derivative of the result with respect to x. The notation f_{yx} has an analogous meaning. The order in which the partial derivatives is taken in part (c) is reversed. The partial derivatives in parts (b) and (c) are referred to as **mixed partial derivatives.**

When $f(x, y) = x \cos y + x^2$, we have the following partial derivatives:

$$f_x(x, y) = \cos y + 2x \text{ and } f_y(x, y) = -x \sin y.$$

The second-order partial derivatives are:

$$f_{xx}(x, y) = 2, \; f_{xy}(x, y) = -\sin y = f_{yx}(x, y), \text{ and } f_{yy}(x, y) = -x \cos y.$$

Note that the mixed partial derivatives $f_{xy}(x, y)$ and $f_{yx}(x, y)$ are equal. This does not always happen, but the mixed partial derivatives will be equal everywhere in an open set S if the mixed second-order partial derivatives are continuous on S.

THEOREM 12.23   Clairaut's Theorem: The Equality of the Mixed Second-Order Partial Derivatives

Let $f(x, y)$ be a function defined on an open subset S of $\mathbb{R}^2$. If the second-order partial derivatives of f are continuous everywhere in S, then $f_{xy}(x, y) = f_{yx}(x, y)$ at every point in S.

We will omit the proof of Clairaut's theorem.

In our earlier example, the function and all of its partial derivatives were continuous everywhere in $\mathbb{R}^2$. Therefore, by Clairaut's theorem the mixed partial derivatives f_{xy} and f_{yx} are guaranteed to be equal, as we computed.

Although we will rarely have the need, we may similarly find third- or even higher-order partial derivatives for functions of two or more variables. We explore a few examples shortly.

Finding a Function When the Partial Derivatives Are Given

Earlier in this section we saw that the function $f(x, y) = x \cos y + x^2$ has the partial derivatives

$$f_x(x, y) = \cos y + 2x \quad \text{and} \quad f_y(x, y) = -x \sin y.$$

Is there a way to reverse this process? That is, given two functions $g(x, y)$ and $h(x, y)$, is it possible to find a function $F(x, y)$ such that $g(x, y) = F_x(x, y)$ and $h(x, y) = F_y(x, y)$? Although we cannot do this for every pair of functions g and h, we can do it when the functions satisfy the following theorem:

THEOREM 12.24

The Existence of a Function with Specified Partial Derivatives

Let $g(x, y)$ and $h(x, y)$ be functions with continuous partial derivatives $g_y(x, y)$ and $h_x(x, y)$, respectively. Then there exists a function $F(x, y)$ such that $F_x(x, y) = g(x, y)$ and $F_y(x, y) = h(x, y)$ if and only if $g_y(x, y) = h_x(x, y)$.

For example, consider the pair of functions $g(x, y) = y^2 e^{xy^2}$ and $h(x, y) = 2xye^{xy^2} - \sin y$. There is a function $F(x, y)$ such that

$$F_x(x, y) = g(x, y) = y^2 e^{xy^2} \text{ and } F_y(x, y) = h(x, y) = 2xye^{xy^2} - \sin y$$

since $g_y(x, y) = (2y + 2xy^3)e^{xy^2} = h_x(x, y)$. In a moment we will outline a procedure for finding F. We first note, however, that for the pair $\hat{g}(x, y) = e^{xy}$ and $\hat{h}(x, y) = e^{xy}$, such a function does *not* exist, since $\hat{g}_y(x, y) = xe^{xy} \neq ye^{xy} = \hat{h}_x(x, y)$. Theorem 12.24 tells us that a function F exists only when functions g and h have the specified relationship.

Given that the functions g and h satisfy the condition in Theorem 12.24, we now outline the steps required to find a function $F(x, y)$. We will use the given pair of functions $g(x, y) = y^2 e^{xy^2}$ and $h(x, y) = 2xye^{xy^2} - \sin y$. To find F, we start by considering the form of a function F whose partial derivative with respect to x is $g(x, y) = F_x(x, y) = y^2 e^{xy^2}$. We integrate $g(x, y)$ with respect to x, treating y like a constant:

$$\int y^2 e^{xy^2} \, dx = e^{xy^2} + q(y) = F(x, y).$$

Note:

▶ Instead of having a simple "constant of integration," we have an unknown function $q(y)$, because when we take the partial derivative of any function of y with respect to x, the derivative is zero.

▶ We can check our work in two ways. First, we may compute $\frac{\partial}{\partial x}(F(x, y))$ to ensure that it would result in the function g that we were given. Second, we will take $\frac{\partial}{\partial y}(F(x, y))$ to ensure that this partial derivative is consistent with the function $h(x, y)$ that we were given. We carry out this step next.

We compute

$$\frac{\partial}{\partial y}(F(x, y)) = \frac{\partial}{\partial y}(e^{xy^2} + q(y)) = 2xye^{xy^2} + q'(y).$$

In order for this partial derivative, $2xye^{xy^2} + q'(y)$, to equal $h(x, y) = 2xye^{xy^2} - \sin y$, we must have $q'(y) = -\sin y$. This happens when $q(y) = \cos y + C$, where C is any constant. Therefore, $F(x, y) = e^{xy^2} + \cos y + C$ is the function we sought. We may check this result by showing that $F_x(x, y) = g(x, y) = y^2 e^{xy^2}$.

To summarize, given functions $g(x, y)$ and $h(x, y)$, there is a function $F(x, y)$ such that $g(x, y) = F_x(x, y)$ and $h(x, y) = F_y(x, y)$ if and only if $g_y(x, y) = h_x(x, y)$. When this is the case, you may find F as follows:

▶ Evaluate $\int g(x, y) \, dx = F(x, y)$. Be sure to include a "function of integration," $q(y)$, as part of the antiderivative.

▶ Compute $\frac{\partial}{\partial y}(F(x, y))$.

▶ Set $\frac{\partial}{\partial y}(F(x, y)) = h(x, y)$, and solve the resulting equation for $q'(y)$.

▶ Finish constructing $F(x, y)$ by finding an antiderivative for $q'(y)$.

▶ Check your work by showing that $F_x(x, y) = g(x, y)$.

Theorem 12.24 may be rephrased in terms of a first-order differential equation. A differential equation of the form

$$g(x, y) + h(x, y)\frac{dy}{dx} = 0$$

is said to be **exact** if $g_y(x, y) = h_x(x, y)$. In this case, the implicitly defined function $F(x, y) + C = 0$, where C is a constant, is the solution of the differential equation. Using the same functions as before, we see that the differential equation

$$y^2 e^{xy^2} + (2xye^{xy^2} - \sin y)\frac{dy}{dx} = 0$$

is exact, since if $g(x, y) = y^2 e^{xy^2}$ and $h(x, y) = 2xye^{xy^2} - \sin y$, then

$$g_y(x, y) = (2y + 2xy^3)e^{xy^2} = h_x(x, y).$$

The solution of this differential equation is the function $F(x, y) = e^{xy^2} + \cos y + C = 0$.

Examples and Explorations

EXAMPLE 1 | **Using the definition to compute partial derivatives**

Use the definition to find the indicated partial derivatives for the following functions:

(a) f_y when $f(x, y) = \dfrac{x}{y}$ **(b)** $\dfrac{\partial g}{\partial z}$ when $g(x, y, z) = \dfrac{z^2}{xy^3}$

SOLUTION

(a) $f_y(x, y) = \lim\limits_{h \to 0} \dfrac{f(x, y + h) - f(x, y)}{h}$ ← the definition of f_y

$= \lim\limits_{h \to 0} \dfrac{\dfrac{x}{y + h} - \dfrac{x}{y}}{h}$ ← the function is $\dfrac{x}{y}$

$= \lim\limits_{h \to 0} -\dfrac{hx}{hy(y + h)}$ ← algebra

$= \lim\limits_{h \to 0} -\dfrac{x}{y(y + h)}$ ← Cancellation Theorem

$= -\dfrac{x}{y^2}$ ← evaluation of the limit

(b) $\dfrac{\partial g}{\partial z} = \lim\limits_{h \to 0} \dfrac{g(x, y, z + h) - g(x, y, z)}{h}$ ← the definition of $\dfrac{\partial g}{\partial z}$

$= \lim\limits_{h \to 0} \dfrac{\dfrac{(z + h)^2}{xy^3} - \dfrac{z^2}{xy^3}}{h}$ ← the function is $\dfrac{z^2}{xy^3}$

$= \lim\limits_{h \to 0} \dfrac{2zh + h^2}{hxy^3}$ ← algebra

$= \lim\limits_{h \to 0} \dfrac{2z + h}{xy^3}$ ← Cancellation Theorem

$= \dfrac{2z}{xy^3}$ ← evaluation of the limit

CHECKING THE ANSWER

We can check our answers by using "shortcuts" for finding derivatives. Since $f(x, y) = xy^{-1}$, to find f_y we treat x as though it were a constant and take the derivative of the resulting function of y. We obtain $f_y(x, y) = -xy^{-2}$, which is equivalent to the derivative we found in part (a). Similarly, if we treat both x and y as constants in the function g and take the derivative of the resulting function of z, we obtain the result we found in part (b).

EXAMPLE 2   **Finding the partial derivatives of a function of two variables at a point and interpreting them graphically**

Let $f(x, y) = \dfrac{x}{y^2} - \dfrac{y}{x}$.

(a) Find the first-order partial derivatives f_x and f_y.

(b) Evaluate $f_x(-1, 2)$ and $f_y(-1, 2)$, and find equations for the lines tangent to the surface in the x and y directions at the point $(-1, 2, f(-1, 2))$.

(c) Find an equation for the plane containing the point $(-1, 2, f(-1, 2))$ and the lines from part (b).

SOLUTION

(a) The first-order partial derivatives are

$$f_x(x, y) = \frac{1}{y^2} + \frac{y}{x^2} \text{ and } f_y(x, y) = -\frac{2x}{y^3} - \frac{1}{x}.$$

(b) At $(-1, 2)$, we have $f(-1, 2) = \frac{7}{4}$, $f_x(-1, 2) = \frac{9}{4}$, and $f_y(-1, 2) = \frac{5}{4}$. These partial derivatives represent the rates of change of the function f in the x direction and y direction, respectively, at the point $(-1, 2)$. Therefore, direction vectors of the lines tangent to the surface at $(-1, 2)$ are $\mathbf{i} + \frac{9}{4}\mathbf{k}$ and $\mathbf{j} + \frac{5}{4}\mathbf{k}$. Thus, the line tangent to the surface at $\left(-1, 2, \frac{7}{4}\right)$ in the x direction is given by the parametric equations

$$x = -1 + t, \ y = 2, z = \frac{7}{4} + \frac{9}{4}t.$$

Similarly, the line tangent to the surface at $\left(-1, 2, \frac{7}{4}\right)$ in the y direction is given by the parametric equations

$$x = -1, \ y = 2 + t, z = \frac{7}{4} + \frac{5}{4}t.$$

(c) To obtain the normal vector for the plane containing the two intersecting lines we found in part (b), we take the cross product of the direction vectors for the lines. That is,

$$\mathbf{N} = \left\langle 0, 1, \frac{5}{4}\right\rangle \times \left\langle 1, 0, \frac{9}{4}\right\rangle = \left\langle \frac{9}{4}, \frac{5}{4}, -1\right\rangle.$$

The equation for the plane orthogonal to this vector and containing the point $\left(-1, 2, \frac{7}{4}\right)$ is

$$\left\langle \frac{9}{4}, \frac{5}{4}, -1\right\rangle \cdot \left\langle x + 1, y - 2, z - \frac{7}{4}\right\rangle = 0,$$

or equivalently, $\frac{9}{4}x + \frac{5}{4}y - z = -\frac{3}{2}$.   □

EXAMPLE 3   **Finding the equation of a plane containing the tangent lines in the x and y directions**

Let $f(x, y)$ be a function with first-order partial derivatives $f_x(a, b)$ and $f_y(a, b)$ at a point (a, b) in the domain of f.

(a) Find equations for the lines tangent to the surface in the x and y directions at the point $(a, b, f(a, b))$.

(b) Find an equation for the plane containing the point $(a, b, f(a, b))$ and the lines from part (a).

SOLUTION

(a) Our analysis follows the work we did in Example 2. The partial derivatives represent the rates of change of the function f in the x and y directions, respectively, at the point (a, b). Therefore, direction vectors of the lines tangent to the surface at (a, b) are $\mathbf{i} + f_x(a, b)\mathbf{k}$ and $\mathbf{j} + f_y(a, b)\mathbf{k}$. Thus, the line tangent to the surface at $(a, b, f(a, b))$ in the x direction is given by the parametric equations

$$x = a + t, \ y = b, \ z = f(a, b) + f_x(a, b)t.$$

Similarly, the line tangent to the surface at $(a, b, f(a, b))$ in the y direction is given by the parametric equations

$$x = a, \ y = b + t, \ z = f(a, b) + f_y(a, b)t.$$

(b) To obtain the normal vector for the plane containing the two intersecting lines we found in part (b), we take the cross product of the direction vectors for the lines. That is,

$$\mathbf{N} = \langle 0, 1, f_y(a, b) \rangle \times \langle 1, 0, f_x(a, b) \rangle = \langle f_x(a, b), f_y(a, b), -1 \rangle.$$

The equation for the plane orthogonal to this vector and containing the point $(a, b, f(a, b))$ is

$$\langle f_x(a, b), f_y(a, b), -1 \rangle \cdot \langle x - a, y - b, z - f(a, b) \rangle = 0,$$

or equivalently, $f_x(a, b)(x - a) + f_y(a, b)(y - b) = z - f(a, b)$.   □

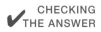
CHECKING THE ANSWER

We may use the general form for the equation of the plane to check our result from Example 2. We had the point $(-1, 2, f(-1, 2)) = \left(-1, 2, \frac{7}{4}\right)$ and partial derivatives $f_x(-1, 2) = \frac{9}{4}$ and $f_y(-1, 2) = \frac{5}{4}$. Therefore, the equation of the plane is

$$\frac{9}{4}(x + 1) + \frac{5}{4}(y - 2) = z - \frac{7}{4},$$

which we see is equivalent to our final answer in Example 2.

EXAMPLE 4   Computing higher order partial derivatives

Let $f(x, y) = \dfrac{x}{y^2} - \dfrac{y}{x}$.

(a) Find all second-order partial derivatives for f, and show that the mixed second-order partial derivatives are equal.

(b) Show that the mixed third-order partial derivatives f_{xxy}, f_{xyx}, and f_{yxx} are all equal.

SOLUTION

(a) We found the first-order partial derivatives $\dfrac{\partial f}{\partial x} = \dfrac{1}{y^2} + \dfrac{y}{x^2}$ and $\dfrac{\partial f}{\partial y} = -\dfrac{2x}{y^3} - \dfrac{1}{x}$ in Example 2. We have

$$\frac{\partial^2 f}{\partial x^2} = \frac{\partial}{\partial x}\left(\frac{1}{y^2} + \frac{y}{x^2}\right) = -\frac{2y}{x^3},$$

$$\frac{\partial^2 f}{\partial y \partial x} = \frac{\partial}{\partial y}\left(\frac{1}{y^2} + \frac{y}{x^2}\right) = -\frac{2}{y^3} + \frac{1}{x^2},$$

$$\frac{\partial^2 f}{\partial x \partial y} = \frac{\partial}{\partial x}\left(-\frac{2x}{y^3} - \frac{1}{x}\right) = -\frac{2}{y^3} + \frac{1}{x^2},$$

$$\frac{\partial^2 f}{\partial y^2} = \frac{\partial}{\partial y}\left(-\frac{2x}{y^3} - \frac{1}{x}\right) = \frac{6x}{y^4}.$$

Note that the partial derivatives $\dfrac{\partial^2 f}{\partial y \partial x}$ and $\dfrac{\partial^2 f}{\partial x \partial y}$ are equal.

(b) We may compute the required third-order partial derivatives by taking the appropriate partial derivatives of the results from part (a).

$$f_{xxy} = \frac{\partial}{\partial y}\left(\frac{\partial^2 f}{\partial x^2}\right) = \frac{\partial}{\partial y}\left(-\frac{2y}{x^3}\right) = -\frac{2}{x^3},$$

$$f_{xyx} = \frac{\partial}{\partial x}\left(\frac{\partial^2 f}{\partial y \partial x}\right) = \frac{\partial}{\partial x}\left(-\frac{2}{y^3} + \frac{1}{x^2}\right) = -\frac{2}{x^3}.$$

Since $f_{yxx} = \frac{\partial}{\partial x}\left(\frac{\partial^2 f}{\partial x \partial y}\right)$ and we already know that $\frac{\partial^2 f}{\partial x \partial y} = \frac{\partial^2 f}{\partial y \partial x}$, we may conclude that $f_{yxx} = -\frac{2}{x^3}$ as well.   □

EXAMPLE 5

Finding a function when the partial derivatives are given

Show that there is a function $F(x, y)$ such that $F_x(x, y) = g(x, y)$ and $F_y(x, y) = h(x, y)$ when

$$g(x, y) = y\cos(xy) \text{ and } h(x, y) = x\cos(xy) + 3y^2.$$

Then find F.

SOLUTION

We use Theorem 12.24 to show that there is such a function. Since the partial derivatives

$$g_y(x, y) = \cos(xy) - xy\sin(xy) = h_x(x, y),$$

Theorem 12.24 guarantees that a function $F(x, y)$ exists.

Now, to find F, we integrate $g(x, y)$ with respect to x:

$$\int y\cos(xy)\,dx = \sin(xy) + q(y) = F(x, y).$$

As we remarked in our discussion earlier in this section, we may check our work by finding both first-order partial derivatives of this function to be sure that $\frac{\partial}{\partial x}(F(x, y)) = g(x, y)$ and $\frac{\partial}{\partial y}(F(x, y))$ is consistent with $h(x, y)$. You should check that $\frac{\partial}{\partial x}(F(x, y)) = g(x, y)$. Checking the consistency of the other partial derivative is part of our procedure. Here,

$$\frac{\partial}{\partial y}(F(x, y)) = \frac{\partial}{\partial y}(\sin(xy) + q(y)) = x\cos(xy) + q'(y).$$

In order for this partial derivative, $x\cos(xy) + q'(y)$, to equal $h(x, y) = x\cos(xy) + 3y^2$, we must have $q'(y) = 3y^2$. This happens when $q(y) = y^3 + C$, where C is any constant. Therefore, $F(x, y) = \sin(xy) + y^3 + C$ is the function we sought.

If we are given a point on the graph of F, we may use that information to evaluate the constant C. For example, here, if we want the graph of F to contain the point $(0, 2, 5)$, then we must have

$$F(0, 2) = \sin(0 \cdot 2) + 2^3 + C = 8 + C = 5.$$

Therefore, $C = -3$, and our function is $F(x, y) = \sin(xy) + y^3 - 3$.   □

CHECKING THE ANSWER

We may check our answer by finding the partial derivatives of our function

$$F(x, y) = \sin(xy) + y^3 + C$$

Since $F_x(x, y) = y\cos(xy) = g(x, y)$ and $F_y(x, y) = x\cos(xy) + 3y^2 = h(x, y)$, our work was correct.

EXAMPLE 6

Solving an exact differential equation

Show that the differential equation

$$y\cos(xy) + (x\cos(xy) + 3y^2)\frac{dy}{dx} = 0$$

is exact. Solve the initial-value problem $F\left(\frac{\pi}{3}, 6\right) = 200$.

SOLUTION

Showing that the differential equation is exact requires the same steps that we performed in Example 5. If we let $g(x, y) = y\cos(xy)$ and $h(x, y) = x\cos(xy) + 3y^2$, then $g_y(x, y) = \cos(xy) - xy\sin(xy) = h_x(x, y)$, so the differential equation is exact. Furthermore, the general solution of the differential equation is $F(x, y) + C = 0$, where $F(x, y)$ is the same function we found in Example 5. That is,

$$F(x, y) = \sin(xy) + y^3 + C = 0.$$

To solve the initial-value problem, we determine the value of C that satisfies $F\left(\frac{\pi}{3}, 6\right) = 200$. When $x = \frac{\pi}{3}$ and $y = 6$, we have $\sin(2\pi) + 6^3 + C = 200$. The unique solution of this equation is $C = -16$. Therefore, the solution to the initial-value problem is

$$\sin(xy) + y^3 - 16 = 0$$

☐

TEST YOUR UNDERSTANDING

▶ What are the definitions of the partial derivatives with respect to x and y for a function $f(x, y)$? How are these definitions similar to the definition for the derivative of a function of a single variable? How are they different?

▶ What are the geometrical interpretations for $f_x(a, b)$ and $f_y(a, b)$ at a point (a, b) in the domain of a function f? When these partial derivatives exist, how can we find the equations of lines tangent to the surface defined by f in the x and y directions? How can we find the equation of the plane containing these lines?

▶ How may the shortcuts for derivatives of a function of a single variable be applied to find partial derivatives of functions of two or three variables?

▶ How are higher order partial derivatives computed? What is meant by mixed second-order partial derivatives? What conditions are sufficient to guarantee that the mixed second-order partial derivatives are equal?

▶ What condition(s) must $g(x, y)$ and $h(x, y)$ satisfy for these functions to equal the first-order partial derivatives of a function $F(x, y)$? That is, when will $g(x, y) = F_x(x, y)$ and $h(x, y) = F_y(x, y)$? What is the procedure for finding F?

EXERCISES 12.3

Thinking Back

▶ *Finding a direction vector for a tangent line:* Find a direction vector for the line tangent to the curve $y = x^3$ when $x = 2$.

▶ *Finding the equation of the plane containing two intersecting lines:* Show that the lines given by $\mathbf{r}_1(t) = \langle 3t - 4, -4t + 1, t \rangle$ and $\mathbf{r}_2(t) = \langle -t + 2, 2t - 9, -2t + 7 \rangle$ intersect, and find the equation of the plane containing the lines.

Concepts

0. *Problem Zero:* Read the section and make your own summary of the material.

1. *True/False:* Determine whether each of the statements that follow is true or false. If a statement is true, explain why. If a statement is false, provide a counterexample.

(a) *True or False:* The partial derivative of a function $f(x, y)$ with respect to x is defined by $\lim\limits_{h \to 0} \frac{f(x+h, y) - f(x, y)}{h}$ if the limit exists.

(b) *True or False:* The partial derivative of a function $g(r, s)$ with respect to s is defined by $\lim\limits_{h \to 0} \frac{g(r, s+h) - g(r, s)}{h}$ if the limit exists.

(c) *True or False:* The second-order partial derivative of a function $f(x, y)$ with respect to x and y is defined by $\lim\limits_{h \to 0} \frac{f(x+h, y+h) - f(x, y)}{h}$ if the limit exists.

(d) *True or False:* If (a, b) is a point in the domain of a function $f(x, y)$ with continuous partial derivatives, then $f_x(a, b) = f_y(a, b)$.

(e) *True or False:* If (a, b) is a point in the domain of a function $f(x, y)$ with continuous second-order partial derivatives, then $f_{xy}(a, b) = f_{yx}(a, b)$.

(f) *True or False:* If (a, b) is a point in the domain of a function $f(x, y)$ at which the second-order partial derivatives exist but are not continuous, then $f_{xy}(a, b) \neq \dfrac{\partial^2 f}{\partial y \partial x}(a, b)$.

(g) *True or False:* If $f(x, y) = |x|y$, then $f_x(x, y)$ at $(0, 0)$ does not exist.

(h) *True or False:* If $f(x, y) = |x|y$, then $f_y(x, y)$ at $(0, 0)$ does not exist.

2. *Examples:* Construct examples of the thing(s) described in the following. Try to find examples that are different than any in the reading.

(a) A function $f(x, y)$ such that $f_x(x, y) = y^2 e^{xy^2}$.

(b) A function $g(x, y)$ such that $\dfrac{\partial g}{\partial y} = xy + e^{4x} \sin 3y$.

(c) A function $F(x, y)$ such that $F_x(x, y) = \sec^2 x + \dfrac{1}{2}y^2 + \dfrac{4}{3}e^{4x} \cos 3y$ and $F_y(x, y) = xy - e^{4x} \sin 3y$.

3. Draw a line on a piece of paper. If the paper is on a horizontal surface, how does the pitch or slope of the line change as you rotate the paper on the tabletop?

4. Draw a line on a piece of paper. If the paper is on an angled surface, how does the pitch or slope of the line change as you rotate the paper on the surface? Why would it *not* make sense to talk about the slope of a plane?

5. Let (x_0, y_0) be a point in the domain of the function $f(x, y)$ at which $f_x(x_0, y_0)$ and $f_y(x_0, y_0)$ both exist. Explain how these partial derivatives may be interpreted as the slopes of lines tangent to the surface defined by $f(x, y)$ at the point (x_0, y_0).

6. Let $f(x)$ be a differentiable function of a single variable x.

(a) What is the relationship between the graph of $f(x)$ and the graph of the function of two variables, $g(x, y) = f(x)$?

(b) For what values of x and y do the first-order partial derivatives of g exist?

(c) What are $\dfrac{\partial g}{\partial x}$ and $\dfrac{\partial g}{\partial y}$? Why do these partial derivatives make sense?

7. Let $f(y)$ be a differentiable function of a single variable y.

(a) What is the relationship between the graph of $f(y)$ and the graph of the function of two variables, $g(x, y) = f(y)$?

(b) For what values of x and y do the first-order partial derivatives of g exist?

(c) What are $\dfrac{\partial g}{\partial x}$ and $\dfrac{\partial g}{\partial y}$? Why do these partial derivatives make sense?

8. The function $f(x, y) = |xy|$ is graphed in the figure that follows. Use the definition of the partial derivatives to show that $f_x(0, 0) = f_y(0, 0) = 0$. What are the equations of the lines tangent to the surface in the x and y directions at $(0, 0, 0)$? What is the equation of the plane containing these two lines?

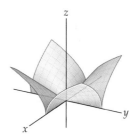

9. The function $g(x, y) = x^2 - y^2$ is graphed in the figure that follows. Use the definition of the partial derivatives to show that $g_x(0, 0) = g_y(0, 0) = 0$. What are the equations of the lines tangent to the surface in the x and y directions at $(0, 0, 0)$? What is the equation of the plane containing these two lines?

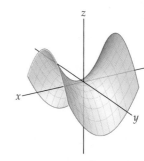

10. Compare the graphs in Exercises 8 and 9. Which of these surfaces do you think has a well-defined tangent plane at $(0, 0, 0)$?

11. Let $f(x, y) = \begin{cases} 0, & \text{if } xy = 0 \\ 1, & \text{if } xy \neq 0. \end{cases}$ Use the definition of the partial derivatives to show that $f_x(0, 0) = f_y(0, 0) = 0$. Explain why this example shows that the existence of the partial derivatives at a point (x_0, y_0) for a function $f(x, y)$ does *not* guarantee that f is continuous at (x_0, y_0).

12. Consider the function f defined in Exercise 11.

(a) Use the definition of the partial derivatives to show that $f_x(x_0, 0)$ exists for every value of x_0 but $f_y(x_0, 0)$ exists only when $x_0 = 0$.

(b) Use the definition of the partial derivatives to show that $f_y(0, y_0)$ exists for every value of y_0 but $f_x(0, y_0)$ exists only when $y_0 = 0$.

13. Let $f(x, y, z) = \begin{cases} 0, & \text{if } xyz = 0 \\ 1, & \text{if } xyz \neq 0. \end{cases}$ Use the definition of the partial derivatives to show that $f_x(0, 0, 0) = f_y(0, 0, 0) = f_z(0, 0, 0) = 0$. Explain why this example shows that the existence of the partial derivatives at a point (x_0, y_0, z_0) for a function $f(x, y, z)$ does *not* guarantee that f is continuous at (x_0, y_0, z_0).

14. Consider the function f defined in Exercise 13. Use the definition of the partial derivatives to show that $f_x(x_0, y_0, 0)$ and $f_y(x_0, y_0, 0)$ exist for every value of x_0 and y_0 but $f_z(x_0, y_0, 0)$ exists only when at least one of x_0 and y_0 is zero.

15. Assume that $f(x, y)$ is a function of two variables with partial derivatives of every order. Assume also that the order in which the partial derivatives are taken is significant.

  (a) How many different second-order partial derivatives does f have?

  (b) How many different third-order partial derivatives does f have?

  (c) How many different nth-order partial derivatives does f have?

16. Assume that $g(x, y, z)$ is a function of three variables with partial derivatives of every order. Assume also the order in which the partial derivatives are taken is significant.

  (a) How many different second-order partial derivatives does g have?

  (b) How many different third-order partial derivatives does g have?

  (c) How many different nth-order partial derivatives does g have?

17. Assume that $f(x, y)$ is a function of two variables with partial derivatives of every order. Assume also that the order in which the partial derivatives are taken is *not* significant.

  (a) How many different second-order partial derivatives does f have?

(b) How many different third-order partial derivatives does f have?

(c) How many different nth-order partial derivatives does f have?

18. Assume that $g(x, y, z)$ is a function of three variables with partial derivatives of every order. Assume also that the order in which the partial derivatives are taken is *not* significant.

  (a) How many different second-order partial derivatives does g have?

  (b) How many different third-order partial derivatives does g have?

  (c) How many different nth-order partial derivatives does g have?

19. Let $f(x, y)$ and $g(x, y)$ be functions of two variables with the property that $\frac{\partial f}{\partial x} = \frac{\partial g}{\partial x}$ for every point $(x, y) \in \mathbb{R}^2$. What is the relationship between f and g?

20. Let $f(x, y)$ and $g(x, y)$ be functions of two variables with the property that $\frac{\partial f}{\partial y} = \frac{\partial g}{\partial y}$ for every point $(x, y) \in \mathbb{R}^2$. What is the relationship between f and g?

21. Let $f(x, y)$ and $g(x, y)$ be functions of two variables with the property that $\frac{\partial f}{\partial x} = \frac{\partial g}{\partial x}$ and $\frac{\partial f}{\partial y} = \frac{\partial g}{\partial y}$ for every point $(x, y) \in \mathbb{R}^2$. What is the relationship between f and g?

22. Let $f(x, y, z)$ and $g(x, y, z)$ be functions of three variables with the property that $\frac{\partial f}{\partial y} = \frac{\partial g}{\partial y}$ for every point $(x, y, z) \in \mathbb{R}^3$. What is the relationship between f and g?

Skills

Use the definition of the partial derivative to find the partial derivatives specified in Exercises 23–26.

23. $f_y(2, 1)$ when $f(x, y) = 9 - x^2 - y^2$

24. $f_x(0, -3)$ and $f_y(0, -3)$ when $f(x, y) = \dfrac{3x}{y^2}$

25. $\dfrac{\partial f}{\partial x}, \dfrac{\partial f}{\partial y}$ and $\dfrac{\partial f}{\partial z}$ when $f(x, y, z) = \dfrac{xy^2}{z}$

26. $\dfrac{\partial g}{\partial x}$ and $\dfrac{\partial g}{\partial y}$ when $g(x, y) = \dfrac{\sqrt{y}}{x+1}$

Find the first-order partial derivatives for the functions in Exercises 27–36.

27. $f(x, y) = e^x \sin(xy)$

28. $g(x, y) = \tan^{-1}(xy^2)$

29. $f(x, y) = x^y$

30. $g(x, y) = \dfrac{\ln(y^2+1)}{x}$

31. $f(x, y) = x \sin y$

32. $g(x, y) = x \cos y$

33. $f(r, \theta) = r \sin \theta$

34. $g(r, \theta) = r \cos \theta$

35. $f(x, y, z) = \dfrac{xy^2}{x+z}$

36. $g(x, y, z) = 2xy + 2xz + 2yz$

In Exercises 37–42, use the partial derivatives you found in Exercises 27–32 and the point (x_0, y_0) specified to (a) find the equation of the line tangent to the surface defined by the function in the x direction, (b) find the equation of the line tangent to the surface defined by the function in the y direction, and (c) find the equation of the plane containing the lines you found in parts (a) and (b).

37. Exercise 27, $\left(0, \dfrac{\pi}{2}\right)$

38. Exercise 28, $(1, 0)$

39. Exercise 29, $(e, 3)$

40. Exercise 30, $(3, 0)$

41. Exercise 31, $\left(2, \dfrac{\pi}{3}\right)$

42. Exercise 32, $(1, \pi)$

In Exercises 43–50, compute all of the second-order partial derivatives for the functions in Exercises 27–34 and show that the mixed partial derivatives are equal.

43. Exercise 27

44. Exercise 28

45. Exercise 29

46. Exercise 30

47. Exercise 31

48. Exercise 32

49. Exercise 33

50. Exercise 34

For the partial derivatives given in Exercises 51–54, find the most general form for a function of two variables, $f(x, y)$, with the given partial derivative.

51. $\dfrac{\partial f}{\partial x} = 0$

52. $\dfrac{\partial f}{\partial y} = 0$

53. $\dfrac{\partial f}{\partial x^2} = 0$

54. $\dfrac{\partial^2 f}{\partial y \partial x} = 0$

For the partial derivatives given in Exercises 55–58, find the most general form for a function of three variables, $f(x, y, z)$, with the given partial derivative.

55. $\dfrac{\partial f}{\partial x} = 0$

56. $\dfrac{\partial f}{\partial y} = 0$

57. $\dfrac{\partial f}{\partial x^2} = 0$

58. $\dfrac{\partial^2 f}{\partial y \partial x} = 0$

For each pair of functions in Exercises 59–62, use Theorem 12.24 to show that there is a function of two variables, $F(x, y)$, such that $\dfrac{\partial F}{\partial x} = g(x, y)$ and $\dfrac{\partial F}{\partial y} = h(x, y)$. Then find F.

59. $g(x, y) = e^x \cos y, \ h(x, y) = -e^x \sin y + 2y$

60. $g(x, y) = \dfrac{1}{xy} - \sin x, \ h(x, y) = -\dfrac{\ln x}{y^2}$

61. $g(x, y) = \dfrac{y}{1 + x^2 y^2}, \ h(x, y) = \dfrac{x}{1 + x^2 y^2}$

62. $g(x, y) = \dfrac{2x}{y}, \ h(x, y) = -\dfrac{x^2}{y^2}$

Solve the exact differential equations in Exercises 63–66.

63. $e^y + (xe^y - 7)\dfrac{dy}{dx} = 0$

64. $y\cos(xy) - 3 + (x\cos(xy) + 2)\dfrac{dy}{dx} = 0$

65. $e^x \ln y + x^3 + \dfrac{e^x}{y}\dfrac{dy}{dx} = 0$

66. $\dfrac{12x^3}{y^5} - \dfrac{15x^4}{y^6}\dfrac{dy}{dx} = 0$

Applications

67. Recall that the volume, V, of a cylinder is given by $V(r, h) = \pi r^2 h$. Find $\dfrac{\partial V}{\partial r}$ and $\dfrac{\partial V}{\partial h}$. What are the physical interpretations of these partial derivatives?

68. The pressure P of an ideal gas is given by $P(n, T, V) = \dfrac{nRT}{V}$, where n is the amount of the gas, R is a constant, T is the temperature of the gas in degrees Kelvin, and V is the volume of the gas. Find $\dfrac{\partial P}{\partial n}$, $\dfrac{\partial P}{\partial V}$, and $\dfrac{\partial P}{\partial T}$. What are the physical interpretations of these partial derivatives?

69. Annie is covering the kayak she is building with fabric. She knows that the shape of the fabric, when stretched across the ribs of her kayak, satisfies the differential equation $f_{xx} + f_{yy} = 0$.

   (a) Annie has no interest in solving that equation; she is just building a kayak. You, however, should show that $\alpha e^{nx} \sin ny$ and $\beta e^{ny} \sin nx$ are solutions for any real numbers n, α, and β.

   (b) Assuming that she needed something to do besides apply the fabric to her kayak, Annie measured and normalized the function describing the ribs near the bow to find that they have the shape given by

$$f(x, 0) = 0.04 \sin\left(\dfrac{\pi x}{2}\right),$$

$$f(x, 1) = 0.04\, e^{\pi/2} \sin\left(\dfrac{\pi x}{2}\right),$$

$$f(0, y) = 0,$$

$$f(1, y) = 0.04\, e^{\pi y/2}.$$

What solution $f(x, y)$ from part (a) satisfies these conditions? Sketch the solution.

70. Alex is modeling traffic patterns at a bottleneck on a freeway as it leaves Denver. He uses a well-known equation $u_x + u_t = 0$, where $u = u(x, t)$ is a scaled traffic density at a point x on the highway at time t.

   (a) Show that $u_n(x, t) = \sin(n(x - t))$ is a solution of the equation for any integer n.

   (b) Alex finds that in heavy traffic at the bottleneck the solution of his equation can look like

$$24 + \sum_{n=1}^{N} \dfrac{\sin(n(x - t))}{n} \quad \text{for some integer } N.$$

Show that this is a solution, and plot it for $N = 8$. Can you tell what happens as N gets large?

Proofs

71. Let $z = (f(x, y))^n$. Show that

$$\dfrac{\partial z}{\partial x} = n(f(x, y))^{n-1}\dfrac{\partial f}{\partial x} \quad \text{and} \quad \dfrac{\partial z}{\partial y} = n(f(x, y))^{n-1}\dfrac{\partial f}{\partial y}.$$

72. Let $z = f(x, y)g(x, y)$. Show that

$$\dfrac{\partial z}{\partial x} = \dfrac{\partial f}{\partial x}g(x, y) + f(x, y)\dfrac{\partial g}{\partial x} \quad \text{and}$$

$$\dfrac{\partial z}{\partial y} = \dfrac{\partial f}{\partial y}g(x, y) + f(x, y)\dfrac{\partial g}{\partial y}.$$

73. Let $z = \dfrac{f(x, y)}{g(x, y)}$. Show that

$$\dfrac{\partial z}{\partial x} = \dfrac{f_x(x, y)g(x, y) - f(x, y)g_x(x, y)}{(g(x, y))^2} \quad \text{and}$$

$$\dfrac{\partial z}{\partial y} = \dfrac{f_y(x, y)g(x, y) - f(x, y)g_y(x, y)}{(g(x, y))^2}.$$

74. Let $p(x)$ be a polynomial of degree n, $q(y)$ be any function of y, and $f(x, y) = p(x)q(y)$. Prove that

$$\frac{\partial^{n+1} f}{\partial x^{n+1}} = 0.$$

75. Let $f(x)$ be a differentiable function of x, $g(y)$ be a differentiable function of y, and $h(x, y) = f(x) + g(y)$. Prove

that

$$\frac{\partial^2 h}{\partial x \partial y} = \frac{\partial^2 h}{\partial y \partial x}.$$

76. Let $f(x)$ be a differentiable function of x, $g(y)$ be a differentiable function of y, and $h(x, y) = f(x)g(y)$. Prove that

$$\frac{\partial^2 h}{\partial x \partial y} = \frac{\partial^2 h}{\partial y \partial x}.$$

Thinking Forward

A chain rule for functions of two variables: Let $f(x, y) = x^2 y^3$, $x = s \cos t$, and $y = s \sin t$.

▶ Find $\dfrac{\partial f}{\partial x}$ and $\dfrac{\partial f}{\partial y}$.

▶ Find $\dfrac{\partial x}{\partial s}$ and $\dfrac{\partial x}{\partial t}$.

▶ Find $\dfrac{\partial y}{\partial s}$ and $\dfrac{\partial y}{\partial t}$.

▶ Substitute the expressions for x and y into $f(x, y)$.

▶ Find $\dfrac{\partial f}{\partial s}$ and $\dfrac{\partial f}{\partial t}$ using the preceding results.

▶ Make a conjecture about the relationship between $\dfrac{\partial f}{\partial s}$, $\dfrac{\partial f}{\partial x}$, $\dfrac{\partial f}{\partial y}$, $\dfrac{\partial x}{\partial s}$, and $\dfrac{\partial y}{\partial s}$.

12.4 DIRECTIONAL DERIVATIVES AND DIFFERENTIABILITY

▶ Directional derivatives

▶ Differentiability of functions of two and three variables

▶ The tangent plane to a function of two variables

The Directional Derivative

In Section 12.3 we defined the partial derivatives for a function of two or three variables. Recall that when $f(x, y)$ is a function of two variables, the partial derivatives $\dfrac{\partial f}{\partial x}$ and $\dfrac{\partial f}{\partial y}$ tell us the rate of change of the function f in the (positive) x and y directions, respectively, when these derivatives exist. How can we compute the rate of change of f in a different direction? To do this we define the ***directional derivative*** of f in the direction of a specified unit vector $\mathbf{u}$.

DEFINITION 12.25

The Directional Derivative of a Function of Two Variables

Let $f(x, y)$ be a function of two variables defined on an open set containing the point (x_0, y_0), and let $\mathbf{u} = \langle \alpha, \beta \rangle$ be a unit vector. The ***directional derivative*** of f at (x_0, y_0) in the direction of $\mathbf{u}$, denoted by $D_{\mathbf{u}} f(x_0, y_0)$, is given by

$$\lim_{h \to 0} \frac{f(x_0 + \alpha h, y_0 + \beta h) - f(x_0, y_0)}{h},$$

provided that this limit exists.

Note that when $\mathbf{u} = \mathbf{i}$ or $\mathbf{u} = \mathbf{j}$, the directional derivative is $\dfrac{\partial f}{\partial x}$ or $\dfrac{\partial f}{\partial y}$, respectively. That is, for a point (x_0, y_0) in the domain of f at which the partial derivatives exist,

$$D_{\mathbf{i}} f(x_0, y_0) = f_x(x_0, y_0) \text{ and } D_{\mathbf{j}} f(x_0, y_0) = f_y(x_0, y_0).$$

Thus, the directional derivative generalizes the partial derivatives. The figure shown next indicates the geometry of the directional derivative. We may visualize $D_{\mathbf{u}}f(x_0, y_0)$ as the slope of the cut edge at the point $(x_0, y_0, f(x_0, y_0))$ when we slice the surface $z = f(x, y)$ with a vertical plane parallel to the vector $\mathbf{u}$ containing the point.

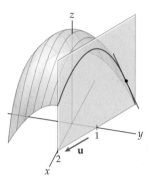

To compute the directional derivative of the function $f(x, y) = \dfrac{x^2}{y}$ at the point $(-1, 2)$ in the direction of the vector $\mathbf{v} = \langle 3, -4 \rangle$, we first need to find a unit vector $\mathbf{u}$ with the same direction as $\mathbf{v}$. We have

$$\mathbf{u} = \frac{1}{\|\mathbf{v}\|}\mathbf{v} = \frac{1}{\sqrt{3^2 + (-4)^2}}\langle 3, -4 \rangle = \left\langle \frac{3}{5}, -\frac{4}{5} \right\rangle.$$

Thus, the directional derivative

$$D_{\mathbf{u}}f(-1, 2) = \lim_{h \to 0} \frac{\dfrac{(-1 + 3h/5)^2}{2 - 4h/5} - \dfrac{(-1)^2}{2}}{h} = \lim_{h \to 0} \frac{-20h + 9h^2}{h(50 - 20h)} = \lim_{h \to 0} \frac{-20 + 9h}{50 - 20h} = -\frac{2}{5}.$$

Geometrically, the number $-\dfrac{2}{5}$ describes the slope of the line tangent to the cut edge of the graph when the surface defined by $f(x, y) = \dfrac{x^2}{y}$ is sliced by the vertical plane parallel to the vector $\mathbf{u} = \left\langle \dfrac{3}{5}, -\dfrac{4}{5} \right\rangle$ and containing the point $(-1, 2)$. We may use all of this information to find the equation of the line tangent to the surface at the point $(-1, 2, f(-1, 2)) = \left(-1, 2, \dfrac{1}{2}\right)$. Here, that equation is

$$x = -1 + \frac{3}{5}t, \ y = 2 - \frac{4}{5}t, \ z = \frac{1}{2} - \frac{2}{5}t.$$

More generally, for a point (x_0, y_0) in the domain of a function $f(x, y)$ and a unit vector $\mathbf{u} = \langle \alpha, \beta \rangle$ for which the directional derivative $D_{\mathbf{u}}f(x_0, y_0)$ exists, the equation of the line tangent to the surface defined by f at the point $(x_0, y_0, f(x_0, y_0))$ will be

$$x = x_0 + \alpha t, \ y = y_0 + \beta t, \ z = f(x_0, y_0) + D_{\mathbf{u}}f(x_0, y_0)t.$$

We may also define the directional derivative for a function of three variables.

DEFINITION 12.26

The Directional Derivative of a Function of Three Variables

Let $f(x, y, z)$ be a function of three variables defined on an open set containing the point (x_0, y_0, z_0), and let $\mathbf{u} = \langle \alpha, \beta, \gamma \rangle$ be a unit vector. The **directional derivative** of f at (x_0, y_0, z_0) in the direction of $\mathbf{u}$, denoted by $D_{\mathbf{u}}f(x_0, y_0, z_0)$, is given by

$$\lim_{h \to 0} \frac{f(x_0 + \alpha h, y_0 + \beta h, z_0 + \gamma h) - f(x_0, y_0, z_0)}{h},$$

provided that this limit exists.

As with a directional derivative for a function of two variables, when **u** is one of the standard basis vectors,

$$D_{\mathbf{i}} f(x_0, y_0, z_0) = f_x(x_0, y_0, z_0),$$
$$D_{\mathbf{j}} f(x_0, y_0, z_0) = f_y(x_0, y_0, z_0),$$
$$D_{\mathbf{k}} f(x_0, y_0, z_0) = f_z(x_0, y_0, z_0).$$

Differentiability

Recall that for a function $y = f(x)$ defined on an open interval containing the point x_0, we defined the derivative of f in terms of the limit of the difference quotient; that is,

$$f'(x_0) = \lim_{h \to 0} \frac{f(x_0 + h) - f(x_0)}{h},$$

provided that the limit exists. Unfortunately, we cannot use a definition analogous to this to define the differentiability of a function of two or more variables. In Exercise 98 of Section 2.2 we asked you to show that if a function $y = f(x)$ is differentiable at x_0 and $\Delta y = f(x_0 + \Delta x) - f(x_0)$, then

$$\Delta y = f'(x_0)\Delta x + \epsilon \Delta x,$$

where ϵ is a function satisfying $\lim_{\Delta x \to 0} \epsilon = 0$. This property tells us that the function f is "locally linear" near x_0 and, furthermore, that we may use the tangent line at $(x_0, f(x_0))$ to approximate the function f on some interval containing x_0. We will use an analogous property to define differentiability for a function of two or three variables.

DEFINITION 12.27

Differentiability for Functions of Two Variables

Let $z = f(x, y)$ be a function of two variables defined on an open set containing the point (x_0, y_0), and let $\Delta z = f(x_0 + \Delta x, y_0 + \Delta y) - f(x_0, y_0)$. The function f is said to be ***differentiable*** at (x_0, y_0) if the partial derivatives $f_x(x_0, y_0)$ and $f_y(x_0, y_0)$ both exist and

$$\Delta z = f_x(x_0, y_0)\Delta x + f_y(x_0, y_0)\Delta y + \epsilon_1 \Delta x + \epsilon_2 \Delta y,$$

where ϵ_1 and ϵ_2 are functions of Δx and Δy, and both go to zero as $(\Delta x, \Delta y) \to (0, 0)$.

In general, it is cumbersome to use Definition 12.27 to show that a function of two variables is differentiable. Fortunately, when the the partial derivatives are continuous at (x_0, y_0), we may use the following theorem:

THEOREM 12.28

The Continuity of the Partial Derivatives Guarantees Differentiability

If $f(x, y)$ is a function of two variables with partial derivatives $\frac{\partial f}{\partial x}$ and $\frac{\partial f}{\partial y}$ that are continuous on some open set containing the point (x_0, y_0), then f is differentiable at (x_0, y_0).

Theorem 12.28 allows us to verify much more readily where a function is differentiable, for most of the common functions we study. For example, consider the function $f(x, y) = \frac{x^2}{y}$

that we discussed earlier in the section. The partial derivatives $f_x(x, y) = \frac{2x}{y}$ and $f_y(x, y) = -\frac{x^2}{y^2}$ of this function are continuous at every point in the domain of f. That is, they are continuous on the set $S = \{(x, y) \mid y \neq 0\}$. Therefore, f is differentiable on S. As this example illustrates, every rational function of two variables will be differentiable everywhere the function is defined, because the only places where the partial derivatives will be discontinuous are the points where the function itself is discontinuous.

It bears repeating that a function with partial derivatives at a point (x_0, y_0) may or may not be differentiable at (x_0, y_0). For example, the function

$$f(x, y) = \begin{cases} 0, & \text{if } x = 0 \text{ or } y = 0 \\ 1, & \text{otherwise} \end{cases}$$

has the property that $f_x(0, 0) = 0 = f_y(0, 0)$. Although these partial derivatives exist, f is not differentiable at the origin, because f does not satisfy Definition 12.27. In fact, the function is not even continuous at the origin.

When a function of two variables, $f(x, y)$, is differentiable at a point (x_0, y_0), we may modify the equality $\Delta z = f_x(x_0, y_0)\Delta x + f_y(x_0, y_0)\Delta y + \epsilon_1 \Delta x + \epsilon_2 \Delta y$ from Definition 12.27 to find the equation of the plane tangent to the surface defined by f at (x_0, y_0). If we let $\Delta x = x - x_0$, $\Delta y = y - y_0$, and $\Delta z = z - f(x_0, y_0)$, we immediately have the equation of the tangent plane when we drop the terms $\epsilon_1 \Delta x$ and $\epsilon_2 \Delta y$.

THEOREM 12.29

Using the Partial Derivatives to Find the Equation of the Tangent Plane

Let $f(x, y)$ be a function of two variables that is differentiable at the point (x_0, y_0). Then the equation of the plane tangent to the surface defined by $f(x, y)$ at (x_0, y_0) is

$$f_x(x_0, y_0)(x - x_0) + f_y(x_0, y_0)(y - y_0) = z - f(x_0, y_0).$$

Before we give the definition of what it means for a function of three variables to be differentiable, recall that the graph of a function of three variables is a three-dimensional object existing in four-dimensional space, $\mathbb{R}^4$. If this graph is sufficiently well behaved, the natural tangent object to the surface is a *hyperplane* in $\mathbb{R}^4$. Perhaps the easiest way to think about a hyperplane is by considering the general equation that defines one. Just as the equation $a_1 x_1 + a_2 x_2 = b$ defines a line in $\mathbb{R}^2$ and the equation $a_1 x_1 + a_2 x_2 + a_3 x_3 = b$ defines a plane in $\mathbb{R}^3$, the equation $a_1 x_1 + a_2 x_2 + a_3 x_3 + a_4 x_4 = b$ defines a hyperplane in $\mathbb{R}^4$.

DEFINITION 12.30

Differentiability for Functions of Three Variables

Let $w = f(x, y, z)$ be a function of three variables defined on an open set containing the point (x_0, y_0, z_0), and let $\Delta w = f(x_0 + \Delta x, y_0 + \Delta y, z_0 + \Delta z) - f(x_0, y_0, z_0)$. The function f is said to be ***differentiable*** at (x_0, y_0, z_0) if the partial derivatives $f_x(x_0, y_0, z_0)$, $f_y(x_0, y_0, z_0)$, and $f_z(x_0, y_0, z_0)$ all exist and

$$\Delta w = f_x(x_0, y_0, z_0)\Delta x + f_y(x_0, y_0, z_0)\Delta y + f_z(x_0, y_0, z_0)\Delta z + \epsilon_1 \Delta x + \epsilon_2 \Delta y + \epsilon_3 \Delta z,$$

where ϵ_1, ϵ_2, and ϵ_3 all go to zero as $(\Delta x, \Delta y, \Delta z) \to (0, 0, 0)$.

The following theorem for functions of three variables is an analog of Theorem 12.28 for functions of two variables:

THEOREM 12.31

> **The Continuity of the Partial Derivatives Guarantees Differentiability**
>
> If $f(x, y, z)$ is a function of three variables with partial derivatives $\frac{\partial f}{\partial x}$, $\frac{\partial f}{\partial y}$, and $\frac{\partial f}{\partial z}$ that are continuous on some open set containing the point (x_0, y_0, z_0), then f is differentiable at (x_0, y_0, z_0).

The hyperplane that may be used to approximate a differentiable function of three variables, $w = f(x, y, z)$, "close to" a point of differentiability (x_0, y_0, z_0) is given by the equation

$$f_x(x_0, y_0, z_0)(x - x_0) + f_y(x_0, y_0, z_0)(y - y_0) + f_z(x_0, y_0, z_0)(z - z_0) = w - f(x_0, y_0, z_0).$$

This equation may be generalized to even higher dimensions.

Examples and Explorations

EXAMPLE 1

Computing directional derivatives

Find the directional derivative of $f(x, y) = \frac{x}{y^2}$ in the direction of the vector $3\mathbf{i} - 2\mathbf{j}$ at the point $(5, -3)$.

SOLUTION

Before we compute any directional derivative, we need to have a unit vector in the correct direction. We divide the vector $3\mathbf{i} - 2\mathbf{j}$ by its magnitude $\sqrt{3^2 + (-2)^2} = \sqrt{13}$ and obtain the unit vector $\mathbf{u} = \frac{3}{\sqrt{13}}\mathbf{i} - \frac{2}{\sqrt{13}}\mathbf{j}$. Now,

$$D_{\mathbf{u}} f(5, -3) = \lim_{h \to 0} \frac{\dfrac{5 + (3/\sqrt{13})h}{(-3 - (2/\sqrt{13})h)^2} - \dfrac{5}{(-3)^2}}{h}$$

$$= \lim_{h \to 0} \frac{1}{h} \cdot \frac{9(5 + (3/\sqrt{13})h) - 5(-3 - (2/\sqrt{13})h)^2}{9(-3 - (2/\sqrt{13})h)^2}$$

$$= \lim_{h \to 0} \frac{1}{h} \cdot \frac{45 + \dfrac{27}{\sqrt{13}}h - 45 - \dfrac{60}{\sqrt{13}}h - \dfrac{20}{13}h^2}{9\left(-3 - \dfrac{2}{\sqrt{13}}h\right)^2} = \lim_{h \to 0} \frac{-\dfrac{33}{\sqrt{13}} - \dfrac{20}{13}h}{9\left(-3 - \dfrac{2}{\sqrt{13}}h\right)^2}$$

$$= -\frac{11}{27\sqrt{13}}. \qquad \square$$

EXAMPLE 2

Using partial derivatives to determine where a function is differentiable

Find the first-order partial derivatives for the functions

(a) $f(x, y) = \dfrac{x^2 \sqrt{y}}{x - 1}$ **(b)** $g(x, y) = \dfrac{\ln x}{\tan^{-1} y}$ **(c)** $P(r, \theta, \phi) = \dfrac{r\theta}{\phi}$

Use the partial derivatives to determine the sets on which the functions are differentiable.

SOLUTION

(a) The partial derivatives of f are

$$\frac{\partial f}{\partial x} = \frac{x(x - 2)\sqrt{y}}{(x - 1)^2} \quad \text{and} \quad \frac{\partial f}{\partial y} = \frac{x^2}{2(x - 1)\sqrt{y}}.$$

These partial derivatives are continuous wherever the denominators are nonzero and $y > 0$. They are continuous on the set $S_1 = \{(x, y) \mid x \neq 1 \text{ and } y > 0\}$, and f is differentiable at every point in S_1.

(b) The partial derivatives of g are

$$\frac{\partial g}{\partial x} = \frac{1}{x \tan^{-1} y} \quad \text{and} \quad \frac{\partial g}{\partial y} = -\frac{\ln x}{(1 + y^2)(\tan^{-1} y)^2}.$$

Since the domain of g consists of just the points in $\mathbb{R}^2$ for which $x > 0$ and $y \neq 0$, both of these partial derivatives are continuous only on the set $S_2 = \{(x, y) \mid x > 0 \text{ and } y \neq 0\}$. Our function g is differentiable at every point in S_2.

(c) Finally, for P we have

$$\frac{\partial P}{\partial r} = \frac{\theta}{\phi}, \quad \frac{\partial P}{\partial \theta} = \frac{r}{\phi}, \quad \text{and} \quad \frac{\partial P}{\partial \phi} = -\frac{r\theta}{\phi^2}.$$

These partial derivatives are continuous everywhere in the domain of P. That is, they are continuous on the set $S_3 = \{(r, \theta, \phi) \mid \phi \neq 0\}$. The function P is differentiable at every point in its domain.   □

EXAMPLE 3 **Finding the equation of the tangent plane**

Find the equation of the tangent plane to the function

(a) $f(x, y) = \dfrac{x^2 \sqrt{y}}{x - 1}$ at $(3, 4)$.   **(b)** $g(x, y) = \dfrac{\ln x}{\tan^{-1} y}$ at $(e, 1)$.

SOLUTION

These are the two functions we discussed in Example 2.

(a) From the work we have already done, we know that the partial derivatives of f are continuous on an open set containing the point $(3, 4)$. At $(3, 4)$, we have $f(3, 4) = 9$, $f_x(3, 4) = \frac{3(3-2)\sqrt{4}}{(3-1)^2} = \frac{3}{2}$, and $f_y(3, 4) = \frac{3^2}{2(3-1)\sqrt{4}} = \frac{9}{8}$. Thus, the equation of the tangent plane is

$$\frac{3}{2}(x - 3) + \frac{9}{8}(y - 4) = z - 9, \quad \text{or equivalently,} \quad 12x + 9y - 8z = 0.$$

(b) At $(e, 1)$, we have $g(e, 1) = \frac{4}{\pi}$, $g_x(e, 1) = \frac{4}{\pi e}$, and $g_y(e, 1) = -\frac{8}{\pi^2}$. Thus, the equation of the tangent plane is

$$\frac{4}{\pi e}(x - e) - \frac{8}{\pi^2}(y - 1) = z - \frac{4}{\pi}.$$ □

? TEST YOUR UNDERSTANDING

▶ What is a directional derivative? What is the geometric significance of a directional derivative for a function of two variables? What is the relationship between a directional derivative and a partial derivative?

▶ Why do we need to use a unit vector when we are computing a directional derivative? What happens if you do not use a unit vector when you compute a directional derivative? How do you compute a directional derivative for a function of two variables? For a function of three variables?

▶ What is the definition of differentiability for a function of two variables? How is the definition of differentiability for a function of two variables related to the definition of differentiability for a function of a single variable?

▶ What is the definition of differentiability for a function of three variables? How could you generalize this definition to a function of four (or more) variables?

▶ When $f(x, y)$ is a differentiable function of two variables at (x_0, y_0), why is

$$f_x(x_0, y_0)(x - x_0) + f_y(x_0, y_0)(y - y_0) = z - f(x_0, y_0)$$

the equation of the tangent plane at (x_0, y_0)?

EXERCISES 12.4

Thinking Back

▶ *Collinear tangent lines:* Show that the left- and right-hand derivatives exist for the function $f(x) = x^3$ at $x = 2$. Show also that the lines with slopes $f'_-(2)$ and $f'_+(2)$ at the point $(2, 8)$ are collinear.

▶ *Noncollinear tangent lines:* Show that the left- and right-hand derivatives exist for the function $f(x) = |x| + 3$ at $x = 0$. Show also that the lines with slopes $f'_-(0)$ and $f'_+(0)$ at the point $(0, 3)$ are not collinear.

Concepts

0. *Problem Zero:* Read the section and make your own summary of the material.

1. *True/False:* Determine whether each of the statements that follow is true or false. If a statement is true, explain why. If a statement is false, provide a counterexample.

(a) *True or False:* If $f(x, y)$ is a function of two variables, then $D_\mathbf{j} f(a, b) = f_y(a, b)$, provided that these derivatives exist at (a, b).

(b) *True or False:* If $\mathbf{u}$ is a unit vector and the directional derivative $D_\mathbf{u} f(a, b)$ exists for a function of two variables, f, then $D_\mathbf{u} f(a, b) = -D_{-\mathbf{u}} f(a, b)$.

(c) *True or False:* If $f(x, y)$ is a function of two variables and $D_\mathbf{u} f(a, b)$ exists for some unit vector $\mathbf{u}$, then $D_\mathbf{v} f(a, b)$ exists for every unit vector $\mathbf{v}$.

(d) *True or False:* If $f(x, y, z)$ is a function of three variables and $\mathbf{v} \in \mathbb{R}^3$ is a nonzero vector, then $D_{\mathbf{v}/\|\mathbf{v}\|} f(a, b, c) = \frac{1}{\|\mathbf{v}\|} D_\mathbf{v} f(a, b, c)$.

(e) *True or False:* If the partial derivatives $f_x(a, b)$ and $f_y(a, b)$ both exist for a function of two variables, $f(x, y)$, then f is differentiable at the point (a, b).

(f) *True or False:* A function of two variables, $f(s, t)$, is differentiable at the point (a, b) if the partial derivatives f_s and f_t are continuous on an open set containing the point (a, b).

(g) *True or False:* If a function of two variables, $f(x, y)$, is differentiable at the point (a, b), then $D_\mathbf{u} f(a, b) = D_\mathbf{v} f(a, b)$ for all unit vectors $\mathbf{u}$ and $\mathbf{v}$.

(h) *True or False:* If $f(x, y)$ is differentiable at the point (a, b), then there is a unit vector $\mathbf{u}$ such that $D_\mathbf{u} f(a, b) \neq 1$.

2. *Examples:* Construct examples of the thing(s) described in the following. Try to find examples that are different than any in the reading.

(a) A function of two variables, $f(x, y)$, such that $D_\mathbf{u} f(0, 0) = 0$ for every unit vector $\mathbf{u} \in \mathbb{R}^2$.

(b) A function of three variables, $f(x, y, z)$, such that $f_x(0, 0, 0) = f_y(0, 0, 0) = f_z(0, 0, 0) = 0$ but f is not differentiable at $(0, 0, 0)$.

(c) A function of three variables, $f(x, y, z)$, such that $D_\mathbf{u} f(0, 0, 0) = 0$ for every unit vector $\mathbf{u} \in \mathbb{R}^3$.

3. How many unit vectors are there in $\mathbb{R}^1$? How many unit vectors are there in $\mathbb{R}^n$ for $n > 1$?

4. Let $\mathbf{u}$ be a unit vector in $\mathbb{R}^2$.

(a) Explain why $-\mathbf{u}$ is a unit vector.

(b) If (a, b) is a point in the domain of the function of two variables, $f(x, y)$, at which $D_\mathbf{u} f(a, b)$ exists, what is the relationship between $D_\mathbf{u} f(a, b)$ and $D_{-\mathbf{u}} f(a, b)$?

5. What is the definition of the directional derivative for a function of two variables, $f(x, y)$? Be sure to include the words "unit vector" in your definition.

6. What is the definition of the directional derivative for a function of three variables, $f(x, y, z)$? Be sure to include the words "unit vector" in your definition.

7. Let $\mathbf{v}$ be a vector in $\mathbb{R}^n$ and let f be a function of n variables. How would we define the directional derivative of f in the direction of a unit vector $\mathbf{u} \in \mathbb{R}^n$ at $\mathbf{v}$?

8. Let $f(x, y) = x + y$ and $\mathbf{u} = \langle \alpha, \beta \rangle$ be a unit vector.

(a) Use the definition of the directional derivative to find $D_\mathbf{u} f(1, 2)$.

(b) Explain why $\langle k\alpha, k\beta \rangle$ is a unit vector only when $|k| = 1$.

(c) Assume that $|k| \neq 1$ and evaluate the limit

$$\lim_{h \to 0} \frac{(1 + (k\alpha)h) + (2 + (k\beta)h) - (1 + 2)}{h}.$$

(d) Use your results from parts (a) and (c) to explain why it is necessary to use a unit vector in the definition of the directional derivative.

9. Let $f(x, y) = xy$ and $\mathbf{u} = \langle \alpha, \beta \rangle$ be a unit vector.

(a) Use the definition of the directional derivative to find $D_\mathbf{u} f(-1, 3)$.

(b) Assume that $|k| \neq 1$ and evaluate

$$\lim_{h \to 0} \frac{(-1 + (k\alpha)h)(3 + (k\beta)h) - (-1 \cdot 3)}{h}.$$

(c) Use your results from parts (a) and (b) to explain why it is necessary to use a unit vector in the definition of the directional derivative.

10. Let $f(x, y) = e^x \sin y$ and $\mathbf{u} = \langle \alpha, \beta \rangle$ be a unit vector.

(a) Use the definition of the directional derivative to find $D_{\mathbf{u}} f(0, \pi)$.

(b) Assume that $|k| \neq 1$ and evaluate

$$\lim_{h \to 0} \frac{e^{k\alpha h} \sin(\pi + k\beta h)}{h}.$$

(*Hint:* $\sin(A + B) = \sin A \cos B + \sin B \cos A$.)

(c) Use your results from parts (a) and (b) to explain why it is necessary to use a unit vector in the definition of the directional derivative.

11. Let $f(x)$ be a function of a single variable. Define the directional derivative of f in the direction of the unit vector $\mathbf{u} = \langle \alpha \rangle$ at a point c. What are the only possible values for α?

12. Use your definition from Exercise 11 to show that the directional derivative of a function of a single variable $f(x)$ at a point c in the direction of $\mathbf{i}$ is $f'(c)$ and that the directional derivative of f at c in the direction of $-\mathbf{i}$ is $-f'(c)$.

13. What does it mean for a function of two variables, $f(x, y)$, to be differentiable at a point (a, b)?

14. If the function $f(x, y)$ is differentiable at a point (a, b), explain why the tangent lines to the graph of f at (a, b) in the x and y directions are sufficient to determine the tangent plane to the surface.

15. Let $\mathbf{u}_1$ and $\mathbf{u}_2$ be two nonparallel unit vectors in $\mathbb{R}^2$. If the function $f(x, y)$ is differentiable at a point (a, b), explain why the tangent lines to the graph of f at (a, b) in the $\mathbf{u}_1$ and $\mathbf{u}_2$ directions are sufficient to determine the tangent plane to the surface.

16. What does it mean for a function of three variables, $f(x, y, z)$, to be differentiable at a point (a, b, c)?

17. If the function $f(x, y, z)$ is differentiable at a point (a, b, c), explain why the tangent lines to the graph of f at (a, b, c) in the x, y, and z directions are sufficient to determine the tangent hyperplane to the surface.

18. Let $\mathbf{u}_1$, $\mathbf{u}_2$, and $\mathbf{u}_3$ be three unit vectors in $\mathbb{R}^3$ that cannot be put into the same plane. If the function $f(x, y, z)$ is differentiable at a point (a, b, c), explain why the tangent lines to the graph of f at (a, b, c) in the $\mathbf{u}_1$, $\mathbf{u}_2$, and $\mathbf{u}_3$ directions are sufficient to determine the tangent hyperplane to the surface.

19. Using Definition 12.30 as a model, provide a definition of differentiability for a function of n variables.

20. Using Theorem 12.31 as a model, provide a conjecture you think would be sufficient to guarantee that a function of n variables is differentiable at a point in its domain.

Skills

In Exercises 21–28, find the directional derivative of the given function at the specified point P and in the direction of the given unit vector $\mathbf{u}$.

21. $f(x, y) = x^2 - y^2$ at $P = (2, 3)$, $\mathbf{u} = \left\langle \frac{\sqrt{2}}{2}, \frac{\sqrt{2}}{2} \right\rangle$

22. $f(x, y) = x^2 - y^2$ at $P = (2, 3)$, $\mathbf{u} = \frac{3}{5}\mathbf{i} - \frac{4}{5}\mathbf{j}$

23. $f(x, y) = \frac{x}{y^2}$ at $P = (-2, 1)$, $\mathbf{u} = \left\langle \frac{\sqrt{10}}{10}, -\frac{3\sqrt{10}}{10} \right\rangle$

24. $f(x, y) = \frac{x}{y^2}$ at $P = (-2, 1)$, $\mathbf{u} = -\frac{5}{13}\mathbf{i} + \frac{12}{13}\mathbf{j}$

25. $f(x, y) = \sqrt{\frac{y}{x}}$ at $P = (4, 9)$, $\mathbf{u} = \left\langle -\frac{\sqrt{17}}{17}, -\frac{4\sqrt{17}}{17} \right\rangle$

26. $f(x, y) = \sqrt{\frac{y}{x}}$ at $P = (4, 9)$, $\mathbf{u} = \frac{15}{17}\mathbf{i} + \frac{8}{17}\mathbf{j}$

27. $f(x, y, z) = x^2 + y^2 - z^3$ at $P = (2, -2, 2)$, $\mathbf{u} = \left\langle \frac{3}{5}, 0, \frac{4}{5} \right\rangle$

28. $f(x, y, z) = x^2 + y^2 - z^3$ at $P = (2, -2, 2)$,

$$\mathbf{u} = \frac{2}{3}\mathbf{i} - \frac{2}{3}\mathbf{j} + \frac{1}{3}\mathbf{k}$$

In Exercises 29–34, find the equation of the line tangent to the surface at the given point P and in the direction of the given unit vector $\mathbf{u}$. Note that these are the same functions, points, and vectors as in Exercises 21–26.

29. $f(x, y) = x^2 - y^2$ at $P = (2, 3)$, $\mathbf{u} = \left\langle \frac{\sqrt{2}}{2}, \frac{\sqrt{2}}{2} \right\rangle$

30. $f(x, y) = x^2 - y^2$ at $P = (2, 3)$, $\mathbf{u} = \frac{3}{5}\mathbf{i} - \frac{4}{5}\mathbf{j}$

31. $f(x, y) = \frac{x}{y^2}$ at $P = (-2, 1)$, $\mathbf{u} = \left\langle \frac{\sqrt{10}}{10}, -\frac{3\sqrt{10}}{10} \right\rangle$

32. $f(x, y) = \frac{x}{y^2}$ at $P = (-2, 1)$, $\mathbf{u} = -\frac{5}{13}\mathbf{i} + \frac{12}{13}\mathbf{j}$

33. $f(x, y) = \sqrt{\frac{y}{x}}$ at $P = (4, 9)$, $\mathbf{u} = \left\langle -\frac{\sqrt{17}}{17}, -\frac{4\sqrt{17}}{17} \right\rangle$

34. $f(x, y) = \sqrt{\frac{y}{x}}$ at $P = (4, 9)$, $\mathbf{u} = \frac{15}{17}\mathbf{i} + \frac{8}{17}\mathbf{j}$

In Exercises 35–38, find the directional derivative of the given function at the specified point P and in the direction of the given vector $\mathbf{v}$.

35. $f(x, y) = x^2 - y^2$ at $P = (3, 3)$, $\mathbf{v} = \langle -1, 5 \rangle$

36. $f(x, y) = \frac{x}{y^2}$ at $P = (9, -3)$, $\mathbf{v} = 2\mathbf{i} + 7\mathbf{j}$

37. $f(x, y) = \sqrt{\frac{y}{x}}$ at $P = (1, 16)$, $\mathbf{v} = \langle 2, -1 \rangle$

38. $f(x, y, z) = x^2 + y^2 - z^3$ at $P = (0, -2, 5)$, $\mathbf{v} = -\mathbf{i} + 3\mathbf{j} + 5\mathbf{k}$

In Exercises 39–42, show that the directional derivative of the given function at the specified point P is zero for every unit vector $\mathbf{u}$.

39. $f(x, y) = xy + 2x - y$ at $P = (1, -2)$

40. $f(x, y) = 3x^2 - 4xy + 2y^2$ at $P = (0, 0)$

41. $f(x, y) = (x + 1)y^2$ at $P = (-1, 0)$

42. $f(x, y, z) = x^2 + y^2 - z^3$ at $P = (0, 0, 0)$

Find all points where the first-order partial derivatives of the functions in Exercises 43–54 are continuous. Then use Theorems 12.28 and 12.31 to determine the sets in which the functions are differentiable.

43. $f(x, y) = x^2 - y^2$

44. $f(x, y) = \dfrac{x}{y^2}$

45. $f(x, y) = \dfrac{x}{x^2 + y^2 - 1}$

46. $f(x, y) = \dfrac{x}{x^2 + y^2}$

47. $f(x, y) = \sin(xy)$

48. $f(x, y) = \cos(xy)$

49. $f(x, y) = \tan(xy)$

50. $f(x, y) = \tan(x + y)$

51. $f(x, y) = \sqrt{\dfrac{y}{x}}$, $x > 0$, $y \geq 0$

52. $f(x, y) = \ln(xy^2)$

53. $f(x, y, z) = x^2 + y^2 - z^3$

54. $f(x, y, z) = \dfrac{x}{y^2 + z^2}$

Use the first-order partial derivatives of the functions in Exercises 55–64 to find the equation of the plane tangent to the graph of the function at the indicated point P. Note that these are the same functions as in Exercises 43–52.

55. $f(x, y) = x^2 - y^2$, $P = (1, -3)$

56. $f(x, y) = \dfrac{x}{y^2}$, $P = (-4, 7)$

57. $f(x, y) = \dfrac{x}{x^2 + y^2}$, $P = (-3, 0)$

58. $f(x, y) = \dfrac{x}{x^2 + y^2 - 1}$, $P = (1, -3)$

59. $f(x, y) = \sin(xy)$, $P = \left(2, \dfrac{\pi}{2}\right)$

60. $f(x, y) = \cos(xy)$, $P = (\pi, -3)$

61. $f(x, y) = \tan(xy)$, $P = \left(1, -\dfrac{\pi}{4}\right)$

62. $f(x, y) = \tan(x + y)$, $P = (0, \pi)$

63. $f(x, y) = \sqrt{\dfrac{y}{x}}$, $P = (1, 9)$

64. $f(x, y) = \ln(xy^2)$, $P = (1, -3)$

Use the first-order partial derivatives of the functions in Exercises 65 and 66 to find the equation of the hyperplane tangent to the graph of the function at the indicated point P. Note that these are the same functions as in Exercises 53 and 54.

65. $f(x, y, z) = x^2 + y^2 - z^3$, $P = (1, -5, 3)$

66. $f(x, y, z) = \dfrac{x}{y^2 + z^2}$, $P = (4, 0, 2)$

Applications

67. Ian is travelling along a glacier on a line directly northeast. The elevation of the glacier in that area is described by the function

$$f(x, y) = 1.2 - 0.2x^2 - 0.3y^2 + 0.1xy - 0.25x,$$

where x, y, and f are given in miles.

(a) If Ian is at the point $(0, 0)$, how steeply is he descending?

(b) In what direction would Ian have to turn in order to contour across (i.e., neither ascend nor descend) the glacier?

68. Alex is laying out a new road that will descend from the mountains near Boulder, Colorado. The local topography is described by the function

$$f(x, y) = 1.1 - 0.2x + 0.05y^2 + 0.1xy.$$

All distances are given in miles. The technical requirements for the road stipulate that the road must be built at a 6% grade. That is, the road will descend 6 feet for every 100 feet it travels horizontally. If Alex's current position is $(-0.5, 0)$ and the road is going generally southeast, in which direction does he need to look to survey the next section of road?

Proofs

69. Let (a, b) be a point in the domain of the function of two variables, $f(x, y)$, and $\mathbf{u}$ be a unit vector for which $D_{\mathbf{u}} f(a, b)$ exists. Prove that $D_{-\mathbf{u}} f(a, b) = -D_{\mathbf{u}} f(a, b)$.

70. Let $f(x, y)$ be a function of two variables and k be a constant. Prove that if f is differentiable at the point (a, b) and $D_{\mathbf{u}} f(a, b) = k$ for every unit vector $\mathbf{u} \in \mathbb{R}^2$, then $k = 0$.

71. Let $f(x, y)$ be a function of two variables. Prove that if f is differentiable at the point (a, b), then there is a unit vector, $\mathbf{u}$, such that $D_{\mathbf{u}} f(a, b) \neq 1$. (*Hint: See Exercise 70.*)

Thinking Forward

▶ *Paraboloid:* Sketch the paraboloid that is the graph of the function $f(x, y) = 4 - x^2 - y^2$. What is the point (a, b) at which the function attains its maximum value? What is the directional derivative $D_{\mathbf{u}} f(a, b)$ for any unit vector $\mathbf{u}$?

▶ *Hyperboloid:* Sketch the hyperboloid that is the graph of the function $g(x, y) = x^2 - y^2$. Does the function g have any extreme values? What is the directional derivative $D_{\mathbf{u}} g(0, 0)$ for any unit vector $\mathbf{u}$?

12.5   THE CHAIN RULE AND THE GRADIENT

▶ The chain rule for functions of two or more variables

▶ The gradient

▶ The gradient and directional derivatives

The Chain Rule

In Theorem 12.29 we saw that if a function of two variables, $f(x, y)$, is differentiable at a point (x_0, y_0) in its domain, then the equation of the plane tangent to the surface is given by

$$f_x(x_0, y_0)(x - x_0) + f_y(x_0, y_0)(y - y_0) = z - f(x_0, y_0).$$

Just as a line tangent to the graph of a function of a single variable may be used to approximate the function close to the point of tangency, we may use the tangent plane to approximate a differentiable function of two variables. If we let $\Delta x = x - x_0$, $\Delta y = y - y_0$, and $\Delta z = z - f(x_0, y_0)$, the previous equation becomes

$$\Delta z = f_x(x_0, y_0)\Delta x + f_y(x_0, y_0)\Delta y.$$

This equation may be interpreted to say that, close to the point of tangency, the difference between the value of the function f and the corresponding value on the tangent plane is approximately $f_x(x_0, y_0)\Delta x + f_y(x_0, y_0)\Delta y$. We will use this property to prove the chain rule for functions of two or more variables.

Recall that when $y = f(x)$ is a function of the single variable x, and x is a function of another variable t, we may find the rate of change of f with respect to t with the chain rule, $\frac{dy}{dt} = \frac{dy}{dx}\frac{dx}{dt}$. If any of our functions is a function of two or more variables, we need to adapt the chain rule to fit the particular functions we have.

THEOREM 12.32

Chain Rule (Version I)

Given functions $z = f(x, y)$, $x = u(t)$, and $y = v(t)$, for all values of t at which u and v are differentiable, and if f is differentiable at $(u(t), v(t))$, then

$$\frac{dz}{dt} = \frac{\partial z}{\partial x}\frac{dx}{dt} + \frac{\partial z}{\partial y}\frac{dy}{dt}.$$

For example, if $z = x^2 \sin y$, $x = t^3$, and $y = e^{5t}$, then

$$\frac{dz}{dt} = \frac{\partial z}{\partial x}\frac{dx}{dt} + \frac{\partial z}{\partial y}\frac{dy}{dt} = (2x \sin y)(3t^2) + (x^2 \cos y)(5e^{5t}).$$

Although this equation is correct, it is better to replace x and y with their respective functions of t and simplify the result:

$$\frac{dz}{dt} = (2(t^3) \sin(e^{5t}))(3t^2) + ((t^3)^2 \cos(e^{5t}))(5e^{5t}) = 6t^5 \sin(e^{5t}) + 5t^6 e^{5t} \cos(e^{5t}).$$

In Example 1 we verify this derivative by first replacing x and y with their respective functions of t and then taking the derivative of the resulting function.

THEOREM 12.33

Chain Rule (Version II)

Given functions $z = f(x, y)$, $x = u(s, t)$, and $y = v(s, t)$, for all values of s and t at which u and v are differentiable, and if f is differentiable at $(u(s, t), v(s, t))$, then

$$\frac{\partial z}{\partial s} = \frac{\partial z}{\partial x}\frac{\partial x}{\partial s} + \frac{\partial z}{\partial y}\frac{\partial y}{\partial s} \quad \text{and} \quad \frac{\partial z}{\partial t} = \frac{\partial z}{\partial x}\frac{\partial x}{\partial t} + \frac{\partial z}{\partial y}\frac{\partial y}{\partial t}.$$

For example, if $z = e^x \cos y$, $x = s^2 t^3$, and $y = \frac{s}{t}$, then

$$\frac{\partial z}{\partial s} = \frac{\partial z}{\partial x}\frac{\partial x}{\partial s} + \frac{\partial z}{\partial y}\frac{\partial y}{\partial s} = (e^x \cos y)(2st^3) + (-e^x \sin y)\left(\frac{1}{t}\right).$$

Again, although this equation is correct, it is better to replace x and y with their respective functions of s and t and simplify the result:

$$\frac{\partial z}{\partial s} = \left(e^{s^2 t^3} \cos\left(\frac{s}{t}\right)\right)(2st^3) + \left(-e^{s^2 t^3} \sin\left(\frac{s}{t}\right)\right)\left(\frac{1}{t}\right) = e^{s^2 t^3}\left(2st^3 \cos\left(\frac{s}{t}\right) - \frac{1}{t}\sin\left(\frac{s}{t}\right)\right).$$

In Example 2 we continue this example by finding $\frac{\partial z}{\partial t}$ and verifying that both of these partial derivatives are correct by replacing x and y with their respective functions of s and t and then taking the partial derivatives of the resulting function.

We now prove Theorem 12.32. The proof of Theorem 12.33 is similar and is left for Exercise 65.

Proof. Let $x = u(t)$, $y = v(t)$, and t_0 be a point in the domains of both u and v, where the functions are differentiable and where f is differentiable at $(x_0, y_0) = (u(t_0), v(t_0))$. By Definition 12.27, at (x_0, y_0) we have

$$\Delta z = \frac{\partial f}{\partial x}\Delta x + \frac{\partial f}{\partial y}\Delta y + \epsilon_1 \Delta x + \epsilon_2 \Delta y,$$

where $\epsilon_1 \to 0$ and $\epsilon_2 \to 0$ as $(\Delta x, \Delta y) \to (0, 0)$. If we divide both sides of this equation by Δt, we obtain

$$\frac{\Delta z}{\Delta t} = \frac{\partial f}{\partial x}\frac{\Delta x}{\Delta t} + \frac{\partial f}{\partial y}\frac{\Delta y}{\Delta t} + \epsilon_1 \frac{\Delta x}{\Delta t} + \epsilon_2 \frac{\Delta y}{\Delta t}.$$

If we let $\Delta x = u(t_0 + \Delta t) - u(t_0)$ and $\Delta y = v(t_0 + \Delta t) - v(t_0)$, then $\Delta x \to 0$ and $\Delta y \to 0$ as $\Delta t \to 0$. Consequently, $\epsilon_1 \to 0$ and $\epsilon_2 \to 0$ as $\Delta t \to 0$. Taking the limit of the quotients, we get:

$$\frac{dz}{dt} = \lim_{\Delta t \to 0} \frac{\Delta z}{\Delta t} = \lim_{\Delta t \to 0}\left(\frac{\partial f}{\partial x}\frac{\Delta x}{\Delta t} + \frac{\partial f}{\partial y}\frac{\Delta y}{\Delta t} + \epsilon_1 \frac{\Delta x}{\Delta t} + \epsilon_2 \frac{\Delta y}{\Delta t}\right)$$

$$= \frac{\partial f}{\partial x}\frac{dx}{dt} + \frac{\partial f}{\partial y}\frac{dy}{dt}. \qquad \blacksquare$$

After examining Theorems 12.32 and 12.33, you may be able to guess that there is a more comprehensive version of the chain rule, which we state in the following theorem:

THEOREM 12.34

Chain Rule (Complete Version)

Given functions $z = f(x_1, x_2, \ldots, x_n)$ and $x_i = u_i(t_1, t_2, \ldots, t_m)$ for $1 \le i \le n$, for all values of $t_1, t_2, \ldots, t_m$ at which each u_i is differentiable, and if f is differentiable at $(u_1(t_1, t_2, \ldots, t_m), u_2(t_1, t_2, \ldots, t_m), \ldots, u_n(t_1, t_2, \ldots, t_m))$, then

$$\frac{\partial z}{\partial t_j} = \frac{\partial z}{\partial x_1}\frac{\partial x_1}{\partial t_j} + \frac{\partial z}{\partial x_2}\frac{\partial x_2}{\partial t_j} + \cdots + \frac{\partial z}{\partial x_n}\frac{\partial x_n}{\partial t_j}, \quad \text{for } 1 \le j \le m.$$

The Gradient

In Section 12.6 we will be finding extrema for functions of two variables. Recall that the extreme values of a function of a single variable occur at the function's critical points (i.e., those points at which the derivative either is zero or does not exist). For functions of two variables, we will use the *gradient* of the function to find extrema.

DEFINITION 12.35 | **The Gradient**

Let $z = f(x, y)$ be a function of two variables. The **gradient** of f is the vector function defined by

$$\nabla f(x, y) = \frac{\partial f}{\partial x}\mathbf{i} + \frac{\partial f}{\partial y}\mathbf{j} = \langle f_x(x, y), f_y(x, y)\rangle.$$

Similarly, if $w = f(x, y, z)$ is a function of three variables, then the **gradient** of f is the vector function defined by

$$\nabla f(x, y, z) = \frac{\partial f}{\partial x}\mathbf{i} + \frac{\partial f}{\partial y}\mathbf{j} + \frac{\partial f}{\partial z}\mathbf{k} = \langle f_x(x, y, z), f_y(x, y, z), f_z(x, y, z)\rangle.$$

The domain of the gradient is the set of all points in the domain of f at which the partial derivatives exist.

The symbol ∇f is read "the gradient of f," "grad f," or "del f."

Recall that when the first-order partial derivatives of f are continuous in a neighborhood of a point P, the function is differentiable at P. Therefore, if the gradient of f is continuous in a neighborhood of a point P, the function is differentiable at P. In particular, a differentiable function of two variables, $f(x, y)$, can have an extreme value only at a point at which the gradient is the zero vector, since such points are the only places where the tangent plane might be horizontal (a necessary, but insufficient, condition for having an extreme). Therefore, our first step for locating extreme values will be to find those places where the gradient is zero. We see that for a function of two or three variables, the gradient plays the role that the derivative plays for functions of a single variable.

Using the Gradient to Compute the Directional Derivative

In Section 12.4 we introduced the directional derivative and defined it in terms of a limit. As we will see in the following theorem, the gradient provides a shortcut for finding the directional derivative when the first-order partial derivatives of the function exist:

THEOREM 12.36 | **The Gradient and the Directional Derivative**

Let $f(x, y)$ be a function of two variables and (x_0, y_0) be a point in the domain of f at which f is differentiable. Then, for every unit vector $\mathbf{u} \in \mathbb{R}^2$,

$$D_{\mathbf{u}} f(x_0, y_0) = \nabla f(x_0, y_0) \cdot \mathbf{u}.$$

Similarly, if $\mathbf{u} \in \mathbb{R}^3$ is a unit vector, $f(x, y, z)$ is a function of three variables, and (x_0, y_0, z_0) is a point in the domain of f at which f is differentiable, then

$$D_{\mathbf{u}} f(x_0, y_0, z_0) = \nabla f(x_0, y_0, z_0) \cdot \mathbf{u}.$$

Recall that in Section 12.4 we used Definition 12.26 to find the directional derivative of the function $f(x, y) = \dfrac{x^2}{y}$ at the point $(-1, 2)$ in the direction of the unit vector $\mathbf{u} = \left\langle \dfrac{3}{5}, -\dfrac{4}{5}\right\rangle$

and saw that $D_{\mathbf{u}}f(-1,2) = -\frac{2}{5}$. We now compute the same derivative, but using Theorem 12.36. First we need the gradient of f:

$$\nabla f(x,y) = \frac{\partial f}{\partial x}\mathbf{i} + \frac{\partial f}{\partial y}\mathbf{j} = \frac{2x}{y}\mathbf{i} - \frac{x^2}{y^2}\mathbf{j}.$$

At $(-1,2)$, $\nabla f(-1,2) = -\mathbf{i} - \frac{1}{4}\mathbf{j}$. Now, using the theorem, we have

$$D_{\mathbf{u}}f(-1,2) = \nabla f(-1,2) \cdot \mathbf{u} = \left(-\mathbf{i} - \frac{1}{4}\mathbf{j}\right) \cdot \left(\frac{3}{5}\mathbf{i} - \frac{4}{5}\mathbf{j}\right) = -\frac{2}{5}.$$

Although the computation using the definition was not terrible, Theorem 12.36 certainly saves time, particularly when the limit in the definition is difficult to work with.

We now prove Theorem 12.36.

Proof. We prove the theorem when f is a function of two variables and leave the case when f is a function of three variables for Exercise 66. By Definition 12.26 we have

$$D_{\mathbf{u}}f(x_0, y_0) = \lim_{h \to 0} \frac{f(x_0 + \alpha h, y_0 + \beta h) - f(x_0, y_0)}{h},$$

where $\mathbf{u} = \langle \alpha, \beta \rangle$. We now define the function $F(h) = f(x_0 + \alpha h, y_0 + \beta h)$ and note that F is a function of the single variable h, because α, β, x_0, and y_0 are all constants. Thus,

$$D_{\mathbf{u}}f(x_0, y_0) = \lim_{h \to 0} \frac{F(h) - F(0)}{h} = F'(0),$$

by the definition of the derivative. If we let $x = x_0 + \alpha \cdot h$ and $y = y_0 + \beta \cdot h$, then, by the chain rule, Theorem 12.32, it follows that

$$F'(h) = \frac{\partial f}{\partial x}\frac{\partial x}{\partial h} + \frac{\partial f}{\partial x}\frac{\partial x}{\partial h} = f_x(x,y)\alpha + f_y(x,y)\beta.$$

Now, when $h = 0$, $x = x_0$ and $y = y_0$. Thus, if we evaluate the equation for $F'(h)$ at $h = 0$, we have

$$F'(0) = D_{\mathbf{u}}f(x_0, y_0) = f_x(x_0, y_0)\alpha + f_y(x_0, y_0)\beta = \nabla f(x_0, y_0) \cdot \mathbf{u},$$

the result we seek. ∎

The gradient of a function $f(x,y)$ gives us information about two important geometric properties of the function. Theorem 12.37 tells us that the gradient ∇f points in the direction in which f increases most rapidly, and Theorem 12.38 says that the gradient at (x_0, y_0) is orthogonal to the level curve $f(x,y) = c_0$, where $c_0 = f(x_0, y_0)$. In the figure that follows, the green gradient vectors are orthogonal to the level curves and point in the direction in which the function increases most rapidly. Generally, that means that they point in the direction in which the level curves are most tightly spaced. If you are a hiker familiar with topographic maps, this is saying that the shortest path to the peak crosses the level curves quickly. It may be a difficult trek, but it does minimize the distance you have to travel.

The gradient vectors are orthogonal to the level curves

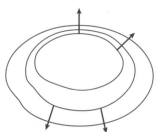

THEOREM 12.37

The Gradient Points in the Direction of Greatest Increase

Let f be a function of two or three variables, and let P be a point in the domain of f at which f is differentiable. Then the gradient of f at P points in the direction in which f increases most rapidly.

Proof. We prove the theorem when f is a function of two variables and leave the case when f is a function of three variables for Exercise 74. By Theorem 12.36,

$$D_{\mathbf{u}} f(x_0, y_0) = \nabla f(x_0, y_0) \cdot \mathbf{u} = \|\nabla f(x_0, y_0)\| \|\mathbf{u}\| \cos \theta,$$

where θ is the angle between the vectors $\nabla f(x_0, y_0)$ and $\mathbf{u}$. Now, since $\mathbf{u}$ is a unit vector, $\|\mathbf{u}\| = 1$. Thus,

$$D_{\mathbf{u}} f(x_0, y_0) = \|\nabla f(x_0, y_0)\| \cos \theta.$$

The quantity on the right is greatest when $\theta = 0$. Therefore, the directional derivative is greatest when $\mathbf{u}$ is parallel to the gradient. ∎

Going back to the function $f(x, y) = \dfrac{x^2}{y}$, we have $\nabla f(-1, 2) = -\mathbf{i} - \dfrac{1}{4}\mathbf{j}$. This vector points in the direction in which f increases most quickly at the point $(-1, 2)$, and the rate of change of f in that direction is the magnitude

$$\left\| \nabla f(-1, 2) \right\| = \left\| -\mathbf{i} - \tfrac{1}{4}\mathbf{j} \right\| = \sqrt{(-1)^2 + \left(-\tfrac{1}{4}\right)^2} = \frac{\sqrt{17}}{4}.$$

The function decreases most rapidly in the opposite direction, $-\nabla f(-1, 2) = \mathbf{i} + \dfrac{1}{4}\mathbf{j}$, and the rate of change of f in this direction is $-\dfrac{\sqrt{17}}{4}$.

THEOREM 12.38

Gradient Vectors are Orthogonal to Level Curves

Let f be a function of two variables, and let (x_0, y_0) be a point in the domain of f at which f is differentiable. If C is the level curve containing the point $c_0 = f(x_0, y_0)$, then $\nabla f(x_0, y_0)$ and C are orthogonal at $f(x_0, y_0)$.

Proof. Let $x = g(t), y = h(t)$ be a parametrization for the level curve C at height $f(x_0, y_0) = c_0$ such that $g(t_0) = x_0$ and $h(t_0) = y_0$. The curve C has equation $f(x, y) = c_0$, or in terms of the parametrization, $F(t) = f(g(t), h(t)) = c_0$. Now, by Theorem 12.32 and the fact that $F(t)$ is a constant function, we have

$$F'(t) = \frac{\partial f}{\partial x}\frac{\partial x}{\partial t} + \frac{\partial f}{\partial y}\frac{\partial y}{\partial t} = 0.$$

In particular, at $t = t_0$

$$F'(t_0) = \nabla f(x_0, y_0) \cdot \langle g'(t_0), h'(t_0) \rangle = 0.$$

But $\langle g'(t_0), h'(t_0) \rangle$ is the tangent vector to the level curve C at $t = t_0$. Therefore, the equation tells us that the gradient at (x_0, y_0) and the level curve are orthogonal at $f(x_0, y_0)$. ∎

Examples and Explorations

EXAMPLE 1

Verifying a derivative (chain rule, version I)

Let $z = x^2 \sin y$, $x = t^3$, and $y = e^{5t}$. Verify that

$$\frac{dz}{dt} = 6t^5 \sin(e^{5t}) + 5t^6 e^{5t} \cos(e^{5t})$$

by replacing x and y with their respective functions of t and then taking the derivative of the resulting function.

SOLUTION

Earlier in the section we used Theorem 12.32 to obtain the derivative shown. Now we express z as a function of t:

$$z = (t^3)^2 \sin(e^{5t}) = t^6 \sin(e^{5t}).$$

Taking the derivative with respect to t, we have

$$\frac{dz}{dt} = 6t^5 \sin(e^{5t}) + 5t^6 e^{5t} \cos(e^{5t}).$$

$\square$

EXAMPLE 2   **Using the chain rule (version II) to find a partial derivative**

Let $z = e^x \cos y$, $x = s^2 t^3$, and $y = \frac{s}{t}$. Earlier in the section we found $\frac{\partial z}{\partial s}$. Now we use Theorem 12.33 to find $\frac{\partial z}{\partial t}$, and then we verify that both of these partial derivatives are correct by replacing x and y with their respective functions of s and t and taking the appropriate partial derivatives of the resulting function.

SOLUTION

By Theorem 12.33,

$$\frac{\partial z}{\partial t} = \frac{\partial z}{\partial x}\frac{\partial x}{\partial t} + \frac{\partial z}{\partial y}\frac{\partial y}{\partial t} = (e^x \cos y)(3s^2 t^2) + (-e^x \sin y)\left(-\frac{s}{t^2}\right).$$

This result is correct, but it is preferable to write the function as a function of just s and t. We use $x = s^2 t^3$ and $y = \frac{s}{t}$ to do so:

$$\frac{\partial z}{\partial t} = \left(e^{s^2 t^3}\cos\left(\frac{s}{t}\right)\right)(3s^2 t^2) + \left(-e^{s^2 t^3}\sin\left(\frac{s}{t}\right)\right)\left(-\frac{s}{t^2}\right) = e^{s^2 t^3}\left(3s^2 t^2 \cos\left(\frac{s}{t}\right) + \frac{s}{t^2}\sin\left(\frac{s}{t}\right)\right).$$

To verify these partial derivatives, we first write z as a function of s and t; that is,

$$z = e^{s^2 t^3}\cos\left(\frac{s}{t}\right).$$

We now find $\frac{\partial z}{\partial s}$ and $\frac{\partial z}{\partial t}$, using the appropriate derivative rules:

$$\frac{\partial z}{\partial s} = 2st^3 e^{s^2 t^3}\cos\left(\frac{s}{t}\right) - \frac{1}{t}e^{s^2 t^3}\sin\left(\frac{s}{t}\right).$$

If we factor this equation, we obtain what we had earlier in the section, namely,

$$\frac{\partial z}{\partial s} = e^{s^2 t^3}\left(2st^3 \cos\left(\frac{s}{t}\right) - \frac{1}{t}\sin\left(\frac{s}{t}\right)\right).$$

Finally,

$$\frac{\partial z}{\partial t} = 3s^2 t^2 e^{s^2 t^3}\cos\left(\frac{s}{t}\right) + \frac{s}{t^2}e^{s^2 t^3}\sin\left(\frac{s}{t}\right).$$

This is equivalent to the result we obtained before.

$\square$

EXAMPLE 3   **Using the chain rule**

If $w = f(x, y, z)$, $x = u(s, t)$, $y = v(s, t)$, and $z = w(s, t)$ are differentiable functions at every point in their domains, what are $\frac{\partial w}{\partial s}$ and $\frac{\partial w}{\partial t}$?

SOLUTION

We have

$$\frac{\partial w}{\partial s} = \frac{\partial w}{\partial x}\frac{\partial x}{\partial s} + \frac{\partial w}{\partial y}\frac{\partial y}{\partial s} + \frac{\partial w}{\partial z}\frac{\partial z}{\partial s} \quad \text{and} \quad \frac{\partial w}{\partial t} = \frac{\partial w}{\partial x}\frac{\partial x}{\partial t} + \frac{\partial w}{\partial y}\frac{\partial y}{\partial t} + \frac{\partial w}{\partial z}\frac{\partial z}{\partial t}.$$

$\square$

| EXAMPLE 4 | **Finding and interpreting the gradient** |

Find the gradient of each of the following functions:

(a) $f(x, y) = e^y \sin\left(\dfrac{x}{y}\right)$ **(b)** $g(x, y, z) = xy^2 - \dfrac{z}{xy^3}$

Then find the direction in which the function f increases most rapidly at the point $(\pi, 1)$ and the rate of change of f in that direction. Finally, find the direction in which g decreases most rapidly at the point $(2, -1, 4)$ and the rate of change of g in that direction.

SOLUTION

(a) The gradient of f is

$$\nabla f(x, y) = \frac{e^y}{y} \cos\left(\frac{x}{y}\right) \mathbf{i} + e^y \left(\sin\left(\frac{x}{y}\right) - \frac{x}{y^2} \cos\left(\frac{x}{y}\right)\right) \mathbf{j}.$$

At $(\pi, 1)$, $\nabla f(\pi, 1) = e \cos(\pi) \mathbf{i} + e(\sin(\pi) - \pi \cos(\pi)) \mathbf{j} = -e\mathbf{i} + \pi e\mathbf{j}$. This is the direction in which f increases most rapidly at $(\pi, 1)$, and the magnitude

$$\|\nabla f(\pi, 1)\| = \|-e\mathbf{i} + \pi e\mathbf{j}\| = \sqrt{e^2 + \pi^2 e^2} = e\sqrt{1 + \pi^2}$$

is the rate of change of f in that direction.

(b) The gradient of g is

$$\nabla g(x, y, z) = \left(y^2 + \frac{z}{x^2 y^3}\right) \mathbf{i} + \left(2xy + \frac{3z}{xy^4}\right) \mathbf{j} - \frac{1}{xy^3} \mathbf{k}.$$

At $(2, -1, 4)$,

$$\nabla g(2, -1, 4) = \left((-1)^2 + \frac{4}{2^2(-1)^3}\right) \mathbf{i} + \left(2 \cdot 2 \cdot (-1) + \frac{3 \cdot 4}{2 \cdot (-1)^4}\right) \mathbf{j} - \frac{1}{2 \cdot (-1)^3} \mathbf{k}$$

$$= 2\mathbf{j} + \frac{1}{2}\mathbf{k}.$$

This is the direction in which g increases most rapidly at $(2, -1, 4)$, so g decreases most rapidly in the opposite direction, $-2\mathbf{j} - \frac{1}{2}\mathbf{k}$. The rate of decrease in this direction is the negative of the magnitude

$$\|\nabla g(2, -1, 4)\| = \sqrt{2^2 + (1/2)^2} = \frac{1}{2}\sqrt{17}.$$

□

| EXAMPLE 5 | **Finding a function given its gradient** |

Find the function $f(x, y)$ whose gradient is $\nabla f(x, y) = (2x + x^2 y^2)e^{xy^2}\mathbf{i} + 2x^3 y e^{xy^2}\mathbf{j}$.

SOLUTION

This is really the same type of problem we saw in Example 5 in Section 12.3. In that example we were given functions $g(x, y)$ and $h(x, y)$ and we used them to find a function f such that $f_x(x, y) = g(x, y)$ and $f_y(x, y) = h(x, y)$. Now we have $g(x, y) = f_x(x, y) = (2x + x^2 y^2)e^{xy^2}$ and $h(x, y) = f_y(x, y) = 2x^3 y e^{xy^2}$. Recall that, to ensure that such a function f existed, we showed that $\frac{\partial g}{\partial y} = \frac{\partial h}{\partial x}$. Here,

$$\frac{\partial}{\partial y}\left((2x + x^2 y^2)e^{xy^2}\right) = (6x^2 y + 2x^3 y^3)e^{xy^2} = \frac{\partial}{\partial x}(2x^3 y e^{xy^2}),$$

so we should be able to find a function f with the given gradient.

Since the function $2x^3ye^{xy^2}$ is slightly simpler than the other function, we choose to integrate this function with respect to y as the first step in finding f:

$$f(x, y) = \int 2x^3ye^{xy^2}\, dy = x^2e^{xy^2} + q(x).$$

(If you are wondering how we integrated this function, try a u-substitution with $u = xy^2$.) Next, to find $q(x)$, we differentiate $f(x, y)$ with respect to x and ensure that the resulting derivative equals $(2x + x^2y^2)e^{xy^2}$. That is,

$$\frac{\partial}{\partial x}(x^2e^{xy^2} + q(x)) = (2x + x^2y^2)e^{xy^2} + q'(x).$$

Thus, we must have $q'(x) = 0$ and $q(x) = C$, where C is a constant. We now have $f(x, y) = x^2e^{xy^2} + C$. ☐

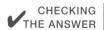
CHECKING THE ANSWER

We may check our answer by showing that for

$$f(x, y) = x^2e^{xy^2} + C, \quad \text{we have} \quad \nabla f(x, y) = (2x + x^2y^2)e^{xy^2}\mathbf{i} + 2x^3ye^{xy^2}\mathbf{j}.$$

EXAMPLE 6   **Showing that gradient vectors are orthogonal to level curves**

Show that the gradient vectors are orthogonal to the level curves for the function $f(x, y) = x^2 + y^2$.

SOLUTION

The graph of $f(x, y) = x^2 + y^2$ is a paraboloid, each of whose level curves is a circle centered at the origin. The gradient is $\nabla f(x, y) = 2x\mathbf{i} + 2y\mathbf{j}$; thus, every gradient vector points directly outward from the origin. For the point (x_0, y_0) the gradient is $\nabla f(x_0, y_0) = 2x_0\mathbf{i} + 2y_0\mathbf{j}$, and a tangent vector to the level curve containing the point (x_0, y_0) is $y_0\mathbf{i} - x_0\mathbf{j}$. The following figure shows the level curves $x^2 + y^2 = c$ for $c = 1, 2, 3$, and 4.

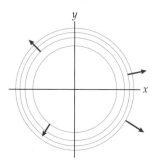

The gradient vector and tangent vector are orthogonal. ☐

TEST YOUR UNDERSTANDING

▶ If f is a function of variables $x_1, x_2, \ldots, x_n$, and each of these variables is a function of $t_1, t_2, \ldots, t_m$, how would we find $\frac{\partial f}{\partial t_1}$? How many summands would there be in this partial derivative?

▶ What is the gradient? What does the gradient tell us about a function?

▶ If f is a function of two variables, what is the relationship between the gradient of f and the level curves of f?

▶ How can the gradient be used to compute the directional derivative of a function?

▶ Given the gradient of a function of two variables, how would we find the function?

EXERCISES 12.5

Thinking Back

▶ *Chain rule:* If f is a function of x and x is a function of t, how is the chain rule used to find the rate of change of f with respect to t?

▶ *Critical points:* What is the definition of a critical point for a function of a single variable? How do we use critical points to locate the extrema of the function?

Concepts

0. *Problem Zero:* Read the section and make your own summary of the material.

1. *True/False:* Determine whether each of the statements that follow is true or false. If a statement is true, explain why. If a statement is false, provide a counterexample.

(a) *True or False:* If $z = f(x, y)$ and $x = u(s, t)$, then $\dfrac{\partial z}{\partial s} = \dfrac{\partial z}{\partial x}\dfrac{\partial x}{\partial s}$.

(b) *True or False:* If $z = f(x, y)$ and $x = u(t)$, then $\dfrac{dz}{dt} = \dfrac{\partial z}{\partial x}\dfrac{dx}{dt}$.

(c) *True or False:* If $z = f(x, y)$, then $\nabla f(x, y) = \dfrac{\partial z}{\partial x} + \dfrac{\partial z}{\partial y}$.

(d) *True or False:* If $z = f(x, y)$ and $\nabla f(1, 3) = \mathbf{0}$, then the graph of f is differentiable at $(1, 3)$.

(e) *True or False:* If $z = f(x, y)$ and $\mathbf{u}$ is a unit vector, then $D_{\mathbf{u}} f(a, b) = \nabla f(a, b) \cdot \mathbf{u}$.

(f) *True or False:* If $f(x, y, z)$ is differentiable at (a, b, c) and $\mathbf{u} \in \mathbb{R}^3$ is a unit vector, then $D_{\mathbf{u}} f(a, b, c) = \nabla f(a, b, c) \cdot \mathbf{u}$.

(g) *True or False:* If $w = f(x, y, z)$ is differentiable at (a, b, c), then $-\nabla f(a, b, c)$ points in the direction in which f is decreasing most rapidly at (a, b, c).

(h) *True or False:* If $z = f(x, y)$ is differentiable at $(-1, 0)$, and if $f(-1, 0) = 4$, then $\nabla f(-1, 0)$ is orthogonal to the level curve $f(x, y) = 4$ at $(-1, 0)$.

2. *Examples:* Construct examples of the thing(s) described in the following. Try to find examples that are different than any in the reading.

(a) A function $z = f(x, y)$ for which $\nabla f(0, 0) = \mathbf{0}$ but f is not differentiable at $(0, 0)$.

(b) A function $z = f(x, y)$ for which $\nabla f(0, 0) = \mathbf{0}$ for every point in $\mathbb{R}^2$.

(c) A function $z = f(x, y)$ and a unit vector $\mathbf{u}$ such that $D_{\mathbf{u}} f(0, 0) \neq \nabla f(0, 0) \cdot \mathbf{u}$.

3. Let $z = e^{-x}(3xy - 4x + y^2)$, $x = \sin t$, and $y = \cos t$.

(a) Find $\dfrac{dz}{dt}$ by using the Chain Rule, Theorem 12.32.

(b) Find $\dfrac{dz}{dt}$ by evaluating $f(x(t), y(t)) = f(\sin t, \cos t)$ and taking the derivative of the resulting function.

(c) Show that your answers from parts (a) and (b) are the same. Which method was easier?

4. Let $z = e^{-xy^2}$, $x = s \sin t$, and $y = s^2 \cos t$.

(a) Find $\dfrac{\partial z}{\partial s}$ by using the Chain Rule, Theorem 12.33.

(b) Find $\dfrac{\partial z}{\partial s}$ by evaluating $f(x(s, t), y(s, t)) = f(s \sin t, s^2 \cos t)$ and taking the partial derivative with respect to s of the resulting function.

(c) Show that your answers from parts (a) and (b) are the same. Which method was easier?

5. Explain why the chain rule from Chapter 2 is a special case of Theorem 12.34 with $n = 1$ and $m = 1$.

6. Explain why Theorem 12.32 is a special case of Theorem 12.34 with $n = 2$ and $m = 1$.

7. Explain why Theorem 12.33 is a special case of Theorem 12.34 with $n = 2$ and $m = 2$.

8. Consider the function $f(x, y) = 2x + 3y$.

(a) Why is the graph of f a plane?

(b) In what direction is f increasing most rapidly at the point $(-1, 4)$?

(c) In what direction is f increasing most rapidly at the point (x_0, y_0)?

(d) Why are your answers to parts (b) and (c) the same?

9. Continue with the function $f(x, y) = 2x + 3y$ from Exercise 8.

(a) What are the level curves of f?

(b) Show that every gradient vector, $\nabla f(x, y)$, is orthogonal to every level curve of f.

10. Consider the function $f(x, y) = ax + by$, where neither a nor b is zero.

(a) Why is the graph of f a plane?

(b) In what direction is f increasing most rapidly at the point $(2, -3)$?

(c) In what direction is f increasing most rapidly at the point (x_0, y_0)?

(d) Why are your answers to parts (b) and (c) the same?

11. Continue with the function $f(x, y) = ax + by$ from Exercise 10.

(a) What are the level curves of f?

(b) Show that every gradient vector, $\nabla f(x, y)$, is orthogonal to every level curve of f.

12. If a function $f(x, y)$ is differentiable at (a, b), explain how to use the gradient $\nabla f(a, b)$ to find the equation of the plane tangent to the graph of f at (a, b).

13. If a function $f(x, y, z)$ is differentiable at (a, b, c), explain how to use the gradient $\nabla f(a, b, c)$ to find the equation of the hyperplane tangent to the graph of f at (a, b, c).

14. Sketch level curves $z = 1, 4, 9,$ and 16 for the function $z = x^2 + y^2$. Include the graphs of three gradient vectors on each level curve. What do you observe?

15. Sketch level curves $z = 9, 16, 21,$ and 24 for the function $z = 25 - x^2 - y^2$. Include the graphs of three gradient vectors on each level curve. What do you observe?

16. Sketch level curves $z = 1, 4, 9,$ and 16 for the function $z = \dfrac{x^2}{4} + \dfrac{y^2}{9}$. Include the graphs of three gradient vectors on each level curve. What do you observe?

17. Sketch level curves $z = -1, 0,$ and 1 for the function $z = x^2 - y^2$. Include the graphs of three gradient vectors on each level curve. What do you observe?

18. When would you have to use the definition of the directional derivative rather than the shortcut $D_{\mathbf{u}} f(x_0, y_0) = \nabla f(x_0, y_0) \cdot \mathbf{u}$?

19. Imagine using a topographic map to plan a hike on a mountain. If your hike stays on a single contour line, what does that mean about the difficulty of the hike? If you intentionally hike perpendicularly to the contour lines, what does *that* mean about the difficulty of the hike?

20. An isotherm is a curve on a weather map connecting points on the map that have the same temperature. From a point on an isotherm, which direction would result in the greatest temperature change?

Skills

Use Theorem 12.32 to find the indicated derivatives in Exercises 21–26. Express your answers as functions of a single variable.

21. $\dfrac{dz}{dt}$ when $z = \sin x \cos y$, $x = e^t$, and $y = t^3$.

22. $\dfrac{dz}{dt}$ when $z = x^3 e^y$, $x = \sin t$, and $y = \cos t$.

23. $\dfrac{dx}{dt}$ when $x = r \cos \theta$, $r = t^2 - 5$, and $\theta = t^3 + 1$.

24. $\dfrac{dy}{dt}$ when $y = r \sin \theta$, $r = t^3$, and $\theta = \sqrt{t}$.

25. $\dfrac{dr}{dt}$ when $r = \sqrt{x^2 + y^2}$, $x = \sqrt{t}$, and $y = t^2$.

26. $\dfrac{d\theta}{dt}$ when $\theta = \tan^{-1}\left(\dfrac{y}{x}\right)$, $x = e^t$, and $y = e^{2t}$.

Use Theorem 12.33 to find the indicated derivatives in Exercises 27–30. Express your answers as functions of two variables.

27. $\dfrac{\partial z}{\partial s}$ when $z = x^2 y^3$, $x = t \sin s$, and $y = s \cos t$.

28. $\dfrac{\partial z}{\partial t}$ when $z = x^2 y^3$, $x = t \sin s$, and $y = s \cos t$.

29. $\dfrac{\partial z}{\partial r}$ when $z = (x^2 + xy)e^y$, $x = r \cos \theta$, and $y = r \sin \theta$.

30. $\dfrac{\partial z}{\partial \theta}$ when $z = (x^2 + xy)e^y$, $x = r \cos \theta$, and $y = r \sin \theta$.

Use Theorem 12.34 to find the indicated derivatives in Exercises 31–36. Be sure to simplify your answers.

31. $\dfrac{dx}{dt}$ when $x = \rho \sin \phi \cos \theta$, $\rho = t^2$, $\phi = t^3$, and $\theta = t^4$.

32. $\dfrac{dx}{dt}$ when $x = \rho \sin \phi \sin \theta$, $\rho = \sqrt{t}$, $\phi = \sqrt[3]{t}$, and $\theta = \sqrt[4]{t}$.

33. $\dfrac{\partial w}{\partial \rho}$ when $w = (x^2 + z)e^y$, $x = \rho \sin \phi \cos \theta$, $y = \rho \sin \phi \sin \theta$, $z = \rho \cos \phi$.

34. $\dfrac{\partial w}{\partial \theta}$ when $w = (x^2 + z)e^y$, $x = \rho \sin \phi \cos \theta$, $y = \rho \sin \phi \sin \theta$, $z = \rho \cos \phi$.

35. $\dfrac{d\rho}{dt}$ when $\rho = \sqrt{x^2 + y^2 + z^2}$, $x = \sqrt{t}$, $y = t^2$, $z = t^3$.

36. $\dfrac{d\theta}{dt}$ when $\theta = \tan^{-1}\left(\dfrac{yz}{x}\right)$, $x = e^t$, $y = e^{2t}$, $z = e^{3t}$.

Find the gradient of the given functions in Exercises 37–42.

37. $z = x^2 \sin y + y \sin x$

38. $z = \tan^{-1}\left(\dfrac{y}{x}\right)$

39. $f(x, y) = \sqrt{x^2 + y^2}$

40. $f(x, y) = y \sec^{-1}\left(\dfrac{1}{x}\right)$

41. $f(x, y, z) = \sqrt{x^2 + y^2 + z^2}$

42. $f(x, y, z) = \cos^{-1}\left(\dfrac{z}{\sqrt{x^2 + y^2 + z^2}}\right)$

In Exercises 43–48:

(a) Find the direction in which the given function increases most rapidly at the specified point.

(b) Find the rate of change of the function in the direction you found in part (a).

(c) Find the direction in which the given function decreases most rapidly at the specified point.

Note: These are the same functions as in Exercises 37–42.

43. $z = x^2 \sin y + y \sin x$ at $\left(\pi, \dfrac{\pi}{2}\right)$

44. $z = \tan^{-1}\left(\dfrac{y}{x}\right)$ at $(1, \sqrt{3})$

45. $f(x, y) = \sqrt{x^2 + y^2}$ at $(2, -3)$

46. $f(x, y) = y \sec^{-1} x$ at $(2, 5)$

47. $f(x, y, z) = \sqrt{x^2 + y^2 + z^2}$ at $(2, -1, -2)$

48. $f(x, y, z) = \cos^{-1}\left(\dfrac{z}{\sqrt{x^2 + y^2 + z^2}}\right)$ at $(\sqrt{10}, -1, 5)$

In Exercises 49–54, find the directional derivative of the given function at the specified point P and in the specified direction $\mathbf{v}$. Note that some of the direction vectors are *not* unit vectors.

49. $z = e^x \tan y$, $P = \left(0, \dfrac{\pi}{4}\right)$, $\mathbf{v} = 3\mathbf{i} - \mathbf{j}$

50. $z = 3x^4 - 5y^2$, $P = (1, -2)$, $\mathbf{v} = \left\langle \dfrac{5}{13}, -\dfrac{12}{13} \right\rangle$

51. $z = \ln\left(\dfrac{x}{y^2}\right)$, $P = (7, 1)$, $\mathbf{v} = -\mathbf{i} - 4\mathbf{j}$

52. $z = \sqrt{x^2 + y^2}$, $P = (-3, 4)$, $\mathbf{v} = \langle 4, -3 \rangle$

53. $w = x \sin y \cos z$, $P = \left(3, \frac{\pi}{4}, -\frac{\pi}{2}\right)$, $v = i - 2j + 3k$

54. $w = \ln\left(\frac{x^2}{yz}\right) + \ln\left(\frac{z}{xy}\right) - \ln\left(\frac{x}{y^2}\right)$, $P = (3, 5, 8)$,

$v = 13i + 21j + 34k$

In Exercises 55–60, find a function of two variables with the given gradient.

55. $\nabla f(x, y) = \left\langle -\frac{3}{x}, \frac{1}{y} \right\rangle$

56. $\nabla f(x, y) = (2x + \cos x \cos y) \, i - \sin x \sin y \, j$

57. $\nabla f(x, y) = -\frac{y}{x^2 + y^2} i + \frac{x}{x^2 + y^2} j$

58. $\nabla f(x, y) = \left\langle y^2 e^{xy^2}, 2xy e^{xy^2} \right\rangle$

59. $\nabla f(x, y) = \left\langle \frac{y}{(x+y)^2}, -\frac{x}{(x+y)^2} \right\rangle$

60. $\nabla f(x, y) = \left(\frac{1}{y} - \frac{y}{x^2}\right) i + \left(\frac{1}{x} - \frac{x}{y^2}\right) j$

Applications

61. Ian is travelling along a glacier. The elevation of the glacier in his area is described by the function

$$f(x, y) = 1.2 - 0.2x^2 - 0.3y^2 + 0.1xy - 0.25x,$$

where x, y, and f are given in miles.

(a) What is the direction that Ian would need to go to descend most steeply if he is at the point $(0.5, -0.5)$?

(b) In what direction would Ian have to turn in order to contour (i.e., , neither ascend nor descend) across the glacier?

62. Emmy is tracking down another source of contamination at Hanford, this time in a warehouse containing numerous 55-gallon drums of waste. Detectors placed throughout the facility have measured a certain amount of radiation at different points. From these measurements, Emmy has constructed a polynomial approximation to the radiation given by

$$f(x, y) = -0.0156x^2 - 0.0392y^2 + 0.0268xy + 1.7x + 20,$$

where the radiation is measured in microrems per hour and the distances x and y are measured in feet from a corner of the warehouse. At what location in the warehouse should Emmy start her search for the contaminated drum?

Proofs

63. Prove that $\|\nabla f(x, y, z)\| = 1$ for every point in the domain of the function $f(x, y, z) = \sqrt{x^2 + y^2 + z^2}$ except the origin.

64. Use at least two methods to prove that $\frac{dr}{dt} = 0$ when $r = \sqrt{x^2 + y^2}$, $x = \alpha \cos t$, and $y = \alpha \sin t$ if α is a constant.

65. Prove Theorem 12.33. That is, show that if $z = f(x, y)$, $x = u(s, t)$, and $y = v(s, t)$, then, for all values of s and t at which u and v are differentiable, and if f is differentiable at $(u(s, t), v(s, t))$, it follows that

$$\frac{\partial z}{\partial s} = \frac{\partial z}{\partial x}\frac{\partial x}{\partial s} + \frac{\partial z}{\partial y}\frac{\partial y}{\partial s} \quad \text{and} \quad \frac{\partial z}{\partial t} = \frac{\partial z}{\partial x}\frac{\partial x}{\partial t} + \frac{\partial z}{\partial y}\frac{\partial y}{\partial t}.$$

66. Prove Theorem 12.36 when f is a function of three variables. That is, show that if (x_0, y_0, z_0) is a point in the domain of $f(x, y, z)$ at which the first-order partial derivatives of f exist, and if $u \in \mathbb{R}^3$ is a unit vector for which the directional derivative $D_u f(x_0, y_0, z_0)$ also exists, then $D_u f(x_0, y_0, z_0) = \nabla f(x_0, y_0, z_0) \cdot u$.

In Exercises 67–72 you will prove several basic properties of the gradient for functions of two variables. In each exercise, assume that f and/or g is differentiable.

67. Prove that $\nabla f = 0$ if f is the constant function $f(x, y) = c$.

68. Prove that $\nabla(\alpha f)(x, y) = \alpha \nabla f(x, y)$, where α is a constant.

69. Prove that

$$\nabla(f(x, y) + g(x, y)) = \nabla f(x, y) + \nabla g(x, y).$$

70. Prove that

$$\nabla(\alpha f(x, y) + \beta g(x, y)) = \alpha \nabla f(x, y) + \beta \nabla g(x, y),$$

where α and β are constants.

71. Prove that

$$\nabla(f(x, y)g(x, y)) = f(x, y)\nabla g(x, y) + g(x, y)\nabla f(x, y).$$

72. Prove that

$$\nabla\left(\frac{f(x, y)}{g(x, y)}\right) = \frac{g(x, y)\nabla f(x, y) - f(x, y)\nabla g(x, y)}{(g(x, y))^2},$$

where $g(x, y) \neq 0$.

73. Analogous properties hold for functions of three variables. What would you have to change in the proofs in Exercises 67–72 to make them work for functions of three variables?

74. Let $f(x, y, z)$ be a function of three variables, and let P be a point in the domain of f at which f is differentiable. Prove that the gradient of f at P points in the direction in which f increases most rapidly. (This is Theorem 12.37.)

Thinking Forward

▶ *The gradient at a maximum:* If a function of two variables, $f(x, y)$, is differentiable at a point (x_0, y_0) where the function has a maximum, what is $\nabla f(x_0, y_0)$?

▶ *The gradient at a minimum:* If a function of three variables, $f(x, y, z)$, is differentiable at a point (x_0, y_0, z_0) where the function has a minimum, what is $\nabla f(x_0, y_0, z_0)$?

12.6   EXTREME VALUES

▶ Local and global extrema of functions of two variables

▶ Critical points of functions of two variables

▶ Using critical points to find the local extrema of functions of two variables

The Gradient at a Local Extremum

Many of the ideas in this section parallel the analogous concepts for functions of a single variable. It would be worthwhile to review the definitions of local extrema and critical points for functions of a single variable. Before proceeding in this section, we also suggest that you think about how the first- and second-derivative tests can be adapted for functions of two variables, if at all. Further, we suggest that when you are done with this section, think about how the concepts presented here might be adapted to analyze functions of three or more variables.

Suppose a function $f(x, y)$ has a local maximum at some point (x_0, y_0). This means that the value of $f(x_0, y_0)$ is greater than or equal to all other "nearby" values of the function. More precisely, we have the following definition:

DEFINITION 12.39

Local and Global Extrema of a Function of Two Variables

(a) A function $f(x, y)$ has a **local** or **relative maximum** at (x_0, y_0) if $f(x_0, y_0) \geq f(x, y)$ for every point (x, y) in some open disk containing (x_0, y_0).

(b) A function $f(x, y)$ has a **local** or **relative minimum** at (x_0, y_0) if $f(x_0, y_0) \leq f(x, y)$ for every point (x, y) in some open disk containing (x_0, y_0).

(c) A function $f(x, y)$ has a **local** or **relative extremum** at (x_0, y_0) if f has either a local maximum or a local minimum at (x_0, y_0).

(d) A function $f(x, y)$ has a **global** or **absolute maximum** at (x_0, y_0) if $f(x_0, y_0) \geq f(x, y)$ for every point (x, y) in the domain of f.

(e) A function $f(x, y)$ has a **global** or **absolute minimum** at (x_0, y_0) if $f(x_0, y_0) \leq f(x, y)$ for every point (x, y) in the domain of f.

(f) A function $f(x, y)$ has a **global** or **absolute extremum** at (x_0, y_0) if f has either a global maximum or a global minimum at (x_0, y_0).

Clearly, every global extremum is also a local extremum. At a local extremum, the plane tangent to the function must be either horizontal or undefined. Consider the following three graphs:

Local maximum with horizontal tangent plane

Local minimum with horizontal tangent plane

Local maximum with no tangent plane

When a function has a horizontal tangent plane at a point P, its gradient at P is zero. Such a point is called a ***stationary point***. When a function is not differentiable at a point, its gradient there is either zero or undefined. Points at which either of these things occur are called ***critical points***.

DEFINITION 12.40

Stationary Points of a Function of Two Variables

A point (x_0, y_0) in the domain of a function $f(x, y)$ is called a ***stationary point*** of f if f is differentiable at (x_0, y_0) and $\nabla f(x_0, y_0) = \mathbf{0}$.

In the first two of the preceding three graphs, the functions have stationary points where the graphs have the horizontal tangent planes. The condition that $f(x, y)$ needs to be differentiable at (x_0, y_0) is necessary, since $\nabla f(x_0, y_0) = \mathbf{0}$ does not ensure that f is differentiable at (x_0, y_0). Consider, for example, the function

$$f(x, y) = \begin{cases} 0, & \text{if } x = 0 \text{ or } y = 0 \\ 1, & \text{otherwise.} \end{cases}$$

In Section 12.4, we saw that $\nabla f(0, 0) = \mathbf{0}$ but that f is *not* differentiable at the origin. Therefore, $(0, 0)$ is not a stationary point for f.

DEFINITION 12.41

Critical Points of a Function of Two Variables

A point (x_0, y_0) in the domain of a function $f(x, y)$ is called a ***critical point*** of f if (x_0, y_0) is a stationary point of f or if f is not differentiable at (x_0, y_0).

Therefore, each of the three graphs shown before has a critical point at its local extremum. In fact, a local extremum can occur only at a critical point.

THEOREM 12.42

Local Extrema Occur at Critical Points

If $f(x, y)$ has a local extremum at (x_0, y_0), then (x_0, y_0) is a critical point of f.

The proof of Theorem 12.42 is similar to the proof of Theorem 3.3 in Section 3.1 and is left for Exercise 56.

The converse of Theorem 12.42 is not true. For example, consider the function $g(x, y) = x^2 - y^2$:

$$g(x, y) = x^2 - y^2$$

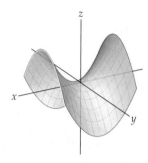

Here we have $\nabla g(0, 0) = \mathbf{0}$, but the function has neither a maximum nor a minimum at $(0, 0)$. This example leads us to the definition of a **saddle point**.

DEFINITION 12.43

Saddle Points of a Function of Two Variables

A point (x_0, y_0) in the domain of a function $f(x, y)$ is called a **saddle point** of f if (x_0, y_0) is a stationary point of f at which there is neither a maximum nor a minimum.

The term "saddle point" comes from the fact that the graphs of the surfaces near the simplest saddle points look like saddles, as we see in the graph of $g(x, y) = x^2 - y^2$. However, the graphs near other saddle points can look far more complicated. For example, the graph of $f(x, y) = x^2 y - y^3$, which also has a saddle point at the origin, is as follows:

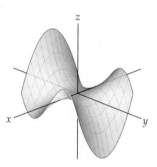

The Second-Derivative Test for Classifying Local Extrema

We will now discuss a method for determining whether a function has a local maximum, local minimum, or saddle point at each stationary point. This test uses the second-order partial derivatives of the function. You may be wondering why we are not starting by discussing a test that uses the first-order partial derivatives. The short answer is that such a test is somewhat impractical. You will explore the reason for that in Exercise 7.

Before we discuss the second-order partial derivative test, we define the ***discriminant*** of a function of two variables.

DEFINITION 12.44

The Hessian and the Discriminant of a Function of Two Variables

Let $f(x, y)$ be a function with continuous second-order partial derivatives on some open set S.

(a) The ***Hessian*** of f is the 2×2 matrix of second-order partial derivatives:

$$H_f = \begin{bmatrix} \dfrac{\partial^2 f}{\partial x^2} & \dfrac{\partial^2 f}{\partial y \partial x} \\[2ex] \dfrac{\partial^2 f}{\partial x \partial y} & \dfrac{\partial^2 f}{\partial y^2} \end{bmatrix}.$$

(b) The ***discriminant*** of f is the determinant of the Hessian. That is,

$$\det(H_f) = \frac{\partial^2 f}{\partial x^2} \frac{\partial^2 f}{\partial y^2} - \left(\frac{\partial^2 f}{\partial x \partial y} \right)^2.$$

Note that since we require that the second-order partial derivatives be continuous to form either the Hessian or the discriminant, the determinant of the Hessian matrix is equivalent to the discriminant because Theorem 12.23 guarantees the equality of the mixed second-order partial derivatives.

The discriminant of a function of two variables is a related function of two variables. For example, when $f(x, y) = x^3 e^{2y}$, we have the first-order partial derivatives

$$\frac{\partial f}{\partial x} = 3x^2 e^{2y} \quad \text{and} \quad \frac{\partial f}{\partial y} = 2x^3 e^{2y}$$

and the second-order partial derivatives

$$\frac{\partial^2 f}{\partial x^2} = 6x e^{2y}, \quad \frac{\partial^2 f}{\partial y^2} = 4x^3 e^{2y}, \quad \text{and} \quad \frac{\partial^2 f}{\partial x \partial y} = 6x^2 e^{2y}.$$

Therefore, the discriminant is

$$\det(H_f(x, y)) = (6x e^{2y})(4x^3 e^{2y}) - (6x^2 e^{2y})^2 = -12x^4 e^{4y}.$$

We finally note that the Hessian may be generalized to functions of three or more variables, although we will not be using the Hessian in these contexts. We will be using the Hessian and the discriminant to analyze the stationary points of a function of two variables. The Hessian of a function of three variables may be used in a similar manner.

THEOREM 12.45

The Second-Order Partial-Derivative Test for Classifying Stationary Points

Let $f(x, y)$ be a function with continuous second-order partial derivatives on some open disk containing the point at (x_0, y_0). If f has a stationary point at (x_0, y_0), then

(a) f has a relative maximum at (x_0, y_0) if $\det(H_f(x_0, y_0)) > 0$ with $f_{xx}(x_0, y_0) < 0$ or $f_{yy}(x_0, y_0) < 0$.

(b) f has a relative minimum at (x_0, y_0) if $\det(H_f(x_0, y_0)) > 0$ with $f_{xx}(x_0, y_0) > 0$ or $f_{yy}(x_0, y_0) > 0$.

(c) f has a saddle point at (x_0, y_0) if $\det(H_f(x_0, y_0)) < 0$.

(d) If $\det(H_f(x_0, y_0)) = 0$, no conclusion may be drawn about the behavior of f at (x_0, y_0).

Before we prove Theorem 12.45, we consider two examples. First, for the function

$$f(x, y) = x^2 + xy + y^2 + 3x - 6y,$$

we have $f_x(x, y) = 2x + y + 3$ and $f_y(x, y) = x + 2y - 6$. When we solve the system of equations

$$2x + y + 3 = 0 \quad \text{and} \quad x + 2y - 6 = 0,$$

we find the only stationary point of f, namely, $(-4, 5)$. The second-order partial derivatives are $f_{xx}(x, y) = 2, f_{yy}(x, y) = 2$, and $f_{xy}(x, y) = 1$. Therefore, $\det(H_f(-4, 5)) = 2 \cdot 2 - 1^2 = 3$. Since $\det(H_f(-4, 5)) > 0, f_{xx}(-4, 5) > 0$, and $f_{yy}(-4, 5) > 0$, it follows by Theorem 12.45 that f has a local minimum at $(-4, 5)$.

Next, if we let $g(x, y) = x^6 + y^6$, then $\nabla g(x, y) = 6x^5 \mathbf{i} + 6y^5 \mathbf{j}$. The point $(0, 0)$ is the only stationary point for g. The second-order partial derivatives are $g_{xx}(x, y) = 30x^4, f_{yy}(x, y) = 30y^4$, and $f_{xy}(x, y) = 0$. Therefore, $\det(H_g(0, 0)) = 0$. Thus, no conclusion can be drawn from Theorem 12.45 about the behavior of g at the origin. However, since $g(0, 0) = 0$ and $g(x, y) > 0$ for every point in $\mathbb{R}^2$ except the origin, we may deduce that g has an absolute minimum of 0 at the origin, without using the theorem.

We will prove Theorem 12.45 in a moment, but perhaps the following remarks will help explain the reasoning in the proof: Recall that the sign of the second derivative of a function of a single variable determines whether a function is concave up or concave down. Similarly, the sign of the second-order directional derivative $D_{\mathbf{u}}(D_{\mathbf{u}} f(x, y)) = D_{\mathbf{u}}^2 f(x, y)$ may be used to determine whether the function is concave up or concave down in the direction of the unit vector $\mathbf{u}$. If $D_{\mathbf{u}}^2 f(x_0, y_0) > 0$ for every unit vector at a stationary point (x_0, y_0), then the function has a local minimum at the stationary point. Similarly, if $D_{\mathbf{u}}^2 f(x_0, y_0) < 0$ for every unit vector at the stationary point, then the function has a local maximum there.

One final note before we turn to the proof: Recall that the discriminant of a quadratic polynomial $y = ax^2 + bx + c$ is the quantity $b^2 - 4ac$ and the discriminant of a function of two variables is given in Definition 12.44. In the proof of Theorem 12.45, we use this term in both contexts.

Proof. In Exercise 55 you will prove that for a function $f(x, y)$ with continuous second-order partial derivatives, the second directional derivative in the direction of the unit vector $\mathbf{u} = \langle \alpha, \beta \rangle$ is given by

$$D_{\mathbf{u}}^2 f(x, y) = \alpha^2 f_{xx}(x, y) + 2\alpha\beta f_{xy}(x, y) + \beta^2 f_{yy}(x, y).$$

Now let (x_0, y_0) be a stationary point of f, and let

$$A = f_{xx}(x_0, y_0), \quad B = f_{xy}(x_0, y_0), \quad \text{and} \quad C = f_{yy}(x_0, y_0).$$

Before proceeding further, note that $AC - B^2$ is the discriminant, $\det(H_f(x_0, y_0))$. We have

$$D_{\mathbf{u}}^2 f(x_0, y_0) = \alpha^2 A + 2\alpha\beta B + \beta^2 C.$$

Now, if $\alpha \neq 0$ and we let $m = \dfrac{\beta}{\alpha}$, then

$$D_{\mathbf{u}}^2 f(x_0, y_0) = \alpha^2 \left(A + 2\frac{\beta}{\alpha} B + \left(\frac{\beta}{\alpha} \right)^2 C \right) = \alpha^2 \left(A + 2mB + m^2 C \right).$$

The polynomial $q(m) = A + 2Bm + Cm^2$ is a quadratic in m and has the same roots as $D_{\mathbf{u}}^2 f(x_0, y_0)$. Now, if the discriminant, $4B^2 - 4AC$, of this *polynomial* is negative, then $q(m)$ has no real roots. Thus, if $AC - B^2 > 0$, then $q(m)$ has no real roots, and the sign of $q(m)$ never changes. In order for $AC - B^2$ to be positive, the signs of A and C must be the same. So, if $A = f_{xx}(x_0, y_0), C = f_{yy}(x_0, y_0)$, and $AC - B^2 = \det(H_f(x_0, y_0))$ are all positive, the function is concave up in every direction at the stationary point (x_0, y_0), and it follows that f has a local minimum at (x_0, y_0). If A and C are both negative with $AC - B^2 > 0$, the function is concave down in every direction at (x_0, y_0), and it follows that f has a local maximum at (x_0, y_0). If $AC - B^2 < 0$, the polynomial $q(m)$

has two real roots, and then the function f is concave up in some directions and concave down in other directions. Therefore, f has a saddle at the stationary point. If $AC - B^2 = 0$, we cannot draw any conclusions about the function. In Exercises 10-12 we ask you to show that a function can have a local maximum, local minimum, or saddle point when $AC - B^2 = 0$. Finally, if $\alpha = 0$, then $\beta \neq 0$, and we may modify the argument by factoring β^2 instead of α^2 out of the equation for $D_{\mathbf{u}}^2 f(x, y)$. ∎

Examples and Explorations

EXAMPLE 1

Finding and analyzing stationary points

Find the local extrema of the function $f(x, y) = x^3 - 3xy - y^3$. Classify each extreme point as a local or global minimum or maximum.

SOLUTION

Our function is a polynomial. Every polynomial in two variables is differentiable at every point in $\mathbb{R}^2$, so the only critical points of f will be stationary points. The gradient of f is

$$\nabla f(x, y) = (3x^2 - 3y)\mathbf{i} + (-3x - 3y^2)\mathbf{j}.$$

To find the stationary points, we look for the points that make the gradient $\mathbf{0}$. We look for solutions of the system of equations

$$3x^2 - 3y = 0 \quad \text{and} \quad -3x - 3y^2 = 0.$$

Using the first equation, we see that $y = x^2$. Substituting x^2 for y in the second equation and simplifying, we obtain $x(x^3 + 1) = 0$. This equation has solutions $x = 0$ and $x = -1$. Since $x^2 = y$, we have the two stationary points $(0, 0)$ and $(-1, 1)$.

Now, the second-order partial derivatives of f are

$$f_{xx}(x, y) = 6x, \quad f_{yy}(x, y) = -6y, \quad \text{and} \quad f_{xy}(x, y) = -3.$$

The discriminant is $\det(H_f(x, y)) = (6x)(-6y) - (-3)^2 = -36xy - 9$. At the stationary points, we have $\det(H_f(0, 0)) = -9$ and $\det(H_f(-1, 1)) = 27$. By Theorem 12.45, since $\det(H_f(0, 0)) < 0$, there is a saddle point at $(0, 0)$, and since $\det(H_f(-1, 1)) > 0$ and $f_{xx}(-1, 1) = -6 < 0$, f has a local maximum at $(-1, 1)$.

Finally, it is good to know if there is an absolute maximum at $(-1, 1)$ or just a relative maximum. We have $f(-1, 1) = 1$, but when $x = 0$, $\displaystyle\lim_{y \to -\infty} f(0, y) = \lim_{y \to -\infty} (-y^3) = \infty$. Therefore, f has just a relative maximum at $(-1, 1)$. □

EXAMPLE 2

Tabulating information about stationary points

Find the local extrema of the function $g(x, y) = 3x^2 y - 3y + y^3$. Classify each extreme point as a local or global minimum or maximum.

SOLUTION

Again, the function is a polynomial, so the only critical points of g will be stationary points. The gradient of g is

$$\nabla g(x, y) = g_x(x, y)\mathbf{i} + g_y(x, y)\mathbf{j} = 6xy\mathbf{i} + (3x^2 - 3 + 3y^2)\mathbf{j}.$$

To find the stationary points, we look for solutions of the system of equations

$$6xy = 0 \quad \text{and} \quad 3x^2 - 3 + 3y^2 = 0.$$

From the first equation, we see that one of x and y must be zero. Substituting $x = 0$ into the second equation and solving for y, we have $y = \pm 1$. Similarly, if $y = 0$, solving for x yields $x = \pm 1$. Thus, we have exactly four stationary points: $(0, 1)$, $(0, -1)$, $(1, 0)$, and $(-1, 0)$.

The second-order partial derivatives of g are

$$g_{xx}(x, y) = 6y, \quad g_{yy}(x, y) = 6y, \quad \text{and} \quad g_{xy}(x, y) = 6x.$$

The discriminant is $\det(H_g(x, y)) = (6y)(6y) - (6x)^2 = 36(y^2 - x^2)$. When there are several stationary points, it is convenient to tabulate the relevant information about the discriminant and second-order partial derivatives in order to classify the stationary points. Accordingly, we construct the following table:

(x_0, y_0)	$\det(H_g(x_0, y_0)) = 36(y_0^2 - x_0^2)$	$g_{xx}(x_0, y_0) = 6y_0$	$g(x_0, y_0)$	conclusion
$(0, 1)$	36	6	-2	minimum
$(0, -1)$	36	-6	2	maximum
$(1, 0)$	-36		0	saddle point
$(-1, 0)$	-36		0	saddle point

It remains to determine whether the function has an absolute minimum at $(0, 1)$ or just a relative minimum, and an absolute maximum at $(0, -1)$ or just a relative maximum. Note that when $x = 0$, we have the limits

$$\lim_{y \to \infty} g(0, y) = \lim_{y \to \infty} (-3y + y^3) = \infty \quad \text{and} \quad \lim_{y \to -\infty} g(0, y) = \lim_{y \to -\infty} (-3y + y^3) = -\infty.$$

Therefore, g has a relative minimum at $(0, 1)$ and a relative maximum at $(0, -1)$. □

EXAMPLE 3 Optimizing a function of two variables

Max is planning to construct a box with five wooden sides and a glass front. If the wood costs $5 per square foot and the glass costs $10 per square foot, what dimensions should Max use for the box to minimize the cost of the materials if the box needs to have a volume of 12 cubic feet?

SOLUTION

Following is a schematic of the box:

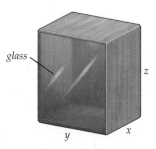

Using the dimensions in the figure and the fact that the volume of the box is to be 12 cubic feet, we have the constraint $V = xyz = 12$ between the variables. The glass front has area yz. Therefore, the cost of the glass will be $10yz$. The back wooden face will cost $5yz$, the left and right sides will each cost $5xz$, and the top and bottom will each cost $5xy$. Thus, in dollars, the cost of the box will be $10xy + 10xz + 15yz$. Now, from our constraint on the volume, we have $z = \dfrac{12}{xy}$, which we may use in our equation for the cost to rewrite our function with just two variables:

$$C(x, y) = 10xy + \frac{120}{y} + \frac{180}{x}, \quad \text{where } x > 0 \quad \text{and} \quad y > 0.$$

To find the stationary points, we find the first-order partial derivatives

$$C_x = 10y - \frac{180}{x^2} \quad \text{and} \quad C_y = 10x - \frac{120}{y^2}.$$

Setting these partial derivatives equal to zero and solving simultaneously, we obtain $x = 3$ and $y = 2$. We also have the second-order partial derivatives

$$C_{xx} = \frac{360}{x^3}, \quad C_{yy} = \frac{240}{y^3}, \quad \text{and} \quad C_{xy} = 10.$$

Thus, the discriminant at $(3, 2)$ is $\det(H_C(3, 2)) = \left(\frac{360}{27}\right)\left(\frac{240}{8}\right) - 10^2 = 300 > 0$, and since $C_{xx}(3, 2) > 0$ as well, this stationary point gives us a minimum cost for the box. These values tell us that the glass should be a square 2 feet on each side and the other dimension of the box should be 3 feet.   □

EXAMPLE 4

Using the derivative to find the distance from a point to a plane

In Theorem 10.39 in Chapter 10 we showed that the distance between a point P and a plane $\mathcal{P}$ is given by $\dfrac{|\mathbf{N} \cdot \overrightarrow{RP}|}{\|\mathbf{N}\|}$, where R is any point in the plane and $\mathbf{N}$ is a normal vector for $\mathcal{P}$. Use the minimization techniques discussed in this section to show that the distance from the point $P(x_0, y_0, z_0)$ to the plane $\mathcal{P}$ with equation $ax + by + cz = d$ is also given by

$$\frac{|ax_0 + by_0 + cz_0 - d|}{\sqrt{a^2 + b^2 + c^2}}.$$

SOLUTION

Let (x, y, z) be an arbitrary point in the plane $\mathcal{P}$. We wish to minimize the distance from P to (x, y, z) given by

$$\text{distance} = \sqrt{(x - x_0)^2 + (y - y_0)^2 + (z - z_0)^2}.$$

However, as we have previously mentioned, if we minimize the square of the distance, we accomplish the same goal. We will minimize

$$\text{distance}^2 = (x - x_0)^2 + (y - y_0)^2 + (z - z_0)^2$$

subject to the constraint that $ax + by + cz = d$. Now, at least one of the coefficients a, b, and c is nonzero. Without loss of generality, we assume that $c \neq 0$, so we have $z = \dfrac{d - ax - by}{c}$. Thus, we will minimize the function

$$D(x, y) = (x - x_0)^2 + (y - y_0)^2 + \left(\frac{d - ax - by}{c} - z_0\right)^2.$$

Here,

$$\nabla D = \left(2(x - x_0) - \frac{2a}{c}\left(\frac{d - ax - by}{c} - z_0\right)\right)\mathbf{i} + \left(2(y - y_0) - \frac{2b}{c}\left(\frac{d - ax - by}{c} - z_0\right)\right)\mathbf{j}.$$

Setting $\nabla D = \mathbf{0}$ and solving for x and y is cumbersome, but the unique stationary point for D is

$$\left(\frac{ad - aby_0 - acz_0 + b^2x_0 + c^2x_0}{a^2 + b^2 + c^2}, \frac{bd - abx_0 - bcz_0 + a^2y_0 + c^2y_0}{a^2 + b^2 + c^2} \right).$$

In Exercise 13, you are asked to provide the details of this result. In Exercise 14, you are also asked to show that this point provides an absolute minimum for the function D. In Exercise 15, you are asked to show that this minimal value is $\frac{(ax_0 + by_0 + cz_0 - d)^2}{a^2 + b^2 + c^2}$, and we have the desired result.   ☐

? TEST YOUR UNDERSTANDING

▶ What are stationary points? What are critical points? What are saddle points?

▶ Where do the local extrema of a function of two variables occur? How do we use the first-order partial derivatives to find local extrema?

▶ How do we use the second-order partial derivatives to classify the stationary points of a function of two variables?

▶ What is the main idea in the proof of Theorem 12.45 that is used to classify the stationary points of a function of two variables?

▶ How could the ideas from this section be adapted to find the extrema of functions of three variables? How would the definitions have to be modified? Could Theorem 12.45 be generalized to handle three variables?

EXERCISES 12.6

Thinking Back

▶ *First-Derivative Test:* Review the first-derivative test for functions of a single variable. Explain how the test works, what conditions a function must satisfy to make the test useful, and when, if ever, the test might fail.

▶ *Second-Derivative Test:* Review the second-derivative test for functions of a single variable. Explain how the test works, what conditions a function must satisfy to make the test useful, and when, if ever, the test might fail.

Concepts

0. *Problem Zero:* Read the section and make your own summary of the material.

1. *True/False:* Determine whether each of the statements that follow is true or false. If a statement is true, explain why. If a statement is false, provide a counterexample.

   (a) *True or False:* If $\nabla f(a, b) = \mathbf{0}$, then f has a stationary point at (a, b).

   (b) *True or False:* If (a, b) is a stationary point of a function $f(x, y)$, then (a, b) is a critical point of f.

   (c) *True or False:* If (a, b) is a critical point of a function $f(x, y)$, then (a, b) is a stationary point of f.

   (d) *True or False:* If (a, b) is a saddle point of a function $f(x, y)$, then (a, b) is a critical point of f.

   (e) *True or False:* If $f(x, y)$ has a local maximum at (a, b), then $f(a, b) > f(x, y)$ for every point in some open disk containing (a, b).

   (f) *True or False:* The function $f(x, y) = \pi$ has an absolute minimum at every point in $\mathbb{R}^2$.

   (g) *True or False:* A function $f(x, y)$ can have an absolute maximum and an absolute minimum at every point in $\mathbb{R}^2$.

   (h) *True or False:* If the graph of $f(x, y)$ is a plane and $D_{\mathbf{u}} f(a, b) \neq 0$ for some point (a, b), then f has no extrema.

2. *Examples:* Construct examples of the thing(s) described in the following. Try to find examples that are different than any in the reading.

   (a) A function of two variables with a local minimum at $(1, -3)$.

   (b) A function of two variables with a point of nondifferentiability at $(4, -5)$.

   (c) A function of two variables with a saddle point at $(0, 0)$.

3. How do you find the critical points of a function of two variables, $f(x, y)$? What is the significance of the critical points?

4. What is a stationary point of a function of two variables, $f(x, y)$? What, if anything, is the difference between a critical point and a stationary point of f?

5. What is a saddle point of a function of two variables, $f(x, y)$?

6. Explain why every saddle point is both a stationary point and a critical point.

7. Review the first-derivative test for a function of a single variable, set forth in Theorem 3.8 from Section 3.2. Explain why it might be difficult to design or implement an analogous test for a function of two or more variables.

8. In the proof of Theorem 12.45 we argued that, for a unit vector $\langle \alpha, \beta \rangle$, if $\alpha = 0$, then $\beta \neq 0$. Explain why by finding the only possible values of β.

9. In the proof of Theorem 12.45 we mentioned that if the quantity $AC - B^2 > 0$, then the signs of A and C are the same. Explain why.

Theorem 12.45 is inconclusive when the discriminant, $\det(H_f)$, is zero at a stationary point. In Exercises 10–12 we ask you to illustrate this fact by analyzing three functions of two variables with stationary points at the origin.

10. Show that the function $f(x, y) = x^4 + y^4$ has a stationary point at the origin. Show that the discriminant $\det(H_f(0,0)) = 0$. Explain why f has an absolute minimum at the origin.

11. Show that the function $g(x, y) = -(x^4 + y^4)$ has a stationary point at the origin. Show that the discriminant $\det(H_g(0,0)) = 0$. Explain why g has an absolute maximum at the origin.

12. Show that the function $h(x, y) = x^3 + y^3$ has a stationary point at the origin. Show that the discriminant $\det(H_h(0,0)) = 0$. Show that there are points arbitrarily close to the origin such that $h(x, y) > 0$. Show that there are points arbitrarily close to the origin such that $h(x, y) < 0$. Explain why all this shows that h has a saddle at the origin.

In Exercises 13–16 we ask you to complete some of the details we omitted in Example 4.

13. Show that the unique solution of the system of equations

$$2(x - x_0) - \frac{2a}{c}\left(\frac{d - ax - by}{c} - z_0\right) = 0 \quad \text{and}$$

$$2(y - y_0) - \frac{2b}{c}\left(\frac{d - ax - by}{c} - z_0\right) = 0$$

is

$$\left(\frac{ad - aby_0 - acz_0 + b^2x_0 + c^2x_0}{a^2 + b^2 + c^2},\right.$$

$$\left.\frac{bd - abx_0 - bcz_0 + a^2y_0 + c^2y_0}{a^2 + b^2 + c^2}\right).$$

14. Use Theorem 12.45 to show that the point

$$\left(\frac{ad - aby_0 - acz_0 + b^2x_0 + c^2x_0}{a^2 + b^2 + c^2},\right.$$

$$\left.\frac{bd - abx_0 - bcz_0 + a^2y_0 + c^2y_0}{a^2 + b^2 + c^2}\right)$$

provides an absolute minimum for the function

$$D(x, y) = (x - x_0)^2 + (y - y_0)^2 + \left(\frac{d - ax - by}{c} - z_0\right)^2.$$

15. Show that the minimal value of

$$D(x, y) = (x - x_0)^2 + (y - y_0)^2 + \left(\frac{d - ax - by}{c} - z_0\right)^2,$$

is $\dfrac{(ax_0 + by_0 + cz_0 - d)^2}{a^2 + b^2 + c^2}$ by evaluating

$$D\left(\frac{ad - aby_0 - acz_0 + b^2x_0 + c^2x_0}{a^2 + b^2 + c^2},\right.$$

$$\left.\frac{bd - abx_0 - bcz_0 + a^2y_0 + c^2y_0}{a^2 + b^2 + c^2}\right).$$

16. Let $\mathcal{P}$ be the plane $ax + by + cz = d$, $\mathbf{N} = \langle a, b, c \rangle$ be the normal vector to $\mathcal{P}$, R be a point on $\mathcal{P}$, and P be the point (x_0, y_0, z_0). Show that the distance formula we derived for computing the distance from point P to plane $\mathcal{P}$ in Chapter 10, $\dfrac{|\mathbf{N} \cdot \overrightarrow{RP}|}{\|\mathbf{N}\|}$, is equivalent to the distance formula we derived in Example 4. That is, show that

$$\frac{|\mathbf{N} \cdot \overrightarrow{RP}|}{\|\mathbf{N}\|} = \frac{|ax_0 + by_0 + cz_0 - d|}{\sqrt{a^2 + b^2 + c^2}}.$$

17. Let P be the point (x_1, y_1, z_1), P_0 be the point (x_0, y_0, z_0), $\mathbf{d}$ be the vector $\langle a, b, c \rangle$, and $\mathcal{L}$ be the line parametrized by $\mathbf{r}(t) = P_0 + t\mathbf{d}$. In Chapter 10 we showed that the distance from P to $\mathcal{L}$ is $\dfrac{\|\mathbf{d} \times \overrightarrow{P_0P}\|}{\|\mathbf{d}\|}$. Explain how to derive this distance formula by minimizing a function of a single variable.

18. Show that $f(x, y) = \sqrt{x^2 + y^2}$ has a critical point at $(0, 0)$. Explain why f has an absolute minimum at $(0, 0)$ and why you cannot use Theorem 12.45 to show this.

19. Every function of a single variable, $f(x) = x^n$, where n is a positive integer greater than 1 will have a critical point at $x = 0$. For which values of n will there be a relative minimum at $x = 0$? For which values of n will there be an inflection point at $x = 0$? Are there any other possibilities for the behavior of the function at $x = 0$?

20. Consider a function of two variables, $f(x, y) = x^m y^n$, where m and n are positive integers. What conditions do m and n have to satisfy in order for f to have a relative minimum at the origin? What conditions do m and n have to satisfy in order for f to have a saddle point at the origin? Are there any other possibilities for the behavior of the function at the origin? It may be helpful to refer to your answers to Exercise 19 before you answer this question.

Skills

In Exercises 21–26, find the discriminant of the given function.

21. $f(x, y) = e^{2x} \cos y$

22. $g(x, y) = e^{-3x} \sin(2y)$

23. $f(s, t) = s^3 e^{t/2}$

24. $g(s, t) = s^2 t e^{-st^2}$

25. $f(\theta, \phi) = \cos \theta \sin \phi$

26. $g(\theta, \phi) = \cos \theta \cos \phi$

In Exercises 27–30, use the result from Example 4 to find the distance from the point P to the given plane.

27. $P = (3, 5, -2)$, $x - 3y + 4z = 8$

28. $P = (-1, 3, 5)$, $5x - 2y - 7z + 2 = 0$

29. $P = (4, 0, -3)$, $12y - 5z = 7$

30. $P = (-3, 2, 6)$, $x = 3y + 5z$

In Exercises 31–52, find the relative maxima, relative minima, and saddle points for the given functions. Determine whether the function has an absolute maximum or absolute minimum as well.

31. $f(x, y) = 3x^2 + 3x + 6y^2 - 7$

32. $g(x, y) = 4x^2 - 8xy + y^2 + 3y + 5$

33. $f(x, y) = x^3 + y^3 - 12x - 3y + 15$

34. $g(x, y) = x^3 + 6x^2 + 6y^2 - 4$

35. $f(x, y) = x^3 - 12xy + y^3$

36. $g(x, y) = x^4 - 8x^2 y + y^4$

37. $f(x, y) = 2x^2 - 2xy + 4y^2 + 3x + 9y - 5$

38. $g(x, y) = 8x^3 - 3xy^2 + 2y^3 - 4x - 2$

39. $f(x, y) = 8xy + \dfrac{1}{x} + \dfrac{1}{y}$

40. $g(x, y) = e^x \cos y$

41. $f(x, y) = x \sin y$

42. $g(x, y) = x^2 + \sin y - 3$

43. $f(x, y) = e^{x^2} \cos y$

44. $g(x, y) = 5x - 4y^2 + x \ln y$, $y > 0$

45. $f(x, y) = e^{xy}$

46. $g(x, y) = \dfrac{1}{x^2 + y^2}$

47. $f(x, y) = \dfrac{1}{x^2 + y^2 - 1}$

48. $g(x, y) = \dfrac{1}{x^2 + y^2 + 1}$

49. $f(x, y) = x^2 y^2$

50. $g(x, y) = xy^2$

51. $f(x, y) = x^3 y^2$

52. $g(x, y) = x^3 y$

Applications

53. Bob plans to build a rectangular wooden box. The plywood he will use for the bottom costs $2 per square foot. The pine for the sides costs $5 per square foot, and the oak for the top costs $7 per square foot. What should the dimensions of the box be to minimize the cost of a box with a volume of 20 cubic feet?

54. Bob plans to build a rectangular wooden box. The wood he will use for the bottom costs $A per square foot. The wood for the sides costs $B per square foot, and the wood for the top costs $C per square foot. What should the dimensions of the box be to minimize the cost of a box with a volume of V cubic feet?

Proofs

55. Prove that for a function $f(x, y)$ with continuous second-order partial derivatives, the second directional derivative in the direction of the unit vector $\mathbf{u} = \langle a, b \rangle$ is given by

$$D_{\mathbf{u}}^2 f(x, y) = a^2 f_{xx}(x, y) + 2ab f_{xx}(x, y) + b^2 f_{yy}(x, y).$$

56. Prove Theorem 12.42. That is, show that if $f(x, y)$ has a local extremum at (x_0, y_0), then (x_0, y_0) is a critical point of f. (*Hint: Consider the functions of a single variable, $g(x) = f(x, y_0)$ and $h(y) = f(x_0, y)$.*)

57. Prove that if the square of the distance from the point (x_0, y_0) to the line with equation $\alpha x + \beta y = \gamma$ is minimized, then the distance from (x_0, y_0) to the line is also minimized.

58. Prove that if the square of the distance from the point (x_0, y_0) to the curve with equation $g(x, y) = 0$ is minimized, then the distance from (x_0, y_0) to the curve is also minimized.

Thinking Forward

▶ *A function without extrema:* What is the domain of the function $f(x, y) = 3x - 4y$? Explain why $f(x, y)$ has no local extrema on its domain.

▶ *Extrema on a closed and bounded set:* How could we find the maximum and minimum values of the function $f(x, y) = 3x - 4y$ if we restrict the domain to the disk $x^2 + y^2 \leq 1$?

12.7   LAGRANGE MULTIPLIERS

▶   Constrained extrema

▶   The method of Lagrange multipliers

▶   The Extreme Value Theorem for functions of two or more variables

Constrained Extrema

In Example 3 of the previous section, we discussed how to construct a box with five sides made of wood and one side made of glass. We saw that if the wood costs \$5 per square foot and the glass costs \$10 per square foot, the cost of constructing the box is $C(x, y) = 10xy + 10xz + 15yz$. In Section 12.6, we discussed how to optimize functions of two independent variables. Fortunately, we were also told that the volume of the box was to be 12 cubic feet, giving the **constraint** equation $12 = xyz$. This constraint equation allowed us to eliminate one of the variables. In the example, we showed that the values $x = 3$ and $y = z = 2$ together minimize the function subject to the constraint. Now, consider the gradient of the function $C(x, y, z) = 10xy + 10xz + 15yz$ at the point $(3, 2, 2)$. Since

$$\nabla C(x, y, z) = (10y + 10z)\mathbf{i} + (10x + 15z)\mathbf{j} + (10x + 15y)\mathbf{k},$$

we have $\nabla C(3, 2, 2) = 40\mathbf{i} + 60\mathbf{j} + 60\mathbf{k}$. In addition, if we write the constraint equation in the form $g(x, y, z) = xyz - 12 = 0$, we may also compute the gradient of this function at the point $(3, 2, 2)$. When we do, we obtain

$$\nabla g(x, y, z) = yz\mathbf{i} + xz\mathbf{j} + xy\mathbf{k} \quad \text{and} \quad \nabla g(3, 2, 2) = 4\mathbf{i} + 6\mathbf{j} + 6\mathbf{k}.$$

Note that $\nabla C(3, 2, 2)$ is a multiple of $\nabla g(3, 2, 2)$. This is not a coincidence; it is a consequence of the main theorem of this section, presented next. The technique we discuss here is called the **method of Lagrange multipliers**, or, more simply, **Lagrange's method**. This technique tells us that the gradient of a function of several variables is always a scalar multiple of the gradient of the constraint equation at a point that optimizes the function.

THEOREM 12.46

The Method of Lagrange Multipliers

Let f and g be functions with continuous first-order partial derivatives. If $f(x, y)$ has a relative extremum at a point (x_0, y_0) subject to the constraint $g(x, y) = 0$, then $\nabla f(x_0, y_0)$ and $\nabla g(x_0, y_0)$ are parallel. Thus, if $\nabla g(x_0, y_0) \neq \mathbf{0}$,

$$\nabla f(x_0, y_0) = \lambda \nabla g(x_0, y_0)$$

for some scalar λ.

Although we have stated Theorem 12.46 in terms of functions of two variables, it is also true if f and g are both functions of three or more variables.

DEFINITION 12.47

Lagrange Multiplier

The scalar λ mentioned in Theorem 12.46 is called a **Lagrange multiplier**.

The following figure shows a schematic illustrating Theorem 12.46.

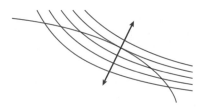

The blue curves are level curves for a function $f(x, y)$. The red curve is the graph of a constraint equation $g(x, y) = 0$. At an extreme of f subject to the constraint, the gradients of both f and g are parallel. These are the green vectors shown in the figure.

Before we prove Theorem 12.46, we again consider our earlier example, to see how Lagrange's method works. We optimize the function $C(x, y, z) = 10xy + 10xz + 15yz$ subject to the constraint $g(x, y, z) = xyz - 12 = 0$. We've seen that

$$\nabla C(x, y, z) = (10y + 10z)\mathbf{i} + (10x + 15z)\mathbf{j} + (10x + 15y)\mathbf{k} \quad \text{and} \quad \nabla g(x, y, z) = yz\mathbf{i} + xz\mathbf{j} + xy\mathbf{k}.$$

By Theorem 12.46, we must then have

$$\nabla C(x, y, z) = \lambda \nabla g(x, y, z)$$

for some constant λ. This equation gives rise to the following system of three equations with four unknowns:

$$10y + 10z = \lambda yz,$$
$$10x + 15z = \lambda xz, \quad \text{and}$$
$$10x + 15y = \lambda xy.$$

We are unable find a unique solution for such a system. We need a fourth equation relating the variables. Fortunately we have the constraint equation $xyz = 12$. The resulting system of four equations in four unknowns is nonlinear, since each of the original three equations has a term that is the product of three variables. Solving a nonlinear system can require numerical tricks. Indeed, many nonlinear systems of equations are impossible to solve exactly. Here, a little perseverance does pay off. It is often helpful to solve each of the equations involving λ for λ and equate the results. When we do this for the preceding system, we obtain

$$\lambda = \frac{10y + 10z}{yz} = \frac{10x + 15z}{xz} = \frac{10x + 15y}{xy},$$

or equivalently,

$$\frac{10}{z} + \frac{10}{y} = \frac{10}{z} + \frac{15}{x} = \frac{10}{y} + \frac{15}{x}.$$

Since $\frac{10}{z} + \frac{15}{x} = \frac{10}{y} + \frac{15}{x}$, we must have $y = z$, and since $\frac{10}{z} + \frac{10}{y} = \frac{10}{z} + \frac{15}{x}$, we must also have $y = z = \frac{2}{3}x$. Now, because $xyz = 12$, it follows that $x = 3$ and $y = z = 2$, the same result we obtained in Section 12.6. Note that we did not solve the system for λ: The actual value of λ is not significant.

We are now ready to prove Theorem 12.46.

Proof. Let $z = f(x, y)$ be the function of two variables that we wish to optimize subject to the constraint $g(x, y) = 0$. Assume that $\nabla g(x, y) \neq \mathbf{0}$ for any point in the domain of g, let C be the curve determined by $g(x, y) = 0$, and let $\mathbf{r}(t) = \langle x(t), y(t) \rangle$ be a parametrization for C. Along the curve C, z may be expressed as a function of the single variable t: $z = f(x(t), y(t))$. By the chain rule, the derivative $\frac{dz}{dt}$ is given by

$$\frac{dz}{dt} = \frac{\partial f}{\partial x}\frac{dx}{dt} + \frac{\partial f}{\partial y}\frac{dy}{dt},$$

or equivalently,

$$\frac{dz}{dt} = f_x(x, y)x'(t) + f_y(x, y)y'(t).$$

But note that $\nabla f(x, y) = f_x(x, y)\mathbf{i} + f_y(x, y)\mathbf{j}$ and $\mathbf{r}'(t) = x'(t)\mathbf{i} + y'(t)\mathbf{j}$ is the tangent function to the curve C. Now, if t_0 is a point that optimizes f subject to the constraint $g(x, y) = 0$, then the derivative $\left.\dfrac{dz}{dt}\right|_{t=t_0} = 0$. So, at t_0, the dot product $\nabla f(x(t_0), y(t_0)) \cdot \mathbf{r}'(t_0) = 0$. Thus, the vectors $\nabla f(x(t_0), y(t_0))$ and $\mathbf{r}'(t_0)$ are orthogonal. But by Theorem 12.38, $\nabla g(x(t_0), y(t_0))$ is also orthogonal to C at $(x(t_0), y(t_0))$. Therefore, $\nabla f(x(t_0), y(t_0))$ and $\nabla g(x(t_0), y(t_0))$ are parallel, since they are orthogonal to the same vector. ■

Optimizing a Function on a Closed and Bounded Set

Recall that the Extreme Value Theorem tells us that every continuous function on a closed interval will have both a maximum value and a minimum value on the interval. We have the following analogous theorem for functions of two variables:

THEOREM 12.48

The Extreme Value Theorem for a Function of Two Variables

Let f be a continuous function of two variables defined on the closed and bounded set S. Then there exist points (x_M, y_M) and (x_m, y_m) in S such that $f(x_M, y_M)$ is the maximum value of f on S and $f(x_m, y_m)$ is the minimum value of f on S.

The proof of Theorem 12.48 lies outside the scope of this text. An analogous theorem also holds when f is a function of more than two variables, and we will use that theorem when appropriate.

Given a continuous function f defined on a closed and bounded set S, we may use the following outline to find those points (x_M, y_M) and (x_m, y_m) which maximize and minimize f on S:

1. Find the stationary points and other critical points of f.
2. Select only those critical points that lie in S.
3. Evaluate the function at each of the critical points found in step 2.
4. Use the method of Lagrange multipliers to locate the points on the boundary of S that maximize and minimize f.
5. Evaluate the function at each of the critical points on the boundary of S.
6. Use the extrema from steps 3 and 5 to find the maximum and minimum values of the function f on S.

Optimizing a Function with Two Constraints

We may extend the ideas of this section to optimize a function of three variables, $w = f(x, y, z)$, subject to two constraints. In this context, the constraints, $g(x, y, z) = 0$ and $h(x, y, z) = 0$, define surfaces in $\mathbb{R}^3$. Assuming that these surfaces intersect in a curve, we would be attempting to find the maximum and minimum values of f on this curve of intersection. To do this, we attempt to solve the system given by

$$\nabla f(x_0, y_0, z_0) = \lambda \nabla h(x_0, y_0, z_0) + \mu \nabla h(x_0, y_0, z_0),$$

where λ and μ are both Lagrange multipliers. (See Example 5.)

Examples and Explorations

EXAMPLE 1   Optimizing a function subject to a constraint

Find the maximum and minimum of the function $f(x, y) = xy$ subject to the constraint $x^2 + 4y^2 = 16$.

SOLUTION

We could solve the equation $x^2 + 4y^2 = 16$ for either x or y and use the resulting equation to rewrite the function f in terms of a single variable. However, we will use Lagrange's method to optimize the function. We start by writing the constraint in the form $g(x, y) = x^2 + 4y^2 - 16 = 0$ and find

$$\nabla f(x, y) = y\mathbf{i} + x\mathbf{j} \quad \text{and} \quad \nabla g(x, y) = 2x\mathbf{i} + 8y\mathbf{j}.$$

We now set up the system of equations determined by $\nabla f(x, y) = \lambda \nabla g(x, y)$, where λ is a constant. Here we have

$$y = 2\lambda x \quad \text{and} \quad x = 8\lambda y.$$

Our third equation relating the variables is the constraint equation $x^2 + 4y^2 = 16$. We solve each of the two equations for λ and equate the results:

$$\frac{y}{2x} = \lambda = \frac{x}{8y}.$$

We may now deduce that $x^2 = 4y^2$. Using this relationship in the constraint equation, we find that $2x^2 = 16$. Therefore, $x = \pm 2\sqrt{2}$, and from this we have $y = \pm\sqrt{2}$. Thus, we have the four pairs of coordinates

$$(2\sqrt{2}, \sqrt{2}), \ (2\sqrt{2}, -\sqrt{2}), \ (-2\sqrt{2}, \sqrt{2}), \quad \text{and} \quad (-2\sqrt{2}, -\sqrt{2}).$$

Finally, we evaluate the function $f(x, y)$ at each of these points and obtain

$$f(2\sqrt{2}, \sqrt{2}) = 4, \ f(2\sqrt{2}, -\sqrt{2}) = -4, \ f(-2\sqrt{2}, \sqrt{2}) = -4, \quad \text{and} \quad f(-2\sqrt{2}, -\sqrt{2}) = 4.$$

The maximum value of f subject to the constraint is 4. This maximum occurs at both $(2\sqrt{2}, \sqrt{2})$ and $(-2\sqrt{2}, -\sqrt{2})$. The minimum value of the function is -4. The minimum occurs at both $(-2\sqrt{2}, \sqrt{2})$ and $(2\sqrt{2}, -\sqrt{2})$.   □

**CHECKING
THE ANSWER**

We may check our answers by evaluating the gradients of f and g at the points we just found and showing that they are indeed parallel. For example, since

$$\nabla f(x, y) = y\mathbf{i} + x\mathbf{j} \quad \text{and} \quad \nabla g(x, y) = 2x\mathbf{i} + 8y\mathbf{j},$$

we have

$$\nabla f(2\sqrt{2}, \sqrt{2}) = \sqrt{2}\mathbf{i} + 2\sqrt{2}\mathbf{j} \quad \text{and} \quad \nabla g(2\sqrt{2}, \sqrt{2}) = 4\sqrt{2}\mathbf{i} + 8\sqrt{2}\mathbf{j}.$$

Note that $\nabla f(2\sqrt{2}, \sqrt{2}) = \frac{1}{4}\nabla g(2\sqrt{2}, \sqrt{2})$. The gradients are indeed parallel when $(x, y) = (2\sqrt{2}, \sqrt{2})$. Similarly, at each of the other three points we found in Example 1, we have $\nabla f(x_0, y_0) = \frac{1}{4}\nabla g(x_0, y_0)$.

EXAMPLE 2   Analyzing the result of an optimization problem

Analyze the result of Example 1 by graphing the level curves of the function $f(x, y) = xy$ corresponding to the constrained extrema, along with the constraint equation $x^2 + 4y^2 = 16$.

SOLUTION

In Example 1 we found that the maximum and minimum of $f(x, y) = xy$ subject to the constraint $x^2 + 4y^2 = 16$ were 4 and -4, respectively. We graph the level curves $xy = 4$ and $xy = -4$ along with the following ellipse, which is the graph of the equation $x^2 + 4y^2 = 16$:

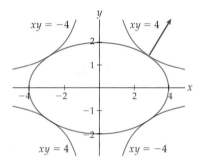

Note that the level curves and the ellipse are tangent at the maxima and minima. In the first and third quadrants, the level curve is the graph of $xy = 4$. This graph intersects the ellipse at $(2\sqrt{2}, \sqrt{2})$ and $(-2\sqrt{2}, -\sqrt{2})$. Since the curves are tangent at these two points, the gradients of the two curves would be orthogonal to the curves at those points. Therefore, the gradients would be parallel at those points. We show only one of these gradients in the preceding figure, because they overlap.

The figure that follows at the left shows the ellipse, together with four additional level curves of the function f. Note that the ellipse is tangent only to the level curves $f(x, y) = 4$ and $f(x, y) = -4$.

$x^2 + 4y^2 = 16$, along with the
level curves $f(x, y) = \pm 2, \pm 4, \pm 6$

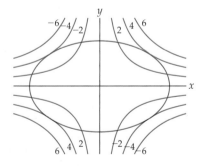

The graph of $f(x, y) = xy$
restricted to $x^2 + 4y^2 = 16$

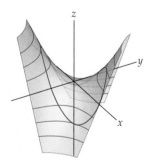

The figure at the right shows the graph of the surface defined by $f(x, y) = xy$, together with the curve on that surface resulting from the constraint $x^2 + 4y^2 = 16$. On this graph, you see the two maxima and two minima that we found.   □

EXAMPLE 3   Understanding the difference between a function and a constraint

Find the maximum and minimum of the function $f(x, y) = x^2 + 4y^2 - 16$ subject to the constraint $xy = 0$.

SOLUTION

Here we have reversed the roles of the function and the constraint given in Example 1. We could proceed as outlined before, but note that since we now have the constraint $xy = 0$, either x or y must be zero. If $x = 0$, then $f(0, y) = 4y^2 - 16$, which has minimum value -16 when $y = 0$, but no maximum value, since the function increases without bound as $y \to \pm\infty$. Similarly, if $y = 0$, then $f(x, 0) = x^2 - 16$. Again, there is a minimum of -16 when $x = 0$ and no maximum value. Therefore, subject to the constraint, the function $f(x, y) = x^2 + 4y^2 - 16$ has a minimum of $-16 = f(0, 0)$ and no maximum value.

Our point here is that if you confuse the function and the constraint, you may obtain vastly different results. Make sure that you understand which function you are trying to optimize and which function is your constraint. □

EXAMPLE 4   Optimizing a function on a closed and bounded domain

Find the maximum and minimum of the function $f(x, y) = x^3 + 9x^2 + 6y^2$ on the region $x^2 + y^2 \le 9$.

SOLUTION

Since f is a polynomial, the only critical points of f are stationary points. We have $\nabla f(x, y) = (3x^2 + 18x)\mathbf{i} + 12y\mathbf{j}$. The stationary points of f will satisfy the equations

$$3x(x + 6) = 0 \quad \text{and} \quad 12y = 0.$$

The only points satisfying these two equations are $(0, 0)$ and $(-6, 0)$. Of these points, only $(0, 0)$ satisfies the inequality $x^2 + y^2 \le 9$, so it is a point where the maximum or minimum may occur. We ignore the other point, since it is outside the domain.

We now consider points on the boundary of the region. These points are given by the constraint $x^2 + y^2 = 9$. We let $g(x, y) = x^2 + y^2 - 9 = 0$. Now, $\nabla g(x, y) = 2x\mathbf{i} + 2y\mathbf{j}$. By Lagrange's method, we have the system of equations $3x^2 + 18x = 2\lambda x$ and $12y = 2\lambda y$. From the second of these equations, we see that either $y = 0$ or $\lambda = 6$. If $y = 0$, then by the constraint equation, $x = \pm 3$. Using $\lambda = 6$ in the first of the equations, we have $3x^2 + 6x = 0$, so either $x = -2$ or $x = 0$. Again using the constraint when $x = -2$, we obtain $y = \pm\sqrt{5}$, and when $x = 0$, we have $y = \pm 3$.

We now have a list of seven points where the maximum and minimum of the function on the given domain may occur:

$$(0, 0), \ (3, 0), \ (-3, 0), \ (-2, \sqrt{5}), \ (-2, -\sqrt{5}), \ (0, 3), \quad \text{and} \quad (0, -3).$$

When we evaluate the function at these seven points, we see that

$$f(0, 0) = 0, \ f(3, 0) = 108, \ f(-3, 0) = 54, \ f(-2, \pm\sqrt{5}) = 58, \quad \text{and} \quad f(0, \pm 3) = 54.$$

We see that the maximum of f on the region is 108 and the minimum of the function on the region is 0. Note that it was *not* necessary to use the second-derivative test to analyze f at $(0, 0)$, since we were looking for the maximum and minimum of the function on the given region and we had to analyze the behavior of f only at the finite number of points we obtained in the previous steps. When we evaluated the function at the seven points we found, the minimum value of the function and the maximum value of the function at those points had to give the two values we wanted.

Although we successfully answered the question posed in this example, in Exercise 13 you will show that the function $f(x, y) = x^3 + 9x^2 + 6y^2$ with domain $\mathbb{R}^2$ has only a relative minimum at $(0, 0)$, not an absolute minimum, and that f has a saddle point at the other stationary point, $(-6, 0)$. □

EXAMPLE 5 Optimizing a function of three variables with two constraints

Find the points in $\mathbb{R}^3$ that are farthest from the origin and that lie on both the cylinder $y^2 + z^2 = 1$ and the plane $x + 2y + 3z = 0$.

SOLUTION

As usual, it is sufficient (and easier) to maximize the square of the distance from the origin, $D(x, y, z) = x^2 + y^2 + z^2$, subject to the constraints $g(x, y, z) = y^2 + z^2 - 1 = 0$ and $h(x, y, z) = x + 2y + 3z = 0$. As the following figure illustrates, geometrically we are looking for the point(s) on the ellipse formed by the intersection of the cylinder and plane farthest from the origin.

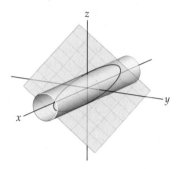

The gradients of the three functions we are interested in are

$$\nabla D(x, y, z) = 2x\mathbf{i} + 2y\mathbf{j} + 2z\mathbf{k},$$
$$\nabla g(x, y, z) = 2y\mathbf{j} + 2z\mathbf{k}, \quad \text{and}$$
$$\nabla h(x, y, z) = \mathbf{i} + 2\mathbf{j} + 3\mathbf{k}.$$

We wish to solve the system $\nabla D(x, y, z) = \lambda \nabla g(x, y, z) + \mu \nabla h(x, y, z)$ subject to $g(x, y, z) = 0$ and $h(x, y, z) = 0$. This leads to the following system of five equations:

$$2x = 0 + \mu,$$
$$2y = \lambda \cdot 2y + \mu \cdot 2,$$
$$2z = \lambda \cdot 2z + \mu \cdot 3,$$
$$y^2 + z^2 = 1, \text{ and}$$
$$x + 2y + 3z = 0.$$

Since $\mu = 2x$, we may eliminate μ to obtain a system with only four equations in four variables:

$$y = \lambda y + 2x, \quad z = \lambda z + 3x, \quad y^2 + z^2 = 1, \quad x + 2y + 3z = 0.$$

A little thought tells us that if either $y = 0$ or $z = 0$, the system does not have a solution. So we may solve the first two of these equations for λ and equate the results. We obtain

$$\frac{y - 2x}{y} = \frac{z - 3x}{z},$$

or equivalently, $x(2z - 3y) = 0$. So, either $x = 0$ or $z = \frac{3}{2}y$.

If $x = 0$, then when we solve the equations $y^2 + z^2 = 1$ and $2y + 3z = 0$ for y and z, we have $(x, y, z) = \left(0, \frac{3}{\sqrt{13}}, -\frac{2}{\sqrt{13}}\right)$ or $(x, y, z) = \left(0, -\frac{3}{\sqrt{13}}, \frac{2}{\sqrt{13}}\right)$. These two points are both 1 unit from the origin.

Now, using $z = \frac{3}{2}y$ we solve the system $y^2 + \left(\frac{3}{2}y\right)^2 = 1$ and $x + 2y + 3 \cdot \frac{3}{2}y = 0$ and obtain the points $\left(-\sqrt{13}, \frac{2}{\sqrt{13}}, \frac{3}{\sqrt{13}}\right)$ and $\left(\sqrt{13}, -\frac{2}{\sqrt{13}}, -\frac{3}{\sqrt{13}}\right)$. These two points are each $\sqrt{14}$ units from the origin. They are the two points on the cylinder and plane that are farthest from the origin. The other two points are the points on the cylinder and plane that are closest to the origin.   ☐

TEST YOUR UNDERSTANDING

▸ What is a constraint equation? Does every continuous function have a maximum value and minimum value subject to a constraint?

▸ What is a Lagrange multiplier? What is the method of Lagrange multipliers? If we try to optimize the function of three variables, $w = f(x, y, z)$, subject to the constraint $g(x, y, z) = 0$, how many equations do we obtain when we use the method of Lagrange multipliers? How many variables are in the system?

▸ What methods do you know to find the maxima and minima of a function of two or more variables subject to a constraint?

▸ What is the Extreme Value Theorem for a function of two variables? Three variables? What is the relationship between the Extreme Value Theorem for a function of a single variable to the Extreme Value Theorem for a function of two or more variables?

▸ Outline the steps you should take to find the maximum value and minimum value of a function of two variables on a closed and bounded set. Would the steps change if you had a function of three variables on a closed and bounded set?

EXERCISES 12.7

Thinking Back

▸ *Optimizing a function of two variables subject to a constraint:* If you wish to find the maximum and minimum of the function $f(x, y) = xy$ subject to the constraint $x^2 + 4y^2 = 16$, you may eliminate one variable of f by solving the constraint equation for either x or y and rewriting f in terms of a single variable. Do this and then use the techniques of Chapter 3 to find the maximum and minimum of the resulting function.

▸ *Optimizing a function of three variables subject to a constraint:* If you wish to find the maximum and minimum of the function $f(x, y, z) = xyz$ subject to the constraint $x^2 + 4y^2 + 9z^2 = 16$, you may eliminate one variable of f by solving the constraint equation for x, y, or z and rewriting f in terms of two variables. Do this and then use the techniques of Section 12.6 to find the maximum and minimum of the resulting function.

Concepts

0. *Problem Zero:* Read the section and make your own summary of the material.

1. *True/False:* Determine whether each of the statements that follow is true or false. If a statement is true, explain why. If a statement is false, provide a counterexample.

(a) *True or False:* Every function of two variables $f(x, y)$ subject to a constraint $g(x, y) = 0$ has both a maximum and a minimum value.

(b) *True or False:* The function $f(x, y, z) = \frac{xy}{y+z}$ has both a minimum and a maximum on the region defined by $x^2 + y^2 + z^2 \leq 1$.

(c) *True or False:* The function $f(x, y, z) = e^{\sin xy} \cos(xy+z)$ has both a minimum and a maximum on the region defined by $x^2 + y^2 + z^2 \leq 1$.

(d) *True or False:* The function $f(x, y) = xy + 2y - 5xy^2$ has both a minimum and a maximum on the region defined by $x^2 + y^2 \geq 1$.

(e) *True or False:* The function $y = x^2$ has both a maximum and a minimum for $x \in (-2, 2)$.

(f) *True or False:* If $f(x_0, y_0)$ is the maximum value of f on a region $\mathcal{R}$, then $f(x_0, y_0)$ is the maximum value of f.

(g) *True or False:* If $f(x_0, y_0)$ is the maximum value of f on its domain, then $f(x_0, y_0)$ is the maximum value of f on every closed and bounded region $\mathcal{R}$.

(h) *True or False:* If $f(x, y)$ is a continuous function and if m and M are the minimum and maximum values of f on its domain, then $m \leq f(x, y) \leq M$ on every closed and bounded region $\mathcal{R}$.

2. *Examples:* Construct examples of the thing(s) described in the following. Try to find examples that are different than any in the reading.

   (a) A function of two variables $f(x, y)$, along with a constraint $g(x, y) = 0$, that could be optimized with the method of Lagrange multipliers.

   (b) A function of two variables $f(x, y)$, along with a constraint $g(x, y) = 0$, that could be optimized by eliminating one of the variables.

   (c) A function of two variables $f(x, y)$ that has neither a maximum nor a minimum value on the set $x^2 + y^2 \leq 4$.

3. What is the distinction between the function $z = f(x, y)$ and the constraint equation $g(x, y) = 0$ when we use the method of Lagrange multipliers?

4. Outline a method for optimizing a function of two variables, $z = f(x, y)$.

5. Describe two methods for optimizing a function $z = f(x, y)$ subject to a constraint.

6. What is meant by a "closed" subset of $\mathbb{R}^2$? What is meant by a "bounded" subset of $\mathbb{R}^2$?

7. Give an example of each of the following:

   (a) A subset of $\mathbb{R}^2$ that is neither closed nor bounded.

   (b) A subset of $\mathbb{R}^2$ that is closed but not bounded.

   (c) A subset of $\mathbb{R}^2$ that is not closed but is bounded.

   (d) A subset of $\mathbb{R}^2$ that is closed and bounded.

8. What is meant by a "closed" subset of $\mathbb{R}^3$? What is meant by a "bounded" subset of $\mathbb{R}^3$?

9. Give an example of each of the following:

   (a) A subset of $\mathbb{R}^3$ that is neither closed nor bounded.

   (b) A subset of $\mathbb{R}^3$ that is closed but not bounded.

   (c) A subset of $\mathbb{R}^3$ that is not closed but is bounded.

   (d) A subset of $\mathbb{R}^3$ that is closed and bounded.

10. Outline the steps required to find the minimum and maximum of a function of two variables on a closed and bounded set $\mathcal{R}$.

11. Given a function of three variables, $w = f(x, y, z)$, and a constraint equation $g(x, y, z) = 0$, how many equations would we obtain if we tried to optimize f by the method of Lagrange multipliers?

12. Given a function of n variables, $z = f(x_1, x_2, \ldots, x_n)$, and a constraint equation, $g(x_1, x_2, \ldots, x_n) = 0$, how many equations would we obtain if we tried to optimize f by the method of Lagrange multipliers?

13. In Example 4 we found that the function $f(x, y) = x^3 + 9x^2 + 6y^2$ has stationary points at $(0, 0)$ and $(-6, 0)$.

   (a) Use the second-derivative test to show that f has a saddle point at $(-6, 0)$.

   (b) Use the second-derivative test to show that f has a relative minimum at $(0, 0)$.

   (c) Use the value of $f(-10, 0)$ to argue that f has a relative minimum at $(0, 0)$, and not an absolute minimum, without using the second-derivative test.

In Exercises 14–17, by considering the function $f(x, y) = x^2 y$ subject to the constraint $x + y = 0$, you will explore a situation in which the method of Lagrange multipliers does not provide an extremum of a function.

14. Show that the only point given by the method of Lagrange multipliers for the function $f(x, y) = x^2 y$ subject to the constraint $x + y = 0$ is $(0, 0)$.

15. Explain why $(0, 0)$ is *not* an extremum of $f(x, y) = x^2 y$ subject to the constraint $x + y = 0$.

16. Why does the method of Lagrange multipliers fail with this function?

17. Optimize $f(x, y) = x^2 y$ subject to the constraint $ax + by = 0$ for nonzero constants a and b. Are there any nonzero values of a and b for which the method of Lagrange multipliers succeeds?

18. Explain how you could use the method of Lagrange multipliers to find the extrema of a function of two variables, $f(x, y)$, subject to the constraint that (x, y) is on the boundary of the rectangle $\mathcal{R}$ defined by $a \leq x \leq b$ and $c \leq y \leq d$.

19. Explain the steps you would take to find the extrema of a function of two variables, $f(x, y)$, if (x, y) is a point in the rectangle $\mathcal{R}$ defined by $a \leq x \leq b$ and $c \leq y \leq d$.

20. Explain how you could use the method of Lagrange multipliers to find the extrema of a function of two variables, $f(x, y)$, subject to the constraint that (x, y) is a point on the boundary of a triangle $\mathcal{T}$ in the xy-plane.

21. Explain the steps you would take to find the extrema of a function of two variables, $f(x, y)$, if (x, y) is a point in a triangle $\mathcal{T}$ in the xy-plane.

22. When you use the method of Lagrange multipliers to find the maximum and minimum of $f(x, y) = x + y$ subject to the constraint $xy = 1$, you obtain two points. Is there a relative maximum at one of the points and a relative minimum at the other? Which is which?

23. Let $f(x, y)$ be a differentiable function such that $\nabla f(x, y) \neq \mathbf{0}$ for every point in the domain of f, and let $\mathcal{R}$ be a closed, bounded subset of $\mathbb{R}^2$. Explain why the maximum and minimum of f restricted to $\mathcal{R}$ occur on the boundary of $\mathcal{R}$.

Skills

In Exercises 24–32, find the maximum and minimum of the function f subject to the given constraint. In each case explain why the maximum and minimum must both exist.

24. $f(x, y) = x + y$ when $x^2 + y^2 = 16$

25. $f(x, y) = x + y$ when $x^2 + 4y^2 = 16$

26. $f(x, y) = xy$ when $x^2 + y^2 = 16$

27. $f(x, y) = xy$ when $x^2 + 4y^2 = 16$

28. $f(x, y, z) = x + y + z$ when $x^2 + y^2 + z^2 = 9$

29. $f(x, y, z) = x + y + z$ when $x^2 + 4y^2 + 16z^2 = 64$

30. $f(x, y, z) = xyz$ when $x^2 + y^2 + z^2 = 9$

31. $f(x, y, z) = xyz$ when $x^2 + 4y^2 + 16z^2 = 64$

32. $f(x, y, z) = xy$ when $x^2 + y^2 + z^2 = 9$

In Exercises 33–38, find the point on the given curve closest to the specified point. Recall that if you minimize the square of the distance, you have minimized the distance as well.

33. $(0, 0)$ and $x^3 + y^3 = 1$

34. $(0, 0)$ and $x^2 + y^3 = 1$

35. $(0, 0)$ and $\sqrt{x} + \sqrt{y} = 1$

36. $(1, 1)$ and $\sqrt{x} + \sqrt{y} = 2$

37. $(0, 0)$ and $x^n + y^n = 1$ for n a positive, odd integer

38. $(0, 0)$ and $x^n + y^n = 1$ where $n > 2$ is a positive, even integer

In Exercises 39–43, find the point on the given surface closest to the specified point.

39. $xyz = 1$ and $(0, 0, 0)$

40. $x + 2y - 3z = 7$ and $(0, 0, 0)$

41. $ax + by + cz = d$ and $(0, 0, 0)$

42. $ax + by + cz = d$ and (α, β, γ)

43. $x + 2y - 3z = 4$ and $(-1, 5, 3)$

In Exercises 44–49, find the maximum and minimum of the given function on the specified region. Also, give the points where the maximum and minimum occur.

44. $f(x, y) = 3x^2 - 5y^2$ on the circular region $x^2 + y^2 \leq 1$

45. $f(x, y) = 3x^2 + 5y^2$ on the square with vertices $(1, 1)$, $(-1, 1)$, $(-1, -1)$, and $(1, -1)$

46. $f(x, y) = x^2 + y$ on the circular region $x^2 + y^2 \leq 4$

47. $f(x, y) = x^2 + y$ on the square with vertices $(1, 0)$, $(0, 1)$, $(-1, 0)$, and $(0, -1)$

48. $f(x, y) = \dfrac{x}{y}$ on the circular region $x^2 + (y - 2)^2 \leq 1$

49. $f(x, y) = \dfrac{x}{y}$ on the rectangle given by $-1 \leq x \leq 1$, $1 \leq y \leq 4$

50. Find the dimensions of the rectangular solid with maximum volume such that all sides are parallel to the coordinate planes; one vertex is at the origin; and the vertex diagonally opposite is in the first octant and on the plane with equation $2x + 5y + 6z = 30$.

51. Find the dimensions of the rectangular solid with maximum volume such that all sides are parallel to the coordinate planes; one vertex is at the origin; and the vertex diagonally opposite is in the first octant and on the plane with equation $ax + by + cz = d$, where a, b, c, and d are positive real numbers.

52. Find the points in $\mathbb{R}^3$ closest to the origin and that lie on the cylinder $x^2 + y^2 = 4$ and the plane $x - 3y + 2z = 6$.

53. Find the points in $\mathbb{R}^3$ closest to the origin and that lie on the cone $z^2 = x^2 + y^2$ and the plane $x + 2y = 6$.

Applications

54. Julia plans to make a cylindrical vase in which the bottom of the vase is to be 0.3 cm thick and the curved, lateral part of the vase is to be 0.2 cm thick. If the vase needs to have a volume of 1 liter, what should its dimensions be to minimize its weight?

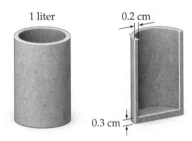

1 liter   0.2 cm

0.3 cm

55. Bob is building a toy chest for his son that will be a rectangular box with an open top. The base will be made from plywood that costs $2 per square foot, and the sides will be made from oak that costs $5 per square foot. What dimensions should Bob make the toy chest if he wants it to have a capacity of 10 cubic feet but wishes to minimize the cost of the wood?

Proofs

56. Let T be a triangle with side lengths a, b, and c. The **semiperimeter** of T is defined to be $s = \frac{1}{2}(a + b + c)$. Heron's formula for the area A of a triangle is

$$A = \sqrt{s(s - a)(s - b)(s - c)}.$$

Use Heron's formula and the method of Lagrange multipliers to prove that, for a triangle with perimeter P, the equilateral triangle maximizes the area.

57. Prove that a square maximizes the area of all rectangles with perimeter P.

58. Prove that if you minimize the square of the distance from the origin to a point (x, y) subject to the constraint $g(x, y) = 0$, you have minimized the distance from the origin to (x, y) subject to the same constraint.

59. Prove that if you minimize the square of the distance from the origin to a point (x, y, z) subject to the constraint

$g(x, y, z) = 0$, then you have minimized the distance from the origin to (x, y, z) subject to the same constraint.

60. Let a and b be nonzero real numbers and let $r > 0$. Prove that the point closest to the origin on the circle $(x - a)^2 + (y - b)^2 = r^2$ is on the line that contains the points $(0, 0)$ and (a, b).

61. Let (α, β) and (a, b) be distinct points in $\mathbb{R}^2$, and let r be a positive real number. Use the result of Exercise 60 to prove that the point closest to (α, β) on the circle $(x - a)^2 + (y - b)^2 = r^2$ is on the line that contains the points (α, β) and (a, b).

62. Let a, b, and c be nonzero real numbers and let $r > 0$. Prove that the point closest to the origin on the sphere $(x - a)^2 + (y - b)^2 + (z - c)^2 = r^2$ is on the line that contains the points $(0, 0, 0)$ and (a, b, c).

63. Let (α, β, γ) and (a, b, c) be distinct points in $\mathbb{R}^3$, and let r be a positive real number. Use the result of Exercise 62 to prove that the point closest to (α, β, γ) on the sphere

$(x-a)^2 + (y-b)^2 + (z-c)^2 = r^2$ is on the line that contains the points (α, β, γ) and (a, b, c).

The **arithmetic mean** of the real numbers $a_1, a_2, \ldots, a_n$ is $\frac{1}{n}(a_1 + a_2 + \cdots + a_n)$. If $a_i > 0$ for $1 \le i \le n$, then the **geometric mean** of $a_1, a_2, \ldots, a_n$ is $(a_1 a_2 \cdots a_n)^{1/n}$. In Exercises 64–66 we ask you to prove that the geometric mean is always less than the arithmetic mean for a set of positive numbers.

64. Use the method of Lagrange multipliers to show that $\sqrt{xy} \le \frac{1}{2}(x + y)$ when x and y are both positive.

65. Use the method of Lagrange multipliers to show that $\sqrt[3]{xyz} \le \frac{1}{3}(x + y + z)$ when x, y, and z are all positive.

66. Use the method of Lagrange multipliers to show that $(a_1 a_2 \cdots a_n)^{1/n} \le \frac{1}{n}(a_1 + a_2 + \cdots + a_n)$ when $a_1, a_2, \ldots, a_n$ are all positive.

Thinking Forward

▶ *A Double Summation:* Let

$$\sum_{i=1}^{m} \sum_{j=1}^{n} f(i, j) = \sum_{i=1}^{m} \left(\sum_{j=1}^{n} f(i, j) \right).$$

Evaluate $\displaystyle\sum_{i=1}^{15} \sum_{j=1}^{10} ij^2$.

▶ *Reordering a Double Summation:* Explain why

$$\sum_{i=1}^{m} \sum_{j=1}^{n} f(i, j) = \sum_{j=1}^{n} \sum_{i=1}^{m} f(i, j).$$

CHAPTER REVIEW, SELF-TEST, AND CAPSTONES

Before you progress to the next chapter, be sure you are familiar with the definitions, concepts, and basic skills outlined here. The capstone exercises at the end bring together ideas from this chapter and look forward to future chapters.

Definitions

Give precise mathematical definitions or descriptions of each of the following concepts that follow. Then illustrate the definition with a graph or an algebraic example.

▶ a *level curve* for $f(x, y)$

▶ a *level surface* for $f(x, y, z)$

▶ an *open disk* in $\mathbb{R}^2$

▶ an *open ball* in $\mathbb{R}^3$

▶ an *open subset* of $\mathbb{R}^2$ or $\mathbb{R}^3$

▶ the *complement* of a subset $\mathbb{R}^2$ or $\mathbb{R}^3$

▶ a *closed subset* of $\mathbb{R}^2$ or $\mathbb{R}^3$

▶ the *boundary* of a subset of $\mathbb{R}^2$ or $\mathbb{R}^3$

▶ a *bounded subset* of $\mathbb{R}^2$ or $\mathbb{R}^3$

▶ the $\epsilon\text{--}\delta$ definition of the statement $\lim_{\mathbf{x} \to \mathbf{a}} f(\mathbf{x}) = L$, where f is a function of two or more variables

▶ the limit of a function of two or three variables *along a path*

▶ what it means in terms of a limit for a function f of two or three variables to be *continuous* at a point in the domain of f

▶ the limit definition of the *partial derivatives* $f_x(x_0, y_0)$ and $f_y(x_0, y_0)$

▶ the limit definition of the *partial derivatives* $f_x(x_0, y_0, z_0)$, $f_y(x_0, y_0, z_0)$, and $f_z(x_0, y_0, z_0)$

▶ the limit definition of the *directional derivative* of a function of two or three variables, f, at a point P in the direction of a unit vector $\mathbf{u}$

▶ the *gradient* of a function f of two or three variables

▶ a function f of two or three variables has a *local minimum* or a *local maximum* at a point P

▶ a function f of two or three variables has a *global minimum* or a *global maximum* at a point P

- a function f of two or three variables has a *stationary point* at a point P
- a function f of two or three variables has a *critical point* at a point P

- a function $f(x, y)$ has a *saddle point* at a point (x_0, y_0)
- the *Hessian* and *discriminant* of a function $f(x, y)$

Theorems

Fill in the blanks to complete each of the following theorem statements:

- If S is a closed subset of $\mathbb{R}^3$, then S^c is _____ subset of $\mathbb{R}^3$.
- Let $f(x, y)$ be a function defined on an open subset S of $\mathbb{R}^2$. If the second-order partial derivatives of a function $f(x, y)$ are _____ in a (an) _____ subset S of $\mathbb{R}^2$, then $f_{xy}(x, y) = f_{yx}(x, y)$ at every point in S.
- If $f(x, y)$ is a function of two variables with the partial derivatives $\dfrac{\partial f}{\partial x}$ and $\dfrac{\partial f}{\partial y}$ that are _____ on _____ containing the point (x_0, y_0), then f is differentiable at (x_0, y_0).
- Let $f(x, y)$ be a function of two variables that is differentiable at the point (x_0, y_0). The equation of the plane tangent to the surface defined by $f(x, y)$ at (x_0, y_0) is _____.

- *The chain rule*: Given functions $z = f(x, y)$, $x = u(s, t)$, and $y = v(s, t)$, for all values of s and t at which u and v are differentiable, and if f is differentiable at $(u(s, t), v(s, t))$, $\dfrac{\partial z}{\partial s} =$ _____ and $\dfrac{\partial z}{\partial t} =$ _____.
- Let $f(x, y)$ be a function of two variables and (x_0, y_0) be a point in the domain of f at which the first-order partial derivatives of f exist. If $\mathbf{u} \in \mathbb{R}^2$ is a _____ for which the directional derivative $D_{\mathbf{u}} f(x_0, y_0)$ also exists, then $D_{\mathbf{u}} f(x_0, y_0) =$ _____.
- If $f(x, y)$ is a function of two or three variables and P is a point in the domain of f at which f is differentiable, then the _____ of f at P points in the direction in which f _____.
- If $f(x, y)$ has a local extremum at (x_0, y_0), then (x_0, y_0) is a _____ of f.

Notation, Algebraic Rules, and Optimization

Notation: Describe the meanings of each of the following mathematical expressions:

For $A \subset \mathbb{R}^2$,

- A^c
- ∂A

- $f_x(x_0, y_0)$
- $f_y(x_0, y_0)$
- $f_z(x_0, y_0, z_0)$

- $\dfrac{\partial w}{\partial x}$
- $\dfrac{\partial w}{\partial y}$
- $\dfrac{\partial w}{\partial z}$

- $D_{\mathbf{u}} f(x_0, y_0)$
- $\nabla f(x, y)$
- $\nabla f(x, y, z)$

Limits of combinations: Fill in the blanks to complete the limit rules. You may assume that $\lim\limits_{\mathbf{x} \to \mathbf{a}} f(\mathbf{x})$ and $\lim\limits_{\mathbf{x} \to \mathbf{a}} g(\mathbf{x})$ exist and that k is a scalar.

- $\lim\limits_{\mathbf{x} \to \mathbf{a}} k f(\mathbf{x}) =$ _____.
- $\lim\limits_{\mathbf{x} \to \mathbf{a}} (f(\mathbf{x}) + g(\mathbf{x})) =$ _____.
- $\lim\limits_{\mathbf{x} \to \mathbf{a}} (f(\mathbf{x}) - g(\mathbf{x})) =$ _____.
- $\lim\limits_{\mathbf{x} \to \mathbf{a}} (f(\mathbf{x}) g(\mathbf{x})) =$ _____.
- $\lim\limits_{\mathbf{x} \to \mathbf{a}} (f(\mathbf{x}) g(\mathbf{x})) =$ _____, provided that _____.

Function optimization: Fill in the blanks to complete optimization facts.

- Let $f(x, y)$ be a function with continuous second-order partial derivatives on some open disk containing the point (x_0, y_0). If (x_0, y_0) is a stationary point in the domain of f, then

  (a) f has a relative maximum at (x_0, y_0) if $\det(H_f(x_0, y_0))$ _____ with $f_{xx}(x_0, y_0)$ _____ or $f_{yy}(x_0, y_0)$ _____.

  (b) f has a relative minimum at (x_0, y_0) if $\det(H_f(x_0, y_0))$ _____ with $f_{xx}(x_0, y_0)$ _____ or $f_{yy}(x_0, y_0)$ _____.

  (c) f has a saddle point at (x_0, y_0) if $\det(H_f(x_0, y_0))$ _____.

  (d) No conclusion may be drawn about the behavior of f at (x_0, y_0) if $\det(H_f(x_0, y_0))$ _____.

- Let f and g be functions with continuous first-order partial derivatives. If $f(x, y)$ has a relative extremum at a point (x_0, y_0) subject to the constraint $g(x, y) = 0$, then _____ and _____ are parallel.

Skill Certification: The Calculus of Multivariable Functions

Level curves: Sketch the level curves $f(x, y) = c$ of the following functions for $c = -3, -2, -1, 0, 1, 2,$ and 3:

1. $f(x, y) = -\dfrac{2y}{x}$

2. $f(x, y) = y^2 - x^2$

3. $f(x, y) = x - \sin^{-1} y$

4. $f(x, y) = x \sec y$

Evaluating limits: Evaluate the following limits, or explain why the limit does not exist.

5. $\lim\limits_{(x,y) \to (1,2)} \dfrac{x^2 + y^2}{x^2 - y^2}$

6. $\lim\limits_{(x,y) \to (4,\pi)} \tan^{-1}\left(\dfrac{y}{x}\right)$

7. $\lim\limits_{(x,y) \to (0,0)} \dfrac{x^3 + y^3}{x^2 + y^2}$

8. $\lim\limits_{(x,y) \to (0,0)} \dfrac{x + y}{x^2 + y^2}$

9. $\lim\limits_{(x,y,z) \to (1,0,-1)} \dfrac{\sin(xy)}{x^2 - y^2 + z^2}$

10. $\lim\limits_{(x,y,z) \to (0,0,0)} \dfrac{x^2 + y^2}{x^2 + y^2 + z^2}$

11. $\lim\limits_{\substack{(x,y) \to (0,0) \\ x=0}} \dfrac{x^2 + y^2}{x^2 - y^2}$

12. $\lim\limits_{\substack{(x,y) \to (0,0) \\ y=0}} \dfrac{x^2 + y^2}{x^2 - y^2}$

Continuity: Find the set of points where the function is continuous.

13. $f(x, y) = \dfrac{x + y}{x - y}$

14. $f(x, y) = \dfrac{x + y}{x^2 + y^2}$

15. $f(x, y) = \ln xy$

16. $f(x, y) = \cos\left(\dfrac{1}{x - y}\right)$

17. $f(x, y, z) = \dfrac{x + y + z}{x - 2y + 3z}$

18. $f(x, y, z) = \dfrac{x + y + z}{x^2 + y^2 + z^2}$

Partial derivatives: Find all first- and second-order partial derivatives for the following functions:

19. $f(x, y) = \dfrac{x + y}{x^2 + y^2}$

20. $f(x, y) = \dfrac{x^2 - y^2}{x^2 + y^2}$

21. $f(x, y) = xye^{x^2}$

22. $f(x, y) = \tan^{-1}\left(\dfrac{y}{x}\right)$

23. $f(x, y, z) = xz^2 e^y$

24. $f(x, y, z) = \ln(x + y + z)$

Directional derivatives: Find the directional derivative of the given function at the specified point P in the direction of the given vector. Note: The given vectors may not be unit vectors.

25. $f(x, y) = \dfrac{y}{x}, P = (4, 3), \mathbf{v} = \langle 2, -3 \rangle$

26. $f(x, y) = \dfrac{y}{x}, P = (4, 3), \mathbf{v} = \langle 3, -2 \rangle$

27. $f(x, y) = \dfrac{x + y}{x^2 + y^2}, P = (1, 2), \mathbf{v} = \langle 3, -2 \rangle$

28. $f(x, y) = \dfrac{xy}{x^2 + y^2}, P = (-2, 1), \mathbf{v} = \left\langle \dfrac{3}{5}, -\dfrac{4}{5} \right\rangle$

29. $f(x, y, z) = xy^2 z^3, P = (0, 0, 0), \mathbf{v} = \langle 1, -2, -1 \rangle$

30. $f(x, y, z) = \sqrt{\dfrac{xy}{z}}, P = (2, 3, 1), \mathbf{v} = \langle 2, 1, -2 \rangle$

Gradients: Find the gradient of the given function, and find the direction in which the function increases most rapidly at the specified point P.

31. $f(x, y) = \dfrac{y}{x}, P = (4, 3)$

32. $f(x, y) = \dfrac{x}{y}, P = (4, 3)$

33. $f(x, y) = x \sin y, P = \left(3, \dfrac{\pi}{2}\right)$

34. $f(x, y) = \tan^{-1}\left(\dfrac{y}{x}\right), P = (-2, 2)$

35. $f(x, y, z) = \dfrac{x^2 y}{z}, P = (0, 3, -1)$

36. $f(x, y, z) = \ln(x + y + z), P = (e, 0, -1)$

Extrema: Find the local maxima, local minima, and saddle points of the given functions.

37. $f(x, y) = 2x^2 + y^2 + y + 5$

38. $f(x, y) = x^3 - 12xy - y^3$

39. $f(x, y) = x^3 + y^3 - 6x^2 + 3y^2 - 4$

40. $f(x, y) = x^4 + y^4 + 4xy$

Capstone Problems

A. *Chain Rule:* Let $z = f(x, y)$, $x = x(s, t)$, and $y = y(t)$.

(a) Find $\dfrac{\partial z}{\partial s}$ and $\dfrac{\partial z}{\partial t}$.

(b) Use the results of part (a) to find $\dfrac{\partial z}{\partial s}$ and $\dfrac{\partial z}{\partial t}$ when $z = x^2 \sin y$, $x = e^{s/t}$, and $y = t^3$. Express your answer in terms of s and t.

B. *Laplace's equation:* A function $f(x, y)$ is said to satisfy **Laplace's equation** if $\dfrac{\partial^2 f}{\partial x^2} + \dfrac{\partial^2 f}{\partial y^2} = 0$.

(a) Show that the function $f(x, y) = \tan^{-1}\left(\dfrac{y}{x}\right)$ satisfies Laplace's equation.

(b) Show that the function $f(x, y) = \ln\left(\dfrac{y}{x}\right)$ does not satisfy Laplace's equation.

C. *Extrema on a closed and bounded set:* Find and classify the maxima, minima, and saddle points of the function $f(x, y) = y^3 - 3x^2 y$ on the square region with vertices $(\pm 3, \pm 3)$.

D. *The Method of Least Squares:* The **method of least squares** is one way to fit a line to a collection of data points. The object of this method is to find the slope m and y-intercept b for a **regression line** $y = mx + b$ that provides the minimal sum for the squares of the vertical distances from the data points to the line, as indicated in the following figure:

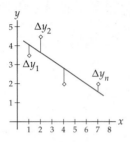

Most scientific calculators have built-in programs to find the equation of the regression line for a collection of data

points (x_1, y_1), (x_2, y_2), ..., (x_n, y_n), but here we ask you to first prove that:

(a)
$$m = \frac{(\sum x_k)(\sum y_k) - n(\sum x_k y_k)}{(\sum x_k)^2 - n(\sum x_k^2)} \quad \text{and}$$

$$b = \frac{1}{n}\left(\sum y_k - m \sum x_k\right),$$

where each of the summations is evaluated from $k = 1$ to $k = n$.

Use the results of part (a) to find the equations of the regression lines for the following collections of data:

(b) $(1, 2)$, $(3, 4)$, $(5, 5)$, $(6, 8)$

(c) $(1, 2)$, $(1, 4)$, $(5, 2)$, $(5, 4)$

CHAPTER 13

Double and Triple Integrals

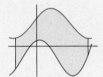

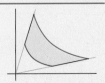

13.1   DOUBLE INTEGRALS OVER RECTANGULAR REGIONS

► Computing volumes bounded by functions of two variables

► Defining double integrals as the limit of a Riemann sum

► Using Fubini's theorem to evaluate double integrals as iterated integrals

Volumes

Consider the following two questions:

► How can we compute the volume of a solid?

► How can we compute the mass of a solid whose density varies from point to point?

The way in which we will answer these two questions is quite similar to the methods we used in single-variable calculus. For example, to find the volume of the solid bounded above by the surface and below by the xy-plane, in the left-hand figure that follows, we subdivide the volume into vertical slices whose volumes we can reasonably approximate, as we see in the middle figure, and then, take a limit as the size of the pieces goes to zero and, simultaneously, the number of pieces goes to infinity. This brief outline summarizes much of what we will do in this chapter.

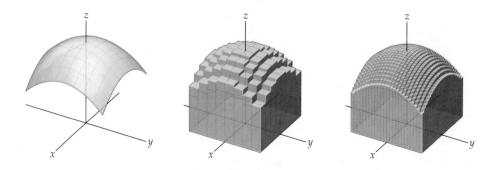

We will begin by reviewing and extending the summation notation we introduced in Chapter 4.

Double and Triple Summations

Recall that

$$\sum_{k=1}^{n} a_k = a_1 + a_2 + a_3 + \cdots + a_{n-1} + a_n.$$

We take this idea a step or two further by introducing double and triple sums.

DEFINITION 13.1   Double Sums and Triple Sums

(a) The *double summation* $\displaystyle\sum_{j=1}^{m}\sum_{k=1}^{n} a_{jk}$ is given by

$$\sum_{j=1}^{m}\sum_{k=1}^{n} a_{jk} = \sum_{j=1}^{m}\left(\sum_{k=1}^{n} a_{jk}\right)$$

$$= \sum_{j=1}^{m}(a_{j1} + a_{j2} + a_{j3} + \cdots + a_{jn})$$

$$= a_{11} + a_{12} + a_{13} + \cdots + a_{1n}$$
$$+ a_{21} + a_{22} + a_{23} + \cdots + a_{2n}$$
$$+ \vdots \quad \vdots \quad \vdots \quad \cdots \quad \vdots$$
$$+ a_{m1} + a_{m2} + a_{m3} + \cdots + a_{mn}.$$

(b) Similarly, the *triple summation* $\displaystyle\sum_{i=1}^{l}\sum_{j=1}^{m}\sum_{k=1}^{n} a_{ijk}$ is given by

$$\sum_{i=1}^{l}\sum_{j=1}^{m}\sum_{k=1}^{n} a_{ijk} = \sum_{i=1}^{l}\left(\sum_{j=1}^{m}\left(\sum_{k=1}^{n} a_{ijk}\right)\right).$$

For example, consider the following double sum:

$$\sum_{j=1}^{2}\sum_{k=1}^{3} j^2 k = (1^2 \cdot 1 + 1^2 \cdot 2 + 1^2 \cdot 3) + (2^2 \cdot 1 + 2^2 \cdot 2 + 2^2 \cdot 3) = 1 + 2 + 3 + 4 + 8 + 12 = 30.$$

We know that, for finite sums, addition obeys both the commutative and associative rules. In Exercises 71 and 72 you will use these properties to prove the following theorem:

THEOREM 13.2   Changing the Order of a Finite Double or Triple Sum

For positive integers l, m, and n,

(a) $\displaystyle\sum_{j=1}^{m}\sum_{k=1}^{n} a_{jk} = \sum_{k=1}^{n}\sum_{j=1}^{m} a_{jk}$

(b) $\displaystyle\sum_{i=1}^{l}\sum_{j=1}^{m}\sum_{k=1}^{n} a_{ijk} = \sum_{j=1}^{m}\sum_{k=1}^{n}\sum_{i=1}^{l} a_{ijk} = \sum_{k=1}^{n}\sum_{i=1}^{l}\sum_{j=1}^{m} a_{ijk}$

Note that there are three additional permutations of the summations in Theorem 13.2(b) equal to those listed.

Double Integrals over Rectangular Regions

Starting with Section 13.2, we will allow more complicated domains, but we begin by considering functions defined on rectangular regions. Let $a < b$ and $c < d$ be real numbers and

$\mathcal{R}$ be the rectangle in the xy-plane defined by

$$\mathcal{R} = \{(x, y) \mid a \le x \le b \text{ and } c \le y \le d\},$$

as shown in the following figure at the left:

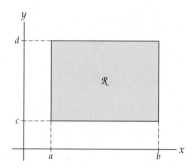
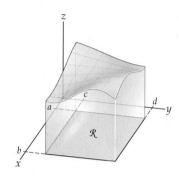

We also assume that $f(x, y) > 0$ is a function defined at every point of $\mathcal{R}$, as we see in the figure on the right. Here is our procedure for finding the volume V of the solid bounded below by $\mathcal{R}$ and bounded above by the graph of f:

▶ We subdivide the interval $[a, b]$ into m equal subintervals, each of width $\Delta x = \frac{b-a}{m}$, and also let $x_j = a + j\Delta x$ for $0 \le j \le m$.

▶ Similarly, we subdivide the interval $[c, d]$ into n equal subintervals, each of width $\Delta y = \frac{d-c}{n}$, and let $y_k = a + k\Delta y$ for $0 \le k \le n$.

▶ The preceding subdivisions partition the rectangle into $m \times n$ congruent rectangles as follows:

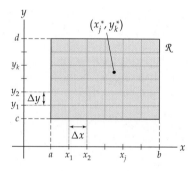

▶ For each $j = 1, 2, \ldots, m$ and $k = 1, 2, \ldots, n$, we select a point (x_j^*, y_k^*) in the subrectangle $\mathcal{R}_{jk} = \{(x, y) \mid x_{j-1} \le x \le x_j \text{ and } y_{k-1} \le y \le y_k\}$. One such point is shown in the previous figure.

▶ We let ΔA be the common area of each subrectangle. That is, $\Delta A = \Delta x \Delta y$.

▶ The product $f(x_j^*, y_k^*)\Delta A$ approximates the volume of the solid bounded below by the rectangle $\mathcal{R}_{jk}$ and above by the graph of f.

▶ When we sum these approximate volumes over all of the subrectangles, we obtain an approximation for the volume of V.

The summations

$$\sum_{j=1}^{m}\sum_{k=1}^{n} f(x_j^*, y_k^*)\Delta A = \sum_{k=1}^{n}\sum_{j=1}^{m} f(x_j^*, y_k^*)\Delta A$$

are both **Riemann sums** for f on $\mathcal{R}$.

We now define Δ to be the length of the diagonal of each subrectangle. That is,

$$\Delta = \sqrt{(\Delta x)^2 + (\Delta y)^2}.$$

Taking the limit as $\Delta \to 0$ ensures that $m \to \infty$ and $n \to \infty$. We use the limit of the Riemann sum as $\Delta \to 0$ to define the volume of the solid V, provided that the limit exists. The following definitions formalize these ideas:

DEFINITION 13.3

Riemann Sums for Functions of Two Variables

Let $a < b$ and $c < d$ be real numbers, let $\mathcal{R}$ be the rectangle defined by

$$\mathcal{R} = \{(x, y) \mid a \le x \le b \text{ and } c \le y \le d\},$$

and let $f(x, y)$ be a function defined on $\mathcal{R}$. The sums

$$\sum_{j=1}^{m} \sum_{k=1}^{n} f(x_j^*, y_k^*)\Delta A = \sum_{k=1}^{n} \sum_{j=1}^{m} f(x_j^*, y_k^*)\Delta A$$

are **Riemann sums** for f on $\mathcal{R}$, where x_j, x_j^*, y_k, y_k^*, Δx, Δy, and ΔA are defined as outlined before.

DEFINITION 13.4

Double Integrals

Let $a < b$ and $c < d$ be real numbers, let $\mathcal{R}$ be the rectangle defined by

$$\mathcal{R} = \{(x, y) \mid a \le x \le b \text{ and } c \le y \le d\},$$

and let $f(x, y)$ be a function defined on $\mathcal{R}$. Provided that the limits exist, the **double integral** of f over $\mathcal{R}$ is

$$\iint_{\mathcal{R}} f(x, y)\, dA = \lim_{\Delta \to 0} \sum_{j=1}^{m} \sum_{k=1}^{n} f(x_j^*, y_k^*)\Delta A = \lim_{\Delta \to 0} \sum_{k=1}^{n} \sum_{j=1}^{m} f(x_j^*, y_k^*)\Delta A,$$

where the double sums are Riemann sums, as outlined in Definition 13.3, and where $\Delta = \sqrt{(\Delta x)^2 + (\Delta y)^2}$. When the limits exist, the function f is said to be **integrable** on $\mathcal{R}$.

Note that as $\Delta \to 0$, the increment of area $\Delta A \to 0$. For every function that is continuous on a rectangular region, the limit will exist and the function will be integrable.

DEFINITION 13.5

The Volume of a Solid and the Signed Volume

Let $a < b$ and $c < d$ be real numbers, let $\mathcal{R}$ be the rectangle defined by

$$\mathcal{R} = \{(x, y) \mid a \le x \le b \text{ and } c \le y \le d\},$$

and let $f(x, y)$ be an integrable function defined on $\mathcal{R}$.

(a) The **signed volume** between the graph of f and the rectangle $\mathcal{R}$ is defined to be the double integral $\iint_{\mathcal{R}} f(x, y)\, dA$.

(b) The **(absolute) volume** between the graph of f and the rectangle $\mathcal{R}$ is defined to be the double integral $\iint_{\mathcal{R}} |f(x, y)|\, dA$.

As a consequence of Definition 13.5, if $f(x, y) \geq 0$ on $\mathcal{R}$, then the double integral $\iint_{\mathcal{R}} f(x, y)\, dA$ represents the volume of the solid bounded above by the graph of f and below by the rectangle $\mathcal{R}$.

In Example 1, we compute the volume of the solid bounded above by the graph of the function $f(x, y) = x^2 y$ on the rectangle

$$\mathcal{R} = \{(x, y) \mid 1 \leq x \leq 3 \text{ and } 2 \leq y \leq 5\}$$

by using Definition 13.4 to evaluate the double integral $\iint_{\mathcal{R}} x^2 y\, dA$. We will see that such computations are rather lengthy and time consuming. Fortunately, we have an alternative.

Iterated Integrals and Fubini's Theorem

As we mentioned, using the definition to evaluate a double integral is a lengthy process. When we can, we will use iterated integrals to evaluate double integrals.

DEFINITION 13.6   Iterated Integrals

Let a, b, c, and d be real numbers. We define the following *iterated integrals*:

(a) $\displaystyle \int_a^b \int_c^d f(x, y)\, dy\, dx = \int_a^b \left(\int_c^d f(x, y)\, dy \right) dx$

(b) $\displaystyle \int_c^d \int_a^b f(x, y)\, dx\, dy = \int_c^d \left(\int_a^b f(x, y)\, dx \right) dy.$

The general idea here is that the subdivisions that are used to construct a Riemann sum for a double integral over a rectangle $\mathcal{R}$ may be added most easily in one of two orders:

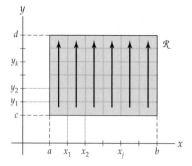 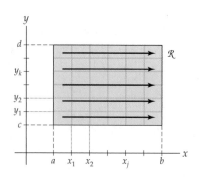

As the figures show, either we sum over the subrectangles, first moving up each column and then moving to the right, or we sum along the rows from left to right and then move up. The first order of summation gives us part (a) of Definition 13.6. The other order gives us the iterated integral in part (b).

We will evaluate an iterated integral after we introduce Fubini's theorem, which tells us that we may use an iterated integral, rather than the definition of the double integral, to evaluate a double integral.

THEOREM 13.7

Fubini's Theorem

Let $a < b$ and $c < d$ be real numbers, let $\mathcal{R}$ be the rectangle defined by

$$\mathcal{R} = \{(x, y) \mid a \le x \le b \text{ and } c \le y \le d\},$$

and let $f(x, y)$ be continuous on $\mathcal{R}$. Then

$$\iint_{\mathcal{R}} f(x, y)\, dA = \int_c^d \int_a^b f(x, y)\, dx\, dy = \int_a^b \int_c^d f(x, y)\, dy\, dx.$$

Proof. A rigorous proof of Fubini's theorem is beyond the scope of this text, but we provide a less formal argument. From the definition of the integral as a limit of a Riemann sum, we have

$$\int_c^d \int_a^b f(x, y)\, dx\, dy = \int_c^d \left(\int_a^b f(x, y)\, dx \right) dy = \lim_{n \to \infty} \sum_{k=1}^n \left(\lim_{m \to \infty} \sum_{j=1}^m f(x_j^*, y_k^*) \Delta x \right) \Delta y,$$

where $\Delta x = \dfrac{b-a}{m}$, $\Delta y = \dfrac{d-c}{n}$, and (x_j^*, y_k^*) is a sample point in the jkth subrectangle. We know that for the finite summations

$$\sum_{k=1}^n \sum_{j=1}^m f(x_j^*, y_k^*) \Delta x \Delta y = \sum_{j=1}^m \sum_{k=1}^n f(x_j^*, y_k^*) \Delta y \Delta x.$$

Fubini's theorem follows from the fact that when f is continuous, the orders of both the limits and the sums may be interchanged:

$$\int_c^d \int_a^b f(x, y)\, dx\, dy = \lim_{n \to \infty} \sum_{k=1}^n \left(\lim_{m \to \infty} \sum_{j=1}^m f(x_j^*, y_k^*) \Delta x \right) \Delta y$$

$$= \lim_{m \to \infty} \sum_{j=1}^m \left(\lim_{n \to \infty} \sum_{k=1}^n f(x_j^*, y_k^*) \Delta y \right) \Delta x$$

$$= \int_a^b \int_c^d f(x, y)\, dy\, dx. \qquad \blacksquare$$

To illustrate Fubini's theorem, let

$$\mathcal{R} = \{(x, y) \mid 1 \le x \le 3 \text{ and } 2 \le y \le 5\}.$$

Then

$$\iint_{\mathcal{R}} x^2 y\, dA = \int_2^5 \int_1^3 x^2 y\, dx\, dy = \int_1^3 \int_2^5 x^2 y\, dy\, dx.$$

For rectangular regions, the level of difficulty of the two iterated integrals we may use to evaluate the double integral is the same. In Section 13.2 we will discuss how to evaluate double integrals over more complicated regions. At that time we will see that, for a more complicated region, one of the two orderings of the iterated integrals can make the evaluation process much simpler. Here we evaluate

$$\iint_{R} x^2 y \, dA = \int_{2}^{5} \int_{1}^{3} x^2 y \, dx \, dy \qquad \leftarrow \text{Fubini's theorem}$$

$$= \int_{2}^{5} \left(\int_{1}^{3} x^2 y \, dx \right) dy \qquad \leftarrow \text{evaluation procedure for the iterated integral}$$

$$= \int_{2}^{5} \left[\frac{1}{3} x^3 y \right]_{x=1}^{x=3} dy \qquad \leftarrow \text{the Fundamental Theorem of Calculus}$$

$$= \int_{2}^{5} \left(\frac{27}{3} y - \frac{1}{3} y \right) dy = \int_{2}^{5} \frac{26}{3} y \, dy \qquad \leftarrow \text{evaluation of the inner antiderivative}$$

$$= \left[\frac{13}{3} y^2 \right]_{y=2}^{y=5} \qquad \leftarrow \text{the Fundamental Theorem of Calculus}$$

$$= \frac{13}{3} (25 - 4) = 91. \qquad \leftarrow \text{evaluation of the outer antiderivative}$$

Note that when we evaluate the inner integral with respect to x, we treat the other variable, y, as a constant. Not until we evaluate the outer integral with respect to y do we consider y as a variable. In Exercise 17 you will use the other order of integration to evaluate this double integral.

Examples and Explorations

EXAMPLE 1   **Using the definition to evaluate a double integral**

Let

$$R = \{(x, y) \mid 1 \leq x \leq 3 \text{ and } 2 \leq y \leq 5\}.$$

Use the definition of the double integral to evaluate $\iint_{R} x^2 y \, dA$.

SOLUTION

Before we evaluate the integral, we note that it represents the volume of the solid bounded above by the graph of $f(x, y) = x^2 y$ and below by the xy-plane on the rectangular region R, since $x^2 y \geq 0$ on R. This solid is shown here:

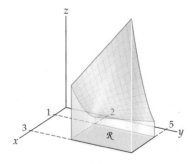

To use Definition 13.5 to evaluate the integral, for our starred points (x_j^*, y_k^*) we will choose $(x_j, y_k) = (1 + j\Delta x, 2 + k\Delta y)$ for each j and k. Thus, the double integral of $x^2 y$ on R is given by

$$\iint_{R} x^2 y \, dA = \lim_{\Delta \to 0} \sum_{j=1}^{m} \sum_{k=1}^{n} (1 + j\Delta x)^2 (2 + k\Delta y) \Delta A.$$

Recall that using the definition to compute a definite integral of a function of a single variable can be quite time consuming. Still, we will use the definition to evaluate this double integral, but will use Fubini's theorem to evaluate other double integrals.

We start working on the Riemann sum. Since the index of the "interior" summation is k,

$$\sum_{j=1}^{m}\sum_{k=1}^{n}(1+j\Delta x)^2(2+k\Delta y)\Delta A = \sum_{j=1}^{m}(1+j\Delta x)^2\,\Delta A\left(\sum_{k=1}^{n}(2+k\Delta y)\right).$$

Now, using properties of summations from Chapter 4, we have

$$\sum_{k=1}^{n}(2+k\Delta y) = 2n + \frac{1}{2}n(n+1)\Delta y.$$

Combining the preceding two equations gives

$$\sum_{j=1}^{m}(1+j\Delta x)^2\,\Delta A\left(\sum_{k=1}^{n}(2+k\Delta y)\right) = \sum_{j=1}^{m}(1+j\Delta x)^2\,\Delta A\left(2n + \frac{1}{2}n(n+1)\Delta y\right).$$

Again using properties of summations from Chapter 4, we obtain

$$\sum_{j=1}^{m}(1+j\Delta x)^2 = \sum_{j=1}^{m}(1+2j\Delta x+j^2(\Delta x)^2) = m + m(m+1)\Delta x + \frac{1}{6}m(m+1)(2m+1)(\Delta x)^2.$$

Incorporating this result into the original Riemann sum, we have

$$\sum_{j=1}^{m}\sum_{k=1}^{n}(1+j\Delta x)^2(2+k\Delta y)\Delta A =$$

$$\left(m + m(m+1)\Delta x + \frac{1}{6}m(m+1)(2m+1)(\Delta x)^2\right)\left(2n + \frac{1}{2}n(n+1)\Delta y\right)\Delta A.$$

But recall that $\Delta x = \frac{b-a}{m} = \frac{2}{m}$, $\Delta y = \frac{d-c}{n} = \frac{3}{n}$ and $\Delta A = \Delta x \Delta y = \frac{6}{mn}$, so the Riemann sum equals

$$\left(m + m(m+1)\frac{2}{m} + \frac{1}{6}m(m+1)(2m+1)\left(\frac{2}{m}\right)^2\right)\left(2n + \frac{1}{2}n(n+1)\frac{3}{n}\right)\frac{6}{mn}$$

$$= 6\left(1 + \frac{2(m+1)}{m} + \frac{2(m+1)(2m+1)}{3m^2}\right)\left(2 + \frac{3(n+1)}{2n}\right).$$

We are finally ready to evaluate the limit of the Riemann sum! We get

$$\iint_{\mathcal{R}} x^2y\,dA = \lim_{\Delta\to 0} 6\left(1 + \frac{2(m+1)}{m} + \frac{2(m+1)(2m+1)}{3m^2}\right)\left(2 + \frac{3(n+1)}{2n}\right)$$

$$= \lim_{m\to\infty}\left(\lim_{n\to\infty} 6\left(1 + \frac{2(m+1)}{m} + \frac{2(m+1)(2m+1)}{3m^2}\right)\left(2 + \frac{3(n+1)}{2n}\right)\right)$$

$$= 6\left(1 + 2 + \frac{4}{3}\right)\left(2 + \frac{3}{2}\right) = 91.$$

□

 **CHECKING THE ANSWER** | Note that we evaluated this double integral earlier in the section by using Fubini's theorem. Of course, we obtained the same result.

EXAMPLE 2 | **Using Fubini's theorem**

Use Fubini's theorem to evaluate the following double integrals:

(a) $\iint_{\mathcal{R}_1} xy\sin(x^2y)\,dA$, where $\mathcal{R}_1 = \{(x,y)\mid 0\le x\le 1 \text{ and } 0\le y\le\pi\}$.

(b) $\iint_{\mathcal{R}_2}(\sin x + e^{-2y})\,dA$, where $\mathcal{R}_2 = \{(x,y)\mid 0\le x\le\pi \text{ and } 0\le y\le 1\}$.

(c) $\iint_{\mathcal{R}_3}\ln y\,dA$, where $\mathcal{R}_3 = \left\{(x,y)\mid 0\le x\le 5 \text{ and } \frac{1}{e}\le y\le e\right\}$.

SOLUTION

(a) Our first integral represents the volume of the solid bounded above by the graph of $xy \sin(x^2 y)$ and below by the xy-plane on the rectangular region $\mathcal{R}_1$, since $xy \sin(x^2 y) \geq 0$ on $\mathcal{R}_1$. Following is a graph of this solid:

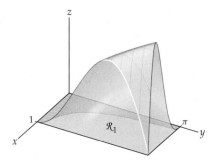

By Fubini's theorem,

$$\iint_{\mathcal{R}_1} xy \sin(x^2 y)\, dA = \int_0^1 \int_0^\pi xy \sin(x^2 y)\, dy\, dx = \int_0^\pi \int_0^1 xy \sin(x^2 y)\, dx\, dy.$$

As we mentioned earlier, for rectangular regions the levels of difficulty of the two iterated integrals given by Fubini's theorem should be quite similar. We choose the second integral here. We begin by evaluating $\int_0^1 xy \sin(x^2 y)\, dx$. Note that when we integrate the "interior" integral with respect to x, we treat y as a constant. Therefore, we may use substitution to find an antiderivative for the function $xy \sin(x^2 y)$ with respect to x. The function $-\frac{1}{2} \cos(x^2 y)$ is such an antiderivative. We may check this by taking the partial derivative $\frac{\partial}{\partial x} \left(-\frac{1}{2} \cos(x^2 y) \right)$ and observing that it is, indeed, $xy \sin(x^2 y)$. Therefore,

$$\int_0^1 xy \sin(x^2 y)\, dx = \left[-\frac{1}{2} \cos(x^2 y) \right]_{x=0}^{x=1} = -\frac{1}{2} (\cos(1^2 y) - \cos(0^2 y)) = \frac{1}{2}(1 - \cos y).$$

We now have

$$\iint_{\mathcal{R}_1} xy \sin(x^2 y)\, dA = \int_0^\pi \int_0^1 xy \sin(x^2 y)\, dx\, dy = \int_0^\pi \frac{1}{2}(1 - \cos y)\, dy.$$

The integral on the right is a definite integral like those we saw in Chapter 4. We have

$$\int_0^\pi \frac{1}{2}(1 - \cos y)\, dy = \left[\frac{1}{2}(y - \sin y) \right]_{y=0}^{y=\pi} = \frac{\pi}{2}.$$

Note that the answer here is a constant. Recall that when $f(x)$ is an integrable function on an interval $[a, b]$, the value of the definite integral $\int_a^b f(x)\, dx$ is a constant. Similarly, when $f(x, y)$ is an integrable function on a rectangle $\mathcal{R}$, the value of the double integral $\iint_{\mathcal{R}} f(x, y)\, dA$ is a constant.

Before we proceed to the other double integrals, we summarize the steps we just used:

▶ Use Fubini's theorem to express the double integral as an iterated integral.

▶ Choose an order of integration. The levels of difficulty of the two iterated integrals given by Fubini's theorem may differ. Choose the order that is easier to evaluate.

▶ Whichever of the two iterated integrals you choose, work on the "interior" integral first, treating the other variable as a constant. (In the example we just completed, the interior variable was x and the exterior variable was y. You may use any of the integration techniques from Chapter 5 in the process.)

▶ When you are done with this interior integration, you should have a new definite integral that involves just the exterior variable. (In the preceding example, the last step involved evaluating a definite integral with respect to y.)

▶ Evaluate the exterior integral.

▶ The result should be a constant.

(b) Our second integral represents the volume of the solid bounded above by the graph of $\sin x + e^{-2y}$ and below by the xy-plane on the rectangular region $\mathcal{R}_2$, since $\sin x + e^{-2y} \geq 0$ on $\mathcal{R}_2$. Following is a graph of this solid:

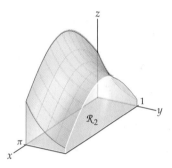

Again by Fubini's theorem,

$$\iint_{\mathcal{R}_2} (\sin x + e^{-2y}) \, dA = \int_0^\pi \int_0^1 (\sin x + e^{-2y}) \, dy \, dx = \int_0^1 \int_0^\pi (\sin x + e^{-2y}) \, dx \, dy.$$

We evaluate the first iterated integral here. We begin by evaluating $\int_0^1 (\sin x + e^{-2y}) \, dy$. We integrate with respect to y, treating x as a constant. Therefore,

$$\int_0^1 (\sin x + e^{-2y}) \, dy = \left[(\sin x)y - \frac{1}{2} e^{-2y} \right]_{y=0}^{y=1} = \sin x + \frac{1 - e^{-2}}{2}.$$

We now integrate this function with respect to x on the interval $[0, \pi]$:

$$\int_0^\pi \left(\sin x + \frac{1 - e^{-2}}{2} \right) dx = \left[-\cos x + \frac{1 - e^{-2}}{2} x \right]_{x=0}^{x=\pi} = 2 + \frac{\pi(1 - e^{-2})}{2}.$$

(c) Our third integral represents a signed volume, since the graph of $\ln y$ takes on both positive and negative values on the rectangular region $\mathcal{R}_3$, as shown in the following figure:

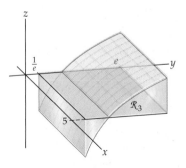

Here we have

$$\iint_{\mathcal{R}_3} \ln y \, dA = \int_0^5 \int_{1/e}^e \ln y \, dy \, dx = \int_{1/e}^e \int_0^5 \ln y \, dx \, dy.$$

We now evaluate the second iterated integral. Note that the function $\ln y$ is constant with respect to x; therefore,

$$\int_0^5 \ln y\, dx = \left[(\ln y)x\right]_{x=0}^{x=5} = 5\ln y.$$

We integrate this function with respect to y on the interval $\left[\frac{1}{e}, e\right]$:

$$\int_{1/e}^e 5\ln y\, dy = 5\left[y\ln y - y\right]_{y=1/e}^{y=e} = 5\left((e\ln e - e) - \left(\frac{1}{e}\ln\frac{1}{e} - \frac{1}{e}\right)\right) = 10e^{-1}. \quad \square$$

EXAMPLE 3

Computing volume

Find the volume of the solid between the graph of the function $f(x, y) = x\sin y$ and the rectangle $\mathcal{R}$,

$$\mathcal{R} = \{(x, y, 0) \mid -1 \le x \le 2 \text{ and } 0 \le y \le \pi\}$$

in the xy-plane.

SOLUTION

Following is a graph of the function f on the rectangular region $\mathcal{R}$:

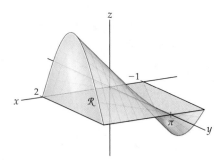

Note that $f(x, y) \ge 0$ on the rectangle $\mathcal{R}_1 = \{(x, y) \mid 0 \le x \le 2 \text{ and } 0 \le y \le \pi\}$ and $f(x, y) \le 0$ on the rectangle $\mathcal{R}_2 = \{(x, y) \mid -1 \le x \le 0 \text{ and } 0 \le y \le \pi\}$. Therefore the volume we want is given by the difference

$$\iint_{\mathcal{R}} |x\sin y|\, dA = \iint_{\mathcal{R}_1} x\sin y\, dA - \iint_{\mathcal{R}_2} x\sin y\, dA.$$

We will use Fubini's theorem to compute both of these double integrals:

$$\iint_{\mathcal{R}_1} x\sin y\, dA = \int_0^2 \int_0^\pi x\sin y\, dy\, dx \quad \text{and} \quad \iint_{\mathcal{R}_2} x\sin y\, dA = \int_{-1}^0 \int_0^\pi x\sin y\, dy\, dx.$$

Both iterated integrals require us to compute the integral

$$\int_0^\pi x\sin y\, dy = \left[-x\cos y\right]_{y=0}^{y=\pi} = 2x.$$

We now finish the evaluation of the two iterated integrals:

$$\iint_{\mathcal{R}_1} x\sin y\, dA = \int_0^2 2x\, dx = \left[x^2\right]_{x=0}^{x=2} = 4 \text{ and } \iint_{\mathcal{R}_2} x\sin y\, dA = \int_{-1}^0 2x\, dx = \left[x^2\right]_{x=-1}^{x=0} = -1.$$

Therefore, the volume $\iint_{\mathcal{R}} |x\sin y|\, dA = 4 - (-1) = 5.$ $\quad \square$

? **TEST YOUR**
UNDERSTANDING

▶ What are double sums and triple sums?

▶ What is a Riemann sum for a function of two variables over a rectangular region?

▶ What is the definition of a double integral?

▶ What is an iterated integral? How are iterated integrals defined? What is the difference between a double integral and an iterated integral?

▶ How is the volume between the graph of a function of two variables $f(x, y)$ and the xy-plane defined? Not every double integral gives a volume. Why? Which double integrals represent volumes? Which do not?

EXERCISES 13.1

Thinking Back

▶ *Using summation formulas:* Verify that the formulas $\sum_{k=1}^{n} k = \frac{n(n+1)}{2}$ and $\sum_{k=1}^{n} k^2 = \frac{n(n+1)(2n+1)}{6}$ hold when $n = 4$ and $n = 7$.

▶ *Using the definition to compute a definite integral:* Evaluate the definite integral $\int_1^4 x^2 \, dx$ as the limit of a right Riemann sum.

Concepts

0. *Problem Zero:* Read the section and make your own summary of the material.

1. *True/False:* Determine whether each of the statements that follow is true or false. If a statement is true, explain why. If a statement is false, provide a counterexample.

(a) *True or False:*
$$\sum_{j=1}^{10} \sum_{k=1}^{15} jk^2 = \sum_{k=1}^{10} \sum_{j=1}^{15} j^2 k$$

(b) *True or False:*
$$\sum_{j=1}^{m} \sum_{k=1}^{n} j^2 k^3 = \left(\sum_{j=1}^{m} j^2\right)\left(\sum_{k=1}^{n} k^3\right)$$

(c) *True or False:*
$$\sum_{j=1}^{m} \sum_{k=1}^{n} e^{jk^2} = \left(\sum_{j=1}^{m} e^j\right)\left(\sum_{k=1}^{n} e^{k^2}\right)$$

(d) *True or False:* If $f(x, y)$ is continuous on the rectangle $\mathcal{R} = \{(x, y) \mid a \leq x \leq b \text{ and } c \leq y \leq d\}$, then $\iint_{\mathcal{R}} f(x, y) \, dA$ exists.

(e) *True or False:* If $f(x, y)$ is continuous on the rectangle $\mathcal{R} = \{(x, y) \mid a \leq x \leq b \text{ and } c \leq y \leq d\}$, then $\iint_{\mathcal{R}} f(x, y) \, dA = \int_a^b \int_c^d f(x, y) \, dx \, dy$.

(f) *True or False:* If $f(x, y)$ is continuous on the rectangle $\mathcal{R} = \{(x, y) \mid a \leq x \leq b \text{ and } c \leq y \leq d\}$, then $\iint_{\mathcal{R}} f(x, y) \, dA = \int_a^b \int_c^d f(x, y) \, dy \, dx$.

(g) *True or False:* If $f(x, y)$ is continuous on the rectangle $\mathcal{R} = \{(x, y) \mid a \leq x \leq b \text{ and } c \leq y \leq d\}$, then $\int_b^a \int_d^c f(x, y) \, dy \, dx = \int_a^b \int_c^d f(x, y) \, dy \, dx$.

(h) *True or False:* If $f(x, y)$ is continuous on the rectangle $\mathcal{R} = \{(x, y) \mid a \leq x \leq b \text{ and } c \leq y \leq d\}$, then $\int_c^d \int_a^b f(x, y) \, dx \, dy = \int_a^b \int_c^d f(x, y) \, dy \, dx$.

2. *Examples:* Construct examples of the thing(s) described in the following. Try to find examples that are different than any in the reading.

(a) Two different sigma notation expressions of the same double sum.

(b) A double summation whose value is zero although none of the summands is zero.

(c) A nonzero function $f(x, y)$ defined on the rectangle $\mathcal{R} = \{(x, y) \mid -2 \leq x \leq 2 \text{ and } -3 \leq y \leq 3\}$ such that $\iint_{\mathcal{R}} f(x, y) \, dA = 0$.

3. Express the sum $3e^4 + 3e^9 + 3e^{16} + 4e^4 + 4e^9 + 4e^{16}$ using double-summation notation.

4. Express the sum $\frac{2}{5} + \frac{2}{6} + \frac{2}{7} + \frac{2}{8} + \frac{4}{5} + \frac{4}{6} + \frac{4}{7} + \frac{4}{8}$ using double-summation notation.

5. How many summands are in $\sum_{j=3}^{13} \sum_{k=5}^{20} \frac{j^2}{e^{jk}}$?

6. How many summands are in $\sum_{j=j_0}^{m} \sum_{k=k_0}^{n} j^k$?

7. How many summands are in $\sum_{i=2}^{15} \sum_{j=3}^{17} \sum_{k=4}^{19} \frac{i}{j+k}$?

8. How many summands are in $\sum_{i=i_0}^{l} \sum_{j=j_0}^{m} \sum_{k=k_0}^{n} k^i j^k$?

9. Discuss the similarities and differences between the definition of the definite integral found in Chapter 4 and the definition of the double integral found in this section.

10. Explain how to construct a Riemann sum for a function of two variables over a rectangular region.

11. Explain how to construct a *midpoint* Riemann sum for a function of two variables over a rectangular region for which each (x_j^*, y_k^*) is the midpoint of the subrectangle

$$\mathcal{R}_{jk} = \{(x, y) \mid x_{j-1} \leq x_j^* \leq x_j \text{ and } y_{k-1} \leq y_k^* \leq y_k\}.$$

Refer to your answer to Exercise 10 or to Definition 13.3.

12. What is the difference between a double integral and an iterated integral?

13. State Fubini's theorem.

14. Explain why using an iterated integral to evaluate a double integral is often easier than using the definition of the double integral to evaluate the integral.

15. Explain how the Fundamental Theorem of Calculus is used in evaluating the iterated integral $\int_a^b \int_c^d f(x, y)\, dy\, dx$.

16. Explain how the Fundamental Theorem of Calculus is used in evaluating the iterated integral $\int_c^d \int_a^b f(x, y)\, dx\, dy$.

17. Earlier in this section, we showed that we could use Fubini's theorem to evaluate the integral $\iint_{\mathcal{R}} x^2 y\, dA$ and we showed that $\int_2^5 \int_1^3 x^2 y\ dx\, dy = 91$. Now evaluate the double integral by evaluating the iterated integral $\int_1^3 \int_2^5 x^2 y\, dy\, dx$.

Explain why it would be difficult to evaluate the double integrals in Exercises 18 and 19 as iterated integrals.

18. $\iint_{\mathcal{R}} e^{xy}\, dA,$

where $\mathcal{R} = \{(x, y) \mid 1 \le x \le 2 \text{ and } 1 \le y \le 3\}$

19. $\iint_{\mathcal{R}} \cos(xy)\, dA,$

where $\mathcal{R} = \left\{(x, y) \mid \dfrac{\pi}{4} \le x \le \dfrac{\pi}{2} \text{ and } \dfrac{\pi}{2} \le y \le \pi\right\}$

20. Use the results of Exercises 18 and 19 to explain why it may not be possible to evaluate a double integral by using an iterated integral.

21. Show that $\int_a^b \int_c^d f(x, y)\, dy\, dx$ does not always equal $\int_c^d \int_a^b f(x, y)\, dx\, dy$ by evaluating these two iterated integrals for $f(x, y) = \dfrac{x - y}{(x+y)^3}$ when $a = c = 0$ and $b = d = 1$. Why does this result not violate Fubini's theorem?

22. Outline the steps required to find the volume of the solid bounded by the graph of a function $f(x, y)$ and the xy-plane for $a \le x \le b$ and $c \le y \le d$.

Skills

Evaluate the sums in Exercises 23–28.

23. $\displaystyle\sum_{j=1}^{3} \sum_{k=1}^{2} j^k$

24. $\displaystyle\sum_{j=1}^{3} \sum_{k=1}^{2} k^j$

25. $\displaystyle\sum_{j=1}^{3} \sum_{k=1}^{4} (3j - 4k)$

26. $\displaystyle\sum_{j=1}^{m} \sum_{k=1}^{n} (j - k)$

27. $\displaystyle\sum_{i=1}^{4} \sum_{j=1}^{3} \sum_{k=1}^{2} ij^2 k^3$

28. $\displaystyle\sum_{i=1}^{l} \sum_{j=1}^{m} \sum_{k=1}^{n} ij^2 k^3$

Use Definition 13.4 to evaluate the double integrals in Exercises 29–32.

29. $\iint_{\mathcal{R}} xy\, dA$

where $\mathcal{R} = \{(x, y) \mid 0 \le x \le 2 \text{ and } 1 \le y \le 4\}$

30. $\iint_{\mathcal{R}} x^2 y\, dA$

where $\mathcal{R} = \{(x, y) \mid -1 \le x \le 0 \text{ and } 0 \le y \le 2\}$

31. $\iint_{\mathcal{R}} xy^3\, dA$

where $\mathcal{R} = \{(x, y) \mid -2 \le x \le 2 \text{ and } -1 \le y \le 1\}$

32. $\iint_{\mathcal{R}} x^2 y^3\, dA$

where $\mathcal{R} = \{(x, y) \mid 1 \le x \le 3 \text{ and } 0 \le y \le 2\}$

Evaluate each of the integrals in Exercises 33–36 as an iterated integral, and then compare your answers with those you found in Exercises 29–32.

33. The integral in Exercise 29.

34. The integral in Exercise 30.

35. The integral in Exercise 31.

36. The integral in Exercise 32.

Evaluate each of the double integrals in Exercises 37–54 as iterated integrals.

37. $\iint_{\mathcal{R}} (3 - x + 4y)\, dA,$

where $\mathcal{R} = \{(x, y) \mid 0 \le x \le 1 \text{ and } -1 \le y \le 3\}$

38. $\iint_{\mathcal{R}} y^2\, dA,$

where $\mathcal{R} = \{(x, y) \mid -3 \le x \le 2 \text{ and } -2 \le y \le 2\}$

39. $\iint_{\mathcal{R}} (2 - 3x^2 + y^2)\, dA,$

where $\mathcal{R} = \{(x, y) \mid -3 \le x \le 2 \text{ and } 3 \le y \le 5\}$

40. $\iint_{\mathcal{R}} (x - e^y)\, dA,$

where $\mathcal{R} = \{(x, y) \mid -3 \le x \le 2 \text{ and } -2 \le y \le 2\}$

41. $\iint_{\mathcal{R}} \sin(x + 2y)\, dA,$

where $\mathcal{R} = \left\{(x, y) \mid 0 \le x \le \pi \text{ and } 0 \le y \le \dfrac{\pi}{2}\right\}$

42. $\iint_{\mathcal{R}} x \sin x \cos y\, dA,$

where $\mathcal{R} = \{(x, y) \mid -3 \le x \le 2 \text{ and } -2 \le y \le 2\}$

43. $\iint_{\mathcal{R}} xe^{xy}\, dA,$

where $\mathcal{R} = \{(x, y) \mid 0 \le x \le 1 \text{ and } 0 \le y \le \ln 5\}$

44. $\iint_{\mathcal{R}} x^2 \cos(xy)\, dA,$

where $\mathcal{R} = \{(x, y) \mid 0 \le x \le \pi \text{ and } 0 \le y \le 1\}$

45. $\iint_{\mathcal{R}} \dfrac{x}{x+y}\, dA,$

where $\mathcal{R} = \{(x, y) \mid 1 \le x \le 4 \text{ and } 0 \le y \le 3\}$

46. $\iint_{\mathcal{R}} x^3 e^{x^2 y}\, dA,$

where $\mathcal{R} = \{(x, y) \mid 0 \le x \le 4 \text{ and } -1 \le y \le 1\}$

47. $\iint_{\mathcal{R}} y \sin x\, dA,$

where $\mathcal{R} = \left\{(x, y) \mid 0 \le x \le \dfrac{\pi}{2} \text{ and } 0 \le y \le 1\right\}$

48. $\iint_{\mathcal{R}} \sin(2x + y)\, dA,$

where $\mathcal{R} = \left\{(x, y) \mid 0 \le x \le \dfrac{\pi}{2} \text{ and } 0 \le y \le \dfrac{\pi}{2}\right\}$

49. $\iint_{\mathcal{R}} y^2 \sin x\, dA,$

where $\mathcal{R} = \{(x, y) \mid 0 \le x \le \pi \text{ and } 0 \le y \le 3\}$

50. $\iint_{\mathcal{R}} xy\sin(x^2)\, dA$,

where $\mathcal{R} = \{(x, y) \mid 0 \le x \le \sqrt{\pi} \text{ and } 0 \le y \le 1\}$

51. $\iint_{\mathcal{R}} y\cos(xy)\, dA$,

where $\mathcal{R} = \left\{(x, y) \mid 0 \le x \le \dfrac{\pi}{2} \text{ and } 0 \le y \le 1\right\}$

52. $\iint_{\mathcal{R}} e^{x+y}\, dA$,

where $\mathcal{R} = \{(x, y) \mid 0 \le x \le 1 \text{ and } 0 \le y \le 1\}$

53. $\iint_{\mathcal{R}} x^2 e^{xy}\, dA$,

where $\mathcal{R} = \{(x, y) \mid 0 \le x \le 1 \text{ and } 0 \le y \le 1\}$

54. $\iint_{\mathcal{R}} \dfrac{dA}{x^2 + 2xy + y^2}$,

where $\mathcal{R} = \{(x, y) \mid 1 \le x \le 2 \text{ and } 0 \le y \le 1\}$

In Exercises 55–58, find the signed volume between the graph of the given function and the xy-plane over the specified rectangle in the xy-plane.

55. $f(x, y) = 3x - 2y^5 + 1$,

where $\mathcal{R} = \{(x, y) \mid -4 \le x \le 6 \text{ and } 0 \le y \le 7\}$

56. $f(x, y) = x^2 - y^3 + 4$,

where $\mathcal{R} = \{(x, y) \mid 0 \le x \le 3 \text{ and } -2 \le y \le 3\}$

57. $f(x, y) = y^3 e^{xy^2}$,

where $\mathcal{R} = \{(x, y) \mid 0 \le x \le 2 \text{ and } -2 \le y \le 3\}$

58. $f(x, y) = xy^3 e^{x^2 y^2}$,

where $\mathcal{R} = \{(x, y) \mid -2 \le x \le -1 \text{ and } -2 \le y \le 0\}$

In Exercises 59–64, find the volume between the graph of the given function and the xy-plane over the specified rectangle.

59. $f(x, y) = xy$,

where $\mathcal{R} = \{(x, y) \mid -2 \le x \le 3 \text{ and } -1 \le y \le 5\}$

60. $f(x, y) = -2x^2 y^3$,

where $\mathcal{R} = \{(x, y) \mid -2 \le x \le 3 \text{ and } -1 \le y \le 5\}$

61. $f(x, y) = \sin x \cos y$,

where $\mathcal{R} = \{(x, y) \mid 0 \le x \le \pi \text{ and } 0 \le y \le \pi\}$

62. $f(x, y) = \dfrac{y}{x} e^y$,

where $\mathcal{R} = \{(x, y) \mid 1 \le x \le e \text{ and } 0 \le y \le 2\}$

63. $f(x, y) = \dfrac{x}{y} + \dfrac{y}{x}$,

where $\mathcal{R} = \{(x, y) \mid 1 \le x \le 3 \text{ and } 1 \le y \le 5\}$

64. $f(x, y) = x^2 y e^{xy}$,

where $\mathcal{R} = \{(x, y) \mid 0 \le x \le 1 \text{ and } 0 \le y \le 2\}$

Use midpoint Riemann sums with the specified numbers of subintervals to approximate the iterated integrals in Exercises 65–68. (See Exercise 11 for the definition of a midpoint Riemann sum for a double integral.)

65. $\int_0^1 \int_0^{3/2} e^{xy}\, dx\, dy$. Let each subrectangle be a square with side length $\dfrac{1}{2}$ unit.

66. $\int_0^{\pi} \int_0^1 \sin(xy)\, dy\, dx$. Use four subrectangles by dividing each of the intervals $[0, \pi]$ and $[0, 1]$ into two equal pieces.

67. $\int_0^1 \int_0^{\pi/2} \cos(xy)\, dx\, dy$. Use six subrectangles by dividing the interval $\left[0, \dfrac{\pi}{2}\right]$ into three equal pieces and the interval $[0, 1]$ into two equal pieces.

68. $\int_0^1 \int_0^1 y\sin(x^2)\, dx\, dy$. Let each subrectangle be a square with side length $\dfrac{1}{3}$ unit.

Applications

69. Emmy oversees the operations of a number of sedimentation tanks, into which a toxic solution is dumped so that the toxic materials will settle out. The base of each tank is a square with side lengths of 80 feet, but she does not know the volume of any of the tanks. She defines a function

$$r(t) = \begin{cases} \dfrac{-(t - 10)^2}{100} + 1, & \text{if } t < 10 \\ 1, & \text{if } t \ge 10. \end{cases}$$

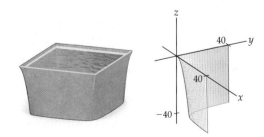

With this function, one-fourth of a tank is described by the surface $-12r(x)r(y)$ for $x \in [0, 40)$, $y \in [0, 40)$. The rest of the tank is the same function reflected about the lines $x = 40$ and $y = 40$ and the point $(40, 40)$. What is the volume of each tank?

70. Leila has been assigned the task of estimating the number of caribou in a rectangular management unit (an area of land) during the summer. She imposes coordinates on the unit, which is 4 miles wide by 5 miles long (i.e., $[0, 4] \times [0, 5]$). She does not have time to commission a count of population, but she knows from a past study that the density of caribou in this region in July is approximated by $d(x, y) = 0.08x^2 y^2 - 0.456x^2 y - 0.08x^2 - 0.328xy^2 + 1.87xy + 0.328x - 0.061y^2 + 0.347y + 0.061$. Roughly how many caribou can be found in the management unit?

Proofs

71. Prove Theorem 13.2(a). That is, prove that $\sum_{j=1}^{m} \sum_{k=1}^{n} a_{jk} = \sum_{k=1}^{n} \sum_{j=1}^{m} a_{jk}$.

72. List the other five orders for the summations in $\sum_{i=1}^{l} \sum_{j=1}^{m} \sum_{k=1}^{n} a_{ijk}$, and prove that all six orders for the summations are equal.

73. Let $a < b$ and $c < d$ be real numbers, and let $\mathcal{R}$ be the rectangle defined by

$$\mathcal{R} = \{(x, y) \mid a \leq x \leq b \text{ and } c \leq y \leq d\}$$

in the xy-plane. If $g(x)$ is continuous on the interval $[a, b]$ and $h(y)$ is continuous on $[c, d]$, use Fubini's theorem to prove that

$$\iint_{\mathcal{R}} g(x)h(y)\, dA = \left(\int_{a}^{b} g(x)\, dx \right)\left(\int_{c}^{d} h(y)\, dy \right).$$

74. Let a and b be positive real numbers, and let

$$\mathcal{R} = \{(x, y) \mid -a \leq x \leq a \text{ and } -b \leq y \leq b\}.$$

Assuming that g and h are continuous on their domains, prove that $\iint_{\mathcal{R}} g(x)h(y)\, dA = 0$ if either g or h is an odd function.

75. Let $a < b$ and $c < d$ be real numbers, and let $\mathcal{R}$ be the rectangle in the xy-plane defined by

$$\mathcal{R} = \{(x, y) \mid a \leq x \leq b \text{ and } c \leq y \leq d\}.$$

Prove that $\iint_{\mathcal{R}} dA = (b-a)(d-c)$. What is the relationship between $\mathcal{R}$ and the product $(b - a)(d - c)$?

Thinking Forward

▶ *Riemann sums for functions of three variables:* Provide a definition of a Riemann sum for a function of three variables. Model your definition on Definition 13.3.

▶ *Triple integrals:* Provide a definition for a triple integral $\iiint_{S} f(x, y, z)\, dV$, where S is a rectangular box in $\mathbb{R}^3$ given by $S = \{(x, y, z) \mid a_1 \leq x \leq a_2, \ b_1 \leq y \leq b_2 \text{ and } c_1 \leq z \leq c_2\}$. Model your definition on Definition 13.4.

13.2 DOUBLE INTEGRALS OVER GENERAL REGIONS

▶ Analyzing bounded regions in the plane

▶ Defining and computing double integrals on general regions

▶ The algebraic properties of double integrals

General Regions in the Plane

In Section 13.1 we defined the double integral of a function of two variables over a rectangular region. We now wish to generalize this concept to define a double integral over a general region Ω (omega) bounded by a simple closed curve—that is, a two-dimensional region in which the boundary may be represented by a finite collection of curves with differentiable parametrizations. Following is an example of such a region:

Unlike definite integrals, which are defined on finite intervals, for double intervals over regions more complicated than a rectangle we have to make an effort to understand the region Ω over which the double integral is defined.

In particular, the regions we use must be subdivided into subregions that fall into one of the following two types:

DEFINITION 13.8

Type I and Type II Regions

(a) Let $y = g_1(x)$ and $y = g_2(x)$ be two functions defined on the interval $[a, b]$ such that $g_1(x) \le g_2(x)$ for every $x \in [a, b]$. The region Ω bounded above by $g_2(x)$, below by $g_1(x)$, on the left by the line $x = a$, and on the right by the line $x = b$ is said to be a ***type I region***.

(b) Let $x = h_1(y)$ and $x = h_2(y)$ be two functions defined on the interval $[c, d]$ such that $h_1(y) \le h_2(y)$ for every $y \in [c, d]$. The region Ω bounded on the left by $h_1(x)$, on the right by $h_2(y)$, below by the line $y = c$, and above by the line $y = d$ is said to be a ***type II region***.

A type I region is shown next at the left, and a type II region next on the right.

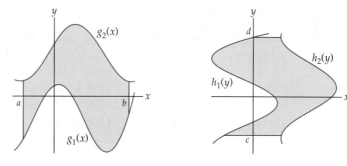

Many regions naturally fall into one or the other of these two categories. Some regions, however, may be considered to be either a type I or a type II region. For example, consider the area in the first quadrant between the functions $y = \sqrt{x}$ and $y = x^3$:

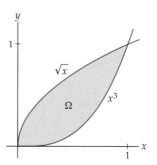

As a type I region, Ω is bounded above by $y = \sqrt{x}$ and below by $y = x^3$ for $x \in [0, 1]$. As a type II region, Ω is bounded on the left by $x = y^2$ and on the right by $x = \sqrt[3]{y}$ for $y \in [0, 1]$.

Many regions need to be subdivided so that their subregions fall more simply into one of the two categories. For example, the region Ω bounded above by the semicircle with equation $y = \sqrt{4 - x^2}$ and below by the quarter circle $y = -\sqrt{4 - x^2}$ on the interval $[-2, 0]$ and by the x-axis on the interval $[0, 2]$ may be subdivided into the two subregions Ω_1 and Ω_2, as pictured next. We see that Ω_1 and Ω_2 are type I regions.

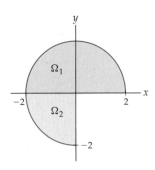

Note, however, that there are always many ways to subdivide such a region. If we wish, we may divide this region into three quarter circles, each of which may be considered to be a type I or a type II region. For example, Ω_2, shown in the preceding figure, is also a type II region, bounded on the left by $x = -\sqrt{4 - y^2}$ and on the right by the y-axis on the interval $-2 \le y \le 0$.

Double Integrals over General Regions

We now expand our definition of the double integral to allow us to compute a double integral over a general region Ω whose boundary is a simple closed curve. Let $f(x, y)$ be defined on Ω. Note that since Ω is bounded, there exists a rectangle $\mathcal{R} = \{(x, y) \mid a \le x \le b$ and $c \le y \le d\}$ such that Ω is a subset of $\mathcal{R}$ (i.e., $\Omega \subseteq \mathcal{R}$).

DEFINITION 13.9

Double Integrals over General Regions

Let Ω be a region in the xy-plane bounded by a simple closed curve, let $f(x, y)$ be a function defined on Ω, and let $\mathcal{R} = \{(x, y) \mid a \le x \le b$ and $c \le y \le d\}$ be a rectangle containing Ω. We define the **double integral** of f over Ω to be

$$\iint_\Omega f(x, y)\, dA = \iint_\mathcal{R} F(x, y)\, dA,$$

where

$$F(x, y) = \begin{cases} f(x, y), & \text{if } (x, y) \in \Omega \\ 0, & \text{if } (x, y) \notin \Omega, \end{cases}$$

provided that the double integral $\iint_\mathcal{R} F(x, y)\, dA$ exists.

As with double integrals defined on rectangular regions, we will try to evaluate double integrals on general regions with Fubini's theorem. To use Fubini's theorem in this context, when we wish to integrate the function $f(x, y)$ over the type I region Ω bounded above by $g_2(x)$, below by $g_1(x)$, on the left by the line $x = a$, and on the right by the line $x = b$, we set up the iterated integral

$$\iint_\Omega f(x, y)\, dA = \int_a^b \int_{g_1(x)}^{g_2(x)} f(x, y)\, dy\, dx.$$

Similarly, to integrate $f(x, y)$ over the type II region bounded on the left by $h_1(x)$, on the right by $h_2(y)$, below by the line $y = c$, and above by the line $y = d$, we set up the iterated integral

$$\iint_\Omega f(x, y)\, dA = \int_c^d \int_{h_1(y)}^{h_2(y)} f(x, y)\, dx\, dy.$$

Note that in each case the limits of the outer integration are constants, and the limits of the inner integration are functions of the outer variable of integration.

For example, let Ω be the following triangular region in the first quadrant:

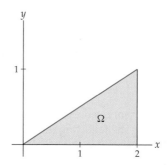

To evaluate the double integral of a function $f(x, y)$ on this region, we may treat Ω either as the type I region bounded below by $y = 0$ and above by $y = \frac{1}{2}x$ for $x \in [0, 2]$ or as the type II region bounded on the left by the function $x = 2y$ and on the right by the function $x = 2$ for $y \in [0, 1]$. Therefore,

$$\iint_{\Omega} f(x, y)\, dA = \int_0^2 \int_0^{x/2} f(x, y)\, dy\, dx = \int_0^1 \int_{2y}^2 f(x, y)\, dx\, dy.$$

Depending upon the function $f(x, y)$, one of these two iterated integrals may be significantly easier to evaluate. When $f(x, y) = \cos(x^2)$, treating Ω as a type I region results in a significantly simpler integral, since the inner integral may be easily evaluated with the Fundamental Theorem of Calculus:

$$\int_0^{x/2} \cos(x^2)\, dy = \left[y \cos(x^2) \right]_{y=0}^{y=x/2} = \frac{1}{2}x \cos(x^2).$$

To finish evaluating the iterated integral, we now use substitution to find an antiderivative for the resulting function. We have

$$\int_0^2 \frac{1}{2}x \cos(x^2)\, dx = \left[\frac{1}{4} \sin(x^2) \right]_{x=0}^{x=2} = \frac{1}{4} \sin 4.$$

By contrast, if we try to evaluate the double integral by using the iterated integral

$$\int_0^1 \int_{2y}^2 \cos(x^2)\, dx\, dy,$$

we are immediately stymied by the fact that the function $\cos(x^2)$ does *not* have a simple antiderivative when we are integrating first with respect to the variable x.

This example illustrates how crucial it is to be able to analyze a given planar region as both a type I region and a type II region. When we write a double integral as an iterated integral, one of the two possible orders of integration can result in a computation that is significantly easier than the other. In such a case, the relatively simple chore of selecting the appropriate order of integration saves a great deal of work integrating.

Algebraic Properties of Double Integrals

The algebraic properties set forth in the next theorem are similar to the algebraic properties of definite integrals we studied in Chapter 4. The proofs of these properties follow from the definition of the double integral and are left for Exercises 67 and 68.

THEOREM 13.10   **Algebraic Properties of the Double Integral**

Let $f(x, y)$ and $g(x, y)$ be integrable functions on the general region Ω, and let $c \in \mathbb{R}$. Then

(a) $\displaystyle\iint_{\Omega} cf(x, y)\, dA = c \iint_{\Omega} f(x, y)\, dA.$

(b) $\displaystyle\iint_{\Omega} (f(x, y) + g(x, y))\, dA = \iint_{\Omega} f(x, y)\, dA + \iint_{\Omega} g(x, y)\, dA.$

As we mentioned earlier in the section, we may also need to decompose a region into subregions. The following theorem tells us that the double integral is well behaved with respect to reasonable subdivisions:

THEOREM 13.11   **Subdividing the Region on Which a Double Integral Is Defined**

Let $f(x, y)$ be an integrable function on the general region Ω. If Ω_1 and Ω_2 are general regions that are subsets of Ω that do not overlap, except possibly on their boundaries, and if $\Omega = \Omega_1 \cup \Omega_2$, then

$$\iint_{\Omega} f(x, y)\, dA = \iint_{\Omega_1} f(x, y)\, dA + \iint_{\Omega_2} f(x, y)\, dA.$$

Example 1 provides an application of Theorem 13.11 that illustrates how to split a region into subregions for integration.

Examples and Explorations

EXAMPLE 1   Expressing a planar region as a type I region and a type II region

Express the double integral $\iint_{\Omega} f(x, y)\, dA$ as an iterated integral over the following region Ω, itself expressed as both a type I region and a type II region:

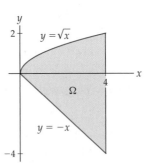

SOLUTION

It is slightly easier to express Ω as type I region. To envision the boundary components of a type I region, it often helps to draw a "typical" vertical strip, as shown in the figure at the left.

Ω as a type I region

Ω as two type II regions

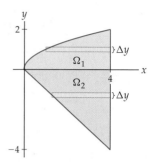

Each vertical strip that crosses the region is bounded below by the function $y = -x$ and above by the function $y = \sqrt{x}$. Such vertical strips would cover Ω as x varies over the interval $[0, 4]$. Thus, to integrate $f(x, y)$ over Ω as a type I region, we have

$$\iint_{\Omega} f(x, y)\, dA = \int_0^4 \int_{-x}^{\sqrt{x}} f(x, y)\, dy\, dx.$$

To understand the region as a type II region, we draw two horizontal strips. Above the x-axis, these strips are bounded on the left by the function $x = y^2$ and on the right by the line $x = 4$. Such strips would cover Ω_1 as y varies over the interval $[0, 2]$, as pictured at the right. Similarly, below the x-axis such strips are bounded on the left by the function $x = -y$ and on the right by the line $x = 4$. They would cover Ω_2 as y varies over the interval $[-4, 0]$. Thus, to integrate a function $f(x, y)$ over Ω as a type II region, we would have

$$\iint_{\Omega} f(x, y)\, dA = \int_{-4}^0 \int_{-y}^4 f(x, y)\, dx\, dy + \int_0^2 \int_{y^2}^4 f(x, y)\, dx\, dy.$$

Note that we have used Theorem 13.11 to break the integral over Ω into the integrals over Ω_1 and Ω_2. Also, observe that for all three iterated integrals shown, the limits of the outer integrals are constants. The limits of the inner integrals are functions of the outer variable, although they may be constant functions.   □

EXAMPLE 2   Drawing a region determined by an iterated integral

Sketch the regions determined by the iterated integrals:

(a) $\displaystyle \int_{-3\pi/4}^{\pi/4} \int_{\cos x}^{\sin x} xy\, dy\, dx$ **(b)** $\displaystyle \int_0^2 \int_0^{\sqrt{2-y}} xe^{x^2}\, dx\, dy$

SOLUTION

(a)   The region described by the first integral is bounded below by the function $y = \cos x$ and above by $y = \sin x$ for values of x in the interval $[-3\pi/4, \pi/4]$. This region is shown here at the left:

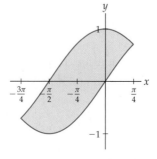

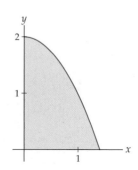

(b) The region described by the second integral is bounded on the left by the line $x = 0$ and on the right by the function $x = \sqrt{2-y}$ for values of y in the interval $[0, 2]$. For convenience, to sketch the function $x = \sqrt{2-y}$, we may rewrite it as $y = 2 - x^2$, where $x \geq 0$. This region is shown at the right. □

EXAMPLE 3   Evaluating an iterated integral

Evaluate the iterated integrals from Example 2.

SOLUTION

To evaluate an iterated integral, we begin by evaluating the inner integral.

(a) Thus, when we evaluate $\int_{-3\pi/4}^{\pi/4} \int_{\cos x}^{\sin x} xy \, dy \, dx$, we start with

$$\int_{\cos x}^{\sin x} xy \, dy = \left[\frac{1}{2}xy^2 \right]_{y=\cos x}^{y=\sin x} = \frac{1}{2}x(\sin^2 x - \cos^2 x) = -\frac{1}{2}x\cos(2x).$$

For the rightmost equality, we used the double-angle identity for the cosine. We now integrate this function on the interval $\left[-\frac{3\pi}{4}, \frac{\pi}{4}\right]$ to complete the evaluation of the iterated integral:

$$\int_{-3\pi/4}^{\pi/4} \left(-\frac{1}{2}x\cos(2x)\right) dx = \left[-\left(\frac{1}{4}x\sin(2x) + \frac{1}{8}\cos(2x)\right)\right]_{-3\pi/4}^{\pi/4} = -\frac{\pi}{4}.$$

Note that we used integration by parts to find the antiderivative of the function.

(b) For the second iterated integral, $\int_0^2 \int_0^{\sqrt{2-y}} xe^{x^2} \, dx \, dy$, we again work first on the inner integral. Here we use substitution to find the antiderivative:

$$\int_0^{\sqrt{2-y}} xe^{x^2} \, dx = \left[\frac{1}{2}e^{x^2}\right]_0^{\sqrt{2-y}} = \frac{1}{2}(e^{2-y} - 1).$$

We now evaluate the outer integral:

$$\int_0^2 \frac{1}{2}(e^{2-y} - 1) \, dy = \left[-\frac{1}{2}(e^{2-y} + y)\right]_0^2 = \frac{1}{2}(e^2 - 3).$$ □

EXAMPLE 4   Reversing the order of integration in an iterated integral

Reverse the order of integration in the iterated integral $\int_0^2 \int_0^{\sqrt{2-y}} xe^{x^2} \, dx \, dy$. Then evaluate the new iterated integral and show that the result is the same as the value obtained in Example 3.

SOLUTION

This is one of the integrals we have been discussing since Example 2. In that example, we drew the region Ω, expressed as a type II region. To reverse the order of integration we must first express Ω as a type I region. Using the work we did in Example 2, we see that the region is bounded above by $y = 2 - x^2$ and below by $y = 0$, where x is in the interval $[0, \sqrt{2}]$. Thus,

$$\int_0^2 \int_0^{\sqrt{2-y}} xe^{x^2} \, dx \, dy = \int_0^{\sqrt{2}} \int_0^{2-x^2} xe^{x^2} \, dy \, dx.$$

Again, working on the inner integral, we have

$$\int_0^{2-x^2} xe^{x^2}\, dy = \left[xe^{x^2}y\right]_0^{2-x^2} = (2x - x^3)e^{x^2}.$$

To finish this example we evaluate the outer integral. Using the techniques of integration by substitution and integration by parts, we obtain

$$\int_0^{\sqrt{2}} (2x - x^3)e^{x^2}\, dx = \left[\frac{1}{2}(3 - x^2)e^{x^2}\right]_0^{\sqrt{2}} = \frac{1}{2}(e^2 - 3).$$

As we have previously mentioned, when we evaluate an iterated integral, sometimes one order of integration leads to a computation that is significantly simpler than the other order of integration. Which order of integration did you find easier for this integral, the one we did in Example 3, or the one we just completed?   □

?   TEST YOUR UNDERSTANDING

▶ What is a general region in the xy-plane? What is a type I region? What is a type II region? Is every general region either a type I region or a type II region?

▶ How is the definition of the double integral over a rectangular region used to define the double integral of a function over a general region? Why does this definition make sense?

▶ How is a region that is neither a type I region nor a type II region decomposed into subregions that are type I or type II?

▶ Which order of integration in an iterated integral is associated with a type I region, and which order of integration in an iterated integral is associated with a type II region? Why?

▶ How can the shape of a region Ω help or hinder in the evaluation of the double integral $\iint_\Omega f(x, y)\, dA$? Does the interval $[a, b]$ ever make a definite integral $\int_a^b g(x)\, dx$ easier or harder to evaluate?

EXERCISES 13.2

Thinking Back

▶ *Finding the area between two curves:* If $g_1(x)$ and $g_2(x)$ are two continuous functions such that $g_1(x) \le g_2(x)$ on the interval $[a, b]$, find a definite integral representing the area between the graphs of the two functions on $[a, b]$.

▶ *Finding the area between two curves:* If $h_1(y)$ and $h_2(y)$ are two continuous functions such that $h_1(y) \le h_2(y)$ on the interval $[c, d]$, find a definite integral representing the area between the graphs of the two functions on $[c, d]$.

Concepts

0. *Problem Zero:* Read the section and make your own summary of the material.

1. *True/False:* Determine whether each of the statements that follow is true or false. If a statement is true, explain why. If a statement is false, provide a counterexample.

(a) *True or False:* Every rectangular region in the plane, with its sides parallel to the coordinate axes, may be considered to be either a type I region or a type II region.

(b) *True or False:* Every rectangular region in the plane may be considered to be a type I region or a type II region.

(c) *True or False:* Every region in the plane can be expressed as either a single type I region or a single type II region.

(d) *True or False:* A general region in the plane with a polygonal boundary can be decomposed into finitely many type I and type II regions.

(e) *True or False:* To evaluate the double integral $\iint_\Omega f(x, y)\, dA$ where Ω is a type I region, you integrate first with respect to y.

(f) *True or False:* If Ω is the set of all points satisfying the inequality $x^2 + y^2 \leq 4$, then $\iint_\Omega f(x, y)\, dA = \int_{-2}^{2} \int_{-\sqrt{4-y^2}}^{\sqrt{4-y^2}} f(x, y)\, dx\, dy$.

(g) *True or False:* If f is a continuous function and $\Omega = \Omega_1 \cup \Omega_2$, then $\iint_\Omega f(x, y)\, dA = \iint_{\Omega_1} f(x, y)\, dA + \iint_{\Omega_2} f(x, y)\, dA$.

(h) *True or False:* If f is a positive continuous function defined on a region Ω, and if $\Gamma \subset \Omega$, then $\iint_\Gamma f(x, y)\, dA \leq \iint_\Omega f(x, y)\, dA$.

2. *Examples:* Construct examples of the thing(s) described in the following. Try to find examples that are different than any in the reading.

(a) A non-rectangular region that can be expressed as a single type I region or as a single type II region.

(b) A rectangular region Ω that can be expressed neither as a single type I region nor as a single type II region.

(c) A region Ω that can be expressed as a single type II region, but that requires two type I regions intersecting only on their boundaries, in order to express Ω simply.

3. Explain the difference between a type I region and a type II region.

4. Let $\mathcal{R} = \{(x, y) \mid a \leq x \leq b \text{ and } c \leq y \leq d\}$ be a rectangular region. Explain why $\mathcal{R}$ is both a type I region and a type II region.

Which of the iterated integrals in Exercises 5–8 could correctly be used to evaluate the double integral $\iint_\mathcal{R} f(x, y)\, dA$, where $f(x, y)$ is a continuous function and $\mathcal{R}$ is the rectangular region bounded by the lines $x = 1$, $x = 4$, $y = 2$, and $y = 6$? For each incorrect integral, how could it be changed to give the correct value?

5. $\displaystyle \int_1^4 \int_2^6 f(x, y)\, dy\, dx$

6. $\displaystyle \int_4^1 \int_6^2 f(x, y)\, dy\, dx$

7. $\displaystyle \int_1^4 \int_2^6 f(x, y)\, dx\, dy$

8. $\displaystyle \int_2^6 \int_1^4 f(x, y)\, dx\, dy$

Which of the iterated integrals in Exercises 9–12 could correctly be used to evaluate the double integral $\iint_\Omega f(x, y)\, dA$, where $f(x, y)$ is a continuous function and Ω is the right triangular region bounded below by the x-axis, on the left by the y-axis, and along the hypotenuse by the line $y = -x + 2$? For each incorrect integral, how could it be changed to give the correct value?

9. $\displaystyle \int_0^2 \int_0^{-x+2} f(x, y)\, dy\, dx$

10. $\displaystyle \int_0^{-x+2} \int_0^2 f(x, y)\, dx\, dy$

11. $\displaystyle \int_0^2 \int_0^{-y+2} f(x, y)\, dx\, dy$

12. $\displaystyle \int_2^0 \int_{-y+2}^0 f(x, y)\, dx\, dy$

13. The following region Ω is bounded by the functions $y = \frac{1}{2}x$ and $y = \sqrt{x}$:

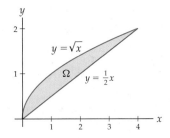

Express Ω as a type I region and as a type II region. Referring to Definition 13.8, if Ω is a type I region, what are a and b? If Ω is a type II region, what are c and d?

14. Explain why the double integral $\iint_\Omega dA$ gives the area of the region Ω. Illustrate your explanation with an example.

15. Let $g_1(x)$ and $g_2(x)$ be two continuous functions such that $g_1(x) \leq g_2(x)$ on the interval $[a, b]$, and let Ω be the region in the xy-plane bounded by g_1 and g_2 on $[a, b]$. Use your answer to Exercise 14 to set up an iterated integral whose value is the area of Ω. How is this iterated integral related to the definite integral you would have used to compute the area of Ω in Chapter 4?

16. Let $h_1(y)$ and $h_2(y)$ be two continuous functions such that $h_1(x) \leq h_2(x)$ on the interval $[c, d]$, and let Ω be the region in the xy-plane bounded by h_1 and h_2 on $[c, d]$. Use your answer to Exercise 14 to set up an iterated integral whose value is the area of Ω. How is this iterated integral related to the definite integral you would have used to compute the area of Ω in Chapter 4?

17. Use the results of Exercises 15 and 16 to find the area of the region Ω shown in Exercise 13.

18. Express the area of the region Ω between the function $f(x) = x^2$ and the x-axis on the interval $[-3, 3]$ as an iterated integral, integrating first with respect to x. Express the area of Ω as a sum of two iterated integrals, integrating first with respect to y in each. Now evaluate your integrals.

19. Express the area of the region Ω between the function $f(x) = x^3$ and the x-axis on the interval $[-3, 3]$ as a sum of two iterated integrals, integrating first with respect to x in each. Express the area of Ω as a sum of two different iterated integrals, integrating first with respect to y. Now evaluate your integrals.

20. When you wish to evaluate the definite integral $\int_a^b f(x)\, dx$ of a continuous function f, the interval $[a, b]$ is never an impediment to using the Fundamental Theorem of Calculus. However, when you wish to evaluate the double integral $\iint_\Omega g(x, y)\, dA$ of a continuous function g over a region Ω, the region can make the evaluation process easier or harder. Why?

Skills

Let $f(x, y)$ be a continuous function. For each region Ω shown in Exercises 21–24, set up one or more (if necessary) iterated integrals to compute $\iint_\Omega f(x, y)\, dA$, (a) where you integrate first with respect to y and (b) where you integrate first with respect to x.

21.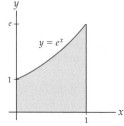

22.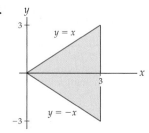

23.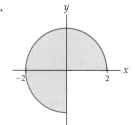

24.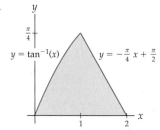

Let $f(x, y)$ be a continuous function. Sketch each region Ω described in Exercises 25–28. Then set up one or more (if necessary) iterated integrals to compute $\iint_\Omega f(x, y)\, dA$, (a) where you integrate first with respect to y and (b) where you integrate first with respect to x.

25. $\Omega = \left\{ (x, y) \mid 0 \le x \le \dfrac{\pi}{4} \text{ and } \sin x \le y \le \cos x \right\}$

26. $\Omega = \{(x, y) \mid -2 \le x \le 2 \text{ and } |x| \le y \le 2\}$

27. $\Omega = \{(x, y) \mid |x| + |y| \le 1\}$

28. $\Omega = \{(x, y) \mid x^2 + y^2 \le 9\}$

In Exercises 29–34, sketch the region determined by the limits of the iterated integrals and then give another iterated integral (or a sum of iterated integrals if necessary) using the opposite order of integration.

29. $\displaystyle\int_1^2 \int_{\ln x}^{e^x} f(x, y)\, dy\, dx$

30. $\displaystyle\int_0^2 \int_x^{-x+4} f(x, y)\, dy\, dx$

31. $\displaystyle\int_0^{\pi/2} \int_0^{\sin y} f(x, y)\, dx\, dy$

32. $\displaystyle\int_0^8 \int_0^{\sqrt[3]{y}} f(x, y)\, dx\, dy$

33. $\displaystyle\int_0^{\pi/2} \int_{\sin y}^1 f(x, y)\, dx\, dy$

34. $\displaystyle\int_0^8 \int_{\sqrt[3]{y}}^2 f(x, y)\, dx\, dy$

In Exercises 35–40, find the volume of the solid bounded above by the given function over the specified region Ω.

35. $f(x, y) = 10 - 2x + y$, with Ω the region from Exercise 21.

36. $f(x, y) = 10 - 2x + y$, with Ω the region from Exercise 22.

37. $f(x, y) = \sqrt{4 - x^2 - y^2}$, with Ω the region from Exercise 23.

38. $f(x, y) = x^2 y$, with Ω the region from Exercise 25.

39. $f(x, y) = \sin x \cos y$, with Ω the region from Exercise 26.

40. $f(x, y) = 1 - |x| - |y|$, with Ω the region from Exercise 27.

Find the volumes of the solids described in Exercises 41–44.

41. The portion of the first octant bounded by the coordinate planes and the plane $3x + 4y + 6z = 12$.

42. The solid bounded above by the plane with equation $2x + 3y - z = 2$ and bounded below by the triangle with vertices $(1, 0, 0)$, $(4, 0, 0)$, and $(0, 2, 0)$.

43. The solid bounded above by the paraboloid with equation $z = 8 - x^2 - y^2$ and bounded below by the rectangle $\mathcal{R} = \{(x, y) \mid 1 \le x \le 2 \text{ and } 0 \le y \le 2\}$ in the xy-plane.

44. The solid bounded above by the hyperboloid with equation $z = x^2 - y^2$ and bounded below by the square with vertices $(2, 2, -4)$, $(2, -2, -4)$, $(-2, -2, -4)$, and $(-2, 2, -4)$.

Evaluate the iterated integrals in Exercises 45–48 by reversing the order of integration. Explain why it is easier to reverse the order of integration than evaluate the given iterated integral.

45. $\displaystyle\int_0^9 \int_{\sqrt{y}}^3 \sqrt{1 + x^3}\, dx\, dy$

46. $\displaystyle\int_0^{\sqrt{\pi}} \int_x^{\sqrt{\pi}} \cos(y^2)\, dy\, dx$

47. $\displaystyle\int_0^{\sqrt{3}} \int_{\tan^{-1} x}^{\pi/3} \sec y\, dy\, dx$

48. $\displaystyle\int_0^1 \int_{\pi/4}^{\cot^{-1} y} \csc x\, dx\, dy$

In Exercises 49–58, sketch the region determined by the iterated integral and then evaluate the integral. For some of these integrals, it may be helpful to reverse the order of integration.

49. $\displaystyle\int_0^3 \int_{x+2}^{2x-3} (x^2 + 3xy)\, dy\, dx$

50. $\displaystyle\int_0^3 \int_{y+2}^{2y-3} (x^2 + 3xy)\, dx\, dy$

51. $\displaystyle\int_4^9 \int_2^{\sqrt{x}} (x^3 + y^2)\, dy\, dx$ **52.** $\displaystyle\int_4^9 \int_2^{\sqrt{y}} (x^2 + 3xy)\, dx\, dy$

53. $\displaystyle\int_0^{\pi/4} \int_{\tan x}^{\sec x} y\, dy\, dx$ **54.** $\displaystyle\int_0^1 \int_{-e^x}^{e^x} \sin(e^x)\, dy\, dx$

55. $\displaystyle\int_0^1 \int_y^1 e^{x^2}\, dx\, dy$ **56.** $\displaystyle\int_0^{\pi/4} \int_0^{\sec x} \sec x\, dy\, dx$

57. $\displaystyle\int_{\pi/4}^{3\pi/4} \int_0^{\csc y} \csc y\, dx\, dy$

58. $\displaystyle\int_0^3 \int_0^{\sqrt{9-y^2}} \sqrt{9 - x^2}\, dx\, dy$

In Exercises 59–62, evaluate the double integral over the specified region.

59. $\iint_\Omega xe^{x^3}\, dA$, where Ω is the triangular region with vertices $(0, 0)$, $(2, 0)$, and $(2, 2)$.

60. $\iint_\Omega xe^{x^3}\, dA$, where Ω is the triangular region in the first quadrant bounded below by the x-axis, bounded above by the line $y = mx$, where $m > 0$, and bounded on the right by the line with equation $x = 1$.

61. $\iint_\Omega x^2 y^3\, dA$, where Ω is the region in the first quadrant bounded by the graphs of the curves $y = x^2$ and $x = y^2$.

62. $\iint_\Omega dA$, where Ω is the region in the first quadrant bounded by the graphs of the curves $y = x^m$ and $y = x^n$, where m and n are distinct positive integers.

Applications

63. Emmy oversees the operation of a sedimentation lagoon that was built and lined using the natural contours of the terrain. The bottom of the lagoon is the part of the surface

$$z = \frac{1}{10}|x| + \frac{1}{10}|y| - 3$$

that lies below the $z = 0$ plane, where all units are in meters and $z = 0$ represents the water level. What is the volume of the lagoon?

64. Leila is designing a new summer range management unit for caribou in the Selkirk Mountains in the Idaho panhandle. The old unit was laid out as a rectangle, which had nothing to do with the behavior of the caribou. The new one is supposed to resemble the actual area in which the caribou live. Leila has used a study which indicates that the density of caribou in this region in July is approximated by $d(x, y) = 0.08x^2y^2 - 0.456x^2y - 0.08x^2 - 0.328xy^2 + 1.87xy + 0.328x - 0.061y^2 + 0.347y + 0.061$. Her proposed southern boundary for the management unit is a mountain ridge that roughly follows the curve $0.0195x^4$, while the northern border is a political boundary at $\frac{x}{4} + 4$. The western boundary is a state line on which she places the y-axis. Roughly how many caribou can be found in the management unit?

Proofs

65. Let $f(x, y)$ be an integrable function on the rectangle $\mathcal{R} = \{(x, y) \mid a \le x \le b \text{ and } c \le y \le d\}$, and let $\alpha \in \mathbb{R}$. Use the definition of the double integral to prove that

$$\iint_\mathcal{R} \alpha f(x, y)\, dA = \alpha \iint_\mathcal{R} f(x, y)\, dA.$$

66. Let $f(x, y)$ and $g(x, y)$ be integrable functions on the rectangle $\mathcal{R} = \{(x, y) \mid a \le x \le b \text{ and } c \le y \le d\}$. Use the definition of the double integral to prove that

$$\iint_\mathcal{R} (f(x, y) + g(x, y))\, dA$$
$$= \iint_\mathcal{R} f(x, y)\, dA + \iint_\mathcal{R} g(x, y)\, dA.$$

67. Prove Theorem 13.10 (a). That is, show that if $f(x, y)$ is an integrable function on the general region Ω and $c \in \mathbb{R}$, then

$$\iint_\Omega \alpha f(x, y)\, dA = \alpha \iint_\Omega f(x, y)\, dA.$$

68. Prove Theorem 13.10 (b). That is, show that if $f(x, y)$ and $g(x, y)$ are integrable functions on the general region Ω, then

$$\iint_\Omega (f(x, y) + g(x, y))\, dA$$
$$= \iint_\Omega f(x, y)\, dA + \iint_\Omega g(x, y)\, dA.$$

69. Let a, b, and c be positive real numbers. Prove that the volume of the pyramid with vertices $(0, 0, 0)$, $(a, 0, 0)$, $(0, b, 0)$, and $(0, 0, c)$ is $\frac{1}{6}abc$.

70. Let a and c be positive real numbers. Prove that the volume of the right-square pyramid with vertices $(a, a, 0)$, $(-a, a, 0)$, $(a, -a, 0)$, $(-a, -a, 0)$, and $(0, 0, c)$ is $\frac{4}{3}a^2c$. (*Hint: Use the result of Exercise 69.*)

Thinking Forward

▶ *Three iterated integrals:* Let $[a_1, a_2]$, $[b_1, b_2]$, and $[c_1, c_2]$ be three closed intervals. Explain why the triple integral

$$\int_{a_1}^{a_2} \int_{b_1}^{b_2} \int_{c_1}^{c_2} dz \, dy \, dx$$

computes the volume of the rectangular solid with length $a_2 - a_1$, width $b_2 - b_1$, and height $c_2 - c_1$.

▶ *Three more iterated integrals:* Evaluate the triple integral

$$\int_0^2 \int_0^{-(3/2)x+3} \int_0^{4-2x-(4/3)y} dz \, dy \, dx,$$

and give a physical interpretation to the integral.

13.3   DOUBLE INTEGRALS IN POLAR COORDINATES

▶ Expressing a double integral with polar coordinates

▶ Finding areas of regions bounded by functions expressed with polar coordinates

▶ Finding volumes of solids bounded by functions expressed with polar coordinates

Polar Coordinates and Double Integrals

In Chapter 9 we saw that every point in the coordinate plane can be expressed with polar coordinates (r, θ), where θ, in radians, measures the counterclockwise rotation from the positive x-axis and r measures the signed distance that the point is from the origin on the line determined by θ. In Chapter 9 we allowed r and θ to be any real numbers. In this section, where we discuss how to use polar coordinates to evaluate double integrals, we will insist that $r \geq 0$ and that θ be a real number in an interval of width 2π, typically $\theta \in [0, 2\pi]$ or $\theta \in [-\pi, \pi]$.

In the first two sections of this chapter we discussed how to use rectangular coordinates to integrate functions of two variables. Here, we extend this idea to functions and regions that are more naturally expressed with polar coordinates. That is, we wish to find

$$\int_{\mathcal{R}} f(r, \theta) \, dA,$$

where f is a function of r and θ and the region $\mathcal{R}$ is also expressed in terms of r and θ. In Section 13.1 we began with a basic rectangular region, which we then partitioned into subrectangles. Here we start with a polar "rectangle" defined by the inequalities

$$0 \leq a \leq r \leq b \quad \text{and} \quad \alpha \leq \theta \leq \beta,$$

as shown in the following figure at the left:

A polar "rectangle"

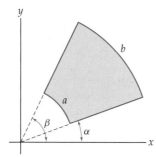

The subdivided rectangle

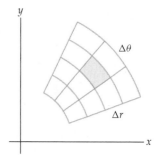

In the right-hand figure, we see the same region divided into "subrectangles." We follow a procedure analogous to that we used in Section 13.1. We assume that $f(r, \theta)$ is a function defined at every point of $\mathcal{R}$. We will outline the steps required to find the volume V of the solid bounded below by $\mathcal{R}$ and bounded above by the graph of f in a moment, but first we need to understand how to compute the area of one of the subregions, shown in the right-hand figure.

The areas of the "subrectangles" depend upon the values of Δr and $\Delta \theta$. Consider the *annulus* with inner radius r_{j-1} and outer radius of r_j, as shown in the following figure at the left:

The annulus with inner and outer radii r_{j-1} and r_j   *A slice of the annulus*

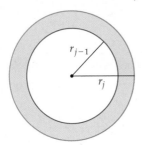 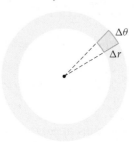

The area of the annulus is given by

$$\pi(r_j^2 - r_{j-1}^2) = \pi(r_j + r_{j-1})(r_j - r_{j-1}) = 2\pi \frac{r_j + r_{j-1}}{2} \Delta r.$$

Note that if we let $\hat{r}_j = \frac{r_j + r_{j-1}}{2}$, then $\hat{r}_j \in [r_{j-1}, r_j]$. Thus, the area of the annulus is $2\pi \hat{r}_j \Delta r$. The area of a "slice" of the annulus corresponding to an angular rotation of $\Delta \theta$ will be

$$(2\pi \hat{r}_j \Delta r)\frac{\Delta \theta}{2\pi} = \hat{r}_j \Delta r \Delta \theta.$$

We are now ready to approximate the volume V of the solid bounded below by $\mathcal{R}$ and above by the graph of f:

▶ We subdivide the interval $[a, b]$ into m equal subintervals, each of width $\Delta r = \frac{b-a}{m}$, and we also let $r_j = a + j\Delta r$ for $0 \le j \le m$.

▶ Similarly, we subdivide the interval $[\alpha, \beta]$ into n equal subintervals, each of width $\Delta \theta = \frac{\beta - \alpha}{n}$, and we let $\theta_k = \alpha + k\Delta \theta$ for $0 \le k \le n$.

▶ The subdivisions we just created partition the polar rectangle into $m \times n$ polar rectangles, like the one shown earlier at the right.

▶ For each $j = 1, 2, \ldots, m$ and $k = 1, 2, \ldots, n$, we select a point (r_j^*, θ_k^*) in the subrectangle $\mathcal{R}_{jk} = \{(r, \theta) \mid r_{j-1} \le r_j^* \le r_j \text{ and } \theta_{k-1} \le \theta_k^* \le \theta_k\}$.

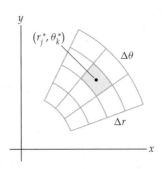

One such point is shown in the preceding figure.

▶ The area of a subregion is given by $\hat{r}_j \Delta r \Delta \theta$, where $\hat{r}_j \in [r_{j-1}, r_j]$.
▶ The product $f(r_j^*, \theta_k^*)\hat{r}_j \Delta r \Delta \theta$ approximates the volume of the solid bounded below by the rectangle $\mathcal{R}_{jk}$ and above by the graph of f.
▶ When we sum these approximate volumes over all of the subregions, we obtain an approximation for the volume of V:

$$\sum_{j=1}^{m} \sum_{k=1}^{n} f(r_j^*, \theta_k^*)\,\hat{r}_j \Delta r \Delta \theta.$$

The limit of this summation provides the iterated integral of the function on the polar region $\mathcal{R}$, namely,

$$\int_{\alpha}^{\beta} \int_{a}^{b} f(r, \theta)r\,dr\,d\theta = \lim_{\Delta \to 0} \sum_{j=1}^{m} \sum_{k=1}^{n} f(r_j^*, \theta_k^*)\,\hat{r}_j \Delta r \Delta \theta,$$

where $\Delta = \sqrt{(\Delta r)^2 + (\Delta \theta)^2}$.

We evaluate this type of iterated integral just as we evaluated iterated integrals expressed with rectangular coordinates. For example, we may prove that the volume of a sphere with radius R is $\frac{4}{3}\pi R^3$ by integrating the function $f(x, y) = \sqrt{R^2 - (x^2 + y^2)}$ over the polar rectangle $\mathcal{R} = \{(r, \theta) \mid 0 \leq r \leq R \text{ and } 0 \leq \theta \leq 2\pi\}$. Before we integrate, however, we must express f in polar coordinates. Recall that $r^2 = x^2 + y^2$. Therefore $z = \sqrt{R^2 - r^2}$. The volume of the top hemisphere is given by the integral

$$\int_{0}^{2\pi} \int_{0}^{R} \sqrt{R^2 - r^2}\, r\,dr\,d\theta.$$

The inner integral is evaluated as

$$\int_{0}^{R} \sqrt{R^2 - r^2}\, r\,dr = \left[-\frac{1}{3}(R^2 - r^2)^{3/2} \right]_0^R = \frac{1}{3}R^3.$$

We now evaluate the outer integral:

$$\int_{0}^{2\pi} \frac{1}{3}R^3 \, d\theta = \left[\frac{1}{3}R^3 \theta \right]_0^{2\pi} = \frac{2}{3}\pi R^3.$$

Therefore the volume of the entire sphere is $\frac{4}{3}\pi R^3$.

Double Integrals in Polar Coordinates over General Regions

We may also use polar coordinates to evaluate an iterated integral over a more general region

$$\Omega = \{(r, \theta) \mid f_1(\theta) \leq r \leq f_2(\theta) \text{ and } \alpha \leq \theta \leq \beta\}.$$

For a polar function $g(r, \theta)$ defined on the region Ω, we have

$$\int_{\alpha}^{\beta} \int_{f_1(\theta)}^{f_2(\theta)} g(r, \theta)\, r\,dr\,d\theta.$$

If $g(r, \theta) \geq 0$ on Ω, then the iterated integral represents the volume of the solid bounded above by the function $g(r, \theta)$ over the region Ω. If $g(r, \theta)$ takes on both positive and negative values on Ω, then the double integral represents the signed volume of the solid between the graph of the function $g(r, \theta)$ and the coordinate plane over the region Ω.

For example, consider the following figures:

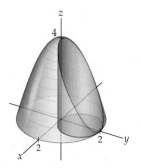
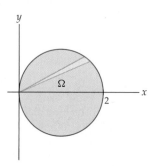

To find the volume of the solid bounded below by the disk whose boundary is the circle with equation $r = 2\cos\theta$ in the coordinate plane and bounded above by the paraboloid $z = 4 - x^2 - y^2$, we evaluate the integral

$$\int_0^\pi \int_0^{2\cos\theta} (4 - r^2)\, r\, dr\, d\theta.$$

Note that:

▶ Our limits of integration for θ are 0 and π, because the entire circle is traced once over this interval.

▶ Our limits of integration for r are 0 and $2\cos\theta$, since every "slice" emanating from the origin has those boundaries, as we see in the right-hand figure.

▶ We have replaced $x^2 + y^2$ with r^2 to express the function with polar coordinates.

▶ If we had preferred, we could have used the expression

$$2\int_0^{\pi/2} \int_0^{2\cos\theta} (4 - r^2)\, r\, dr\, d\theta$$

to evaluate the volume, since both the circular region and the surface are symmetric with respect to the xz-plane.

To evaluate either of the preceding integrals, we begin with the inner integration:

$$\int_0^{2\cos\theta} (4 - r^2)\, r\, dr = \left[2r^2 - \frac{1}{4}r^4\right]_0^{2\cos\theta} = 8\cos^2\theta - 4\cos^4\theta.$$

To finish the computation of the volume, we have

$$\int_0^\pi (8\cos^2\theta - 4\cos^4\theta)\, d\theta = \left[\frac{5}{2}\theta + \frac{5}{2}\sin\theta\cos\theta - \sin\theta\cos^3\theta\right]_0^\pi = \frac{5}{2}\pi.$$

Every double integral expressed with rectangular coordinates may also be expressed with polar coordinates. Recall that we may use the equations

$$x = r\cos\theta \quad \text{and} \quad y = r\sin\theta$$

to change from rectangular coordinates to polar coordinates. Therefore, the double integral

$$\iint_\Omega f(x, y)\, dA$$

may also be expressed as

$$\iint_\Omega f(r\cos\theta, r\sin\theta)\, r\, dr\, d\theta.$$

The value of making this transformation depends upon the region Ω and the particular function f we are trying to integrate. If Ω has a more "natural" expression in polar coordinates, or if f has a simpler antiderivative in polar coordinates, the change to polar coordinates may make the evaluation of the integral considerably easier. We look at such an integral in Example 4.

Examples and Explorations

EXAMPLE 1

Finding the area bounded by a polar rose

Use a double integral to calculate the area bounded by the curve $r = \cos 4\theta$.

SOLUTION

As with all area computations, we must understand the region whose area we are trying to compute. When we can, we will use the symmetry of the region to simplify our work. In Chapter 9 we saw that the graph of an equation of the form $r = \cos n\theta$ or $r = \sin n\theta$ is a polar rose when $n > 1$ is an integer. In addition, if n is even, the figure has $2n$ petals; if n is odd, the figure has n petals. The graph of this curve is as follows:

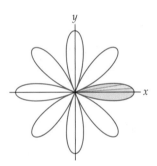

To compute the area bounded by the curve, we will use the symmetry of the rose. Note that we have shaded one of the eight petals of the rose. The values of θ for which the graph passes through the origin correspond to the roots of $\cos 4\theta$ (i.e., odd multiples of $\pi/8$). A sample radial slice of each petal starts at the origin and ends on the curve. The shaded region corresponds to the values of θ in the interval $[-\pi/8, \pi/8]$. The double integral whose value is the area of the shaded region is

$$\int_{-\pi/8}^{\pi/8} \int_0^{\cos 4\theta} r \, dr \, d\theta.$$

Evaluating the inner integral first, we obtain

$$\int_0^{\cos 4\theta} r \, dr = \left[\frac{1}{2} r^2 \right]_0^{\cos 4\theta} = \frac{1}{2} \cos^2 4\theta.$$

Now, evaluating the outer integral, we have

$$\int_{-\pi/8}^{\pi/8} \frac{1}{2} \cos^2 4\theta \, d\theta = \left[\frac{1}{4} \theta + \frac{1}{16} \sin 4\theta \cos 4\theta \right]_{-\pi/8}^{\pi/8} = \frac{\pi}{16}.$$

Since the area of the shaded region is one-eighth of the region bounded by the polar rose, the area bounded by the rose is $\frac{\pi}{2}$ square units.

| EXAMPLE 2 | Finding the area between two polar curves |

Use a double integral to calculate the area of the region in the polar plane that is inside both the circle $r = 3\cos\theta$ and the cardioid $r = 1 + \cos\theta$.

SOLUTION

By the symmetry of the graphs, half of the region whose area we wish to calculate is the shaded portion of the following figure:

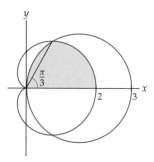

We will compute the area of the top half of the region and then multiply by 2. Solving the equation $3\cos\theta = 1 + \cos\theta$, we see that the two curves intersect when $\theta = \frac{\pi}{3}$. In the figure, we have drawn the ray $\theta = \frac{\pi}{3}$. The iterated integral that represents the area of the portion of the shaded region for values of $\theta \le \frac{\pi}{3}$ is

$$\int_0^{\pi/3} \int_0^{1+\cos\theta} r\, dr\, d\theta,$$

and the iterated integral that represents the area of the portion of the shaded region for $\frac{\pi}{3} \le \theta \le \frac{\pi}{2}$ is

$$\int_{\pi/3}^{\pi/2} \int_0^{3\cos\theta} r\, dr\, d\theta.$$

In Exercise 12 you will show that the values of these two integrals are $\frac{1}{4}\pi + \frac{9}{16}\sqrt{3}$ and $\frac{3}{8}\pi - \frac{9}{16}\sqrt{3}$, respectively. Therefore, the area of the shaded region is $\frac{5}{8}\pi$ and the value of the area we wish to compute is $\frac{5}{4}\pi$ square units.   □

| EXAMPLE 3 | Using a double integral to prove the area formula for the circle |

Use a double integral to prove that the area of a circle with radius R is πR^2.

SOLUTION

We will compute the area of the circle whose equation is $r = R$. The area is given by the iterated integral

$$\int_0^{2\pi} \int_0^R r\, dr\, d\theta.$$

The inner integral is

$$\int_0^R r\, dr = \left[\frac{1}{2}r^2\right]_0^R = \frac{1}{2}R^2.$$

We now evaluate the outer integral:

$$\int_0^{2\pi} \frac{1}{2} R^2 \, d\theta = \left[\frac{1}{2} R^2 \theta \right]_0^{2\pi} = \pi R^2.$$

In Exercises 63 and 64 you will prove this result twice more by finding the area of the circles with equations $r = 2R \cos \theta$ and $r = 2R \sin \theta$.

□

EXAMPLE 4

Changing an integral from rectangular to polar coordinates

Evaluate the integral $\displaystyle\int_0^2 \int_{\sqrt{3}x}^{\sqrt{16-x^2}} (x^2 + y^2) \, dy \, dx.$

SOLUTION

We may try to evaluate the integral as it is written. To do this, we evaluate the inner integral first:

$$\int_{\sqrt{3}x}^{\sqrt{16-x^2}} (x^2 + y^2) \, dy = \left[x^2 y + \frac{1}{3} y^3 \right]_{\sqrt{3}x}^{\sqrt{16-x^2}}$$

$$= \left(x^2 \sqrt{16 - x^2} + \frac{1}{3} (16 - x^2)^{3/2} \right) - \left(\sqrt{3} x^3 + \frac{1}{3} 3\sqrt{3} x^3 \right)$$

$$= x^2 \sqrt{16 - x^2} + \frac{1}{3} (16 - x^2)^{3/2} - 2\sqrt{3} x^3.$$

Now, to finish the problem, we would need to evaluate the integral

$$\int_0^2 \left(x^2 \sqrt{16 - x^2} + \frac{1}{3} (16 - x^2)^{3/2} - 2\sqrt{3} x^3 \right) dx.$$

With diligence, we would use trigonometric substitution to find an antiderivative for the integrand, but we present an alternative.

We may replace the original integrand $x^2 + y^2$ with its polar coordinate equivalent r^2. We shall similarly see that the region described by the limits of the integral may be expressed with polar coordinates. These are two essential things to consider when you are deciding between using rectangular and polar coordinates. The region in question is bounded above by the graph of the function $y = \sqrt{16 - x^2}$ and bounded below by the line whose equation is $y = \sqrt{3}x$, where $x \in [0, 2]$. Here is the region:

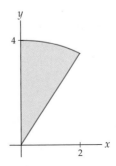

This region is a sector of the circle $r = 4$, where $\frac{\pi}{3} \leq \theta \leq \frac{\pi}{2}$. We may now rewrite the integral in terms of polar coordinates. Whenever we convert from rectangular to polar coordinates, we must replace either $dy\,dx$ or $dx\,dy$ by $r\,dr\,d\theta$. As we already mentioned, in this example we replace $x^2 + y^2$ with r^2. The limits for the integration with respect to r extend from 0 to 4, and the limits for θ extend from $\frac{\pi}{3}$ to $\frac{\pi}{2}$. Therefore we have

$$\int_0^2 \int_{\sqrt{3}x}^{\sqrt{16-x^2}} (x^2 + y^2)\, dy\, dx = \int_{\pi/3}^{\pi/2} \int_0^4 r^3\, dr\, d\theta.$$

We first evaluate the inner integral:

$$\int_0^4 r^3\, dr = \left[\frac{1}{4}r^4\right]_0^4 = 64.$$

We now finish by evaluating the outer integral:

$$\int_{\pi/3}^{\pi/2} 64\, d\theta = \left[64\theta\right]_{\pi/3}^{\pi/2} = 64\left(\frac{\pi}{2} - \frac{\pi}{3}\right) = \frac{32}{3}\pi.$$

□

EXAMPLE 5

Using polar coordinates to compute a volume

Find the volume of the solid bounded above by the sphere $x^2 + y^2 + z^2 = 8$ and bounded below by the cone with equation $z = \sqrt{x^2 + y^2}$.

SOLUTION

The sphere and cone intersect along a circle. If we rewrite the equations in the forms $z^2 = 8 - x^2 - y^2$ and $z^2 = x^2 + y^2$ and equate the results, we see that $x^2 + y^2 = 4$. The disk bounded by the circle defined by this equation is the region over which we need to integrate the difference of the two functions. The graph that follows at the left shows the entire hemisphere and cone, along with the disk in the xy-plane, over which we will integrate. The open figure depicted at the right shows a quarter of the left-hand figure.

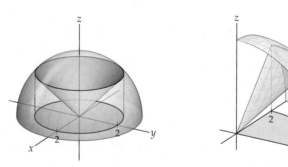

To find the volume of the solid, we will integrate the difference of the two functions over the indicated circle in the xy-plane. In polar coordinates, the equations of the top hemisphere and the cone are $z = \sqrt{8 - r^2}$ and $z = r$, respectively. The equation of the circle of intersection of the sphere and cone projected onto the xy-plane is $r = 2$. Therefore, the iterated integral representing the volume is

$$\int_0^{2\pi} \int_0^2 \left(\sqrt{8 - r^2} - r\right) r\, dr\, d\theta.$$

We now evaluate the iterated integral by first working on the inner integral:

$$\int_0^2 \left(r\sqrt{8 - r^2} - r^2\right) dr = \left[-\frac{1}{3}(8 - r^2)^{3/2} - \frac{1}{3}r^3\right]_0^2 = \frac{16}{3}\left(\sqrt{2} - 1\right).$$

We next integrate the rightmost quantity with respect to θ over the interval $[0, 2\pi]$:

$$\int_0^{2\pi} \frac{16}{3}(\sqrt{2} - 1)\, d\theta = \frac{32}{3}\pi(\sqrt{2} - 1).$$

This is the volume of the region we wished to find.   □

✔ **CHECKING THE ANSWER** | The volume we were asked to find is the volume of the solid of revolution created when the region in the first quadrant bounded above by the graph of $y = \sqrt{8 - x^2}$ and below by the line $y = x$ is rotated about the y-axis, as we illustrate in the following figures:

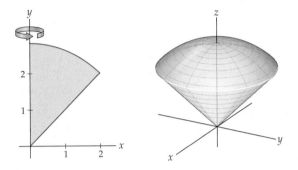

We may evaluate the volume of this region using either the washer method or the shell method, both of which we discussed in Chapter 6. Using the shell method, we have

$$2\pi \int_0^2 x\left(\sqrt{8 - x^2} - x\right) dx.$$

You should check that the value of this integral is also $\frac{32}{3}\pi(\sqrt{2} - 1)$.

? TEST YOUR UNDERSTANDING

▶ How are polar coordinates used to express double integrals?

▶ What are the formulas needed to convert from rectangular to polar coordinates? What are the formulas needed to convert from polar to rectangular coordinates?

▶ How are iterated integrals used to calculate areas between functions that are expressed in polar coordinates? How can a single integral be used to express such areas?

▶ How are iterated integrals used to calculate the volumes bounded by functions that are expressed in polar coordinates?

▶ How is an integral that is expressed in rectangular coordinates rewritten in terms of polar coordinates? When would such rewriting be advantageous? When would it be disadvantageous? How would an integral that is expressed in polar coordinates be expressed in terms of rectangular coordinates?

EXERCISES 13.3

Thinking Back

Each of the integral expressions that follow represents the area of a region in the plane bounded by a function expressed in polar coordinates. Use the ideas from this section and from Chapter 9 to sketch the regions, and then evaluate each integral.

▶ $\dfrac{1}{2}\displaystyle\int_0^\pi \cos^2 3\theta\, d\theta$

▶ $4\displaystyle\int_0^{\pi/4} \cos^2 2\theta\, d\theta$

▶ $\displaystyle\int_0^{2\pi/3} \left(\frac{1}{2} + \cos\theta\right)^2 d\theta - \int_\pi^{4\pi/3} \left(\frac{1}{2} + \cos\theta\right)^2 d\theta$

▶ $\dfrac{1}{2}\displaystyle\int_{\pi/6}^{5\pi/6} \left((3\sin\theta)^2 - (1 + \sin\theta)^2\right) d\theta$

Concepts

0. *Problem Zero:* Read the section and make your own summary of the material.

1. *True/False:* Determine whether each of the statements that follow is true or false. If a statement is true, explain why. If a statement is false, provide a counterexample.

(a) *True or False:* To approximate the area of the region in the polar plane bounded by the function $r = f(\theta)$ and the rays $\theta = \alpha$ and $\theta = \beta$, we can use a sum of areas of sectors of circles.

(b) *True or False:* Suppose we subdivide the interval of angles $\theta \in \left[\frac{\pi}{4}, \frac{\pi}{2}\right]$ into four equal subintervals. Then $\Delta\theta = \frac{\pi}{16}$.

(c) *True or False:* If $f(\theta) \geq 0$ on the interval $[\alpha, \beta]$, then the area of the region in the polar plane bounded by the function $r = f(\theta)$ and the rays $\theta = \alpha$ and $\theta = \beta$ is given by the iterated integral $\int_\alpha^\beta \int_0^{f(\theta)} dr\, d\theta$.

(d) *True or False:* If $f(\theta) \geq 0$ on the interval $[\alpha, \beta]$, then the volume of the solid bounded above by the function $g(r, \theta) = r^2$, over the region in the polar plane bounded by the function $r = f(\theta)$ and the rays $\theta = \alpha$ and $\theta = \beta$, is given by the iterated integral $\int_\alpha^\beta \int_0^{f(\theta)} r^2 \, dr\, d\theta$.

(e) *True or False:* Since the graph of $r = 2\cos 2\theta$ is a circle with radius 1, the value of the integral $\int_0^{2\pi} \int_0^{2\cos\theta} r\, dr\, d\theta$ is π.

(f) *True or False:* The polar equation $r = \sin 4\theta$ is traced twice as θ varies from 0 to 2π.

(g) *True or False:*
$$\int_0^\pi \int_1^5 (x^3 + y^3)^2 \, dy\, dx = \int_0^\pi \int_1^5 (\theta^3 + r^3)^2 \, dr\, d\theta.$$

(h) *True or False:*
$$\int_0^2 \int_0^{\sqrt{4-x^2}} (x^2 + y^2)^3 \, dy\, dx = \int_0^{\pi/2} \int_0^2 r^7 \, dr\, d\theta.$$

2. *Examples:* Construct examples of the thing(s) described in the following. Try to find examples that are different than any in the reading.

(a) An iterated integral that represents the area of a circle with radius R express with polar coordinates.

(b) An iterated integral using polar coordinates that represents the volume of a sphere with radius R.

(c) An iterated integral in rectangular coordinates that would be easier to evaluate by using polar coordinates.

3. Let $\alpha < \beta$ and $a < b$. In polar coordinates, a polar rectangle is bounded by the two rays $\theta = \alpha$ and $\theta = \beta$ and the two circles $r = a$ and $r = b$. Sketch a polar rectangle and explain why this is the basic region for integration in the polar coordinate plane.

4. When we use rectangular coordinates to approximate the area of a region, we subdivide the region into vertical strips and approximate the area by using a sum of areas of rectangles. Explain why we use a wedge (i.e., a sector of a circle), and not a rectangle, when we employ polar coordinates to compute an area.

5. In this section we described a method for approximating the volume bounded by a function f above a polar region bounded by two rays $\theta = \alpha$ and $\theta = \beta$ and between two circles $r = a$ and $r = b$. The method employed a "subdivide, approximate, and add" strategy that involved using some general notation. Draw a carefully labeled picture that illustrates the roles of $\Delta\theta$, θ_k^*, Δr, and r_j^* for one approximating region.

6. In Section 9.4 we showed that the area in the polar coordinate plane bounded by the function $r = f(\theta)$ on the interval $[\alpha, \beta]$ is given by the integral $\frac{1}{2} \int_\alpha^\beta (f(\theta))^2 \, d\theta$. In this section we discussed how to use the iterated integral $\int_\alpha^\beta \int_0^{f(\theta)} r\, dr\, d\theta$ to compute the same area. Explain why the value of these two integrals is the same.

7. Let $0 < f_1(\theta) < f_2(\theta)$ on the interval $[\alpha, \beta]$. What does the integral $\int_\alpha^\beta \int_{f_1(\theta)}^{f_2(\theta)} dr\, d\theta$ represent in a rectangular θr-coordinate system? What does the integral represent in a polar coordinate system?

8. Why do we require that $0 \leq \beta - \alpha \leq 2\pi$ when we are trying to find the volume of a solid bounded above by the graph of a function $z = g(r, \theta)$, over a region bounded by the polar functions $r = f_1(\theta)$ and $r = f_2(\theta)$, where $f_1(\theta) \leq f_2(\theta)$ on the interval $[\alpha, \beta]$? If $0 \leq \beta - \alpha \leq 2\pi$, does that condition ensure that the integral
$$\int_\alpha^\beta \int_{f_1(\theta)}^{f_2(\theta)} g(r, \theta)\, r\, dr\, d\theta$$
will represent the volume we want?

9. Consider the three-petaled polar rose defined by $r = \cos 3\theta$. Explain why the iterated integral $\int_0^{2\pi} \int_0^{\cos 3\theta} r\, dr\, d\theta$ calculates *twice* the area bounded by the petals of this rose.

10. Explain how the symmetries of the graphs of polar functions can be used to simplify area calculations.

11. Give a geometric explanation why
$$n \int_0^{2\pi/n} \int_0^R r\, dr\, d\theta = \pi R^2$$
for any positive real number R and any positive integer n. Would the equation also hold for nonintegral values of n?

12. Complete Example 2 by showing that
$$\int_0^{\pi/3} \int_0^{1+\cos\theta} r\, dr\, d\theta = \frac{1}{4}\pi + \frac{9}{16}\sqrt{3}$$
and
$$\int_{\pi/3}^{\pi/2} \int_0^{3\cos\theta} r\, dr\, d\theta = \frac{3}{8}\pi - \frac{9}{16}\sqrt{3}.$$

In Exercises 13–20, we explore the relationship between the shell method for finding volumes of solids of revolution discussed in Chapter 6 and the method of double integrals using polar coordinates.

13. Sketch a function $z = f(x)$ in the xz-plane such that $f(x) \geq 0$ on the interval $[0, b]$. Use the shell method to find an integral that represents the volume of the solid of revolution obtained when the region bounded above by the graph of f and bounded below by the x-axis on the interval $[0, b]$ is rotated about the z-axis.

14. Explain why the function $z = g(x, y) = f(\sqrt{x^2 + y^2})$ is the equation of the surface obtained when the graph of f is rotated about the z-axis. Sketch the surface obtained when your function from Exercise 13 is rotated about the z-axis.

15. Use the techniques of Section 13.2 to obtain an iterated integral that employs rectangular coordinates to represent the volume of the solid that is bounded above by the graph of the function g from Exercise 14 and below by the xy-plane over the circular disk $x^2 + y^2 \leq b^2$.

16. Use the techniques of this section to obtain an iterated integral that employs polar coordinates to represent the volume of the solid discussed in Exercise 15.

17. Show that the integrals from Exercises 13 and 16 evaluate to the same quantity.

18. Let $0 < a < b$. Use the shell method to find an integral that represents the volume of the solid of revolution obtained when the region bounded above by the graph of f and bounded below by the x-axis on the interval $[a, b]$ is rotated about the z-axis.

19. Use the techniques of this section to obtain an iterated integral that employs polar coordinates to represent the volume of the solid bounded above by the graph of the function g from Exercise 14 and below by the xy-plane over the annulus $a^2 \leq x^2 + y^2 \leq b^2$.

20. Show that the integrals from Exercises 18 and 19 evaluate to the same quantity.

Skills

Each of the integrals or integral expressions in Exercises 21–28 represents the area of a region in the plane. Use polar coordinates to sketch the region and evaluate the expressions.

21. $\displaystyle\int_0^{2\pi} \int_0^{1+\sin\theta} r\,dr\,d\theta$

22. $\displaystyle 2\int_0^{\pi} \int_0^{1+\cos\theta} r\,dr\,d\theta$

23. $\displaystyle 2\int_{-\pi/2}^{\pi/2} \int_0^{2-\sin\theta} r\,dr\,d\theta$

24. $\displaystyle 2\int_0^{\pi/2} \int_0^{\sin\theta} r\,dr\,d\theta$

25. $\displaystyle 2\int_0^{\pi/2} \int_0^{\sin 3\theta} r\,dr\,d\theta$

26. $\displaystyle \int_0^{2\pi} \int_0^{2+\sin 4\theta} r\,dr\,d\theta$

27. $\displaystyle 2\int_{-\pi/4}^{\pi/2} \int_0^{(\sqrt{2}/2)+\sin\theta} r\,dr\,d\theta - 2\int_{-\pi/2}^{-\pi/4} \int_0^{(\sqrt{2}/2)+\sin\theta} r\,dr\,d\theta$

28. $\displaystyle 2\int_0^{2\pi/3} \int_0^{(1/2)+\cos\theta} r\,dr\,d\theta - 2\int_{\pi}^{4\pi/3} \int_0^{(1/2)+\cos\theta} r\,dr\,d\theta$

In Exercises 29–38, find an iterated integral in polar coordinates that represents the area of the given region in the polar plane and then evaluate the integral.

29. The region enclosed by the spiral $r = \theta$ and the x-axis on the interval $0 \leq \theta \leq \pi$.

30. The region inside one loop of the lemniscate $r^2 = \sin 2\theta$.

31. The region between the two loops of the limaçon $r = 1 + \sqrt{2}\cos\theta$.

32. The region between the two loops of the limaçon $r = \sqrt{3} - 2\sin\theta$.

33. The region inside the cardioid $r = 3 - 3\sin\theta$ and outside the cardioid $r = 1 + \sin\theta$.

34. The region where the two cardioids $r = 3 - 3\sin\theta$ and $r = 1 + \sin\theta$ overlap.

35. The region inside the circle $x^2 + y^2 = 1$ and to the right of the vertical line $x = \dfrac{1}{2}$.

36. One loop of the curve $r = 4\sin 3\theta$.

37. The region bounded by the limaçon $r = 1 + k\sin\theta$, where $0 < k < 1$. Explain why it makes sense for the area to approach π as $k \to 0$.

38. The graph of the polar equation $r = \sec\theta - 2\cos\theta$ is called a **strophoid**. Graph the strophoid and find the area bounded by the loop of the graph.

Each of the integrals or integral expressions in Exercises 39–46 represents the volume of a solid in $\mathbb{R}^3$. Use polar coordinates to describe the solid, and evaluate the expressions.

39. $\displaystyle 2\int_0^{2\pi} \int_0^4 r\sqrt{16 - r^2}\,dr\,d\theta$

40. $\displaystyle 2\int_0^{2\pi} \int_0^R r\sqrt{R^2 - r^2}\,dr\,d\theta$

41. $\displaystyle \int_0^{2\pi} \int_0^2 (4r - r^3)\,dr\,d\theta$

42. $\displaystyle \int_0^{\pi} \int_0^R (R^2 r - r^3)\,dr\,d\theta$

43. $\displaystyle \int_0^{2\pi} \int_0^3 (6r - 2r^2)\,dr\,d\theta$

44. $\displaystyle h\int_0^{2\pi} \int_0^R \left(r - \frac{r^2}{R}\right)\,dr\,d\theta$

45. $\displaystyle \int_{-\pi/4}^{5\pi/4} \int_0^{(\sqrt{2}/2)+\sin\theta} r\,dr\,d\theta$

46. $\displaystyle 2\int_{-\pi/4}^{\pi/2} \int_0^{(\sqrt{2}/2)+\sin\theta} r\,dr\,d\theta - 2\int_{-\pi/2}^{-\pi/4} \int_0^{(\sqrt{2}/2)+\sin\theta} r\,dr\,d\theta$

In Exercises 47–56, use polar coordinates to find an iterated integral that represents the volume of the solid described and then find the volume of the solid.

47. The region enclosed by the paraboloids $z = x^2 + y^2$ and $z = 16 - x^2 - y^2$.

48. The region enclosed by the paraboloids $z = x^2 + y^2$ and $z = R^2 - x^2 - y^2$.

49. The region between the cone with equation $z = \sqrt{x^2 + y^2}$ and the unit sphere centered at the origin.

50. The region between the cone with equation $z = \sqrt{x^2 + y^2}$ and the sphere centered at the origin and with radius R.

51. The region bounded above by the unit sphere centered at the origin and bounded below by the plane $z = \dfrac{3}{5}$.

52. The region bounded above by the unit sphere centered at the origin and bounded below by the plane $z = h$ where $0 \le h \le 1$.

53. The region between two spheres with radius 1 if each passes through the center of the other.

54. The region between two spheres with radius R if each passes through the center of the other.

55. The region bounded below by the graph of the cone with equation $z = \sqrt{x^2 + y^2}$ and bounded above by the plane $z = h$, where $h > 0$.

56. The region bounded below by the graph of the cone with equation $z = \sqrt{x^2 + y^2}$ and bounded above by the plane $z = 8 - \dfrac{x}{2}$.

Sketch the region of integration for each of integrals in Exercises 57–60, and then evaluate the integral by converting to polar coordinates.

57. $\displaystyle\int_0^{3\sqrt{2}/2} \int_x^{\sqrt{9-x^2}} dy\,dx$

58. $\displaystyle\int_0^1 \int_{(\sqrt{3}/3)y}^{\sqrt{4-y^2}} \sqrt{x^2 + y^2}\,dx\,dy$

59. $\displaystyle\int_0^4 \int_0^{\sqrt{16-x^2}} e^{x^2} e^{y^2}\,dy\,dx$

60. $\displaystyle\int_{1/2}^1 \int_{-\sqrt{1-x^2}}^{\sqrt{1-x^2}} \left(2\ln x + \ln\left(1 + \left(\dfrac{y}{x}\right)^2\right)\right) dy\,dx$

Applications

61. Leila has been assigned the task of determining the risk to elk herds from the wolf population in a certain region of Idaho. She needs to find out the number of wolves near two distinct elk herds, one centered at the origin and the other 12 miles due north. She estimates the density of wolves in the region as 0.08 wolf per square mile. How many wolves would she expect to find within 12 miles of both herds?

62. Emmy needs to determine the volume of a sedimentation tank. The tank is circular with a radius of 75 feet as viewed from above, with a small concrete island that contains circulation and monitoring equipment in the center. The island has a radius of 7 feet. The depth of the tank at any distance r from the center is $d(r) = 0.15 \times 10^{-13} r^8 - 15$, where the surface of the solution in the tank is at depth zero. What is the volume of the tank?

Proofs

63. Use a double integral to prove that the area of the circle with radius R and equation $r = 2R\cos\theta$ is πR^2.

64. Use a double integral to prove that the area of the circle with radius R and equation $r = 2R\sin\theta$ is πR^2.

65. Use a double integral in polar coordinates to prove that the volume of a sphere with radius R is $\dfrac{4}{3}\pi R^3$.

66. Let h and R be positive real numbers. Explain why the region bounded above by the graph of the function $f(x, y) = h - \dfrac{h}{R}\sqrt{x^2 + y^2}$ and below by the xy-plane is a cone with height h and radius R. Use a double integral with polar coordinates to prove that the volume of this cone is $\dfrac{1}{3}\pi R^2 h$.

67. Use a double integral with polar coordinates to prove that the area of a sector with central angle ϕ in a circle of radius R is given by $A = \dfrac{1}{2}\phi R^2$.

68. Use a double integral with polar coordinates to prove that the area enclosed by one petal of the polar rose $r = \cos 3\theta$ is the same as the area enclosed by one petal of the polar rose $r = \sin 3\theta$.

69. Use a double integral with polar coordinates to prove that the combined area enclosed by all of the petals of the polar rose $r = \cos 2n\theta$ is the same for every positive integer n.

70. Use a double integral with polar coordinates to prove that the combined area enclosed by all of the petals of the polar rose $r = \sin(2n + 1)\theta$ is the same for every positive integer n.

Thinking Forward

▶ *A triple integral using cylindrical coordinates:* Show that the triple integral
$$\int_0^{2\pi} \int_0^R \int_0^h r\,dz\,dr\,d\theta = \pi R^2 h.$$
Explain why this integral gives the volume of a right circular cylinder with radius R and height h.

▶ *A triple integral using spherical coordinates:* Evaluate the triple integral
$$\int_0^\pi \int_0^{2\pi} \int_0^R \rho^2 \sin\phi\,d\rho\,d\theta\,d\phi.$$
Note that this integral gives the volume of a sphere with radius R.

13.4   APPLICATIONS OF DOUBLE INTEGRALS

▶ Using an iterated integral to find the mass of an object in the plane

▶ Finding the center of mass and centroid of a planar region

▶ Using double integrals to define the moment of inertia and the radius of gyration

The Mass of a Planar Region

In Section 6.4 we saw that the mass of a linear rod parallel to the x-axis with varying density $\rho(x) > 0$ is given by the integral

$$\int_a^b \rho(x)\,dx.$$

In the current section we first discuss how to find the mass of a region lying in the xy-plane when the density is a function of both variables x and y. Recall that in Sections 13.1 and 13.2 we discussed a strategy for computing the volume of a solid bounded above by a positive function $f(x, y)$ over a rectangular region in the plane and over a more general region in $\mathbb{R}^2$, respectively. We employ the same strategy here to find the mass of an object in the plane when the density of the object is given by the function $\rho(x, y)$, where $\rho(x, y) > 0$ for every point in the domain, as indicated in the following figure.

Region in the plane with variable density $\rho(x, y)$

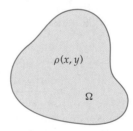

The technique of subdividing, approximating, and adding is precisely what is required here to approximate the mass of a planar region whose density is given by $\rho(x, y)$. Therefore, in the limit, to find the exact value of the mass of such a region in the plane, we may use the double integral

$$\text{Mass} = \iint_\Omega \rho(x, y)\,dA.$$

Depending upon the density function ρ and the region Ω, we may choose to evaluate the double integral as either a type I or type II region, using rectangular coordinates, or we may choose to use polar coordinates. For example, consider the triangular region Ω with vertices $(1, 1)$, $(2, 0)$, and $(2, 3)$. If the density at every point in Ω is proportional to the distance the point is from the y-axis, then $\rho(x, y) = kx$, where $k > 0$ is a constant of proportionality. It is somewhat easier to treat Ω as a type I region. The equations of the boundary segments are shown in the following figure:

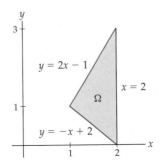

Therefore, the integral

$$\int_1^2 \int_{-x+2}^{2x-1} kx \, dy \, dx$$

represents the mass of the region. In Exercise 15 you will show that the value of this integral is $\frac{5}{2}k$. In Example 2, we discuss finding the mass of a region when the use of polar coordinates provides a simpler computation.

The Center of Mass and First Moments

Consider a system of two point masses on the x-axis: mass m_1 at point x_1 and mass m_2 at point x_2. The **center of mass**, $\bar{x}$, of this simple system may be found with the formula:

$$\bar{x} = \frac{m_1 x_1 + m_2 x_2}{m_1 + m_2}.$$

For example, if a mass of 3 grams is located at $x = 2$ and a mass of 6 grams is located at $x = 5$, the center of mass of this system is found at $\bar{x} = \frac{3 \cdot 2 + 6 \cdot 5}{3 + 6} = 4$, as shown in the following figure:

This value conforms to our experience that the center of mass of the system should be closer to the larger mass.

Similarly, for n point masses $m_1, m_2, m_3, \ldots, m_n$, at the x-coordinates $x_1, x_2, x_3, \cdots, x_n$, respectively, the center of mass of this system is given by

$$\bar{x} = \frac{m_1 x_1 + m_2 x_2 + \cdots + m_n x_n}{m_1 + m_2 + \cdots + m_n} = \frac{\sum_{k=1}^n m_k x_k}{\sum_{k=1}^n m_k}.$$

We may generalize this idea to point masses in two (or three) dimensions:

Thus, for point masses, $m_1, m_2, m_3, \ldots, m_n$, at the coordinate pairs (x_1, y_1), (x_2, y_2), $(x_3, y_3), \ldots, (x_n, y_n)$, the center of mass of this system is given by the point $(\bar{x}, \bar{y})$, where

$$\bar{x} = \frac{\sum_{k=1}^n m_k x_k}{\sum_{k=1}^n m_k} \quad \text{and} \quad \bar{y} = \frac{\sum_{k=1}^n m_k y_k}{\sum_{k=1}^n m_k}.$$

Note that the computations for $\bar{x}$ and $\bar{y}$ are independent of each other; $\bar{x}$ depends only on the masses and their x-coordinates and $\bar{y}$ depends only on the masses and their y-coordinates.

We now turn our attention to finding the center of mass of a planar region with a density function $\rho(x, y)$. Such a region is called a **lamina** (*plural* laminæ). We employ our usual strategy of subdividing the region into small pieces, approximating the mass of each small subregion by a single value of the density function over the subregion and then adding all those values. In this context, we may approximate the x- and y-coordinates of the center of mass independently. The following figure provides a schematic:

A lamina Ω

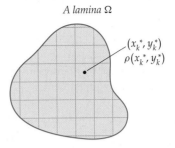

The approximate mass of each subregion is given by its (approximate) density $\rho(x_k^*, y_k^*)$ times its area ΔA. In the denominators that follow, we sum over all of the subregions. In the numerators, we need the extra factors x_k^* and y_k^*, as we are trying to approximate $\bar{x}$ and $\bar{y}$, respectively. Thus, $\bar{x}$ and $\bar{y}$ may be approximated by

$$\bar{x} \approx \frac{\sum_{k=1}^{n} x_k^* \rho(x_k^*, y_k^*) \Delta A}{\sum_{k=1}^{n} \rho(x_k^*, y_k^*) \Delta A} \quad \text{and} \quad \bar{y} \approx \frac{\sum_{k=1}^{n} y_k^* \rho(x_k^*, y_k^*) \Delta A}{\sum_{k=1}^{n} \rho(x_k^*, y_k^*) \Delta A}.$$

In the limit, as the mesh of our grid goes to zero, we obtain the exact coordinates of $\bar{x}$ and $\bar{y}$:

$$\bar{x} = \frac{\iint_{\Omega} x \rho(x, y) \, dA}{\iint_{\Omega} \rho(x, y) \, dA} \quad \text{and} \quad \bar{y} = \frac{\iint_{\Omega} y \rho(x, y) \, dA}{\iint_{\Omega} \rho(x, y) \, dA}.$$

We use the integrals in the numerators in the following definition:

DEFINITION 13.12

First Moments About the x- and y-axes

Let Ω be a lamina in the xy-plane in which the density at each point is given by the continuous function $\rho(x, y)$.

(a) The ***first moment*** of the mass in Ω about the y-axis is

$$M_y = \iint_{\Omega} x \rho(x, y) \, dA.$$

(b) The ***first moment*** of the mass in Ω about the x-axis is

$$M_x = \iint_{\Omega} y \rho(x, y) \, dA.$$

Thus, if we let $m = \iint_{\Omega} \rho(x, y) \, dA$, then m is the mass of the lamina Ω and the coordinates of the center of mass are

$$\bar{x} = \frac{M_y}{m} \quad \text{and} \quad \bar{y} = \frac{M_x}{m}.$$

The first moments are also known as the ***linear moments***. Upon first glance, you might think that there are errors in the definitions of the first moment. We see, for example, that $M_y = \iint_{\Omega} x \rho(x, y) \, dA$. To help remember this notation, recall that the factor x in the integrand measures the distance from the y-axis—hence the designation M_y. The first moments M_y and M_x provide measures of the distribution of the mass with respect to the y- and x-axes, respectively. It is possible to generalize these concepts to measure the distributions about other vertical and horizontal lines. For example, to measure the distribution of the mass about the vertical line $x = x_0$, we would evaluate the integral $\iint_{\Omega} (x - x_0) \rho(x, y) \, dA$. We will not be using these more general moments in this text.

Returning to our earlier example, we again consider the triangular region Ω with vertices $(1, 1)$, $(2, 0)$, and $(2, 3)$. Recall that the density at every point in Ω was proportional to the point's distance from the y-axis. At that time, we computed the mass of the lamina as

$$m = \int_1^2 \int_{-x+2}^{2x-1} kx \, dy \, dx = \frac{5}{2}k,$$

where k is a constant of proportionality. The integrals we need for the first moments require only minor adjustments:

$$M_y = \int_1^2 \int_{-x+2}^{2x-1} kx^2 \, dy \, dx = \frac{17}{4}k,$$

and

$$M_x = \int_1^2 \int_{-x+2}^{2x-1} kxy \, dy \, dx = \frac{27}{8}k.$$

Therefore,

$$\bar{x} = \frac{M_y}{m} = \frac{(17/4)k}{(5/2)k} = \frac{17}{10} \quad \text{and} \quad \bar{y} = \frac{M_x}{m} = \frac{(27/8)k}{(5/2)k} = \frac{27}{20}.$$

Note that the center of mass, $(17/20, 27/20)$, is located within the region Ω. When a region is convex, the center of mass lies within the region. Physically, when the center of mass lies within the region, the lamina may be balanced at that point.

When the density of a two-dimensional region Ω is constant, the center of mass is also called the **centroid** of Ω. In Exercise 20, you will explain why the centroid relates only to the geometry of the region and not to its mass.

Moments of Inertia

The moment of inertia of an object measures how difficult it is to change the angular momentum of the object. The moment of inertia of a spinning wheel is what makes it difficult for the wheel to stop when it is rotating. The integrals required to calculate the moments of inertia about the x- and y-axes are quite similar to those needed for the first moments.

DEFINITION 13.13

Moments of Inertia

Let Ω be a lamina in the xy-plane for which the density at each point is given by the continuous function $\rho(x, y)$.

(a) The **moment of inertia about the y-axis** of the lamina is

$$I_y = \iint_\Omega x^2 \rho(x, y) \, dA.$$

(b) The **moment of inertia about the x-axis** of the lamina is

$$I_x = \iint_\Omega y^2 \rho(x, y) \, dA.$$

(c) The **moment of inertia about the origin** of the lamina, also known as the **polar moment**, is

$$I_o = \iint_\Omega (x^2 + y^2) \rho(x, y) \, dA.$$

The moments of inertia are also known as the **second moments**. We may use these moments of inertia to compute the **radius of gyration** about the y-axis, the x-axis, and the origin. If the lamina has mass m, these are

$$R_y = \sqrt{\frac{I_y}{m}}, \quad R_x = \sqrt{\frac{I_x}{m}}, \quad \text{and} \quad R_0 = \sqrt{\frac{I_0}{m}},$$

respectively. The radii of gyration are the radial distances at which the mass of the lamina could be concentrated without changing its rotational inertia. We compute the moments of inertia and radii of gyration for a region in Examples 4 and 5.

Probability Distributions

We've seen that when we integrate a density function ρ over a laminar region Ω, we obtain the mass of the lamina. Similarly, if we know a population density over a planar region, we may compute the population. We provide such a computation in Example 6. We may also use an iterated integral to find the probability of an event when we know a **joint probability distribution function**. Recall that a continuous probability distribution function on an interval $[a, b]$ is a positive-valued function f defined on $[a, b]$ such that $\int_a^b f(x)\, dx = 1$. A joint probability distribution function of two variables on a subset $X \subset \mathbb{R}^2$ is a positive-valued function $f : X \to [0, \infty)$ such that $\iint_X f(x, y)\, dA = 1$. To compute the probability of an event $Z \subset X$, we evaluate the integral $\iint_Z f(x, y)\, dA$. Joint probability distribution functions of more than two variables are defined in an analogous manner, but we shall not be using them in this text.

Examples and Explorations

EXAMPLE 1   Finding the centroid of a triangular region

Find the centroid of the triangular region with vertices $(1, 1)$, $(2, 0)$, and $(2, 3)$.

SOLUTION

This is the same triangular region we used earlier in the section, but here we assume that the density ρ is constant:

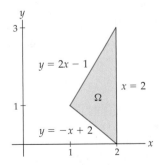

Since ρ is constant, we may factor it out of the integral. In our computation for $\bar{x}$ we have

$$\bar{x} = \frac{\iint_\Omega x \rho\, dA}{\iint_\Omega \rho\, dA} = \frac{\rho \iint_\Omega x\, dA}{\rho \iint_\Omega dA} = \frac{\iint_\Omega x\, dA}{\iint_\Omega dA}.$$

Similarly, we would have

$$\bar{y} = \frac{\iint_{\Omega} y \, dA}{\iint_{\Omega} dA}.$$

The integrals in the rightmost denominators each evaluate to the area of Ω. For this example, we may use either the indicated double integral or, since Ω is a triangle, the area formula for triangles to compute the area of Ω. We cannot avoid setting up an iterated integral, however, because we need to evaluate the integrals in the numerators. Here we need to evaluate the following three related integrals:

$$\text{Area} = \int_1^2 \int_{-x+2}^{2x-1} dy \, dx, \quad M_y = \int_1^2 \int_{-x+2}^{2x-1} x \, dy \, dx, \quad \text{and} \quad M_x = \int_1^2 \int_{-x+2}^{2x-1} y \, dy \, dx.$$

These integrals evaluate to

$$\text{Area} = \frac{3}{2}, \quad M_y = \frac{5}{2}, \quad \text{and} \quad M_x = 2.$$

Therefore, we have

$$\bar{x} = \frac{M_y}{\text{Area of } \Omega} = \frac{5/2}{3/2} = \frac{5}{3} \quad \text{and} \quad \bar{y} = \frac{M_x}{\text{Area of } \Omega} = \frac{2}{3/2} = \frac{4}{3}.$$

Thus, the centroid of the triangle is $(5/3, 4/3)$. Note that this point is inside the triangle. As we mentioned when we computed the center of mass in our earlier example, if a region Ω is convex, the centroid must be located within Ω. $\square$

EXAMPLE 2   **Using polar coordinates to find a mass**

Find the mass of the semicircular lamina whose boundary on the left is given by the line $x = 1$ and on the right is given by $(x - 1)^2 + y^2 = 1$ if the density at every point is proportional to its distance from the origin.

SOLUTION

The lamina we have described is shown here:

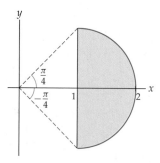

Since the density is proportional to the distance from the origin, it may be expressed easily in polar coordinates as a function of the radial distance r. That is, $\rho(r) = kr$, where $k > 0$ is a constant of proportionality. The equation of the circle in polar coordinates is $r = 2\cos\theta$, and the vertical line $x = 1$ in polar coordinates is given by $r = \sec\theta$. Finally, we use $r \, dr \, d\theta$ for our increment of area, dA. Therefore, the mass of the plate is given by

$$\int_{-\pi/4}^{\pi/4} \int_{\sec\theta}^{2\cos\theta} kr^2 \, dr \, d\theta.$$

To evaluate this iterated integral, we first evaluate the inner integral:

$$\int_{\sec\theta}^{2\cos\theta} kr^2\, dr = \left[\frac{1}{3}kr^3\right]_{\sec\theta}^{2\cos\theta} = \frac{1}{3}k(8\cos^3\theta - \sec^3\theta).$$

We may now use the techniques of Chapter 5 to evaluate the outer integral:

$$\int_{-\pi/4}^{\pi/4} \frac{1}{3}k(8\cos^3\theta - \sec^3\theta)\, d\theta = \frac{1}{9}k(17\sqrt{2} + 3\ln(\sqrt{2} - 1)).$$

You will show the details of this integration in Exercise 22. □

EXAMPLE 3

Using polar coordinates to find a center of mass

Find the center of mass of the semicircular lamina from Example 2.

SOLUTION

To start, we note that the y-coordinate of the center of mass is $\bar{y} = 0$, since the region Ω and the density function are symmetric with respect to the x-axis. Now, we have already computed the mass of the plate, using the integral

$$\int_{-\pi/4}^{\pi/4} \int_{\sec\theta}^{2\cos\theta} kr^2\, dr\, d\theta.$$

We still need to calculate M_y, the first moment of the mass about the y-axis. For M_y, all we have to do is introduce x into the preceding integrand. Since we are using polar coordinates, we will use the fact that $x = r\cos\theta$ instead. Therefore,

$$M_y = \int_{-\pi/4}^{\pi/4} \int_{\sec\theta}^{2\cos\theta} kr^3 \cos\theta\, dr\, d\theta = \frac{1}{60}k(157\sqrt{2} + 15\ln(\sqrt{2} - 1)).$$

You will show the details of this integration in Exercise 23. Thus, the x-coordinate of the center of mass is given by

$$\frac{M_y}{\text{mass}} = \bar{x} = \frac{\frac{k}{60}(157\sqrt{2} + 15\ln(\sqrt{2} - 1))}{\frac{k}{9}(17\sqrt{2} + 3\ln(\sqrt{2} - 1))} = \frac{471\sqrt{2} + 45\ln(\sqrt{2} - 1)}{340\sqrt{2} + 60\ln(\sqrt{2} - 1)} \approx 1.4638.$$

Again, as a quick check on the reasonableness of the answer, the point $(\bar{x}, \bar{y}) \approx (1.4638, 0)$ is within the region Ω. □

 CHECKING THE ANSWER

In Example 3, we used the symmetry of the region Ω and the density function to claim that $\bar{y} = 0$. If we wish, we may evaluate an iterated integral to compute M_x. Here, since the variable of integration on the inner integral is r, we may factor the integral as follows:

$$M_x = \int_{-\pi/4}^{\pi/4} \int_{\sec\theta}^{2\cos\theta} kr^3 \sin\theta\, dr\, d\theta = \left(\int_{-\pi/4}^{\pi/4} \sin\theta\, d\theta\right)\left(\int_{\sec\theta}^{2\cos\theta} kr^3\, dr\right).$$

Now, since sine is an odd function and the interval $[-\pi/4, \pi/4]$ is symmetric with respect to the origin, it follows that $\int_{-\pi/4}^{\pi/4} \sin\theta\, d\theta = 0$, and therefore M_x and $\bar{y}$ are both zero, as we reasoned before.

EXAMPLE 4

Finding the moments of inertia for a triangular region

Find the moments of inertia for the triangular lamina Ω in the xy-plane and with vertices $(1, 1)$, $(2, 0)$, and $(2, 3)$ if the density at every point in Ω is proportional to the distance the point is from the y-axis.

SOLUTION

This is the same region we used in previous examples. We have $\rho(x, y) = kx$, where $k > 0$ is a constant of proportionality. Earlier in the section we saw that

$$M_y = \int_1^2 \int_{-x+2}^{2x-1} kx^2 \, dy \, dx \quad \text{and} \quad M_x = \int_1^2 \int_{-x+2}^{2x-1} kxy \, dy \, dx.$$

The integrals we need for I_y and I_x require the extra factors x and y, respectively. In Exercises 18 and 19 you will show that the values of these integrals are

$$I_y = \int_1^2 \int_{-x+2}^{2x-1} kx^3 \, dy \, dx = \frac{147}{20}k \quad \text{and} \quad I_x = \int_1^2 \int_{-x+2}^{2x-1} kxy^2 \, dy \, dx = \frac{28}{5}k.$$

From Definition 13.13, we see that $I_o = I_y + I_x$. Here we have $I_o = \frac{147}{20}k + \frac{28}{5}k = \frac{259}{20}k$. ☐

EXAMPLE 5   Finding the radii of gyration for a triangular region

Find the radii of gyration about the x-axis, y-axis, and origin for the triangular lamina Ω with vertices $(1, 1)$, $(2, 0)$, and $(2, 3)$ if the density at every point in Ω is proportional to the the distance the point is from the y-axis.

SOLUTION

In Example 4 we computed the moments of inertia for this region, and earlier in this section we saw that the mass of Ω is $m = \frac{5}{2}k$. The radii of gyration with respect to the y-axis, x-axis, and origin are, respectively,

$$R_y = \sqrt{\frac{I_y}{m}} = \sqrt{\frac{(147/20)k}{(5/2)k}} = \sqrt{\frac{147}{50}}, \quad R_x = \sqrt{\frac{I_x}{m}} = \sqrt{\frac{(28/5)k}{(5/2)k}} = \frac{\sqrt{56}}{5}, \quad \text{and}$$

$$R_o = \sqrt{\frac{I_o}{m}} = \sqrt{\frac{(259/20)k}{(5/2)k}} = \sqrt{\frac{259}{50}}.$$

☐

EXAMPLE 6   Determining a population from a population density

A biologist is culturing a population of bacteria in a circular petri dish with a radius of 5 centimeters. She introduces bacteria into the dish and incubates the dish for 24 hours. At the end of that time she estimates that the colony of bacteria in the dish contains 100 bacteria per square centimeter at the center of the dish, with the population decreasing linearly to the edge of the dish. She estimates the population density to be 10 bacteria per square centimeter at the edge. Approximately how many bacteria are in the dish?

SOLUTION

This problem is nearly identical to those in which we were given a (mass) density function ρ for a laminar region Ω. To find the total mass, we used an iterated integral to integrate the density function ρ over the region Ω. We will do the same here to find the population of bacteria. We start by imposing a polar coordinate system on the circular petri dish, with the center of the dish at the origin. Since there are 100 bacteria per square centimeter at the origin and the density decreases linearly to 10 bacteria per square centimeter at the edge of the dish, the function $\rho(r, \theta) = 100 - 18r$ gives us the population density at every point in the petri dish for $r \in [0, 5]$. Therefore, the population of bacteria in the dish is given by the iterated integral

$$\int_0^{2\pi} \int_0^5 (100 - 18r)\, r\, dr\, d\theta.$$

The value of this integral is $1000\pi \approx 3100$ bacteria. Note that we have rounded our answer, since the biologist clearly approximated her values for the two population densities.   □

? TEST YOUR UNDERSTANDING

▶ Given a system of point masses in the plane, how is the center of mass of the system computed? How is this idea used to find the center of mass of a lamina Ω whose density function is $\rho(x, y)$?

▶ What is the centroid of a region Ω? What is the relationship between the centroid of a region and its center of mass? Why is the centroid of Ω independent of the mass of Ω?

▶ How is the moment of inertia of a region Ω with respect to the x- and y-axes defined? How are these definitions related to the definitions of the first moments of Ω with respect to the x- and y-axes? What is meant by the polar moment?

▶ How are the radii of gyration with respect to the x- and y-axes, R_x, and R_y defined for a region Ω? What integrals need to be computed to find R_x and R_y?

▶ How are the integrals for masses, first moments, and moments of inertia related? Which of the three integrals is easiest to step up? Given one of these integrals, how would you modify it to set up the other integrals? When would rectangular coordinates be easier to use for these integrals? When would polar coordinates be easier to use?

EXERCISES 13.4

Thinking Back

▶ *Mass of a rod:* Suppose a thin rod with a radius of 1 centimeter and a length of 20 centimeters has a varying density such that the density of the rod x centimeters from the left end is given by the function $\rho(x) = 10 + 0.01x^2$ grams per cubic centimeter. Find the mass of the rod.

▶ *More questions about the rod:* How far from the left end of the rod is the center of mass in the previous problem? How far from the left end of the rod is the moment of inertia? How far from the left end of the rod is the radius of gyration?

Concepts

0. *Problem Zero:* Read the section and make your own summary of the material.

1. *True/False:* Determine whether each of the statements that follow is true or false. If a statement is true, explain why. If a statement is false, provide a counterexample.

(a) *True or False:* The center of mass of a lamina, Ω, in the xy-plane is a point in Ω.

(b) *True or False:* The centroid of a region, Ω, in the xy-plane is a point in Ω.

(c) *True or False:* The center of mass of a circular region is the center of the circle for every density function $\rho(x, y)$.

(d) *True or False:* The centroid of a circular region is the center of the circle.

(e) *True or False:* The first moment of the mass in a lamina Ω about the x-axis is given by $M_x = \iint_\Omega x\, \rho(x, y)\, dA$, where $\rho(x, y)$ is the lamina's density function.

(f) *True or False:* If M_x and I_x are the first and second moments, respectively, for a lamina in the xy-plane, then $xM_x = I_x$.

(g) *True or False:* If I_x, I_y, and I_o are the moments of inertia of a lamina about the x-axis, y-axis, and origin, respectively, then $I_o = I_x + I_y$.

(h) *True or False:* If m is the mass of a lamina, I_y is the moment of inertia of the lamina about the y-axis, and R_y is the radius of gyration about the y-axis, then $R_y^2 = I_y / m$.

2. *Examples:* Construct examples of the thing(s) described in the following. Try to find examples that are different than any in the reading.

   (a) A lamina Ω in the xy-plane and a density function $\rho(x,y)$ such that the center of mass of Ω and the centroid of Ω are the same.

   (b) A lamina Ω in the xy-plane and a non-constant density function $\rho(x,y)$ such that the center of mass of Ω and the centroid of Ω are the same.

   (c) A lamina Ω in the xy-plane such that the center of mass is *not* in Ω.

Identify the quantities determined by the integral expressions in Exercises 3–11. If x and y are both measured in centimeters and $\rho(x,y)$ is a density function in grams per square centimeter, give the units of the expression.

3. $\displaystyle\iint_{\Omega} dA$

4. $\displaystyle\iint_{\Omega} x\,dA$ and $\displaystyle\iint_{\Omega} y\,dA$

5. $\displaystyle\frac{\iint_{\Omega} x\,dA}{\iint_{\Omega} dA}$ and $\displaystyle\frac{\iint_{\Omega} y\,dA}{\iint_{\Omega} dA}$

6. $\displaystyle\iint_{\Omega} \rho(x,y)\,dA$

7. $\displaystyle\iint_{\Omega} x\,\rho(x,y)\,dA$ and $\displaystyle\iint_{\Omega} y\,\rho(x,y)\,dA$

8. $\displaystyle\frac{\iint_{\Omega} x\,\rho(x,y)\,dA}{\iint_{\Omega} \rho(x,y)\,dA}$ and $\displaystyle\frac{\iint_{\Omega} y\,\rho(x,y)\,dA}{\iint_{\Omega} \rho(x,y)\,dA}$

9. $\displaystyle\iint_{\Omega} x^2\rho(x,y)\,dA$ and $\displaystyle\iint_{\Omega} y^2\rho(x,y)\,dA$

10. $\displaystyle\iint_{\Omega} (x^2+y^2)\rho(x,y)\,dA$

11. $\displaystyle\sqrt{\frac{\iint_{\Omega} x^2\,\rho(x,y)\,dA}{\iint_{\Omega} \rho(x,y)\,dA}}$ and $\displaystyle\sqrt{\frac{\iint_{\Omega} y^2\,\rho(x,y)\,dA}{\iint_{\Omega} \rho(x,y)\,dA}}$

Throughout this section we computed several integrals relating to the triangular region Ω with vertices $(1,1)$, $(2,0)$, and $(2,3)$. In Exercises 12–19 you are asked to provide the details of those computations.

12. Show that the area of Ω is $\frac{3}{2}$ by using the area formula for triangles and by evaluating the integral

$$\int_1^2 \int_{-x+2}^{2x-1} dy\,dx.$$

13. Show that when the density of the region is constant, the first moment about the y-axis is

$$M_y = \int_1^2 \int_{-x+2}^{2x-1} x\,dy\,dx = \frac{5}{2}.$$

14. Show that when the density of the region is constant, the first moment about the x-axis is

$$M_x = \int_1^2 \int_{-x+2}^{2x-1} y\,dy\,dx = 2.$$

15. Show that when the density of the region is proportional to the distance from the y-axis, the mass of Ω is given by

$$\int_1^2 \int_{-x+2}^{2x-1} kx\,dy\,dx = \frac{5}{2}k.$$

16. Show that when the density of the region is proportional to the distance from the y-axis, the first moment about the y-axis is

$$M_y = \int_1^2 \int_{-x+2}^{2x-1} kx^2\,dy\,dx = \frac{17}{4}k.$$

17. Show that when the density of the region is proportional to the distance from the y-axis, the first moment about the x-axis is

$$M_x = \int_1^2 \int_{-x+2}^{2x-1} kxy\,dy\,dx = \frac{27}{8}k.$$

18. Show that when the density of the region is proportional to the distance from the y-axis, the moment of inertia about the y-axis is

$$I_y = \int_1^2 \int_{-x+2}^{2x-1} kx^3\,dy\,dx = \frac{147}{20}k.$$

19. Show that when the density of the region is proportional to the distance from the y-axis, the first moment about the x-axis is

$$I_x = \int_1^2 \int_{-x+2}^{2x-1} kxy^2\,dy\,dx = \frac{28}{5}k.$$

20. Explain why the location of the centroid relates only to the geometry of the region and not its mass.

21. Find the moments of inertia about the x- and y-axes for the semicircular lamina described in Example 2. Assume that the density at every point is proportional to the distance of the point from the origin.

22. Complete Example 2 by showing that

$$\int_{-\pi/4}^{\pi/4} \frac{k}{3}(8\cos^3\theta - \sec^3\theta)\,d\theta$$

$$= \frac{k}{9}(17\sqrt{2} + 3\ln(\sqrt{2}-1)).$$

23. Complete Example 3 by showing that

$$\int_{-\pi/4}^{\pi/4} \int_{\sec\theta}^{2\cos\theta} kr^3\cos\theta\,dr\,d\theta$$

$$= \frac{k}{60}(157\sqrt{2} + 15\ln(\sqrt{2}-1)).$$

Skills

In Exercises 24–30, let T be the triangular region with vertices $(0, 0)$, $(1, 1)$, and $(1, -1)$.

24. Find the centroid of T.

25. If the density at each point in T is proportional to the point's distance from the y-axis, find the mass of T.

26. If the density at each point in T is proportional to the point's distance from the y-axis, find the center of mass of T.

27. If the density at each point in T is proportional to the point's distance from the y-axis, find the moments of inertia about the x- and y-axes. Use these answers to find the radii of gyration of T about the x- and y-axes.

28. If the density at each point in T is proportional to the point's distance from the x-axis, find the mass of T.

29. If the density at each point in T is proportional to the point's distance from the x-axis, find the center of mass of T.

30. If the density at each point in T is proportional to the point's distance from the x-axis, find the moments of inertia about the x- and y-axes. Use these answers to find the radii of gyration of T about the x- and y-axes.

In Exercises 31–37, let T_2 be the triangular region with vertices $(1, 0)$, $(2, 1)$, and $(2, -1)$. Note that T_2 is a translation of the triangle we used in Exercises 24–30.

31. Find the centroid of T_2.

32. If the density at each point in T_2 is proportional to the point's distance from the y-axis, find the mass of T_2.

33. If the density at each point in T_2 is proportional to the point's distance from the y-axis, find the center of mass of T_2.

34. If the density at each point in T_2 is proportional to the point's distance from the y-axis, find the moments of inertia about the x- and y-axes. Use these answers to find the radii of gyration of T_2 about the x- and y-axes.

35. If the density at each point in T_2 is proportional to the square of the point's distance from the y-axis, find the mass of T_2.

36. If the density at each point in T_2 is proportional to the square of the point's distance from the y-axis, find the center of mass of T_2.

37. If the density at each point in T_2 is proportional to the square of the point's distance from the y-axis, find the moments of inertia about the x- and y-axes. Use these answers to find the radii of gyration of T_2 about the x- and y-axes.

In Exercises 38–44, let R be the rectangular region with vertices $(0, 0)$, $(b, 0)$, $(0, h)$, and (b, h).

38. Find the centroid of R.

39. If the density at each point in R is proportional to the point's distance from the y-axis, find the mass of R.

40. If the density at each point in R is proportional to the point's distance from the y-axis, find the center of mass of R.

41. If the density at each point in R is proportional to the point's distance from the y-axis, find the moments of inertia about the x- and y-axes. Use these answers to find the radii of gyration of R about the x- and y-axes.

42. If the density at each point in R is proportional to the square of the point's distance from the y-axis, find the mass of R.

43. If the density at each point in R is proportional to the square of the point's distance from the y-axis, find the center of mass of R.

44. If the density at each point in R is proportional to the square of the point's distance from the y-axis, find the moments of inertia about the x- and y-axes. Use these answers to find the radii of gyration of R about the x- and y-axes.

In Exercises 45–51, let $C = \{(x, y) \mid x^2 + y^2 \le 1\}$.

45. Find the centroid of C.

46. If the density at each point in C is proportional to the point's distance from the y-axis, find the mass of C.

47. If the density at each point in C is proportional to the point's distance from the y-axis, find the center of mass of C.

48. If the density at each point in C is proportional to the point's distance from the y-axis, find the moments of inertia about the x- and y-axes. Use these answers to find the radii of gyration of C about the x- and y-axes.

49. If the density at each point in C is proportional to the point's distance from the origin, find the mass of C.

50. If the density at each point in C is proportional to the point's distance from the origin, find the center of mass of C.

51. If the density at each point in C is proportional to the point's distance from the origin, find the moments of inertia about the x-axis, the y-axis, and the origin. Use these answers to find the radii of gyration of C about the x-axis, the y-axis, and the origin.

In Exercises 52–58, let $S = \{(x, y) \mid x^2 + y^2 \le 1 \text{ and } x \ge 0\}$.

52. Find the centroid of S.

53. If the density at each point in S is proportional to the point's distance from the y-axis, find the mass of S.

54. If the density at each point in S is proportional to the point's distance from the y-axis, find the center of mass of S.

55. If the density at each point in S is proportional to the point's distance from the y-axis, find the moments of inertia about the x- and y-axes. Use these answers to find the radii of gyration of S about the x- and y-axes.

56. If the density at each point in S is proportional to the point's distance from the origin, find the mass of S.

57. If the density at each point in S is proportional to the point's distance from the origin, find the center of mass of S.

58. If the density at each point in S is proportional to the point's distance from the origin, find the moments of inertia about the x-axis, the y-axis, and the origin. Use these answers to find the radii of gyration of S about the x-axis, the y-axis, and the origin.

59. Let Ω be a lamina in the xy-plane. Suppose Ω is composed of two non-overlapping laminæ Ω_1 and Ω_2, as follows:

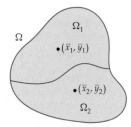

Show that if the masses and centers of masses of Ω_1 and Ω_2 are m_1 and m_2, and $(\bar{x}_1, \bar{y}_1)$ and $(\bar{x}_2, \bar{y}_2)$, respectively, then the center of mass of Ω is $(\bar{x}, \bar{y})$, where

$$\bar{x} = \frac{m_1 \bar{x}_1 + m_2 \bar{x}_2}{m_1 + m_2} \quad \text{and} \quad \bar{y} = \frac{m_1 \bar{y}_1 + m_2 \bar{y}_2}{m_1 + m_2}.$$

60. Let Ω be a lamina in the xy-plane. Suppose Ω is composed of n non-overlapping laminæ $\Omega_1, \Omega_2, \ldots, \Omega_n$. Show that if the masses of these laminæ are $m_1, m_2, \ldots, m_n$ and the centers of masses are $(\bar{x}_1, \bar{y}_1), (\bar{x}_2, \bar{y}_2), \ldots, (\bar{x}_n, \bar{y}_n)$, then the center of mass of Ω is $(\bar{x}, \bar{y})$, where

$$\bar{x} = \frac{\sum_{k=1}^{n} m_k \bar{x}_k}{\sum_{k=1}^{n} m_k} \quad \text{and} \quad \bar{y} = \frac{\sum_{k=1}^{n} m_k \bar{y}_k}{\sum_{k=1}^{n} m_k}.$$

Use the results of Exercises 59 and 60 to find the centers of masses of the laminæ in Exercises 61–67.

61. In the following lamina, all angles are right angles and the density is constant:

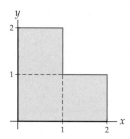

62. Use the lamina from Exercise 61, but assume that the density is proportional to the distance from the x-axis.

63. In the following lamina, all angles are right angles and the density is constant:

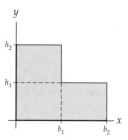

64. In the following lamina, all angles are right angles and the density is constant:

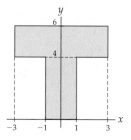

65. Use the lamina from Exercise 64, but assume that the density is proportional to the distance from the x-axis.

66. The lamina in the figure that follows is bounded above by the lines with equations $y = x + 2a$ and $y = -x + 2a$ and below by the x-axis on the interval $-a \leq x \leq a$. The density of the lamina is constant.

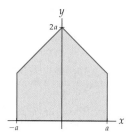

67. Use the lamina from Exercise 66, but assume that the density is proportional to the distance from the x-axis.

Applications

68. Leila has been required to track down a grizzly that attacked a person. The bear once had a radio collar, but has lost it. From the old data, Leila knows that the bear's range is roughly a region bounded by the x-axis, the y-axis, and the line $y = 4.3 - 0.8x$. The bear is equally likely to be found in any part of that region. The probability density function for the bear's location is $f(x, y) =$

0.0873. There is a trail that lies roughly along the y-axis. How far must Leila's team bushwhack from the trail in order for her team to have a 50% chance of finding the bear on the first day of the search?

69. In Exercise 68, what is the expected distance from the road at which the bear is likely to be found? (*Hint: Compute the first moment in the x-direction for the distribution.*)

Proofs

Recall that a **median** of a triangle is a segment connecting a vertex of a triangle to the midpoint of the opposite side. Let T be the triangle with vertices $(0, 0)$, $(a, 0)$, and (c, d). In Exercises 70–72, prove the given statements.

70. The medians of triangle T are **concurrent**; that is, all three medians intersect at the same point, P.

71. Use the integral definition for the centroid to show that the centroid of T is point P from Exercise 70.

72. Prove that the centroid of triangle T is two-thirds of the way from each vertex to the opposite side.

73. Prove that the centroid of a circle is the center of the circle.

74. Recall that an annulus is the region between two concentric circles. Prove that the centroid of an annulus is the common center of the two circles.

Thinking Forward

▶ *A triple iterated integral:* Let α, β, γ, δ, ϵ, and ζ be real numbers. Evaluate the triple iterated integral

$$\int_\alpha^\beta \int_\gamma^\delta \int_\epsilon^\zeta dz\, dy\, dx.$$

What does this integral represent?

▶ *A triple iterated integral of a density function:* Let α, β, γ, δ, ϵ, and ζ be real numbers, and let $\rho(x, y, z)$ be a function giving the density at each point of a three-dimensional rectangular solid. What does the triple integral

$$\int_\alpha^\beta \int_\gamma^\delta \int_\epsilon^\zeta \rho(x, y, z)\, dz\, dy\, dx.$$

represent?

13.5   TRIPLE INTEGRALS

▶ Generalizing integration to functions of three variables

▶ Computing triple integrals defined on rectangular solids and on more general regions

▶ Computing mass, center of mass, and moments of inertia of three-dimensional solids

Triple Integrals over Rectangular Solids

In Section 13.1 we defined the double integral of a function of two variables over a rectangular region in the plane. We will be building upon Definition 13.4 to define the triple integral of a function of three variables over a region in $\mathbb{R}^3$ that is a rectangular solid, such as the following one:

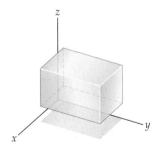

Later in this section we will extend our definition to more general regions in $\mathbb{R}^3$. We will define the triple integral in terms of the limit of a Riemann sum, just as we did for the double integral. Rather than belabor the details, we quickly outline the steps needed to

define a triple integral. If you have questions, then you should reread Section 13.1, which discusses the steps more completely for double integrals. We start with Riemann sums for a function of three variables.

DEFINITION 13.14

Riemann Sums for a Function of Three Variables

Let $a_1 < a_2, b_1 < b_2$, and $c_1 < c_2$ be real numbers, let $\mathcal{R}$ be the rectangular solid defined by

$$\mathcal{R} = \{(x, y, z) \mid a_1 \leq x \leq a_2, \ b_1 \leq y \leq b_2, \ \text{and } c_1 \leq z \leq c_2\},$$

and let $f(x, y, z)$ be a function defined on $\mathcal{R}$. The sum

$$\sum_{i=1}^{l} \sum_{j=1}^{m} \sum_{k=1}^{n} f(x_i^*, y_j^*, z_k^*) \Delta V$$

is a **Riemann Sum** for f on $\mathcal{R}$, where

$$\Delta x = \frac{a_2 - a_1}{l}, \quad \Delta y = \frac{b_2 - b_1}{m}, \quad \Delta z = \frac{c_2 - c_1}{n}, \quad \Delta V = \Delta x \Delta y \Delta z,$$

$$x_i = a_1 + i\Delta x \text{ for } 0 \leq i \leq l, \ y_j = b_1 + j\Delta y \text{ for } 0 \leq j \leq m, \ z_k = c_1 + k\Delta z \text{ for } 0 \leq k \leq n,$$

$$\text{and } x_i^* \in [x_{i-1}, x_i], \ y_j^* \in [y_{j-1}, y_j], \ z_k^* \in [z_{k-1}, z_k].$$

Note that the nested summations in Definition 13.14 may be done in any of $3! = 6$ orders. The order we give in the definition is just one of those six, and we could have used any of the other five instead.

We now define Δ to be the length of the space diagonal of each rectangular solid. That is,

$$\Delta = \sqrt{(\Delta x)^2 + (\Delta y)^2 + (\Delta z)^2}.$$

When we take the limit $\Delta \to 0$, this ensures that $l \to \infty$, $m \to \infty$, and $n \to \infty$ simultaneously. We use the limit of the Riemann sum as $\Delta \to 0$ to define the triple integral, provided that the limit exists.

DEFINITION 13.15

Triple Integrals

Let $a_1 < a_2, b_1 < b_2$, and $c_1 < c_2$ be real numbers, let $\mathcal{R}$ be the rectangular solid defined by

$$\mathcal{R} = \{(x, y, z) \mid a_1 \leq x \leq a_2, \ b_1 \leq y \leq b_2, \ \text{and } c_1 \leq z \leq c_2\},$$

and let $f(x, y, z)$ be a function defined on $\mathcal{R}$. Provided that the limit exists, the **triple integral** of f over $\mathcal{R}$ is

$$\iiint_{\mathcal{R}} f(x, y, z) \, dV = \lim_{\Delta \to 0} \sum_{i=1}^{l} \sum_{j=1}^{m} \sum_{k=1}^{n} f(x_i^*, y_j^*, z_k^*) \Delta V,$$

where the triple sum in the equation is a Riemann sum as outlined in Definition 13.14, and where $\Delta = \sqrt{(\Delta x)^2 + (\Delta y)^2 + (\Delta z)^2}$. When the triple integral exists on $\mathcal{R}$, the function f is said to be **integrable** on $\mathcal{R}$.

Note that as $\Delta \to 0$, the increment of volume $\Delta V \to 0$ as well. The summations in Definition 13.15 can be ordered in five other ways to obtain the same value for the triple integral.

At this point you may be wondering what the value of a particular triple integral represents about the function $f(x, y, z)$. As we have seen before, the answer depends upon the context. For example, if f is the constant function $f(x, y, z) = 1$, the triple integral $\iiint_{\mathcal{R}} dV$ would be a complicated way to compute the volume of the rectangular solid $\mathcal{R}$. If this were the only application of triple integration, we would not be bothering with the concept. However, recall that in Section 13.4 we showed that the double integral $\int_a^b \int_c^d \rho(x, y)\, dy\, dx$ gives the mass of the laminar rectangle $\mathcal{R} = \{(x, y) \mid a \le x \le b \text{ and } c \le y \le d\}$ if $\rho(x, y)$ is the density function for the lamina. Similarly, if $\rho(x, y, z) \ge 0$ is a density function of a rectangular solid $\mathcal{R}$, the triple integral $\iiint_{\mathcal{R}} \rho(x, y, z)\, dV$ gives the mass of the solid. Later in this section we will see how to use triple integrals to find the center of mass and moments of inertia for three-dimensional regions also. These will be our primary applications of triple integration.

Iterated Integrals and Fubini's Theorem

As we saw in Section 13.1, it is usually simpler to evaluate a double integral by using an iterated integral rather than the definition. The situation here is similar.

DEFINITION 13.16

Iterated Triple Integrals

Let a_1, a_2, b_1, b_2, c_1, and c_2 be real numbers. The *iterated triple integral* is

$$\int_{a_1}^{a_2} \int_{b_1}^{b_2} \int_{c_1}^{c_2} f(x, y, z)\, dz\, dy\, dx = \int_{a_1}^{a_2} \left(\int_{b_1}^{b_2} \left(\int_{c_1}^{c_2} f(x, y, z)\, dz \right) dy \right) dx.$$

Again, there are five other orderings we could use to construct an iterated triple integral. For example,

$$\int_{b_1}^{b_2} \int_{c_1}^{c_2} \int_{a_1}^{a_2} f(x, y, z)\, dx\, dz\, dy = \int_{b_1}^{b_2} \left(\int_{c_1}^{c_2} \left(\int_{a_1}^{a_2} f(x, y, z)\, dx \right) dz \right) dy.$$

For a region that is a rectangular solid, the different orderings for the integrals usually do not significantly change the level of difficulty in a problem. The ordering shown in Definition 13.16 is the one we will try first in most problems.

We have a version of Fubini's theorem that tells us that we may use an iterated triple integral, rather than the definition of the triple integral, to evaluate a triple integral.

THEOREM 13.17

Fubini's Theorem for Triple Integrals

Let a_1, a_2, b_1, b_2, c_1, and c_2 be real numbers, and let $\mathcal{R}$ be the rectangular solid defined by

$$\mathcal{R} = \{(x, y, z) \mid a_1 \le x \le a_2,\ b_1 \le y \le b_2,\ \text{and } c_1 \le z \le c_2\}.$$

If $f(x, y, z)$ is continuous on $\mathcal{R}$, then

$$\iiint_{\mathcal{R}} f(x, y, z)\, dV = \int_{a_1}^{a_2} \int_{b_1}^{b_2} \int_{c_1}^{c_2} f(x, y, z)\, dz\, dy\, dx$$

and is also equal to any of the other five possible orderings for the iterated triple integral.

The proof of this version of Fubini's theorem is again beyond the scope of the text, but here is an example illustrating how it is used: Let

$$\mathcal{R} = \{(x, y, z) \mid 1 \leq x \leq 3, \ 2 \leq y \leq 6, \text{ and } 4 \leq z \leq 7\}.$$

Assume that the density at each point of $\mathcal{R}$ is proportional to the distance of the point from the xy-plane. That is, $\rho(x, y, z) = kz$, where k is a constant of proportionality. We may use Theorem 13.17 to find the mass of $\mathcal{R}$. The mass of the solid is given by

$$\iiint_{\mathcal{R}} kz \, dV = \int_1^3 \int_2^6 \int_4^7 kz \, dz \, dy \, dx.$$

Here we have

$$\iiint_{\mathcal{R}} kz \, dV = \int_1^3 \int_2^6 \int_4^7 kz \, dz \, dy \, dx \qquad \leftarrow \text{Fubini's theorem}$$

$$= \int_1^3 \left(\int_2^6 \left(\int_4^7 kz \, dz \right) dy \right) dx \qquad \leftarrow \text{evaluation procedure for the iterated integral}$$

$$= \int_1^3 \left(\int_2^6 \left[\frac{1}{2} kz^2 \right]_{z=4}^{z=7} dy \right) dx \qquad \leftarrow \text{the Fundamental Theorem of Calculus}$$

$$= \int_1^3 \left(\int_2^6 \left(\frac{1}{2} k(49 - 16) \right) dy \right) dx \qquad \leftarrow \text{evaluation of the innermost antiderivative}$$

$$= \int_1^3 \left[\frac{33}{2} ky \right]_{y=2}^{y=6} dx = \int_1^3 66k \, dx \qquad \leftarrow \text{evaluation of the next integral}$$

$$= \left[66kx \right]_{x=1}^{x=3} = 132k. \qquad \leftarrow \text{evaluation of the outer integral}$$

If we had used any of the five other orders of integration to evaluate the triple integral, we would have obtained the same result.

Triple Integrals over General Regions

We now expand our definition of the triple integral to allow us to compute a triple integral over more general regions. The general regions we will use are of three types. First, consider a region Ω_{xy} in the xy-plane and two functions $z = g_1(x, y)$ and $z = g_2(x, y)$ defined on Ω_{xy} such that $g_1(x, y) \leq g_2(x, y)$ for every point in Ω_{xy}. We define the three-dimensional region Ω to be the set

$$\Omega = \{(x, y, z) \mid g_1(x, y) \leq z \leq g_2(x, y) \text{ for } (x, y) \in \Omega_{xy}\}.$$

For example, in the following figure, Ω_{xy} is the circle in the xy-plane, with radius 2, and centered at the origin, and the solid shown is bounded below by the plane with equation $g_1(x, y) = \frac{1}{2}x - \frac{1}{2}y - 2$ and is bounded above by the paraboloid with equation $g_2(x, y) = 4 - x^2 - y^2$:

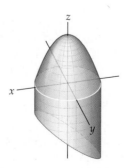

The other two types of regions we will be using are defined in a similar manner. In one, we start with a region Ω_{xz} in the xz-plane and consider two functions $y = h_1(x, z)$ and $y = h_2(x, z)$ such that $h_1(x, z) \leq h_2(x, z)$ for every $(x, z) \in \Omega_{xz}$. In the other, we start with a region Ω_{yz} in the yz-plane and consider two functions $x = k_1(y, z)$ and $x = k_2(y, z)$ such that $k_1(y, z) \leq k_2(y, z)$ for every $(y, z) \in \Omega_{yz}$.

DEFINITION 13.18

Triple Integrals over General Regions

Let Ω be a general region in $\mathbb{R}^3$ of one of the three types just described, and let

$$\mathcal{R} = \{(x, y, z) \mid a_1 \leq x \leq a_2, \ b_1 \leq y \leq b_2, \text{ and } c_1 \leq z \leq c_2\}$$

be a rectangular solid containing Ω. We define the **triple integral** of f over Ω to be

$$\iiint_\Omega f(x, y, z)\, dV = \iiint_\mathcal{R} F(x, y, z)\, dV,$$

where

$$F(x, y, z) = \begin{cases} f(x, y, z), & \text{if } (x, y, z) \in \Omega \\ 0, & \text{if } (x, y, z) \notin \Omega \end{cases}$$

provided that the triple integral $\iiint_\mathcal{R} F(x, y, z)\, dV$ exists.

We will use Fubini's theorem to evaluate triple integrals on such general regions. To use Fubini's theorem in this context, when we wish to integrate the function $f(x, y, z)$ over a region Ω bounded below by $g_1(x, y)$ and above by $g_2(x, y)$ over a region Ω_{xy} in the xy-plane, we evaluate

$$\iiint_\Omega f(x, y, z)\, dV = \iint_{\Omega_{xy}} \int_{g_1(x,y)}^{g_2(x,y)} f(x, y, z)\, dz\, dA.$$

Once we have evaluated the inner integral $\int_{g_1(x,y)}^{g_2(x,y)} f(x, y, z)\, dz$, we evaluate the remaining double integral, using the techniques we discussed in Sections 13.2 and 13.3, depending upon the particular integral we are facing. We evaluate integrals over the other general regions in an analogous manner.

For example, to evaluate the triple integral of the function $f(x, y, z) = 5x - 3y$ over the tetrahedral region Ω bounded by the coordinate planes and the plane containing the points $(2, 0, 0)$, $(0, 3, 0)$, and $(0, 0, 6)$, we first try to visualize Ω. The region, which lies in the first octant, is shown here at the left:

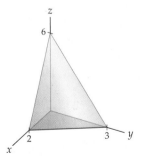

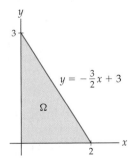

Because this region is relatively simple, we may consider it as belonging to *any* of the three general types we outlined earlier. We will consider it to be of the first type. The equation of the plane containing the three specified points as a function of x and y is

$g_2(x, y) = -3x - 2y + 6$. Since Ω is bounded below by the xy-plane, we use $g_1(x, y) = 0$. Therefore,

$$\iiint_\Omega (5x - 3y)\, dV = \iint_{\Omega_{xy}} \int_0^{-3x-2y+6} (5x - 3y)\, dz\, dA,$$

where Ω_{xy} is the triangular region shown at the right. Here it is particularly easy to find Ω_{xy}, since it is the bottom face of the tetrahedron. More generally, however, Ω_{xy} is the projection of Ω onto the xy-plane. To evaluate the double integral over Ω_{xy} we may treat this region as either a type I region or a type II region, as described in Section 13.2. For this example, we will treat Ω_{xy} as a type I region. Thus we have

$$\iint_{\Omega_{xy}} \int_0^{-3x-2y+6} (5x - 3y)\, dz\, dA = \int_0^2 \int_0^{-(3/2)x+3} \int_0^{-3x-2y+6} (5x - 3y)\, dz\, dy\, dx.$$

Before we evaluate this triple integral, notice that the limits of the innermost integral are functions of the variables of the outer two integrals (although one of them is a constant function), the limits of the middle integral are functions of the variable of the outermost integral (although, again, one of them is a constant function), and the limits of the outermost integral are constants. When we complete the evaluation of the iterated integral, we will see that our answer is also a constant. These are properties you should *always* have when you set up and evaluate an iterated triple integral. Now,

$$\int_0^2 \int_0^{-(3/2)x+3} \int_0^{-3x-2y+6} (5x - 3y)\, dz\, dy\, dx$$

$$= \int_0^2 \int_0^{-(3/2)x+3} \Big[(5x - 3y)z \Big]_{z=0}^{z=-3x-2y+6}\, dy\, dx \qquad \leftarrow \text{the Fundamental Theorem of Calculus}$$

$$= \int_0^2 \int_0^{-(3/2)x+3} (-15x^2 - xy + 6y^2 + 30x - 18y)\, dy\, dx \qquad \leftarrow \text{evaluation and multiplication}$$

$$= \int_0^2 \Big[-15x^2 y - \frac{1}{2}xy^2 + 2y^3 + 30xy - 9y^2 \Big]_{y=0}^{y=-(3/2)x+3}\, dx \qquad \leftarrow \text{the Fundamental Theorem of Calculus}$$

$$= \int_0^2 \Big(\frac{117}{8}x^3 - \frac{261}{4}x^2 + \frac{171}{2}x - 27 \Big)\, dx \qquad \leftarrow \text{evaluation and simplification}$$

$$= \Big[\frac{117}{32}x^4 - \frac{87}{4}x^3 + \frac{171}{4}x^2 - 27x \Big]_{x=0}^{x=2} = \frac{3}{2}. \qquad \leftarrow \text{the Fundamental Theorem of Calculus}$$

In this example, we see that the evaluation of even a relatively simple triple integral over a simple region can necessitate a lengthy computation. None of the steps in this computation are terribly difficult, but there are certainly many points at which small mistakes can lead to an incorrect result. Care must be taken in setting up and evaluating triple integrals, as we will see in the examples that follow.

Applications of Triple Integration

The applications we discussed in Section 13.4 generalize to three-dimensional space in a natural way. We have already mentioned that if we know the density function $\rho(x, y, z)$ for a region $\Omega \subset \mathbb{R}^3$, then $\iiint_\Omega \rho(x, y, z)\, dV$ represents the mass of Ω. The definitions for the first and second moments and for the center of mass of a three-dimensional region are closely related to the definitions we made for laminæ in $\mathbb{R}^2$.

DEFINITION 13.19

First Moments About the Coordinate Planes

Let Ω be a region in $\mathbb{R}^3$ in which the density at each point is given by the continuous function $\rho(x, y, z)$.

(a) The *first moment* of the mass in Ω with respect to the yz-plane is

$$M_{yz} = \iiint_{\Omega} x\,\rho(x, y, z)\,dV.$$

(b) The *first moment* of the mass in Ω with respect to the xz-plane is

$$M_{xz} = \iiint_{\Omega} y\,\rho(x, y, z)\,dV.$$

(c) The *first moment* of the mass in Ω with respect to the xy-plane is

$$M_{xy} = \iiint_{\Omega} z\,\rho(x, y, z)\,dV.$$

Just as we did with laminæ, the center of mass of a region $\Omega \subset \mathbb{R}^3$ is computed from the first moments and the mass of Ω. If the mass of Ω is $m = \iiint_{\Omega} \rho(x, y, z)\,dV$, then the center of mass of Ω is $(\bar{x}, \bar{y}, \bar{z})$, where

$$\bar{x} = \frac{M_{yz}}{m}, \quad \bar{y} = \frac{M_{xz}}{m}, \quad \text{and} \quad \bar{z} = \frac{M_{xy}}{m}.$$

We may also define moments of inertia about the three coordinate axes and use them to compute the radius of gyration about each axis.

DEFINITION 13.20

Moments of Inertia About the Coordinate Axes

Let Ω be a region in $\mathbb{R}^3$ in which the density at each point is given by the continuous function $\rho(x, y, z)$.

(a) The *moment of inertia of Ω about the x-axis* is

$$I_x = \iiint_{\Omega} (y^2 + z^2)\,\rho(x, y, z)\,dV.$$

(b) The *moment of inertia of Ω about the y-axis* is

$$I_y = \iiint_{\Omega} (x^2 + z^2)\,\rho(x, y, z)\,dV.$$

(c) The *moment of inertia of Ω about the z-axis* is

$$I_z = \iiint_{\Omega} (x^2 + y^2)\,\rho(x, y, z)\,dV.$$

To compute the *radii of gyration* about the x-, y-, and z-axes, we use

$$R_x = \sqrt{I_x/m}, \quad R_y = \sqrt{I_y/m}, \quad \text{and} \quad R_z = \sqrt{I_z/m},$$

respectively. These are the radial distances at which the mass of the solid could be concentrated without changing its rotational inertia about the specified axis. We will provide examples of all of these computations in the examples that follow.

Examples and Explorations

EXAMPLE 1

Finding the volume of a rectangular parallelepiped

Set up the six iterated triple integrals that could be used to find the volume of the rectangular parallelepiped defined by

$$\mathcal{R} = \{(x, y, z) \mid -1 \leq x \leq 3,\ 2 \leq y \leq 4,\ \text{and } 1 \leq z \leq 6\}.$$

Evaluate one of the triple integrals to show that the volume of the parallelepiped is the product of its length, width, and height.

SOLUTION

The following rectangular solid measures 4 units in the x direction, 2 units in the y direction, and 5 units in the z direction:

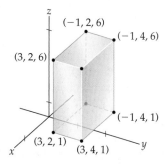

The volume of this solid is $4 \times 2 \times 5 = 40$ cubic units.

The first iterated integral we could use to compute the volume is

$$\text{Volume} = \iiint_{\mathcal{R}} dV = \int_{-1}^{3} \int_{2}^{4} \int_{1}^{6} dz\, dy\, dx.$$

This order of integration is typically the first one we try. We will discuss the reasons for this in Example 3. Here we note that the innermost integration is with respect to z, the middle integration is with respect to y, and the outer integration is with respect to x. Therefore, the limits on the inner, middle, and outer integrals are the values that correspond to those variables from the definition of the parallelepiped. To obtain the other five possible iterated integrals, we permute the differentials dx, dy, and dz, moving the limits of integration in the corresponding way. Thus, we also have

$$\text{Volume} = \int_{2}^{4} \int_{-1}^{3} \int_{1}^{6} dz\, dx\, dy = \int_{-1}^{3} \int_{1}^{6} \int_{2}^{4} dy\, dz\, dx$$

$$= \int_{1}^{6} \int_{-1}^{3} \int_{2}^{4} dy\, dx\, dz = \int_{2}^{4} \int_{1}^{6} \int_{-1}^{3} dx\, dz\, dy = \int_{1}^{6} \int_{2}^{4} \int_{-1}^{3} dx\, dy\, dz.$$

We now evaluate the triple integral, using the first ordering we provided. Note that we work from the innermost integral to the outermost integral:

$$\int_{-1}^{3} \int_{2}^{4} \int_{1}^{6} dz\, dy\, dx = \int_{-1}^{3} \int_{2}^{4} [z]_{1}^{6} dy\, dx = \int_{-1}^{3} \int_{2}^{4} 5\, dy\, dx$$

$$= \int_{-1}^{3} [5y]_{2}^{4}\, dx = \int_{-1}^{3} 10\, dx$$

$$= [10x]_{-1}^{3} = 40.$$

Our answers agree: The volume of $\mathcal{R}$ is 40 cubic units. As we mentioned earlier in the section, we would not have bothered to define triple integration if this were the only application. For example, in Example 2 we find the mass of the solid $\mathcal{R}$, assuming that the density is a nonconstant function.   □

EXAMPLE 2   Finding the mass of a rectangular parallelepiped

Find the mass of the solid $\mathcal{R}$ from Example 1, assuming that the density at every point of $\mathcal{R}$ is proportional to the distance of the point from the xz-plane.

SOLUTION

Instead of integrating the constant function $f(x, y, z) = 1$ to find the volume of $\mathcal{R}$, here we integrate the density function $\rho(x, y, z) = ky$, where k is a constant of proportionality, to find the mass of the solid. Almost all of the work involved in setting up the iterated triple integral was done in Example 1; the only change is the integrand. We choose the same order of variables that we used before:

$$
\begin{aligned}
\text{Mass} &= \iiint_{\mathcal{R}} ky \, dV = \int_{-1}^{3} \int_{2}^{4} \int_{1}^{6} ky \, dz \, dy \, dx \\
&= \int_{-1}^{3} \int_{2}^{4} \left[kyz \right]_{1}^{6} dy \, dx = \int_{-1}^{3} \int_{2}^{4} 5ky \, dy \, dx \\
&= \int_{-1}^{3} \left[\tfrac{5}{2} ky^2 \right]_{2}^{4} dx = \int_{-1}^{3} 30k \, dx \\
&= \left[30kx \right]_{-1}^{3} = 120k.
\end{aligned}
$$

Note again that there are five other orders we could have used for the iterated integral. For this particular problem, the levels of difficulty for the integrations would be similar.   □

EXAMPLE 3   Finding the volume of a pyramid

Use a triple integral to find the volume of the pyramid $\mathcal{P}$ whose base is the square with vertices $(1, 0, 0)$, $(0, 1, 0)$, $(-1, 0, 0)$, and $(0, -1, 0)$ and whose top vertex is $(0, 0, 1)$.

SOLUTION

The pyramid $\mathcal{P}$ is shown in the following figure at the left:

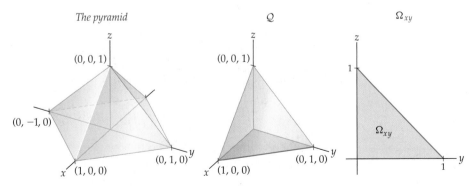

By the symmetry of $\mathcal{P}$, the pyramid $\mathcal{Q}$ in the middle figure has a volume that is one-quarter of the volume of $\mathcal{P}$. We will find the volume of $\mathcal{Q}$ and then multiply by 4 to obtain the volume of $\mathcal{P}$. The oblique planar face of $\mathcal{Q}$ has equation $x + y + z = 1$. We mentioned in Example 1 that we would explain the reason for typically choosing the order of integration specified by $dz \, dy \, dx$. When we have a function of two variables, we tend to think of z as a function of x and y. Similarly, when we have a function of a single variable, we tend to think of y as a function x. This is the convention we have grown accustomed to and that may cause us to prefer it. Notice that in the iterated integral we evaluated in Example 2, the upper limit on the innermost integral is a function of two variables. We preferred to express the equation of the oblique planar face as a function of x and y. Similarly, when we provide the limits on the middle integral, we need to express one or both boundaries as functions of a single variable. When we do so, we prefer to express the limits on this integral as functions of x. The projection of Q onto the xy-plane gives the region Ω_{xy} shown at the right. The right-hand boundary of Ω_{xy} may be expressed either as $y = -x + 1$ or as $x = -y + 1$. We prefer the former, which is the upper limit that we use on the middle integral. In Example 6 we will use one of the other orders of integration to construct an iterated triple integral, but for now the volume of Q is given by the following iterated triple integral:

$$\text{Volume of } Q = \int_0^1 \int_0^{-x+1} \int_0^{-x-y+1} dz\, dy\, dx = \int_0^1 \int_0^{-x+1} [z]_0^{-x-y+1} dy\, dx$$

$$= \int_0^1 \int_0^{-x+1} (-x - y + 1) dy\, dx = \int_0^1 \left[-xy - \frac{1}{2}y^2 + y \right]_0^{-x+1} dx$$

$$= \int_0^1 \left(\frac{1}{2}x^2 - x + \frac{1}{2} \right) dx = \left[\frac{1}{6}x^3 - \frac{1}{2}x^2 + \frac{1}{2}x \right]_0^1 = \frac{1}{6} \text{ cubic units.}$$

Multiplying this answer by 4, we find that the volume of $\mathcal{P}$ is $\frac{2}{3}$ cubic unit.   □

 **CHECKING THE ANSWER**

Recall that the volume of a pyramid is $\frac{1}{3}$(area of the base)(height). Each side of the square base of pyramid $\mathcal{P}$ has length $\sqrt{2}$. The height of $\mathcal{P}$ is 1. Therefore, the volume of $\mathcal{P}$ is $\frac{1}{3}\sqrt{2} \cdot \sqrt{2} \cdot 1 = \frac{2}{3}$ cubic unit, as we just found.

EXAMPLE 4

Finding the center of mass of a pyramid

Find the center of mass of the pyramid $\mathcal{P}$ from Example 3, assuming that the density of the pyramid is uniform.

SOLUTION

By the symmetries of $\mathcal{P}$ and its density function, the first moments of the mass with respect to the yz- and xz-planes are both zero. Therefore, if we let $(\bar{x}, \bar{y}, \bar{z})$ be the center of mass of $\mathcal{P}$, then both $\bar{x}$ and $\bar{y}$ are zero. To find $\bar{z}$, we could compute both the mass of $\mathcal{P}$ and the first moment M_{xy} of $\mathcal{P}$ when the density function $\rho(x, y, z) = k$, where k is a constant. However, note that the z-coordinates for the centers of mass of the pyramids $\mathcal{P}$ and Q are equal. Although the x- and y-coordinates of the centers of mass of $\mathcal{P}$ and Q would be different, we already know that $\bar{x} = \bar{y} = 0$ for $\mathcal{P}$. Therefore, using our work from Example 3, we find that the mass of Q is given by

$$m = \int_0^1 \int_0^{-x+1} \int_0^{-x-y+1} k\, dz\, dy\, dx = \frac{1}{6}k.$$

The moment of inertia of Q with respect to the xy-plane is

$$M_{xy} = \int_0^1 \int_0^{-x+1} \int_0^{-x-y+1} kz\, dz\, dy\, dx = \int_0^1 \int_0^{-x+1} \left[\frac{1}{2}kz^2 \right]_0^{-x-y+1} dy\, dx$$

$$= \int_0^1 \int_0^{-x+1} \frac{1}{2}k(-x - y + 1)^2 dy\, dx = \int_0^1 \left[-\frac{1}{6}k(-x - y + 1)^3 \right]_0^{-x+1} dx$$

$$= \int_0^1 \frac{1}{6}k(-x + 1)^3 dx = \left[-\frac{1}{24}k(-x + 1)^4 \right]_0^1 = \frac{1}{24}k.$$

Therefore, the z-coordinate of the center of mass of both $\mathcal{P}$ and $\mathcal{Q}$ is

$$\bar{z} = \frac{M_{xy}}{m} = \frac{k/24}{k/6} = \frac{1}{4},$$

and the center of mass of $\mathcal{P}$ is $(0, 0, 1/4)$. Note that the point $(0, 0, 1/4)$ lies inside the pyramid. This shows that our result is plausible.   □

EXAMPLE 5

Finding the moments of inertia and radii of gyration for a pyramid

Find the moments of inertia about the coordinate axes for the pyramid $\mathcal{Q}$ from Example 3, assuming that its density is uniform. Use those moments to find the radius of gyration about each coordinate axis.

SOLUTION

In Example 4 we saw that the mass of $\mathcal{Q}$ is

$$m = \int_0^1 \int_0^{-x+1} \int_0^{-x-y+1} k \, dz \, dy \, dx = \frac{1}{6}k.$$

To find the moment of inertia about the x-axis, we modify this integral by including the factor $(y^2 + z^2)$ in the integrand. Thus, we have

$$I_x = \int_0^1 \int_0^{-x+1} \int_0^{-x-y+1} k(y^2 + z^2) \, dz \, dy \, dx.$$

The evaluation procedure for this integral is quite similar to those in Examples 3 and 4. In Exercise 15 you will show that $I_x = \frac{1}{30}k$.

To find the radius of gyration about the x-axis, we compute $R_x = \sqrt{\frac{I_x}{m}}$, where m is the mass of $\mathcal{Q}$. We get

$$R_x = \sqrt{\frac{I_x}{m}} = \sqrt{\frac{(1/30)k}{(1/6)k}} = \frac{\sqrt{5}}{5}.$$

Now, to find the moments of inertia about the y- and z-axes, and the corresponding radii of gyration, we could evaluate the integrals

$$I_y = \int_0^1 \int_0^{-x+1} \int_0^{-x-y+1} k(x^2 + z^2) \, dz \, dy \, dx, \quad I_z = \int_0^1 \int_0^{-x+1} \int_0^{-x-y+1} k(x^2 + y^2) \, dz \, dy \, dx,$$

and follow that by calculating $R_y = \sqrt{I_y/m}$ and $R_z = \sqrt{I_z/m}$. You will do exactly that in Exercises 16 and 17. Here we ask you to convince yourself that pyramid $\mathcal{Q}$ has rotational symmetry about the line given by the parametric equations

$$x = t, \ y = t, \ z = t$$

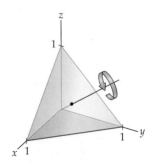

A rotation of $\frac{\pi}{3}$ about this line takes each coordinate axis to one of the others, depending upon the direction of the rotation. In the rotation shown in the preceding figure, the x-axis would rotate onto the y-axis, etc. This means that the moments of inertia and the radii of gyration for all three coordinate axes must be equal. That is,

$$I_x = I_y = I_z = \frac{1}{30}k \quad \text{and} \quad R_x = R_y = R_z = \frac{\sqrt{5}}{5}.$$

□

EXAMPLE 6   Finding the mass of a slice of a cylinder

Find the mass of the slice of the right circular cylinder $x^2 + z^2 = 4$ bounded on the left by the xz-plane and on the right by the plane with equation $x - y + z = -4$ if the density at each point in the cylinder is proportional to the distance of the point from the xz-plane.

SOLUTION

The region we have described is shown here:

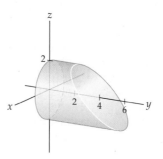

The density function we need to integrate is $\rho(x, y, z) = ky$, where k is a positive constant of proportionality. Let $\mathcal{C}$ represent the slice of the cylinder under consideration. We need to evaluate the triple integral

$$\iiint_{\mathcal{C}} ky \, dV.$$

No matter which order of integration we settle upon, the function ρ will be relatively simple to integrate. The order we will use depends upon the characteristics of the region. In particular, we will consider the projection of $\mathcal{C}$ onto the coordinate planes and decide which is simplest. Since the cylinder is symmetric with respect to the y-axis, the projection of $\mathcal{C}$ onto the xz-plane will be the circle with radius 2, centered at the origin of that plane. Let $\mathcal{C}_{xz}$ be this projection.

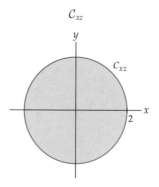

The projections of $\mathcal{C}$ onto the other coordinate planes would be more difficult to analyze; thus we will integrate *first* with respect to y. The limits on this integral are $y = 0$ and

$y = x + z + 4$, the equations of the planes that form the boundaries of C. We now have

$$\iiint_C ky \, dV = \iint_{C_{xz}} \int_0^{x+z+4} ky \, dy \, dA.$$

We next turn our attention to setting up iterated integrals to evaluate the double integral over C_{xz}. The region C_{xz} is bounded above by the semicircle with equation $z = \sqrt{4 - x^2}$ and below by the semicircle with equation $z = -\sqrt{4 - x^2}$. Thus, we have

$$\iiint_C ky \, dV = \int_{-2}^{2} \int_{-\sqrt{4-x^2}}^{\sqrt{4-x^2}} \int_0^{x+z+4} ky \, dy \, dz \, dx.$$

The evaluation of this integral is left for you in Exercise 18. As we see in this example, the main difficulty in evaluating a triple integral often lies in analyzing the region of integration. When you evaluate the preceding integral, you will see that it is a chore, but you will still be using only basic integration techniques that were first introduced in Chapter 5.   □

TEST YOUR UNDERSTANDING

▶ What is a Riemann sum for a function of three variables? How many different orderings are possible for the nested summations? Does the order of the summations matter?

▶ How is the triple integral for a function f over a rectangular solid defined? How is this definition similar to the definition of the double integral of a function over a rectangular region? How is it different?

▶ What is an iterated triple integral? What is Fubini's theorem? How are iterated integrals used to compute triple integrals? How many choices are possible when you order the variables of integration in an iterated triple integral?

▶ What are the three types of three-dimensional regions we discussed in this section? What are the orders of integration associated with each of these different regions?

▶ What applications of triple integration did we discuss in this section? How are integrals for masses, first moments of mass, and moments of inertia related to each other?

EXERCISES 13.5

Thinking Back

▶ *Using iterated double integrals to compute area:* Let Ω be the region in the first quadrant bounded by the x-axis and the lines with equations $x = 1$ and $y = x$. Set up iterated double integrals that represent the area of Ω in which Ω is treated first as a type I region and then as a type II region.

▶ *Using iterated double integrals to compute mass:* Let Ω be the first-quadrant region bounded by the y-axis and the lines with equations $y = x$ and $y = 2 - x$. If the density at each point of Ω is given by the function $\rho(x, y)$, set up iterated double-integral expressions that represent the mass of Ω in which Ω is treated first as a type I region and then as a type II region.

Concepts

0. *Problem Zero:* Read the section and make your own summary of the material.

1. *True/False:* Determine whether each of the statements that follow is true or false. If a statement is true, explain why. If a statement is false, provide a counterexample.

(a) *True or False:* If $f(x, y, z)$ is a continuous function on the rectangular solid defined by $\mathcal{R} = \{(x, y, z) \mid a_1 \leq x \leq a_2, \ b_1 \leq y \leq b_2, \ \text{and} \ c_1 \leq z \leq c_2\}$, then $\iiint_{\mathcal{R}} f(x, y, z) \, dV = \int_{a_1}^{a_2} \int_{b_1}^{b_2} \int_{c_1}^{c_2} f(x, y, z) \, dx \, dy \, dz.$

(b) *True or False:* If $f(x, y, z)$ is a continuous function on the rectangular solid defined by $\mathcal{R} = \{(x, y, z) \mid a_1 \leq x \leq a_2, \ b_1 \leq y \leq b_2, \ \text{and} \ c_1 \leq z \leq c_2\}$, then $\iiint_{\mathcal{R}} f(x, y, z) \, dV = \int_{a_1}^{a_2} \int_{c_1}^{c_2} \int_{b_1}^{b_2} f(x, y, z) \, dy \, dz \, dx.$

(c) *True or False:* If $f(x, y, z)$ is a continuous function of three variables and $\Omega = \Omega_1 \cup \Omega_2$ is a subset of $\mathbb{R}^3$, then $\iiint_{\Omega} f(x, y, z) \, dV = \iiint_{\Omega_1} f(x, y, z) \, dV + \iiint_{\Omega_2} f(x, y, z) \, dV.$

(d) *True or False:* If Ω is a bounded subset of $\mathbb{R}^3$, then there is a rectangular solid $\mathcal{R}$ with its sides parallel to the coordinate planes such that $\Omega \subseteq \mathcal{R}$.

(e) *True or False:* If f is a positive continuous function defined on a region Ω and $\Gamma \subseteq \Omega$, then $\iiint_\Gamma f(x, y, z)\, dV \le \iiint_\Omega f(x, y, z)\, dV$.

(f) *True or False:* If $\rho(x, y, z)$ gives the density at every point of a region Ω, then the first moment of the mass in Ω with respect to the xy-plane is $M_{xy} = \iiint_\Omega xy\, \rho(x, y, z)\, dV$.

(g) *True or False:* If $\rho(x, y, z)$ gives the density at every point of a region Ω, then the moment of inertia of Ω about the x-axis is $I_x = \iiint_\Omega (y^2 + z^2)\, \rho(x, y, z)\, dV$.

(h) *True or False:* If $g_1(x, y) \le g_2(x, y)$ on the square region $\mathcal{R} = \{(x, y) \mid 0 \le x \le 2 \text{ and } 0 \le y \le 2\}$, then the iterated integral $\int_{g_1(x,y)}^{g_2(x,y)} \int_0^2 \int_0^2 dV$ represents the volume of the solid bounded below by $g_1(x, y)$ and above by $g_2(x, y)$ on $\mathcal{R}$.

2. *Examples:* Construct examples of the thing(s) described in the following. Try to find examples that are different than any in the reading.

   (a) An iterated triple integral over a rectangular solid such that the integral gives the mass of the solid.

   (b) An iterated triple integral over a non-rectangular solid such that the integral gives the volume of the solid.

   (c) An iterated triple integral over a solid such that the integral gives the moment of inertia of the solid about the y-axis.

3. Explain why
$$\sum_{i=1}^l \sum_{j=1}^m \sum_{k=1}^n ij^2 k^3 = \sum_{j=1}^m \sum_{k=1}^n \sum_{i=1}^l ij^2 k^3.$$

4. Explain how to construct a Riemann sum for a function of three variables over a rectangular solid.

5. Explain how to construct a *midpoint* Riemann sum for a function of three variables over a rectangular solid for which each (x_i^*, y_j^*, z_k^*) is the midpoint of the subsolid $\mathcal{R}_{ijk} = \{(x, y, z) \mid x_{i-1} \le x_i^* \le x_i, y_{j-1} \le y_j^* \le y_j, \text{ and } z_{k-1} \le z_k^* \le z_k\}$. Refer either to your answer to Exercise 4 or to Definition 13.14.

6. Discuss the similarities and differences between the definition of the double integral found in Section 13.1 and the definition of the triple integral found in this section.

7. What is the difference between a triple integral and an *iterated* triple integral?

8. Let $f(x, y, z)$ be a continuous function of three variables, let
$$\Omega_{xy} = \{(x, y) \mid a \le x \le b \text{ and } h_1(x) \le y \le h_2(x)\}$$
be a set of points in the xy-plane, and let
$$\Omega = \{(x, y, z) \mid (x, y) \in \Omega_{xy} \text{ and } g_1(x, y) \le z \le g_2(x, y)\}$$
be a set of points in 3-space. Find an iterated triple integral equal to the the triple integral $\iiint_\Omega f(x, y, z)\, dV$. How

would your answer change if
$$\Omega_{xy} = \{(x, y) \mid a \le y \le b \text{ and } h_1(y) \le x \le h_2(y)\}?$$

9. Let $f(x, y, z)$ be a continuous function of three variables, let
$$\Omega_{yz} = \{(y, z) \mid a \le y \le b \text{ and } h_1(y) \le z \le h_2(y)\}$$
be a set of points in the yz-plane, and let
$$\Omega = \{(x, y, z) \mid (y, z) \in \Omega_{yz} \text{ and } g_1(y, z) \le x \le g_2(y, z)\}$$
be a set of points in 3-space. Find an iterated triple integral equal to the triple integral $\iiint_\Omega f(x, y, z)\, dV$. How would your answer change if
$$\Omega_{yz} = \{(y, z) \mid a \le z \le b \text{ and } h_1(z) \le y \le h_2(z)\}?$$

10. Let $f(x, y, z)$ be a continuous function of three variables, let
$$\Omega_{xz} = \{(x, z) \mid a \le x \le b \text{ and } h_1(x) \le z \le h_2(x)\}$$
be a set of points in the xz-plane, and let
$$\Omega = \{(x, y, z) \mid (x, z) \in \Omega_{xz} \text{ and } g_1(x, z) \le y \le g_2(x, z)\}$$
be a set of points in 3-space. Find an iterated triple integral equal to the triple integral $\iiint_\Omega f(x, y, z)\, dV$. How would your answer change if
$$\Omega_{xz} = \{(x, z) \mid a \le z \le b \text{ and } h_1(z) \le x \le h_2(z)\}?$$

11. Let $\rho(x, y, z)$ be a density function defined on the rectangular solid $\mathcal{R}$ where
$$\mathcal{R} = \{(x, y, z) \mid -1 \le x \le 3,\ 0 \le y \le 2,\ \text{and } 2 \le z \le 7\}.$$
Set up iterated integrals representing the mass of $\mathcal{R}$, using all six distinct orders of integration.

12. Let $\rho(x, y, z)$ be a density function defined on the rectangular solid $\mathcal{R}$ where
$$\mathcal{R} = \{(x, y, z) \mid a_1 \le x \le a_2, b_1 \le y \le b_2,\ \text{and } c_1 \le z \le c_2\}.$$
Set up iterated integrals representing the mass of $\mathcal{R}$, using all six distinct orders of integration.

13. Let $\rho(x, y, z)$ be a density function defined on the tetrahedron Ω with vertices $(0, 0, 0)$, $(2, 0, 0)$, $(0, 4, 0)$, and $(0, 0, 3)$. Set up iterated integrals representing the mass of Ω, using all six distinct orders of integration.

14. Let $\rho(x, y, z)$ be a density function defined on the tetrahedron Ω with vertices $(0, 0, 0)$, $(a, 0, 0)$, $(0, b, 0)$, and $(0, 0, c)$, where a, b, and c are positive real numbers. Set up iterated integrals representing the mass of Ω, using all six distinct orders of integration.

In Exercises 15–17, evaluate the three integrals that can be used to find the moments of inertia for the pyramid $\mathcal{Q}$ described in Example 5 and then use those values to find the radii of gyration about the coordinate axes. Recall that the mass of $\mathcal{Q}$ is $\frac{1}{6}k$.

15. Show that
$$I_x = \int_0^1 \int_0^{-x+1} \int_0^{-x-y+1} k(y^2 + z^2)\, dz\, dy\, dx = \frac{1}{30}k.$$

16. Show that

$$I_y = \int_0^1 \int_0^{-x+1} \int_0^{-x-y+1} k(x^2 + z^2)\, dz\, dy\, dx = \frac{1}{30}k.$$

Use your answer to show that $R_y = \dfrac{\sqrt{5}}{5}$.

17. Show that

$$I_z = \int_0^1 \int_0^{-x+1} \int_0^{-x-y+1} k(x^2 + y^2)\, dz\, dy\, dx = \frac{1}{30}k.$$

Use your answer to show that $R_z = \dfrac{\sqrt{5}}{5}$.

18. Complete Example 6 by evaluating the iterated integral

$$\int_{-2}^2 \int_{-\sqrt{4-x^2}}^{\sqrt{4-x^2}} \int_0^{x+z+4} ky\, dy\, dz\, dx.$$

Identify the quantities determined by the integral expressions in Exercises 19–24. If x, y, and z are all measured in centimeters and $\rho(x, y, z)$ is a density function in grams per cubic centimeter on the three-dimensional region Ω, give the units of the expression.

19. $\displaystyle\iiint_\Omega dV$

20. $\displaystyle\iiint_\Omega \rho(x, y, z)\, dV$

21. $\displaystyle\iiint_\Omega x\,\rho(x, y, z)\, dV$

22. $\dfrac{\iiint_\Omega z\,\rho(x, y, z)\, dV}{\iiint_\Omega \rho(x, y, z)\, dV}$

23. $\displaystyle\iiint_\Omega (x^2 + z^2)\rho(x, y, z)\, dV$

24. $\sqrt{\dfrac{\iiint_\Omega (x^2 + z^2)\,\rho(x, y, z)\, dV}{\iiint_\Omega \rho(x, y, z)\, dV}}$

Skills

Evaluate the iterated integrals in Exercises 25–30.

25. $\displaystyle\int_0^1 \int_2^4 \int_{-1}^5 (x + yz^2)\, dx\, dy\, dz$

26. $\displaystyle\int_1^3 \int_{-1}^5 \int_0^3 xy^2z\, dz\, dy\, dx$

27. $\displaystyle\int_0^1 \int_0^{6-y} \int_0^{3-3y-(1/2)z} (y - z)\, dx\, dz\, dy$

28. $\displaystyle\int_0^1 \int_0^{3-3y} \int_0^{6y} \frac{x}{y}\, dz\, dx\, dy$

29. $\displaystyle\int_0^{\pi/2} \int_0^{\cos y} \int_0^{\sin y} (2x + y)\, dz\, dx\, dy$

30. $\displaystyle\int_0^1 \int_0^3 \int_0^{\sin^{-1} x} (x + y)\, dz\, dy\, dx$

Evaluate the triple integrals in Exercises 31–34 over the specified rectangular solid regions.

31. $\displaystyle\iiint_\mathcal{R} x^2yz\, dV$, where

$\mathcal{R} = \{(x, y, z) \mid -2 \le x \le 1,\ 0 \le y \le 3,\ \text{and } 1 \le z \le 5\}$

32. $\displaystyle\iiint_\mathcal{R} (x + 2y + 3z)\, dV$, where

$\mathcal{R} = \{(x, y, z) \mid 0 \le x \le 4,\ 1 \le y \le 5,\ \text{and } 2 \le z \le 7\}$

33. $\displaystyle\iiint_\mathcal{R} z\sin x \cos y\, dV$, where

$\mathcal{R} = \left\{(x, y, z) \mid 0 \le x \le \pi,\ \dfrac{3\pi}{2} \le y \le 2\pi,\ \text{and } 1 \le z \le 3\right\}$

34. $\displaystyle\iiint_\mathcal{R} \ln(xyz^2)\, dV$, where

$\mathcal{R} = \{(x, y, z) \mid 1 \le x \le 3,\ 1 \le y \le e,\ \text{and } 1 \le z \le 2\}$

Describe the three-dimensional region expressed in each iterated integral in Exercises 35–44.

35. $\displaystyle\int_{-2}^4 \int_2^6 \int_0^5 f(x, y, z)\, dy\, dx\, dz$

36. $\displaystyle\int_{-1}^5 \int_{-3}^2 \int_4^8 f(x, y, z)\, dz\, dx\, dy$

37. $\displaystyle\int_0^3 \int_0^3 \int_0^{3-y} f(x, y, z)\, dz\, dy\, dx$

38. $\displaystyle\int_0^2 \int_0^3 \int_{4x/3}^4 f(x, y, z)\, dz\, dx\, dy$

39. $\displaystyle\int_0^3 \int_0^{1-y/3} \int_0^{2-(2/3)y-2z} f(x, y, z)\, dx\, dz\, dy$

40. $\displaystyle\int_0^2 \int_0^{1-3x/2} \int_{2x+(4/3)y-4}^0 f(x, y, z)\, dz\, dy\, dx$

41. $\displaystyle\int_{-3}^3 \int_{-\sqrt{9-x^2}}^{\sqrt{9-x^2}} \int_{-\sqrt{9-x^2-y^2}}^{\sqrt{9-x^2-y^2}} f(x, y, z)\, dz\, dy\, dx$

42. $\displaystyle\int_{-3}^3 \int_{-\sqrt{9-x^2}}^{\sqrt{9-x^2}} \int_{-3}^3 f(x, y, z)\, dz\, dy\, dx$

43. $\displaystyle\int_{-3}^3 \int_{-\sqrt{9-y^2}}^{\sqrt{9-y^2}} \int_0^{9-y^2-z^2} f(x, y, z)\, dx\, dz\, dy$

44. $\displaystyle\int_{-3}^3 \int_{-\sqrt{9-z^2}}^{\sqrt{9-z^2}} \int_0^{3-\sqrt{x^2+z^2}} f(x, y, z)\, dy\, dx\, dz$

In Exercises 45–52, rewrite the indicated integral with the specified order of integration.

45. Exercise 35 with the order $dx\,dy\,dz$.

46. Exercise 36 with the order $dy\,dz\,dx$.

47. Exercise 37 with the order $dx\,dy\,dz$.

48. Exercise 38 with the order $dx\,dy\,dz$.

49. Exercise 39 with the order $dz\,dy\,dx$.

50. Exercise 40 with the order $dx\,dy\,dz$.

51. Exercise 41 with the order $dy\,dx\,dz$.

52. Exercise 42 with the order $dy\,dx\,dz$.

Find the masses of the solids described in Exercises 53–56.

53. The first-octant solid bounded by the coordinate planes and the plane $3x + 4y + 6z = 12$ if the density at each point is proportional to the distance of the point from the xz-plane.

54. The solid bounded above by the plane with equation $2x + 3y - z = 2$ and bounded below by the triangle with vertices $(1, 0, 0)$, $(4, 0, 0)$, and $(0, 2, 0)$ if the density at each point is proportional to the distance of the point from the xy-plane.

55. The solid bounded above by the paraboloid with equation $z = 8 - x^2 - y^2$ and bounded below by the rectangle $\mathcal{R} = \{(x, y, 0) \mid 1 \leq x \leq 2 \text{ and } 0 \leq y \leq 2\}$ in the xy-plane if the density at each point is proportional to the square of the distance of the point from the origin.

56. The solid bounded above by the hyperboloid with equation $z = x^2 - y^2$ and bounded below by the square with vertices $(2, 2, -4)$, $(2, -2, -4)$, $(-2, -2, -4)$, and $(-2, 2, -4)$ if the density at each point is proportional to the distance of the point from the plane with equation $z = -4$.

Applications

In Exercises 57–60, let $\mathcal{R}$ be the rectangular solid defined by $\mathcal{R} = \{(x, y, z) \mid 0 \leq x \leq 4,\ 0 \leq y \leq 3,\ 0 \leq z \leq 2\}$.

57. Assume that the density of $\mathcal{R}$ is uniform throughout.

(a) Without using calculus, explain why the center of mass is $(2, 3/2, 1)$.

(b) Verify that the center of mass is $(2, 3/2, 1)$, using the appropriate integral expressions.

58. Assume that the density of $\mathcal{R}$ is uniform throughout, and find the moment of inertia about the x-axis and the radius of gyration about the x-axis.

59. Assume that the density at each point in $\mathcal{R}$ is proportional to the distance of the point from the xy-plane.

(a) Without using calculus, explain why the x- and y-coordinates of the center of mass are $\bar{x} = 2$ and $\bar{y} = \dfrac{3}{2}$, respectively.

(b) Use an appropriate integral expression to find the z-coordinate of the center of mass.

60. Assuming that the density at each point in $\mathcal{R}$ is proportional to the distance of the point from the xy-plane, find the moment of inertia about the x-axis and the radius of gyration about the x-axis.

In Exercises 61–64, let $\mathcal{R}$ be the rectangular solid defined by $\mathcal{R} = \{(x, y, z) \mid 0 \leq a_1 \leq x \leq a_2,\ 0 \leq b_1 \leq y \leq b_2,\ 0 \leq c_2 \leq z \leq c_2\}$.

61. Assume that the density of $\mathcal{R}$ is uniform throughout.

(a) Without using calculus, explain why the center of mass is $\left(\dfrac{a_1 + a_2}{2}, \dfrac{b_1 + b_2}{2}, \dfrac{c_1 + c_2}{2} \right)$.

(b) Verify that $\left(\dfrac{a_1 + a_2}{2}, \dfrac{b_1 + b_2}{2}, \dfrac{c_1 + c_2}{2} \right)$ is the center of mass by using the appropriate integral expressions.

62. Assume that the density of $\mathcal{R}$ is uniform throughout, and find the moment of inertia about the x-axis and the radius of gyration about the x-axis.

63. Assume that the density at each point in $\mathcal{R}$ is proportional to the distance of the point from the yz-plane.

(a) Without using calculus, explain why the y- and z-coordinates of the center of mass are $\bar{y} = \dfrac{b_1 + b_2}{2}$ and $\bar{z} = \dfrac{c_1 + c_2}{2}$, respectively.

(b) Use an appropriate integral expression to find the x-coordinate of the center of mass.

64. Assuming that the density at each point in $\mathcal{R}$ is proportional to the distance of the point from the yz-plane, find the first moment of inertia about the x-axis and the radius of gyration about the x-axis.

Let a, b, and c be positive real numbers. In Exercises 65–68, let $\mathcal{T}$ be the tetrahedron with vertices $(0, 0, 0)$, $(a, 0, 0)$, $(0, b, 0)$, and $(0, 0, c)$.

65. Assume that the density at each point in $\mathcal{T}$ is uniform throughout.

(a) Find the x-coordinate of the center of mass of $\mathcal{T}$.

(b) Explain how to use your answer from part (a) to find the y- and z-coordinates of the center of mass without doing any other computations.

66. Assume that the density of $\mathcal{T}$ is uniform throughout. Set up the integrals required to find the moment of inertia about the x-axis and the radius of gyration about the x-axis.

67. Assume that the density at each point in $\mathcal{T}$ is proportional to the distance of the point from the xz-plane. Set up the integral expressions required to find the center of mass of $\mathcal{T}$.

68. Assuming that the density at each point in $\mathcal{T}$ is proportional to the distance of the point from the yz-plane, set up the integrals required to find the first moment of inertia about the x-axis and the radius of gyration about the x-axis.

Proofs

69. Let $\mathcal{R} = \{(x, y, z) \mid a_1 \leq x \leq a_2, \ b_1 \leq y \leq b_2, \text{ and } c_1 \leq z \leq c_2\}$. If $\alpha(x)$, $\beta(y)$, and $\gamma(z)$ are integrable on the intervals $[a_1, a_2]$, $[b_1, b_2]$, and $[c_1, c_2]$, respectively, use Fubini's theorem to prove that

$$\iiint_{\mathcal{R}} \alpha(x)\beta(y)\gamma(z) \, dV$$
$$= \left(\int_{a_1}^{a_2} \alpha(x) \, dx \right)\left(\int_{b_1}^{b_2} \beta(y) \, dy \right)\left(\int_{c_1}^{c_2} \gamma(z) \, dz \right).$$

70. Let a, b, and c be positive real numbers, and let $\mathcal{R} = \{(x, y, z) \mid -a \leq x \leq a, \ -b \leq y \leq b, \text{ and } -c \leq z \leq c\}$. Prove that $\iiint_{\mathcal{R}} \alpha(x)\beta(y)\gamma(z) \, dV = 0$ if any of α, β, and γ is an odd function.

71. Let $\mathcal{R} = \{(x, y, z) \mid a_1 \leq x \leq a_2, \ b_1 \leq y \leq b_2, \text{ and } c_1 \leq z \leq c_2\}$. Prove that

$$\iiint_{\mathcal{R}} dV = (a_2 - a_1)(b_2 - b_1)(c_2 - c_1).$$

What is the relationship between $\mathcal{R}$ and the product $(a_2 - a_1)(b_2 - b_1)(c_2 - c_1)$?

72. Let $f(x, y, z)$ be an integrable function on the rectangular solid $\mathcal{R} = \{(x, y, z) \mid a_1 \leq x \leq a_2, \ b_1 \leq y \leq b_2, \text{ and } c_1 \leq z \leq c_2\}$, and let $\kappa \in \mathbb{R}$. Use the definition of the triple integral to prove that

$$\iiint_{\mathcal{R}} \kappa f(x, y, z) \, dV = \kappa \iiint_{\mathcal{R}} f(x, y, z) \, dV.$$

73. Let $f(x, y, z)$ and $g(x, y, z)$ be integrable functions on the rectangular solid $\mathcal{R} = \{(x, y, z) \mid a_1 \leq x \leq a_2, \ b_1 \leq y \leq b_2, \text{ and } c_1 \leq z \leq c_2\}$. Use the definition of the triple integral to prove that

$$\iiint_{\mathcal{R}} (f(x, y, z) + g(x, y, z)) \, dV$$
$$= \iiint_{\mathcal{R}} f(x, y, z) \, dV + \iiint_{\mathcal{R}} g(x, y, z) \, dV.$$

Thinking Forward

▶ *Cylindrical coordinates:* When we use polar coordinates in the xy-plane and the usual z-coordinate, we are using *cylindrical coordinates*. Let Ω_{xy} be a region in the xy-plane such that for each point (r, θ) in Ω_{xy}, $h_1(\theta) \leq r \leq h_2(\theta)$ and $\alpha \leq \theta \leq \beta$. Then, if $g_1(x, y) \leq z \leq g_2(x, y)$ for each (x, y) in Ω_{xy}, explain why

$$\iiint_{\Omega} f(x, y, z) \, dV = \iint_{\Omega_{xy}} \int_{g_1(x,y)}^{g_2(x,y)} f(x, y, z) \, dz \, dA$$
$$= \int_{\alpha}^{\beta} \int_{h_1(\theta)}^{h_2(\theta)} \int_{g_1(r\cos\theta, r\sin\theta)}^{g_2(r\cos\theta, r\sin\theta)} f(r\cos\theta, r\sin\theta, z) \, r \, dz \, dr \, d\theta.$$

▶ *Cylindrical coordinates:* Use the results of the previous exercise to explain why the triple integral

$$\int_0^{2\pi} \int_0^3 \int_0^{9-r^2} kr^2 \, dz \, dr \, d\theta$$

represents the mass of a solid bounded below by the xy-plane and bounded above by the paraboloid with equation $z = 9 - x^2 - y^2$ if the density at every point in the solid is proportional to the distance of the point from the origin. Then determine the mass of the solid.

13.6 INTEGRATION WITH CYLINDRICAL AND SPHERICAL COORDINATES

▶ Generalizing polar coordinates to three dimensions

▶ Converting between rectangular coordinates, cylindrical coordinates, and spherical coordinates

▶ Constructing triple integrals with cylindrical and spherical coordinates

Cylindrical Coordinates

In this section we will discuss two generalizations of polar coordinates to three-dimensional space: *cylindrical coordinates* and *spherical coordinates*. We will also discuss how these coordinate systems may be used in triple integrals. We will see that the cylindrical coordinate system uses two linear measures and one angular measure to locate each point in $\mathbb{R}^3$ and the spherical coordinate system uses one linear measure and two angular measures to locate points.

In the cylindrical coordinate system, we use polar coordinates in the xy-plane and the usual z-coordinate to locate positions off of the xy-plane. As we see in the following figure, given a point P with rectangular coordinates (x, y, z), r measures the distance from the projection of P in the xy-plane to the origin and θ measures the angular rotation from the x-axis to the ray containing the origin and the point $(x, y, 0)$:

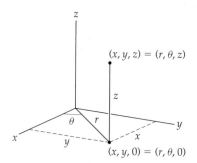

That is, if (x, y, z) is a point in $\mathbb{R}^3$ given in rectangular coordinates, then, in cylindrical coordinates we have (r, θ, z), where

$$r^2 = x^2 + y^2, \quad \tan\theta = \frac{y}{x}, \quad \text{and} \quad z = z.$$

Note that the values for r and θ are computed just as they were when we were discussing polar coordinates in the plane. Since our primary motivation in this chapter is integration, we will typically choose $r \geq 0$ and restrict values of θ to an interval of length 2π, usually either $[0, 2\pi]$ or $[-\pi, \pi]$.

When we have the coordinates of a point (r, θ, z) in cylindrical coordinates, the rectangular coordinates are (x, y, z), where

$$x = r\cos\theta, \quad y = r\sin\theta, \quad \text{and} \quad z = z.$$

From Section 13.5, we know that, for a three-dimensional region, Ω, that is bounded below by the function $g_1(x, y)$ and above by $g_2(x, y)$ on the domain Ω_{xy}, the triple integral of the function $f(x, y, z)$ is

$$\iiint_\Omega f(x, y, z)\, dV = \iint_{\Omega_{xy}} \int_{g_1(x,y)}^{g_2(x,y)} f(x, y, z)\, dz\, dA.$$

In Section 13.5 we discussed how to use rectangular coordinates to evaluate the exterior double integral over the region Ω_{xy}. However, in Section 13.3 we saw that the increment of area dA is given by $r\, dr\, d\theta$ in polar coordinates. Therefore, to use cylindrical coordinates to evaluate the triple integral of f, we replace each occurrence of x in f, g_1, and g_2 with $r\cos\theta$ and each occurrence of y with $r\sin\theta$ and we use $r\, dz\, dr\, d\theta$ in place of the increment of volume dV. We also need to express the region Ω_{xy} in terms of *polar* coordinates, as we did in Section 13.3.

For example, we will compute the mass of the region Ω bounded above by the paraboloid with equation $g(x, y) = 4 - x^2 - y^2$ and bounded below by the xy-plane, assuming that the density at each point is proportional to the distance of the point from the xy-plane. The triple integral we wish to evaluate is $\iiint_\Omega kz\, dV$. Using rectangular coordinates, we could evaluate

$$\int_{-2}^{2} \int_{-\sqrt{4-x^2}}^{\sqrt{4-x^2}} \int_{0}^{4-x^2-y^2} kz\, dz\, dy\, dx.$$

However, using cylindrical coordinates, we will have a simpler iterated integral. With cylindrical coordinates, the equation of the paraboloid is $z = 4 - r^2$. The projection of the solid onto the xy-plane is a circle with radius 2 and centered at the origin. The equation of this circle is $r = 2$ for $\theta \in [0, 2\pi]$. Finally, our increment of volume is $dV = r\, dz\, dr\, d\theta$. Therefore, the mass may be computed by

$$\int_0^{2\pi} \int_0^2 \int_0^{4-r^2} kzr\, dz\, dr\, d\theta = \int_0^{2\pi} \int_0^2 \left[\frac{1}{2}krz^2\right]_0^{4-r^2} dr\, d\theta$$

$$= \int_0^{2\pi} \int_0^2 \frac{1}{2}kr(4 - r^2)^2 dr\, d\theta$$

$$= -\int_0^{2\pi} \left[\frac{1}{12}k(4 - r^2)^3\right]_0^2 d\theta$$

$$= \int_0^{2\pi} \frac{16}{3}k\, d\theta = \left[\frac{16}{3}k\theta\right]_0^{2\pi} = \frac{32\pi}{3}k.$$

Every iterated triple integral requires three integrations, so we expect the computation to require at least that much work, but as we see in this example, using the cylindrical coordinate system here provides a much simpler path toward the solution. Evaluating the equivalent iterated triple integral in rectangular coordinates would be even more laborious.

Consider using cylindrical coordinates to evaluate triple integrals when the projection of the region of integration Ω onto the xy-plane, Ω_{xy}, has a "natural" expression in polar coordinates. This will always occur when Ω displays rotational symmetry about the z-axis.

Spherical Coordinates

Let (x, y, z) be the rectangular coordinates of a point P in $\mathbb{R}^3$. As we just discussed, we may also express the coordinates of P as cylindrical coordinates (r, θ, z), where

$$r^2 = x^2 + y^2, \quad \tan\theta = \frac{y}{x}, \quad \text{and} \quad z = z.$$

In the spherical system, the coordinates of a point P in $\mathbb{R}^3$ are given by (ρ, θ, ϕ), where ρ is the distance that P is from the origin, θ is the same angle we used in the cylindrical system, and $\phi \in [0, \pi]$ is the angle that the ray from the origin through P makes with the positive z-axis, as we see in the figure that follows at the left. The figure at the right shows the relationships between the spherical, cylindrical, and rectangular coordinates.

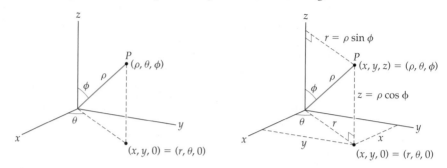

The angles θ and ϕ will be measured in radians. We will always assume that $0 \leq \phi \leq \pi$ and that θ lies in some interval of length 2π, typically $\theta \in [0, 2\pi]$ or $\theta \in [-\pi, \pi]$. In the figures that follow, we show a sphere, a plane, and a cone. These are the graphs obtained when ρ, θ, and ϕ, respectively, are constant.

$$\rho = \rho_0 \qquad\qquad \theta = \theta_0 \qquad\qquad \phi = \phi_0$$

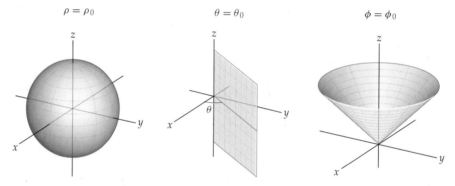

We summarize the relationships between the variables of our three-dimensional coordinate systems in the following theorem:

THEOREM 13.21

Converting Between the Three-Dimensional Coordinate Systems

Let P be a point in $\mathbb{R}^3$ with coordinates (x, y, z) in the rectangular coordinate system, (r, θ, z) in the cylindrical coordinate system, and (ρ, θ, ϕ) in the spherical coordinate system.

(a) The cylindrical coordinates and rectangular coordinates for P are related by the following equations:

$$r = \sqrt{x^2 + y^2}, \quad \tan\theta = \frac{y}{x}, \quad \text{and} \quad z = z$$
$$x = r\cos\theta, \quad y = r\sin\theta, \quad \text{and} \quad z = z.$$

(b) The cylindrical coordinates and spherical coordinates for P are related by the following equations:

$$\rho = \sqrt{r^2 + z^2}, \quad \theta = \theta, \quad \text{and} \quad \tan\phi = \frac{r}{z}$$
$$r = \rho\sin\phi, \quad \theta = \theta, \quad \text{and} \quad z = \rho\cos\phi.$$

(c) The rectangular coordinates and spherical coordinates for P are related by the following equations:

$$\rho = \sqrt{x^2 + y^2 + z^2}, \quad \tan\theta = \frac{y}{x}, \quad \text{and} \quad \cos\phi = \frac{z}{\sqrt{x^2 + y^2 + z^2}}$$
$$x = \rho\sin\phi\cos\theta, \quad y = \rho\sin\phi\sin\theta, \quad \text{and} \quad z = \rho\cos\phi.$$

For example, if the rectangular coordinates of a point P are $(1, \sqrt{3}, 2\sqrt{3})$, then

$$r = \sqrt{x^2 + y^2} = \sqrt{1^2 + (\sqrt{3})^2} = 2, \; z = 2\sqrt{3}, \text{ and } \tan\theta = \frac{y}{x} = \frac{\sqrt{3}}{1} = \sqrt{3}.$$

Therefore, $\theta = \frac{\pi}{3}$ and the cylindrical coordinates for P are $\left(2, \frac{\pi}{3}, 2\sqrt{3}\right)$. Now that we have the cylindrical coordinates for P, we may use either those or the rectangular coordinates for P to find the spherical coordinates for P. We'll use the cylindrical coordinates. We have

$$\rho = \sqrt{r^2 + z^2} = \sqrt{2^2 + (2\sqrt{3})^2} = 4, \; \theta = \frac{\pi}{3}, \quad \text{and} \quad \tan\phi = \frac{r}{z} = \frac{2}{2\sqrt{3}} = \frac{1}{\sqrt{3}}.$$

Therefore, since $\phi \in [0, \pi]$, we have $\phi = \frac{\pi}{6}$. So the spherical coordinates for P are $\left(4, \frac{\pi}{3}, \frac{\pi}{6}\right)$.

Integration with Spherical Coordinates

Recall that to integrate a function $f(r, \theta)$ over a region $\Omega \subset \mathbb{R}^2$, we use $dA = r \, dr \, d\theta$ for our increment of area. That is,

$$\iint_\Omega f(r, \theta) \, dA = \iint_\Omega f(r, \theta) \, r \, dr \, d\theta.$$

Similarly, as we saw earlier in this section, when we integrate a function $f(r, \theta, z)$ over a region $\Omega \subset \mathbb{R}^3$, we use $dV = r \, dz \, dr \, d\theta$ for our increment of volume and obtain

$$\iiint_\Omega f(r, \theta, z) \, dV = \iiint_\Omega f(r, \theta, z) \, r \, dz \, dr \, d\theta.$$

When we use spherical coordinates, we must also adapt our increment of volume. Here, we will have $dV = \rho^2 \sin\phi \, d\rho \, d\theta \, d\phi$. To see this, start by considering an increment of volume in spherical coordinates determined by $\Delta\rho = \rho_i - \rho_{i-1}$, $\Delta\theta = \theta_j - \theta_{j-1}$, and $\Delta\phi = \phi_k - \phi_{k-1}$. When $\Delta\rho$, $\Delta\theta$, and $\Delta\phi$ are small, this region is nearly a rectangular solid, as shown next. Thus, its volume may be approximated by the product of its three dimensions.

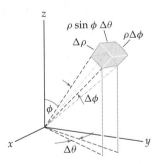

The measure of the side of this solid determined by $\Delta\rho$ is just that linear measure. The approximate measure of the side determined by $\Delta\phi$ is $\rho\Delta\phi$, since it is the arc of the circle of radius ρ subtended by the angle $\Delta\phi$. The projection of the third side into the xy-plane is shown in the figure. This projection is an arc of a circle with radius r (in cylindrical coordinates) subtended by the angle $\Delta\theta$. Therefore, the measure of the projection is $r\Delta\theta$. However, since $r = \rho\sin\phi$, the approximate measure of the third side is $r\Delta\theta = \rho\sin\phi\Delta\theta$. Multiplying these three factors together, we have $\Delta V = \rho^2 \sin\phi \, \Delta\rho \, \Delta\theta \, \Delta\phi$. When we take the limit as $\Delta\rho$, $\Delta\theta$, and $\Delta\phi$ all tend to zero, we have the volume differential $dV = \rho^2 \sin\phi \, d\rho \, d\theta \, d\phi$. Therefore, when we have a function $f(\rho, \theta, \phi)$ given in spherical coordinates and defined on a three-dimensional region Ω, the integral of f on Ω is

$$\iiint_\Omega f(\rho, \theta, \phi) \, \rho^2 \sin\phi \, d\rho \, d\theta \, d\phi.$$

Typically, the spherical coordinate system is used when the boundaries of a solid are spheres or parts of spheres. For example, let $\mathcal{E}$ be the portion of the unit sphere centered at the origin in the first octant, as shown in the following figure:

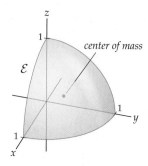

We will find the center of mass of $\mathcal{E}$, assuming that it has uniform density k. To find the x-coordinate of the center of mass, $\bar{x}$, we start by computing the first moment with respect to the yz-plane. We have

$$M_{yz} = \iiint_{\mathcal{E}} kx\, dV = \int_0^{\pi/2} \int_0^{\pi/2} \int_0^1 k\rho^3 \sin^2\phi \cos\theta\, d\rho\, d\theta\, d\phi,$$

since $x = \rho \sin\phi \cos\theta$ and $dV = \rho^2 \sin\phi\, d\rho\, d\theta\, d\phi$. In Exercise 18 you will show that $M_{yz} = \frac{1}{16}\pi k$. Similarly, the mass, m, of $\mathcal{E}$ is given by the integral

$$m = \iiint_{\mathcal{E}} k\, dV = \int_0^{\pi/2} \int_0^{\pi/2} \int_0^1 k\rho^2 \sin\phi\, d\rho\, d\theta\, d\phi.$$

However, we may circumvent this computation because the value of that integral has to be one-eighth of the volume of a sphere times the (uniform) density k. That is, $m = \frac{1}{8}\left(\frac{4}{3}\pi\right)k = \frac{1}{6}\pi k$. In Exercise 19 we ask that you evaluate the integral to obtain this result. Therefore, $\bar{x} = \frac{M_{yz}}{m} = \frac{3}{8}$. Because of the symmetry of $\mathcal{E}$, the y- and z-coordinates of the center of mass must also be $\frac{3}{8}$. So the center of mass of $\mathcal{E}$ is $\left(\frac{3}{8}, \frac{3}{8}, \frac{3}{8}\right)$. (In Exercises 20 and 21, you are asked to show that M_{xy} and M_{xz} both equal $\frac{1}{16}\pi k$ to confirm this result.)

Examples and Explorations

EXAMPLE 1

Finding the volume of a region between a cylinder and a sphere

Use rectangular, cylindrical, and spherical coordinates to set up triple integrals representing the volume inside the sphere with equation $x^2 + y^2 + z^2 = 4$ but outside the cylinder with equation $x^2 + y^2 = 1$.

SOLUTION

The sphere is centered at the origin and has radius 2. The z-axis is the axis of symmetry for the cylinder with radius 1. The graph is as follows:

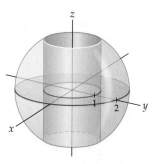

Let Ω represent the region. Since Ω is symmetric with respect to the z-axis, when we use rectangular coordinates we will integrate first with respect to z. The projection of Ω onto the xy-plane is the annulus Ω_{xy}, shown here:

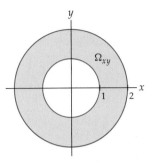

It would require several subdivisions to represent Ω_{xy} as a sum of type I and type II regions. However, we can use rectangular coordinates to represent the volume of Ω by setting up a triple integral giving the volume of the entire sphere and then subtracting a triple integral that represents the portion of the volume of the sphere inside the cylinder. That is, the volume of Ω is

$$\int_{-2}^{2}\int_{-\sqrt{4-x^2}}^{\sqrt{4-x^2}}\int_{-\sqrt{4-x^2-y^2}}^{\sqrt{4-x^2-y^2}} dz\,dy\,dx - \int_{-1}^{1}\int_{-\sqrt{1-x^2}}^{\sqrt{1-x^2}}\int_{-\sqrt{4-x^2-y^2}}^{\sqrt{4-x^2-y^2}} dz\,dy\,dx.$$

Note that it is more natural to represent Ω_{xy} with polar coordinates than with rectangular coordinates. This is why it will be easier to represent the volume of Ω with cylindrical coordinates than it was with rectangular coordinates. In cylindrical coordinates, the equations of the upper and lower hemispheres are $z = \sqrt{4-r^2}$ and $z = -\sqrt{4-r^2}$, respectively. We must also remember to include the factor r in the integrand when we use cylindrical coordinates. Therefore, the triple integral representing the volume of Ω is

$$\int_{0}^{2\pi}\int_{1}^{2}\int_{-\sqrt{4-r^2}}^{\sqrt{4-r^2}} r\,dz\,dr\,d\theta.$$

Lastly, we will use spherical coordinates to set up a triple integral for the volume. In spherical coordinates, the equation of the sphere is $\rho = 2$. Since $x = \rho \sin\phi \cos\theta$ and $y = \rho \sin\phi \sin\theta$, the equation of the cylinder $x^2 + y^2 = 1$ in spherical coordinates is

$$(\rho \sin\phi \cos\theta)^2 + (\rho \sin\phi \sin\theta)^2 = 1.$$

We ask you to show in Exercise 22 that this equation simplifies to $\rho = \csc\phi$. To set up the iterated integral, we also need the correct interval of values for ϕ. The sphere and cylinder intersect when $\csc\phi = 2$. The values of ϕ satisfying this relationship within the interval $[0, \pi]$ are $\phi = \dfrac{\pi}{6}$ and $\phi = \dfrac{5\pi}{6}$. Recall that when we integrate with spherical coordinates, we must include the factor $\rho^2 \sin\phi$ in the integrand. Therefore, the volume of Ω is given by

$$\int_{\pi/6}^{5\pi/6}\int_{0}^{2\pi}\int_{\csc\phi}^{2} \rho^2 \sin\phi\,d\rho\,d\theta\,d\phi. \qquad\square$$

EXAMPLE 2

Finding the volume of a region between a cone and a sphere

Use rectangular, cylindrical, and spherical coordinates to set up triple integrals representing the volume bounded below by the cone with equation $z = \sqrt{x^2 + y^2}$ and bounded above by the unit sphere centered at the origin.

SOLUTION

The z-axis is the axis of symmetry for the cone. Its vertex is at the origin and it opens upward. Here is the graph:

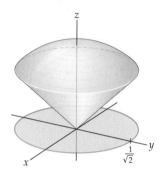

Let Ω represent the region. Since Ω is symmetric with respect to the z-axis, when we use rectangular coordinates we will integrate first with respect to z. The cone and sphere intersect at height $z = \dfrac{1}{\sqrt{2}}$ in a circle with radius $\dfrac{1}{\sqrt{2}}$. Therefore, the projection of Ω onto the xy-plane is a circle Ω_{xy} with radius $\dfrac{1}{\sqrt{2}}$ and centered at the origin. The equation for this circle is $x^2 + y^2 = \dfrac{1}{2}$.

The region Ω_{xy} may be treated as either a type I or type II region. We will treat it as a type I region. Using rectangular coordinates, we find that the volume of Ω is

$$\int_{-1/\sqrt{2}}^{1/\sqrt{2}} \int_{-\sqrt{(1/2)-x^2}}^{\sqrt{(1/2)-x^2}} \int_{\sqrt{x^2-y^2}}^{\sqrt{1-x^2-y^2}} dz\, dy\, dx.$$

We will now use an iterated integral with cylindrical coordinates to represent the volume of Ω. In cylindrical coordinates, the equation of the upper hemisphere of the unit sphere is $z = \sqrt{1 - r^2}$ and the equation of the cone is $z = r$. We include the factor r in the integrand when we use cylindrical coordinates, and we obtain the integral

$$\int_{0}^{2\pi} \int_{0}^{1/\sqrt{2}} \int_{r}^{\sqrt{1-r^2}} r\, dz\, dr\, d\theta$$

for the volume of Ω.

Finally, we will use spherical coordinates to set up a triple integral for the volume. In spherical coordinates, the equation of the sphere is $\rho = 1$ and the equation of the cone is $\phi = \dfrac{\pi}{4}$. Recall that when we integrate with spherical coordinates, we must include the factor $\rho^2 \sin \phi$ in the integrand. Therefore, the volume of Ω is given by

$$\int_{0}^{\pi/4} \int_{0}^{2\pi} \int_{0}^{1} \rho^2 \sin \phi\, d\rho\, d\theta\, d\phi.$$

□

? TEST YOUR UNDERSTANDING

▶ How do you convert between rectangular and polar coordinates in $\mathbb{R}^2$? How do you convert between rectangular and cylindrical coordinates in $\mathbb{R}^3$? How do you convert between rectangular and spherical coordinates? How do you convert between cylindrical and spherical coordinates?

▶ What do x, y, and z represent in rectangular coordinates? What range of values is used for x, y, and z in rectangular coordinates? If α, β, and γ are constants, what are the graphs of $x = \alpha$, $y = \beta$, and $z = \gamma$ in rectangular coordinates?

▶ What do r, θ, and z represent in cylindrical coordinates? What range of values is used for r in cylindrical coordinates? What range of values is used for θ in cylindrical coordinates? What range of values is used for z in cylindrical coordinates? If α, β, and γ are constants, what are the graphs of $r = \alpha$, $\theta = \beta$, and $z = \gamma$ in cylindrical coordinates?

▶ What do ρ, θ, and ϕ represent in spherical coordinates? What range of values is used for ρ in spherical coordinates? What range of values is used for θ in spherical coordinates? What range of values is used for ϕ in spherical coordinates? If α, β, and γ are constants, what are the graphs of $\rho = \alpha$, $\theta = \beta$, and $\phi = \gamma$ in spherical coordinates?

▶ Under what conditions should you use rectangular, cylindrical, or spherical coordinates in evaluating triple integrals?

EXERCISES 13.6

Thinking Back

▶ *Graphing with polar coordinates:* Use polar coordinates in $\mathbb{R}^2$ to graph each of the following equations:

▶ $r = 3$
▶ $\theta = \dfrac{\pi}{3}$
▶ $r = 2\cos\theta$
▶ $r = 1 + 2\sin\theta$

▶ *Rectangular versus polar coordinates:* When is it easier to use the rectangular coordinate system in $\mathbb{R}^2$? When is it easier to use the polar coordinate system?

▶ *Integrating with polar coordinates:* Let Ω be a region in $\mathbb{R}^2$. Provide a double integral that represents the area of Ω when you integrate with polar coordinates.

Concepts

0. *Problem Zero:* Read the section and make your own summary of the material.

1. *True/False:* Determine whether each of the statements that follow is true or false. If a statement is true, explain why. If a statement is false, provide a counterexample.

(a) *True or False:* With cylindrical coordinates, the graph of the function $z = \sqrt{2r\cos\theta - r^2}$ is a hemisphere.

(b) *True or False:* With spherical coordinates, the graph of $\phi = \dfrac{\pi}{2}$ is the xy-plane.

(c) *True or False:* With spherical coordinates, for every value of $a \in (0, \pi)$ the intersection of the graphs of $\phi = a$ and $\rho = 1$ is a circle with radius 1.

(d) *True or False:* The graph of $z = r$ in cylindrical coordinates is the same as the graph of $\phi = \dfrac{\pi}{4}$ in spherical coordinates.

(e) *True or False:* If f is an integrable function of three variables on a region Ω, then

$$\iiint_\Omega f(x, y, z)\, dz\, dy\, dx$$

$$= \iiint_\Omega f(r, \theta, z)\, dz\, dr\, d\theta.$$

(f) *True or False:* If f is an integrable function of three variables on a region Ω, then

$$\iiint_\Omega f(x, y, z)\, dz\, dy\, dx$$

$$= \iiint_\Omega f(x(r, \theta), y(r, \theta), z)\, r\, dz\, dr\, d\theta.$$

(g) *True or False:* The integral

$$\int_0^{2\pi} \int_0^{2\pi} \int_0^{R} \rho^2 \sin\phi\, d\rho\, d\theta\, d\phi = \frac{4}{3}\pi R^3.$$

(h) *True or False:* The rectangular, cylindrical, and spherical coordinate systems are the only coordinate systems in $\mathbb{R}^3$.

2. *Examples:* Construct examples of the thing(s) described in the following. Try to find examples that are different than any in the reading.

(a) A region in $\mathbb{R}^3$ that is most easily expressed with rectangular coordinates.

(b) A region in $\mathbb{R}^3$ that is most easily expressed with cylindrical coordinates.

(c) A region in $\mathbb{R}^3$ that is most easily expressed with spherical coordinates.

3. What are the graphs of the constant functions $x = x_0$, $y = y_0$, and $z = z_0$ in the rectangular coordinate system?

4. What are the graphs of the constant functions $r = r_0$, $\theta = \theta_0$, and $z = z_0$ in the cylindrical coordinate system?

5. What are the graphs of the constant functions $\rho = \rho_0$, $\theta = \theta_0$, and $\phi = \phi_0$ in the spherical coordinate system?

6. For what values of $\alpha \in [0, \pi]$ in the spherical coordinate system is the graph $\phi = \alpha$ *not* a cone?

Fill in the blanks for the conversion formulas in Exercises 7–12.

7. To convert from rectangular to cylindrical coordinates:
$r = \underline{\hspace{1cm}}$, $\theta = \underline{\hspace{1cm}}$, $z = \underline{\hspace{1cm}}$.

8. To convert from cylindrical to rectangular coordinates:
$x =$ _____, $y =$ _____, $z =$ _____.

9. To convert from rectangular to spherical coordinates:
$\rho =$ _____, $\theta =$ _____, $\phi =$ _____.

10. To convert from spherical to rectangular coordinates:
$x =$ _____, $y =$ _____, $z =$ _____.

11. To convert from cylindrical to spherical coordinates:
$\rho =$ _____, $\theta =$ _____, $\phi =$ _____.

12. To convert from spherical to cylindrical coordinates:
$r =$ _____, $\theta =$ _____, $z =$ _____.

13. What are the six forms used to express the volume increment dV when you use rectangular coordinates to evaluate a triple integral? How do you decide which order to use?

14. The volume increment $dV =$ _____ when you use cylindrical coordinates to evaluate a triple integral. Why is this the standard order of integration for cylindrical coordinates?

15. The volume increment $dV =$ _____ when you use spherical coordinates to evaluate a triple integral. Why is this the standard order of integration for spherical coordinates?

16. What geometric conditions do you look for when you are deciding which coordinate system to use in $\mathbb{R}^3$?

17. What geometric conditions do you look for when you are deciding which coordinate system to use when you are evaluating a triple integral?

In Exercises 18–21, we ask you to confirm a result from earlier in the section. Let $\mathcal{E}$ be the portion of the unit sphere centered at the origin in the first octant. Assume that $\mathcal{E}$ has uniform density k.

18. Show that the first moment of $\mathcal{E}$ is $M_{yz} = \frac{1}{16}\pi k$.

19. Show that the mass of $\mathcal{E}$ is $\frac{1}{6}\pi k$ by evaluating the integral

$$\iiint_{\mathcal{E}} k \, dV = \int_0^{\pi/2} \int_0^{\pi/2} \int_0^1 k\rho^2 \sin\phi \, d\rho \, d\theta \, d\phi.$$

20. Set up the appropriate triple integral with spherical coordinates to show that $M_{xy} = \frac{1}{16}\pi k$.

21. Set up the appropriate triple integral with spherical coordinates to show that $M_{xz} = \frac{1}{16}\pi k$.

22. From Example 1, recall that $x^2 + y^2 = 1$ is the equation of the cylinder with radius 1, whose axis of symmetry is the z-axis. Show that the equation of this cylinder in spherical coordinates is $\rho = \csc\phi$.

Skills

Find the coordinates specified in Exercises 23–28.

23. Give the cylindrical and spherical coordinates for the point with rectangular coordinates $(1, 0, 0)$.

24. Give the cylindrical and spherical coordinates for the point with rectangular coordinates $(-6, 6, 6)$.

25. Give the rectangular and spherical coordinates for the point with cylindrical coordinates $(\sqrt{48}, \pi/3, 4)$.

26. Give the rectangular and spherical coordinates for the point with cylindrical coordinates $(4, \pi/3, 6)$.

27. Give the rectangular and cylindrical coordinates for the point with spherical coordinates $(8, \pi/2, \pi)$.

28. Give the rectangular and cylindrical coordinates for the point with spherical coordinates $(6, \pi/4, \pi/4)$.

Describe the graphs of the equations in Exercises 29–38 in $\mathbb{R}^3$, and provide alternative equations in the specified coordinate systems.

29. Change $x = 4$ to the cylindrical and spherical systems.

30. Change $z = x + y$ to the cylindrical and spherical systems.

31. Change $r = 2$ to the rectangular and spherical systems.

32. Change $r = 4\sin\theta$ to the rectangular and spherical systems.

33. Change $\theta = \frac{\pi}{2}$ to the rectangular system.

34. Change $\theta = \alpha$, where α is a constant, to the rectangular system.

35. Change $\rho = 2$ to the rectangular and cylindrical systems.

36. Change $\phi = \frac{\pi}{4}$ to the rectangular and spherical systems.

37. Change $\phi = \frac{\pi}{2}$ to the rectangular and cylindrical systems.

38. Change $\phi = \frac{3\pi}{4}$ to the rectangular and cylindrical systems.

The iterated integrals in Exercises 39–42 use cylindrical coordinates. Describe the solids determined by the limits of integration.

39. $\displaystyle \int_0^{2\pi} \int_0^3 \int_0^r f(r, \theta, z)\, r \, dz \, dr \, d\theta$

40. $\displaystyle \int_0^\pi \int_1^2 \int_0^{r^2} f(r, \theta, z)\, r \, dz \, dr \, d\theta$

41. $\displaystyle \int_0^\pi \int_0^{2\sin\theta} \int_0^{\sqrt{16-r^2}} f(r, \theta, z)\, r \, dz \, dr \, d\theta$

42. $\displaystyle \int_0^{\pi/2} \int_0^1 \int_0^{\sqrt{1-r^2}} f(r, \theta, z)\, r \, dz \, dr \, d\theta$

The iterated integrals in Exercises 43–46 use spherical coordinates. Describe the solids determined by the limits of integration.

43. $\displaystyle\int_{\pi/2}^{\pi}\int_{0}^{2\pi}\int_{0}^{2} f(\rho,\theta,\phi)\rho^2\sin\phi\,d\rho\,d\theta\,d\phi$

44. $\displaystyle\int_{0}^{\pi/2}\int_{0}^{\pi/2}\int_{0}^{1} f(\rho,\theta,\phi)\rho^2\sin\phi\,d\rho\,d\theta\,d\phi$

45. $\displaystyle\int_{0}^{\pi/4}\int_{0}^{2\pi}\int_{0}^{3\sec\theta} f(\rho,\theta,\phi)\rho^2\sin\phi\,d\rho\,d\theta\,d\phi$

46. $\displaystyle\int_{\pi/4}^{\pi/2}\int_{0}^{2\pi}\int_{0}^{3\csc\theta} f(\rho,\theta,\phi)\rho^2\sin\phi\,d\rho\,d\theta\,d\phi$

Use a triple integral with either cylindrical or spherical coordinates to find the volumes of the solids described in Exercises 47–56.

47. The region inside both the sphere with equation $x^2+y^2+z^2=4$ and the cylinder with equation $x^2+(y-1)^2=1$.

48. The region inside the cylinder with equation $x^2+(y-1)^2=1$, bounded below by the xy-plane and bounded above by the cone with equation $z=\sqrt{x^2+y^2}$.

49. The region bounded below by the xy-plane, bounded above by the sphere with radius 2 and centered at the origin, and outside the cylinder with equation $x^2+y^2=1$.

50. The region bounded below by the plane with equation $z=c$ and bounded above by the sphere with equation $x^2+y^2+z^2=R^2$, where c and R are constants such that $0<c<R$.

51. The region bounded above by the plane with equation $z=x$ and bounded below by the paraboloid with equation $z=x^2+y^2$.

52. The region bounded above by the sphere with equation $\rho=2$ and bounded below by the cone with equation $\phi=\dfrac{\pi}{3}$.

53. The region in the next figure which is bounded below by the xy-plane, bounded above by the hyperboloid with equation $x^2+y^2-z^2=1$, and inside the cylinder with equation $x^2+y^2=5$.

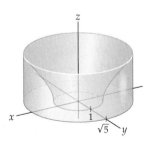

54. The first-octant region bounded above by the sphere with equation $\rho=R$ and bounded below by the plane with equation $z=x+y$.

55. The region bounded above by the sphere with equation $\rho=R$ and bounded below by the cone with equation $\phi=\alpha$. Explain why the volume should be zero if $\alpha=0$ and $\dfrac{4}{3}\pi R^3$ if $\alpha=\pi$.

56. The region bounded above by the sphere with equation $\rho=R$ and bounded below by the plane with equation $z=b$, where $0\le b\le R$. Explain why the volume should be $\dfrac{2}{3}\pi R^3$ if $b=0$ and zero if $b=R$. Use that result to find the volume of the region when $-R\le b<0$.

Find the specified quantities for the solids described in Exercises 57–66.

57. The mass of the region from Exercise 47, assuming that the density at every point is proportional to the square of the point's distance from the z-axis.

58. The mass of the region from Exercise 48, assuming that the density at every point is proportional to the square of the point's distance from the xy-plane.

59. The center of mass of the region from Exercise 49, assuming that the density at every point is proportional to the point's distance from the z-axis.

60. The center of mass of the region from Exercise 50, assuming that the density at every point is proportional to the point's distance from the xy-plane.

61. The mass of the region from Exercise 51, assuming that the density at every point is proportional to the square of the point's distance from the xy-plane.

62. The moment of inertia about the z-axis of the region from Exercise 52, assuming that the density at every point is inversely proportional to the point's distance from the z-axis.

63. The moment of inertia about the z-axis of the region from Exercise 53, assuming that the density at every point is inversely proportional to the point's distance from the z-axis.

64. The mass of the region from Exercise 54, assuming that the density at every point is proportional to the point's distance from the xy-plane.

65. The center of mass of the region from Exercise 55, assuming that the density of the region is constant.

66. The moment of inertia about the xy-plane of the region from Exercise 56, assuming that the density of the region is constant.

Applications

67. Emmy is responsible for a tank that is circular when viewed from above. The tank has a radius of 75 feet. An island with a radius of 8 feet lies at the center of the tank and contains monitoring equipment. Emmy knows that the sides of the tank are vertical but the bottom of the tank is the surface of a sphere of radius 150 feet, forming a bowl at the bottom. At the edge of the tank, the depth is 8 feet. What is the volume of the tank?

68. Annie is a sea-kayaking guide who finds herself with time on her hands in the winter. She is building a wood-and-fabric kayak for herself, but is concerned about its capacity for cargo. The outside surface of the kayak satisfies the equation

$$r = \frac{(49 - l^2)(2 - \sin\theta)}{147},$$

where the coordinates are cylindrical, r is the distance from a line joining the tips of the boat, and l is the horizontal distance from the center.

(a) What is the volume of the kayak?

(b) When Annie sits in a kayak, the seat and her body fill about 5 cubic feet of the boat. The stringers, which are supports inside that run the length of the kayak, take up space, so the interior has a radius that is 1.5 inches smaller than what you calculated in part (a). Approximately how much cargo space will she actually have in the kayak?

Proofs

69. Let a be a constant. Prove that the equation of the plane $x = a$ is $r = a\sec\theta$ in cylindrical coordinates.

70. Let b be a constant. Prove that the equation of the plane $y = b$ is $r = b\csc\theta$ in cylindrical coordinates.

71. Let a be a constant. Prove that the equation of the plane $x = a$ is $\rho = a\csc\phi\sec\theta$ in spherical coordinates.

72. Let b be a constant. Prove that the equation of the plane $y = b$ is $r = b\csc\phi\csc\theta$ in spherical coordinates.

73. Let R be the radius of the base of a cone and h be the height of the cone. Use cylindrical coordinates to set up

and evaluate a triple integral proving that the volume of the cone is $\frac{1}{3}\pi R^2 h$.

74. Let R be the radius of a sphere. Use cylindrical coordinates to set up and evaluate a triple integral proving that the volume of the sphere is $\frac{4}{3}\pi R^3$.

75. Repeat Exercise 73, but using spherical coordinates.

76. Repeat Exercise 74, but using spherical coordinates.

Thinking Forward

A non-standard coordinate system in $\mathbb{R}^2$: Imagine a coordinate system in $\mathbb{R}^2$ in which the coordinate axes u and v are *not* perpendicular, as shown in the figure.

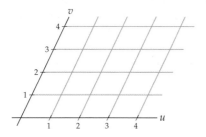

▶ Explain why every point in $\mathbb{R}^2$ has unique coordinates in this coordinate system.

▶ What is the area of each of the small parallelograms in the figure?

▶ If **a** is a vector parallel to the u-axis and **b** is a vector parallel to the v-axis, what is the area of the parallelogram determined by **a** and **b**?

A non-standard coordinate system in $\mathbb{R}^3$: Imagine a three-dimensional analog to the coordinate system described in the previous problem, except that now there are three coordinate axes meeting at an origin but the three axes are not mutually perpendicular.

▶ What is the volume of each parallelepiped determined by a unit change in the u, v, and w directions?

▶ If **a**, **b**, and **c** are vectors parallel to the u-, v-, and w-axes, respectively, what is the volume of the parallelepiped determined by **a**, **b**, and **c**?

13.7   JACOBIANS AND CHANGE OF VARIABLES

▶ Using nonstandard coordinate systems in $\mathbb{R}^2$

▶ Using nonstandard coordinate systems in $\mathbb{R}^3$

▶ Using Jacobians to construct double and triple integrals

Change of Variables in $\mathbb{R}^2$

The first integration technique that we discussed in Chapter 5 was integration by substitution. At that time we saw that

$$\int_a^b f(g(x))g'(x)\,dx = \int_c^d f(u)\,du,$$

where $u = g(x)$, $c = g(a)$, and $d = g(b)$. The point of changing the variable in that context was to obtain an integral $\int_c^d f(u)\,du$ that was simpler to evaluate than the original integral. We studied related phenomena when we introduced polar coordinates, cylindrical coordinates, and spherical coordinates. Some integrals are easier to evaluate when we represent them in those coordinate systems. No matter how comfortable we may feel with the rectangular coordinate system, analyzing an integral with an alternative coordinate system often makes the evaluation of a double or triple integral simpler. For example, we discussed how to evaluate a double integral $\iint_\Omega g(x, y)\,dA$ when we analyzed Ω as a type I region or a type II region in Section 13.2, but then, in Section 13.3, we extended our discussion by analyzing Ω in terms of polar coordinates. In Section 13.3 we saw that

$$\iint_\Omega g(x, y)\,dA = \int_\alpha^\beta \int_{f_1(\theta)}^{f_2(\theta)} g(r\cos\theta, r\sin\theta)\, r\,dr\,d\theta$$

when the boundary components of Ω are determined by the polar functions $r = f_1(\theta)$ and $r = f_2(\theta)$ for $\theta \in [\alpha, \beta]$. In the current section we will learn how to generalize this idea so that we can use other coordinate systems to analyze double and triple integrals.

Typically we start with some region Ω in $\mathbb{R}^2$. We wish to find a function $T : \Omega \to \Omega'$ such that Ω' is a simpler subset of $\mathbb{R}^2$. Such functions are called **transformations**. In particular, we will require the transformations that we use in this section to be both one-to-one and differentiable at every point in the interior of Ω. As the following schematic illustrates, we consider the points in Ω to be given by the rectangular coordinates (x, y) and the points in our target to be given by coordinates (u, v):

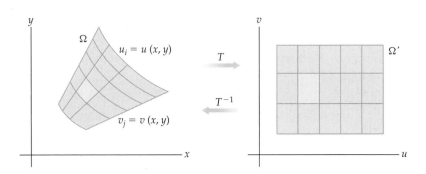

The goal is to determine a transformation T such that $T(\Omega) = \Omega'$ may be more simply analyzed in the new coordinate system. In our schematic, the image set Ω' is a rectangle, but the image set does not need to be that simple.

The coordinates u and v are both functions of x and y. That is,

$$u = u(x, y) \quad \text{and} \quad v = v(x, y).$$

Since we require T to be one-to-one on the interior of Ω, it is also invertible. Therefore, the x- and y-coordinates in the interior of Ω are functions of u and v; that is,

$$x = x(u, v) \quad \text{and} \quad y = y(u, v).$$

The grid lines in Ω in the previous figure at the left are level curves of the form $u_i = u(x, y)$ for $1 \leq i \leq m$ and $v_j = v(x, y)$ for $1 \leq j \leq n$. Optimally, the boundaries of Ω are also level curves. If we let $\Delta u = u_{i+1} - u_i$ and $\Delta v = v_{j+1} - v_j$, then when the mesh of the grid is relatively small, each subregion, such as the highlighted region in the left-hand figure, may be approximated by a parallelogram. We expand this subregion in the following figure:

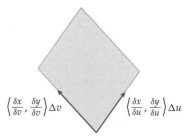

$\left(\frac{\delta x}{\delta v}, \frac{\delta y}{\delta v} \right) \Delta v$ $\left(\frac{\delta x}{\delta u}, \frac{\delta y}{\delta u} \right) \Delta u$

We may normalize the vectors $\left(\frac{\partial x}{\partial u}, \frac{\partial y}{\partial u} \right)$ and $\left(\frac{\partial x}{\partial v}, \frac{\partial y}{\partial v} \right)$ at a typical point (x, y) so that they both have unit length. In this case, the sides of the parallelogram have length $\left(\frac{\partial x}{\partial u}, \frac{\partial y}{\partial u} \right) \Delta u$ and $\left(\frac{\partial x}{\partial v}, \frac{\partial y}{\partial v} \right) \Delta v$. As we know from Chapter 10, the area of this parallelogram is given by the magnitude of the cross product

$$\left(\left(\frac{\partial x}{\partial u}, \frac{\partial y}{\partial u}, 0 \right) \Delta u \right) \times \left(\left(\frac{\partial x}{\partial v}, \frac{\partial y}{\partial v}, 0 \right) \Delta v \right) = \left(\left(\frac{\partial x}{\partial u}, \frac{\partial y}{\partial u}, 0 \right) \times \left(\frac{\partial x}{\partial v}, \frac{\partial y}{\partial v}, 0 \right) \right) \Delta u \Delta v.$$

Recall that the cross product

$$\left(\frac{\partial x}{\partial u}, \frac{\partial y}{\partial u}, 0 \right) \times \left(\frac{\partial x}{\partial v}, \frac{\partial y}{\partial v}, 0 \right) = \det \begin{bmatrix} \mathbf{i} & \mathbf{j} & \mathbf{k} \\ \frac{\partial x}{\partial u} & \frac{\partial y}{\partial u} & 0 \\ \frac{\partial x}{\partial v} & \frac{\partial y}{\partial v} & 0 \end{bmatrix} = \left(\frac{\partial x}{\partial u} \frac{\partial y}{\partial v} - \frac{\partial x}{\partial v} \frac{\partial y}{\partial u} \right) \mathbf{k}.$$

Furthermore, the quantity $\frac{\partial x}{\partial u} \frac{\partial y}{\partial v} - \frac{\partial x}{\partial v} \frac{\partial y}{\partial u}$ is the determinant of the 2×2 matrix:

$$\det \begin{bmatrix} \frac{\partial x}{\partial u} & \frac{\partial y}{\partial u} \\ \frac{\partial x}{\partial v} & \frac{\partial y}{\partial v} \end{bmatrix} = \frac{\partial x}{\partial u} \frac{\partial y}{\partial v} - \frac{\partial x}{\partial v} \frac{\partial y}{\partial u}.$$

We use this determinant in the next definition.

DEFINITION 13.22

The Jacobian for a Transformation in $\mathbb{R}^2$

Let Ω and Ω' be subsets of $\mathbb{R}^2$. If the transformation $T : \Omega \to \Omega'$ has a differentiable inverse with $x = x(u, v)$ and $y = y(u, v)$, then we define the *Jacobian* of the transformation T, denoted by $\frac{\partial(x,y)}{\partial(u,v)}$, to be the determinant of the matrix of partial derivatives; that is

$$\frac{\partial(x, y)}{\partial(u, v)} = \det \begin{bmatrix} \dfrac{\partial x}{\partial u} & \dfrac{\partial y}{\partial u} \\ \dfrac{\partial x}{\partial v} & \dfrac{\partial y}{\partial v} \end{bmatrix} = \frac{\partial x}{\partial u}\frac{\partial y}{\partial v} - \frac{\partial x}{\partial v}\frac{\partial y}{\partial u}.$$

Using Definition 13.22, we see that the area of the parallelogram we were discussing is

$$\left| \frac{\partial(x, y)}{\partial(u, v)} \right| \Delta u \Delta v.$$

Now, to approximate the integral of a function $g(x, y)$ over a region Ω, we may use the Riemann sum

$$\sum_{i=1}^{m} \sum_{j=1}^{n} g(x_i, y_j) \Delta A = \sum_{i=1}^{m} \sum_{j=1}^{n} g(x(u_i, v_j), y(u_i, v_j)) \left| \frac{\partial(x, y)}{\partial(u, v)} \right| \Delta u \Delta v.$$

When we take the limit of this Riemann sum as $m \to \infty$ and $n \to \infty$, we have

$$\iint_{\Omega} g(x, y)\, dA = \iint_{\Omega'} g(x(u, v), y(u, v)) \left| \frac{\partial(x, y)}{\partial(u, v)} \right| du\, dv.$$

If the preceding iterated integral is simple enough, we may use it to evaluate the double integral.

Our formulas for converting between rectangular and polar coordinates define a transformation if we omit the origin and insist that $\Omega \in [0, 2\pi)$. Recall that to convert from rectangular to polar coordinates, we use

$$r = \sqrt{x^2 + y^2} \quad \text{and} \quad \tan \theta = \frac{y}{x}.$$

To convert from polar coordinates to rectangular coordinates, we use

$$x = r \cos \theta \quad \text{and} \quad y = r \sin \theta.$$

Thus, subject to the restrictions we mentioned, the Jacobian of the transformation is

$$\frac{\partial(x, y)}{\partial(r, \theta)} = \det \begin{bmatrix} \dfrac{\partial x}{\partial r} & \dfrac{\partial y}{\partial r} \\ \dfrac{\partial x}{\partial \theta} & \dfrac{\partial y}{\partial \theta} \end{bmatrix} = \det \begin{bmatrix} \cos \theta & \sin \theta \\ -r \sin \theta & r \cos \theta \end{bmatrix} = r(\cos^2 \theta + \sin^2 \theta) = r.$$

This is the result we expect, since we know that

$$\iint_{\Omega} g(x, y)\, dA = \iint_{\Omega'} g(r \cos \theta, r \sin \theta) \left| \frac{\partial(x, y)}{\partial(r, \theta)} \right| dr\, d\theta = \iint_{\Omega'} g(r \cos \theta, r \sin \theta)\, r\, dr\, d\theta.$$

That is, the factor r we introduced in order to use polar coordinates to evaluate a double integral is just the Jacobian of the transformation.

To find a change of variable in order to simplify an integral $\iint_{\Omega} f(x, y)\, dA$, we determine invertible and differentiable functions $u = u(x, y)$ and $v = v(x, y)$ that allow us to analyze Ω more easily. To use Definition 13.22 to find the Jacobian we must also find the inverse of the transformation, which requires us to determine the functions $x = x(u, v)$ and $y = y(u, v)$.

Rather than go through this step, we may use the fact that $\frac{\partial(x,y)}{\partial(u,v)} \frac{\partial(u,v)}{\partial(x,y)} = 1$. The proof of this equation is left for Exercise 60.

Change of Variables in $\mathbb{R}^3$

The situation is quite similar in three dimensions. As we have already seen, a three-dimensional region may be easier to analyze with a coordinate system other than the rectangular system in $\mathbb{R}^3$. Rather than derive the analogous iterated triple integral in as great a detail as we did for the two-dimensional case, we will summarize the changes. We start with some region Ω initially expressed with rectangular coordinates in $\mathbb{R}^3$. A function $T : \Omega \to \Omega'$ is a transformation in $\mathbb{R}^3$. We require our transformations in $\mathbb{R}^3$ to be both invertible and differentiable in the interior of Ω. We consider the points in Ω to be given by the rectangular coordinates (x, y, z) and the points in our target to be given by coordinates (u, v, w).

The coordinates u, v, and w are all functions of x, y, and z. That is,

$$u = u(x, y, z), \quad v = v(x, y, z), \quad \text{and} \quad w = w(x, y, z).$$

Since T is invertible, it follows that x, y, and z are each functions of u, v, and w:

$$x = x(u, v, w), \quad y = y(u, v, w), \quad \text{and} \quad z = z(u, v, w).$$

We use level surfaces to subdivide Ω:

$$u_i = u(x, y, z) \text{ for } 1 \le i \le l, v_j = v(x, y, z) \text{ for } 1 \le j \le m, \text{ and } w_k = w(x, y, z) \text{ for } 1 \le k \le n.$$

Optimally, the boundaries of Ω are level surfaces.

We subdivide Ω into small pieces, each of which may be approximated with a small parallelepiped, as follows:

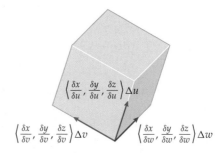

The volume of the parallelepiped is given by the triple scalar product

$$\left(\left(\left\langle \frac{\partial x}{\partial u}, \frac{\partial y}{\partial u}, \frac{\partial z}{\partial u} \right\rangle \Delta u \right) \times \left(\left\langle \frac{\partial x}{\partial v}, \frac{\partial y}{\partial v}, \frac{\partial z}{\partial v} \right\rangle \Delta v \right) \right) \cdot \left(\left\langle \frac{\partial x}{\partial w}, \frac{\partial y}{\partial w}, \frac{\partial z}{\partial w} \right\rangle \Delta w \right) =$$

$$\left(\left\langle \frac{\partial x}{\partial u}, \frac{\partial y}{\partial u}, \frac{\partial z}{\partial u} \right\rangle \times \left\langle \frac{\partial x}{\partial v}, \frac{\partial y}{\partial v}, \frac{\partial z}{\partial v} \right\rangle \right) \cdot \left\langle \frac{\partial x}{\partial w}, \frac{\partial y}{\partial w}, \frac{\partial z}{\partial w} \right\rangle \Delta u \Delta v \Delta w.$$

Recall that the triple scalar product is equal to the determinant of an appropriate 3×3 matrix. We use this matrix to define the Jacobian of a transformation in $\mathbb{R}^3$.

DEFINITION 13.23

The Jacobian for a Transformation in $\mathbb{R}^3$

Let Ω and Ω' be subsets of $\mathbb{R}^3$. If the transformation $T : \Omega \to \Omega'$ has a differentiable inverse with $x = x(u, v, w)$, $y = y(u, v, w)$, and $z = z(u, v, w)$, then we define the ***Jacobian*** of the transformation T, denoted by $\frac{\partial(x,y,z)}{\partial(u,v,w)}$, to be the determinant of the matrix of partial derivatives; that is,

$$\frac{\partial(x, y, z)}{\partial(u, v, w)} = \det \begin{bmatrix} \dfrac{\partial x}{\partial u} & \dfrac{\partial y}{\partial u} & \dfrac{\partial z}{\partial u} \\[2mm] \dfrac{\partial x}{\partial v} & \dfrac{\partial y}{\partial v} & \dfrac{\partial z}{\partial v} \\[2mm] \dfrac{\partial x}{\partial w} & \dfrac{\partial y}{\partial w} & \dfrac{\partial z}{\partial w} \end{bmatrix}.$$

Thus, we may express the volume of the parallelepiped as $\left|\frac{\partial(x,y,z)}{\partial(u,v,w)}\right| \Delta u \Delta v \Delta w$. Using this expression in a Riemann sum and taking the appropriate limit, we see that

$$\iiint_\Omega g(x, y, z)\, dV = \iiint_{\Omega'} g(x(u, v, w), y(u, v, w), z(u, v, w)) \left|\frac{\partial(x, y, z)}{\partial(u, v, w)}\right| du\, dv\, dw.$$

If the preceding iterated integral is simple enough, we may use it to evaluate the triple integral.

The procedure just given is consistent with the evaluation procedure we established when we introduced spherical coordinates in Section 13.6. From Theorem 13.21, we know that, given the spherical coordinates (ρ, θ, ϕ) of a point, we may obtain the rectangular coordinates of the point with

$$x = \rho \sin \phi \cos \theta, \quad y = \rho \sin \phi \sin \theta, \quad \text{and} \quad z = \rho \cos \phi.$$

In Exercise 52 you are asked to compute the Jacobian of this transformation and show that

$$\frac{\partial(x, y, z)}{\partial(\rho, \theta, \phi)} = \det \begin{bmatrix} \dfrac{\partial x}{\partial \rho} & \dfrac{\partial y}{\partial \rho} & \dfrac{\partial z}{\partial \rho} \\[2mm] \dfrac{\partial x}{\partial \theta} & \dfrac{\partial y}{\partial \theta} & \dfrac{\partial z}{\partial \theta} \\[2mm] \dfrac{\partial x}{\partial \phi} & \dfrac{\partial y}{\partial \phi} & \dfrac{\partial z}{\partial \phi} \end{bmatrix} = -\rho^2 \sin \phi.$$

Therefore, we again find that

$$\iiint_\Omega g(x, y, z)\, dV = \iiint_{\Omega'} g(x(\rho, \theta, \phi), y(\rho, \theta, \phi), z(\rho, \theta, \phi)) \left|\frac{\partial(x, y, z)}{\partial(\rho, \theta, \phi)}\right| d\rho\, d\theta\, d\phi$$

$$= \iiint_{\Omega'} g(x(\rho, \theta, \phi), y(\rho, \theta, \phi), z(\rho, \theta, \phi)) \rho^2 \sin \phi\, d\rho\, d\theta\, d\phi.$$

Examples and Explorations

EXAMPLE 1

Evaluating a double integral by changing the variable

Evaluate the double integral $\iint_\Omega \frac{1}{(x+y)^2}\, dA$, where Ω is the trapezoidal region in the first quadrant bounded by the lines $x + y = 1$ and $x + y = 4$ and the x- and y-axes.

SOLUTION

Following is the region Ω:

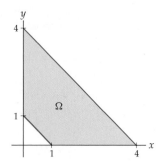

It would be possible to decompose Ω into two type I regions or two type II regions, but we will use a change of variable that allows us to evaluate the integral with a single iterated integral. The difficulty here is that, relative to the x- and y-axes, the region is mildly complicated. If one of our coordinate axes were parallel to the parallel sides of the trapezoid and the other were perpendicular to those sides, we would be able to treat Ω more simply. Fortunately, we may accomplish this aim with the change of variable

$$u = x + y \quad \text{and} \quad v = y - x.$$

Note that the level curves for $u = x + y$ are lines with slope -1 and the level curves for v are lines with slope 1, as we suggested. With this transformation, the equation of the lines $x + y = 1$ and $x + y = 4$ are expressed more simply as $u = 1$ and $u = 4$, respectively. Solving the system $u = x + y$ and $v = y - x$ simultaneously, we may express x and y as the following functions of u and v:

$$x = \frac{1}{2}(u - v) \quad \text{and} \quad y = \frac{1}{2}(u + v).$$

Using these functions, we see that the x-axis ($y = 0$) has equation $v = -u$ and the y-axis ($x = 0$) has equation $v = u$. Therefore, under the transformation, the graph of Ω' in the uv-plane is as follows:

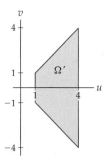

Our new variables make it easier to treat Ω' as a type I region. Before we set up the appropriate double integral, we need the Jacobian of the transformation. It is

$$\frac{\partial(x, y)}{\partial(u, v)} = \det \begin{bmatrix} \dfrac{\partial x}{\partial u} & \dfrac{\partial y}{\partial u} \\ \dfrac{\partial x}{\partial v} & \dfrac{\partial y}{\partial v} \end{bmatrix} = \det \begin{bmatrix} \dfrac{1}{2} & \dfrac{1}{2} \\ -\dfrac{1}{2} & \dfrac{1}{2} \end{bmatrix} = \frac{1}{2}.$$

Now, the double integral will be

$$\iint_{\Omega} \frac{1}{(x + y)^2}\, dA = \frac{1}{2} \int_{1}^{4} \int_{-u}^{u} \frac{1}{u^2}\, dv\, du.$$

Note that because the Jacobian is a constant for this transformation, we were able to factor it out of the iterated integral. The most difficult part of the process is finding the appropriate change of variable to simplify Ω and, therefore, the double integral. The transformation we used was *not* the only one we could have used, but it sufficed to simplify the integral. We leave the evaluation of the iterated integral to Exercise 22. □

EXAMPLE 2 **Evaluating another double integral by changing the variable**

Evaluate the double integral $\iint_{\Omega_2} \frac{x}{y}\, dA$, where Ω_2 is the first-quadrant region bounded by the lines $y = \frac{1}{4}x$ and $y = 4x$ and the curves $y = \frac{1}{x}$ and $y = \frac{4}{x}$.

SOLUTION
Following is the region:

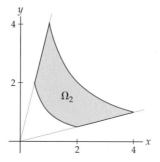

It would be possible to decompose Ω_2 into three type I regions or three type II regions, but we will use a change of variable that allows us to evaluate the integral with a single iterated integral. Note that the curves $y = \frac{1}{x}$ and $y = \frac{4}{x}$ may also be expressed as $xy = 1$ and $xy = 4$, respectively. If we let $u = xy$, then we may express these two curves as $u = 1$ and $u = 4$. The lines $y = \frac{1}{4}x$ and $y = 4x$ may be expressed as $\frac{y}{x} = \frac{1}{4}$ and $\frac{y}{x} = 4$, respectively. If we let $v = \frac{y}{x}$, we may express these lines as $v = \frac{1}{4}$ and $v = 4$. Thus, in the uv-plane, the graph of Ω_2 is transformed to the rectangle Ω'_2, as shown in the following figure:

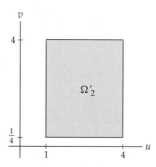

We will treat Ω'_2 as a type I region, although we could just as easily treat it as a type II region.

Before we set up the appropriate double integral, we need the Jacobian of the transformation. Here we will use the fact that $\frac{\partial(x,y)}{\partial(u,v)} \frac{\partial(u,v)}{\partial(x,y)} = 1$. We have

$$\frac{\partial(u,v)}{\partial(x,y)} = \det \begin{bmatrix} \dfrac{\partial u}{\partial x} & \dfrac{\partial v}{\partial x} \\[2mm] \dfrac{\partial u}{\partial y} & \dfrac{\partial v}{\partial y} \end{bmatrix} = \det \begin{bmatrix} y & -\dfrac{y}{x^2} \\[2mm] x & \dfrac{1}{x} \end{bmatrix} = \frac{2y}{x}.$$

Now, since $\frac{y}{x} = v$, it follows that $\frac{\partial(u,v)}{\partial(x,y)} = 2v$. Thus, $\frac{\partial(x,y)}{\partial(u,v)} = \frac{1}{2v}$.

Alternatively, we could solve the system

$$u = xy \quad \text{and} \quad v = \frac{y}{x}$$

for x and y. In Exercise 25, you will show that

$$x = \sqrt{\frac{u}{v}} = u^{1/2}\,v^{-1/2} \quad \text{and} \quad y = \sqrt{uv} = u^{1/2}\,v^{1/2}$$

and to use these equations to show that $\frac{\partial(x,y)}{\partial(u,v)} = \frac{1}{2v}$. Either way, we need to express the Jacobian in terms of the new variables, here u and v.

We are now ready to rewrite the double integral in terms of the new variables. Since $v = \frac{y}{x}$, we have $\frac{x}{y} = \frac{1}{v}$; thus, the double integral will be

$$\iint_{\Omega_2} \frac{x}{y}\,dA = \frac{1}{2}\int_1^4 \int_{1/4}^4 \frac{1}{v^2}\,dv\,du.$$

The factor $\frac{1}{2}$ in front of the integral was originally a factor of the Jacobian. Again, the most difficult part of this process is finding the appropriate change of variable to simplify Ω_2 and, therefore, the double integral. We leave the evaluation of the iterated integral to Exercise 26.   □

EXAMPLE 3

Finding the area of an elliptical annulus

Find the area between the ellipses with equations $4x^2 + 9y^2 = 36$ and $4x^2 + 9y^2 = 144$.

SOLUTION

Let $\mathcal{S}$ represent this annulus, shown in the following figure:

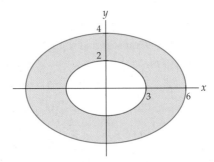

If this were a circular annulus, it would be quite easy to find its area. Fortunately, using a suitably chosen transformation we can obtain a circular annulus. Let $u = 2x$ and $v = 3y$. The equations of the ellipses are $4x^2 + 9y^2 = 36$ and $4x^2 + 9y^2 = 144$, or equivalently, $(2x)^2 + (3y)^2 = 36$ and $(2x)^2 + (3y)^2 = 144$. Now, using the transformation, we have

$$u^2 + v^2 = 36 \quad \text{and} \quad u^2 + v^2 = 144.$$

That is, $\mathcal{S}'$ is a circular ellipse with inner radius 6 and outer radius 12. In order to find the Jacobian of the transformation, we solve for x and y in terms of u and v. Here we have $x = \frac{1}{2}u$ and $y = \frac{1}{3}v$. Thus, the Jacobian of the transformation is

$$\frac{\partial(x,y)}{\partial(u,v)} = \det \begin{bmatrix} \dfrac{\partial x}{\partial u} & \dfrac{\partial y}{\partial u} \\[2mm] \dfrac{\partial x}{\partial v} & \dfrac{\partial y}{\partial v} \end{bmatrix} = \det \begin{bmatrix} \dfrac{1}{2} & 0 \\[2mm] 0 & \dfrac{1}{3} \end{bmatrix} = \frac{1}{6}.$$

The area of the ellipse is given by the double integral

$$\iint_S dA = \iint_{S'} \left| \frac{\partial(x, y)}{\partial(u, v)} \right| dA = \iint_{S'} \frac{1}{6} dA.$$

In our previous examples, we used rectangular coordinates to evaluate the transformed integrals, but here it is simpler to use polar coordinates. We obtain

$$\iint_S dA = \frac{1}{6} \int_0^{2\pi} \int_6^{12} r \, dr \, d\theta = 18\pi.$$ □

CHECKING THE ANSWER

We could also have computed the area of this annulus by means of the area formula for an ellipse, or

$$\text{Area} = \pi ab,$$

where a and b are the lengths of the semimajor and semiminor axes of the ellipse. In this example, the area of the larger ellipse is $4 \cdot 6\pi = 24\pi$ and the area of the smaller ellipse is $2 \cdot 3\pi = 6\pi$. Therefore, the area of the annulus is 18π, as we saw before.

? TEST YOUR UNDERSTANDING

▶ What is the integration-by-substitution formula for evaluating a definite integral? Why is this formula consistent with the change-of-variable formula that uses Jacobians for evaluating a double integral? What is the "Jacobian" in the integration-by-substitution formula?

▶ How is a Jacobian related to an area in $\mathbb{R}^2$? How is it related to a volume in $\mathbb{R}^3$?

▶ What is the Jacobian when you convert from rectangular coordinates to polar coordinates in $\mathbb{R}^2$? The Jacobians for the transformations from rectangular coordinates to polar coordinates in $\mathbb{R}^2$ and from rectangular coordinates to cylindrical coordinates in $\mathbb{R}^3$ are equal. Why does this make sense?

▶ What is the Jacobian for the transformation from rectangular coordinates to spherical coordinates in $\mathbb{R}^3$?

▶ What do we look for when we are considering a coordinate system to help us evaluate a double or triple integral?

EXERCISES 13.7

Thinking Back

▶ *Integration by substitution:* What is the integration-by-substitution formula for definite integrals? How is it derived?

▶ *Integrating with polar coordinates:* Let Ω be a region in $\mathbb{R}^2$. Give a double integral that represents the area of Ω when you integrate with polar coordinates.

▶ *Integrating with cylindrical coordinates:* Let Ω be a region in $\mathbb{R}^3$. Give a triple integral that represents the volume of Ω when you integrate with cylindrical coordinates.

▶ *Integrating with spherical coordinates:* Let Ω be a region in $\mathbb{R}^3$. Give a triple integral that represents the volume of Ω when you integrate with spherical coordinates.

Concepts

0. *Problem Zero:* Read the section and make your own summary of the material.

1. *True/False:* Determine whether each of the statements that follow is true or false. If a statement is true, explain why. If a statement is false, provide a counterexample.

(a) *True or False:* There is a transformation in $\mathbb{R}^2$ that takes a circle to a rectangle.

(b) *True or False:* There is a transformation in $\mathbb{R}^2$ that takes a parallelogram to a square.

(c) *True or False:* There is a transformation in $\mathbb{R}^3$ that takes a cylinder to a rectangular solid.

(d) *True or False:* There is a transformation in $\mathbb{R}^3$ that takes a sphere to a rectangular solid.

(e) *True or False:* If $u = 2x + y$ and $v = x - 2y$, then $x = \frac{2}{5}u + \frac{1}{5}v$ and $y = \frac{1}{5}u - \frac{2}{5}v$.

(f) *True or False:* If $u = 2x + y$ and $v = x - 2y$, then $\frac{\partial(x,y)}{\partial(u,v)} = \frac{1}{5}$.

(g) *True or False:* Given a rectangular region $\mathcal{R}$ in the xy-plane and a square region $\mathcal{R}'$ in the uv-plane, there is a transformation taking $\mathcal{R}$ to $\mathcal{R}'$.

(h) *True or False:* Given a rectangular region $\mathcal{R}$ in the xy-plane and a square region $\mathcal{R}'$ in the uv-plane, there is a unique transformation taking $\mathcal{R}$ to $\mathcal{R}'$.

2. *Examples:* Construct examples of the thing(s) described in the following. Try to find examples that are different than any in the reading.

(a) A transformation $T : \mathbb{R}^2 \to \mathbb{R}^2$ whose Jacobian is a positive constant.

(b) A transformation $T : \mathbb{R}^2 \to \mathbb{R}^2$ whose Jacobian varies.

(c) A transformation $T : \mathbb{R}^3 \to \mathbb{R}^3$ whose Jacobian is a negative constant.

3. Let T be a transformation in $\mathbb{R}^2$. How is the Jacobian of the transformation related to the change in the increment of area, ΔA?

Let $\mathcal{R}$ be the rectangle in the Cartesian coordinate system with vertices $(1, 0)$, $(3, 2)$, $(2, 3)$, and $(0, 1)$. Use $\mathcal{R}$ to answer Exercises 4–9.

4. Set up and evaluate iterated double integrals equal to $\iint_{\mathcal{R}} x^2 y \, dA$, treating $\mathcal{R}$ as a union of type I regions.

5. Set up and evaluate iterated double integrals equal to $\iint_{\mathcal{R}} x^2 y \, dA$, treating $\mathcal{R}$ as a union of type II regions.

6. Show that the boundaries of $\mathcal{R}$ are the lines with equations $x + y = 1$, $x + y = 5$, $x - y = 1$, and $x - y = -1$.

(a) Use these equations to explain why the change of variables $u = x + y$ and $v = x - y$ results in the simpler system of equations in the uv-coordinate system.

(b) Sketch the graph of the transformed rectangle $\mathcal{R}'$ in the uv-coordinate system.

7. Use your transformation from Exercise 6 to express x and y as functions of u and v, and find the Jacobian of the transformation.

8. Find the Jacobian of the transformation from Exercise 6, using the fact that $\frac{\partial(x,y)}{\partial(u,v)} \frac{\partial(u,v)}{\partial(x,y)} = 1$.

9. Set up and evaluate a single iterated double integral equal to $\iint_{\mathcal{R}} x^2 y \, dA$ in which you integrate with respect to u and v rather than with respect to x and y.

Let a and b be constants. Use the transformation $T : \mathbb{R}^2 \to \mathbb{R}^2$ defined by $u = x + a$ and $v = y + b$ to answer Exercises 10–13.

10. What does this transformation do to the point with co-ordinates (x_0, y_0)? What does the transformation do to a vertical line $x = x_0$? What does the transformation do to a horizontal line $y = y_0$?

11. What does this transformation do to a region Ω? In particular, if $a = 3$, $b = 4$, and Ω is the rectangle defined by the inequalities $0 \le x \le 1$ and $0 \le y \le 2$, what is the image of Ω under the given transformation?

12. What are x and y as functions of u and v? What is the Jacobian of this transformation? How does your answer conform to the answer to Exercise 3?

13. Explain why this transformation would or would not be useful for evaluating a double integral.

Let a and b be positive constants. Use the transformation $T : \mathbb{R}^2 \to \mathbb{R}^2$ defined by $u = ax$ and $v = by$ to answer Exercises 14–17.

14. What does this transformation do to the point with co-ordinates (x_0, y_0)? What does the transformation do to a vertical line $x = x_0$? What does the transformation do to a horizontal line $y = y_0$?

15. What does this transformation do to a region Ω? In particular, if $a = 3$, $b = \frac{1}{2}$, and Ω is the square defined by the inequalities $0 \le x \le 1$ and $0 \le y \le 1$, what is the image of Ω under the given transformation?

16. What are x and y as functions of u and v? What is the Jacobian of this transformation? How does your answer conform to the answer to Exercise 3?

17. If we allow either a or b to be negative, how would that change your answers to Exercises 14 and 16?

Let $\theta \in [0, 2\pi)$. Use the transformation $T : \mathbb{R}^2 \to \mathbb{R}^2$ defined by $u = (\sin\theta)x - (\cos\theta)y$ and $v = (\cos\theta)x + (\sin\theta)y$ to answer Exercises 18 and 19.

18. What does this transformation do to a region Ω? In particular, if $\theta = \frac{\pi}{6}$ and Ω is the rectangle defined by the inequalities $0 \le x \le 1$ and $0 \le y \le 2$, what is the image of Ω under the given transformation?

19. What are x and y as functions of u and v? What is the Jacobian of this transformation? How does your answer conform to the answer to Exercise 3?

20. Let $T : \mathbb{R}^3 \to \mathbb{R}^3$ be a transformation. How is the Jacobian of the transformation related to the change of the increment of volume, ΔV?

Exercises 21–23 continue the work started in Example 1.

21. Evaluate the double integral $\iint_{\Omega} \frac{1}{(x+y)^2} \, dA$ without using a change of variables. Note that you will need to treat Ω as the union of two type I regions or two type II regions.

22. Complete Example 1 by evaluating the iterated integral $\frac{1}{2} \int_1^4 \int_{-u}^{u} \frac{1}{u^2} \, dv \, du$.

23. Find the areas of the trapezoid Ω and the transformed trapezoid Ω' in Example 1. What is the relationship between these two areas and the Jacobian of the transformation?

Exercises 24–26 continue the work started in Example 2.

24. We mentioned that to evaluate the double integral $\iint_{\Omega_2} \frac{x}{y} \, dA$ without using a change of variables, you would need to decompose Ω_2 into a union of three type I regions or three type II regions. What are those decompositions?

25. Show that if $u = xy$ and $v = \frac{y}{x}$, then

$$x = \sqrt{\frac{u}{v}} = u^{1/2} v^{-1/2} \quad \text{and} \quad y = \sqrt{uv} = u^{1/2} v^{1/2}.$$

Use these formulas to show that $\frac{\partial(x,y)}{\partial(u,v)} = \frac{1}{2v}$.

26. Complete Example 2 by evaluating the iterated integral

$$\iint_{\Omega_2} \frac{x}{y} \, dA = \frac{1}{2} \int_1^4 \int_{1/4}^4 \frac{1}{v^2} \, dv \, du.$$

Skills

In Exercises 27–32, functions $x = x(u, v)$ and $y = y(u, v)$ are given that determine transformations from an xy-coordinate system to a uv-coordinate system in $\mathbb{R}^2$. Use these functions to determine a region in the xy-plane that has the image specified for the given values of u and v, and find the Jacobian of the transformation.

27. $x = u - v$ and $y = u + v$ for $0 \le u \le 2$ and $0 \le v \le 1$

28. $x = 3u + 4v$ and $y = 4u - 3v$ for $1 \le u \le 3$ and $1 \le v \le 5$

29. $x = \frac{u}{v}$ and $y = uv$ for $1 \le u \le 2$ and $1 \le v \le 3$

30. $x = u^2 + v^2$ and $y = u^2 - v^2$ for $0 \le u \le 4$ and $0 \le v \le 4$

31. $x = u \sin v$ and $y = u \cos v$ for $0 \le u \le 2$ and $0 \le v \le \pi$

32. $x = u \sec v$ and $y = u \tan v$ for $0 \le u \le 2$ and $0 \le v \le \frac{\pi}{4}$

For each double integral in Exercises 33–38, (a) sketch the region Ω, (b) use the specified transformation to sketch the transformed region, and (c) use the transformation to evaluate the integral.

33. $\iint_\Omega (x + y)^2 \, dA$, where Ω is the trapezoid with vertices $(1, 3)$, $(3, 1)$, $(9, 3)$, and $(3, 9)$. Use the transformation given by $u = x + y$ and $v = x - y$.

34. Evaluate the integral from Exercise 33, but use the transformation given by $u = x + y$ and $v = 3x - y$.

35. Evaluate the integral from Exercise 33, but use the transformation given by $u = x + y$ and $v = \frac{x}{y}$.

36. $\iint_\Omega xy \, dA$, where Ω is the trapezoid with vertices $(2, 3)$, $(3, 2)$, $(5, 3)$, and $(3, 5)$. Use the transformation from Exercise 33.

37. Evaluate the double integral from Exercise 36, using the transformation given by $u = x + y$ and $v = 2y - x$.

38. $\iint_\Omega \left(\frac{y^3}{x} - xy \right) dA$, where Ω is the region in the first and second quadrants that is bounded above by the hyperbola $y^2 - x^2 = 12$, bounded below by the hyperbola $y^2 - x^2 = 3$, and bounded on the left and right by the lines $y = -2x$ and $y = 2x$, respectively. Use the transformation given by $u = \frac{y}{x}$ and $v = y^2 - x^2$.

Evaluate the double integrals in Exercises 39–48. Use suitable transformations as necessary.

39. $\iint_\Omega (x - y)^3 \, dA$, where Ω is the parallelogram with vertices $(0, 0)$, $(3, 0)$, $(5, 2)$, and $(2, 2)$.

40. $\iint_\Omega xy \, dA$, where Ω is the parallelogram from Exercise 39.

41. $\iint_\Omega \frac{2y - x}{3x + y + 1} \, dA$, where Ω is the parallelogram with vertices $(0, 0)$, $(2, 1)$, $(1, 4)$, and $(-1, 3)$.

42. $\iint_\Omega (3x^2 - 5xy - 2y^2) \, dA$, where Ω is the parallelogram from Exercise 41.

43. $\iint_\Omega \left(\frac{x^2}{y^2} + x^2 y^2 \right) dA$, where Ω is the following region:

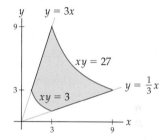

44. $\iint_\Omega xy^3 \, dA$, where Ω is the region from Exercise 43.

45. $\iint_\Omega \frac{y^2}{x^3} \, dA$, where Ω is the following region:

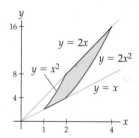

46. $\iint_\Omega \frac{y + xy}{x^2} \, dA$, where Ω is the region from Exercise 45.

47. $\iint_\Omega \frac{x^3}{y^3} \, dA$, where Ω is the following region:

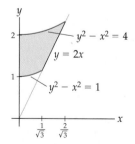

48. $\iint_\Omega x^2 \left(1 - \frac{1}{y^2} \right) dA$, where Ω is the region from Exercise 47.

Applications

49. Joe plans to build a swimming pool in his backyard in the shape of parallelogram to fit the contours of his property in the dimensions shown in the following figure:

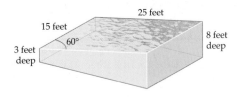

The shallow end of the pool will start at a depth of 3 feet and will increase linearly to a depth of 8 feet on the opposite side of the pool. How much water will the pool hold?

50. Joe also plans to build a goldfish pond in the front of his house in the shape of an ellipse with a 5-foot major axis and a 3-foot minor axis. The water in the pond will be deepest at the center of the pond. The depth of the water, in feet, at any point in the pond will be given by the function $d(x, y) = 2\sqrt{\frac{x^2}{25} + \frac{y^2}{9}} - 2$, where $(x, y) = (0, 0)$ represents the center of the surface of the pond. How much water will the pond hold?

Proofs

51. The formulas for converting from cylindrical coordinates to rectangular coordinates are

$$x = r \cos\theta, \quad y = r \sin\theta, \quad \text{and} \quad z = z.$$

Prove that the Jacobian $\frac{\partial(x,y,z)}{\partial(r,\theta,z)} = r$.

52. The formulas for converting from spherical coordinates to rectangular coordinates are

$$x = \rho \sin\phi \cos\theta, \quad y = \rho \sin\phi \sin\theta, \quad \text{and} \quad z = \rho \cos\phi.$$

Prove that the Jacobian $\frac{\partial(x,y,z)}{\partial(\rho,\theta,\phi)} = -\rho^2 \sin\phi$.

Let $\alpha, \beta, \gamma,$ and δ be constants. A transformation $T : \mathbb{R}^2 \to \mathbb{R}^2$, where

$$x = \alpha u + \beta v \quad \text{and} \quad y = \gamma u + \delta v,$$

is called a ***linear transformation*** of $\mathbb{R}^2$. Use this transformation to answer Exercises 53–55.

53. Prove that a linear transformation takes a line $ax + by = c$ in the xy-plane to a line in the uv-plane if the Jacobian of the transformation is nonzero.

54. Prove that there is a linear transformation that takes a line in the xy-plane to a point in the uv-plane if the Jacobian of the transformation is zero.

55. Assuming that the Jacobian is nonzero, find expressions for u and v as functions of x and y.

56. Let $\alpha_i, \beta_i,$ and γ_i be constants for $i = 1, 2,$ and 3. A transformation $T : \mathbb{R}^3 \to \mathbb{R}^3$ defined by

$$x = \alpha_1 u + \beta_1 v + \gamma_1 w,$$
$$y = \alpha_2 u + \beta_2 v + \gamma_2 w, \quad \text{and}$$
$$z = \alpha_3 u + \beta_3 v + \gamma_3 w.$$

is called a ***linear transformation*** of $\mathbb{R}^3$. Prove that this transformation takes a plane $ax + by + cz = d$ in the xyz-coordinate system to a plane in the uvw-coordinate system if the Jacobian of the transformation is nonzero.

57. Use a change of variables to prove that the area of the ellipse with equation $\left(\frac{x}{a}\right)^2 + \left(\frac{y}{b}\right)^2 = 1$ is πab.

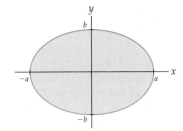

58. Use a change of variables to prove that the volume of the ellipsoid with equation $\left(\frac{x}{a}\right)^2 + \left(\frac{y}{b}\right)^2 + \left(\frac{z}{c}\right)^2 = 1$ is $\frac{4}{3}\pi abc$.

59. Prove the following chain rule for Jacobians: If x and y are differentiable functions of u and v, and if u and v are differentiable functions of s and t, then

$$\frac{\partial(x,y)}{\partial(s,t)} = \frac{\partial(x,y)}{\partial(u,v)} \frac{\partial(u,v)}{\partial(s,t)}.$$

60. Let Ω and Ω' be subsets of $\mathbb{R}^2$. Use the results of Exercise 59 to prove that if a transformation $T : \Omega \to \Omega'$ is invertible, and if both T and T^{-1} are differentiable, then

$$\frac{\partial(x,y)}{\partial(u,v)} \frac{\partial(u,v)}{\partial(x,y)} = 1.$$

Thinking Forward

Determinants of 3×3 *matrices:* Let

$$A = \begin{bmatrix} a_{11} & a_{12} & a_{13} \\ a_{21} & a_{22} & a_{23} \\ a_{31} & a_{32} & a_{33} \end{bmatrix}.$$

The determinant of A, $\det(A)$, may be defined in terms of determinants of 2×2 matrices as follows:

$$\det(A) = a_{11} \det \begin{bmatrix} a_{22} & a_{23} \\ a_{32} & a_{33} \end{bmatrix}$$

$$- a_{12} \det \begin{bmatrix} a_{21} & a_{23} \\ a_{31} & a_{33} \end{bmatrix} + a_{13} \det \begin{bmatrix} a_{21} & a_{22} \\ a_{31} & a_{32} \end{bmatrix}.$$

▶ Show that the previous definition of $\det(A)$ provides the same result as the procedure we used to find the determinant of a 3×3 matrix earlier in this section.

▶ Use the preceding definition to find

$$\det \begin{bmatrix} 1 & 3 & 4 \\ -2 & 5 & -1 \\ -4 & 2 & -3 \end{bmatrix}.$$

▶ In an analogous fashion, find a recursion formula for computing the determinant of a 4×4 matrix in terms of determinants of 3×3 matrices.

CHAPTER REVIEW, SELF-TEST, AND CAPSTONES

Before you progress to the next chapter, be sure you are familiar with the definitions, concepts, and basic skills outlined here. The capstone exercises at the end bring together ideas from this chapter and look forward to the next chapter.

Definitions

Give precise mathematical definitions or descriptions of each of the concepts that follow. Then illustrate the definition with a graph or an algebraic example.

▶ a *Riemann sum* for a function $z = f(x, y)$ defined on a rectangle $\mathcal{R} = \{(x, y) \mid a \le x \le b$ and $c \le y \le d\}$

▶ the *double integral* of f over
$$\mathcal{R} = \{(x, y) \mid a \le x \le b \text{ and } c \le y \le d\}$$

▶ the *double integral* of f over a general region
$$\Omega \subseteq \{(x, y) \mid a \le x \le b \text{ and } c \le y \le d\}$$

▶ the *Jacobian* of a transformation $T : \Omega \to \Omega'$

Theorems

Fill in the blanks to complete each of the following theorem statements:

▶ *Fubini's theorem:* Let $a < b$ and $c < d$ be real numbers, and let $\mathcal{R}$ be the rectangle defined by

$$\mathcal{R} = \{(x, y) \mid a \le x \le b \text{ and } c \le y \le d\}.$$

If $f(x, y)$ is continuous on $\mathcal{R}$, then $\iint_{\mathcal{R}} f(x, y)\, dA =$ _____.

▶ If $f(x, y)$ is an integrable function and $c \in \mathbb{R}$, then $\iint_{\Omega} cf(x, y)\, dA =$ _____.

▶ If $f(x, y)$ and $g(x, y)$ are integrable functions, then $\iint_{\Omega} (f(x, y) + g(x, y))\, dA =$ _____.

▶ Let $f(x, y)$ be an integrable function on the general region Ω. If Ω_1 and Ω_2 are general regions that are subsets of _____, and if $\Omega = \Omega_1 \cup \Omega_2$, then $\iint_{\Omega} f(x, y)\, dA =$ _____.

Notation and Algebraic Rules

Notation: Give the meanings of each of the following mathematical expressions:

▶ $\displaystyle \sum_{j=1}^{m} \sum_{k=1}^{n} a_{jk}$

▶ $\displaystyle \sum_{i=1}^{l} \sum_{j=1}^{m} \sum_{k=1}^{n} a_{ijk}$

▶ $\displaystyle \iint_{\mathcal{R}} f(x, y)\, dA$

▶ $\displaystyle \int_{a}^{b} \int_{g_1(x)}^{g_2(x)} f(x, y)\, dy\, dx$

▶ $\displaystyle \int_{c}^{d} \int_{h_1(y)}^{h_2(y)} f(x, y)\, dx\, dy$

▶ $\displaystyle \iiint_{\mathcal{R}} f(x, y, z)\, dV$

▶ $\displaystyle \int_{a}^{b} \int_{h_1(x)}^{h_2(x)} \int_{g_1(x,y)}^{g_2(x,y)} f(x, y, z)\, dz\, dy\, dx.$

▶ $\dfrac{\partial(x, y)}{\partial(u, v)}$

▶ $\dfrac{\partial(x, y, z)}{\partial(u, v, w)}$

Converting between coordinate systems: Provide the conversion formulas between each of the coordinate systems that follow. Include a sketch indicating the reason for each formula.

▶ Given rectangular coordinates x and y, the polar coordinates are $r =$ _____ and $\tan \theta =$ _____.

▶ Given polar coordinates r and θ, the rectangular coordinates are $x = $ _____ and $y = $ _____.

▶ Given rectangular coordinates x, y, and z, the cylindrical coordinates are $r = $ _____, $\tan\theta = $ _____, and $z = $ _____.

▶ Given cylindrical coordinates r, θ, and z, the rectangular coordinates are $x = $ _____, $y = $ _____, and $z = $ _____.

▶ Given rectangular coordinates x, y, and z, the spherical coordinates are $\rho = $ _____, $\theta = $ _____, and $\phi = $ _____.

▶ Given spherical coordinates ρ, $\tan\theta$, and $\cos\phi$, the spherical coordinates are $x = $ _____, $y = $ _____, and $z = $ _____.

▶ Given cylindrical coordinates r, θ, and z, the spherical coordinates are $\rho = $ _____, $\theta = $ _____, and $\tan\phi = $ _____.

▶ Given spherical coordinates ρ, θ, and ϕ, the cylindrical coordinates are $r = $ _____, $\theta = $ _____, and $z = $ _____.

First and Second Moments: Let $\rho(x, y)$ be a continuous density function for a lamina Ω in the xy-plane. Provide iterated integrals or iterated integral expressions that could be used to compute each of the following quantities:

▶ the mass of Ω

▶ the first moment of the mass in Ω about the y-axis

▶ the first moment of the mass in Ω about the x-axis

▶ the center of mass of Ω

▶ the moment of inertia of the mass in Ω about the y-axis

▶ the moment of inertia of the mass in Ω about the x-axis

▶ the moment of inertia of the mass in Ω about the origin

▶ the radius of gyration of the mass in Ω about the y-axis

▶ the radius of gyration of the mass in Ω about the x-axis

▶ the radius of gyration of the mass in Ω about the origin

Skill Certification: Integrating Functions of Two and Three Variables ———

Using the definition to evaluate a double integral: Evaluate the given double integrals as a limit of a Riemann sum. For each integral, let $\mathcal{R} = \{(x, y) \mid 0 \le x \le 2 \text{ and } 1 \le y \le 4\}$.

1. $\displaystyle\iint_{\mathcal{R}} (x + 2y)\, dA$ 2. $\displaystyle\iint_{\mathcal{R}} (xy^2)\, dA$

Evaluating a double integral as an iterated integral: Use Fubini's theorem to evaluate the given double integrals. For each integral, show that you obtain the same result when you integrate using both possible orders of integration when $\mathcal{R} = \{(x, y) \mid 0 \le x \le 2 \text{ and } 1 \le y \le 4\}$.

3. $\displaystyle\iint_{\mathcal{R}} (x + 2y)\, dA$ 4. $\displaystyle\iint_{\mathcal{R}} (xy^2)\, dA$

Evaluating iterated integrals: Sketch the region determined by the limits of the given iterated integrals, and then evaluate the integrals.

5. $\displaystyle\int_0^1 \int_{x^2}^{\sqrt{x}} x^2 y^3\, dy\, dx$ 6. $\displaystyle\int_0^2 \int_0^{\sqrt{4-x^2}} y^3\, dy\, dx$

7. $\displaystyle\int_0^1 \int_{-\sqrt{1-y^2}}^{\sqrt{1-y^2}} \frac{x}{y+1}\, dx\, dy$ 8. $\displaystyle\int_1^2 \int_0^{1/x} \sqrt{xy}\, dy\, dx$

9. $\displaystyle\int_1^4 \int_0^{1/y} \frac{x}{y}\, dx\, dy$ 10. $\displaystyle\int_0^2 \int_0^y x\sqrt{y^2 - x^2}\, dx\, dy$

Reversing the order of integration: Sketch the region determined by the limits of the given iterated integrals, and then evaluate the integrals by reversing the order of integration.

11. $\displaystyle\int_0^4 \int_{\sqrt{x}}^2 y \cos x\, dy\, dx$ 12. $\displaystyle\int_0^{\sqrt{\pi}} \int_x^{\sqrt{\pi}} \sin y^2\, dy\, dx$

13. $\displaystyle\int_0^9 \int_{\sqrt{y}}^3 \frac{1}{1+x^3}\, dx\, dy$ 14. $\displaystyle\int_0^{16} \int_{\sqrt[4]{x}}^2 \frac{1}{1+y^5}\, dy\, dx$

15. $\displaystyle\int_0^4 \int_y^4 e^{y/x}\, dx\, dy$ 16. $\displaystyle\int_0^1 \int_{\sqrt{x}}^1 e^{x/y^2}\, dy\, dx$

Using polar coordinates to evaluate iterated integrals: Sketch the region determined by the limits of the given iterated integrals, and then evaluate the integrals.

17. $\displaystyle\int_0^{\pi/2} \int_0^3 r^2\, dr\, d\theta$ 18. $\displaystyle\int_{\pi/2}^{3\pi/2} \int_1^4 r^{3/2}\, dr\, d\theta$

19. $\displaystyle\int_0^{\pi/2} \int_0^{2\sin\theta} r^3\, dr\, d\theta$ 20. $\displaystyle\int_0^{2\pi} \int_0^{1+\sin\theta} r\, dr\, d\theta$

Using polar coordinates to evaluate iterated integrals: Evaluate the given iterated integrals by converting them to polar coordinates. Include a sketch of the region.

21. $\displaystyle\int_0^2 \int_0^{\sqrt{4-y^2}} e^{x^2+y^2}\, dx\, dy$

22. $\displaystyle\int_0^4 \int_0^{\sqrt{4y-y^2}} \frac{1}{\sqrt{x^2+y^2}}\, dx\, dy$

23. $\displaystyle\int_{-3}^3 \int_{-\sqrt{9-x^2}}^{\sqrt{9-x^2}} \frac{x+2y}{x^2+y^2}\, dy\, dx$

24. $\displaystyle\int_{-5}^0 \int_{-\sqrt{25-x^2}}^0 \frac{3}{(4+x^2+y^2)^3}\, dy\, dx$

Evaluating triple integrals: Each of the triple integrals that follows represents the volume of a solid. Sketch the solid and evaluate the integral.

25. $\displaystyle\int_0^2 \int_1^4 \int_{-2}^3 dx\, dy\, dz$ 26. $\displaystyle\int_1^5 \int_{-2}^0 \int_{-1}^3 dz\, dy\, dx$

27. $\displaystyle\int_0^4 \int_0^{3-(3/4)x} \int_0^{2-(1/2)x-(2/3)y} dz\, dy\, dx$

28. $\displaystyle\int_0^3 \int_0^{2-2y/3} \int_0^{4-(4/3)y-2z} dx\, dz\, dy$

29. $\displaystyle\int_{-2}^{2}\int_{0}^{\sqrt{4-x^2}}\int_{0}^{y} dz\, dy\, dx$

30. $\displaystyle\int_{0}^{2\pi}\int_{0}^{2}\int_{0}^{5} r\, dz\, dr\, d\theta$

31. $\displaystyle\int_{0}^{2\pi}\int_{0}^{3}\int_{0}^{\sqrt{9-r^2}} r\, dz\, dr\, d\theta$

32. $\displaystyle\int_{0}^{2\pi}\int_{0}^{4}\int_{r}^{4} r\, dz\, dr\, d\theta$

33. $\displaystyle\int_{0}^{2\pi}\int_{0}^{4}\int_{0}^{4-r} r\, dz\, dr\, d\theta$

34. $\displaystyle\int_{0}^{\pi}\int_{0}^{2}\int_{0}^{r\sin\theta} r\, dz\, dr\, d\theta$

35. $\displaystyle\int_{0}^{\pi}\int_{0}^{2\pi}\int_{0}^{5} \rho^2 \sin\phi\, d\rho\, d\theta\, d\phi$

36. $\displaystyle\int_{0}^{\pi/4}\int_{0}^{2\pi}\int_{0}^{4\sec\theta} \rho^2 \sin\phi\, d\rho\, d\theta\, d\phi$

Capstone Problems

A. Show that

$$\int_{0}^{1}\int_{0}^{1} \frac{x-y}{(x+y)^3}\, dy\, dx \neq \int_{0}^{1}\int_{0}^{1} \frac{x-y}{(x+y)^3}\, dx\, dy.$$

Explain why this does not contradict Fubini's theorem.

B. A hole with radius $\dfrac{R}{2}$ is drilled though the center of a sphere with radius R, as shown in the following figure. Find the volume of the resulting solid. *(Hint: You may assume that the equation of the sphere is $x^2+y^2+z^2 = R^2$ and the equation of the cylinder through the sphere is $x^2 + y^2 = (R/2)^2$.)*

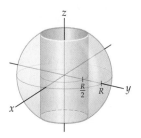

C. Find the center of mass and moment of inertia about the z-axis for the solid from Problem B, assuming that it has a uniform density.

D. Evaluate the integral

$$\iint_{T} \ln\left(\frac{y-x}{x+y}\right) dA,$$

where T is the first-quadrant triangular region with vertices $(2, 3)$, $(2, 4)$, and $(3, 4)$.

Vector Analysis

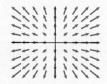

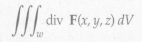

14.1   VECTOR FIELDS

▶ An overview of vector analysis

▶ Assigning a vector to each point in Euclidean space

▶ Conservative vector fields and their potential functions

Calculus and Vector Analysis

In this final chapter, we unify and generalize our understanding of integration and the Fundamental Theorem of Calculus. We will generalize the original notion of integration that was introduced in Chapter 4 to enable us to integrate over curves other than the x-axis, extend the integration results of Chapter 13 to allow us to perform double integration over regions other than regions in the xy-plane, and see new multivariate extensions of the Fundamental Theorem of Calculus. The basic ideas will be the same, but the context in which we implement them will be more subtle and more varied. Now would be an excellent time to review what you have studied about the integration of functions of one variable, vector-valued functions, and multivariate functions.

Here are two examples of the types of questions we will address in this chapter. First, consider a fence whose base is a straight line, like the x-axis, and another fence whose base is along an undulating hill:

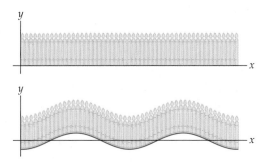

Both fences clearly have well-defined areas. To find the area of the first, we integrate

$$A = \int h(x)\, dx,$$

where $h(x)$ is a function that tells us the height of the fence above the x-axis. The area of the second fence is not quite this integral, since the curve of its base will distort the fence. Still, the fence height and its base appear smooth, so its area ought to be expressible as a one-variable integration of something, where we integrate along the green curve in the figure. The resulting integral will combine ideas we used for computing arc lengths and areas in previous chapters. The resulting integral will have the form

$$A = \int h(t)\varphi(t)\, dt,$$

where $\varphi(t)$ is a factor that accounts for the bending of the base.

Similarly, in Section 6.4 we discussed work W as the product of force F and distance d, or $W = Fd$. From this equation, we saw how to choose an appropriate integral to calculate the work done in pumping water out the top of a tank, opposite the downward force of gravity. If we wanted to know the work required to move an object along a more complicated path, and opposed to a variable force, we would expect that this, too, could be expressed as an integral, perhaps now with multiple inputs.

In Section 14.2 we will see how to solve both of these problems. For now, we introduce a new mathematical object that is necessary for the solution.

Vector Fields

One immediate question that confronts us when considering how to generalize the Fundamental Theorem of Calculus to several dimensions is the issue of what, exactly, we are trying to integrate. As we saw in the preceding example we might be interested in finding the area of a fence whose base is curved; we might also be interested in the work done by that fence as it stays upright in a windstorm, when whirling wind exerts forces of different strengths and in different directions on the fence at different points. This idea of varying forces associated with points in space or points in the plane motivates the notion of a *vector field*.

A vector field is a function that assigns a vector to each point in its domain. Thus, a vector field in the plane has inputs that are points in the xy-plane and outputs that are vectors in the xy-plane; a vector field in $\mathbb{R}^3$ has inputs that are points in $\mathbb{R}^3$ and outputs that are vectors in $\mathbb{R}^3$.

DEFINITION 14.1

Vector Field

A ***vector field in*** $\mathbb{R}^2$ is a function $\mathbf{F}(x, y)$ with domain $D \subseteq \mathbb{R}^2$ and whose outputs are vectors in $\mathbb{R}^2$ of the form

$$\mathbf{F}(x, y) = \langle F_1(x, y), F_2(x, y) \rangle$$

for each point (x, y) in D.

Similarly, a ***vector field in*** $\mathbb{R}^3$ is a function $\mathbf{G}(x, y, z)$ with domain $D \subseteq \mathbb{R}^3$ and whose outputs are vectors in $\mathbb{R}^3$ of the form

$$\mathbf{G}(x, y, z) = \langle G_1(x, y, z), G_2(x, y, z), G_3(x, y, z) \rangle$$

for each point (x, y, z) in D.

In this chapter, the only vector fields we are interested in are those whose domains are well-behaved enough to support vector analysis. Thus, for the remainder of our discussion of vector fields, we will always suppose that every vector field has a domain D that is open, connected, and simply connected. This means that D is open in the sense of Chapter 12; that, for any two points P and Q in D, there is a path from P to Q that lies in D; and that any loop in D can be smoothly contracted to a point in D. In general, a two-dimensional region is simply connected if it does not contain any holes.

A region in $\mathbb{R}^2$ that is simply connected   *A region in $\mathbb{R}^2$ that is not simply connected*

A region in $\mathbb{R}^3$ that is simply connected   *A region in $\mathbb{R}^3$ that is not simply connected*

A three-dimensional region may have holes and yet still be simply connected, as shown in the following sequence of figures:

We will also usually assume that the vector field in question is sufficiently smooth so that the component functions are continuous and have continuous first partial derivatives.

The notion of a vector field in $\mathbb{R}^n$ may of course be defined analogously for any natural number n, but we will not need such vector fields in this text.

Following are pictures of the vector fields $\mathbf{F}(x, y) = 2\mathbf{i} + 0\mathbf{j}$ and $\mathbf{G}(x, y) = \mathbf{i} - \mathbf{j}$, where at each point (x, y) we draw the vector $\langle F_1(x, y), F_2(x, y) \rangle$:

$\mathbf{F}(x, y) = 2\mathbf{i} + 0\mathbf{j}$ $\mathbf{G}(x, y) = \mathbf{i} - \mathbf{j}$

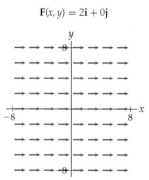 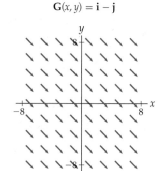

Note that these diagrams show only some of the vectors. A true representation would look like a solid block of color, because every point in the plane would have a vector associated with it. The following vector fields $\mathbf{F}(x, y) = \langle x, y \rangle$ and $\mathbf{G}(x, y) = \langle \sin x, \sin y \rangle$ are slightly more interesting:

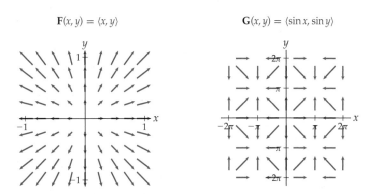

$$\mathbf{F}(x, y) = \langle x, y \rangle \qquad\qquad \mathbf{G}(x, y) = \langle \sin x, \sin y \rangle$$

Note that in the vector field $\mathbf{F}$ the magnitudes of the vectors in the codomain increase as we move away from the origin and have constant magnitude on concentric circles about the origin. For example, $\|\mathbf{F}(x, y)\| = 1$ if and only if (x, y) lies on the unit circle. In addition, all vectors emanating from points near the origin point away from the origin. In such a situation the origin is called a ***source***. In all of our vector fields, the vectors are drawn so that their lengths are representative of their magnitudes.

In the vector field $\mathbf{G}$, all vectors emanating from points near the origin again point away from the origin, so the origin is a source. In addition because of the periodicity of the sine function, all coordinate pairs of the form $(2k\pi, 2n\pi)$, where k and n are both integers, are also sources. Note, however, that all of the vectors that emanate from points surrounding (π, π) point towards (π, π). Such a point is called a ***sink***. Again by the periodicity of the sine, all points of the form $((2k+1)\pi, (2n+1)\pi)$, where k and n are both integers, are also sinks.

Conservative Vector Fields

Some vector fields, such as $\mathbf{F}(x, y) = \langle x, y \rangle$, can be written as the gradient of a function. For example, we might have

$$\mathbf{F}(x, y) = \nabla f(x, y), \quad \text{where} \quad f(x, y) = \frac{x^2}{2} + \frac{y^2}{2} \quad \text{and} \quad \nabla f = \frac{\partial f}{\partial x}\mathbf{i} + \frac{\partial f}{\partial y}\mathbf{j}$$

This is one of the infinitely many possible choices for the function $f(x, y)$. It is also true that

$$\mathbf{F}(x, y) = \nabla g(x, y), \quad \text{where} \quad g(x, y) = \frac{x^2}{2} + \frac{y^2}{2} + \alpha,$$

for any choice of the constant α.

If a vector field can be expressed as a gradient, then it has many mathematically desirable attributes, including being easy to integrate along curves, a property we will see in the next section. Such fields are known as *conservative vector fields*.

DEFINITION 14.2

Conservative Vector Field

A ***conservative vector field*** $\mathbf{F}$ is a vector field that can be written as the gradient of some function f. That is,

$$\mathbf{F}(x, y) = \nabla f(x, y) = \frac{\partial f}{\partial x}\mathbf{i} + \frac{\partial f}{\partial y}\mathbf{j}$$

if $f(x, y)$ is a function of two variables, or

$$\mathbf{F}(x, y, z) = \nabla f(x, y, z) = \frac{\partial f}{\partial x}\mathbf{i} + \frac{\partial f}{\partial y}\mathbf{j} + \frac{\partial f}{\partial z}\mathbf{k}$$

if $f(x, y, z)$ is a function of three variables.

In either case, any function f whose gradient is equal to $\mathbf{F}$ is called a ***potential function*** for $\mathbf{F}$.

Recall that if a function $g(x, y)$ has continuous first and second partial derivatives, then

$$\frac{\partial^2 g}{\partial y \partial x} = \frac{\partial^2 g}{\partial x \partial y}.$$

So, to verify that $\mathbf{F}(x, y) = \langle F_1(x, y), F_2(x, y) \rangle$ is a conservative vector field with a potential function $f(x, y)$ (with continuous second partial derivatives), it is sufficient to check that

$$\frac{\partial F_1}{\partial y} = \frac{\partial F_2}{\partial x},$$

since equality of these first partial derivatives of the components of $\mathbf{F}$ means that the mixed second-order partial derivatives of f are equal. In fact, a vector field

$$\mathbf{F}(x, y) = \langle F_1(x, y), F_2(x, y) \rangle,$$

whose first partial derivatives $\frac{\partial F_1}{\partial y}$ and $\frac{\partial F_2}{\partial x}$ are both continuous on an open subset of $\mathbb{R}^2$, is conservative if and only if $\frac{\partial F_1}{\partial y} = \frac{\partial F_2}{\partial x}$.

The situation is similar for vector fields in $\mathbb{R}^3$, except there are more mixed partial derivatives to check. Note that potential functions are not unique. As we saw earlier, if $f(x, y)$ is a potential function for a conservative vector field $\mathbf{F}(x, y)$, then so is $f(x, y) + \alpha$ for any constant α.

In some cases it will be enough to know that a vector field is conservative; in others, it will be important to find a potential function for the field. By thinking through the definitions of conservative vector fields and potential functions, we can see that the question is one of recovering a function from its partial derivatives. That is easy to do by integrating, but there is a subtle point to consider.

For example, let

$$\mathbf{G}(x, y, z) = 6xz\mathbf{i} + \cos y\mathbf{j} + 3x^2\mathbf{k}.$$

By checking all six mixed second partial derivatives, we find that $\mathbf{G}$ is indeed conservative. Now, we want a potential function $g(x, y, z)$ whose gradient is $\mathbf{G}$. Any g that satisfies

$$\frac{\partial g}{\partial x} = 6xz, \qquad \frac{\partial g}{\partial y} = \cos y, \qquad \frac{\partial g}{\partial z} = 3x^2$$

will work. We let $G_1(x, y, z) = 6xz$, $G_2(x, y, z) = \cos y$, and $G_3(x, y, z) = 3x^2$ and begin by integrating G_1 with respect to x:

$$\int G_1(x, y, z)\, dx = \int 6xz\, dx = 3x^2z + \alpha = g(x, y, z).$$

But the gradient of this function is not $\mathbf{G}(x, y, z)$. Here we have

$$\nabla g(x, y, z) = 6xz\mathbf{i} + 0\mathbf{j} + 3x^2\mathbf{k}.$$

We are missing a term that will give the correct partial derivative for y. Since the terms of g that do not involve x are constants with respect to differentiation by x, we can fix the problem by integrating all the terms of G_2 that do not involve x, integrating the terms of G_3 that involve neither x nor y, and adding the results to the function g that we have already found. This gives

$$\int G_2(x, y, z)\, dy = \int \cos y\, dy = \sin y + \alpha.$$

Since none of the terms of G_3 involve only z, when we add $\sin y + \alpha$ to our previous function and set the constant term to zero, we obtain

$$g(x, y, z) = 3x^2 z + \sin y.$$

You may check that

$$\nabla g(x, y, z) = \langle 6xz, \cos y, 3x^2 \rangle = \mathbf{G}(x, y, z).$$

More generally, to find the potential functions $f(x, y)$ and $g(x, y, z)$ for the respective conservative vector fields

$$\mathbf{F}(x, y) = \langle F_1(x, y), F_2(x, y) \rangle \quad \text{and} \quad \mathbf{G}(x, y, z) = \langle G_1(x, y, z), G_2(x, y, z), G_3(x, y, z) \rangle,$$

we construct

$$f(x, y) = \int F_1(x, y) \, dx + B \quad \text{and} \quad g(x, y, z) = \int G_1(x, y, z) \, dx + B + C,$$

where B is the integral with respect to y of the terms in $F_2(x, y)$ or $G_2(x, y, z)$ that have no x factor and C is the integral with respect to z of the terms in $G_3(x, y, z)$ that have no x or y factor.

It is also possible, and sometimes more convenient, to perform these integrations in other orders. Consider the complexities of the component functions when you are deciding.

Examples and Explorations

EXAMPLE 1

Drawing vector fields in the plane

Sketch or use graphing software to create plots of the following vector fields in $\mathbb{R}^2$:

(a) $\mathbf{F}(x, y) = \langle x, -y \rangle$

(b) $\mathbf{G}(x, y) = \dfrac{y}{x^2 + y^2} \mathbf{i} + \dfrac{x}{x^2 + y^2} \mathbf{j}$

SOLUTION

Following are the two vector fields:

$\mathbf{F}(x, y) = \langle x, -y \rangle$

$\mathbf{G}(x, y) = \dfrac{y}{x^2 + y^2} \mathbf{i} + \dfrac{x}{x^2 + y^2} \mathbf{j}$

Each vector in the field is computed using the component functions. For example, $\mathbf{F}(1, 1/2) = \langle 1, -1/2 \rangle$. Our graphs are drawn with software, but they may be drawn by hand. □

EXAMPLE 2

Drawing vector fields in $\mathbb{R}^3$

Sketch or use graphing software to create plots of the given vector fields in $\mathbb{R}^3$. Identify any sources or sinks in each vector field.

(a) $\mathbf{F}(x, y, z) = \langle x, y, z \rangle$

(b) $\mathbf{G}(x, y, z) = y\mathbf{i} - x\mathbf{j} + z\mathbf{k}$

SOLUTION

Three-dimensional vector fields can be a little difficult to understand, even when we use graphing software. As an aid, it may be useful to visualize the analogous vector field one dimension lower.

(a) The vector field $\mathbf{F}(x, y) = \langle x, y \rangle$ that we drew earlier in the section is a two-dimensional analog for the field shown next. Each vector points away from the origin, their magnitudes increase as you move away from the origin, and the magnitudes are constant on spheres centered at the origin. Note that to determine any particular vector in the field, we evaluate the function at a point. For example, $\mathbf{F}(1, 2, 3) = \langle 1, 2, 3 \rangle$.

$$\mathbf{F}(x, y, z) = \langle x, y, z \rangle$$

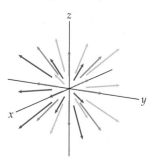

Note that the origin is a source in this vector field, since all vectors emanating from points near the origin point away from it.

(b) We first sketch the two-dimensional vector field $\mathbf{G}(x, y) = y\mathbf{i} - x\mathbf{j}$ shown next at the left. We see that the vectors are tangent to concentric circles centered at the origin and that their magnitudes increase as you move away from the origin. In the three-dimensional vector field $\mathbf{G}(x, y, z) = y\mathbf{i} - x\mathbf{j} + z\mathbf{k}$ shown at the right, the $\mathbf{k}$-components of the vectors increase as you move away from the xy-plane as well.

$$\mathbf{G}(x, y) = y\mathbf{i} - x\mathbf{j} \qquad\qquad \mathbf{G}(x, y, z) = y\mathbf{i} - x\mathbf{j} + z\mathbf{k}$$

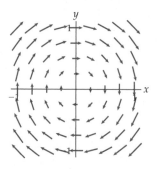
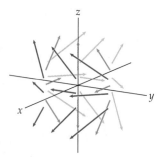

This vector field has no sources or sinks.

EXAMPLE 3   **Determining whether a vector field is conservative**

Determine whether the vector fields from Example 1 are conservative.

(a) $\mathbf{F}(x, y) = \langle x, -y \rangle$ **(b)** $\mathbf{G}(x, y) = \dfrac{y}{x^2 + y^2}\mathbf{i} + \dfrac{x}{x^2 + y^2}\mathbf{j}$

SOLUTION

A vector field $\mathbf{F}(x, y) = \langle F_1(x, y), F_2(x, y) \rangle$ is conservative if and only if

$$\frac{\partial F_1}{\partial y} = \frac{\partial F_2}{\partial x}$$

when $\frac{\partial F_1}{\partial y}$ and $\frac{\partial F_2}{\partial x}$ are both continuous on some open subset of $\mathbb{R}^2$.

(a) For the vector field $\mathbf{F}(x, y) = \langle x, -y \rangle$,

$$\frac{\partial F_1}{\partial y} = \frac{\partial}{\partial y}(x) = 0 \quad \text{and} \quad \frac{\partial F_2}{\partial x} = \frac{\partial}{\partial x}(-y) = 0.$$

Since these partial derivatives are the same, $\mathbf{F}(x, y)$ is indeed conservative.

(b) For the vector field $\mathbf{G}(x, y) = \frac{y}{x^2+y^2}\mathbf{i} + \frac{x}{x^2+y^2}\mathbf{j}$,

$$\frac{\partial G_1}{\partial y} = \frac{\partial}{\partial y}\left(\frac{y}{x^2 + y^2}\right) = \frac{x^2 - y^2}{(x^2 + y^2)^2} \quad \text{and} \quad \frac{\partial G_2}{\partial x} = \frac{\partial}{\partial x}\left(\frac{x}{x^2 + y^2}\right) = \frac{y^2 - x^2}{(x^2 + y^2)^2}.$$

Since $\frac{\partial G_1}{\partial y} \neq \frac{\partial G_2}{\partial x}$ and these partial derivatives are continuous on every open subset of $\mathbb{R}^2$ not containing the origin, $\mathbf{G}(x, y)$ is not conservative. $\square$

EXAMPLE 4   **Finding a potential function for a vector field in $\mathbb{R}^2$**

Find a potential function for the vector field $\mathbf{F}(x, y) = \langle x, -y \rangle$ from Example 1.

SOLUTION

Since

$$\mathbf{F}(x, y) = \langle x, -y \rangle,$$

$$f(x, y) = \int x \, dx + B = \frac{x^2}{2} + \alpha + B,$$

where α is an arbitrary constant and B is the integral with respect to y of the terms in $F_2(x, y)$ in which the factor x does not appear. In this case, that is all of $F_2(x, y)$, so

$$B = \int (-y) \, dy = \frac{-y^2}{2} + \beta,$$

where β is an arbitrary constant. Setting the constants equal to zero since they do not affect the gradient of $f(x, y)$, we have

$$f(x, y) = \frac{x^2}{2} - \frac{y^2}{2}.$$ $\square$

CHECKING THE ANSWER

We can verify this solution by computing ∇f directly:

$$\nabla f(x, y) = f_x(x, y)\mathbf{i} + f_y(x, y)\mathbf{j} = x\mathbf{i} - y\mathbf{j} = \mathbf{F}(x, y).$$

EXAMPLE 5   **Finding a potential function for a vector field in $\mathbb{R}^3$**

Find a potential function for

$$\mathbf{G}(x, y, z) = \left(\frac{y}{x^2 + 1} + z\right)\mathbf{i} + \tan^{-1}x\,\mathbf{j} + (x + \cos z)\,\mathbf{k}.$$

SOLUTION

We start by integrating $G_1(x, y, z)$. We have

$$g(x, y, z) = \int \left(\frac{y}{x^2 + 1} + z\right) dx + B + C = y\tan^{-1}x + xz + \alpha + B + C,$$

where α is an arbitrary constant, B is the integral with respect to y of the terms in $G_2(x, y, z)$ in which the factor x does not appear, and C is the integral with respect to z of the terms in $G_3(x, y, z)$ in which neither the factor x nor the factor y appears. In this case $B = 0$, because the only term in $G_2(x, y, z)$ is $\tan^{-1}x$ and this part of the function $g(x, y, z)$ is already recovered by the integral of $G_1(x, y, z)$. For C, however, there is a term with no x- or y-component, so

$$C = \int \cos z\, dz = \sin z + \beta,$$

where β is an arbitrary constant. Combining these results and setting the arbitrary constants to zero, we see that

$$g(x, y, z) = y\tan^{-1}x + xz + \sin z$$

is a potential function for $\mathbf{G}(x, y, z)$.   ☐

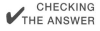
CHECKING THE ANSWER

We can verify this solution by computing $\nabla g(x, y, z)$ directly:

$$\nabla g(x, y, z) = \langle g_x, g_y, g_z \rangle = \left(\frac{y}{x^2 + 1} + z\right)\mathbf{i} + \tan^{-1}x\,\mathbf{j} + (x + \cos z)\,\mathbf{k} = \mathbf{G}(x, y, z).$$

? TEST YOUR UNDERSTANDING

▶ Which curve is longer, the curve $y = 0$ for $0 \le x \le \pi$ or the curve $y = \sin x$ for $0 \le x \le \pi$? Why? With the answer to this question in mind, which of the fences pictured in the illustration on page 1086 has a greater area? Why?

▶ Which of the vector fields pictured in this section would carry an object caught within towards the origin? Why?

▶ If $f(x, y, z)$ is a potential function for $\mathbf{F}(x, y, z)$, then what is $\nabla f(x, y, z)$?

▶ What does it mean to say that a vector field $\mathbf{F}(x, y)$ is conservative? How would you show that a vector field failed to be conservative?

▶ Given a conservative vector field, how do we find its potential function? How does the procedure for finding a potential function for a conservative vector field in the plane differ from the procedure for finding a potential function for a conservative vector field in $\mathbb{R}^3$?

EXERCISES 14.1

Thinking Back

▶ *Work as an integral of force and distance:* Find the work done in moving an object along the x-axis from the origin to $x = \dfrac{\pi}{2}$ if the force acting on the object at a given value of x is $F(x) = x \sin x$.

Vector geometry: Use properties of vectors to obtain each of the specified quantities.

▶ Find a unit vector that points in the direction of $\mathbf{F}(3, 3, \sqrt{13})$ for the vector field

$$\mathbf{F} = z(x - y)^2 \mathbf{i} + \sqrt{y}\,\mathbf{j} + \mathbf{k}.$$

▶ Find the equation of a line through the origin and parallel to $\mathbf{F}(3, 3, \sqrt{13})$.

Calculus of vector-valued functions: Calculate each of the following.

▶ $\dfrac{d}{dt}(\mathbf{r}(t))$, where $\mathbf{r}(t) = 3\cos^2 t\,\mathbf{i} + 5t\mathbf{j} + \dfrac{t}{t^2 + 1}\mathbf{k}$.

▶ $\int \mathbf{r}(t)\,dt$, where $\mathbf{r}(t) = e^t\mathbf{i} + t^3\mathbf{j} - 4\mathbf{k}$.

Concepts

0. *Problem Zero:* Read the section and make your own summary of the material.

1. *True/False:* Determine whether each of the statements that follow is true or false. If a statement is true, explain why. If a statement is false, provide a counterexample.

  (a) *True or False:* A vector field is a function whose outputs are scalars.

  (b) *True or False:* A vector field is a function whose outputs are vectors.

  (c) *True or False:* A vector field is a function whose inputs are scalars.

  (d) *True or False:* A vector field in $\mathbb{R}^3$ is a function whose inputs are points in $\mathbb{R}^3$.

  (e) *True or False:* A conservative vector field has infinitely many potential functions.

  (f) *True or False:* Every vector field $\mathbf{F}(x, y)$ is the gradient of some function $f(x, y)$.

  (g) *True or False:* If two functions have the same gradient, they are the same function.

  (h) *True or False:* Work is the integral of force times distance.

2. *Examples:* Construct examples of the thing(s) described in the following. Try to find examples that are different than any in the reading.

  (a) A vector field in $\mathbb{R}^2$ and another in $\mathbb{R}^3$.

  (b) A conservative vector field.

  (c) A vector field that is not conservative.

3. What are the inputs of a vector field in the Cartesian plane?

4. What are the inputs of a vector field in $\mathbb{R}^3$?

5. What are the outputs of a vector field in the Cartesian plane?

6. What are the outputs of a vector field in $\mathbb{R}^3$?

7. What does it mean to say that a vector field is conservative?

8. Do the vectors in the range of $\mathbf{F}(x, y) = x\mathbf{i} + y\mathbf{j}$ point towards or away from the origin?

9. What is the difference between the graphs of

$$\mathbf{G}(x, y) = \mathbf{i} + \mathbf{j} \text{ and } \mathbf{F}(x, y) = 2\mathbf{i} + 2\mathbf{j}?$$

10. What is the difference between the graphs of

$$\mathbf{G}(x, y, z) = -\mathbf{i} - \mathbf{j} - \mathbf{k} \text{ and } \mathbf{F}(x, y, z) = \mathbf{i} + \mathbf{j} + \mathbf{k}?$$

11. What is the difference between the graphs of

$$\mathbf{G}(x, y) = x\mathbf{i} + y\mathbf{j} \text{ and } \mathbf{F}(x, y) = -x\mathbf{i} + -y\mathbf{j}?$$

12. What is the difference between the graphs of

$$\mathbf{G}(x, y, z) = 2\mathbf{i} - 3\mathbf{j} + z\mathbf{k} \text{ and } \mathbf{F}(x, y, z) = -2\mathbf{i} + 3\mathbf{j} - z\mathbf{k}?$$

13. Consider the vector field $\mathbf{F}(x, y, z) = \langle yz, xz, xy \rangle$. Find a vector field $\mathbf{G}(x, y, z)$ with the property that, for all points in $\mathbb{R}^3$, $\mathbf{G}(x, y, z) = 2\mathbf{F}(x, y, z)$.

14. What is the difference between vector fields

$$\mathbf{F}(x, y) = \langle x^2, y^2 \rangle \text{ and } \mathbf{G}(x, y) = \langle x^2, y^2, 0 \rangle?$$

15. How would you show that a given vector field in $\mathbf{R}^2$ is not conservative?

16. How would you show that a given vector field in $\mathbf{R}^3$ is not conservative?

Skills

In Exercises 17–24, find a potential function for the given vector field.

17. $\mathbf{F}(x, y) = \langle 3x^2 \cos y, -x^3 \sin y \rangle$

18. $\mathbf{F}(x, y) = \langle e^y \sec^2 x, e^y \tan x \rangle$

19. $\mathbf{G}(x, y) = \langle 5x^4 + y, x - 12y^3 \rangle$

20. $\mathbf{G}(x, y) = \mathbf{i} - \mathbf{j}$

21. $\mathbf{F}(x, y, z) = yz\mathbf{i} + xz\mathbf{j} + xy\mathbf{k}$

22. $\mathbf{F}(x, y, z) = e^{y^2}\mathbf{i} + (2xye^{y^2} + \sin z)\mathbf{j} + y\cos z\,\mathbf{k}$

23. $\mathbf{G}(x, y, z) = \cos y\mathbf{i} + (\sin z - x\sin y)\mathbf{j} + y\cos z\mathbf{k}$

24. $\mathbf{G}(x, y, z) = \langle ye^{xy+z}, xe^{xy+z}, e^{xy+z} \rangle$

Sketch the vector fields in Exercises 25–32.

25. $\mathbf{F}(x, y) = 2\mathbf{i} + 0\mathbf{j}$
26. $\mathbf{F}(x, y) = 0\mathbf{i} + 2\mathbf{j}$
27. $\mathbf{F}(x, y) = \mathbf{i} + \mathbf{j}$
28. $\mathbf{F}(x, y) = 2\mathbf{i} + 2\mathbf{j}$
29. $\mathbf{F}(x, y) = \mathbf{i} - \mathbf{j}$
30. $\mathbf{F}(x, y) = -\mathbf{i} + \mathbf{j}$
31. $\mathbf{F}(x, y) = x\mathbf{i} + 2y\mathbf{j}$
32. $\mathbf{F}(x, y) = -2x\mathbf{i} - 3y\mathbf{j}$

Show that the vector fields in Exercises 33–40 are not conservative.

33. $\mathbf{F}(x, y) = \langle xy, -y \rangle$

34. $\mathbf{F}(x, y) = \langle x^2 + y^2, \cos y \rangle$

35. $\mathbf{G}(x, y) = \dfrac{1}{x^2 + y}\mathbf{i} + \dfrac{y}{x}\mathbf{j}$

36. $\mathbf{G}(x, y) = y\mathbf{i} - x\mathbf{j}$

37. $\mathbf{F}(x, y, z) = 2\mathbf{i} - z\mathbf{j} + e^{yz}\mathbf{k}$

38. $\mathbf{F}(x, y, z) = \tan(yz)\mathbf{i} + (xz\sec^2(yz) - 2)\mathbf{j} + 4z^3\mathbf{k}$

39. $\mathbf{G}(x, y, z) = \langle 3, yz, z + 12 \rangle$

40. $\mathbf{G}(x, y, z) = \langle e^y + z, xe^y + z, x + y \rangle$

Determine whether or not each of the vector fields in Exercises 41–48 is conservative. If the vector field is conservative, find a potential function for the field.

41. $\mathbf{F}(x, y) = e^y\mathbf{i} + \sin y\mathbf{j}$

42. $\mathbf{F}(x, y) = \left(\tan^{-1} y, \dfrac{x}{1 + y^2} \right)$

43. $\mathbf{G}(x, y) = \langle 2x + y\cos(xy), x\cos(xy) - 1 \rangle$

44. $\mathbf{G}(x, y) = yx^2\mathbf{i} + e^y\mathbf{j}$

45. $\mathbf{F}(x, y, z) = \langle ye^{2z} + 1, xe^{2z}, 2xye^{2z} \rangle$

46. $\mathbf{F}(x, y, z) = \mathbf{i} + 2\mathbf{j} - 3\mathbf{k}$

47. $\mathbf{G}(x, y, z) = (z - y)\mathbf{i} - xy\mathbf{j} + (xz + y)\mathbf{k}$

48. $\mathbf{G}(x, y, z) = \langle \sin(yz), \cos(yz), x^2 \rangle$

Applications

49. Construct a vector field to describe each of the situations that follow. There may be more than one choice of constant that gives an accurate answer. (*Hint: Examples from the section and earlier exercises may be useful, as well as plotting a few vectors from a given field.*)

    (a) A thin film of water flowing across a flat plate from right to left at a constant rate.

    (b) A thin film of liquid flowing across a flat plate diagonally from top left to bottom right at a constant rate.

    (c) A thin film of liquid flowing on a flat plate rotating counterclockwise about the origin.

50. Write a vector field to describe each of the flows that follow. As in Exercise 49, there may be more than one choice of constant that gives an accurate answer.

    (a) A thin film of liquid flowing across a flat plate from top to bottom at a constant rate.

    (b) A thin film of liquid flowing across a flat plate in the direction of $3\mathbf{i} + 2\mathbf{j}$ at a constant rate.

    (c) Fluid rotating clockwise around the origin.

51. Annie uses a vector field to model the current in a channel in the San Juan Islands in Washington State. She has imposed a coordinate system centered at the origin of the area she is interested in, and she is looking at the region $[-1, 1] \times [-1, 1]$. The velocity of the current is given by $\langle 0.9 - x^2 + 0.5y^2, xy \rangle$.

    (a) Sketch the vector field.

    (b) Is the vector field conservative? If so, what is an equation of a potential function? Explain the relationship between the potential function and the current.

    (c) Can you tell where the land that influences the current lies?

Proofs

52. Prove that a conservative vector field has infinitely many potential functions.

53. Prove that if two functions $f(x, y)$ and $g(x, y)$ have the same gradient, then they differ by at most a constant.

Thinking Forward

Integration of vector fields: In order to answer the questions posed at the beginning of this section, we ought to be able to use the "subdivide, approximate, and add" strategy from Chapter 6. The results—area and work—should be scalars.

▶ How might the integral to find the arc length of $\mathbf{r}(t)$ from $t = a$ to $t = b$ in Section 11.4 be modified to produce the area of a fence whose base is $\mathbf{r}(t)$ and whose height is given by $h(t) = h(x(t), y(t))$?

▶ Write down some possible interpretations of what integrating a vector field along a curve to find work could mean, bearing in mind that the result should be a scalar.

14.2   LINE INTEGRALS

▶ Defining the integral of a multivariate function along a curve

▶ Defining the integral of a vector field along a curve

▶ Using the Fundamental Theorem of Line Integrals

Integrals Along Curves in Space

In this section, we address both of the problems posed at the beginning of the chapter: how to find the area of a fence whose base is not the x-axis and how to compute work along a curve in space. We will also see an extension of the Fundamental Theorem of Calculus.

Since both of our motivating questions have to do with curves in space, we recall the discussion in Section 11.4 of parametrized curves in $\mathbb{R}^2$ and $\mathbb{R}^3$. In general, in order for us to be able to compute the arc length of a curve C, we need the curve to be described by a vector function $\mathbf{r}(t) = \langle x(t), y(t), z(t) \rangle$ such that $\mathbf{r} : [a, b] \rightarrow C$ is one-to-one, $\|\mathbf{r}'(t)\| \neq 0$ for all $t \in [a, b]$, and each of the component functions $x(t)$, $y(t)$, and $z(t)$ is differentiable and has a continuous first derivative. Throughout the rest of the chapter, we will call such curves and parametrizations *smooth*. For a smooth curve C with smooth parametrization $\mathbf{r}(t)$ and endpoints $\mathbf{r}(a)$, $\mathbf{r}(b)$, Theorem 11.18 gives the arc length $\mathcal{L}$ of C as

$$\mathcal{L}(a, b) = \int_a^b \|\mathbf{r}'(t)\| \, dt.$$

Line Integrals of Multivariate Functions

Our process for computing the area of a fence whose base is a smooth curve C uses the same ideas we found in Definition 4.9, when we defined the definite integral. When we computed the area under a curve on an interval, we used the "subdivide, approximate, and add" strategy. We saw that the area is the limit of the sum of areas of approximating subrectangles. More formally,

$$A = \lim_{n \to \infty} \sum_{k=1}^{n} (\text{height of } k\text{th rectangle} \times \text{width of } k\text{th rectangle}).$$

In Chapter 4 this equation became

$$A = \lim_{n \to \infty} \sum_{k=1}^{n} f(x_k^*) \Delta x = \int_a^b f(x) \, dx.$$

Here we want to say the same thing, adjusting for the fact that our curve C is smoothly parametrized by $\mathbf{r}(t)$ for $a \le t \le b$. So now the width of an approximating subrectangle is given by

$$\|\mathbf{r}'(t)\| = \sqrt{\left(\frac{dx}{dt}\right)^2 + \left(\frac{dy}{dt}\right)^2 + \left(\frac{dz}{dt}\right)^2},$$

and the height is given by $f(t_k^*)$, where, as before, t_k^* is an arbitrary value of t in the kth subdivision of $[a, b]$, as seen in the following figure:

Approximating the area of a function defined on a curve

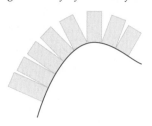

Combining these insights, we define the integral of a multivariate function along a curve in space.

DEFINITION 14.3

Line Integral of a Multivariate Function

Let C be a curve in $\mathbb{R}^3$ with a smooth parametrization $\mathbf{r}(t) = \langle x(t), y(t), z(t) \rangle$ for $t \in [a, b]$. Then the ***integral of*** $f(x, y, z)$ ***along*** C is

$$\int_C f(x, y, z) ds = \int_a^b f(x(t), y(t), z(t)) \| \mathbf{r}'(t) \| \, dt.$$

The definition for curves in $\mathbb{R}^2$ is analogous. They are the same as curves in $\mathbb{R}^3$, except that all z-components are zero. Also, note the new notation:

$$ds = \| \mathbf{r}'(t) \| \, dt.$$

The integral along C is sometimes referred to as the ***integral with respect to arc length***.

Line integrals of multivariate functions act like the single-variable integrals of Chapter 4 because, after the substitutions for $f(x, y, z)$ and $\| \mathbf{r}'(t) \|$ in terms of t, they *are* single-variable integrals. Note that if $f(x, y, z)$ is constantly equal to 1, the line integral returns just the arc length of C. This situation is parallel to that for single-variable integrals: The interval $[a, b]$ has length $b - a$, and

$$\int_a^b 1 \, dx = b - a.$$

Line integrals are additive, like single-variable integrals. That is, if $C = C_1 \cup C_2$, where C_1 and C_2 overlap in a single point, then

$$\int_C f(x, y, z) \, ds = \int_{C_1} f(x, y, z) \, ds + \int_{C_2} f(x, y, z) \, ds.$$

This property is particularly useful when a natural parametrization of C by $\mathbf{r}(t)$ has $\mathbf{r}'(t) = \mathbf{0}$ at a point. We can then rewrite the integral along C as a sum of two integrals that do not have this problem. If a curve C is not itself smooth, but can be written as a finite sum of smooth curves, we say that C is ***piecewise smooth***.

Another point of similarity between line integrals and single-variable integrals is that the direction of integration matters. Recall from Chapter 4 that

$$\int_a^b f(x) \, dx = - \int_b^a f(x) \, dx.$$

The same is true for the integral along a curve C with endpoints P and Q. If $\mathbf{r}_1(t)$ is a smooth parametrization of C that starts at P and ends at Q, and if $\mathbf{r}_2(t)$ is a smooth parametrization of C that starts at Q and ends at P, then the integral along C using $\mathbf{r}_1(t)$ will be the negative of the integral along C using $\mathbf{r}_2(t)$.

Line Integrals of Vector Fields

In Chapter 6 we discussed the concept of *work*. We now generalize this concept to include work done by a variable force acting along a curve in space. We represent the force acting in each of the **i**, **j**, and **k** directions as a vector field whose component functions vary with location in space:

$$\mathbf{F}(x, y, z) = F_1(x, y, z)\mathbf{i} + F_2(x, y, z)\mathbf{j} + F_3(x, y, z)\mathbf{k}.$$

Here $F_1(x, y, z)$ measures the action of $\mathbf{F}(x, y, z)$ in the **i** direction, $F_2(x, y, z)$ measures the action of $\mathbf{F}(x, y, z)$ in the **j** direction, and $F_3(x, y, z)$ measures the action of $\mathbf{F}(x, y, z)$ in the **k** direction.

To compute work done by a vector field along a curve, we follow our usual strategy of subdividing, approximating, and adding, followed by taking a limit as the size of the subdivisions approaches zero. In this case, the limit will be the integral of the vector field along the curve. Subdividing the curve uses the position vector $\langle x_n - x_{n-1}, y_n - y_{n-1}, z_n - z_{n-1}\rangle$ to approximate the curve between points (x_n, y_n, z_n) and $(x_{n-1}, y_{n-1}, z_{n-1})$. To compute the action of the vector field $\mathbf{F}(x, y, z)$ on this small piece of the curve, $F_1(x_n, y_n, z_n)$ is used to approximate F_1, the action of $\mathbf{F}(x, y, z)$ in the **i** direction for that subdivision. The components $F_2(x_n, y_n, z_n)$ and $F_3(x_n, y_n, z_n)$ are similarly used in the **j** and **k** directions. This approximation is a constant force in a fixed direction, so we multiply $F_1(x_n, y_n, z_n) \cdot (x_n - x_{n-1})$ to get the action of $\mathbf{F}(x, y, z)$ in the **i** direction. Also, summing in the **j** and **k** directions, we have

$$W \approx F_1(x_n, y_n, z_n) \cdot (x_n - x_{n-1}) + F_2(x_n, y_n, z_n) \cdot (y_n - y_{n-1}) + F_3(x_n, y_n, z_n) \cdot (z_n - z_{n-1})$$

as the approximate work done by $\mathbf{F}(x, y, z)$ on the nth subdivided piece of the curve. Adding the action along the n pieces of the subdivided curve, we obtain

$$\sum_{n=1}^{M} (F_1(x_n, y_n, z_n) \cdot (x_n - x_{n-1}) + F_2(x_n, y_n, z_n) \cdot (y_n - y_{n-1}) + F_3(x_n, y_n, z_n) \cdot (z_n - z_{n-1})).$$

As M approaches infinity, the change in x, y, and z becomes $\mathbf{r}'(t)$ and $\mathbf{F}(x_n, y_n, z_n)$ becomes the value of the field at a point; thus, the limit of the summation becomes

$$\int_a^b \left(F_1(x, y, z) \cdot \frac{\partial}{\partial x}\mathbf{r}(t) + F_2(x, y, z) \cdot \frac{\partial}{\partial y}\mathbf{r}(t) + F_3(x, y, z) \cdot \frac{\partial}{\partial z}\mathbf{r}(t) \right) dt$$

$$= \int_a^b \mathbf{F}(x, y, z) \cdot \mathbf{r}'(t)\, dt = \int_C \mathbf{F}(x, y, z) \cdot d\mathbf{r}.$$

This result motivates our definition of the line integral of a vector field.

DEFINITION 14.4   **Line Integral of a Vector Field**

Suppose that C is a smooth curve in $\mathbb{R}^3$ with a smooth parametrization $\mathbf{r}(t)$ for $t \in [a, b]$ and with a vector field $\mathbf{F}(x, y, z) = F_1\mathbf{i} + F_2\mathbf{j} + F_3\mathbf{k}$ whose component functions are each continuous on C and whose domain is open, connected, and simply connected. Then the *line integral of* $\mathbf{F}(x, y, z)$ *along* C is

$$\int_C \mathbf{F}(x, y, z) \cdot d\mathbf{r} = \int_a^b (F_1(x, y, z)x'(t) + F_2(x, y, z)y'(t) + F_3(x, y, z)z'(t))\, dt.$$

Two alternative ways of writing this integral are

$$\int_C \mathbf{F}(x, y, z) \cdot d\mathbf{r} = \int_C (F_1\, dx + F_2\, dy + F_3\, dz) = \int_C \mathbf{F} \cdot \mathbf{T}\, ds,$$

where $\mathbf{T}$ is the unit tangent vector. You are encouraged to check that all of the representations really are equivalent by carrying out the relevant substitutions and simplifications. For example, since the unit tangent vector is $\mathbf{T} = \dfrac{\mathbf{r}'(t)}{\|\mathbf{r}'(t)\|}$ and the differential is $ds = \|\mathbf{r}'(t)\|\,dt$, upon cancellation we see that $\int_C \mathbf{F} \cdot \mathbf{T}\,ds$ is equivalent to the line integral given in Definition 14.4.

As with the line integrals of multivariate functions discussed earlier in the section, changing the direction of travel along C changes the sign. If $\mathbf{r}(t)$ and $\mathbf{q}(t)$ are smooth parametrizations that traverse the curve C in opposite directions, then

$$\int_C \mathbf{F}(x, y, z) \cdot d\mathbf{r} = -\int_C \mathbf{F}(x, y, z) \cdot d\mathbf{q}.$$

The Fundamental Theorem of Line Integrals

Line integrals of vector fields can be complicated to compute in practice, depending on the curve C and the choice of $\mathbf{r}(t)$ to describe it. In the case of conservative vector fields, however, the fact that $\mathbf{F}(x, y, z)$ is a gradient allows us to reduce the line integral of the vector field to subtraction.

THEOREM 14.5

The Fundamental Theorem of Line Integrals

Let C be a smooth curve that is the graph of the vector function $\mathbf{r}(t)$ defined on the interval $[a, b]$ with $P = \mathbf{r}(a)$ and $Q = \mathbf{r}(b)$. If $\mathbf{F}(x, y, z)$ is a conservative vector field with $\mathbf{F}(x, y, z) = \nabla f(x, y, z)$ on an open, connected, and simply connected domain containing the curve C, then

$$\int_C \mathbf{F}(x, y, z) \cdot d\mathbf{r} = f(Q) - f(P).$$

Before we prove Theorem 14.5, note that $P = \mathbf{r}(a)$ and $Q = \mathbf{r}(b)$. These equations represent a common abuse of notation, because P and Q are points, but $\mathbf{r}(a)$ and $\mathbf{r}(b)$ are vectors. By this we mean that if P is the point (x_0, y_0, z_0), then $\mathbf{r}(a) = \langle x_0, y_0, z_0 \rangle$ and $f(P) = f(x_0, y_0, z_0)$.

Proof. If $\mathbf{F}(x, y, z) = \nabla f(x, y, z)$ and $\mathbf{r}(t)$ is a smooth parametrization of C with $\mathbf{r}(a) = P$ and $\mathbf{r}(b) = Q$, and if $\|\mathbf{r}'(t)\| \neq 0$ for all $t \in [a, b]$, then

$$\mathbf{F}(x, y, z) = \frac{\partial f}{\partial x}\mathbf{i} + \frac{\partial f}{\partial y}\mathbf{j} + \frac{\partial f}{\partial z}\mathbf{k} \qquad \text{and} \qquad \mathbf{r}'(t) = \frac{dx}{dt}\mathbf{i} + \frac{dy}{dt}\mathbf{j} + \frac{dz}{dt}\mathbf{k}.$$

Therefore,

$$\mathbf{F} \cdot d\mathbf{r} = \left(\frac{\partial f}{\partial x}\frac{dx}{dt} + \frac{\partial f}{\partial y}\frac{dy}{dt} + \frac{\partial f}{\partial z}\frac{dz}{dt} \right) dt.$$

By the chain rule, we have

$$\mathbf{F} \cdot d\mathbf{r} = \frac{d}{dt}(f(x(t), y(t), z(t)))\,dt = \frac{d}{dt}(f(t))\,dt.$$

Therefore,

$$\int_C \mathbf{F} \cdot d\mathbf{r} = \int_a^b \frac{d}{dt}(f(t))\,dt$$
$$= f(x(b), y(b), z(b)) - f(x(a), y(a), z(a))$$
$$= f(Q) - f(P).$$ ∎

Theorem 14.5 is very useful, since it allows us to avoid finding an explicit parametrization of C and many important vector fields in the sciences are conservative. In Exercise 65, you will show that a consequence of this theorem is that line integrals of conservative vector fields depend only on the endpoints of curves, not on the curves themselves. Such integrals are therefore said to be *independent of path*.

Examples and Explorations

EXAMPLE 1 | Area of a fence whose base is a curve in the plane

Find the area of a wall whose base is the part of the circle of radius 2 centered at the origin, lying in the first quadrant, and whose height at point (x, y) is given by $f(x, y) = 2x + y$.

SOLUTION

Following is an illustration of the wall, along with the plane defined by $z = 2x + y$:

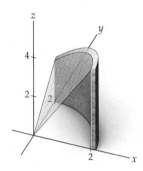

The quarter-circle base of the wall may be parametrized by $\mathbf{r}(t) = \langle 2\cos t, 2\sin t \rangle$, where $0 \leq t \leq \frac{\pi}{2}$, so

$$\|\mathbf{r}'(t)\| = \sqrt{4\sin^2 t + 4\cos^2 t} = 2.$$

We then substitute into the two-variable version of Definition 14.3:

$$\int_C f(x, y)\, ds = \int_a^b f(x(t), y(t)) \|\mathbf{r}'(t)\|\, dt$$

$$= \int_0^{\pi/2} (2(2\cos t) + 2\sin t) \cdot 2\, dt$$

$$= \left[8\sin t - 4\cos t \right]_0^{\pi/2} = 12.$$

As usual, when we are computing an area, it is important to remember that integration gives the *signed* area; thus, if $f(x, y)$ were negative at any point on C, we would need to compensate by integrating $|f(x, y)|$.   ☐

EXAMPLE 2 | Integrating a multivariate function along a curve in $\mathbb{R}^3$

Find

$$\int_C f(x, y, z)\, ds,$$

where C is the line segment from $(1, 4, 2\sqrt{3})$ to $(3, 7, 4\sqrt{3})$ and

$$f(x, y, z) = x + y + z.$$

SOLUTION

Recall that the line segment from the point (a, b, c) to the point (d, e, f) may be parametrized by

$$\mathbf{r}(t) = \langle a + t(d - a), b + t(e - b), c + t(f - c) \rangle, \quad \text{for } 0 \leq t \leq 1.$$

In this case we have

$$\mathbf{r}(t) = \langle 1 + 2t, 4 + 3t, 2\sqrt{3} + 2\sqrt{3}\,t \rangle \text{ for } t \in [0, 1].$$

So,

$$\mathbf{r}'(t) = \langle 2, 3, 2\sqrt{3} \rangle, \qquad \|\mathbf{r}'(t)\| = 5,$$

and on the curve,

$$f(x(t), y(t), z(t)) = 5 + 2\sqrt{3} + (5 + 2\sqrt{3})t.$$

Substituting gives

$$\int_C f(x, y, z)\, ds = \int_0^1 (5 + 2\sqrt{3} + (5 + 2\sqrt{3})t)5\, dt$$

$$= 5\left[(5 + 2\sqrt{3})t + \frac{1}{2}(5 + 2\sqrt{3})t^2\right]_0^1$$

$$= \frac{75}{2} + 15\sqrt{3}.$$

$\square$

EXAMPLE 3 **Computing line integrals of vector fields**

Find the line integral $\int_C \mathbf{F} \cdot d\mathbf{r}$ for each of the following vector fields along the given curve:

(a) $\mathbf{F}(x, y) = \langle -y, x \rangle$, where C_1 is the straight segment from $(3, 0)$ to $(8, \pi)$.

(b) $\mathbf{F}(x, y, z) = \langle e^y, xe^z, y^2 \rangle$, where C_2 is the curve given by $\mathbf{r}(t) = \langle t, 2, t^2 \rangle$, for $-2 \le t \le 2$.

SOLUTION

(a) The curve C_1 is smoothly parametrized by

$$\mathbf{r}(t) = \langle 3 + 5t, \pi t \rangle, \qquad 0 \le t \le 1.$$

Thus, $\mathbf{r}'(t) = \langle 5, \pi \rangle$. So,

$$\int_{C_1} \mathbf{F}(x, y) \cdot d\mathbf{r} = \int_0^1 \langle -\pi t, 3 + 5t \rangle \cdot \langle 5, \pi \rangle dt$$

$$= \int_0^1 3\pi\, dt = 3\pi.$$

(b) For C_2 and $\mathbf{F}(x, y, z)$,

$$\mathbf{r}(t) = \langle t, 2, t^2 \rangle, \qquad \text{for} \qquad -2 \le t \le 2.$$

Therefore, $\mathbf{r}'(t) = \langle 1, 0, 2t \rangle$. So,

$$\int_{C_2} \mathbf{F}(x, y, z) \cdot d\mathbf{r} = \int_{-2}^2 \langle e^2, te^{t^2}, 4 \rangle \cdot \langle 1, 0, 2t \rangle\, dt$$

$$= \int_{-2}^2 (e^2 + 8t)\, dt$$

$$= [e^2 t + 4t^2]_{-2}^2$$

$$= (2e^2 + 16) - (-2e^2 + 16) = 4e^2.$$

$\square$

EXAMPLE 4   **Applying the Fundamental Theorem of Line Integrals**

Use the Fundamental Theorem of Line Integrals to evaluate the following:

(a) The line integral

$$\int_{C_1} \mathbf{F}(x, y) \cdot d\mathbf{r},$$

where $\mathbf{F}(x, y) = \left\langle 6x^2 \ln(y+1), \frac{2x^3}{y+1} \right\rangle$ and C_1 is the straight segment from $(1, 1)$ to $(3, 7)$.

(b) The line integral

$$\int_{C_2} \mathbf{G}(x, y, z) \cdot d\mathbf{r},$$

where $\mathbf{G}(x, y, z) = \langle yze^{xyz}, xze^{xyz}, xye^{xyz} \rangle$ and C_2 is the circular helix about the z-axis given by

$$\mathbf{r}(t) = \langle \cos t, \sin t, t \rangle, \text{ for } t \in \left[\frac{\pi}{4}, \frac{3\pi}{4} \right].$$

SOLUTION

(a) The Fundamental Theorem of Line Integrals makes it quite easy to carry out these integrations; it is enough to verify that the vector field in question is conservative, compute the potential function, substitute according to the theorem, and subtract. In the case of the first field,

$$\frac{\partial}{\partial y}(6x^2 \ln(y+1)) = \frac{6x^2}{y+1} = \frac{\partial}{\partial x}\left(\frac{2x^3}{y+1} \right),$$

so $\mathbf{F}$ is conservative. Integrating, we find a potential function for $\mathbf{F}$:

$$f(x, y) = 2x^3 \ln(y+1).$$

By the Fundamental Theorem,

$$\int_{C_1} \mathbf{F}(x, y) \cdot d\mathbf{r} = f(3, 7) - f(1, 1) = 54 \ln 8 - 2 \ln 2 = 162 \ln 2 - 2 \ln 2 = 160 \ln 2.$$

(b) In the case of the second field, it is easier to integrate $g_1(x, y, z) = yze^{xyz}$ with respect to x and check that the resulting function can be adjusted as necessary to produce a potential function for $\mathbf{G}$. Integration yields

$$g(x, y, z) = e^{xyz}; \qquad \mathbf{G}(x, y, z) = \nabla g(x, y, z).$$

Therefore, $g(x(t), y(t), z(t)) = e^{t \cos t \sin t}$. Evaluating $\mathbf{r}(\pi/4)$ and $\mathbf{r}(3\pi/4)$ gives us $(\sqrt{2}/2, \sqrt{2}/2, \pi/4)$ as the starting point of C_2 and $(-\sqrt{2}/2, \sqrt{2}/2, 3\pi/4)$ as the endpoint of C_2. Then

$$\int_{C_2} \mathbf{G}(x, y, z) \cdot d\mathbf{r} = g\left(-\frac{\sqrt{2}}{2}, \frac{\sqrt{2}}{2}, \frac{3\pi}{4} \right) - g\left(\frac{\sqrt{2}}{2}, \frac{\sqrt{2}}{2}, \frac{\pi}{4} \right) = e^{-3\pi/8} - e^{\pi/8}.$$

□

EXAMPLE 5   **Evaluating line integrals of vector functions along closed curves**

Integrate each of the following vector fields along the specified closed curve C:

(a) $\mathbf{F}(x, y) = (y+2)\mathbf{i} + (x+3)\mathbf{j}$, where C is the unit circle in the plane, traversed counterclockwise starting and ending at $(1, 0)$.

(b) $\mathbf{G}(x, y) = y\mathbf{i} - y^2\mathbf{j}$, where C is the square with vertices $(1, 0)$, $(0, 1)$, $(-1, 0)$, and $(0, -1)$, traversed counterclockwise starting and ending at $(1, 0)$.

SOLUTION

(a) The vector field $\mathbf{F}(x, y)$ is conservative, since $\frac{\partial}{\partial y}(y + 2) = 1 = \frac{\partial}{\partial x}(x + 3)$, so we can use the Fundamental Theorem of Line Integrals. A potential function for $\mathbf{F}(x, y)$ is

$$f(x, y) = 2x + xy + 3y.$$

The unit circle is smoothly parametrized by $\mathbf{r}(t) = (\cos t)\mathbf{i} + (\sin t)\mathbf{j}$ for $t \in [0, 2\pi]$. So, $\mathbf{r}'(t) = \langle -\sin t, \cos t \rangle$. Since we are integrating over a closed curve, we expect to end where we started. As we know, when we evaluate $\mathbf{r}(t)$ at $t = 0$ and $t = 2\pi$, we start and stop at $(1, 0)$. Applying the Fundamental Theorem gives

$$\int_C \mathbf{F}(x, y) \cdot d\mathbf{r} = f(1, 0) - f(1, 0) = 0.$$

The zero value we obtain here is not a coincidence: In Exercise 64 you will prove that the integral of *any* conservative vector field along any closed curve is equal to zero.

(b) The vector field $\mathbf{G}(x, y)$ is not conservative, since $\frac{\partial}{\partial y}(y) = 1$ but $\frac{\partial}{\partial x}(y^2) = 0$. Because the vector field is not conservative, we may not use the Fundamental Theorem of Line Integrals. For this, piecewise-smooth, simple closed curve, we will perform a separate integration for each of the sides of the square, as indicated in the following figure:

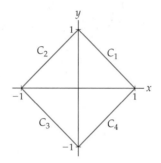

We parametrize each side of the square as follows:

- For side C_1 we use $x = 1 - t$ and $y = t$, where $0 \le t \le 1$.
- For C_2 we use $x = 1 - t$ and $y = 2 - t$, where $1 \le t \le 2$.
- For C_3 we use $x = -3 + t$ and $y = 2 - t$, where $2 \le t \le 3$.
- Finally, for C_4 we use $x = -3 + t$ and $y = -4 + t$, where $3 \le t \le 4$.

To evaluate the line integral, we have

$$\int_C \mathbf{G}(x, y) \cdot d\mathbf{r} = \int_{C_1} \mathbf{G}(x, y) \cdot d\mathbf{r} + \int_{C_2} \mathbf{G}(x, y) \cdot d\mathbf{r} + \int_{C_3} \mathbf{G}(x, y) \cdot d\mathbf{r} + \int_{C_4} \mathbf{G}(x, y) \cdot d\mathbf{r}.$$

On C_1 we have $\mathbf{G}(x(t), y(t)) = \langle t, -t^2 \rangle$ and $d\mathbf{r} = \langle -1, 1 \rangle dt$. Therefore,

$$\int_{C_1} \mathbf{G}(x, y) \cdot d\mathbf{r} = \int_0^1 \langle t, -t^2 \rangle \cdot \langle -1, 1 \rangle \, dt$$

$$= \int_0^1 (-t - t^2) \, dt$$

$$= \left[-\frac{1}{2}t^2 - \frac{1}{3}t^3 \right]_0^1 = -\frac{5}{6}.$$

On C_2 we have $\mathbf{G}(x(t), y(t)) = \langle 2 - t, -(2 - t)^2 \rangle$ and $d\mathbf{r} = \langle -1, -1 \rangle\, dt$. Hence,

$$\int_{C_2} \mathbf{G}(x, y) \cdot d\mathbf{r} = \int_1^2 \langle 2 - t, -(2 - t)^2 \rangle \cdot \langle -1, -1 \rangle\, dt$$

$$= \int_1^2 (t - 2 + (2 - t)^2)\, dt$$

$$= \left[\frac{1}{2}t^2 - 2t - \frac{1}{3}(2 - t)^3 \right]_1^2 = -\frac{1}{6}.$$

The integrals for sides C_3 and C_4 are similar. You should check that the values of these integrals are $-\frac{1}{6}$ and $-\frac{5}{6}$, respectively. Therefore,

$$\int_C \mathbf{G}(x, y) \cdot d\mathbf{r} = -\frac{5}{6} - \frac{1}{6} - \frac{1}{6} - \frac{5}{6} = -2$$

☐

| EXAMPLE 6 | **Computing the work done along a curve** |

Find the work done by the vector field $\mathbf{F}(x, y) = y^2\mathbf{i} - x^2\mathbf{j}$ along the curve parametrized by $x = t^2$, $y = t^3$, for $0 \le t \le 2$.

SOLUTION

To find the work, we compute the line integral $\int_C \mathbf{F}(x, y) \cdot d\mathbf{r}$, where $\mathbf{r}(t) = \langle t^2, t^3 \rangle$. Here we have $\mathbf{F}(x(t), y(t)) = \langle t^6, -t^4 \rangle$ and $d\mathbf{r} = \langle 2t, 3t^2 \rangle\, dt$. Thus,

$$\int_C \mathbf{F}(x, y) \cdot d\mathbf{r} = \int_0^2 \langle t^6, -t^4 \rangle \cdot \langle 2t, 3t^2 \rangle\, dt$$

$$= \int_0^2 (2t^7 - 3t^6)\, dt = \left[\frac{1}{4}t^8 - \frac{3}{7}t^7 \right]_0^2 = \frac{64}{7}.$$

☐

? TEST YOUR UNDERSTANDING

▶ How does the integral of a multivariate function along a smooth curve resemble a single-variable integral from Chapter 4? Why does it do so?

▶ What does ds stand for in terms of $\mathbf{r}(t)$?

▶ What does $d\mathbf{r}$ stand for in terms of $\mathbf{r}(t)$?

▶ Why does the integral of any conservative vector field along any smooth or piecewise-smooth curve depend only on the endpoints and not on the curve?

▶ Why does the integral of any conservative vector field along any smooth or piecewise-smooth closed curve always equal zero?

EXERCISES 14.2

Thinking Back

Average value: Review the average value formula from Section 4.6. Use the formula to compute the average value of the following functions.

▶ $g(x) = x^2 + x + 2$ on $[1, 5]$

▶ $f(x) = \cos x$ on $[\pi, 2\pi]$

Arc length: Review the discussion of arc length in Sections 6.3 and 11.4. Compute the length of the following curves.

▶ The plane curve $\langle t, t^2 \rangle$ from $t = 4$ to $t = 6$

▶ The circular helix given by $\mathbf{r}(t) = \cos z\mathbf{i} + \sin z\mathbf{j} + z\mathbf{k}$ from $z = 0$ to $z = 2\pi$

Concepts

0. *Problem Zero:* Read the section and make your own summary of the material.

1. *True/False:* Determine whether each of the statements that follow is true or false. If a statement is true, explain why. If a statement is false, provide a counterexample.

   (a) *True or False:* The integral of a multivariate function along a smooth curve C produces a scalar.

   (b) *True or False:* The integral of a vector field $\mathbf{F}(x, y)$ along a smooth curve C produces a vector.

   (c) *True or False:* A smooth parametrization $\mathbf{r}(t)$ must have $\mathbf{r}'(t) \neq 0$ for all points on the curve.

   (d) *True or False:* If a curve is not itself smooth, but can be written as a finite sum of smooth curves, it is possible to integrate either a multivariate function or a vector field along the curve by integrating over each smooth piece separately.

   (e) *True or False:* If a vector field is conservative, it cannot be integrated along any curve.

   (f) *True or False:* The Fundamental Theorem of Line Integrals applies to the integral of a conservative vector field along a smooth curve.

   (g) *True or False:* Every integral of a vector field along a smooth curve can be evaluated with the Fundamental Theorem of Line Integrals.

   (h) *True or False:* The integral of a vector field along a smooth closed curve is always zero.

2. *Examples:* Construct examples of the thing(s) described in the following. Try to find examples that are different than any in the reading.

   (a) A smooth parametrization of the straight line between two points in $\mathbb{R}^3$.

   (b) A smooth curve C that is not a straight line, and a smooth parametrization of C.

   (c) A vector field $\mathbf{F}(x, y)$ such that the integral of $\mathbf{F}(x, y)$ around the unit circle is zero.

3. Write, but do not evaluate, the line integral $\int_C f(x, y, z)\, ds$ as an integral explicitly in terms of t if $f(x, y, z) = e^{xyz}$ and $\mathbf{r}(t) = t\mathbf{i} + t^2\mathbf{j} + t^4\mathbf{k}$ for $a \leq t \leq b$.

4. Make a chart of all the new notation in this section, including what each item means in terms you already understand.

For each integral in Exercises 5–8, give the vector field that is being integrated.

5. $\displaystyle\int_C (x + y)\, dx + xy\, dy$

6. $\displaystyle\int_C \cos(x^2 y)\, dx - 3e^y\, dy$

7. $\displaystyle\int_C xy^2\, dx + (xy - z)\, dy + \cos y\, dz$

8. $\displaystyle\int_C xyz\, dx + \sin(zy)\, dy + (z - x)\, dz$

9. Write the first alternative form of $\int_C \mathbf{F}(x, y) \cdot d\mathbf{r}$ for $\mathbf{F}(x, y)$ and C_1 from Example 3. (This alternative form is described immediately after Definition 14.4 and is used in Exercises 5–8.)

10. Write the first alternative form of $\int_C \mathbf{G}(x, y, z) \cdot d\mathbf{r}$ for $\mathbf{G}(x, y, z)$ and C_2 from Example 3. (This alternative form is described immediately after Definition 14.4 and is used in Exercises 5–8.)

11. Translate ds into terms of $\mathbf{r}(t)$.

12. Translate $d\mathbf{r}$ into terms of $\mathbf{r}(t)$.

13. Write $\int_C \mathbf{F}(x, y) \cdot d\mathbf{r}$ explicitly as an integral of t, where $\mathbf{F}(x, y) = \langle x^2 y, x - y \rangle$ and $\mathbf{r}(t) = \langle 2t, e^t \rangle$ for $a \leq t \leq b$.

14. Write $\int_C \mathbf{F}(x, y.z) \cdot d\mathbf{r}$ explicitly as an integral of t, where $\mathbf{F}(x, y, z) = \langle zy - x, xz - y, xy - z \rangle$ and $\mathbf{r}(t) = \langle t, \cos t, \sin t \rangle$ for $a \leq t \leq b$.

15. Give a smooth parametrization $\mathbf{r}(t)$ for the unit circle, starting and ending at $(1, 0)$ and travelling in the counterclockwise direction.

16. Give a smooth parametrization $\mathbf{r}(t)$ for the unit circle, starting and ending at $(1, 0)$ and travelling in the clockwise direction.

17. Give a smooth parametrization $\mathbf{r}(t)$ for the circular helix with radius 2 and centered about the y-axis, starting at $(2, 0, 0)$ and completing two full revolutions.

18. Give a smooth parametrization $\mathbf{r}(t)$ for the straight line between the points $(\pi, e, 1)$ and $(3, 7, 13)$.

19. In Chapter 4 we defined the ***average value*** of an integrable function f on the interval $[a, b]$ to be $\frac{1}{b-a} \int_a^b f(x)\, dx$.

    Generalize this expression to a formula that will give the average value of $f(x, y, z)$ along a smooth curve C parametrized by $\mathbf{r}(t)$ for $a \leq t \leq b$.

20. Use your intuition from Exercise 19 to propose a formula for the average value of $\mathbf{F}(x, y, z)$ along a smooth curve C parametrized by $\mathbf{r}(t)$ for $a \leq t \leq b$.

Skills

In Exercises 21–28, evaluate the multivariate line integral of the given function over the specified curve.

21. $g(x, y) = x$, with C the graph of $y = x^2$ in the xy-plane from the origin to $(2, 4)$.

22. $f(x, y) = e^{xy}$, with C the line with equation $x = 3y$ for $-1 \leq y \leq 1$.

23. $f(x, y) = x^2 + y^2$, with C the unit circle traversed counterclockwise.

24. $f(x, y) = x^2 + y^2$, with C the unit circle traversed *clockwise*.

25. $f(x, y, z) = e^{x+y+z}$, with C the straight line segment from the origin to $(1, 2, 3)$.

26. $g(x, y, z) = xyz$, with C the curve parametrized by $\mathbf{r}(t) = \left\langle \frac{\sqrt{2}}{3} t^3, t^2, t \right\rangle$ for $1 \le t \le 4$.

27. $f(x, y, z) = e^{x^2 + y + z^2}$, with C the circular helix of radius 1, centered about the z-axis, and parametrized by $\mathbf{r}(t) = \langle \cos t, \sin t, t \rangle$ from height 0 to π.

28. $f(x, y, z) = e^{x^2 + y^2 + z^2}$, with C the circular helix of radius 1, centered about the y-axis, and parametrized by $\mathbf{r}(t) = \langle \cos t, t, \sin t \rangle$ for $\pi \le y \le 3\pi$.

Evaluate each of the vector field line integrals in Exercises 29–36 over the indicated curves.

29. $\mathbf{F}(x, y) = \mathbf{i} - \mathbf{j}$, with C the curve with equation $y - x^2 = 1$ for $5 \le x \le 10$.

30. $\mathbf{F}(x, y) = y\,\mathbf{i} - x\,\mathbf{j}$, with C the circle of radius 2, centered at the origin, and traversed counterclockwise starting at $(2, 0)$.

31. $\mathbf{F}(x, y)$ and C are as in Exercise 29, but C is traversed in the reverse direction, from $x = 10$ to $x = 5$.

32. $\mathbf{F}(x, y, z) = 2^z\,\mathbf{i} + \ln x\,\mathbf{j} + xz\,\mathbf{k}$, with C the curve parametrized by $\langle t, \ln t, t \rangle$ for $1 \le t \le e^2$.

33. $\mathbf{F}(x, y) = y \sin xy\,\mathbf{i} + x \cos xy\,\mathbf{j}$, with C the cardioid given by $r = 1 + 2\cos\theta$ from $\theta = 0$ to $\theta = 2\pi$.

34. $\mathbf{F}(x, y, z) = yz\,\mathbf{i} + x\,\mathbf{j} + z^2\,\mathbf{k}$, with C the straight line segment from the origin to $(1, 0, 4)$.

35. $\mathbf{F}(x, y, z) = -x^2 y\,\mathbf{i} + x^3\,\mathbf{j} + y^2\,\mathbf{k}$, with C the circular helix parametrized by $x = \cos t$, $y = \sin t$, $z = t$, for $2\pi \le t \le 3\pi$.

36. $\mathbf{F}(x, y, z) = \frac{2}{1-z}\,\mathbf{i} + \mathbf{j} - \frac{1}{z^2 + 1}\,\mathbf{k}$, with C the curve parametrized by $x = \ln(1 + t)$, $y = \ln(1 - t^2)$, $z = t$, for $3 \le t \le 7$.

Use the Fundamental Theorem of Line Integrals, if applicable, to evaluate the integrals in Exercises 37–44. Otherwise, show that the vector field is not conservative.

37. $\mathbf{F}(x, y) = \left\langle \frac{1}{x} + \ln y, \frac{1}{y} + \frac{x}{y} \right\rangle$, with C the straight line segment from (π, e) to $(1, \pi)$.

38. $\mathbf{F}(x, y) = \langle 2^x \ln 2, 2y \rangle$, with C the straight line segment from $(3, 7)$ to $(0, 1)$.

39. $\mathbf{F}(x, y, z) = \langle -z, 1, x \rangle$, with C the circular helix given by $x = \cos t$, $y = t$, $z = \sin t$, for $0 \le t \le 2\pi$.

40. $\mathbf{F}(x, y) = \left\langle \frac{xe^x}{(x+y+4)^2}, \frac{-e^x}{(x+y+4)^2} \right\rangle$, with C the cardioid given by $r = (1 + 2\sin\theta)$ from $\theta = 0$ to $\theta = \pi$.

41. $\mathbf{F}(x, y, z) = yz^{xy} \ln z\,\mathbf{i} + xz^{xy} \ln z\,\mathbf{j} + \frac{z^{xy}}{z}\,\mathbf{k}$, with C any curve from $(0, 0, 1)$ to $(2, 16, 3)$.

42. $\mathbf{F}(x, y, z) = (\cos(zy) - \sin(zy))\,\mathbf{i} - (xz \sin(zy) + xz \cos(zy))\,\mathbf{j} - (xy \sin(zy) + xy \cos(zy))\,\mathbf{k}$, with C any curve from $\left(5, \frac{\pi}{2}, \frac{\pi}{3} \right)$ to $(2, 0, 2\pi)$.

43. $\mathbf{F}(x, y, z) = z \sin x\,\mathbf{i} + x \ln(y + 4)\,\mathbf{j} + y\,\mathbf{k}$, with C the unit circle traversed counterclockwise starting at the point $(1, 0)$.

44. $\mathbf{F}(x, y, z) = (z^3 + 1)\,\mathbf{i} + x \cos z\,\mathbf{j} + xye^{xyz}\,\mathbf{k}$, with C the curve parametrized by $\mathbf{r}(t) = t^2\mathbf{i} + t^3\mathbf{j} - t\mathbf{k}$ for $0 \le t \le 4$.

Evaluate the line integrals in Exercises 45–50.

45. $\int_C \mathbf{F}(x, y) \cdot d\mathbf{r}$, where $\mathbf{F}(x, y) = -y\,\mathbf{i} + x\,\mathbf{j}$ and C is the spiral $x = t \cos t$, $y = t \sin t$, for $\pi \le t \le 2\pi$.

46. $\int_C \mathbf{F}(x, y, z) \cdot d\mathbf{r}$, where
$$\mathbf{F}(x, y, z) = \frac{y^2}{z}\mathbf{i} + \frac{2xy}{z}\mathbf{j} - \frac{xy^2}{z^2}\mathbf{k}$$
and C is the portion of the conical helix $x = t \cos t$, $y = t \sin t$, $z = t$, for $t = \pi$ to $t = \frac{3\pi}{2}$.

47. $\int_C f(x, y)\, ds$, where $f(x, y) = (x^2 + y^2)^{1/2}$ and C is the curve parametrized by $x = t \sin t$, $y = t \cos t$, for $0 \le t \le 4\pi$.

48. $\int_C \mathbf{F}(x, y, z) \cdot d\mathbf{r}$, where
$$\mathbf{F}(x, y, z) = \frac{\ln(z + 4)}{x}\mathbf{i} + xyz\,\mathbf{j} + y\,\mathbf{k}$$
and C is the curve parametrized by $x = 4t$, $y = t^2$, $z = 1 - t$, for $0 \le t \le 1$.

49. $\int_C \mathbf{F}(x, y, z) \cdot d\mathbf{r}$, where
$$\mathbf{F}(x, y, z) = (yze^{xyz} + 2)\,\mathbf{i} + (xze^{xyz} - 1)\,\mathbf{j} + xye^{xyz}\,\mathbf{k}$$
and C is the curve of intersection of the surface $z = \sqrt{x^2 + y^2}$ and the plane $z - x + y = 10$.

50. $\int_C f(x, y, z)\, ds$, where $f(x, y, z) = (x + z)3^y$ and C is the curve parametrized by $x = 2 - t$, $y = 4t$, $z = t + 5$, for $t = 1$ to $t = 4$.

Applications

In Exercises 51–54, assume that a thin wire in space follows the given curve C and that the density of the wire at any point on C is given by $\rho(x, y, z)$. Find the mass of the wire by computing the line integral of the density function along the specified curve.

51. C is the portion of the unit circle that lies in the first quadrant, and $\rho(x, y) = \ln(e^2 \sqrt{x^2 + y^2})$.

52. C is the straight line through $(5, -10)$ and $(0, 2)$, and $\rho(x, y) = e^{x+y}$.

53. C is parametrized by $\mathbf{r}(t) = \left\langle 2t + 1, 10 - t, \frac{t^2}{2} \right\rangle$ from $t = 0$ to $t = 1$, and $\rho(x, y, z) = (x + 2y)\sqrt{2z + 5}$.

54. C is the conical helix parametrized by $\mathbf{r}(t) = \langle t \cos t, t \sin t, t \rangle$ from $t = 0$ to $t = 3\pi$, and $\rho(x, y, z) = \sqrt{x^2 + y^2 + z^2}$.

In Exercises 55–58, find the work done by the given vector field **F** along the specified curve C.

55. $F(x, y) = y\mathbf{i} - x\mathbf{j}$, where C is the portion of the circle of radius 2, centered at the origin, and in the first quadrant, traversed counterclockwise from $(2, 0)$ to $(0, 2)$.

56. $F(x, y) = xe^y\mathbf{i} + 2y\mathbf{j}$, where C is the curve $y = \ln x$ from $(e, 1)$ to $(e^2, 2)$.

57. $F(x, y, z) = x\ln y\mathbf{i} + z\mathbf{j} + z^2\mathbf{k}$, where C is the curve parametrized by $\langle t, e^t, e^{-t}\rangle$ from $(1, e, 1/e)$ to $(\ln 2, 2, 1/2)$.

58. $F(x, y, z) = yz\mathbf{i} + xz\mathbf{j} + xy\mathbf{k}$, where C is the curve that is on the hyperbolic saddle and is parametrized by $\langle \cos t, \sin t, \cos^2 t - \sin^2 t\rangle$ for $0 \le t \le \dfrac{\pi}{4}$.

For a function $f(x, y)$ defined along a curve C in the plane, we may define the **average value** of f on C to be the integral of f along C, divided by the length of C:

$$f_{\text{avg}}(C) = \frac{\int_C f(x, y)\, ds}{\int_C ds}.$$

A similar definition applies to functions of three variables. (See also Exercise 19.) In Exercises 59 and 60, use these definitions to find the average value of the specified function along the given curve.

59. $f(x, y) = \sin^{-1} y \cos^{-1} x$, with C the portion of the unit circle that lies in the first quadrant.

60. $g(x, y, z) = xze^{xyz}$, with C the straight line segment from the origin to the point $(1, 1, 1)$.

61. If C is a smooth closed curve in the plane, the **flux** of a vector field $F(x, y)$ across C measures the net flow out of the region enclosed by C and is defined to be

$$\int_C F(x, y) \cdot \mathbf{n}\, ds,$$

where $\mathbf{n}$ is a unit vector that is perpendicular to C and points in the "outward" direction. Find the flux of $F(x, y) = xe^{x^2+y^2}\mathbf{i} + ye^{x^2+y^2}\mathbf{j}$ across the boundary of the unit circle. (On the unit circle, $\mathbf{n} = \langle x, y\rangle$.)

62. Ian is planning a trip into the Wind River Range of mountains in Wyoming. He will carry ice gear, rock gear, and backpacking gear to a total of 80 pounds. He will follow a trail so that his position is given by

$$\langle 0.7t^2 + 1, 1.5t(2 - t), 0.1t^3\rangle \text{ for } t \in [0, 1].$$

Distances are given in miles. How much work does Ian have to do to carry his pack on this trail?

63. Annie is making a crossing from one island to another, with a strong current in the channel. Her path is given by $\mathbf{r}(t) = \langle t, 0.2t(1.2 - t)\rangle$ for $t \in [0, 1.2]$, while the velocity of the current is $V = \langle 0, 0.9x(1.2 - x)\rangle$, with the times in hours and the speeds in miles per hour.

(a) Is the velocity of the current a conservative vector field?

(b) Evaluate the line integral of the velocity of the current along the curve of Annie's paddling path.

(c) What is the physical significance of the line integral?

Proofs

64. Prove that if **F** is a conservative vector field, then the line integral of **F** along any smooth closed curve C is zero.

65. Prove that if $F(x, y, z)$ is a conservative vector field, then line integrals of $F(x, y, z)$ are independent of path.

Thinking Forward

Integration across a surface: In the next section, we will use the results of this section to enable us to deal with integrals on surfaces. We already know some of the basics, developed in Chapter 11.

▶ Our original computation of the arc length of a space curve in Chapter 11 used the "subdivide, approximate, and add" strategy and then took a limit, to end up with $\int \|\mathbf{r}(t)\|\, dt = \int ds$ as the integral used to compute arc length. How could this construction be modified to compute the area of a surface S that is not part of the xy-plane?

▶ What should be added to the answer to the previous problem to compute the average value of a function $f(x(t), y(t), z(t))$ across a smooth surface S?

▶ How could the integral of a vector field along a curve that we developed in this section be generalized to integrate a vector field across a surface?

14.3   SURFACES AND SURFACE INTEGRALS

▶ Using parametric equations to define surfaces

▶ Computing the integral of a multivariate function on a surface

▶ Using an integral to compute the flow through a surface

Parametrized Surfaces

In this section we will address the question of how to compute double integrals over surfaces that do not lie in a coordinate plane, and we will see how to integrate a vector field across a surface. In essence, this section generalizes double integrals in the same way that the previous section generalized integrals of functions of one variable. Surface area in the context of integrating over surfaces plays a role analogous to that of arc length in the context of line integrals: In both cases, the basic size of the object is calculated.

The surfaces we are most familiar with are those that are the graphs of functions of two variables, $z = f(x, y)$. One such surface, given by $z = x^2 - y^2$ for $x \in [-1, 3]$ and $y \in [-1, 3]$, is the following:

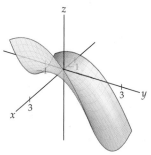

As we have done in other contexts, we will move back and forth between expressing a surface S in terms of points and in terms of position vectors.

Some surfaces cannot be expressed in the form $z = f(x, y)$ but may still be obtained by distorting a portion of a plane while preserving its two-dimensionality. In general, a surface is the image of a region of the uv-plane under a change of variables like those we studied in Section 13.7. Just as a curve C is called a parametrized curve when described by $\mathbf{r}(t) = x(t)\mathbf{i} + y(t)\mathbf{j} + z(t)\mathbf{k}$, a surface S given by

$$\mathbf{r}(u, v) = x(u, v)\mathbf{i} + y(u, v)\mathbf{j} + z(u, v)\mathbf{k},$$

for points $(u, v) \in D$, where D is a region in the uv-plane, is known as a **_parametrized surface_**. Typically D is a relatively simple subset of the uv-plane, such as a rectangle or a circle. For instance, consider the half of the unit sphere in $\mathbb{R}^3$ that lies on the nonnegative side of the xz-plane:

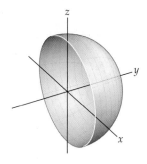

This surface can be parametrized by

$$\mathbf{r}(u, v) = \langle \sin u \sin v, \cos u, \sin u \cos v \rangle, \text{ where } u \in \left[0, \frac{\pi}{2}\right] \text{ and } v \in [0, 2\pi].$$

Since we are interested in doing calculus, we want to deal only with smooth surfaces and parametrizations. A parametrization

$$\mathbf{r}(u, v) = x(u, v)\mathbf{i} + y(u, v)\mathbf{j} + z(u, v)\mathbf{k} \text{ for } (u, v) \in D$$

is **smooth** if it is differentiable, if it has a continuous first derivative, and if $\nabla \mathbf{r}(u, v) \neq \mathbf{0}$ for all $(u, v) \in D$. A surface S is smooth if it has a smooth parametrization. We will also work only with surfaces that are **orientable**; that is, surfaces for which there is a consistent choice of normal vector $\mathbf{n}$. Planes and spheres have this property. The Möbius band is an example of a **nonorientable** surface, since there is no consistent choice of normal vector.

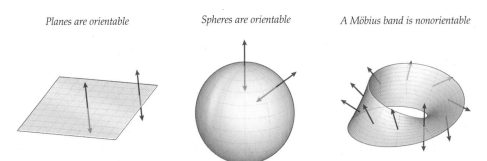

Planes are orientable   *Spheres are orientable*   *A Möbius band is nonorientable*

Nonorientable surfaces will not be discussed in this book, although they are interesting in their own right.

Surface Area and Surface Integrals of Multivariate Functions

If we restrict our attention to smooth surfaces, we can use the "subdivide, approximate, and add" strategy first to approximate surface area and then to take the limit of increasingly fine approximations. This limit will result in an integration that yields the exact surface area.

Just as we approximated curves by their tangent vectors, we approximate surfaces by their tangent planes. Recall from Section 12.3 that the vectors $\langle 1, 0, f_x(x_0, y_0) \rangle$ and $\langle 0, 1, f_y(x_0, y_0) \rangle$ both lie in the plane tangent to the graph of $z = f(x, y)$ at the point $(x_0, y_0, f(x_0, y_0))$.

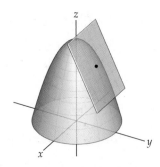

Since these vectors lie in the tangent plane, the parallelogram they determine is a piece of the tangent plane. We will use such parallelograms to approximate the area of the surface.

Starting with the familiar case, suppose we are interested in the area of the surface S described by

$$x\mathbf{i} + y\mathbf{j} + z(x, y)\mathbf{k}$$

for $(x, y) \in D \subseteq \mathbb{R}^2$, with $z(x, y)$ smooth on D. By partitioning D into n subregions and choosing (x_n, y_n) in each region, we can use the area of the parallelogram with sides $\langle 1, 0, z_x(x_n, y_n)\rangle \Delta x$ and $\langle 0, 1, z_y(x_n, y_n)\rangle \Delta y$,

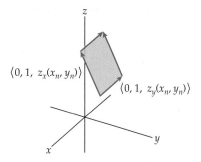

as shown in the preceding figure, to approximate the area of the part of the surface that lies over the nth subregion:

$$A_n \approx \left\| \langle 1, 0, z_x(x_n, y_n)\rangle \Delta x \times \langle 0, 1, z_y(x_n, y_n)\rangle \Delta y \right\|$$

$$\approx \sqrt{(z_x(x_n, y_n))^2 + (z_y(x_n, y_n))^2 + 1}\, \Delta x \Delta y.$$

Adding approximations over the subregions, we have

$$\text{Surface area} \approx \sum_{n=1}^{M} \sqrt{(z_x(x_n, y_n))^2 + (z_y(x_n, y_n))^2 + 1}\, \Delta x \Delta y.$$

The limit as the sizes of the subregions approach zero gives the exact surface area:

$$\text{Surface area} = \int_D \sqrt{(z_x)^2 + (z_y)^2 + 1}\, dA$$

$$= \iint_S \sqrt{(z_x)^2 + (z_y)^2 + 1}\, dy\, dx.$$

If instead S is a general surface parametrized by $\mathbf{r}(u, v) = \langle x(u, v), y(u, v), z(u, v)\rangle$, then the approximation is done with the formulas

$$\mathbf{r}_u(u, v) = x_u\mathbf{i} + y_u\mathbf{j} + z_u\mathbf{k} \quad \text{and} \quad \mathbf{r}_v(u, v) = x_v\mathbf{i} + y_v\mathbf{j} + z_v\mathbf{k},$$

both of which lie in the plane tangent to S at the point $(x(u, v), y(u, v), z(u, v))$. Substituting $\|\mathbf{r}_u \times \mathbf{r}_v\|$ for $\|\langle 1, 0, z_x(x_n, y_n)\rangle \times \langle 0, 1, z_y(x_n, y_n)\rangle\|$ in the approximation for A_n, and dA for $\Delta x \Delta y$ yields

$$\text{Surface area} = \int_D \|\mathbf{r}_u(x, y, z) \times \mathbf{r}_v(x, y, z)\|\, dA.$$

This equation motivates our definition of the area of a surface and simplifies some notation as well.

DEFINITION 14.6   Surface Area

The **surface area** of a smooth surface S is

$$\int_S 1 \, dS.$$

(a) If S is given by $z = f(x, y)$ for $(x, y) \in D \subseteq \mathbb{R}^2$, then

$$dS = \sqrt{\left(\frac{\partial z}{\partial x}\right)^2 + \left(\frac{\partial z}{\partial y}\right)^2 + 1} \; dA.$$

(b) If S is parametrized by $\mathbf{r}(u, v)$ for $(u, v) \in D$, then

$$dS = \|\mathbf{r}_u \times \mathbf{r}_v\| \, dA$$
$$= \|\mathbf{r}_u \times \mathbf{r}_v\| \, du \, dv.$$

Once we know how to find the area of a surface, it is easy to see what we ought to do to integrate a function $f(x, y, z)$ on that surface: multiply the area integrand dS by $f(x, y, z)$, and integrate. Then the integral of a function $f(x, y, z)$ over a surface S is

$$\int_S f(x, y, z) \, dS$$

The motivation for this definition of the integral of a function over a surface is the same as our motivation for the integral of a function along a curve. In each case, the integrand becomes

$$(\text{value of function}) \times (\text{size of region}),$$

where the size of the region is a length or an area, depending on the dimension. Thus, our definition of the integral of a multivariate function over a surface is as follows:

DEFINITION 14.7   Surface Integral of a Multivariate Function

The **integral of** $f(x, y, z)$ **over a smooth surface** S is

$$\int_S f(x, y, z) \, dS = \iint_D f(x(u, v), y(u, v), z(u, v)) \|\mathbf{r}_u \times \mathbf{r}_v\| \, dA$$

$$= \iint_D f(x(u, v), y(u, v), z(u, v)) \|\mathbf{r}_u \times \mathbf{r}_v\| \, du \, dv,$$

where D is the domain of the smooth parametrization of S by $\mathbf{r}(u, v)$.

This definition is more tricky in practice than it appears, since S may be quite complex. One of the attractive features of the Divergence Theorem, which appears later in this chapter, is that it provides an alternative way of evaluating some surface integrals.

The Flux of a Vector Field Across a Surface

Integrating a vector field over a surface is similar to integrating a vector field along a curve, although in the case of a surface there is a conceptual change beyond the increase in dimension. The interaction of vector field and surface of most general interest is the *flux* through the surface; that is, the measurement of what is transferred through the surface by the action of the vector field. As an example, consider the passage of fluid through a membrane. If we know the rates and directions of flow at every point on the surface, we should be able to find the total flow through the surface by integrating the rates of flow at every point on the surface. We are thus summing the action of vector field $\mathbf{F}(x, y, z)$ through the surface, in the direction normal to the surface at that point. Because we are interested only in the direction that is perpendicular, and not a magnitude, we will represent the normal direction by a unit normal vector $\mathbf{n}$. Because we have restricted our attention to surfaces that are orientable, any surface we consider has a continuous and well-defined unit normal vector. We need only to decide in which direction we want to travel across the surface. For instance, the vectors $\mathbf{k}$ and $-\mathbf{k}$ are both unit vectors perpendicular to the xy-plane; to evaluate the flux of a vector field through the xy-plane, as shown in the figure that follows, we would need to know which direction we were interested in, $\mathbf{k}$ or $-\mathbf{k}$.

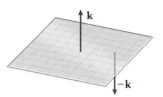

Now that we have resolved to count the activity of $\mathbf{F}(x, y, z)$ in the direction normal to S, our usual strategy of "subdivide, approximate, and add" will involve subdividing the surface into small subregions, then choosing an arbitrary point (x_n, y_n, z_n) in each subregion, and approximating the total flow through that region by computing $\mathbf{F}(x_n, y_n, z_n) \cdot \mathbf{n}$, where $\mathbf{n}$ is the unit normal vector at (x_n, y_n, z_n).

DEFINITION 14.8

Flux of a Vector Field Across a Surface

If $\mathbf{F}(x, y, z)$ is a continuous vector field defined on an oriented surface S given by $\mathbf{r}(u, v)$ for $(u, v) \in D$ with unit normal vector $\mathbf{n}$, then the *flux* of $\mathbf{F}(x, y, z)$ through S is

$$\int_S \mathbf{F}(x, y, z) \cdot \mathbf{n}\, dS.$$

(a) If S is the graph of $z = z(x, y)$, then

$$\int_S \mathbf{F}(x, y, z) \cdot \mathbf{n}\, dS = \iint_D (\mathbf{F}(x, y, z) \cdot \mathbf{n}) \|z_x \times z_y\|\, dy\, dx.$$

(b) If S is parametrized by u and v, then

$$\int_S \mathbf{F}(x, y, z) \cdot \mathbf{n}\, dS = \iint_D (\mathbf{F}(x, y, z) \cdot \mathbf{n}) \|\mathbf{r}_u \times \mathbf{r}_v\|\, du\, dv.$$

Note that $\mathbf{n}$, being a unit normal vector to S, satisfies

$$\mathbf{n} = \pm \frac{\mathbf{r}_u \times \mathbf{r}_v}{\|\mathbf{r}_u \times \mathbf{r}_v\|}.$$

We might be tempted to cancel the magnitudes and simply integrate $\mathbf{F} \cdot (\mathbf{r}_u \times \mathbf{r}_v)$, but this is correct only if $\mathbf{r}_u \times \mathbf{r}_v$ happens to point in the desired direction.

Examples and Explorations

EXAMPLE 1

Computing surface areas

Find the areas of the following surfaces:

(a) The portion of the plane $z = 3x + 5y$ that lies above the rectangle $[0, 1] \times [1, 5]$ in the xy-plane.

(b) The unit sphere.

SOLUTION

(a) The portion of the plane we seek is the graph of the function $f(x, y) = 3x + 5y$ with $f_x = 3$ and $f_y = 5$. Then

$$dS = \sqrt{9 + 25 + 1} \, dA = \sqrt{35} \, dA.$$

So,

$$\int_S dS = \int_1^5 \int_0^1 \sqrt{35} \, dx \, dy = 4\sqrt{35}.$$

(b) The unit sphere is parametrized by

$$\mathbf{r}(u, v) = \langle \cos u \sin v, \sin u \sin v, \cos v \rangle \quad \text{for } u \in [0, 2\pi] \text{ and } v \in [0, \pi].$$

Thus,

$$\mathbf{r}_u = \langle -\sin u \sin v, \cos u \sin v, 0 \rangle \quad \text{and} \quad \mathbf{r}_v = \langle \cos u \cos v, \sin u \cos v, -\sin v \rangle.$$

You should verify that

$$\|\mathbf{r}_u \times \mathbf{r}_v\| = \sin v.$$

So,

$$\int_S dS = \int_0^{2\pi} \int_0^{\pi} \sin v \, dv \, du = 4\pi.$$

$\square$

 **CHECKING THE ANSWER**

(a) The surface here is a parallelogram whose four vertices are $(0, 1, f(0, 1)) = (0, 1, 5)$, $(1, 1, f(1, 1)) = (1, 1, 8)$, $(1, 5, f(1, 5)) = (1, 5, 28)$, and $(0, 5, f(0, 5)) = (0, 5, 25)$. This parallelogram is congruent to the parallelogram determined by the vectors $\langle 1, 0, 3 \rangle$ and $\langle 0, 4, 20 \rangle$. As we saw in Section 10.4, the area of the latter parallelogram is $\|\langle 1, 0, 3 \rangle \times \langle 0, 4, 20 \rangle\|$. You may check that the value of this norm is also $4\sqrt{35}$.

(b) The surface area of a sphere with radius R is given by $4\pi R^2$. Since the radius was 1 in our example, we obtained the correct result.

EXAMPLE 2

Integrating a multivariate function on a surface

Integrate the indicated functions over the accompanying surfaces:

(a) The function $g(x, y, z) = 4x^2 + 4y^2 + 1$ over the portion of the surface S defined by $z = 1 - x^2 - y^2$ and that lies above the unit disk in the xy-plane.

(b) The function $h(x, y, z) = e^{x-2z}$, where S is the surface parametrized by

$$\mathbf{r}(u, v) = \langle 2u + 3v, 7v, u + 5v \rangle \quad \text{for } 0 \leq u \leq \pi \text{ and } 2 \leq v \leq 4.$$

SOLUTION

(a) This surface is the graph of z as a function of x and y, where $\frac{\partial z}{\partial x} = -2x$ and $\frac{\partial z}{\partial y} = -2y$, so we have

$$dS = \sqrt{4x^2 + 4y^2 + 1}\, dA$$

and

$$\int_S g(x, y, z)\, dS = \iint_{\text{unit disk}} (4x^2 + 4y^2 + 1)\sqrt{4x^2 + 4y^2 + 1}\, dA.$$

It is easier to evaluate this integral with polar coordinates. Recall that $x^2 + y^2 = r^2$ and $dA = r\, dr\, d\theta$. Since we are integrating above the unit disk, we have

$$\int_0^{2\pi} \int_0^1 (4r^2 + 1)\sqrt{4r^2 + 1}\, r\, dr\, d\theta = \int_0^{2\pi} \int_0^1 r(4r^2 + 1)^{3/2}\, dr\, d\theta$$

$$= \int_0^{2\pi} \frac{1}{20}\big[(4r^2 + 1)^{5/2}\big]_{r=0}^{r=1}\, d\theta$$

$$= \int_0^{2\pi} \frac{5^{5/2} - 1}{20}\, d\theta = \frac{\pi(25\sqrt{5} - 1)}{10}.$$

(b) This surface is parametrized by $\mathbf{r}(u, v) = \langle 2u + 3v, 7v, u + 5v \rangle$. We have

$$\mathbf{r}_u = \langle 2, 0, 1 \rangle \quad \text{and} \quad \mathbf{r}_v = \langle 3, 7, 5 \rangle.$$

Then

$$dS = \|\mathbf{r}_u \times \mathbf{r}_v\|dA = \|\langle -7, -7, 14 \rangle\|\, dA = 7\sqrt{6}\, dA.$$

We rewrite h in terms of the parameters u and v, giving

$$h(x(u, v), y(u, v), z(u, v)) = e^{(2u+3v)-2(u+5v)} = e^{-7v}.$$

So,

$$\int_S h(x, y, z)\, dS = 7\sqrt{6} \int_D e^{-7v}\, dA$$

$$= 7\sqrt{6} \int_0^{\pi} \int_2^4 e^{-7v}\, dv\, du$$

$$= 7\sqrt{6} \int_0^{\pi} \Big[-\frac{1}{7}e^{-7v}\Big]_2^4\, du$$

$$= \sqrt{6} \int_0^{\pi} (e^{-14} - e^{-28})\, du = \frac{\sqrt{6}\pi(e^{14} - 1)}{e^{28}}.$$
□

EXAMPLE 3   The mass of a lamina

Recall that a lamina is a two-dimensional object with a density function ρ and that the mass is the integral of the density function over the region occupied by the lamina.

Find the mass of each lamina:

(a) S is the portion of the cone $z = \sqrt{x^2 + y^2}$ that lies above the disk of radius 3 in the xy-plane and centered at the origin. S has the constant density function $\rho(x, y, z) = k$.

(b) S is the same surface, but with density function $\rho(x, y, z) = x^2 + y^2 + z^2$.

SOLUTION

(a) This surface, shown next, is a graph of z in terms of x and y, with

$$\frac{\partial z}{\partial x} = \frac{x}{\sqrt{x^2 + y^2}} \quad \text{and} \quad \frac{\partial z}{\partial y} = \frac{y}{\sqrt{x^2 + y^2}}.$$

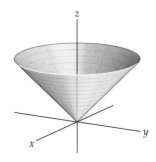

So,

$$dS = \sqrt{\frac{x^2}{x^2 + y^2} + \frac{y^2}{x^2 + y^2} + 1} \, dA = \sqrt{2} \, dA.$$

The mass of this lamina is given by

$$\int_S k \, dS = \int_D k\sqrt{2} \, dA,$$

where D is the circle with radius 3 and centered at the origin. Because of the symmetry of the lamina, it is convenient to integrate with polar coordinates. We have

$$\int_0^{2\pi} \int_0^3 k\sqrt{2} \, r \, dr \, d\theta = \int_0^{2\pi} \left[\frac{k\sqrt{2}r^2}{2} \right]_0^3 \, d\theta$$

$$= \int_0^{2\pi} \frac{9k\sqrt{2}}{2} \, d\theta = 9k\sqrt{2}\pi.$$

(b) In this case the differential, dS, is the same as before, but

$$\rho(x, y, z) = x^2 + y^2 + z^2 = x^2 + y^2 + \left(\sqrt{x^2 + y^2} \right)^2 = 2x^2 + 2y^2,$$

since $z = \sqrt{x^2 + y^2}$. So,

$$\int_S \rho(x, y, z) \, dS = \int_D (2x^2 + 2y^2)\sqrt{2} \, dA$$

$$= \int_0^{2\pi} \int_0^3 2r^2\sqrt{2} \, r \, dr \, d\theta$$

$$= \int_0^{2\pi} \int_0^3 2\sqrt{2}r^3 \, dr \, d\theta$$

$$= \int_0^{2\pi} \left[\frac{\sqrt{2}}{2}r^4 \right]_0^3 \, d\theta$$

$$= \int_0^{2\pi} \frac{81\sqrt{2}}{2} \, d\theta = 81\sqrt{2}\pi.$$

EXAMPLE 4

Finding the flux through a surface

Find the fluxes of

(a) $\mathbf{F}(x, y, z) = yz\mathbf{i} + 3x\cos z\mathbf{j} - 4y^2x\mathbf{k}$ through the rectangle $1 \le y \le 10$ and $2 \le z \le 4$ in the yz-plane in the positive x direction.

(b) $\mathbf{F}(x, y, z) = x\mathbf{i} + y\mathbf{j} + z\mathbf{k}$ through the upper half of the unit sphere in the direction of the unit normal vector with positive z-component.

SOLUTION

(a) This surface is just a piece of the yz-plane. It is parametrized by

$$y\mathbf{j} + z\mathbf{k} \qquad 1 \le y \le 10,\ 2 \le z \le 4.$$

Because of the orientation, $\mathbf{n} = \mathbf{i}$, so

$$\mathbf{F}(x, y, z) \cdot \mathbf{n} = (yz\mathbf{i} + 3x\cos z\mathbf{j} - 4y^2x\mathbf{k}) \cdot \mathbf{i} = yz.$$

The flux of $\mathbf{F}(x, y, z)$ through S is

$$\int_S \mathbf{F} \cdot \mathbf{n}\, dS = \int_1^{10} \int_2^4 yz\, dz\, dy = 297.$$

(b) Here the surface is given by

$$z = \sqrt{1 - x^2 - y^2}.$$

The choice of $\mathbf{n}$ should have a positive z-component. We recall from Chapter 12 that if $z = f(x, y)$, then the vector

$$\left\langle 1, 0, \frac{\partial z}{\partial x} \right\rangle \times \left\langle 0, 1, \frac{\partial z}{\partial y} \right\rangle = \left\langle -\frac{\partial z}{\partial x}, -\frac{\partial z}{\partial y}, 1 \right\rangle$$

is normal to the surface. This vector already has a positive z-component, so all we have to do is scale it to form a unit vector. We have

$$\mathbf{v} = \left\langle -\frac{\partial z}{\partial x}, -\frac{\partial z}{\partial y}, 1 \right\rangle = \left\langle \frac{x}{\sqrt{1 - x^2 - y^2}}, \frac{y}{\sqrt{1 - x^2 - y^2}}, 1 \right\rangle$$

perpendicular to the surface. This equation yields

$$\frac{\mathbf{v}}{\|\mathbf{v}\|} = \mathbf{n} = \left\langle x, y, \sqrt{1 - x^2 - y^2} \right\rangle$$

as the desired unit normal. We can also use that information to compute

$$dS = \sqrt{\frac{x^2}{1 - x^2 - y^2} + \frac{y^2}{1 - x^2 - y^2} + 1}\, dA = \frac{1}{\sqrt{1 - x^2 - y^2}}\, dA.$$

If we let D be the unit disk in the xy-plane and centered at the origin, then the flux through S in the positive z direction is

$$\int_S \mathbf{F}(x, y, z) \cdot \mathbf{n}\, dS = \int_D \left(\langle x, y, z \rangle \cdot \left\langle x, y, \sqrt{1 - x^2 - y^2} \right\rangle \right) \frac{1}{\sqrt{1 - x^2 - y^2}}\, dA$$

$$= \int_D \left(\left\langle x, y, \sqrt{1 - x^2 - y^2} \right\rangle \cdot \left\langle x, y, \sqrt{1 - x^2 - y^2} \right\rangle \right) \frac{1}{\sqrt{1 - x^2 - y^2}}\, dA$$

$$= \int_D \frac{1}{\sqrt{1 - x^2 - y^2}}\, dA$$

$$= \int_0^{2\pi} \int_0^1 \frac{r}{\sqrt{1 - r^2}}\, dr\, d\theta$$

$$= \int_0^{2\pi} \left[-(1 - r^2)^{1/2} \right]_0^1\, d\theta$$

$$= \int_0^{2\pi} 1\, d\theta = 2\pi.$$

$\square$

? TEST YOUR UNDERSTANDING

▶ How was the "subdivide, approximate, and add" approach used in motivating the definition of surface area?

▶ What is measured by $\int_S dS$?

▶ What is measured by $\int_S f(x, y, z)\, dS$?

▶ Why does this section refer only to vector fields in $\mathbb{R}^3$?

▶ What is measured by $\int_S \mathbf{F} \cdot \mathbf{n}\, dS$?

EXERCISES 14.3

Thinking Back

▶ *Area:* Finding the area of a region in the xy-plane is one of the motivating applications of integration. It is also a special case of the surface area calculation developed in this section. Find the area of the region in the xy-plane bounded by the curves $y = x^2$ and $x = \sqrt{y}$.

▶ *Double integrals:* The double integrals of Chapter 13 were our first exposure to surface integrals, on the very well-behaved surface of the xy-plane. Evaluate $f(x, y) = x^2 - y^2$ on the region bounded by the x-axis and the curve defined by $y = 4 - x^2$.

▶ *Average value:* Recall that the average value of a function of two variables $f(x, y)$ on a region Ω is given by the quotient

$$\frac{\iint_\Omega f(x, y)\, dA}{\iint_\Omega dA}.$$

Compute the average value of $g(x, y) = xy$ on the first-quadrant region bounded by $y = x$ and $y = x^2$.

Concepts

0. *Problem Zero:* Read the section and make your own summary of the material.

1. *True/False:* Determine whether each of the statements that follow is true or false. If a statement is true, explain why. If a statement is false, provide a counterexample.

  (a) *True or False:* The result of integrating a vector field over a surface is a vector.

  (b) *True or False:* The result of integrating a function over a surface is a scalar.

  (c) *True or False:* For a region R in the xy-plane, $dS = dA$.

  (d) *True or False:* In computing $\int_S f(x, y, z)\, dS$, the direction of the normal vector is irrelevant.

  (e) *True or False:* If $f(x, y, z)$ is defined on an open region containing a smooth surface S, then $\int_S f(x, y, z)\, dS$ measures the flow through S in the positive z direction determined by $f(x, y, z)$.

  (f) *True or False:* If $\mathbf{F}(x, y, z)$ is defined on an open region containing a smooth surface S, then $\int_S \mathbf{F}(x, y, z) \cdot \mathbf{n}\, dS$ measures the flow through S in the direction of $\mathbf{n}$ determined by the field $\mathbf{F}(x, y, z)$.

  (g) *True or False:* In computing

$$\int_S \mathbf{F}(x, y, z) \cdot \mathbf{n}\, dS,$$

the direction of the normal vector is irrelevant.

  (h) *True or False:* In computing

$$\int_S \mathbf{F}(x, y, z) \cdot \mathbf{n}\, dS$$

with $\mathbf{n}$ pointing in the correct direction, we could use a scalar multiple of $\mathbf{n}$, since the length will cancel in the dS term.

2. *Examples:* Construct examples of the thing(s) described in the following. Try to find examples that are different than any in the reading.

  (a) Two different surfaces with the same area. (Try to make these very different, not just shifted copies of each other.)

  (b) Let S be the surface parametrized by

$$\mathbf{r}(u, v) = x(u, v)\mathbf{i} + y(u, v)\mathbf{j} + z(u, v)\mathbf{k}.$$

  Give two different unit normal vectors to S at the point $r(u_0, v_0)$.

  (c) A smooth surface that can be smoothly parametrized as $\mathbf{r}(x, z) = \langle x, f(x), z \rangle$.

3. Why do surface integrals of multivariate functions *not* include an $\mathbf{n}$ term, whereas surface integrals of vector fields *do* include this term?

4. Make a chart of all the new notation, definitions, and theorems in this section, including what each new item means in terms you already understand.

5. Compute a general formula for dS for any plane $ax + by + cz = k$ if $c \neq 0$.

6. Relate the integrand $\mathbf{F} \cdot \mathbf{n} \, dS$ to the discussions of work in Sections 6.4 and 14.2.

7. Give a smooth parametrization of the upper half of the unit sphere in terms of x and y.

8. Compute dS for your parametrization in Exercise 7.

9. Give a smooth parametrization, in terms of u and v, of the sphere of radius k and centered at the origin.

10. Compute dS for your parametrization in Exercise 9.

11. Given a smooth surface S described as a function $z = f(x, y)$, calculate the upwards-pointing normal vector for S.

12. Give a smooth parametrization for a "generalized cylinder" S, given by extending the curve $y = x^2$ upwards and downwards from $z = -2$ to $z = 3$.

13. Compute $\mathbf{n}$ for the surface S in Exercise 12.

14. Generalize your answer to Exercise 12 to give a parametrization and a normal vector for the extension of any differentiable plane curve $y = f(x)$ through $a \leq z \leq b$.

15. Use what you know about average value from previous sections to propose a formula for the average value of a multivariate function $f(x, y, z)$ on a smooth surface S.

16. Use what you know about averages to propose a formula for the average rate of flux of a vector field $\mathbf{F}(x, y, z)$ through a smooth surface S in the direction of $\mathbf{n}$.

17. If S is parametrized by $\mathbf{r}(u, v)$, why is $dS = \|\mathbf{r}_u \times \mathbf{r}_v\| \, du \, dv$ the correct factor to use to account for distortion of area?

18. Examine the parametrization of surfaces and the computation of dS in light of the discussion of change of variables and the Jacobian in Section 13.7. How are these ideas related?

19. Give a formula for a normal vector to the surface S determined by $x = f(y, z)$, where $f(y, z)$ is a function with continuous partial derivatives.

20. Give a formula for a normal vector to the surface S determined by $y = g(x, z)$, where $g(x, z)$ is a function with continuous partial derivatives.

Skills

Find the areas of the given surfaces in Exercises 21–26.

21. S is the portion of the plane with equation $y - z = \frac{\pi}{2}$ that lies above the rectangle determined by $0 \leq x \leq 4$ and $3 \leq y \leq 6$.

22. S is the portion of the plane with equation $x = y + z$ that lies above the region in the xy-plane that is bounded by $y = x$, $y = 5$, $y = 10$, and the y-axis.

23. S is the portion of the saddle surface determined by $z = x^2 - y^2$ that lies above and/or below the annulus in the xy-plane determined by the circles with radii $\frac{\sqrt{3}}{2}$ and $\sqrt{2}$ and centered at the origin.

24. S is the portion of the surface determined by $x = 9 - y^2 - z^2$ that lies on the positive side of the yz-plane (i.e., where $x \geq 0$).

25. S is the portion of the surface parametrized by $\mathbf{r}(u, v) = \langle 3u - v, v + u, v - u \rangle$ whose preimage (the domain in the uv-plane) is the unit square $[0, 1] \times [0, 1]$.

26. S is the lower branch of the hyperboloid of two sheets $z^2 = x^2 + y^2 + 1$ that lies below the annulus determined by $1 \leq r \leq 2$ in the xy-plane.

Integrate the given function over the accompanying surface in Exercises 27–34.

27. $f(x, z) = e^{-(x^2 + z^2)}$, where S is the unit disk centered at the point $(0, 2, 0)$ and in the plane $y = 2$.

28. $f(x, y, z) = x^2 - y + 3z$, where S is the portion of the plane with equation $2x - 6y + 3z = 1$ whose preimage in the xz-plane is the region bounded by the coordinate axes and the lines with equations $z = 4$ and $x = z$.

29. $f(x, y, z) = xyz^2$, where S is the portion of the cone $\sqrt{3}\,z = \sqrt{x^2 + y^2}$ that lies within the sphere of radius 4 and centered at the origin.

30. $f(x, y, z) = e^z$, where S is the portion of the unit sphere in the first octant.

31. $f(x, y, z) = \frac{y}{x}\sqrt{4z^2 + 1}$, where S is the portion of the paraboloid $z = x^2 + y^2$ that lies above the rectangle determined by $1 \leq x \leq e$ and $0 \leq y \leq 2$ in the xy-plane.

32. $f(x, y, z) = e^{\sqrt{8y^2 - 4z + 1}}(4z + 8x^2 + 1)^{-1/2}$, where S is the portion of the saddle surface with equation $z = y^2 - x^2$ that lies above the region in the xy-plane bounded by the line with equation $y = x$ in the first quadrant, the line with equation $y = -x$ in the second quadrant, and the unit circle.

33. $f(x, y, z) = x - z + y^2$, where S is given by

$$\mathbf{r}(u, v) = (u + v)\,\mathbf{i} + 2\sqrt{u^2 + v^2}\,\mathbf{j} + (u - v)\,\mathbf{k}$$

on the region in the uv-plane bounded by the graphs of $v = u$ and $v = u^2$.

34. $f(x, y, z) = \dfrac{1}{x \ln(z + 2y)}$, where S is the surface given by $\mathbf{r}(u, v) = 2u\,\mathbf{i} + (u + v)\,\mathbf{j} + (2u - 2v)\,\mathbf{k}$ for $3 \leq u \leq 7$ and $2 \leq v \leq 10$.

Find the flux of the given vector field through the given surface in Exercises 35–42.

35. $\mathbf{F}(x, y, z) = \cos(xyz)\,\mathbf{i} + \mathbf{j} - yz\,\mathbf{k}$, where S is the portion of the surface with equation $z = y^3 - y^2$ that lies above and/or below the rectangle determined by $-3 \leq x \leq 2$ and $-1 \leq y \leq 1$ in the xy-plane, with $\mathbf{n}$ pointing in the positive z direction.

36. $\mathbf{F} = \left\langle x \ln(xz), 5z, \dfrac{1}{y^2 + 1} \right\rangle$, where S is the region of the plane with equation $12x - 9y + 3z = 10$, where $2 \leq x \leq 3$ and $5 \leq y \leq 10$, with $\mathbf{n}$ pointing upwards.

37. $\mathbf{F}(x, y, z) = -xz\,\mathbf{i} - yz\,\mathbf{j} + z^2\,\mathbf{k}$, where S is the cone with equation $z = \sqrt{x^2 + y^2}$ between $z = 2$ and $z = 4$, with $\mathbf{n}$ pointing outwards.

38. $\mathbf{F}(x, y, z) = \langle yz, xz, xy \rangle$, where S is the portion of the saddle determined by $z = x^2 - y^2$ that lies above the region in the xy-plane bounded by the x-axis and the parabola with equation $y = 1 - x^2$.

39. $\mathbf{F}(x, y, z) = z\cos yz\,\mathbf{j} + z\sin yz\,\mathbf{k}$, where S is the portion of the plane with equation $2x - 8y - 10z = 42$ that lies on the positive side of the rectangle with corners $(0, -\pi, 0)$, $(0, \pi, 0)$, $(0, \pi, \pi)$, and $(0, -\pi, \pi)$ in the yz-plane.

40. $\mathbf{F}(x, y, z) = \mathbf{i} + \mathbf{j} + \mathbf{k}$, where S is the lower half of the unit sphere, with $\mathbf{n}$ pointing outwards.

41. $\mathbf{F}(x, y, z) = 5\,\mathbf{i} + 13\,\mathbf{j} + 2\,\mathbf{k}$, where S is the unit sphere, with $\mathbf{n}$ pointing outwards.

42. $\mathbf{F}(x, y, z) = -y\,\mathbf{i} + x\,\mathbf{j} - e^{yz}\,\mathbf{k}$, where S is the cylinder with equation $x^2 + y^2 = 9$ from $z = 2$ to $z = 4$, with $\mathbf{n}$ pointing outwards.

Evaluate the integrals in Exercises 43–48.

43. Find $\int_S 1\,dS$, where S is the portion of the surface determined by $z = x^2 - \sqrt{3}y$ that lies above the region in the xy-plane bounded by the x-axis and the lines with equations $y = 2x$ and $x = 3$.

44. Find $\int_S \mathbf{F}(x, y, z) \cdot \mathbf{n}\,dS$ if
$$\mathbf{F}(x, y, z) = \frac{\ln(x^2 + y^2 + 1)}{z + 3}\,\mathbf{i} + \frac{y}{y + 1}\,\mathbf{j} + e^{z^2}\,\mathbf{k},$$
where S is the portion of the sphere with radius 2, centered at the origin, and that lies below the plane with equation $z = -\sqrt{2}$, with $\mathbf{n}$ pointing outwards.

45. Find the integral of $f(x, y, z) = z^3 + z(x^2 + 2^y)$ on the portion of the unit sphere that lies in the first octant, above the rectangle $\left[0, \frac{1}{2}\right] \times \left[0, \frac{1}{3}\right]$ in the xy-plane.

46. Find $\int_S 1\,dS$, where S is the portion of the surface with equation $x = e^{yz} - e^{-yz}$ that lies on the positive side of the circle of radius 3 and centered at the origin in the yz-plane.

47. Find $\int_S \mathbf{F}(x, y, z) \cdot \mathbf{n}\,dS$ if
$$\mathbf{F}(x, y, z) = 2xz\,\mathbf{i} + 2yz\,\mathbf{j} - 18\,\mathbf{k}$$
and S is the portion of the hyperboloid $x^2 + y^2 - 9 = z^2$ that lies between the planes $z = -4$ and $z = 0$, with $\mathbf{n}$ pointing outwards.

48. Find the integral of $f(x, y, z) = x - y - z$ on the portion of the plane with equation $10x - \sqrt{33}y + 36z = 30$ with $2 \leq x \leq 7$ and $1 \leq z \leq 2$.

Applications

Find the masses of the laminæ in Exercises 49 and 50.

49. The lamina occupies the region of the hyperbolic saddle with equation $z = x^2 - y^2$ that lies above and/or below the disk of radius 2 about the origin in the xy-plane where the density is uniform.

50. The lamina occupies the region of the hyperboloid with equation $z^2 + 4 = x^2 + y^2$ that lies above and/or below the disk of radius 5 about the origin in the xy-plane, and the density function, $\rho(x, y, z)$, is proportional to distance from the origin.

Find the flux of the given vector field through a permeable membrane described by surface S in Exercises 51 and 52 .

51. $\mathbf{F}(x, y, z) = y\,\mathbf{i} + x\,\mathbf{j} + \mathbf{k}$, where S is the paraboloid with equation $z = 4x^2 + y^2 + 1$ that lies above the annulus determined by $1 \leq x^2 + y^2 \leq 4$ in the xy-plane.

52. $\mathbf{F}(x, y, z) = -z\,\mathbf{i} + y\,\mathbf{j} + x\,\mathbf{k}$, where S is the surface with equation $y = \cos z$ for $1 \leq x \leq 5$ and $0 \leq z \leq \frac{\pi}{2}$.

53. Suppose that an electric field is given by
$$\mathbf{E} = 2y\,\mathbf{i} + 2xy\,\mathbf{j} + yz\,\mathbf{k}.$$
Compute the flux $\int_S \mathbf{E} \cdot \mathbf{n}\,dA$ of the field through the unit cube $[0, 1] \times [0, 1] \times [0, 1]$.

54. Suppose that an electric field is given by
$$\mathbf{E} = \frac{K}{(x^2 + y^2 + z^2)^{3/2}}(x\,\mathbf{i} + y\,\mathbf{j} + z\,\mathbf{k}),$$
where K is a constant. Show that the flux of the field through any sphere centered at the origin is constant.

Mac owns a farm in the Palouse, a large agricultural region in the northwestern United States. One field on his farm can be modeled as the surface $s(x, y) = 0.24\sqrt{x^2 + y^2}$ on the square $[0, 0.25] \times [0, 0.25]$, where all distances are given in miles. Use this information to answer Exercises 55 and 56.

55. What is the actual area of Mac's field?

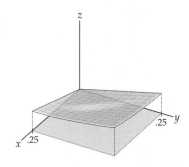

56. Mac is concerned with evapotranspiration from his land: the process wherein moisture from the soil evaporates, or is taken up by plants and then transpired into the air. The problem is that certain parts of his field tend to be drier than others, and since the whole field is tilted so as to face south, the entire field dries out more quickly than if it were flat. The rate at which the field dries out depends on how much solar radiation hits Mac's land (i.e., a flux of radiation into the field). On a hot day in July at noon, the rays of the sun hit the earth parallel to the line $\langle 0, t, -3t \rangle$ for $t \in \mathbb{R}$. Compare the solar flux through this field with that through the horizontal rectangle $[0, 0.25] \times [0, 0.25]$ in the xy-plane that underlies it.

Proofs

57. Show that reversing the orientation of a surface S reverses the sign of $\int_S \mathbf{F}(x, y, z) \cdot \mathbf{n} \, dS$.

58. Show that the two definitions of dS in Definition 14.6 are equivalent, by showing that if S is a surface described by $\mathbf{r}(x, y) = (x, y, z(x, y))$ for $(x, y) \in D$, then

$$\|\mathbf{r}_x(x, y) \times \mathbf{r}_y(x, y)\| = \sqrt{\left(\frac{\partial z}{\partial x}\right)^2 + \left(\frac{\partial z}{\partial y}\right)^2 + 1}.$$

59. Let R be a simply connected region in the xy-plane. Show that the portion of the paraboloid with equation $z = x^2 + y^2$ determined by R has the same area as the portion of the saddle with equation $z = x^2 - y^2$ determined by R.

60. Let a, b, and c be nonzero constants. Find a general formula for the area of the portion of the plane with equation $ax + by + cz = k$ that lies above a rectangle $[\alpha, \beta] \times [\gamma, \delta]$ in the xy-plane.

Thinking Forward

▶ *Integrating vector fields over three-dimensional regions:* Let W be a three-dimensional region in $\mathbb{R}^3$. If a vector field measures the movement of a gas or a fluid in space, what should the integral of a field over W measure?

▶ *Comparing double integrals and surface integrals:* Think about how surface integrals compared and contrasted with double integrals in the plane. What changes and what is the same in these two cases?

▶ *Computations in $\mathbb{R}^4$:* In $\mathbb{R}^3$, a surface is determined by (1) its normal vector at every point and (2) one point on the surface. An intuitive way to think of this is that, in $\mathbb{R}^3$, $\mathbf{n}$ tells us which direction is not occupied by the surface because it is perpendicular to the surface. It can be difficult to visualize, but the same procedure can lifted to solids in $\mathbb{R}^4$. How might we write an integral to determine the volume of a three-dimensional region W in $\mathbb{R}^4$? How might we integrate over surfaces in $\mathbb{R}^4$?

14.4   GREEN'S THEOREM

▶ Defining the del operator, ∇

▶ Understanding the relationships between the divergence and the curl

▶ Extending the Fundamental Theorem of Calculus into the plane

The Fundamental Theorem of Calculus and Green's Theorem

In this section, and in the two that follow, we generalize the spirit of the Fundamental Theorem of Calculus in a new way. Recall that the Fundamental Theorem says that, under appropriate conditions,

$$\int_a^b f(x)\, dx = F(b) - F(a),$$

where $F(x)$ is an antiderivative of $f(x)$. One way of interpreting this theorem is that it relates the behavior of the function $F(x)$ on the boundary of a set in $\mathbb{R}$ to the behavior of the derivative of $F(x)$ in the interior of the region. The region in question is the interval $[a, b]$; its boundary is the set containing points a and b, and its interior is the open interval (a, b).

Given a function $g(x, y)$ defined on a bounded region $\mathcal{R}$ of the plane, it is reasonable, therefore, to ask if there is some relationship between $\int_{\mathcal{R}} g(x, y)\, dA$ and an operation on the curve C that is the boundary of $\mathcal{R}$. Since the boundary curve is more complicated than the two endpoints of an interval in $\mathbb{R}$, we expect that this operation will be more complicated; we shall see that it is an integration. Similarly, we can hope that there is a relationship between $f(x, y, z)$ defined on a bounded region $\mathcal{R}$ in $\mathbb{R}^3$ and an operation on the surface S that forms the boundary of $\mathcal{R}$. As in the case of line and surface integrals, we will sometimes find that vector fields are relevant to the discussion.

The Del Operator, the Divergence, and the Curl

The theorems of the next three subsections are improved by compact notation involving two quantities associated with vector fields: the *divergence* and the *curl*. To state these quantities, it is convenient and traditional to define a vector operator: the *del operator*, ∇.

DEFINITION 14.9

The Del Operator, ∇

The **del operator**, ∇, is the vector of operations

$$\nabla = \frac{\partial}{\partial x}\mathbf{i} + \frac{\partial}{\partial y}\mathbf{j} + \frac{\partial}{\partial z}\mathbf{k}$$

or, in $\mathbb{R}^2$,

$$\nabla = \frac{\partial}{\partial x}\mathbf{i} + \frac{\partial}{\partial y}\mathbf{j}.$$

This operator is used in dot or cross products with vector fields. For example, if

$$\mathbf{F}(x, y, z) = (x + y)\mathbf{i} + e^{xyz}\mathbf{j} + (x - y + z^2)\mathbf{k},$$

then

$$\nabla \cdot \mathbf{F}(x, y, z) = \frac{\partial}{\partial x}(x + y) + \frac{\partial}{\partial y}e^{xyz} + \frac{\partial}{\partial z}(x - y + z^2)$$

$$= 1 + xze^{xyz} + 2z$$

and

$$\nabla \times \mathbf{F}(x,y,z) = \left(\frac{\partial}{\partial x}\mathbf{i} + \frac{\partial}{\partial y}\mathbf{j} + \frac{\partial}{\partial z}\mathbf{k} \right) \times ((x+y)\mathbf{i} + e^{xyz}\mathbf{j} + (x-y+z^2)\mathbf{k})$$

$$= \left(\frac{\partial}{\partial y}(x-y+z^2) - \frac{\partial}{\partial z}(e^{xyz}) \right)\mathbf{i} + \left(\frac{\partial}{\partial z}(x+y) - \frac{\partial}{\partial x}(x-y+z^2) \right)\mathbf{j}$$

$$+ \left(\frac{\partial}{\partial x}(e^{xyz}) - \frac{\partial}{\partial y}(x+y) \right)\mathbf{k}$$

$$= (-1 - xye^{xyz})\mathbf{i} - \mathbf{j} + (yze^{xyz} - 1)\mathbf{k}.$$

For a vector field in $\mathbb{R}^3$, the operation $\nabla \cdot \mathbf{F}(x,y,z)$ produces a scalar that is the sum of $\mathbf{F}$'s directional derivatives in each of the $\mathbf{i}$, $\mathbf{j}$, and $\mathbf{k}$ directions. If $\mathbf{F}(x,y,z)$ represents the flow of a gas or fluid, then the sign of this scalar corresponds to whether the gas or fluid is compressing ($\nabla \cdot \mathbf{F}(x,y,z)$ is negative) or expanding ($\nabla \cdot \mathbf{F}(x,y,z)$ is positive). The scalar is known as the *divergence* of the vector field $\mathbf{F}(x,y,z)$.

DEFINITION 14.10   Divergence of a Vector Field

The ***divergence*** of vector field $\mathbf{F}$ is the dot product of ∇ and $\mathbf{F}$.

(a) In $\mathbb{R}^2$, if $\mathbf{F}(x,y) = F_1(x,y)\mathbf{i} + F_2(x,y)\mathbf{j}$, then

$$\text{div } \mathbf{F}(x,y) = \nabla \cdot \mathbf{F}(x,y) = \frac{\partial F_1}{\partial x} + \frac{\partial F_2}{\partial y}.$$

(b) In $\mathbb{R}^3$, if $\mathbf{F}(x,y,z) = F_1(x,y,z)\mathbf{i} + F_2(x,y,z)\mathbf{j} + F_3(x,y,z)\mathbf{k}$, then

$$\text{div } \mathbf{F}(x,y,z) = \nabla \cdot \mathbf{F}(x,y,z) = \frac{\partial F_1}{\partial x} + \frac{\partial F_2}{\partial y} + \frac{\partial F_3}{\partial z}.$$

The divergence at a point of a vector field measures the magnitude of the rate of change of the vector field away from a source or towards a sink.

In a similar vein, the cross product $\nabla \times \mathbf{F}$ is called the *curl* of the field. At a point in the vector field, the curl corresponds physically to the circulation or rotation of the field at that point. This interpretation of the curl is motivated by applications in engineering and physics.

DEFINITION 14.11   Curl of a Vector Field

The ***curl*** of a vector field $\mathbf{F}(x,y,z) = F_1(x,y,z)\mathbf{i} + F_2(x,y,z)\mathbf{j} + F_3(x,y,z)\mathbf{k}$ is the cross product of ∇ with $\mathbf{F}(x,y,z)$:

$$\text{curl } \mathbf{F} = \nabla \times \mathbf{F}(x,y,z) = \left(\frac{\partial F_3}{\partial y} - \frac{\partial F_2}{\partial z} \right)\mathbf{i} + \left(\frac{\partial F_1}{\partial z} - \frac{\partial F_3}{\partial x} \right)\mathbf{j} + \left(\frac{\partial F_2}{\partial x} - \frac{\partial F_1}{\partial y} \right)\mathbf{k}.$$

If $\mathbf{F}(x,y) = (F_1(x,y), F_2(x,y))$ is a vector field in $\mathbb{R}^2$, we define the curl of $\mathbf{F}$ to be curl $(F_1(x,y), F_2(x,y), 0)$. So,

$$\text{curl } \mathbf{F} = \text{curl } (F_1(x,y), F_2(x,y), 0) = \nabla \times \langle F_1(x,y), F_2(x,y), 0 \rangle$$

$$= \left(\frac{\partial F_2}{\partial x} - \frac{\partial F_1}{\partial y} \right)\mathbf{k}.$$

Using the same field as before, namely, $\mathbf{F}(x,y,z) = (x+y)\mathbf{i} + e^{xyz}\mathbf{j} + (x-y+z^2)\mathbf{k}$, we have

$$\text{div } \mathbf{F} = 1 + xze^{xyz} + 2z,$$

which is a scalar because it is the result of a dot product, and

$$\text{curl } \mathbf{F} = (-1 - xye^{xyz})\mathbf{i} - \mathbf{j} + (yze^{xyz} - 1)\mathbf{k},$$

which is a vector because it is the result of a cross product.

Recall that the cross product is not commutative: $\nabla \times \mathbf{F} \neq \mathbf{F} \times \nabla$. In Chapter 10, we saw that the cross product may be computed as the determinant of a 3×3 matrix. We may use that approach to find the curl as well:

$$\nabla \times \mathbf{F}(x, y, z) = \nabla \times \langle F_1(x, y, z), F_2(x, y, z), F_3(x, y, z) \rangle$$

$$= \det \begin{bmatrix} \mathbf{i} & \mathbf{j} & \mathbf{k} \\ \dfrac{\partial}{\partial x} & \dfrac{\partial}{\partial y} & \dfrac{\partial}{\partial z} \\ F_1(x, y, z) & F_2(x, y, z) & F_3(x, y, z) \end{bmatrix}.$$

There is a special relationship between the divergence and curl of a field, on the one hand, and the curl of a conservative vector field, on the other:

THEOREM 14.12   **Divergence of Curl and Curl of a Gradient**

(a) If $\mathbf{F} = \langle F_1(x, y, z), F_2(x, y, z) \rangle$ is a vector field in $\mathbb{R}^2$ or $\mathbf{F} = \langle F_1(x, y, z), F_2(x, y, z), F_3(x, y, z) \rangle$ is a vector field in $\mathbb{R}^3$, for which F_1, F_2, and F_3 have continuous second-order partial derivatives, then

$$\text{div}(\text{curl } \mathbf{F}) = \nabla \cdot (\nabla \times \mathbf{F}) = 0.$$

(b) For a multivariate function f in $\mathbb{R}^2$ or $\mathbb{R}^3$ with continuous second partial derivatives,

$$\text{curl } \nabla f = \nabla \times (\nabla f) = \mathbf{0}.$$

In each case proofs may be obtained by using the equality of mixed partial derivatives. We ask you to prove these identities in Exercises 55 and 56.

Green's Theorem

Our next extension of the Fundamental Theorem of Calculus is in the plane. Consider a region $\mathcal{R}$ of the plane whose boundary is a smooth, simple closed curve C, oriented counterclockwise.

Green's Theorem asserts that integrating along the boundary curve C is equal to integrating a double integral over the interior of $\mathcal{R}$. Recall the notation from Section 14.2: For a vector field $\mathbf{F}(x, y) = \langle F_1(x, y), F_2(x, y) \rangle$ and a curve $\mathbf{r}(t) = \langle x(t), y(t) \rangle$, for $a \leq t \leq b$,

$$\int_C \mathbf{F}(x, y) \cdot d\mathbf{r} = \int_C F_1(x, y)\, dx + F_2(x, y)\, dy = \int_a^b \left(F_1(x, y)\frac{dx}{dt} + F_2(x, y)\frac{dy}{dt} \right) dt.$$

THEOREM 14.13

Green's Theorem

Let $\mathbf{F}(x, y) = \langle F_1(x, y), F_2(x, y) \rangle$ be a vector field defined on a region R in the plane whose boundary is a smooth or piecewise-smooth, simple closed curve C. If $\mathbf{r}(t)$ is a parametrization of C in the counterclockwise direction (as viewed from the positive z-axis), then

$$\int_C \mathbf{F}(x, y) \cdot d\mathbf{r} = \int_C F_1(x, y)\, dx + F_2(x, y)\, dy = \iint_R \left(\frac{\partial F_2}{\partial x} - \frac{\partial F_1}{\partial y} \right) dA.$$

Proof. A thorough proof of Green's Theorem is beyond the scope of this text. However, we will sketch the proof for a region that can be described both as the set of points lying between smooth functions $g_1(x)$ and $g_2(x)$, for $a \leq x \leq b$, with $g_1(a) = g_2(a)$ and $g_1(b) = g_2(b)$, and also as the set of points lying between smooth functions $h_1(y)$ and $h_2(y)$, for $c \leq y \leq d$, with $h_1(c) = h_2(c)$ and $h_1(d) = h_2(d)$.

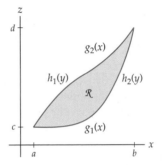

To prove Green's Theorem for this region, let C be the boundary curve traversed in the counterclockwise direction. Then $C = C_1 \cup C_2$, where $C_1 = \mathbf{r}_1(t) = (t, g_1(t))$ for $a \leq t \leq b$ and $C_2 = \mathbf{r}_2(t) = (t, g_2(t))$ for $a \leq t \leq b$. (Since we are going around the boundary of R counterclockwise, our path along C_2 will be in the negative direction.)

Fix a vector field $\mathbf{F}(x, y) = F_1(x, y)\mathbf{i} + F_2(x, y)\mathbf{j}$. Then, for a given $x \in [a, b]$,

$$\int_{g_1(x)}^{g_2(x)} \frac{\partial F_1}{\partial y}\, dy = \left[F_1(x, y) \right]_{g_1(x)}^{g_2(x)} = F_1(x, g_2(x)) - F_1(x, g_1(x)).$$

So,

$$\int_a^b \int_{g_1(x)}^{g_2(x)} \frac{\partial F_1}{\partial y}\, dy\, dx = \int_a^b F_1(x, g_2(x)) - F_1(x, g_1(x))\, dx$$

$$= -\int_b^a F_1(x, g_2(x))\, dx - \int_a^b F_1(x, g_1(x))\, dx = -\int_C F_1(x, y)\, dx.$$

A parallel argument shows that

$$\int_c^d \int_{h_1(y)}^{h_2(y)} \frac{\partial F_2}{\partial x}\, dx\, dy = \int_c^d F_2(h_2(y), y) - F_2(h_1(y), y)\, dy$$

$$= \int_c^d F_2(h_2(y), y)\, dy + \int_d^c F_2(h_1(y), y)\, dy = \int_C F_2(x, y)\, dy.$$

Combining the two results gives the desired equality:

$$\int_C \mathbf{F}(x, y) \cdot d\mathbf{r} = \iint_R \left(\frac{\partial F_2}{\partial x} - \frac{\partial F_1}{\partial y} \right) dA.$$

Green's Theorem can be used to evaluate line integrals over simple closed curves with reverse (clockwise) parametrizations. Since reversing the direction of travel in a parametrization of a curve changes the sign of the resulting line integral, we can multiply both sides of the preceding equation by -1 to obtain a formula for clockwise-traversed curves. (See Example 5.)

The relationship between interior double integrals and boundary line integrals given by Green's Theorem also applies to regions S that are not themselves bounded by smooth or piecewise-smooth simple closed curves, provided that S can be written as a sum of such regions. For instance, if Q is the annular region $Q = \{(x, y) \mid 4 \leq x^2 + y^2 \leq 16\}$ shown next at the left, we may represent this region as the union of $\mathcal{R}_1$ and $\mathcal{R}_2$, where each of the subregions has a piecewise-smooth simple closed boundary, as in the figure at the right:

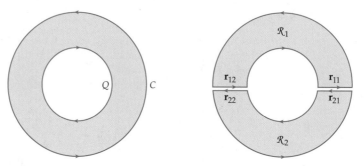

Moreover, if both $\mathcal{R}_1$ and $\mathcal{R}_2$ have parametrizations with counterclockwise orientations, as indicated in the two figures, then the line integrals along the pairs $\mathbf{r}_{11}$, $\mathbf{r}_{21}$ and $\mathbf{r}_{12}$, $\mathbf{r}_{22}$ will cancel in the total summation, since we are integrating the same vector field $\mathbf{F}(x, y)$ along the same curve in opposite directions. So these curves can be disregarded in our treatment of Q. Looking at the curves that form the boundary of Q, we see that it is important that the inner curve D be parametrized in the *clockwise* direction, because of how its components must be parametrized as boundaries of $\mathcal{R}_1$ and $\mathcal{R}_2$. Taking all this into account, we may conclude that

$$\int_C \mathbf{F}(x, y) \cdot d\mathbf{r} + \int_D \mathbf{F}(x, y) \cdot d\mathbf{r} = \iint_Q \left(\frac{\partial F_2}{\partial y} - \frac{\partial F_1}{\partial x} \right) dA,$$

where C is the outer edge traversed counterclockwise and D is the inner edge traversed clockwise.

Green's Theorem can be restated in terms of both the divergence and the curl.

THEOREM 14.14

Green's Theorem: Curl and Divergence Expressions

Let $\mathcal{R}$ be a region in the plane to which Green's Theorem applies, with smooth boundary curve C oriented in the counterclockwise direction by $\mathbf{r}(t) = \langle (x(t), y(t) \rangle$ for $a \leq t \leq b$, with vector field $\mathbf{F}(x, y) = \langle (F_1(x, y), F_2(x, y) \rangle$ defined on $\mathcal{R}$.

(a) Green's Theorem, curl form:
A unit vector perpendicular to the xy-plane and thus to the region R in the positive direction is just $\mathbf{n} = \mathbf{k}$. So we can rewrite Green's Theorem as

$$\int_C \mathbf{F}(x, y) \cdot d\mathbf{r} = \iint_{\mathcal{R}} \left(\frac{\partial F_2}{\partial x} - \frac{\partial F_1}{\partial y} \right) dA = \iint_{\mathcal{R}} \text{curl } \mathbf{F} \cdot \mathbf{k} \, dA.$$

(b) Green's Theorem, divergence form:
If we restrict our attention to the plane, we see that a unit vector that lies in the xy-plane and is perpendicular to the curve C is given by

$$\mathbf{n} = \frac{y'(t)}{\sqrt{(x'(t))^2 + (y'(t))^2}} \mathbf{i} + \frac{-x'(t)}{\sqrt{(x'(t))^2 + (y'(t))^2}} \mathbf{j}.$$

Then Green's Theorem is equivalent to the statement

$$\int_C \mathbf{F}(x, y) \cdot \mathbf{n} \, ds = \iint_{\mathcal{R}} \text{div } \mathbf{F} \, dA.$$

Note that in Theorem 14.14 (a) we treat $\mathbf{F}$ as a vector field in $\mathbb{R}^3$ whose $\mathbf{k}$-component is zero, since the integrand on the right-hand side of the equation is the $\mathbf{k}$-component of the curl of a vector field in $\mathbb{R}^3$.

Proof.

(a) The curl form of Green's Theorem is the direct result of labeling the terms in the original statement of the theorem.

(b) The divergence form requires a little manipulation:

$$\int_C \mathbf{F}(x, y) \cdot \mathbf{n} \, ds$$

$$= \int_a^b \left(F_1(x, y) \frac{y'(t)}{\sqrt{(x'(t))^2 + (y'(t))^2}} + F_2(x, y) \frac{-x'(t)}{\sqrt{(x'(t))^2 + (y'(t))^2}} \right) \sqrt{(x'(t))^2 + (y'(t))^2} \, dt$$

$$= \int_C F_1(x, y) \, dy - F_2(x, y) \, dx = \int_C -F_2(x, y) \, dx + F_1(x, y) \, dy.$$

Applying Green's Theorem to this last line integral gives

$$\int_C -F_2(x, y) \, dx + F_1(x, y) \, dy = \iint_{\mathcal{R}} \left(\frac{\partial F_1}{\partial x} + \frac{\partial F_2}{\partial y} \right) dA$$

$$= \iint_{\mathcal{R}} \operatorname{div} \mathbf{F} \, dA.$$

These equivalent formulations do not affect our application of Green's Theorem; rather, they foreshadow the generalization of the theorem to regions that are surfaces other than the xy-plane, to Stokes' Theorem, and to a higher dimensional analog called the Divergence Theorem. There are various statements of Green's Theorem, including the divergence and curl forms, throughout mathematical literature.

Examples and Explorations

EXAMPLE 1

Computing the divergence and the curl

(a) Compute the divergence of $\mathbf{F}(x, y) = x^2 \mathbf{i} + xy \mathbf{j}$.

(b) Compute the divergence and curl of $\mathbf{G}(x, y, z) = \dfrac{1}{xyz} \mathbf{i} + e^{xyz} \mathbf{j} + \dfrac{1}{z^2 + 1} \mathbf{k}$.

SOLUTION

(a) We use Definition 14.10 to find the divergence of $\mathbf{F}$:

$$\operatorname{div} \mathbf{F}(x, y) = \nabla \cdot \mathbf{F}(x, y)$$

$$= \frac{\partial F_1}{\partial x} + \frac{\partial F_2}{\partial y}$$

$$= 2x + x = 3x.$$

(b) We again use Definition 14.10 to find the divergence of $\mathbf{G}$, or

$$\mathbf{G}(x, y, z) = G_1(x, y, z)\mathbf{i} + G_2(x, y, z)\mathbf{j} + G_3(x, y, z)\mathbf{k} = \frac{1}{xyz}\mathbf{i} + e^{xyz}\mathbf{j} + \frac{1}{z^2 + 1}\mathbf{k}.$$

We have

$$\operatorname{div} \mathbf{G}(x, y, z) = \nabla \cdot \mathbf{G}(x, y, z)$$

$$= \frac{\partial G_1}{\partial x} + \frac{\partial G_2}{\partial y} + \frac{\partial G_3}{\partial z}$$

$$= -\frac{yz}{(xyz)^2} + xze^{xyz} - \frac{2z}{(z^2 + 1)^2}.$$

We now use Definition 14.11 to compute the curl:

$$\text{curl } \mathbf{G}(x, y, z) = \nabla \times \mathbf{G}(x, y, z)$$

$$= \left(\frac{\partial G_3}{\partial y} - \frac{\partial G_2}{\partial z} \right) \mathbf{i} + \left(\frac{\partial G_1}{\partial z} - \frac{\partial G_3}{\partial x} \right) \mathbf{j} + \left(\frac{\partial G_2}{\partial x} - \frac{\partial G_1}{\partial y} \right) \mathbf{k}$$

$$= (0 - xye^{xyz}) \mathbf{i} + \left(-\frac{xy}{(xyz)^2} - 0 \right) \mathbf{j} + \left(yze^{xyz} + \frac{xz}{(xyz)^2} \right) \mathbf{k}$$

$$= -xye^{xyz} \mathbf{i} - \frac{xy}{(xyz)^2} \mathbf{j} + \left(yze^{xyz} + \frac{xz}{(xyz)^2} \right) \mathbf{k}.$$

□

EXAMPLE 2   Verifying Green's Theorem

Verify Green's Theorem for the unit disk in the xy-plane with boundary curve C, traversed counterclockwise, where $\mathbf{F}(x, y) = x\mathbf{i} + 2x\mathbf{j}$.

SOLUTION

Let R be the unit disk and C be the boundary curve parametrized by

$$\mathbf{r}(t) = \langle \cos t, \sin t \rangle, \quad \text{for } 0 \le t \le 2\pi.$$

So, $\frac{d\mathbf{r}}{dt} = \langle -\sin t, \cos t \rangle$, for $0 \le t \le 2\pi$. We also express the vector field $\mathbf{F}$ in terms of the parameter t:

$$\mathbf{F}(x(t), y(t)) = \langle x(t), 2x(t) \rangle = \langle \cos t, 2 \cos t \rangle.$$

Then

$$\int_C \mathbf{F}(x, y) \cdot d\mathbf{r} = \int_0^{2\pi} (-\cos t \sin t + 2 \cos^2 t) \, dt.$$

You should verify that $\frac{1}{2} \cos^2 t + \sin t \cos t + t$ is an antiderivative of $-\cos t \sin t + 2 \cos^2 t$. Therefore,

$$\int_0^{2\pi} (-\cos t \sin t + 2 \cos^2 t) \, dt = \left[\frac{1}{2} \cos^2 t + \sin t \cos t + t \right]_0^{2\pi} = 2\pi.$$

We will now perform the double integration given by Green's Theorem. Since $\mathbf{F}(x, y) = x\mathbf{i} + 2x\mathbf{j}$, we have $\frac{\partial F_2}{\partial x} = 2$ and $\frac{\partial F_1}{\partial y} = 0$. Therefore,

$$\iint_R \left(\frac{\partial F_2}{\partial x} - \frac{\partial F_1}{\partial y} \right) dA = \iint_R 2 \, dA$$

$$= \int_0^{2\pi} \int_0^1 2 \, r \, dr \, d\theta$$

$$= \int_0^{2\pi} \left[r^2 \right]_0^1 \, d\theta = 2\pi.$$

As expected, our results agree.

□

EXAMPLE 3   Computing with Green's Theorem

Evaluate $\int_C \mathbf{F}(x, y) \cdot d\mathbf{r}$, where $\mathbf{F}(x, y) = xy\mathbf{i} + xy^3\mathbf{j}$ and C is the boundary of the rectangle in the xy-plane and whose sides are the coordinate axes and the lines $y = 2$ and $x = 3$.

SOLUTION

This problem shows some of the strength of Green's Theorem. Parametrizing the four distinct pieces of C would not be difficult in this case, but it would be tedious. Meanwhile, the region $\mathcal{R}$ that C encloses is very easy to describe. Applying Green's Theorem, we have

$$\int_C \mathbf{F}(x, y) \cdot d\mathbf{r} = \iint_R \left(\frac{\partial F_2}{\partial x} - \frac{\partial F_1}{\partial y} \right) dA$$

$$= \int_0^3 \int_0^2 (y^3 - x) dy\, dx$$

$$= \int_0^3 \left[\frac{1}{4} y^4 - xy \right]_0^2 dx$$

$$= \int_0^3 (4 - 2x)\, dx$$

$$= \left[4x - x^2 \right]_0^3 = 3.$$

EXAMPLE 4

Using Green's Theorem with a conservative vector field

Use Green's Theorem to evaluate

$$\int_C \mathbf{F}(x, y) \cdot d\mathbf{r},$$

where C is the boundary of the first-quadrant region $\mathcal{R}$ that lies between the curves $y = x^2$ and $y = x$ and where the conservative vector field $\mathbf{F}(x, y) = \nabla f(x, y)$, in which $f(x, y) = e^{xy}$.

SOLUTION

Computing the partial derivatives of $f(x, y)$, we have

$$\mathbf{F}(x, y) = ye^{xy}\mathbf{i} + xe^{xy}\mathbf{j}.$$

Then,

$$\int_C \mathbf{F}(x, y) \cdot d\mathbf{r} = \int_R \left(\frac{\partial F_2}{\partial x} - \frac{\partial F_1}{\partial y} \right) dA$$

$$= \int_R ((e^{xy} + xye^{xy}) - (e^{xy} + xye^{xy}))\, dA = 0.$$

Green's Theorem, together with the equality of the second-order mixed partial derivatives, gives us another proof of the Fundamental Theorem of Line Integrals for smooth or piecewise-smooth simple closed curves. If $\mathbf{F}(x, y) = F_1(x, y)\mathbf{i} + F_2(x, y)\mathbf{j}$ is conservative, then, by the equality of the mixed partial derivatives,

$$\frac{\partial F_2}{\partial x} = \frac{\partial F_1}{\partial y}.$$

So, for a conservative vector field and a smooth or piecewise-smooth simple closed curve C that is the boundary of a region $\mathcal{R}$,

$$\int_C \mathbf{F}(x, y) \cdot d\mathbf{r} = \int_R \left(\frac{\partial F_2}{\partial x} - \frac{\partial F_1}{\partial y} \right) dA = \int_{\mathcal{R}} 0\, dA = 0.$$

CHECKING THE ANSWER

Applying the Fundamental Theorem of Line Integrals to

$$\int_C \mathbf{F}(x, y) \cdot d\mathbf{r}$$

directly, where C is parametrized by $\mathbf{r}(t)$ for $a \le t \le b$, gives

$$\int_C \mathbf{F}(x, y) \cdot d\mathbf{r} = f(x(b), y(b)) - f(x(a), y(a)) = 0,$$

since the curve's being closed entails that $(x(b), y(b)) = (x(a), y(a))$.

EXAMPLE 5 | **Using Green's Theorem for curves with reverse orientation**

Use Green's Theorem to evaluate $\int_C \mathbf{F}(x, y) \cdot d\mathbf{r}$, where $\mathbf{F}(x, y) = y\mathbf{i} + 2x\mathbf{j}$ and C is the boundary of the region bounded by the x-axis and the curve $y = 1 - x^2$, traversed in the clockwise direction.

SOLUTION

From Section 14.2, we know that reversing the direction of travel along a curve changes the sign of the line integrals along that curve. Here, since $\mathbf{F}(x, y) = y\mathbf{i} + 2x\mathbf{j}$, we have

$$\frac{\partial}{\partial x}(2x) - \frac{\partial}{\partial y}(y) = 2 - 1 = 1.$$

So, in this case, Green's Theorem implies that

$$\int_C \mathbf{F}(x, y) \cdot d\mathbf{r} = -\iint_R 1 \, dA$$

$$= -\int_{-1}^{1} \int_0^{1-x^2} 1 \, dy \, dx$$

$$= \int_{-1}^{1} x^2 - 1 \, dx$$

$$= \left[\frac{1}{3}x^3 - x \right]_{-1}^{1} = -\frac{4}{3}.$$

□

EXAMPLE 6 | **Analyzing water flow in a river channel**

The Los Angeles River is channelized for much of its length. The channelized portion of the river flows in concrete banks with uniform width and slope and runs straight from north to south. Therefore, the velocity vector of the flow is $\mathbf{F} = 0\mathbf{i} + f(y)\mathbf{j}$. Suppose we measure the rate of flow at two points that are close to one another.

(a) One day the flow is found to have speed 0.4 feet per second at the top and bottom edges of a rectangle $\mathcal{R}$ with boundary C enclosing the river. Find $\int_C \mathbf{F} \cdot d\mathbf{r}$.

(b) Suppose that another day we find that the rate of flow at the lower edge of the rectangle is slightly higher than the flow at the top. Use the divergence form of Green's Theorem to explain what unusual phenomenon is occurring in Los Angeles that day.

SOLUTION

(a) Since the flow at the top and bottom is the same, the edge vectors have opposite signs, and the x-component of the velocity field is zero, it follows that the integral is zero.

(b) We have

$$0 < \int_C \mathbf{F} \cdot \mathbf{n} \, ds = \iint_R \frac{\partial f}{\partial y} \, dA.$$

But since the river is channelized, $\frac{\partial f}{\partial y} > 0$ at some points in $\mathcal{R}$. This means that the flow is increasing, so it must be raining.

□

? TEST YOUR UNDERSTANDING

▶ What is the definition of the divergence of a vector field? What does the divergence measure?

▶ What is the definition of the curl of a vector field? What does the curl measure?

▶ How is Green's Theorem similar to the Fundamental Theorem of Calculus from Chapter 4?

▶ How can Green's Theorem be used to compute line integrals around smooth or piecewise-smooth simple closed curves that are parametrized in the *clockwise* direction?

▶ In our examples, we used Green's Theorem to reduce line integrals to double integrals (though the reverse direction is also sometimes useful, as we will see in the exercises). In what way is this use different from our use of the Fundamental Theorem of Calculus from Chapter 4?

EXERCISES 14.4

Thinking Back

Remembering the Fundamental Theorem of Calculus: Answer the following questions about the Fundamental Theorem of Calculus.

▶ Consider the "region" in $\mathbb{R}^1$ given by the interval $[a, b]$. What is the boundary of this region? What is its interior?

▶ Suppose that $g(x)$ is a differentiable function on $[a, b]$. Express $g(b) - g(a)$ in terms of a function on the interior of $[a, b]$.

▶ Express $\int_a^b f(x)\, dx$ in terms of something on the boundary of $[a, b]$.

Intractable Integrals: One of the virtues of Green's Theorem—in both directions—is that it gives us a way to rewrite difficult integrals in a form we hope will be easier to manipulate.

▶ Give an example of a differentiable function $f(x, y)$ that is not obviously integrable.

Concepts

0. *Problem Zero:* Read the section and make your own summary of the material.

1. *True/False:* Determine whether each of the statements that follow is true or false. If a statement is true, explain why. If a statement is false, provide a counterexample.

(a) *True or False:* The del operator, ∇, converts vectors into scalars.

(b) *True or False:* The del operator, ∇, measures the rotation of a vector field.

(c) *True or False:* The divergence of a vector field is a scalar.

(d) *True or False:* The curl of a vector field is a vector.

(e) *True or False:* The curl of a gradient vector field is **0**.

(f) *True or False:* Both the Fundamental Theorem of Calculus (Theorem 4.24) and Green's Theorem relate the integral of a function on a (mathematically well-behaved) region to a quantity measured on the boundary of that region.

(g) *True or False:* The curl of a vector field measures how much the field is compressing or expanding.

(h) *True or False:* The conclusion of Green's Theorem does not depend on the direction of parametrization of the boundary curve in question.

2. *Examples:* Construct examples of the thing(s) described in the following. Try to find examples that are different than any in the reading.

(a) A smooth simple closed curve parametrized in the counterclockwise direction.

(b) A smooth simple closed curve parametrized in the clockwise direction.

(c) A simple closed curve that is not smooth, but is piecewise smooth, parametrized in the counterclockwise direction.

3. In what sense is the integrand of the double integral in Green's Theorem the antiderivative of the vector field?

4. Make a chart of all the new notation, definitions, and theorems in this section, including what each new item means in terms you already understand.

5. Give two examples of quantities that may be computed by $\int_C \mathbf{F} \cdot d\mathbf{r}$.

6. Give three examples of quantities that may be computed by $\iint_{\mathcal{R}} G(x, y)\, dA$.

7. Explain how your answer to Exercise 6 is relevant to the discussion of Green's Theorem.

8. Draw the annular region bounded by $r = 1$ and $r = 2$. Divide it into two regions, each of which is suitable for the application of Green's Theorem.

9. If the velocity of a flow of a gas at a point (x, y, z) is represented by $\mathbf{F}$ and the gas is compressing at that point, what does this imply about the divergence of $\mathbf{F}$ at the point?

10. Give an example of a field with negative divergence at the origin.

11. If the velocity of a flow of a gas at a point (x, y, z) is represented by $\mathbf{F}$ and the gas is expanding at that point, what does this imply about the divergence of $\mathbf{F}$ at the point?

12. Give an example of a field with positive divergence at $(1, 0, \pi)$.

13. Use the vector field $\mathbf{F}(x, y) = x^2 e^y \mathbf{i} + \cos x \sin y \mathbf{j}$ and Green's Theorem to write the line integral of $\mathbf{F}(x, y)$ about the unit circle, traversed counterclockwise, as a double integral. Do not evaluate the integral.

14. Use the same vector field as in Exercise 13 together with the divergence form of Green's Theorem to write the line integral of $\mathbf{F}(x, y)$ about the unit circle as a double integral. Do not evaluate the integral.

15. Use the same vector field as in Exercise 13, and compute the $\mathbf{k}$-component of the curl of $\mathbf{F}(x, y)$.

16. Use the curl form of Green's Theorem to write the line integral of $\mathbf{F}(x, y)$ about the unit circle as a double integral. Do not evaluate the integral.

Skills

Compute the divergence of the vector fields in Exercises 17–22.

17. $\mathbf{F}(x, y) = yx^2\mathbf{i} - x\cos y\,\mathbf{j}$

18. $\mathbf{G}(x, y) = \langle x\cos(xy), y\cos(xy)\rangle$

19. $\mathbf{G}(x, y, z) = yz^2\mathbf{i} - x\sin z\mathbf{j} + x^2e^{xy}\mathbf{k}$

20. $\mathbf{G}(x, y, z) = (x^2 + y - z)\mathbf{i} - 2y\cos z\mathbf{j} + e^{x^2+y^2+z^2}\mathbf{k}$

21. $\mathbf{F}(x, y, z) = \sin^{-1}(xy)\mathbf{i} + \ln(y + z)\mathbf{j} + \dfrac{1}{(2x + 3y + 5z + 1)}\mathbf{k}$

22. $\mathbf{F}(x, y, z) = xe^{yz}\mathbf{i} + ye^{xz}\mathbf{j} + ze^{xy}\mathbf{k}$

Compute the curl of the vector fields in Exercises 23–28.

23. $\mathbf{F}(x, y, z) = \sin^{-1}(xy)\mathbf{i} + \ln(y + z)\mathbf{j} + \dfrac{1}{(2x + 3y + 5z + 1)}\mathbf{k}$

24. $\mathbf{G}(x, y, z) = (x^2 + y - z)\mathbf{i} - 2y\cos z\mathbf{j} + e^{x^2+y^2+z^2}\mathbf{k}$

25. $\mathbf{F}(x, y, z) = xe^{yz}\mathbf{i} + ye^{xz}\mathbf{j} + ze^{xy}\mathbf{k}$

26. $\mathbf{F}(x, y) = -4x^2y\mathbf{i} + 4xy^2\mathbf{j}$

27. $\mathbf{F}(x, y) = \cos(x + y)\mathbf{i} + \sin(x - y)\mathbf{j}$

28. $\mathbf{G}(x, y, z) = (2y + 3z)\mathbf{i} + 2xy^2\mathbf{j} + (zx - y)\mathbf{k}$

Use Green's Theorem to evaluate the integrals in Exercises 29–34.

29. Find $\int_C \mathbf{F} \cdot d\mathbf{r}$, where $\mathbf{F}(x, y) = -4x^2y\mathbf{i} + 4xy^2\mathbf{j}$ and C is the unit circle traversed counterclockwise.

30. Find $\int_C \mathbf{F} \cdot d\mathbf{r}$, where $\mathbf{F}(x, y) = (y^3 + 3x^2y)\mathbf{i} + 2x\mathbf{j}$ and C is the boundary of the half of the unit circle that lies above the x-axis, traversed counterclockwise.

31. Find $\int_C \mathbf{F} \cdot d\mathbf{r}$, where $\mathbf{F}(x, y) = (x + 2y)\mathbf{i} + (x - 2y)\mathbf{j}$ and C is the boundary of the region bounded by the curves $x = y^2, x = 4$, traversed counterclockwise.

32. Find $\int_C \mathbf{F} \cdot d\mathbf{r}$, where $\mathbf{F}(x, y) = y\ln(x+1)^2\mathbf{i} + \mathbf{j}$ and C is the boundary of the square with vertices at $(0, 0)$, $(0, 2)$, $(2, 0)$, and $(2, 2)$, traversed counterclockwise.

33. Find $\int_C \mathbf{F} \cdot d\mathbf{r}$, where

$$\mathbf{F}(x, y) = xe^{x^2+y^2}\mathbf{i} + (10\sin y + x)\mathbf{j}$$

and C is the boundary of the region in the third quadrant bounded by $y = x$ and the x-axis, for $x \in [-1, 0]$, traversed counterclockwise.

34. Find $\int_C \mathbf{F} \cdot d\mathbf{r}$, where $\mathbf{F}(x, y) = xy\mathbf{i} + 4(x - y)\mathbf{j}$ and C is the boundary of the region bounded by the curves $x = 1, x = e, y = e^x$, and $y = \ln x$, traversed counterclockwise.

Verify Green's Theorem by working Exercises 35–42 in pairs. In each pair, evaluate the desired integrals first directly and then using the theorem.

35. Directly compute (i.e., without using Green's Theorem) $\int_C \mathbf{F}(x, y) \cdot d\mathbf{r}$, where $\mathbf{F}(x, y) = y^2\mathbf{i} - xy\mathbf{j}$ and C is the unit circle traversed counterclockwise.

36. Use Green's Theorem to evaluate the line integral in Exercise 35.

37. Directly compute (i.e., without using Green's Theorem) $\int_C \mathbf{F}(x, y) \cdot d\mathbf{r}$, where

$$\mathbf{F}(x, y) = (2^x + y)\mathbf{i} + (2x - y)\mathbf{j}$$

and C is the triangle described by $x = 0, y = 0$, and $y = 1 - x$, traversed counterclockwise.

38. Use Green's Theorem to evaluate the line integral in Exercise 37.

39. Directly compute (i.e., without using Green's Theorem) $\iint_R (3y - 3x)\,dA$, where R is the portion of the disk of radius 2, centered at the origin, and lying above the x-axis.

40. Use Green's Theorem to evaluate the double integral in Exercise 39.

41. Directly compute (i.e., without using Green's Theorem) $\iint_R (e^x + e^y)\,dA$, where R is the region bounded by the lines $x = 1, x = \ln 2, y = 0$, and $y = 2$.

42. Use Green's Theorem to evaluate the double integral in Exercise 41.

Evaluate the integrals in Exercises 43–46 directly or using Green's Theorem.

43. $\int_C \mathbf{F} \cdot d\mathbf{r}$, where $\mathbf{F}(x, y) = y2^x\mathbf{i} + xy^e\mathbf{j}$ and C is the boundary of the square with vertices at $(0, 0)$, $(0, 1)$, $(1, 0)$, and $(1, 1)$, traversed clockwise.

44. $\int_C \mathbf{F}(x, y) \cdot d\mathbf{r}$, where $\mathbf{F}(x, y) = \frac{1}{2}x^2y^3\mathbf{i} + xy\mathbf{j}$ and C is the circle of radius 3, centered at the origin, traversed clockwise.

45. $\iint_R (2xe^{x^2+y^2} + 2ye^{-(x^2+y^2)})\,dA$, where R is the annulus bounded by $r = 1$ and $r = 2$.

46. $\iint_R (3xy - 4x^2y)\,dA$, where R is the unit disk.

Applications

47. Find the work done by the vector field

$$\mathbf{F}(x, y) = (\cos(x^2) + 4xy^2)\mathbf{i} + (2^y - 4x^2y)\mathbf{j}$$

in moving an object around the unit circle, starting and ending at $(1, 0)$.

48. Find the work done by the vector field

$$\mathbf{F}(x, y) = x^3y^2\mathbf{i} + (y - x)\mathbf{j}$$

in moving an object around the triangle with vertices $(1, 1)$, $(2, 2)$, and $(3, 1)$, starting and ending at $(2, 2)$.

49. Find the work done by the vector field

$$F(x, y) = \frac{xe^{\sqrt{x^2+y^2}}}{\sqrt{x^2 + y^2}}\mathbf{i} + \frac{ye^{\sqrt{x^2+y^2}}}{\sqrt{x^2 + y^2}}\mathbf{j}$$

in moving an object around the unit circle, starting and ending at $(1, 0)$.

50. Find the work done by the vector field

$$\mathbf{F}(x, y) = (\cos x - 3ye^x)\mathbf{i} + \sin x \sin y\mathbf{j}$$

in moving an object around the periphery of the rectangle with vertices $(0, 0)$, $(2, 0)$, $(2, \pi)$, and $(0, \pi)$, starting and ending at $(2, \pi)$.

51. The current through a certain passage of the San Juan Islands in Washington State is given by

$$F = \langle F_1(x, y), F_2(x, y)\rangle = \langle 0, 1.152 - 0.8x^2\rangle.$$

Consider a disk $\mathcal{R}$ of radius 1 mile and centered on this region. Denote the boundary of the disk by $\partial\mathcal{R}$.

(a) Compute $\displaystyle\iint_{\mathcal{R}}\left(\frac{\partial F_2}{\partial x} - \frac{\partial F_1}{\partial y}\right) dA$.

(b) Show that

$$\int_{\partial\mathcal{R}} F \cdot n\, ds = \int_0^{2\pi} (1.152 - 0.8\cos^2\theta)\sin\theta\, d\theta.$$

Conclude that Green's Theorem is valid for the current in this area of the San Juan Islands.

(c) What do the integrals from Green's Theorem tell us about this region of the San Juan Islands?

52. Emmy has to examine how well a waste tank with radius 50 feet is stirred. The current at the edge of the tank moves at $\frac{\pi}{4}$ radians per minute in a counterclockwise direction. Emmy knows that if the vector field of the current velocity is $\mathbf{v} = \langle v_1(x, y), v_2(x, y)\rangle$, then the average value of $\frac{\partial v_2}{\partial x} - \frac{\partial v_1}{\partial y}$ in the tank must be at least 1.5 in order to get adequate mixing in the tank. Does she need to speed up stirring for this tank?

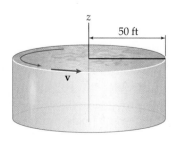

53. Traffic can be described by fluid properties such as density and momentum; hence it is usually modeled as a continuum. Consider a straight road with multiple lanes going due north. Fast traffic uses the lanes farther left, and slower traffic uses the ones to the right. Denote the velocity of traffic at any point of the road by $\mathbf{v}(x, y) = \langle 0, v_2(x, y)\rangle$. Also; that is, assume that the speed in any particular lane is constant; that is, $\frac{\partial v_2}{\partial y} = 0$.

(a) What can we say about the curl $\frac{\partial v_2}{\partial x} - \frac{\partial v_1}{\partial y}$ at any point on the road? What does it mean for that quantity to be small? What if it is large? Does this quantity have anything to do with traffic safety?

(b) Suppose that $v_2(x, y) = 75 - \alpha x$ in miles per hour, where x is measured in miles. Compute both sides of Green's Theorem for the traffic flow over a segment of highway described by $S = [0, 0.0113] \times [0, 0.5]$ with boundary C. In other words, verify that

$$\int_C \mathbf{v}(x, y) \cdot d\mathbf{r} = \iint_S \left(\frac{\partial v_2}{\partial x} - \frac{\partial v_1}{\partial y}\right) dx\, dy.$$

(c) What is the significance of either of the integrals from part (b) for the traffic? Comment on whether the integral on the left or the integral on the right would be easier for traffic engineers to calculate from measurements.

54. Traffic engineers use road tubes to measure the number and speed of cars passing a line in the road. Suppose traffic engineers set road tubes on a freeway whose inner edge lies along the circle $x^2 + y^2 = 0.2$ and whose outer edge is $x^2 + y^2 = 0.2113$. The traffic is flowing counterclockwise in the first quadrant of the circles. The road tubes lie along the lines $y = 0$ and $x = 0$, and the measurements show that the speed of cars along the inner circle is 69 mph while along the outside circle it is 61 mph. Denote the stretch of road in the first quadrant by S with oriented boundary C, and denote the velocity of the cars at any point by $\mathbf{v}(x, y)$.

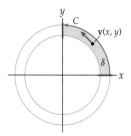

(a) Compute $\int_C \mathbf{v}(x, y) \cdot d\mathbf{r}$. What is the significance of the sign of this quantity?

(b) Use Green's Theorem to compute the average curl of the traffic on this stretch of road:

$$\frac{1}{A}\iint_S \left(\frac{\partial v_2}{\partial x} - \frac{\partial v_1}{\partial y}\right) dx\, dy.$$

Here, A is the area of the roadway.

Proofs

55. Let $\mathbf{F}$ be a vector field in $\mathbb{R}^2$ or $\mathbb{R}^3$. Prove that div curl $\mathbf{F} = 0$. (This is Theorem 14.12, part (a).)

56. Let f be a function of two or three variables. Prove that curl $\nabla f = \mathbf{0}$. (This is Theorem 14.12, part (b).)

57. Prove that, for any conservative vector field $\mathbf{F}(x, y)$,

$$\iint_R \left(\frac{\partial F_2}{\partial x} - \frac{\partial F_1}{\partial y} \right) dA = 0$$

for any simply connected region $R \subset \mathbb{R}^2$ whose boundary is smooth or piecewise smooth.

58. Give an example of a vector field defined on a simply connected region $\mathcal{R} \subset \mathbb{R}^2$ whose boundary is smooth or piecewise smooth, that is not a conservative field, but whose integral over the unit disk does have the property that

$$\iint_{\mathcal{R}} \left(\frac{\partial F_2}{\partial x} - \frac{\partial F_1}{\partial y} \right) dA = 0.$$

Your example shows that the converse of Exercise 57 fails.

Thinking Forward

▶ *Generalizing Green's Theorem:* How might we generalize Green's Theorem to two-dimensional regions that are surfaces in $\mathbb{R}^3$, rather than patches in the xy-plane? What sort of statement do you expect?

▶ *Another generalization of Green's Theorem:* How might we generalize Green's Theorem to higher dimensions? What sort of relationship might we hope to find between a triple integral over a (reasonably well-behaved) three-dimensional region of space and a double integral over the surface that is the boundary of this region?

14.5   STOKES' THEOREM

▶ Generalizing Green's Theorem to smooth surfaces other than the plane

▶ Expressing Stokes' Theorem in terms of the curl

▶ Computations using Stokes' Theorem

A Generalization of Green's Theorem

The goal of this section is to generalize Green's Theorem to regions that are two dimensional but that do not lie in the xy-plane. For this to work, we will need our surfaces to be well behaved enough that they act like a small portion of the xy-plane that floated off into space and were deformed only in a smooth (or piecewise-smooth) way, as shown in the following figure:

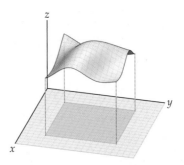

Recall from Section 14.3 that we restrict our attention to surfaces that are smooth or piecewise smooth. In these circumstances, Green's Theorem still holds, provided that we take account of the distortion of the surface.

Stokes' Theorem

The generalized form of Green's Theorem known as *Stokes' Theorem* is traditionally written in terms of the curl of a vector field. Recall that the curl form of Green's Theorem states that, under appropriate restrictions on C and $\mathcal{R}$,

$$\int_C \mathbf{F}(x, y) \cdot d\mathbf{r} = \iint_{\mathcal{R}} \text{curl } \mathbf{F}(x, y) \cdot \mathbf{k} \, dA.$$

In the context of a general surface, what matters is that $\mathbf{k}$ is a unit vector normal to $\mathcal{R}$ and pointing in the direction from which C was parametrized counterclockwise. Generalizing gives Stokes' Theorem:

THEOREM 14.15

Stokes' Theorem

Let S be a smooth or piecewise-smooth oriented surface with a smooth or piecewise-smooth boundary curve C. Suppose that S has an (oriented) unit normal vector $\mathbf{n}$ and that C has a parametrization that traverses C in the counterclockwise direction with respect to $\mathbf{n}$. If $\mathbf{F}(x, y, z) = F_1(x, y, z)\mathbf{i} + F_2(x, y, z)\mathbf{j} + F_3(x, y, z)\mathbf{k}$ is a vector field on an open region containing S, then

$$\int_C \mathbf{F}(x, y, z) \cdot d\mathbf{r} = \iint_S \text{curl } \mathbf{F}(x, y, z) \cdot \mathbf{n} \, dS.$$

If S is a region of the xy-plane, then the unit normal is $\mathbf{k}$ and the right-hand side of the preceding equation is

$$\iint_S \left(\frac{\partial F_2}{dx} - \frac{\partial F_1}{dy} \right) dA,$$

which is Green's Theorem. Another way to think of this is that Green's Theorem is a special case of Stokes' Theorem for regions in the xy-plane.

Proof. A thorough proof of Stokes' Theorem is beyond the scope of this course; however, we do provide a sketch.

To sketch the proof, first consider a plane P in space, other than the xy-plane. Since P is a plane, we could relabel points and treat the new plane as the xy-plane. This would involve a change of variables using the techniques of Chapter 13. Green's Theorem will hold here, since the relabeled plane behaves exactly like the old xy-plane in every respect. (For an example, consider substituting the xz-plane for the xy-plane. It is not too difficult to figure out what Green's Theorem should be in the xz-plane.)

Now consider an oriented smooth surface S in space. Since S is smooth, we can seek to understand it by subdividing it into many small subregions, each of which is closely approximated by the tangent plane at a point in the subregion. On this tiny piece of tangent plane, Green's Theorem applies. With a careful trace through definitions and notation, we can show that when we modify Green's Theorem to account for the plane no longer being the xy-plane, the new integrand will be curl $\mathbf{F}(x, y, z) \cdot \mathbf{n} \, dS$, as is required by Stokes' Theorem.

Next we recall the discussion of Green's Theorem for regions that are sums of regions bounded by simple closed curves, like the annular region discussed in the previous section. As the following figure indicates, we approximate the surface with small portions of approximating tangent planes. It takes a bit more work to show that the line integrals around adjoining pieces of these approximating tangent planes cancel as they do for the annular region. (The subtlety in this case is that the *planes* are different.)

A smooth surface approximated with portions of tangent planes

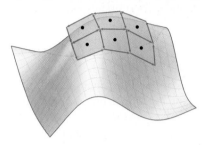

Once this is done, we resort to our usual plan of taking a limit of approximations to find the exact value. The outline of the proof of Stokes' Theorem follows the same "subdivide, approximate, and add" strategy that we have used in previous integration results. The major steps are as follows:

▶ Show that Green's Theorem can be applied in any plane and that when it is, the integrand is as desired.

▶ Subdivide S into subregions, and argue that S is well approximated by gluing lots of subdivision tangent planes together.

▶ Argue that applying Green's Theorem to all these tiny tangent planes cancels all shared borders, leaving only the perimeter.

▶ Approximate the true integral by calculating the sum of Green's Theorem on tangent planes.

▶ Add up the subregion approximations, and take a limit as the subregions become arbitrarily small. Addition becomes integration in the limit.   ■

Examples and Explorations

EXAMPLE 1   **Verifying Stokes' Theorem**

Verify Stokes' Theorem for the vector fields

(a) $\mathbf{F}(x, y, z) = yz\mathbf{i} + xz\mathbf{j} + xy\mathbf{k}$ and   **(b)** $\mathbf{G}(x, y, z) = -y\mathbf{i} + x\mathbf{j} + e^z\mathbf{k}$

on the surface defined by $S = \{(x, y, z) \mid z = 1 - x^2 - y^2, \; x^2 + y^2 \le 1\}$, with outward unit normal vector.

SOLUTION

The graph of S is the part of the downwards-opening paraboloid $z = 1 - x^2 - y^2$ that lies above the xy-plane, with boundary curve, C, equal to the unit circle in the xy-plane.

To verify Stokes' Theorem, we separately compute

$$\iint_S \text{curl } \mathbf{F}(x, y, z) \cdot \mathbf{n} \, dS \quad \text{and} \quad \int_C \mathbf{F}(x, y, z) \cdot d\mathbf{r}$$

and compare the results.

(a) For $\mathbf{F}(x, y, z) = yz\mathbf{i} + xz\mathbf{j} + xy\mathbf{k}$,

$$\text{curl } \mathbf{F}(x, y, z) = \nabla \times \mathbf{F}(x, y, z) = \mathbf{0}.$$

This equation implies that

$$\iint_S \text{curl } \mathbf{F}(x, y, z) \cdot \mathbf{n} \, dS = 0.$$

Turning to the line integral of $\mathbf{F}(x, y, z)$ about the boundary of S, we observe that $\mathbf{F}(x, y, z)$ is conservative:

$$\mathbf{F}(x, y, z) = \nabla f, \quad \text{where} \quad f(x, y, z) = xyz.$$

By the Fundamental Theorem of Line Integrals,

$$\int_C \mathbf{F}(x, y, z) \cdot d\mathbf{r} = 0,$$

because C is closed. (Note that Theorem 14.12 also explains why curl $\mathbf{F}(x, y, z) = \mathbf{0}$, since $\mathbf{F}(x, y, z)$ is conservative.)

(b) Recall that, for a surface S given by $z = f(x, y)$, a normal vector is $\mathbf{N} = \left(-\dfrac{\partial z}{\partial x}, -\dfrac{\partial z}{\partial y}, 1 \right)$. This vector in general needs to be scaled before it becomes a unit normal vector, and it may point in the opposite of the desired direction, depending on orientation of S. In this case,

$$\mathbf{N} = \langle 2x, 2y, 1 \rangle,$$

which does point in the desired direction, away from the z-axis. The scaling will cancel in the integration.

Combining the above with the fact that curl $\mathbf{G}(x, y, z) = \text{curl } (-y\mathbf{i} + x\mathbf{j} + e^z\mathbf{k}) = 2\mathbf{k}$, we have

$$\iint_S \text{curl } \mathbf{G}(x, y, z) \cdot \mathbf{N} \, dS = \iint 2 \, dA$$
$$= \int_0^{2\pi} \int_0^1 2 \, r \, dr \, d\theta$$
$$= \int_0^{2\pi} \left[r^2 \right]_0^1 \, d\theta$$
$$= \int_0^{2\pi} 1 \, d\theta = 2\pi.$$

To compute the line integral $\int_C \mathbf{G}(x, y, z) \cdot d\mathbf{r}$ directly, we first note that C is parametrized by

$$\mathbf{r}(t) = \cos t \, \mathbf{i} + \sin t \, \mathbf{j} + 0\mathbf{k} \quad \text{for} \quad 0 \le t \le 2\pi$$

with

$$d\mathbf{r} = \langle -\sin t, \cos t, 0 \rangle \, dt \quad \text{for} \quad 0 \le t \le 2\pi.$$

On the curve C,

$$\mathbf{G}(x, y, z) = -y\mathbf{i} + x\mathbf{j} + e^z\mathbf{k} = -\sin t \, \mathbf{i} + \cos t \, \mathbf{j} + 1\mathbf{k}.$$

Substituting gives

$$\int_C \mathbf{G}(x, y, z) \cdot d\mathbf{r} = \int_0^{2\pi} (\sin^2 t + \cos^2 t) \, dt = \int_0^{2\pi} 1 \, dt = 2\pi.$$

Again, we have verified Stokes' Theorem.

EXAMPLE 2   **Using Stokes' Theorem for a graph $y = f(x, z)$**

Use Stokes' Theorem to compute $\int_C \mathbf{F}(x, y, z) \cdot d\mathbf{r}$, where

$$\mathbf{F}(x, y, z) = xyz\mathbf{i} + (y - 2)\mathbf{j} + yz\mathbf{k}$$

and C is the boundary of the region of the plane

$$x + y + z = 4$$

that lies to the right of the triangular region $0 \le z \le 4,\ 0 \le x \le z$ in the xz-plane, oriented in the positive y direction.

SOLUTION

The surface shown next is the graph of $y = 4 - x - z$.

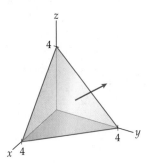

Modifying our previous calculation of $\mathbf{N}$ for surfaces $z = f(x, y)$, we have the normal vector

$$\mathbf{N} = \left\langle -\frac{\partial y}{\partial x}, 1, -\frac{\partial y}{\partial z} \right\rangle = \langle 1, 1, 1 \rangle.$$

Computing curl $\mathbf{F}(x, y, z)$ yields:

$$\text{curl } \mathbf{F}(x, y, z) = z\mathbf{i} + xy\mathbf{j} - xz\mathbf{k}.$$

Then,

$$\iint_S \text{curl } \mathbf{F}(x, y, z) \cdot \mathbf{N}\, dS = \int_0^2 \int_0^z (z + xy - xz)\, dx\, dz$$

$$= \int_0^2 \int_0^z (z + x(4 - x - z) - xz)\, dx\, dz$$

$$= \int_0^2 \left[xz + \frac{1}{2}(4 - 2z)x^2 - \frac{1}{3}x^3 \right]_0^z dz$$

$$= \int_0^2 \left(3z^2 - \frac{4}{3}z^3 \right) dz$$

$$= \left[z^3 - \frac{1}{3}z^4 \right]_0^2 = \frac{8}{3}. \qquad \square$$

EXAMPLE 3   **Using Stokes' Theorem for an arbitrary surface**

Use Stokes' Theorem to evaluate the integral of the vector field

$$\mathbf{F}(x, y, z) = \left\langle e^{xyz}, -xy^2z, xyz^2 \right\rangle$$

around the curve C given by $y^2 + z^2 = 9$ in the plane $x = 5$ and traversed in the counterclockwise direction when viewed from the right (i.e., where $x > 5$).

SOLUTION

Here any smooth orientable surface whose boundary is C will do. We make the simple choice of a disk of radius 3 in the plane $x = 5$ and with center $(5, 0, 0)$. This plane is parallel to the yz-plane. Hence, a unit normal vector is $\mathbf{i}$.

Evaluating $\mathbf{F}(x, y, z)$, we have

$$\text{curl } \mathbf{F} = \nabla \times \langle e^{xyz}, -xy^2z, xyz^2 \rangle$$

$$= \langle xz^2 + xy^2, xye^{xyz} - yz^2, -y^2z - xze^{xyz} \rangle.$$

If we let D be the disk with radius 3, centered on the x-axis, and in the plane $x = 5$, substituting according to Stokes' Theorem gives

$$\int_C \mathbf{F}(x, y, z) \cdot d\mathbf{r} = \iint_S \text{curl } \mathbf{F}(x, y, z) \cdot \mathbf{n} \, dS = \iint_D (xz^2 + xy^2) \, dA$$

$$= \iint_D x(z^2 + y^2) \, dA.$$

To finish the integration, we will use polar coordinates. Here, $r^2 = z^2 + y^2$ and $dA = r \, dr \, d\theta$. Furthermore, since $x = 5$, we have

$$\iint_D x(z^2 + y^2) \, dA = \int_0^{2\pi} \int_0^3 5r^2 \, r \, dr \, d\theta = \int_0^{2\pi} \left[\frac{5}{4}r^4 \right]_0^3 \, d\theta$$

$$= \int_0^{2\pi} \frac{405}{4} \, d\theta = \frac{405\pi}{2}.$$

□

? TEST YOUR UNDERSTANDING

▶ What is the relationship between Green's Theorem and Stokes' Theorem?

▶ What is the relationship between Stokes' Theorem and the Fundamental Theorem of Calculus?

▶ Why are both sides of the equation in Stokes' Theorem potentially more complicated than they are in either the Fundamental Theorem of Calculus or Green's Theorem?

▶ Suppose that S_1 and S_2 are two different smooth oriented surfaces with a common boundary curve C and a common orientation relative to C. (That is, travelling counterclockwise along C with respect to S_1 is the same direction as travelling counterclockwise with respect to S_2.) Let $\mathbf{F}(x, y, z)$ be a vector field that is defined on an open set containing both surfaces. What does Stokes' Theorem imply about

$$\iint_{S_1} \mathbf{F}(x, y, z) \cdot \mathbf{n_1} \, dS \quad \text{and} \quad \iint_{S_2} \mathbf{F}(x, y, z) \cdot \mathbf{n_2} \, dS$$

if, for example, S_1 is the hemisphere that is the graph of $z = \sqrt{1 - x^2 - y^2}$ and S_2 is the paraboloid that is the graph of $z = 4(x^2 + y^2 - 1)$, as shown in the figures that follow? Both of these graphs are bounded by the unit circle in the xy-plane.

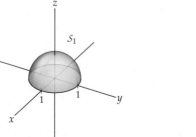
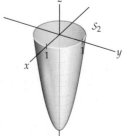

▶ In the examples, why are $\mathbf{n}$ and dS never explicitly separated? (It may be useful to look at Section 14.3.)

EXERCISES 14.5

Thinking Back

Antecedents: Stokes' Theorem is an immediate generalization of Green's Theorem, but it is related to other theorems as well. For instance, it can also be viewed as a two-dimensional version of the Fundamental Theorem of Line Integrals.

▶ Review the Fundamental Theorem of Line Integrals in Section 14.2. For conservative vector fields, what does the theorem imply about $\int_{C_1} \mathbf{F} \cdot d\mathbf{r}$ and $\int_{C_2} \mathbf{F} \cdot d\mathbf{r}$ if C_1

and C_2 are two smooth curves with common initial and terminal points and $\mathbf{F}$ is smooth?

▶ Review the Fundamental Theorem of Calculus in Chapter 4. What parts of the Stokes' Theorem equation correspond to the interval of integration, the integrand, and the antiderivative?

Concepts

0. *Problem Zero:* Read the section and make your own summary of the material.

1. *True/False:* Determine whether each of the statements that follow is true or false. If a statement is true, explain why. If a statement is false, provide a counterexample.

   (a) *True or False:* Stokes' Theorem asserts that the flux of a vector field through a smooth surface with a smooth boundary is equal to the line integral of this field about the boundary of the surface.

   (b) *True or False:* Stokes' Theorem can be interpreted as a generalization of Green's Theorem.

   (c) *True or False:* Stokes' Theorem applies only to conservative vector fields.

   (d) *True or False:* Stokes' Theorem is always used as a way to evaluate difficult surface integrals.

   (e) *True or False:* Stokes' Theorem can be interpreted as a generalization of the Fundamental Theorem of Line Integrals.

   (f) *True or False:* If $\mathbf{F}(x, y, z)$ is a conservative vector field, then Stokes' Theorem and Theorem 14.12 together give an alternative proof of the Fundamental Theorem of Line Integrals for simple closed curves.

   (g) *True or False:* Stokes' Theorem can be interpreted as a generalization of the Fundamental Theorem of Calculus.

   (h) *True or False:* Stokes' Theorem can be used to evaluate surface area.

2. *Examples:* Construct examples of the thing(s) described in the following. Try to find examples that are different than any in the reading.

   (a) A smooth surface with a smooth boundary.

   (b) A surface that is not smooth, but that has a smooth boundary.

   (c) A surface that is smooth, but does not have a smooth boundary.

3. Write two different normal vectors for a smooth surface S given by $(x, y, g(x, y))$ at the point $(x_0, y_0, g(x_0, y_0))$.

4. Make a chart of all the new notation, definitions, and theorems in this section, including what each new thing means in terms you already understand.

5. Suppose that S_1 is the upper half of the unit sphere, with outwards-pointing normal $\mathbf{n}_1$, and S_2 is a balloon-shaped surface whose boundary is the unit circle and

whose orientation leads to counterclockwise parametrization of the unit circle. If $\mathbf{F}(x, y, z)$ is a smooth vector field defined on a region large enough to include both surfaces, what is the relationship between $\iint_{S_1} \text{curl } \mathbf{F} \cdot \mathbf{n}_1 \, dS$ and $\iint_{S_2} \text{curl } \mathbf{F} \cdot \mathbf{n}_2 \, dS$?

6. In what way is Green's Theorem a special case of Stokes' Theorem?

7. If curl $\mathbf{F}(x, y, z) \cdot \mathbf{n}$ is constantly equal to 1 on a smooth surface S with a smooth boundary curve C, then Stokes' Theorem can reduce the integral for the surface area to a line integral. State this integral.

8. In what way is Stokes' Theorem a generalization of the Fundamental Theorem of Line Integrals?

9. Why is the orientation of S important to the statement of Stokes' Theorem? What will change if the orientation is reversed?

10. Draw a picture of a smooth surface with subdivisions and locally approximating tangent planes.

11. Give an example of a vector field whose orientation does not affect the outcome of Stokes' Theorem.

12. In your own words, explain how the "subdivide, approximate, and add" strategy from Chapter 4 is applied in the sketch of the proof of Stokes' Theorem.

13. Why is dA in Green's Theorem replaced by dS in Stokes' Theorem?

14. Given a smooth surface S with boundary curve the unit circle traversed counterclockwise, rewrite

$$\iint_S \text{curl } \mathbf{F}(x, y, z) \cdot \mathbf{n} \, dS$$

as a double integral in the plane.

15. Given an integral of the form $\int_C \mathbf{F} \cdot d\mathbf{r}$, what considerations would lead you to evaluate the integral with Stokes' Theorem?

16. Given an integral of the form $\iint_S \text{curl } \mathbf{F}(x, y, z) \cdot \mathbf{n} \, dS$, what considerations would lead you to evaluate the integral with Stokes' Theorem?

17. Why does the statement of Stokes' Theorem require that the surface S be smooth or piecewise smooth? What, if anything, goes wrong if this condition is not met?

18. Why does the statement of Stokes' Theorem require that the boundary curve C be smooth or piecewise smooth? What, if anything, goes wrong if this condition is not met?

Skills

Evaluate the integrals in Exercises 19–32. In some cases, Stokes' Theorem will help; in other cases, it may be preferable to evaluate the integrals directly.

19. $\int_C \mathbf{F}(x, y, z) \cdot d\mathbf{r}$, where C is the boundary of the triangle in the plane $y = 2$ and with vertices $(1, 2, 0)$, $(0, 2, 0)$, and $(0, 2, 1)$, with the normal vector pointing in the positive y direction and

$$\mathbf{F}(x, y, z) = 3yz\mathbf{i} + e^x\mathbf{j} + x^2 z\mathbf{k}.$$

20. $\int_C \mathbf{F}(x, y, z) \cdot d\mathbf{r}$, where C is in the plane $z = 12 - x - y$, with upwards-pointing normal vector, and is the boundary of the region that lies above the square in the xy-plane and with vertices $(3, 5, 0)$, $(3, 7, 0)$, $(4, 5, 0)$, and $(4, 7, 0)$, and where

$$\mathbf{F}(x, y, z) = (y - z)\mathbf{i} + 2x\mathbf{j} + 5xz\mathbf{k}.$$

21. $\iint_S \operatorname{curl} \mathbf{F}(x, y, z) \cdot \mathbf{n}\, dS$, where S is the portion of the plane $x + y - z = 0$ with upwards-pointing normal vector and

$$\mathbf{F}(x, y, z) = yze^{xyz}\mathbf{i} + xze^{xyz}\mathbf{j} + xye^{xyz}\mathbf{k}.$$

22. $\iint_S \operatorname{curl} \mathbf{F}(x, y, z) \cdot \mathbf{n}\, dS$, where S is the portion of the paraboloid $z = x^2 + y^2$ that lies above the square $-1 \le x \le 1$, $0 \le y \le 2$ in the xy-plane with upwards-pointing normal vector and

$$\mathbf{F}(x, y, z) = x^3 yz\mathbf{i} + xy^2 z\mathbf{j} + yz\mathbf{k}.$$

23. $\int_C \mathbf{F}(x, y, z) \cdot d\mathbf{r}$, where C is the closed curve in the plane $y = x$ and formed by the curves $x = z$ and $x^2 = z$, traversed counterclockwise with respect to normal vector $\mathbf{n} = \langle 1, -1, 0 \rangle$, and where

$$\mathbf{F}(x, y, z) = 7xy\mathbf{i} - z\mathbf{j} + 3xyz\mathbf{k}.$$

24. $\int_C \mathbf{F}(x, y, z) \cdot d\mathbf{r}$, where C is the boundary of the region in the plane $z = 2x - y + 10$ and that lies above the curves $x = 1$, $x = 2$, and $y = e^x$, and where

$$\mathbf{F}(x, y, z) = (3x + y)\mathbf{i} + (y - 2z)\mathbf{j} + (2 + 3z)\mathbf{k}.$$

25. $\iint_S \operatorname{curl} \mathbf{F}(x, y, z) \cdot \mathbf{n}\, dS$, where S is the cap of the unit sphere that lies below the xy-plane and inside the cylinder $x^2 + y^2 = \frac{1}{9}$ with outwards-pointing normal vector and where

$$\mathbf{F}(x, y, z) = -yz^2\mathbf{i} + xz^2\mathbf{j} + 3^{-xyz}\mathbf{k}.$$

26. $\iint_S \operatorname{curl} \mathbf{F}(x, y, z) \cdot \mathbf{n}\, dS$, where S is the portion of the hyperbolic paraboloid $z = x^2 - y^2$ that lies inside the elliptical cylinder $4x^2 + 9y^2 = 36$ with upwards-pointing normal vector and $\mathbf{F}(x, y, z) = (1 - yz\sin(xyz))\mathbf{i} - (1 + xz\sin(xyz))\mathbf{j} + (1 - xy\sin(xyz))\mathbf{k}$.

27. $\int_C \mathbf{F}(x, y, z) \cdot d\mathbf{r}$, where C is the curve in the plane $x - y + z = 20$ and that lies above the curves $y = 4$ and $y = x^2$ in the xy-plane, traversed counterclockwise with respect to $\mathbf{n} = \langle 1, -1, 1 \rangle$, and where $\mathbf{F}(x, y, z) = (2x - 3y + 4z)\mathbf{i} + (5x + y - z)\mathbf{j} + (x + 4y + 2z)\mathbf{k}$.

28. $\int_C \mathbf{F}(x, y, z) \cdot d\mathbf{r}$, where C is the curve on the paraboloid $z = x^2 + y^2$ that lies above the unit circle, traversed counterclockwise with respect to the outwards-pointing normal vector, and where

$$\mathbf{F}(x, y, z) = (3x + y - z)\mathbf{i} + (4y - 2z)\mathbf{j} + (x - 3z)\mathbf{k}.$$

29. $\iint_S \operatorname{curl} \mathbf{F}(x, y, z) \cdot \mathbf{n}\, dS$, where S is the portion of the surface $y = \sqrt{4 - x^2 - z^2}$ that lies between $y = 4$ and $y = \sqrt{3}$ with normal vector pointing in the positive y direction and where

$$\mathbf{F}(x, y, z) = (-4z - xz^2)\mathbf{i} + \sin(xyz)\mathbf{j} + (4x + x^2 z)\mathbf{k}.$$

30. $\iint_S \operatorname{curl} \mathbf{F}(x, y, z) \cdot \mathbf{n}\, dS$, where S is the portion of the surface $z = e^{x^2 + y^2}$ that lies above the plane $z = \frac{1}{4}$ with upwards-pointing normal vector and where

$$\mathbf{F}(x, y, z) = \ln(2x + 1)\mathbf{i} + 2^{y+1}\mathbf{j} + xe^y\mathbf{k}.$$

31. $\int_C \mathbf{F} \cdot d\mathbf{r}$, where C is the intersection of the surface $z = e^{-(x^2 + y^2)}$ and the cylinder $x^2 + y^2 = 9$ and where

$$\mathbf{F}(x, y, z) = \langle 3x + 3, 4x + \ln(y^2 + 1) - z, 2x + y \rangle.$$

32. $\iint_S \operatorname{curl} \mathbf{F}(x, y, z) \cdot \mathbf{n}\, dS$, where S is the bottom sheet of the hyperboloid $z^2 = x^2 + y^2 + 1$ that lies above the plane $z = 2\sqrt{2}$ with outwards-pointing normal vector.

The hypothesis of Stokes' Theorem requires that the vector field $\mathbf{F}$ be defined and continuously differentiable on an open set containing the surface S bounded by C, that C be simple, smooth, and closed, and that S be oriented and smooth. For each of the situations in Exercises 33–36, show that Stokes' Theorem does not apply.

33. $\mathbf{F}(x, y, z) = \left\langle \frac{x}{x^2 + y^2}, -\frac{y}{x^2 + y^2}, z^2 \right\rangle$, and C is the unit circle.

34. $\mathbf{F}(x, y, z) = \langle \ln(xy + 1) + 5^x 3^y 2^z, 4xz^2 \rangle$, and C is the boundary of the square in the plane $z = 6$ and with vertices at $(2, 0, 6)$, $(-2, 0, 6)$, $(2, 4, 6)$, and $(-2, 4, 6)$.

35. S is the portion of the cone $z = \sqrt{x^2 + y^2}$ inside the cylinder $x^2 + y^2 = k^2$ for some $k \ge 0$.

36. S is the pyramid with vertices at $(0, 0, 6)$, $(2, 0, 0)$, $(-2, 0, 0)$, $(0, 3, 0)$, and $(0, -3, 0)$.

Applications

For a given vector field $\mathbf{F}(x, y, z)$ and simple closed curve C, traversed counterclockwise to a chosen normal vector $\mathbf{n}$, the **circulation** of $\mathbf{F}(x, y, z)$ around C measures the rotation of the fluid about C in the direction counterclockwise to the aforementioned chosen normal vector and is defined to be $\int_C \mathbf{F}(x, y, z) \cdot d\mathbf{r}$. Find the circulation of the given vector field around C in Exercises 37 and 38.

37. $\mathbf{F}(x, y, z) = \mathbf{i} + \mathbf{j} + \mathbf{k}$, and C is the curve of intersection of the plane $\theta = \frac{\pi}{4}$ or $\theta = \frac{5\pi}{4}$ and the unit sphere.

38. $\mathbf{F}(x, y, z) = \langle 3y, -x, e^{x+y} \rangle$, and C is the intersection of the cone $z = \sqrt{x^2 + y^2}$ and the unit sphere.

Consider once again the notion of the rotation of a vector field. If a vector field $\mathbf{F}(x, y, z)$ has curl $\mathbf{F} = \mathbf{0}$ at a point P, then the field is said to be **irrotational** at that point. Show that the fields in Exercises 39–42 are irrotational at the given points.

39. $\mathbf{F}(x, y, z) = \langle -\sin x, 3y^3, 4z + 12 \rangle$, $P = (2, 3, 4)$.

40. $\mathbf{F}(x, y, z) = \langle y + z, x + z, x + y \rangle$, $P = (1, -1, 1)$.

41. $\mathbf{F}(x, y, z) = \langle 2xyz, x^2z, x^2y \rangle$, $P = (1, 1, 1)$.

42. $\mathbf{F}(x, y, z) = \nabla f(x, y, z)$, where $f(x, y, z)$ is a function defined on $\mathbb{R}^3$ and has continuous first and second partial derivatives, and where $P = (x_0, y_0, z_0)$.

Find the work done by the given vector field moving around the curve in the indicated direction in Exercises 43 and 44.

43. $\mathbf{F}(x, y, z) = yz\mathbf{i} + 2xz\mathbf{j} + xy\mathbf{k}$, and C is the curve formed by the intersection of the plane $12x + 2y - z = 15$ and the cylinder $y^2 + z^2 = 4$, traversed counterclockwise with respect to the normal vector $\mathbf{n} = \langle 12, 2, -1 \rangle$.

44. $\mathbf{F}(x, y, z) = \left\langle \dfrac{2x}{x^2 + y^2 + z^2 + 1}, \dfrac{2y}{x^2 + y^2 + z^2 + 1}, \dfrac{2z}{x^2 + y^2 + z^2 + 1} \right\rangle$, and C is the curve created by the intersection of the plane $z = 4$ with the surface $x^2 + 2y - z = 0$, when $y \geq 0$, together with the line segment connecting the points $(2, 0, 4)$ and $(-2, 0, 4)$, traversed counterclockwise with respect to the normal vector $\mathbf{k}$.

45. The current through a certain region of the San Juan Islands in Washington State is given by $\mathbf{F} = \langle 0, 1.152 - 0.8x^2 \rangle$. Consider a disk R of radius 1 mile centered on this region. Denote the boundary of the disk by ∂R.

  (a) Compute $\iint_R \nabla \times \mathbf{F} \cdot \mathbf{n} \, dA$.

  (b) Show that

$$\iint_{\partial R} \mathbf{F} \cdot d\mathbf{r} = \int_0^{2\pi} (1.152 - 0.8\cos^2\theta)\sin\theta \, d\theta = 0.$$

Conclude that Stokes' Theorem is valid for the current in this region of the San Juan Islands.

  (c) What do the integrals from Stokes' Theorem tell us about this region of the San Juan Islands?

Proofs

46. (a) Use Stokes' Theorem and the Fundamental Theorem of Line Integrals to show that

$$\iint_S \text{curl } \mathbf{F}(x, y, z) \cdot \mathbf{n} \, dS = 0$$

for any conservative vector field $\mathbf{F}$.

  (b) Without using Stokes' Theorem, show that

$$\iint_S \text{curl } \mathbf{F}(x, y, z) \cdot \mathbf{n} \, dS = 0$$

for any conservative vector field $\mathbf{F}$.

47. Show that if C is a smooth curve in a plane $z = r$, where r is an arbitrary constant, that is parallel to the xy-plane, and if $\mathbf{F}(x, y, z) = \langle F_1(x, y, z), F_2(x, y, z), F_3(x, y, z) \rangle$, then $\int_C \mathbf{F} \cdot d\mathbf{r}$ does not depend on F_3.

48. State and prove a version of Exercise 47 for smooth curves in planes of the form $x = r$ and of the form $y = r$.

49. Let C be a smooth simple closed curve in the plane $ax + by + cz = k$ with nonzero constants a, b, c, and k, and

let $\mathbf{F}(x, y, z) = \langle F_1(x, y, z), F_2(x, y, z), F_3(x, y, z) \rangle$ be a vector field whose component functions are linear. Show that $\int_C \mathbf{F} \cdot d\mathbf{r}$ depends only on the area enclosed by C.

50. Show that $\int_C \mathbf{F} \cdot d\mathbf{r} = 0$ does not imply that $\mathbf{F}$ is conservative, by (1) computing $\int_C \mathbf{F} \cdot d\mathbf{r}$ for C the unit circle and $\mathbf{F}(x, y, z) = x\mathbf{i} - y\mathbf{j} + 2^z \ln(z + 1)\mathbf{k}$ and (2) showing that $\mathbf{F}$ is not conservative but has zero curl.

51. Suppose S_1 and S_2 are two smooth surfaces with the same smooth boundary curve C and that traversing the boundary in the counterclockwise direction determined by the orientation of the surfaces describes the same direction of travel along C. Moreover, let $\mathbf{F}(x, y, z)$ be a smooth vector field defined on a region containing both surfaces. Show that

$$\iint_{S_1} \text{curl } \mathbf{F}(x, y, z) \cdot \mathbf{n} \, dS = \iint_{S_2} \text{curl } \mathbf{F}(x, y, z) \cdot \mathbf{n} \, dS$$

In this case, the integral is **surface independent**, a property analogous to the path independence of line integrals for conservative vector fields.

Thinking Forward

Generalizing to Higher Dimensions: Stokes' Theorem is about a surface inside of $\mathbb{R}^3$. But the essence of the theorem has to do with integrating over a surface, and does not use any properties that $\mathbb{R}^3$ has which $\mathbb{R}^4$ does not.

  ▶ How would a version of Stokes' Theorem for $\mathbb{R}^4$ need to change to accommodate the extra dimension?

  ▶ If we take the perspective that Stokes' Theorem is a two-dimensional version of the Fundamental Theorem of Line Integrals, it is natural to ask what a three-dimensional version of Stokes' Theorem would look like. Make a conjecture about a three-dimensional version of Stokes' Theorem. (Note that the previous question asks about a surface in $\mathbb{R}^4$, while this question is about a three-dimensional region in $\mathbb{R}^4$.)

14.6   THE DIVERGENCE THEOREM

▶   Extending the Fundamental Theorem of Calculus to three-dimensional regions

▶   The Divergence Theorem

▶   Justifying the Divergence Theorem

The Divergence Theorem

Our final theorem of vector analysis presents the Divergence Theorem, another generalization of Green's Theorem and, in turn, of the Fundamental Theorem of Calculus. Here the essential boundary-to-interior relationship is extended to three-dimensional solids, whose boundaries are surfaces.

To set up the Divergence Theorem, we will need to discuss bounded three-dimensional regions in space. The sort of solids we are interested in are regions $W \subseteq \mathbb{R}^3$ whose boundaries are smooth (or piecewise-smooth), simple closed oriented surfaces. Intuitively, we want the boundary of W to be a reasonably well-behaved surface with a choice of inward and outward normal vector. Recall our requirement that all vector fields we use in this chapter be smooth; that is, that their component functions be continuous and have continuous first derivatives.

To see how Green's Theorem can be scaled up in dimension, let W be a reasonably well-behaved region in $\mathbb{R}^3$. For example, W might be described as

$$\{(x, y, z) \mid a \le x \le b, \ g_1(x) \le y \le g_2(x), \ f_1(x, y) \le z \le f_2(x, y)\}$$

or analogous solids for each of the other five permutations of x, y, and z.

Recall the divergence form of Green's Theorem, which reads, under the usual assumptions about $\mathbf{F}(x, y, z)$, C, and R,

$$\int_C \mathbf{F}(x, y) \cdot \mathbf{n} \, ds = \iint_R \operatorname{div} \mathbf{F} \, dA.$$

If W is as described, then Green's Theorem applies to every slice of W that is parallel to the xy-plane. On the basis of past experience, it is reasonable to suppose that subdividing the height range into small subintervals, using a choice from within each subinterval to approximate, and adding the approximations will, in the limit, yield the correct value.

On the left-hand side of the preceding equation, we have an integral of line integrals, which should result in a two-dimensional, or surface, integral. On the right-hand side of the equation, we have a double integral, which should result in a volume.

The Divergence Theorem says that this is in essence what happens:

THEOREM 14.16   **Divergence Theorem**

Suppose W is a region in $\mathbb{R}^3$ bounded by a smooth or piecewise-smooth closed oriented surface $\mathcal{S}$. If $\mathbf{F}(x, y, z)$ is defined on an open region containing W, then

$$\iiint_W \operatorname{div} \mathbf{F}(x, y, z) \, dV = \iint_{\mathcal{S}} \mathbf{F}(x, y, z) \cdot \mathbf{n} \, d\mathcal{S}$$

where $\mathbf{n}$ is the outwards unit normal vector.

Proof. Again, a thorough proof of this theorem is beyond the scope of this book. We sketch a direct proof of the Divergence Theorem for very well-behaved regions in $\mathbb{R}^3$.

To prove the Divergence Theorem for the type of region W that we have described, it suffices to prove that

$$\iiint_W \frac{\partial F_1}{\partial x}\, dV = \iint_S F_1(x,y,z)\mathbf{i}\cdot\mathbf{n}\, dS,$$

$$\iiint_W \frac{\partial F_2}{\partial y}\, dV = \iint_S F_2(x,y,z)\mathbf{j}\cdot\mathbf{n}\, dS, \text{ and}$$

$$\iiint_W \frac{\partial F_3}{\partial z}\, dV = \iint_S F_3(x,y,z)\mathbf{k}\cdot\mathbf{n}\, dS$$

where $\mathbf{n} = \langle n_1, n_2, n_3\rangle$ and $\mathbf{F} = \langle F_1, F_2, F_3\rangle$. We prove the last equation for regions W as we have described them. Such regions can be understood as the regions between $S_1 = \{(x,y,z) \mid z = f_1(x,y)\}$ and $S_2 = \{(x,y,z) \mid z = f_2(x,y)\}$ for points (x,y) in the region D of the xy-plane described by $a \le x \le b, g_1(x) \le y \le g_2(x)$. Then

$$\iint_S F_3(x,y,z)\mathbf{k}\cdot\mathbf{n}\, dS = \iint_{S_2} F_3(x,y,z)\mathbf{k}\cdot\mathbf{n}\, dS + \iint_{S_1} F_3(x,y,z)\mathbf{k}\cdot\mathbf{n}\, dS.$$

We let $M = \left\| \left\langle \frac{\partial F_2}{\partial x}, \frac{\partial F_2}{\partial y}, -1\right\rangle \right\|$. On S_2, $\mathbf{n}$ has a positive $\mathbf{k}$-component, so $\mathbf{n} = \frac{1}{M}\left\langle -\frac{\partial F_2}{\partial x}, -\frac{\partial F_2}{\partial y}, 1\right\rangle$, and on S_1, $\mathbf{n}_1$ has a negative $\mathbf{k}$-component, so $\mathbf{n}_1 = \frac{1}{M}\left\langle \frac{\partial F_2}{\partial x}, \frac{\partial F_2}{\partial y}, -1\right\rangle$, and $\mathbf{n}_1 = -\mathbf{n}$. Substituting, we have

$$\iint_S F_3(x,y,z)\,\mathbf{k}\cdot\mathbf{n}\, dS = \iint_{S_2} F_3(x,y,z)\,\mathbf{k}\cdot\mathbf{n}\, dS - \iint_{S_1} F_3(x,y,z)\,\mathbf{k}\cdot\mathbf{n}\, dS$$

$$= \iint_S \left(\int_{f_1(x,y)}^{f_2(x,y)} \frac{\partial F_3}{\partial z}\, dz\right) dS = \iiint_W \frac{\partial F_3}{\partial z}\, dV. \qquad \blacksquare$$

One way to give a physical intuition for the Divergence Theorem is to recall that the divergence of a vector field in $\mathbb{R}^3$ measures whether or not a gas represented by the field is expanding or compressing. If a gas occupies a region of space, and it is expanding, then the left-hand side of the Divergence Theorem—the integral of div $\mathbf{F}$—will be positive. At the same time, if a gas in a region is expanding, we would expect that it will have a positive flux flowing out through its boundary. This is part of what the Divergence Theorem asserts: Since both sides of the equation give the same value, in particular they have the same sign. The finer detail in the theorem keeps track of the exact relationship.

The Divergence Theorem can be extended to relate surface and volume integrals for solids bounded by more than one closed surface. This procedure is similar to that for Green's Theorem and, as in that case, requires some bookkeeping to make sure that the theorem is being applied correctly. An example is given in Example 3.

A Summary of Theorems About Vector Analysis

Having arrived at the end of a chapter on vector analysis, which is in turn at the end of a three-semester calculus sequence, it is appropriate to examine some of the most salient results side by side, to appreciate their similarities as well as their differences arising from context. A summary of the chief results of this chapter and their precursors follows. You may find it a useful exercise to reconsider the notation, background assumptions, and context pertaining to each theorem.

Theorem Name	Integral Statement	Reference
Fundamental Theorem of Calculus	$\int_a^b f(x)\,dx = F(b) - F(a)$	Theorem 4.24
Fundamental Theorem of Line Integrals	$\int_C \nabla f(x, y, z) \cdot d\mathbf{r} = f(Q) - f(P)$	Theorem 14.5
Green's Theorem	$\iint_R \left(\frac{\partial F_2}{\partial x} - \frac{\partial F_1}{\partial y} \right) dA = \int_C \mathbf{F}(x, y) \cdot d\mathbf{r}$	Theorem 14.13
Stokes' Theorem	$\iint_S \operatorname{curl} \mathbf{F}(x, y, z) \cdot \mathbf{n}\,dS = \int_C \mathbf{F}(x, y, z) \cdot d\mathbf{r}$	Theorem 14.15
Divergence Theorem	$\iiint_W \operatorname{div} \mathbf{F}(x, y, z)\,dV = \iint_S \mathbf{F}(x, y, z) \cdot \mathbf{n}\,dS$	Theorem 14.16

Examples and Explorations

EXAMPLE 1

Verifying the Divergence Theorem

Verify the conclusion of the Divergence Theorem for the vector field

$$\mathbf{F}(x, y, z) = x^2 \mathbf{i} + y^2 \mathbf{j} + z^2 \mathbf{j}$$

with region $\mathcal{R}$, the unit ball centered at the origin.

SOLUTION

To compute

$$\iint_S \mathbf{F}(x, y, z) \cdot \mathbf{n}\,dS,$$

recall that $\mathcal{S}$, the unit sphere that is the boundary of $\mathcal{R}$, is parametrized by

$$\mathbf{r}(\phi, \theta) = (\cos\theta \sin\phi, \sin\theta \sin\phi, \cos\phi), \quad \text{for } 0 \leq \phi \leq \pi \text{ and } 0 \leq \theta \leq 2\pi.$$

Then

$$\mathbf{r}_\phi(\phi, \theta) = \langle \cos\theta \cos\phi, \sin\theta \cos\phi, -\sin\phi \rangle \text{ and } \mathbf{r}_\theta(\phi, \theta) = \langle -\sin\theta \sin\phi, \cos\theta \sin\phi, 0 \rangle.$$

The cross product,

$$\mathbf{r}_\phi \times \mathbf{r}_\theta = \langle \cos\theta \sin^2\phi, \sin\theta \sin^2\phi, \cos\phi \sin\phi \rangle,$$

and $\|\mathbf{r}_\phi \times \mathbf{r}_\theta\| = \sin\phi$. For all values of ϕ and θ, the vector $\mathbf{r}_\phi \times \mathbf{r}_\theta$ points away from the origin and thus in the direction of the outwards unit normal. Therefore, we let

$$\mathbf{n} = \frac{\mathbf{r}_\phi \times \mathbf{r}_\theta}{\sin\phi} = \langle \cos\theta \sin\phi, \sin\theta \sin\phi, \cos\phi \rangle.$$

On $\mathcal{S}$,

$$\mathbf{F}(x, y, z) = \langle \cos^2\theta \sin^2\phi, \sin^2\theta \sin^2\phi, \cos^2\phi \rangle.$$

So,

$$\mathbf{F}(x, y, z) \cdot \mathbf{n} = \langle \cos^2\theta \sin^2\phi, \sin^2\theta \sin^2\phi, \cos^2\phi \rangle \cdot \langle \cos\theta \sin\phi, \sin\theta \sin\phi, \cos\phi \rangle$$
$$= \cos^3\theta \sin^3\phi + \sin^3\theta \sin^3\phi + \cos^3\phi.$$

Substituting, we have

$$\iint_S \mathbf{F}(x,y,z) \cdot \mathbf{n} \, dS = \int_0^{2\pi} \int_0^{\pi} (\cos^3 \theta \sin^3 \phi + \sin^3 \theta \sin^3 \phi + \cos^3 \phi) \, d\phi \, d\theta$$

$$= \int_0^{2\pi} \left[(\cos^3 \theta + \sin^3 \theta) \left(-\frac{1}{3} \sin^2 \phi \cos \phi - \frac{2}{3} \cos \phi \right) - \frac{1}{3} \cos^2 \phi \sin \phi - \frac{2}{3} \phi \right]_0^{\pi} d\theta$$

$$= \int_0^{2\pi} \frac{4}{3} (\cos^3 \theta + \sin^3 \theta) \, d\theta$$

$$= \frac{4}{3} \left[\sin \theta - \frac{1}{3} \sin^3 \theta - \cos \theta + \frac{1}{3} \cos^3 \theta \right]_0^{2\pi} = 0.$$

To compute the triple integral given by the Divergence Theorem, we find

$$\text{div } \mathbf{F}(x,y,z) = 2x + 2y + 2z.$$

Since $\mathcal{R}$ is the unit sphere, this integral is most convenient to evaluate in spherical coordinates. Converting div $\mathbf{F}(x,y,z)$ into spherical coordinates gives

$$\iiint_R \text{div } \mathbf{F}(x,y,z) \, dV$$

$$= 2 \int_0^{2\pi} \int_0^{\pi} \int_0^1 (\rho \cos \theta \sin \phi + \rho \sin \theta \sin \phi + \rho \cos \phi) \, \rho^2 \sin \phi \, d\rho \, d\phi \, d\theta$$

$$= 2 \int_0^{2\pi} \int_0^{\pi} \int_0^1 \rho^3 (\cos \theta \sin^2 \phi + \sin \theta \sin^2 \phi + \cos \phi \sin \phi) \, d\rho \, d\phi \, d\theta$$

$$= 2 \int_0^{2\pi} \int_0^{\pi} \left[\frac{1}{4} \rho^4 (\cos \theta \sin^2 \phi + \sin \theta \sin^2 \phi + \cos \phi \sin \phi) \right]_0^1 d\phi \, d\theta$$

$$= \frac{1}{2} \int_0^{2\pi} \int_0^{\pi} (\cos \theta \sin^2 \phi + \sin \theta \sin^2 \phi + \cos \phi \sin \phi) \, d\phi \, d\theta$$

$$= \frac{1}{2} \int_0^{2\pi} \left[(\cos \theta + \sin \theta) \left(\frac{1}{2} \phi - \frac{1}{4} \sin 2\phi \right) + \frac{1}{2} \sin^2 \phi \right]_0^{\pi} d\theta$$

$$= \frac{\pi}{4} \int_0^{2\pi} (\cos \theta + \sin \theta) \, d\theta$$

$$= \frac{\pi}{4} \left[\sin \theta - \cos \theta \right]_0^{2\pi} = 0.$$

EXAMPLE 2 **Using the Divergence Theorem**

Use the Divergence Theorem to evaluate the following flux integrals:

(a)

$$\iint_S \mathbf{F}(x,y,z) \cdot \mathbf{n} \, dS,$$

where

$$\mathbf{F}(x,y,z) = x\mathbf{i} + ze^{xz}\mathbf{j} + xyz\mathbf{k}$$

and S is the surface of the rectangular solid bounded by the planes $x = 0$, $x = 5$, $y = 0$, $y = 3$, $z = 0$, and $z = 7$.

(b)

$$\iint_S \mathbf{G}(x,y,z) \cdot \mathbf{n} \, dS$$

where

$$\mathbf{G}(x,y,z) = (e^{yz} - x^2 yz)\mathbf{i} + xy^2 z\mathbf{j} + 2z\mathbf{k}$$

and S is the surface of the ellipsoid $x^2 + y^2 + 4z^2 = 16$.

SOLUTION

(a) Computing the divergence of $\mathbf{F}(x, y, z)$, we find that

$$\text{div } \mathbf{F} = 1 + xy.$$

Applying the Divergence Theorem allows us to rewrite the integral:

$$\iint_S \mathbf{F}(x, y, z) \cdot \mathbf{n} \, dS = \iiint_R (1 + xy) \, dz \, dy \, dx$$

$$= \int_0^5 \int_0^3 \int_0^7 (1 + xy) \, dz \, dy \, dx$$

$$= \int_0^5 \int_0^3 (7 + 7xy) \, dy \, dx$$

$$= \int_0^5 \left[7y + \frac{7}{2}xy^2 \right]_0^3$$

$$= \int_0^5 \left(21 + \frac{63}{2}x \right) dx$$

$$= \left[21x + \frac{63}{4}x^2 \right]_0^5 = \frac{1995}{4}.$$

(b) Computing the divergence of $\mathbf{G}(x, y, z)$, we find that

$$\text{div } \mathbf{G} = -2xyz + 2xyz + 2 = 2.$$

Applying the Divergence Theorem allows us to rewrite the surface integral as a volume integral, which we evaluate by means of cylindrical coordinates.

$$\iint_S \mathbf{G} \cdot \mathbf{n} \, dS = \iiint_R \text{div } \mathbf{G} \, dV$$

$$= \int_0^{2\pi} \int_0^4 \int_{-\sqrt{4-(r^2/4)}}^{\sqrt{4-(r^2/4)}} 2 \, r \, dz \, dr \, d\theta$$

$$= \int_0^{2\pi} \int_0^4 [2rz]_{-\sqrt{4-(r^2/4)}}^{\sqrt{4-(r^2/4)}} \, dr \, d\theta$$

$$= \int_0^{2\pi} \int_0^4 4r\sqrt{4 - \frac{1}{4}r^2} \, dr \, d\theta$$

$$= \int_0^{2\pi} \left[-\frac{16}{3}\left(4 - \frac{1}{4}r^2 \right)^{3/2} \right]_0^4 d\theta$$

$$= \int_0^{2\pi} \frac{128}{3} \, d\theta$$

$$= \frac{256\pi}{3}.$$

☐

EXAMPLE 3   **Using the Divergence Theorem for a region bounded by two surfaces**

Use the Divergence Theorem to evaluate

$$\iint_{S_1} \mathbf{F}(x, y, z) \cdot \mathbf{n} \, dS - \iint_{S_2} \mathbf{F}(x, y, z) \cdot \mathbf{n} \, dS,$$

where

$$\mathbf{F}(x, y, z) = x\mathbf{i} + y\mathbf{j} + z\mathbf{k},$$

S_1 is the surface of the sphere of radius 4 centered at the origin, and S_2 is the unit sphere.

SOLUTION

The region is bounded by two concentric spheres:

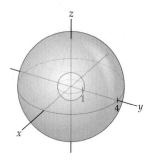

Computing div $\mathbf{F}$, we find that div $\mathbf{F} = 1 + 1 + 1 = 3$. Now, applying the Divergence Theorem gives

$$\iint_{S_1} \mathbf{F}(x, y, z) \cdot \mathbf{n}\, dS - \iint_{S_2} \mathbf{F}(x, y, z) \cdot \mathbf{n}\, dS = \int_0^{2\pi} \int_0^{\pi} \int_1^4 3\rho^2 \sin\phi\, d\rho\, d\phi\, d\theta$$

$$= \int_0^{2\pi} \int_0^{\pi} \left[\rho^3 \sin\phi\right]_1^4 d\phi\, d\theta$$

$$= \int_0^{2\pi} \int_0^{\pi} 63 \sin\phi\, d\phi\, d\theta$$

$$= \int_0^{2\pi} \left[-63 \cos\phi\right]_0^{\pi} d\theta$$

$$= \int_0^{2\pi} 126\, d\theta = 252\pi.$$

☐

EXAMPLE 4   **Calculating an electric field intensity by using Gauss' Law**

Gauss' Law is a method for calculating an electric field intensity $\mathbf{E}$ generated by simple charge distributions. Given some charge density $\rho(\mathbf{x})$ in a region Ω enclosed by a surface S, Gauss' Law states that the electric field at any point $\mathbf{x} \in \Omega$ satisfies

$$\nabla \cdot \mathbf{E}(\mathbf{x}) = \frac{\rho(\mathbf{x})}{\epsilon_0},$$

where ϵ_0 is a constant called the ***permittivity of free space***. Integrate both sides of this equation and apply the Divergence Theorem to establish the following integral form of Gauss' Law:

$$\iint_S \operatorname{div} \mathbf{E}\, dA = \frac{1}{\epsilon_0} \iiint_\Omega \rho\, dV.$$

SOLUTION

Integrating both sides of the the equation, we see that

$$\iiint_\Omega \nabla \cdot \mathbf{E}\, dV = \frac{1}{\epsilon_0} \iiint_\Omega \rho\, dV.$$

Applying the Divergence Theorem to the left side and noting that the boundary of the region Ω is S gives

$$\iint_S \operatorname{div} \mathbf{E}\, dA = \frac{1}{\epsilon_0} \iiint_\Omega \rho\, dV.$$

☐

TEST YOUR
? UNDERSTANDING

▶ What is the relationship between the Divergence Theorem and the Fundamental Theorem of Calculus?

▶ What relationship exists among the Divergence Theorem, Green's Theorem, and Stokes' Theorem?

▶ If we wish to use the Divergence Theorem to compute the flux of a vector field flowing into a closed bounded region $W \subseteq \mathbb{R}^3$, how does the theorem need to be adapted?

▶ In Example 1, what properties of the vector field caused the outcome to be zero?

▶ How do the change-of-variables factors arise in each of the integrals in Example 1?

EXERCISES 14.6

Thinking Back

A Long Look Back: Here at end of the chapter, which is in turn the end of the course, and moreover the end of a multisemester calculus odyssey, it is appropriate to reflect on the unifying questions, insights, and results that characterize the subject.

▶ How do the definitions of differentiation and integration that we have made throughout this text rely on the earlier idea of a limit that we defined in Chapter 1?

▶ How does the strategy of subdivision, approximation, and addition interact with limits in the justifications of the theorems given in Chapter 14?

▶ Are there any questions about integration that you had before reading this chapter that are now resolved? What are they?

▶ Are there any questions about integration that you had before reading this chapter that you still have? What are they?

▶ Do you have any new questions about integration now that had not occurred to you before reading this chapter? What are they?

Concepts

0. *Problem Zero:* Read the section and make your own summary of the material.

1. *True/False:* Determine whether each of the statements that follow is true or false. If a statement is true, explain why. If a statement is false, provide a counterexample.

   (a) *True or False:* The Fundamental Theorem of Line Integrals, Green's Theorem, Stokes' Theorem, and the Divergence Theorem bear a family resemblance to the Fundamental Theorem of Calculus and can in some way be interpreted as generalizations of that theorem.

   (b) *True or False:* The vector field **F** in the statement of the Divergence Theorem must satisfy the condition we have required throughout this chapter that the component functions of **F** be continuous with continuous derivatives on an open region that contains the region in which we wish to apply the theorem.

   (c) *True or False:* The Divergence Theorem is interesting only as a tool for evaluating difficult flux integrals.

   (d) *True or False:* It is often, though not always, easier to integrate over a region in $\mathbb{R}^3$ than to evaluate a surface integral.

   (e) *True or False:* The Divergence Theorem is a generalization of the technique of integration by parts.

   (f) *True or False:* The Divergence Theorem is a consequence of Stokes' Theorem.

   (g) *True or False:* The Divergence Theorem can be used to calculate the flux through a region that is bounded by two distinct smooth simple closed surfaces.

   (h) *True or False:* The Divergence Theorem is a generalization of Green's Theorem.

2. *Examples:* Construct examples of the thing(s) described in the following. Try to find examples that are different than any in the reading.

   (a) A simple closed surface that is smooth.

   (b) A simple closed surface that is not smooth, but is piecewise smooth.

   (c) A simple closed surface that is neither smooth nor piecewise smooth.

3. Does the Divergence Theorem apply to surfaces that are not closed?

4. Make a chart of all the new notation, definitions, and theorems in this section, including what each new item means in terms you already understand.

5. Give an intuitive explanation of why the paraboloid $z = x^2 + y^2$ is smooth but the surface of the rectangular solid bounded by the coordinate planes and the planes $x = 14$, $y = 5$, and $z = 32$ is only piecewise smooth.

6. Give an intuitive explanation of why the surface of a tetrahedron is piecewise smooth but the surface of the cone $z = \sqrt{x^2 + y^2}$ is not.

7. (a) Let $\mathbf{F}(x, y, z) = x\mathbf{i} + y\mathbf{j} + z\mathbf{k}$, and let S be the unit sphere. Make a sketch of the field. Without evaluating the integral, do you expect that $\iint_S \mathbf{F} \cdot \mathbf{n}\, dS$ will be positive, negative, or zero? Why?

(b) Let $\mathbf{F}(x, y, z) = 3\mathbf{i} + 7\mathbf{j} + 9\mathbf{k}$, and let S be the unit sphere. Make a sketch of the field. Without evaluating the integral, do you expect that $\iint_S \mathbf{F} \cdot \mathbf{n}\, dS$ will be positive, negative, or zero? Why?

(c) Use the Divergence Theorem to check your guesses in parts (a) and (b).

8. Recall from Section 14.4 that the divergence of a vector field is related to the expansion or compression of a gas whose motion is represented by that field. If a gas whose motion is represented by a vector field is expanding (or compressing) in a region of space, what effect should that have on the flux of the vector field out of (or into) a closed region W? Making some additional assumptions about the direction of the vectors in $\mathbf{F}$ may help you to think about this situation.

9. If $\mathbf{F}(x, y, z) = \nabla f(x, y, z)$ is a conservative vector field, what will div $\mathbf{F}$ be?

10. Give an example of a conservative vector field whose divergence is not uniformly equal to zero in $\mathbb{R}^3$.

11. Give an example of a conservative vector field whose divergence is uniformly equal to zero in $\mathbb{R}^3$.

12. Give an example of a non-conservative vector field whose divergence is uniformly equal to zero in $\mathbb{R}^3$.

13. Give an example of a non-conservative vector field whose divergence is never equal to zero in $\mathbb{R}^3$.

14. Considering your answers to Exercises 10–13, what do you conjecture is true or not true about conservative vector fields in the context of the Divergence Theorem?

Decide whether or not each of the integrals in Exercises 15–20 can be evaluated by means of the Divergence Theorem. (You do not need to evaluate the integrals.)

15. $\iint_S \mathbf{F}(x, y, z) \cdot \mathbf{n}\, dS$, where S is the cone $y = \sqrt{x^2 + z^2}$ between $y = 0$ and $y = 4$ and where

$$\mathbf{F}(x, y, z) = xyz\mathbf{i} + 2^{xyz}\mathbf{j} + \frac{x}{y^2 + 1}\mathbf{k}.$$

16. $\iint_S \mathbf{F}(x, y, z) \cdot \mathbf{n}\, dS$, where S is the surface of the tetrahedron bounded by the three coordinate planes and the plane $3x + 2y + 6z = 12$ and where

$$\mathbf{F}(x, y, z) = z\sin y\mathbf{i} - x\cos z\mathbf{j} + z\cos y\mathbf{k}.$$

17. $\iint_S \mathbf{F}(x, y, z) \cdot \mathbf{n}\, dS$, where S is the unit sphere and $\mathbf{F}(x, y, z) = \langle e^{\sqrt{x^2+y^2+z^2}}, \ln(x^2 y^2 z^2), 5x^3 z \rangle$.

18. $\iint_S \mathbf{F}(x, y, z) \cdot \mathbf{n}\, dS$, where S is the portion of the surface $x = z^3$ that lies between the planes $y = 13$ and $y = 42$ and where

$$\mathbf{F}(x, y, z) = \left\langle \cos(xy), \frac{1}{y^2 z^2 + 1}, x + y \right\rangle.$$

19. $\iint_S \mathbf{F}(x, y, z) \cdot \mathbf{n}\, dS$, where S is the portion of the hyperboloid of two sheets $x^2 + y^2 + 1 = z^2$ that lies between the planes $z = 5$ and $z = 10$ and where

$$\mathbf{F}(x, y, z) = \langle x^2 - 2y + z^2, 4x + z - 2, 2^{xyz} \rangle.$$

20. $\iint_S \mathbf{F}(x, y, z) \cdot \mathbf{n}\, dS$, where S is the surface of the torus parametrized by

$$x = (3 + \cos v)\cos u, \ \ y = (3 + \cos v)\sin u, \ \ z = \sin v$$

and where $\mathbf{F}(x, y, z) = (3x - z)\mathbf{i} - x^2 z\mathbf{j} + xy^7\mathbf{k}$.

Skills

In Exercises 21–24, compute the divergence of the given vector field.

21. $\mathbf{F}(x, y, z) = x\mathbf{i} + y\mathbf{j} + z\mathbf{k}$

22. $\mathbf{F}(x, y, z) = yz\cos z\mathbf{i} + (z - x)\mathbf{j} + e^x y\mathbf{k}$

23. $\mathbf{F}(x, y, z) = xe^{xyz}\mathbf{i} + ye^{xyz}\mathbf{j} + ze^{xyz}\mathbf{k}$

24. $\mathbf{F}(x, y, z) = \cos x\mathbf{i} - y\sin zy\mathbf{j} + \cos z\mathbf{k}$

In Exercises 25–40, evaluate the integral

$$\iint_S \mathbf{F}(x, y, z) \cdot \mathbf{n}\, dS$$

for the specified function $\mathbf{F}(x, y, z)$ and the given surface S. In each integral, $\mathbf{n}$ is the outwards-pointing normal vector.

25. $\mathbf{F}(x, y, z) = 4x^3 yz\mathbf{i} + 6x^2 y^2 z\mathbf{j} + 6x^2 yz^2\mathbf{k}$, and S is the surface of the first-octant cube with side length π and with one vertex at the origin.

26. $\mathbf{F}(x, y, z) = xy^2\mathbf{i} + y(z - 3x)\mathbf{j} + 4xyz\mathbf{k}$, and S is the surface of the region W bounded by the planes $y = 0$, $y = z$, $z = 3$, $x = 0$, and $x = 4$.

27. $\mathbf{F}(x, y, z) = e^z x\sin y\mathbf{i} + e^z \cos y\mathbf{j} + e^x \tan^{-1} y\mathbf{k}$, and S is the surface of the region W that lies within the unit sphere and above the plane $z = \frac{\sqrt{2}}{2}$.

28. $\mathbf{F}(x, y, z) = x^3\mathbf{i} + y^3\mathbf{j} + z^3\mathbf{k}$, and S is the sphere of radius 3 and centered at the origin.

29. $\mathbf{F}(x, y, z) = 15xz^2\mathbf{i} + 15yx^2\mathbf{j} + 15y^2 z\mathbf{k}$, and S is the surface of the lower half of the unit sphere, along with the unit circle in the plane.

30. $\mathbf{F}(x, y, z) = 4xy\mathbf{i} - yz\mathbf{j} + z^3 x\mathbf{k}$, and S is the surface of the region W that lies within the cone $z = \sqrt{x^2 + y^2}$ and between the planes $z = 1$ and $z = 4$.

31. $\mathbf{F}(x, y, z) = \langle xz, yz, xyz \rangle$, and S is the surface of the cylinder with equation $x^2 + y^2 = 9$ for $-2 \leq z \leq 2$.

32. $\mathbf{F}(x, y, z) = \langle e^{x^2+y^2+z^2}, 3y - z, ye^x \rangle$, and S is the unit sphere.

33. $\mathbf{F}(x, y, z) = \sin y \cos z\,\mathbf{i} + yz^2\mathbf{j} + zx^2\mathbf{k}$, and S is the surface of the region W bounded by the paraboloid $y = x^2 + z^2$ and the planes $y = 1$ and $y = 4$.

34. $\mathbf{F}(x, y, z) = \langle 2x - y, yz - ye^{2x}, x^2 + yz \rangle$, and S is the surface of the region that lies within $x = y^2 + z^2$ and $x = 2z + 1$.

35. $\mathbf{F}(x, y, z) = \left\langle x^2y^2, 2xy, \dfrac{2xz^3}{3} \right\rangle$, and S is the surface of the region that lies within the hyperboloid with equation $x^2 = z^2 + y^2 + 1$ and between the planes $x = 1$ and $x = 4$.

36. $\mathbf{F}(x, y, z) = x2^z \ln 2\,\mathbf{i} + (x - y + z)\mathbf{j} + (2y + 4z)\mathbf{k}$, and S is the surface of the first-octant region W bounded by the coordinate planes and the plane $x + y + 2z = 2$.

37. $\mathbf{F}(x, y, z) = \left\langle \dfrac{xe}{2}, zx - ye, \dfrac{3ez}{2} + \ln(x^2y^2z^2 + 1) \right\rangle$, and S is the surface of the sphere of radius 3 and centered at $(2, 1, 3)$.

38. $\mathbf{F}(x, y, z) = \langle e^{x^2+z^2}, 4y, e^{x^2+z^2} \rangle$, and S is the surface of the right circular cylinder of radius 1 whose axis is the line $\langle 2, t, 2 \rangle$ that runs parallel to the y-axis. The left and right sides of the cylinder lie in the planes $y = 3$ and $y = 0$, respectively.

39. $\mathbf{F}(x, y, z) = 8x\,\mathbf{i} - 13y\,\mathbf{j} + (13z - 12e^y)\mathbf{k}$, and S is the surface of the region W that lies above the region $|y| \leq |x|, y \geq 0$ in the xy-plane, within the unit circle, and below the saddle $z = x^2 - y^2$.

40. $\mathbf{F}(x, y, z) = 4\cos z + 2x\,\mathbf{i} + (3y - z)\mathbf{j} + 12z\,\mathbf{k}$, and S is the surface of the region W bounded below by $z = x^2 + y^2$ and above by the sphere $x^2 + y^2 + z^2 = 2$ centered at the origin.

In Exercises 41–44, find the fluxes of the vector fields through the given surfaces in the direction of the outwards-pointing normal vector.

41. $\mathbf{F}(x, y, z) = \langle xe^y, \ln(xyz), xyz^2 \rangle$, and S is the surface of the rectangular solid with vertices $(1, 1, 1)$, $(1, 5, 1)$, $(1, 1, 4)$, $(1, 5, 4)$, $(7, 1, 1)$, $(7, 5, 1)$, $(7, 1, 4)$, and $(7, 5, 4)$.

42. $\mathbf{F}(x, y, z) = 2xz\,\mathbf{i} + 4xy\,\mathbf{j} - 8zy\,\mathbf{k}$, and S is the surface determined by $y = 4 - x^2, y \geq 0, 0 \leq z \leq 5$.

43. $\mathbf{F}(x, y, z) = \langle y\cos z, 3y, \sin(xy) \rangle$, and S is the surface of the pyramid with the square base in the xy-plane and with vertices $(1, 1, 0)$, $(-1, 1, 0)$, $(1, -1, 0)$, $(-1, -1, 0)$ and apex $(0, 0, 4)$.

44. $\mathbf{F}(x, y, z) = \left\langle x^2y^2z^2, z - x - y, \dfrac{y}{z+1} \right\rangle$, and S is the surface of the region bounded by $z = x^2, x = 0$, and $4x + 2z = 4$.

Applications

45. The current through a certain region of the San Juan Islands in Washington State is given by $\mathbf{F} = \langle 0, 1.152 - 0.8x^2 \rangle$. Consider a disk R of radius 1 centered on this region. Denote the circle that comprises the boundary of the disk by ∂R.

 (a) Compute $\iint_R \nabla \cdot \mathbf{F}\, dA$.

 (b) Show that $\int_{\partial R} \mathbf{F} \cdot \mathbf{n}\, ds = 0$. Conclude that the Divergence Theorem is valid for the current in this region of the San Juan Islands.

 (c) What do the values of the integrals from the Divergence Theorem tell us about this region of the San Juan Islands?

46. Suppose that an electric field is given by $\mathbf{E} = \langle 2y, 2xy, yz \rangle$. Use the Divergence Theorem to compute the flux $\iint_S \mathbf{E}\, d\mathbf{A}$ of the field through the surface of the unit cube $[0, 1] \times [0, 1] \times [0, 1]$.

47. Consider a straight road with multiple lanes going due north. The road is 0.0113 mile wide. Denote the velocity of traffic at any point of the road by $\mathbf{v}(x, y) = \langle 0, v_2(x, y) \rangle$. Traffic engineers are using road tubes along the lines $y = 0$ and $y = 1$ to monitor the speed of traffic.

 (a) Suppose that the road tubes indicate that $v_2(x, 1) = v_2(x, 0) - 5$. Compute

$$\int_C \mathbf{v} \cdot \mathbf{n}\, ds < 0,$$

 and discuss what this inequality means for traffic in this stretch of road.

 (b) Compute the average divergence

$$\frac{1}{A} \iint_S \nabla \cdot \mathbf{v}\, dx\, dy$$

 of $\mathbf{v}$ on this stretch, where A is the area of the roadway. What is the significance of the average divergence as regards the traffic?

Proofs

48. Show that if $\mathbf{F}(x, y, z) = \text{curl } \mathbf{G}(x, y, z)$ for some smooth vector field $\mathbf{G}$, and if S is a smooth or piecewise-smooth simple closed surface, then

$$\iint_S \mathbf{F} \cdot \mathbf{n}\, dS = 0.$$

49. Show that if $\mathbf{F}$ is conservative, it does not follow that $\iint_S \mathbf{F}(x, y, z) \cdot \mathbf{n}\, dS = 0$, by using the Divergence Theorem

to evaluate $\iint_S \mathbf{F}(x, y, z) \cdot \mathbf{n}\, dS$, where $\mathbf{F}(x, y, z) = \nabla f(x, y, z)$, $f(x, y, z) = x^2 + y^2 + z^2$, and S is the unit sphere with outwards-pointing normal vector.

50. Suppose that $\mathbf{F}(x, y, z) = \text{curl } \mathbf{G}(x, y, z)$ for some smooth vector field $\mathbf{G}$ and W is a region in $\mathbb{R}^3$ that is bounded by a smooth simple closed surface. Use the table of integral theorems on page 1145 to equate $\iiint_W \text{div } \mathbf{F}(x, y, z)\, dV$ with a line integral.

51. Suppose S is any smooth or piecewise-smooth simple closed surface enclosing a region W in $\mathbb{R}^3$, and let

$$\mathbf{F}(x, y, z) = \langle ax + b, cy + d, ez + f \rangle,$$

where a, b, c, d, e, and f are scalars and a, b, and c are not all zero. Find a condition for the relationship of a, b, and c that will guarantee that $\iint_S \mathbf{F} \cdot \mathbf{n}\, dS$ is positive for any choice of smooth or piecewise-smooth simple closed surface S in $\mathbb{R}^3$.

52. Let $\mathbf{F}(x, y, z)$ and S be as in Exercise 51. Find a condition for the relationship of a, b, and c which will guarantee that $\iint_S \mathbf{F} \cdot \mathbf{n}\, dS = 0$ for any choice of smooth or piecewise-smooth simple closed surface S in $\mathbb{R}^3$.

53. Show that if $\iint_S \mathbf{F}(x, y, z) \cdot \mathbf{n}\, dS = 0$, it does not follow that $\mathbf{F}$ is conservative.

Thinking Forward

Calculus in $\mathbb{R}^4$ and beyond: The results of this chapter have illustrated several ways that the Fundamental Theorem of Calculus generalizes to two and three dimensions. Earlier chapters have shown us how to integrate and find partial derivatives of functions taking any finite number of inputs. It is reasonable to ask, then, if the Fundamental Theorem has analogs in $\mathbb{R}^n$, $n \geq 4$.

▶ What do you think the Divergence Theorem for $\mathbb{R}^4$ might say? What is the guiding intuition for a Divergence Theorem in $\mathbb{R}^n$?

▶ As in the case of generalizing Green's Theorem to $\mathbb{R}^3$, there is more than one way to try to generalize the Divergence Theorem to $\mathbb{R}^4$ and beyond. What is a way to generalize the Divergence Theorem that is different from the preceding answer you gave.

Calculus for functions from $\mathbb{R}^n$ to $\mathbb{R}^m$: Vector fields can be construed as functions from, say, $\mathbb{R}^3$ to $\mathbb{R}^3$, both multivariate and vector-valued. There are many applications for functions of the general form

$$\mathbf{f}(x_1, \ldots, x_n) = \langle f_1(x_1, \ldots, x_n), \ldots f_m(x_1, \ldots, x_n) \rangle,$$

where n and m are natural numbers. Consider $f : \mathbb{R}^4 \to \mathbb{R}^5$. There are $4 \times 5 = 20$ partial derivatives of f, but if f has smooth component functions, we can still expect to do calculus.

▶ What might an integral of f represent? How might we handle the partial derivatives of f in a matrix?

Analysis: This book is an introduction to, among other things, the theory of integration. But there are more questions to ask.

▶ Is there a way to integrate a function with dense discontinuities—for example, the function whose output is 1 if the input is irrational and zero otherwise? This function cannot be integrated by the methods of Chapter 4, but there is a way to expand the notion of integration that will allow us to integrate the function on finite intervals. This and other questions are addressed in analysis courses.

CHAPTER REVIEW, SELF-TEST, AND CAPSTONES

Be sure you are familiar with the definitions, concepts, and basic skills outlined here. The capstone exercises at the end bring together ideas from this chapter.

Definitions

Give precise mathematical definitions or descriptions of each of the concepts that follow. Then illustrate the definition with a graph or an algebraic example.

▶ a *vector field*

▶ a *conservative vector field*

▶ a *potential function*

▶ a *smooth parametrization*

▶ the *integral* of a function of two or three variables along a curve C

▶ the *line integral* of a vector field $\mathbf{F}(x, y, z)$ along a curve C

▶ the *surface area* of a smooth surface S

▶ The *integral* of $f(x, y, z)$ over a smooth surface S

▶ the *flux* of a vector field across a surface

▶ the *divergence* of a vector field

▶ the *curl* of a vector field

Theorems

Fill in the blanks to complete each of the following theorem statements:

▶ *The Fundamental Theorem of Line Integrals:* If $\mathbf{F}(x, y, z)$ is a _____ vector field with $\mathbf{F}(x, y, z) = \nabla f(x, y, z)$ on an _____, _____, and _____ domain containing the curve C with endpoints P and Q, then

$$\int_C \mathbf{F}(x, y, z) \cdot d\mathbf{r} = \underline{\qquad}.$$

▶ For any vector field $\mathbf{F}$ in $\mathbb{R}^2$ or $\mathbb{R}^3$,

$$\operatorname{div}(\operatorname{curl} \mathbf{F}) = \underline{\qquad} = \underline{\qquad}.$$

▶ For multivariate functions f in $\mathbb{R}^2$ or $\mathbb{R}^3$ with continuous second partial derivatives,

$$\operatorname{curl} \nabla f = \underline{\qquad} = \underline{\qquad}.$$

▶ *Green's Theorem:* Let $\mathbf{F}(x, y) = \langle F_1(x, y), F_2(x, y) \rangle$ be a vector field defined on a region R in the xy-plane whose boundary is a _____ curve C. If $\mathbf{r}(t)$ is a parametrization of C in the counterclockwise direction (as viewed from the positive z-axis), then

$$\int_C \mathbf{F}(x, y) \cdot d\mathbf{r} = \int_C \underline{\qquad} dx + \underline{\qquad} dy$$

$$= \iint_R (\underline{\qquad} - \underline{\qquad})\, dA.$$

▶ *Stokes' Theorem:* Let S be a _____ or _____ surface with a _____ or _____ boundary curve C. Suppose that S has an (oriented) unit normal vector $\mathbf{n}$ and that C has a parametrization which traverses C in the _____ direction with respect to $\mathbf{n}$. If $\mathbf{F}(x, y, z) = F_1(x, y, z)\mathbf{i} + F_2(x, y, z)\mathbf{j} + F_3(x, y, z)\mathbf{k}$ is a vector field on an open region containing S, then

$$\int_C \mathbf{F}(x, y, z) \cdot d\mathbf{r} = \iint_S \underline{\qquad} dS.$$

▶ *Divergence Theorem:* Suppose W is a region in $\mathbb{R}^3$ bounded by a _____ or _____ surface S. If $\mathbf{F}(x, yz)$ is defined on an open region containing W, then

$$\iiint_W \operatorname{div} \mathbf{F}(x, y, z)\, dV = \iint_S \underline{\qquad} dS$$

where $\mathbf{n}$ is the outwards _____.

Notation and Integration Rules

Notation: Describe the meanings of each of the following mathematical expressions.

▶ ∇ ▶ $\operatorname{div} \mathbf{F}(x, y, z)$ ▶ $\operatorname{curl} \mathbf{F}(x, y, z)$

Integration Theorems: Assuming that the necessary hypotheses are present for each of the following integration theorems, provide an alternative method for evaluating the given integral.

▶ *The Fundamental Theorem of Calculus:*

$$\int_a^b f(x)\, dx = \underline{\qquad\qquad}$$

▶ *The Fundamental Theorem of of Line Integrals:*

$$\int_C \nabla f(x, y, z) \cdot d\mathbf{r} = \underline{\qquad\qquad}$$

▶ *Green's Theorem:*

$$\iint_R \left(\frac{\partial f_2}{\partial x} - \frac{\partial f_1}{\partial y} \right) dA = \underline{\qquad\qquad}$$

▶ *Stokes' Theorem:*

$$\iint_S \operatorname{curl} \mathbf{F}(x, y, z) \cdot \mathbf{n}\, dS = \underline{\qquad\qquad}$$

▶ *Divergence Theorem:*

$$\iiint_W \operatorname{div} \mathbf{F}(x, y, z)\, dV = \underline{\qquad\qquad}$$

Skill Certification: Vector Calculus

Potential Functions: Find a potential function for each vector field.

1. $\mathbf{F}(x, y) = \langle 3x^2 y^2, 2x^3 y \rangle$

2. $\mathbf{F}(x, y) = \dfrac{y^2}{x}\mathbf{i} + 2y \ln x \mathbf{j}$

3. $\mathbf{G}(x, y, z) = xze^{y^2}\mathbf{i} + 2xyze^{y^2}\mathbf{j} + xe^{y^2}\mathbf{k}$

4. $\mathbf{G}(x, y, z) = \langle 2xy^2 ze^{x^2 yz}, (1 + x^2 yz)e^{x^2 yz}, x^2 y^2 e^{x^2 yz} \rangle$

Conservative Vector Fields: Determine whether the vector fields that follow are conservative. If the field is conservative, find a potential function for it.

5. $\mathbf{F}(x, y) = \dfrac{2x}{y}\mathbf{i} + \dfrac{x^2}{y^2}\mathbf{j}$

6. $\mathbf{F}(x, y) = \langle (1 + xy)e^{xy}, e^{xy} \rangle$

7. $\mathbf{F}(x, y) = \langle y^3 e^{xy^2}, (1 + 2xy^2)e^{xy^2} \rangle$

8. $\mathbf{F}(x, y, z) = \left\langle \frac{y}{z}, \frac{x}{z}, -\frac{xy}{z^2} \right\rangle$

9. $\mathbf{G}(x, y, z) = \cos y \sin z \mathbf{i} - x \sin y \sin z \mathbf{j} - x \cos y \cos z \mathbf{k}$

10. $\mathbf{G}(x, y, z) = y^2 z \mathbf{i} + 2xyz \mathbf{j} + xy^2 \mathbf{k}$

Line Integrals: Evaluate the line integral of the given function over the specified curve.

11. $\int_C \mathbf{F}(x, y) \cdot d\mathbf{r}$, where $\mathbf{F}(x, y) = 7y^2 \mathbf{i} - 3xy \mathbf{j}$ and C is the line segment $x = t$, $y = 3t$ for $0 \le t \le 1$.

12. $\int_C \mathbf{F}(x, y) \cdot d\mathbf{r}$, where $\mathbf{F}(x, y) = \langle 8x^2 y, -9xy^2 \rangle$ and C is the curve parametrized by $x = t^2$, $y = t^3$ for $0 \le t \le 1$.

13. $\int_C \mathbf{F}(x, y) \cdot d\mathbf{r}$, where $\mathbf{F}(x, y) = \frac{x}{x^2 + y^2} \mathbf{i} - \frac{y}{x^2 + y^2} \mathbf{j}$ and C is the unit circle centered at the origin.

14. $\int_C \mathbf{F}(x, y, z) \cdot d\mathbf{r}$, where $\mathbf{F}(x, y, z) = yz \mathbf{i} + xz \mathbf{j} + xy \mathbf{k}$ and C is the curve parametrized by $x = t^2$, $y = t^{-2}$, $z = t$ for $1 \le t \le 3$.

15. $\int_C \mathbf{F}(x, y, z) \cdot d\mathbf{r}$, where $\mathbf{F}(x, y, z) = \langle e^x, e^y, e^z \rangle$ and C is the line segment from $(0, 0, 0)$ to $(1, 1, 1)$.

16. $\int_C \mathbf{F}(x, y, z) \cdot d\mathbf{r}$, where $\mathbf{F}(x, y, z) = \langle e^x, e^y, e^z \rangle$ and C is the curve parametrized by $x = t$, $y = t^2$, $z = t^3$ for $0 \le t \le 1$. Compare your answer with that of Exercise 15.

Surface Area: Find the area of each of the following surfaces.

17. The portion of the plane $2x + 3y + 4z = 12$ that lies in the first octant.

18. The portion of the plane $2x + 3y + 4z = 12$ that lies inside the cylinder with equation $x^2 + y^2 = 9$.

19. The portion of the sphere with equation $x^2 + y^2 + z^2 = 16$ that lies inside the cylinder with equation $x^2 + y^2 = 9$.

20. The portion of the paraboloid with equation $z = 9 - x^2 - y^2$ that lies above the xy-plane.

Flux: Find the outward flux of the given vector field through the specified surface.

21. $\mathbf{F}(x, y, z) = \langle x, y, z \rangle$ and S is the portion of the cone with equation $z = \sqrt{x^2 + y^2}$ for $z \le 3$.

22. $\mathbf{F}(x, y, z) = \langle x, y, z \rangle$ and S is the sphere with equation $x^2 + y^2 + z^2 = 9$.

23. $\mathbf{F}(x, y, z) = xyz \mathbf{i}$ and S is the portion of the plane with equation $x + y + z = 4$ in the first octant.

24. $\mathbf{F}(x, y, z) = x^2 \mathbf{i} + y^2 \mathbf{j} + z^2 \mathbf{k}$ and S is the top half of the unit sphere centered at the origin.

Divergence and Curl: Find the divergence and curl of the following vector fields.

25. $\mathbf{F}(x, y) = (3x + 4y) \mathbf{i} + (x - 5y) \mathbf{j}$

26. $\mathbf{F}(x, y) = x^2 y \mathbf{i} + xy^3 \mathbf{j}$

27. $\mathbf{F}(x, y, z) = (x - y) \mathbf{i} + (y - z) \mathbf{j} + (z - x) \mathbf{k}$

28. $\mathbf{F}(x, y, z) = (y^2 - z^2) \mathbf{i} + (z^2 - x^2) \mathbf{j} + (x^2 - y^2) \mathbf{k}$

Green's Theorem: Use Green's Theorem to evaluate the integral $\int_C \mathbf{F} \cdot d\mathbf{r}$ for the given vector field and curve.

29. $\mathbf{F}(x, y) = (y^2 + 1) \mathbf{i} + 2xy \mathbf{j}$ and C is the circle with equation $x^2 + y^2 = 9$, traversed counterclockwise.

30. $\mathbf{F}(x, y) = \langle e^x \cos y, e^x \sin y \rangle$ and C is the ellipse with equation $3x^2 + 4y^2 = 25$, traversed counterclockwise.

31. $\mathbf{F}(x, y) = xy \mathbf{j}$ and C is the square with vertices $(\pm 3, \pm 3)$, traversed counterclockwise.

32. $\mathbf{F}(x, y) = y \mathbf{i} - x \mathbf{j}$ and C is the square with vertices $(\pm 3, \pm 3)$, traversed counterclockwise.

Stokes' Theorem: Evaluate the integrals that follow. In some cases Stokes' Theorem will help; in other cases it may be preferable to evaluate the integrals directly.

33. $\int_C \mathbf{F}(x, y, z) \cdot d\mathbf{r}$, where

$$\mathbf{F}(x, y, z) = (y^2 - x^2) \mathbf{i} + (z^2 - y^2) \mathbf{j} + (x^2 - y^2) \mathbf{k},$$

C is the triangle with vertices $(2, 0, 0)$, $(0, 6, 0)$, and $(0, 0, 3)$, and $\mathbf{n}$ points upwards.

34. $\int_C \mathbf{F}(x, y, z) \cdot d\mathbf{r}$, where $\mathbf{F}(x, y, z) = \langle y^3, -4z, 4x^2 \rangle$, C is the circle in the xy-plane parametrized by $x = 2 \cos t$, $y = 2 \sin t$, and $\mathbf{n}$ points upwards.

35. $\iint_S \operatorname{curl} \mathbf{F}(x, y, z) \cdot \mathbf{n} \, dS$, where

$$\mathbf{F}(x, y, z) = \langle 5y, 5x, z^2 \rangle,$$

S is the portion of the unit sphere $x^2 + y^2 + z^2 = 1$ above the plane $z = \frac{1}{2}$, and $\mathbf{n}$ is the upwards-pointing normal vector.

36. $\iint_S \operatorname{curl} \mathbf{F}(x, y, z) \cdot \mathbf{n} \, dS$, where

$$\mathbf{F}(x, y, z) = \langle x^3 yz, xy^2 z, yz \rangle,$$

S is the portion of the hyperboloid $z = x^2 - y^2 + 1$ that lies above the square with vertices $(\pm 1, \pm 1)$, and $\mathbf{n}$ is the upwards-pointing normal vector.

Divergence Theorem: Evaluate the integral

$$\iint_S \mathbf{F}(x, y, z) \cdot \mathbf{n} \, dS$$

for the specified vector field $\mathbf{F}$ and surface S. In each case, $\mathbf{n}$ is the outwards-pointing normal vector.

37. $\mathbf{F}(x, y, z) = x^2 y \mathbf{i} + y^2 z \mathbf{j} + xz^2 \mathbf{k}$ and S is the hemisphere bounded above by $x^2 + y^2 + z^2 = 1$ and bounded below by the xy-plane.

38. $\mathbf{F}(x, y, z) = x^2 \mathbf{i} + y^2 \mathbf{j} + z^2 \mathbf{k}$ and S is the surface of the rectangular solid described by the inequalities $0 \le x \le 3$, $0 \le y \le 2$, and $0 \le z \le 1$.

39. $\mathbf{F}(x, y, z) = x^3 z \mathbf{i} + xy \mathbf{j} + 4yz \mathbf{k}$ and S is the boundary of the frustum of the cone with equation $x = \sqrt{y^2 + z^2}$ between the planes $x = 2$ and $x = 5$.

40. $\mathbf{F}(x, y, z) = xy \mathbf{i} + xyz \mathbf{j} + yz \mathbf{k}$ and S is the boundary of the cylinder with equation $x^2 + z^2 = 4$ for $0 \le y \le 5$.

Capstone Problems

A. Let $\mathbf{F} = \mathbf{F}(x, y, z)$ and $\mathbf{G} = \mathbf{G}(x, y, z)$. Prove that

$$\nabla(\mathbf{F} \cdot \mathbf{G}) = (\mathbf{F} \cdot \nabla)\mathbf{G} + (\mathbf{G} \cdot \nabla)\mathbf{F}$$
$$+ \mathbf{F} \times (\nabla \times \mathbf{G}) + \mathbf{G} \times (\nabla \times \mathbf{F}).$$

B. The **Laplacian** of a function f of two or three variables is the divergence of the gradient. That is,

$$\nabla^2 f = \nabla \cdot (\nabla f).$$

The **vector Laplacian** of a vector field

$$\mathbf{F}(x, y, z) = \langle F_1(x, y, z), F_2(x, y, z), F_3(x, y, z) \rangle$$

is

$$\nabla^2 \mathbf{F} = \langle \nabla^2 F_1, \nabla^2 F_2, \nabla^2 F_3 \rangle.$$

Prove that, for a vector field $\mathbf{F} = \mathbf{F}(x, y, z)$,

$$\nabla \times (\nabla \times \mathbf{F}) = \nabla(\nabla \cdot \mathbf{F}) - \nabla^2 \mathbf{F}.$$

C. Show that $\int_C \mathbf{F} \cdot d\mathbf{r} = \pi \rho^2$ when $\mathbf{F}(x, y) = -y\mathbf{i} + x\mathbf{j}$ and C is the circle with radius ρ centered at the origin and traversed counterclockwise.

   (a) Formulate a conjecture for the value of the integral $\int_C \mathbf{F} \cdot d\mathbf{r}$, where $\mathbf{F}(x, y) = -y\mathbf{i} + x\mathbf{j}$ and C is any simple closed curve.

   (b) Prove your conjecture.

   (c) Use your conjecture to find the area of the ellipse with parametrization

$$x = a \cos t \quad \text{and} \quad y = b \sin t.$$

D. Let $\mathbf{F} = \mathbf{F}(x, y, z)$ and $\mathbf{G} = \mathbf{G}(x, y, z)$ be vector fields defined on a region R enclosed by an orientable surface S with outwards unit normal vector $\mathbf{n}$. Prove that if

$$\nabla \cdot \mathbf{F} = \nabla \cdot \mathbf{G} \quad \text{and} \quad \nabla \times \mathbf{F} = \nabla \times \mathbf{G}$$

on R and $\mathbf{F} \cdot \mathbf{n} = \mathbf{G} \cdot \mathbf{n}$ on S, then $\mathbf{F} = \mathbf{G}$ on S.

Chapter 9

Section 9.1

1. T, F, F, F, F, F, T, F.

3. (a) To the right and up. (b) To the right and down. (c) To the left and up. (d) To the left and down.

5. Up and to the left when $t < 0$, up and to the right when $t > 0$.

7. Up and to the right

9. (i) $y = x^2 - 1$, (ii) (a) the right half of the parabola, (b) the left half of the parabola; (iii) motion to the right as t increases for (a), and motion to the left as t increases for (b).

11. The graphs are the portions of $y = \sin x$ to the right of the y-axis. The direction of motion for both is to the right.

13. There is a horizontal tangent line at a value, t_0, where $\frac{dy}{dt} = 0$ and $\frac{dx}{dt} \neq 0$ or when $\lim_{t \to t_0} \frac{dy/dt}{dx/dt} = 0$. There is a vertical tangent line at value t_0 where $\frac{dy}{dt} \neq 0$ and $\frac{dx}{dt} = 0$ or when $\lim_{t \to t_0} \frac{dy/dt}{dx/dt} = \pm\infty$.

15. Since the sine function is periodic and $-1 \leq \sin t \leq 1$ for all values of t, the values of x and y will also be bounded; $0 \leq x \leq 2$ and $-4 \leq y \leq -3$.

17.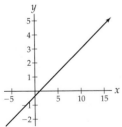

19.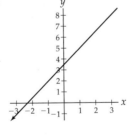

21.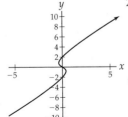

23.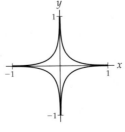

25. $y = 3\left(\frac{x+1}{2}\right)^2 + 5$

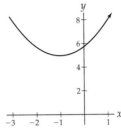

27. $y = x$

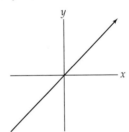

29. $\left(\frac{x}{3}\right)^2 + \left(\frac{y}{4}\right)^2 = 1$

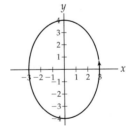

31. $x^2 - y^2 = 1$

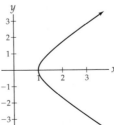

33. $x^2 - y^2 = 1$, same graph as in Exercise 31

35. $x^2 + y^2 = 1$, when $t = \pi$, $(x, y) = (-1, 0)$, $t = \frac{3\pi}{2}$, $(x, y) = (0, -1)$ and $t = 2\pi$, $(x, y) = (1, 0)$

37. $x = -3\sin(2\pi t)$, $y = 3\cos(2\pi t)$

39. $x = \dfrac{2t}{1+t^2}$, $x = \dfrac{1-t^2}{1+t^2}$

41. $y - 2 = \dfrac{3}{2}(x+3)$

43. $y - \dfrac{9}{4} = -3\left(x - \dfrac{1}{4}\right)$

45. 0

47. 8

49. $\dfrac{2}{27}\left(10\sqrt{10} - 1\right)$

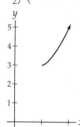

51. 6

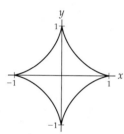

53. $e - \dfrac{1}{e}$

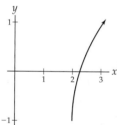

55. $x = 1 + 5t$, $y = -3 + 10t$, $t \in [0, 1]$

57. $x = 1 - 4t$, $y = 4 + t$, $t \in [0, 1]$

59. $x = \pi$, $y = 3 + 5t$, $t \in [0, 1]$

61. $5\sqrt{5}$

63. $x = (k+r)\cos\theta - k\cos\left(\dfrac{(k+r)\theta}{k}\right)$,
$y = (k+r)\sin\theta - r\sin\left(\dfrac{(k+r)\theta}{k}\right)$

65. $\int_0^{2\pi} \sqrt{\cos^2 t + 9\sin^2 t}\, dt \approx 13.4$

67. (a) $x(t) = 1.932t$, $y(t) = 0.552t^3 - 1.287t^2 + 0.518t$,
(b) 0.232 miles, (c) 3.044 miles

69. When $x = t$ and $y = f(t)$, we have
$\int_a^b \sqrt{(x'(t))^2 + (y'(t))^2}\, dt = \int_a^b \sqrt{1 + (f'(t))^2}\, dt$.

71. $\int_a^b \sqrt{((kf)'(t))^2 + ((kg)'(t))^2}\, dt =$
$k\int_a^b \sqrt{(f'(t))^2 + (g'(t))^2}\, dt = k$ times the arc length of
the second set of parametric equations. The arc length
of the final set of parametric equations is the same as
the length of the second set of parametric equations.

73. $\dfrac{dy}{dx} = \dfrac{\sin\theta}{1 - \cos\theta}$. By L'Hôpital's rule
$\lim\limits_{\theta \to 2k\pi^+} \dfrac{\sin\theta}{1 - \cos\theta} = \lim\limits_{\theta \to 2k\pi^+} \dfrac{\cos\theta}{\sin\theta} = \infty$ and
$\lim\limits_{\theta \to 2k\pi^-} \dfrac{\sin\theta}{1 - \cos\theta} = \lim\limits_{\theta \to 2k\pi^-} \dfrac{\cos\theta}{\sin\theta} = -\infty$. Thus, there is a
vertical tangent line at each even multiple of π.

Section 9.2

1. T, F, F, F, T, T, T, T.

3. The value of θ in the polar coordinates for a point P
measures the angular rotation of P from the x-axis.
Adding an integer rotation of 2π doesn't change the
location of the point.

5. The point $\left(8, -\dfrac{\pi}{3}\right)$ is in the fourth quadrant. The point
$(-4, 4\sqrt{3})$ is in the second quadrant.

7. The graphs of $r = c$ and $r = -c$ are the same for every
real number c.

9. $\alpha = k\dfrac{\pi}{2}$ for any integer k.

11. Every point, (r, θ), is on a ray in the first quadrant when
$r > 0$ and $0 < \theta < \dfrac{\pi}{2}$. The points in the third quadrant
are given by the inequalities $r > 0$ and $\pi < \theta < \dfrac{3\pi}{2}$.

13. $(1, 2k\pi)$ and $(-1, (2k+1)\pi)$ for every integer k.

15. $(1, (2k+1)\pi)$ and $(-1, 2k\pi)$ for every integer k.

17. $\left(\dfrac{3\sqrt{3}}{2}, \dfrac{3}{2}\right)$, $\left(-\dfrac{3\sqrt{3}}{2}, -\dfrac{3}{2}\right)$, $\left(\dfrac{3\sqrt{3}}{2}, -\dfrac{3}{2}\right)$, $\left(-\dfrac{3\sqrt{3}}{2}, \dfrac{3}{2}\right)$

19. $(0, 5)$, $(0, -5)$, $(0, 5)$, $(0, 5)$

21. $\left(\dfrac{\sqrt{2}}{2}, \dfrac{\sqrt{2}}{2}\right)$, $\left(\sqrt{2}, \sqrt{2}\right)$, $\left(\dfrac{3\sqrt{2}}{2}, \dfrac{3\sqrt{2}}{2}\right)$, $\left(2\sqrt{2}, 2\sqrt{2}\right)$

23. $(0, -1)$, $(0, -2)$, $(0, -3)$, $(0, -4)$

25. $(1, 2k\pi)$ and $(-1, (2k+1)\pi)$ for any integer k.

27. $\left(2, \dfrac{\pi}{2} + 2\pi k\right)$ and $\left(-2, -\dfrac{\pi}{2} + 2\pi k\right)$ for any integer k

29. $\left(4\sqrt{3}, \dfrac{\pi}{6} + 2\pi k\right)$ and $\left(-4\sqrt{3}, \dfrac{7\pi}{6} + 2\pi k\right)$ for any
integer k.

31. $\left(3\sqrt{2}, -\dfrac{\pi}{4} + 2\pi k\right)$ and $\left(-3\sqrt{2}, \dfrac{3\pi}{4} + 2\pi k\right)$ for any
integer k.

33. $y = x$ **35.** $y = \dfrac{\sqrt{3}}{3}x$

37. $x^2 + \left(y - \dfrac{5}{2}\right)^2 = \left(\dfrac{5}{2}\right)^2$ **39.** $y = 6$

41. $(x^2 + y^2)^{3/2} = 2xy$ **43.** $\pm(x^2 + y^2)^{3/2} = x$.

45. $x\tan\sqrt{x^2 + y^2} = y$ **47.** $(x^2 + y^2)^2 = y^3$

49. $\theta = 0$.   **51.** $\theta = \dfrac{\pi}{4}$.

53. $r = -3 \csc \theta$.

55. $\tan \theta = m$

57. $x^2 + y^2 = k^2$

59. For $k = 1$ and $k = 7$, $y = \frac{\sqrt{3}}{3}x$. For $k = 2$ and $k = 8$, $y = \sqrt{3}x$. For $k = 3$ and $k = 9$, $x = 0$. For $k = 4$ and $k = 10$, $y = -\sqrt{3}x$. For $k = 5$ and $k = 11$, $y = -\frac{\sqrt{3}}{3}x$. For $k = 6$ and $k = 12$, $y = 0$.

61. (a) Assuming the arms of the cam remain symmetrically placed, each $\frac{3}{4}$ inch arm must fit in a half inch of the crack. Thus, we require $0.75 \sin \theta = \frac{1}{2}$, so $\theta = \sin^{-1}\frac{2}{3}$. (b) This causes the cam to push even harder against the walls of the crack.

63. Since $\csc \theta = \frac{r}{y}$, we have, $r = k\frac{r}{y}$, or equivalently $y = k$.

65. We have $a = r - br\cos\theta$. In rectangular coordinates this can be written $\sqrt{x^2 + y^2} = a + bx$. Squaring both sides, completing the square, and simplifying, we may obtain $\left(x - \frac{ab}{1-b^2}\right)^2 + \frac{y^2}{1-b^2} = \left(\frac{a}{1-b^2}\right)^2$ which is the equation of an ellipse if $0 < b < 1$. If $b > 1$ the equation gives a hyperbola.

67. Since $\cos \theta = \frac{x}{r}$, we have $r = a\frac{x}{r}$, or equivalently $r^2 = x^2 + y^2 = ax$. When you complete the square for this equation you obtain $\left(x - \frac{a}{2}\right)^2 + y^2 = \left(\frac{a}{2}\right)^2$. So the graph of the equation is a circle with center $\left(\frac{a}{2}, 0\right)$ and radius $\frac{a}{2}$.

69. Since $\cos \theta = \frac{x}{r}$ and $\sin \theta = \frac{y}{r}$, we have $r = k\frac{y}{r} + l\frac{x}{r}$, or equivalently $r^2 = x^2 + y^2 = lx + ky$. When you complete the square for this equation you obtain $\left(x - \frac{l}{2}\right)^2 + \left(y - \frac{k}{2}\right)^2 = \frac{k^2 + l^2}{4}$. So the graph of the equation is a circle with center $\left(\frac{l}{2}, \frac{k}{2}\right)$ and radius $\frac{\sqrt{k^2 + l^2}}{2}$.

Section 9.3

1. F, T, F, T, F, T, T, T.

3. The θ-intercepts of the graph in the θr-plane correspond to the places where the polar graph passes through the pole. The location of the graph in the θr-plane corresponds to the quadrant in which the polar graph is drawn.

5. (a) All θ-intercepts of the graph in the θr-plane correspond to the places where the polar graph passes through the pole. (b) The points on the θ-axis correspond to the points on the horizontal axis in the polar plane. (c) All points on the vertical lines $\theta = (2k+1)\frac{\pi}{2}$ correspond to the points on the vertical axis in the polar plane.

7. (a) $\left(2, -\frac{\pi}{3}\right)$ (b) $\left(2, \frac{2\pi}{3}\right)$ (c) $(-3, 5)$ (d) $\left(2, \frac{4\pi}{3}\right)$ (e) $(-3, -5)$

9. (a) $r = 2\cos\theta$ (b) $r = 2\sin\theta$ (c) $r = \sin 2\theta$ (d) $r^2 = \sin 2\theta$ (e) $r = \sin 2\theta$ (f) not possible

11. A cardioid, a limaçon, and a lemniscate.

13. $r = \cos 5\theta$: symmetry about the horizontal axis and rotational symmetry, five petals. $r = \sin 8\theta$: symmetry about the x- and y-axes, symmetry about the origin, rotational symmetry, sixteen petals.

15. Since $\cos(-2\theta) = \cos 2\theta$ the graph will be symmetrical with respect to the horizontal axis.

17.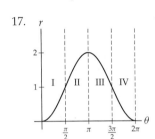

19.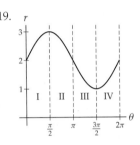

21.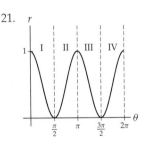

23.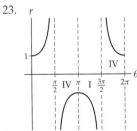

25.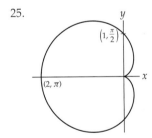

27.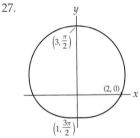

29.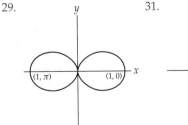

31.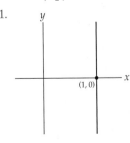

33. Use the interval $[0, 10\pi]$.

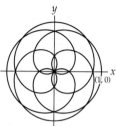

35. Use the intervals $\left[0, \frac{\pi}{2}\right) \cup \left(\frac{\pi}{2}, \frac{3\pi}{2}\right) \cup \left(\frac{3\pi}{2}, 2\pi\right]$.

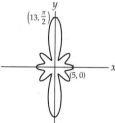

37. Use the interval $[0, 24\pi]$.

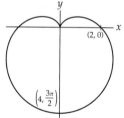

39.

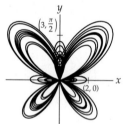

41. (a) $\frac{\pi}{4} + k\pi$ for every integer k. (c) The curves intersect at $(0, 0)$ and $\left(\frac{\sqrt{2}}{2}, \frac{\pi}{4}\right)$.

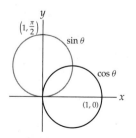

43. (a) $\frac{3\pi}{4} + k\pi$ for every integer k. (b) Graph is shown below. (c) The curves intersect at $(0, 0)$, $\left(1 + \frac{\sqrt{2}}{2}, \frac{3\pi}{4}\right)$ and $\left(1 - \frac{\sqrt{2}}{2}, \frac{7\pi}{4}\right)$.

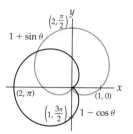

45. (a) $\frac{\pi}{4} + \frac{k\pi}{2}$ for $k = 0, 1, 2$ and 3. (b) Graph is shown below. (c) The curves intersect at $(0, 0)$, $\left(\frac{1}{2}, \frac{\pi}{4} + \frac{k\pi}{2}\right)$ for every integer k.

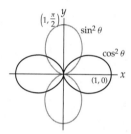

47. They are the mirror images of each other.

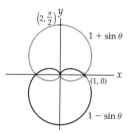

49. When $0 \le b \le \frac{3}{2}$ the limaçon will be convex. When $\frac{3}{2} < b \le 3$ there will be a dimple. When $b = 3$ the curve is a cardioid.

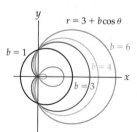

51. (a) $x = \frac{1}{2} + \frac{1}{2}\cos t$, $y = \frac{1}{2}\sin t$, $t \in \mathbb{R}$,
(b) $\left(x - \frac{1}{2}\right)^2 + y^2 = \frac{1}{4}$, (c) $r = \cos\theta$

53.

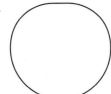

55. We will prove the statement for $f(\theta) = \sin n\theta$. The argument for $f(\theta) = \cos n\theta$ is similar. Let (r, θ) be a point on the graph of $f(\theta) = \sin n\theta$. That is, $r = \sin n\theta$. If n is a positive odd integer, then by the angle sum identity for sine we have $\sin(n(\theta + \pi)) = -\sin(n\theta)$. Therefore the point $(-r, \sin(n(\theta + \pi)) = (-r, -\sin(n\theta))$ has the same location as (r, θ). Thus, the graph traced on the interval $[\pi, 2\pi]$ is the same as the graph traced on the interval $[0, \pi]$.

57. Let (r, θ) be a point on the graph of $f(\theta) = \sin n\theta$ where n is even. We have $f(-\theta) = \sin(-n\theta) = \sin(2\pi - n\theta) = \sin(n\theta)$. Therefore, the point $(r, -\theta)$ is also on the graph. Thus, the graph is symmetrical with respect to the x-axis.

59. Let (r, θ) be a point on the graph of $f(\theta) = \cos n\theta$, where n is even. It suffices to show that $(r, \pi - \theta)$ is also on the graph. We have $f(\pi - \theta) = \cos(n(\pi - \theta)) = \cos(n\theta)$. Therefore, the point $(r, \pi - \theta)$ is also on the graph.

Section 9.4

1. T, T, F, T, F, F, T, F.

3. For a function $r = f(\theta)$ in polar coordinates, θ is an angular rotation. A small change $\Delta\theta$ determines a wedge-shaped slice of the region.

5.

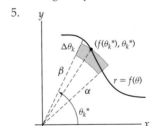

7. To compute an area we need $\beta > \alpha$ for the limits of integration in the integral. So that we don't compute a portion of the area twice, we need $\beta - \alpha < 2\pi$.

9. It is often simpler to compute a known fraction of the total area and then finish by multiplying by the appropriate factor. Review Example 2 to see how this may be done.

11. $x = t, y = f(t)$

13. The graph of the equation $r = \sin\theta$ is a circle with radius $\frac{1}{2}$. Since he integrated on the interval $[0, 2\pi]$ the circle is traced twice, so he got the correct answer.

15. The expression computes the area inside the two circles below. The area is $\frac{\pi - 2}{8}$.

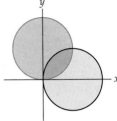

17. $\dfrac{1}{2}\displaystyle\int_0^\pi \theta^2\, d\theta = \dfrac{\pi^3}{6}$

19. $\displaystyle\int_0^{3\pi/4} (1 + \sqrt{2}\cos\theta)^2\, d\theta - \int_{3\pi/4}^{\pi} (1 + \sqrt{2}\cos\theta)^2\, d\theta = 3 + \pi$

21. $\displaystyle\int_{-\pi/2}^{\pi/6} ((3 - 3\sin\theta)^2 - (1 + \sin\theta)^2)\, d\theta = 8\pi + 9\sqrt{3}$

23. $\displaystyle\int_0^{\pi/3} (1 - \sec^2\theta)\, d\theta = \dfrac{\pi}{3} - \dfrac{\sqrt{3}}{4}$

25. $\displaystyle\int_0^{\pi/4} (\sec\theta - 2\cos\theta)^2\, d\theta = 2 - \dfrac{\pi}{2}$

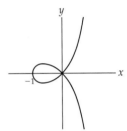

27. $\displaystyle\int_0^{2\pi} \sqrt{2}\, e^\theta\, d\theta = \sqrt{2}(e^{2\pi} - 1)$

29. $\displaystyle\int_0^{2\pi} \sqrt{1 + \alpha^2}\, e^{\alpha\theta}\, d\theta = \dfrac{\sqrt{1 + \alpha^2}}{\alpha}(e^{2\pi\alpha} - 1)$

31. $2\displaystyle\int_0^{\pi/4} \sqrt{4\sin^2 2\theta + \cos^2 2\theta}\, d\theta \approx 2.422$

33. $2\displaystyle\int_0^{\pi/8} \sqrt{16\sin^2 4\theta + \cos^2 4\theta}\, d\theta \approx 2.145$

35. $\displaystyle\int_{7\pi/6}^{11\pi/6} \sqrt{(1 + 2\sin\theta)^2 + 4\cos^2\theta}\, d\theta \approx 2.682$

37. $\dfrac{3\sqrt{3} + \pi}{4}$

39. $\dfrac{3\pi}{2}$

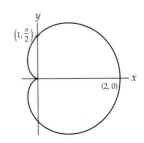

41. 1

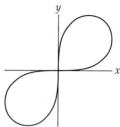

43. $\dfrac{9\pi}{2}$

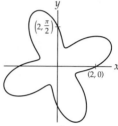

45. The area is $\dfrac{1}{2}\int_0^{2\pi} a^2\,d\theta = \pi a^2$.

47. The circumference is $\int_0^{2\pi} a\,d\theta = 2\pi a$.

49. $a^2\left(\dfrac{2\pi}{3} - \dfrac{\sqrt{3}}{2}\right)$

51. The length of the central cross section of the kayak is about 6.15 feet or about 1.87 meters. Annie should be able to use just two widths of the fabric.

53. $A = \dfrac{1}{2}\displaystyle\int_{\theta_0}^{\theta_0+\phi} r^2\,d\theta = \dfrac{1}{2}\phi r^2$.

55. For any positive integer n, the area is given by
$n\int_0^{2\pi/n}\cos^2(2n\theta)\,d\theta = \dfrac{\pi}{2}$.

57. The length of the curve is given by the improper integral $\displaystyle\int_1^\infty \sqrt{\dfrac{1}{\theta^2} + \dfrac{1}{\theta^4}}\,d\theta$ which diverges to ∞.

Section 9.5

1. T, T, F, T, F, T, F, T.

3. See Definition 9.15

5. See Definition 9.19

7. Since a hyperbola is the set of points in the plane for which the difference between the distances of the foci is a constant, and there are infinitely many differences which can be obtained in this way, there are infinitely many different hyperbolas with the same foci.

9. (a) $\dfrac{\sqrt{A^2-B^2}}{A}$, (b) $\dfrac{\sqrt{B^2-A^2}}{B}$, (c) Since $\sqrt{A^2 - B^2} < A$ and $\sqrt{B^2 - A^2} < B$, (d) It becomes more circular. (e) It elongates and flattens.

11. $y = \pm\dfrac{B}{e}$

13.

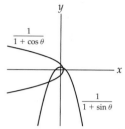

Both are parabolas with a focus at the origin; $r = \dfrac{1}{1+\cos\theta}$ opens to the left, the other opens down. All parabolas have eccentricity one.

15.

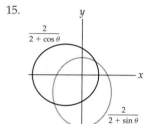

Both are ellipses with a focus at the origin; $r = \dfrac{2}{2+\cos\theta}$ has its major axis on the x-axis, the other has its major axis on the y-axis. Each has eccentricity $\dfrac{1}{2}$.

17. $\dfrac{\dfrac{\alpha}{\beta}}{\dfrac{\beta+\gamma\sin\theta}{}} = \dfrac{\dfrac{\gamma}{\beta}\dfrac{\alpha}{\gamma}}{1+\dfrac{\gamma}{\beta}\sin\theta}$

19. The equation is equivalent to $\dfrac{(x+1)^2}{8} - \dfrac{(y+3)^2}{16} = 1$. Its graph is a hyperbola with center $(-1, -3)$ opening to the left and right.

21. The equation is equivalent to $\dfrac{(y-4)^2}{25} - \dfrac{4(x+1)^2}{25} = 1$. Its graph is a hyperbola with center $(-1, 4)$ opening up and down.

23. $y = \dfrac{1}{2}x^2 + \dfrac{1}{2}$

25. $y = -\dfrac{1}{4}x^2 - 7$

27. $y = \dfrac{1}{2(y_1-y_0)}(x - x_1)^2 + \dfrac{y_0+y_1}{2}$

29. $y = -\dfrac{1}{8}x^2 + 2$

31. $y = -\dfrac{1}{2\alpha}x^2 + \dfrac{\alpha}{2}$

33. $\dfrac{x^2}{9} + \dfrac{(y-1)^2}{4} = 1$

35. $\dfrac{4x^2}{9} + \dfrac{y^2}{9} = 1$

37. $\dfrac{x^2}{\alpha^2} + \dfrac{y^2}{2\alpha^2} = 1$

39. $\dfrac{x^2}{6} - \dfrac{y^2}{30} = 1$

41. $\dfrac{y^2}{8} - \dfrac{x^2}{8} = 1$

43. $\dfrac{(y-5)^2}{4} - \dfrac{(x-3)^2}{12} = 1$

45.

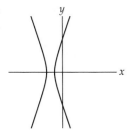

47.

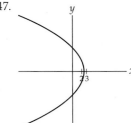

49.

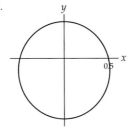

51.

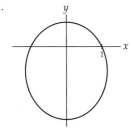

53. (a) $\dfrac{x^2}{1.00000011^2} + \dfrac{y^2}{0.9998606552^2} = 1,$

(b) $r = \dfrac{1.00000011}{1 + 0.0167\cos\theta}$

55. Model your argument on the proof that the distance from any point on the graph of the curve to the point $(0, \sqrt{A^2 - B^2})$ is $D_1 = \dfrac{A^2 - Cx}{A^2}$.

57. Model your proof on the analogous argument for an ellipse by showing that the distance from any point (x, y) on the curve in the first quadrant to the focus $(\sqrt{A^2 + B^2}, 0)$ is $D_1 = \dfrac{Cx - A^2}{A}$ and the distance from (x, y) to the focus $(-\sqrt{A^2 + B^2}, 0)$ is $D_2 = \dfrac{Cx + A^2}{A}$, where $C = \sqrt{A^2 + B^2}$. Thus, the difference of the distances is $2A$ for every point on the curve.

59. For an ellipse with equation $\dfrac{x^2}{A^2} + \dfrac{y^2}{B^2} = 1$ where $A > B > 0$, the distance between the foci is $2\sqrt{A^2 - B^2}$ and the distance between the vertices is $2A$. We have $\dfrac{2\sqrt{A^2 - B^2}}{2A} = e$. The situation when $0 < A < B$ is similar, as is the computation for a hyperbola.

61. Let $P = (x, y)$ be a point in the first quadrant on the hyperbola defined by $\dfrac{x^2}{A^2} - \dfrac{y^2}{B^2} = 1$. The coordinates of the focus on the positive x-axis is $F = (\sqrt{A^2 + B^2}, 0)$ and the equation of the directrix which intersects the positive x-axis is $x = \dfrac{A}{e}$. Let D be the point on the directrix closest to P. Then $DP = x - \dfrac{A}{e}$ and $FP = \sqrt{(x - \sqrt{A^2 + B^2})^2 + y^2}$. Use the fact that $y^2 = \dfrac{B^2}{A^2}x^2 - B^2$ to show that $\dfrac{FP}{DP} = e$.

For answers to odd-numbered Skill Certification exercises in the Chapter Review, please visit the Book Companion Web Site at www.whfreeman.com/tkcalculus.

Chapter 10

Section 10.1

1. F, T, F, F, F, T, T, T.

3. (a) $y = 5$ is a line parallel to the x-axis; $x = -3$ is a line parallel to the y-axis (b) $y = 5$ is a plane parallel to the xz-plane; $x = -3$ is a plane parallel to the yz-plane

5. $(1, 1, 1)$, $(1, 1, -1)$, $(1, -1, 1)$, $(1, -1, -1)$, $(-1, 1, 1)$, $(-1, 1, -1)$, $(-1, -1, 1)$, and $(-1, -1, -1)$

7. The set of all points in $\mathbb{R}^3$ equidistant from a given point (x_0, y_0, z_0).

9. (a) A circle with radius 2 centered at the origin, (b) a right circular cylinder with radius 2 centered on the z-axis.

11. $(-5, 6, -7)$

13. $(3, -7, 6)$

15. There are three possible exchanges: x with y, x with z and y with z. We will discuss the first of these. The others are similar. If the placement of x, y, and z form a right-hand coordinate system, then the axes can be oriented to be superimposed on the graph below left. Switching the labels on the x and y axes we obtain the left-hand system below right. Exchanging two pairs of the labels will give another right-handed system.

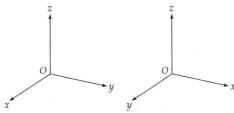

17. The distance between the points $(a_1, a_2, \ldots, a_n)$ and $(b_1, b_2, \ldots, b_n)$ is
$$\sqrt{(a_1 - b_1)^2 + (a_2 - b_2)^2 + \cdots + (a_n - b_n)^2}.$$

19. (a) $x > 0, y > 0, z > 3$,
(b) $(x - 1)^2 + (y - 2)^2 + (z - 4)^2 < 25$,
(c) $(x - 2)^2 + (y + 3)^2 < 25$

21. $\sqrt{130}$

23. $\sqrt{14}$

25. $\sqrt{26}$

27. $\sqrt{51}$

29. $(x-3)^2 + (y+2)^2 + (z-5)^2 = 25$

31. $(x-2)^2 + (y-5)^2 + (z+7)^2 = 49$

33. $(x-2)^2 + (y-5)^2 + (z+7)^2 = 78$

35. $(x-2)^2 + (y+8)^2 + z^2 = 66$

37. $\left(x-\frac{3}{2}\right)^2 + (y-1)^2 + \left(z-\frac{5}{2}\right)^2 = \frac{53}{2}$

39. Center $\left(0, -\frac{3}{2}, \frac{5}{2}\right)$, radius $\frac{\sqrt{22}}{2}$

41.
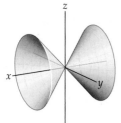

43.
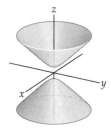

45.
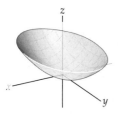

47.
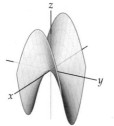

49. $3 - 2\sqrt{2}$

51. Use the distance formula to show that the length of each side of the triangle is $\sqrt{8}$.

53. Use the distance formula to show that the lengths of the sides of the triangle are 7, 14 and $\sqrt{245}$. Since $7^2 + 14^2 = (\sqrt{245})^2$, the triangle is right. The area is 49 square units.

55. $\left(x-\frac{1}{2}\right)^2 + \left(y-\frac{1}{2}\right)^2 + \left(z-\frac{1}{2}\right)^2 = \frac{3}{4}$ and $\left(x-\frac{1}{6}\right)^2 + \left(y-\frac{1}{6}\right)^2 + \left(z-\frac{1}{6}\right)^2 = \frac{3}{4}$

57. $\frac{4}{3}$

59. The compass readings give the slopes of lines from her boat to the respective peaks. Thus her boat is on the line $(y-4.0) = \tan(75\pi/180)(x-7.9)$ from Constitution Peak, and it is on the line $(y+3.4) = \tan\left(\frac{-10\pi}{180}\right)(x-9.3)$. Those lines intersect at $(6.07, -2.83)$. Thus, she is 6.7 miles from Deer Harbor.

61. $a = \pm 3.81$

63. Consider the regular tetrahedron with vertices $(1, 0, 0)$, $(0, 1, 0)$, $(0, 0, 1)$ and $(1, 1, 1)$ from Exercise 54. The midpoint of the segment connecting the midpoints of the opposite edges is the point $\left(\frac{1}{2}, \frac{1}{2}, \frac{1}{2}\right)$.

Section 10.2

1. T, F, F, F, T, T, T, T.

3. $\overrightarrow{AC}$

5. **0**

7. **0**

9. Algebraically, you add them "componentwise." Geometrically, you place the initial point of a vector $\mathbf{w}$ at the terminal point of a vector $\mathbf{v}$. The vector $\mathbf{v} + \mathbf{w}$ is the vector that has the same initial point as $\mathbf{v}$ and the same terminal point as $\mathbf{w}$.

11. $\langle 2a, 2b, 2c \rangle$

13. $(-1, 5, -1)$

15. $\left(\frac{5\sqrt{14}}{14}, 3 + \frac{10\sqrt{14}}{14}, -2 + \frac{15\sqrt{14}}{14}\right)$

17. (a) The set of all position vectors whose terminal points are on the sphere of radius 4 centered at the origin. (b) The set of all position vectors whose terminal points are in or on the sphere of radius 4 centered at the origin. (c) The set of all vectors whose initial point is (a, b, c) and whose terminal point is on the sphere of radius 4 centered at (a, b, c).

19. Use vectors with 4 (or n) components. Algebraically, you still add vectors component by component and when you multiply a vector by a scalar you multiply each component by that scalar.

21. The set of all position vectors whose terminal points are on the sphere of radius 5 centered at the origin.

23. $\mathbf{u} + \mathbf{v} = \langle 8, -4 \rangle$; $\mathbf{u} - \mathbf{v} = \langle -4, -8 \rangle$

25. $\mathbf{u} + \mathbf{v} = \langle -2, 4 \rangle$; $\mathbf{u} - \mathbf{v} = \langle 4, -8 \rangle$

27. $\mathbf{u} + \mathbf{v} = \langle 3, -8, 13 \rangle$; $\mathbf{u} - \mathbf{v} = \langle -1, 0, -1 \rangle$

29. $\langle -6, -8 \rangle$

31. $\langle 0, -5, -7 \rangle$

33. 5

35. $\frac{\sqrt{3}}{3}$

37. $\|\mathbf{v}\| = 5$, $\left\langle \frac{3}{5}, -\frac{4}{5} \right\rangle$

39. $\|\mathbf{v}\| = \frac{\sqrt{34}}{15}$, $\frac{15}{\sqrt{34}} \left\langle \frac{1}{5}, \frac{1}{3} \right\rangle$

41. $\|\mathbf{v}\| = \sqrt{3}$, $\frac{1}{\sqrt{3}} \langle 1, 1, 1 \rangle$

43. $\frac{5}{\sqrt{14}} \langle 3, 1, 2 \rangle$

45. $\frac{2}{\sqrt{117}} \langle 8, -7, 2 \rangle$

47. $\frac{7}{\sqrt{53}} \langle 1, 4, 6 \rangle$

49. $\frac{7}{\sqrt{53}} \langle 1, 4, 6 \rangle$

51. $-\frac{3}{\sqrt{14}} \langle 1, -2, 3 \rangle$

53. The magnitude of the force in the rope on the left is $50(3\sqrt{2} - \sqrt{6})$ pounds and the magnitude of the force in the rope on the right is $100(\sqrt{3} - 1)$ pounds.

55. (a) $\langle 0, -2 \rangle$, (b) $\langle \sqrt{2}, -\sqrt{2} \rangle$, (c) $\langle \sqrt{3} - 2, -\sqrt{3} \rangle$

57. $\mathbf{u} + \mathbf{v} = \langle u_1, u_2, u_3 \rangle + \langle v_1, v_2, v_3 \rangle =$
$\langle u_1 + v_1, u_2 + v_2, u_3 + v_3 \rangle =$
$\langle v_1 + u_1, v_2 + u_2, v_3 + u_3 \rangle = \langle v_1, v_2, v_3 \rangle + \langle u_1, u_2, u_3 \rangle =$
$\mathbf{v} + \mathbf{u}.$

59. $c(\mathbf{u} + \mathbf{v}) = c \left(\langle u_1, u_2, u_3 \rangle + \langle v_1, v_2, v_3 \rangle \right) =$
$c \left(\langle u_1 + v_1, u_2 + v_2, u_3 + v_3 \rangle \right) =$
$\langle c(u_1 + v_1), c(u_2 + v_2), c(u_3 + v_3) \rangle =$
$\langle cu_1 + cv_1, cu_2 + cv_2, cu_3 + cv_3 \rangle = \langle cu_1, cu_2, cu_3 \rangle +$
$\langle cv_1, cv_2, cv_3 \rangle = c \langle u_1, u_2, u_3 \rangle + c \langle v_1, v_2, v_3 \rangle = c\mathbf{u} + c\mathbf{v}.$

61. $(c + d)\mathbf{v} = (c + d)\langle v_1, v_2, v_3 \rangle = \langle (c+d)v_1, (c+d)v_2, (c+d)v_3 \rangle = \langle cv_1 + dv_1, cv_2 + dv_2, cv_3 + dv_3 \rangle = \langle cv_1, cv_2, cv_3 \rangle + \langle dv_1, dv_2, dv_3 \rangle = c\langle v_1, v_2, v_3 \rangle + d\langle v_1, v_2, v_3 \rangle = c\mathbf{v} + d\mathbf{v}.$

63. Let $\mathbf{u} = \langle u_1, u_2 \rangle$ and $\mathbf{v} = \langle v_1, v_2 \rangle$ be two position vectors. The vertices of the parallelogram determined by $\mathbf{u}$ and $\mathbf{v}$ are $(0, 0)$, (u_1, u_2), (v_1, v_2) and $(u_1 + v_1, u_2 + v_2)$. The midpoint of both opposite pairs of vertices is $\left(\frac{u_1 + v_1}{2}, \frac{u_2 + v_2}{2} \right)$.

Section 10.3

1. T, F, F, F, T, T, T, F.

3. See Definition 10.13.

5. When the angle between the sides with lengths a and b is right, the Law of Cosines reduces to the Pythagorean theorem since $\cos 90° = 0$.

7. (a) Assuming none of the constants are zero, we need $\frac{\alpha}{a} = \frac{\beta}{b} = \frac{\gamma}{c}$; (b) $a\alpha + b\beta + c\gamma = 0.$

9. (a) Let $\mathbf{u} = \mathbf{v} = \mathbf{i}$ and $\mathbf{w} = \mathbf{i} + \mathbf{j}$; (b) $\mathbf{u}$ must be orthogonal to $\mathbf{v} - \mathbf{w}$

11. The two vectors must be parallel and point in the same direction.

13. Any position vector that lies in the yz-plane.

15. Any vector of the form $\langle 1, v_2, v_3 \rangle$.

17. (a) $\langle 1, m \rangle$, (b) slope $-\frac{1}{m}$, $\langle -m, 1 \rangle$,
(c) $\langle 1, m \rangle \cdot \langle -m, 1 \rangle = -m + m = 0.$

19. (a) If there were a scalar c such that $c\mathbf{v}_1 = \mathbf{v}_2$, in order for the first components to be equal we would need $c = 3$, but $3\mathbf{v}_1 = \langle 3, -6 \rangle \neq \mathbf{v}_2$. (b) $2\mathbf{v}_1 + 1\mathbf{v}_2 = \langle 5, 1 \rangle.$

21. $-1, \cos^{-1} \left(-\frac{1}{\sqrt{1711}} \right)$

23. $38, \cos^{-1} \left(\frac{38}{\sqrt{2170}} \right)$

25. $\text{proj}_{\mathbf{u}} \mathbf{v} = \mathbf{0}, \text{comp}_{\mathbf{u}} \mathbf{v} = 0$, and the component of $\mathbf{v}$ orthogonal to $\mathbf{u}$ is $\mathbf{v}$

27. $\text{comp}_{\mathbf{u}} \mathbf{v} = -2\sqrt{14}; \text{proj}_{\mathbf{u}} \mathbf{v} = \mathbf{v}$; and the component of $\mathbf{v}$ orthogonal to $\mathbf{u}$ is $\mathbf{0}$

29. $\text{proj}_{\mathbf{u}} \mathbf{v} = \mathbf{0}, \text{proj}_{\mathbf{v}} \mathbf{u} = \mathbf{0}$

31. $\text{proj}_{\mathbf{u}} \mathbf{v} = -\frac{5}{27} \langle 1, -5, -1 \rangle, \text{proj}_{\mathbf{v}} \mathbf{u} = \langle 0, -5, 0 \rangle$

33. (a) -1; (b) $\cos^{-1} \left(-\frac{1}{2\sqrt{2451}} \right)$; (c) $-\frac{1}{38}\mathbf{u}$

35. (a) 24; (b) $\cos^{-1} \left(\frac{24\sqrt{2}}{50} \right)$; (c) $\frac{24}{25}\mathbf{u}$

37. $\frac{48}{83}\sqrt{166}$

39. $\frac{\pi}{2}$

41. $\cos^{-1} \left(\frac{\sqrt{6}}{3} \right)$

43. $\cos \alpha = \frac{1}{\sqrt{14}}, \alpha = \cos^{-1} \left(\frac{1}{\sqrt{14}} \right), \cos \beta = \frac{2}{\sqrt{14}},$
$\beta = \cos^{-1} \left(\frac{2}{\sqrt{14}} \right), \cos \gamma = \frac{3}{\sqrt{14}}, \gamma = \cos^{-1} \left(\frac{3}{\sqrt{14}} \right)$

45. $\cos \alpha = -\frac{1}{\sqrt{18}}, \alpha = \cos^{-1} \left(-\frac{1}{\sqrt{18}} \right), \cos \beta = \frac{1}{\sqrt{18}},$
$\beta = \cos^{-1} \left(\frac{1}{\sqrt{18}} \right), \cos \gamma = -\frac{4}{\sqrt{18}}, \gamma = \arccos \left(-\frac{4}{\sqrt{18}} \right)$

47. Let $\mathbf{v} = \langle a, b, c \rangle$. From their definitions, $\cos \alpha = \frac{a}{\|\mathbf{v}\|},$
$\cos \beta = \frac{b}{\|\mathbf{v}\|}$ and $\cos \gamma = \frac{c}{\|\mathbf{v}\|}$. Thus,
$\|\mathbf{v}\| ((\cos \alpha)\mathbf{i} + (\cos \beta)\mathbf{j} + (\cos \gamma)\mathbf{k}) = a\mathbf{i} + b\mathbf{j} + c\mathbf{k} = \mathbf{v}.$

49. $\cos \gamma = \pm \frac{\sqrt{2}}{2}$

51. $\cos \beta = \pm \frac{3}{4}$

53. (a) $\alpha_i = \cos^{-1} \left(\frac{1}{\sqrt{n}} \right)$; (b) $\lim_{n \to \infty} \cos^{-1} \left(\frac{1}{\sqrt{n}} \right) = \frac{\pi}{2}$

55. (a) $\tan^{-1} \left(\frac{1}{4} \right)$, (b) $4\sqrt{4/17} \approx 1.9$, (c) slightly more than 1 hour.

57. If $\mathbf{v} = \langle a, b, c \rangle$, then $\mathbf{v} \cdot \mathbf{i} = a$, $\mathbf{v} \cdot \mathbf{j} = b$ and $\mathbf{v} \cdot \mathbf{k} = c$. Thus, $(\mathbf{v} \cdot \mathbf{i})\mathbf{i} + (\mathbf{v} \cdot \mathbf{j})\mathbf{j} + (\mathbf{v} \cdot \mathbf{k})\mathbf{k} = a\mathbf{i} + b\mathbf{j} + c\mathbf{k} = \mathbf{v}.$

59. Taking the absolute value of each side of $\mathbf{u} \cdot \mathbf{v} = \|\mathbf{u}\| \|\mathbf{v}\| \cos \theta$, we see that $|\mathbf{u} \cdot \mathbf{v}| = \|\mathbf{u}\| \|\mathbf{v}\| |\cos \theta|$. Since $|\cos \theta| \leq 1$ we have our result. In order to have the equality, the two vectors must be parallel.

61. If either $\mathbf{u}$ or $\mathbf{v}$ is $\mathbf{0}$, then $\mathbf{u} \cdot \mathbf{v} = 0$ and $\mathbf{u}$ and $\mathbf{v}$ are orthogonal, since $\mathbf{0}$ is orthogonal to every vector. If neither $\mathbf{u}$ nor $\mathbf{v}$ is $\mathbf{0}$, then $\mathbf{u} \cdot \mathbf{v} = \|\mathbf{u}\| \|\mathbf{v}\| \cos \theta$ by Theorem 10.15, where θ is the angle between the position vectors $\mathbf{u}$ and $\mathbf{v}$. Since neither $\mathbf{u}$ nor $\mathbf{v}$ is $\mathbf{0}$, the dot product is only zero if $\cos \theta = 0$, that is, if $\mathbf{u}$ and $\mathbf{v}$ are orthogonal.

63. Assume that $\mathbf{v} = \mathbf{v}_{\|} + \mathbf{v}_{\perp}$ and $\mathbf{v} = \mathbf{v}'_{\|} + \mathbf{v}'_{\perp}$ are two sums such that $\mathbf{v}_{\|}$ and $\mathbf{v}'_{\|}$ are both parallel to $\mathbf{u}$ and $\mathbf{v}_{\perp}$ and $\mathbf{v}'_{\perp}$ are both orthogonal to $\mathbf{u}$. Then $\mathbf{v}_{\|} + \mathbf{v}_{\perp} = \mathbf{v}'_{\|} + \mathbf{v}'_{\perp}$, so $\mathbf{v}_{\|} - \mathbf{v}'_{\|} = \mathbf{v}'_{\perp} - \mathbf{v}_{\perp}$. The difference on the left will also be parallel to $\mathbf{u}$ and the difference on the right will also be orthogonal to $\mathbf{u}$. Thus, they are orthogonal to each other. The only way for this to happen is if they are both $\mathbf{0}$. Therefore, the two decompositions are the same.

65. Let $\mathbf{u}$ and $\mathbf{v}$ be two nonparallel position vectors. The diagonals of the parallelogram determined by $\mathbf{u}$ and $\mathbf{v}$ are the vectors $\mathbf{u} + \mathbf{v}$ and $\mathbf{u} - \mathbf{v}$. The dot product $(\mathbf{u}+\mathbf{v}) \cdot (\mathbf{u}-\mathbf{v}) = \mathbf{u} \cdot \mathbf{u} - \mathbf{u} \cdot \mathbf{v} + \mathbf{v} \cdot \mathbf{u} + \mathbf{v} \cdot \mathbf{v} = \|\mathbf{u}\|^2 - \|\mathbf{v}\|^2$, since the dot product is commutative. If the parallelogram is a rhombus, then $\|\mathbf{u}\| = \|\mathbf{v}\|$. Thus, the dot product of the vectors forming the diagonals is zero, so they are perpendicular. Conversely, if the diagonals of the parallelogram are perpendicular, we must have $\|\mathbf{u}\|^2 - \|\mathbf{v}\|^2 = 0$ and $\|\mathbf{u}\| = \|\mathbf{v}\|$, so the parallelogram is a rhombus.

Section 10.4

1. F, F, T, F, T, T, T, F.

3. See Definition 10.25.

5. The cross product, $\mathbf{u} \times \mathbf{v}$, is orthogonal to both $\mathbf{u}$ and $\mathbf{v}$. The magnitude of $\mathbf{u} \times \mathbf{v}$ is $\|\mathbf{u}\|\|\mathbf{v}\| \sin \theta$ where θ is the angle between $\mathbf{u}$ and $\mathbf{v}$. The vectors $\mathbf{u}, \mathbf{v}$ and $\mathbf{u} \times \mathbf{v}$ form a right-hand triple.

7. The dot product is zero if and only if the vectors are orthogonal.

9. Embed the vectors in $\mathbb{R}^3$ as $\langle 1, 2, 0 \rangle$ and $\langle 3, -1, 0 \rangle$. The cross product $\langle 1, 2, 0 \rangle \times \langle 3, -1, 0 \rangle = \langle 0, 0, -7 \rangle$. The norm of this vector is 7, which is the area of the parallelogram in square units.

11. They are orthogonal.

13. Let $\mathbf{u} = \langle 1, 0, 0 \rangle$, $\mathbf{v} = \langle 2, 1, 1 \rangle$ and $\mathbf{w} = \langle 4, 1, 1 \rangle$. If $\mathbf{u} \times \mathbf{v} = \mathbf{u} \times \mathbf{w}$, then $\mathbf{u}$ is parallel to $\mathbf{v} - \mathbf{w}$.

15. As position vectors, the three vectors lie in the same plane.

17. The cross product is not associative, so we must specify whether we mean $(\mathbf{u} \times \mathbf{v}) \times \mathbf{w}$ or $\mathbf{u} \times (\mathbf{v} \times \mathbf{w})$.

19. At least one of them is $\mathbf{0}$.

21. They are parallel.

23. $\langle 23, -4, 14 \rangle$, $\langle -23, 4, -14 \rangle$

25. $\langle -46, 3, -22 \rangle$, $\langle -42, -30, -38 \rangle$

27. $\sqrt{213}$

29. 106 cubic units. No, they form a left-handed triple.

31. $\langle 25, 35, -10 \rangle$, $\langle -25, -35, 10 \rangle$

33. $0, 0$

35. 0

37. (a) $\langle -24, 7, 15 \rangle$ (b) $\pm \dfrac{1}{5\sqrt{34}} \langle -24, 7, 15 \rangle$, (c) $\dfrac{5}{2}\sqrt{34}$.

39. (a) $\langle 20, -9, 30 \rangle$; (b) $\pm \dfrac{1}{\sqrt{1381}} \langle 20, -9, 30 \rangle$; (c) $\dfrac{\sqrt{1381}}{2}$

41. (a) $\mathbf{k}$; (b) $\pm \mathbf{k}$; (c) 26

43. $\dfrac{\sqrt{850}}{2}$

45. (a) The lengths of the sides of a quadrilateral do not determine the quadrilateral's area. For example, a square with side lengths 1 unit has an area of 1 square unit, but a rhombus with side lengths 1 and opposite interior angles of $30°$ has area $\dfrac{\sqrt{3}}{2}$. (b) If you know the side lengths and the length of one diagonal you can decompose the quadrilateral into two triangles. (c) The side lengths adjacent to the angle whose measure is known determine a unique triangle. You may find the length of the third side of the triangle, which is a diagonal of the quadrilateral and use the reasoning from (b).

47. $\dfrac{3}{4}\left(\sqrt{759} + \sqrt{2079}\right)$

49. $16\sqrt{759} + 6\sqrt{70}$

51. (a) Parallelogram, (b) 16 square units

53. (a) Not a parallelogram, (b) $\dfrac{49}{2}$ square units

55. (a) $\overrightarrow{PQ} = \overrightarrow{SR}$ but P, Q, R and S are not collinear, (b) parallelogram, (b) $7\sqrt{254}$ square units

57. (a) $\overrightarrow{PQ}$ and $\overrightarrow{SR}$ are parallel, (b) not a parallelogram, (b) $\dfrac{7}{2}\sqrt{146}$ square units

59. $4\sqrt{2\sqrt{2} - 2} \approx 3.64$ cubic centimeters

61. 30 foot-pounds

63. If $A = \begin{bmatrix} a & b & c \\ d & e & f \\ g & h & i \end{bmatrix}$, then $\det A = aei + bfg + cdh - afh - bdi - ceg$. Show that when you exchange any two rows of A, you obtain the negative of this sum.

65. If $\mathbf{v} = \langle a, b, c \rangle$ and α is a scalar, then $\mathbf{v} \times (\alpha \mathbf{v}) = (\alpha bc - \alpha bc)\mathbf{i} + (\alpha ac - \alpha ac)\mathbf{j} + (\alpha ab - \alpha ab)\mathbf{k} = \mathbf{0}$.

67. Let $\mathbf{u} = \langle u_1, u_2, u_3 \rangle$ and $\mathbf{v} = \langle v_1, v_2, v_3 \rangle$. Show that $c(\mathbf{u} \times \mathbf{v})$, $(c\mathbf{u}) \times \mathbf{v}$, and $\mathbf{u} \times (c\mathbf{v})$ are all equal $\langle c(u_2v_3 - u_3v_2), c(u_3v_1 - u_1v_3), c(u_1v_2 - u_2v_1) \rangle$.

69. Let $\mathbf{u} = \langle u_1, u_2, u_3 \rangle$ and $\mathbf{v} = \langle v_1, v_2, v_3 \rangle$, the triple scalar product $\mathbf{v} \cdot (\mathbf{u} \times \mathbf{v}) = \det \begin{bmatrix} v_1 & v_2 & v_3 \\ u_1 & u_2 & u_3 \\ v_1 & v_2 & v_3 \end{bmatrix}$. Show that this determinant is zero.

71. (a) The volume of the parallelepiped determined by $\mathbf{u}, \mathbf{v}$ and $\mathbf{w}$ is the absolute value of the determinant of the matrix in which the rows are the components of $\mathbf{u}, \mathbf{v}$ and $\mathbf{w}$. Since each component is an integer, the determinant and its absolute value are integers. (b) The area of the parallelogram determined by $\mathbf{i}$ and $\mathbf{j}$ is 1. The area of the parallelogram determined by $\mathbf{i}$ and $\mathbf{j} + \mathbf{k}$ is $\sqrt{2}$.

73. Divide the left and right sides of the equations $\|\mathbf{u} \times \mathbf{v}\| = \|\mathbf{u}\|\|\mathbf{v}\| \sin \theta$ and $\mathbf{u} \cdot \mathbf{v} = \|\mathbf{u}\|\|\mathbf{v}\| \cos \theta$ to obtain the result.

75. If $\mathbf{u} \times \mathbf{v} = \mathbf{u} \times \mathbf{w}$, then $\mathbf{u}$ is parallel to $\mathbf{v} - \mathbf{w}$ and if $\mathbf{u} \cdot \mathbf{v} = \mathbf{u} \cdot \mathbf{w}$, then $\mathbf{u}$ is orthogonal to $\mathbf{v} - \mathbf{w}$. Thus, $\mathbf{v} - \mathbf{w}$ is orthogonal to itself. This can only happen if $\mathbf{v} - \mathbf{w} = \mathbf{0}$.

77. Let $\mathbf{r} = \langle r_1, r_2, r_3 \rangle$, $\mathbf{s} = \langle s_1, s_2, s_3 \rangle$, $\mathbf{u} = \langle u_1, u_2, u_3 \rangle$ and $\mathbf{v} = \langle v_1, v_2, v_3 \rangle$. Show that both sides of the equality are equal to $r_1s_2u_1v_2 + r_1s_3u_1v_3 + r_2s_1u_2v_1 + r_2s_3u_2v_3 + r_3s_2u_3v_2 + r_3s_2u_3v_3 - (r_1s_2u_2v_1 + r_1s_3u_3v_1 + r_2s_1u_1v_2 + r_2s_3u_3v_2 + r_3s_1u_1v_3 + r_3s_2u_2v_3)$.

79. If $\mathbf{r}, \mathbf{s}, \mathbf{u}$, and $\mathbf{v}$ all lie in some plane $\mathcal{P}$, then the cross products $\mathbf{r} \times \mathbf{s}$ and $\mathbf{u} \times \mathbf{v}$ are both orthogonal to $\mathcal{P}$. Therefore, these two vectors are parallel and the cross product of two parallel vectors is $\mathbf{0}$.

Section 10.5

1. T, F, T, F, T, F, F, F.

3. A linear equation in three variables is an equation that can be written in the form $Ax + By + Cz = D$, where A, B, C, and D are real numbers. An example is $2x - 3y + \pi z = 12$. The graph of every linear equation in three variables is a plane.

5. Let $P = (a, b, c)$ and $Q = (\alpha, \beta, \gamma)$. The direction vector for the line is $\pm \langle a - \alpha, b - \beta, c - \gamma \rangle$. We will choose the positive sign for this vector, and we may choose either point P or point Q; we will choose Q. The equation of the line as a vector function is $\mathbf{r}(t) = \langle \alpha + (a - \alpha)t, \beta + (b - \beta)t, \gamma + (c - \gamma)t \rangle$.

7. If the slopes of the two lines are equal, the lines are either parallel or identical. Two lines in $\mathbb{R}^2$ that aren't parallel or identical intersect in a unique point.

9. (a) $\mathbf{r}_1(t) = \langle -1 + 2t, 3 - 4t, 7 + 9t \rangle$
   (b) $\mathbf{r}_2(t) = \langle -1 + 4t, 3 - 8t, 7 + 18t \rangle$
   (c) $x = 2t + 5, y = -4t - 9, z = 9t + 34$

11. $\left(0, \frac{52}{3}, \frac{19}{3}\right)$; $\left(\frac{52}{7}, 0, \frac{27}{7}\right)$; $(19, -27, 0)$

13. $(0, -9, 1)$; $\left(\frac{9}{2}, 0, -17\right)$; $\left(\frac{1}{4}, -\frac{17}{2}, 0\right)$

15. $(-7, 0, 5)$. The line is parallel to the y-axis.

17. (a) $x = 2 + 7t, y = 3 - 5t, z = 2t$. (b) $\frac{x-2}{7} = \frac{y-3}{-5} = \frac{z}{2}$.

19. (a) $\mathbf{r}(t) = \langle 4, 3 - 5t, t \rangle$; (b) $x = 4, \frac{y-3}{-5} = z$

21. In slope-intercept form, $y = -3x + 5$. Next, $x = 2t$ and $y = 5 - 6t$. Eliminate the parameter t to obtain the previous equation.

23. (a) $\mathbf{r}(t) = \langle t, 2t, -4t \rangle$; (b) $x = t, y = 2t, z = -4t$;
   (c) $x = \frac{y}{2} = -\frac{z}{4}$

25. (a) $\mathbf{r}(t) = \langle 2t - 1, 3, 4t + 7 \rangle$;
   (b) $x = 2t - 1, y = 3, z = 4t + 7$; (c) $y = 3, \frac{x+1}{2} = \frac{z-7}{4}$

27. (a) $\mathbf{r}(t) = \langle 2t + 3, 5t + 1 \rangle$; (b) $x = 2t + 3, y = 5t + 1$;
   (c) $\frac{x-3}{2} = \frac{y-1}{5}$

29. (a) $\mathbf{r}(t) = \langle 4t, -t, 6t \rangle$; (b) $x = 4t, y = -t, z = -6t$;
   (c) $\frac{x}{4} = -y = \frac{z}{6}$

31. (a) $\mathbf{r}(t) = \langle 8t - 4, 11, 2t \rangle$; (b) $x = 8t - 4, y = 11, z = 2t$;
   (c) $y = 11, \frac{x+4}{8} = \frac{z}{2}$

33. (a) $\mathbf{r}(t) = \langle 3t + 1, -t + 6 \rangle$; (b) $x = 3t + 1, y = -t + 6$;
   (c) $\frac{x-1}{3} = -y + 6$

35. (a) $\mathbf{r}(t) = \langle (x_1 - x_0)t + x_0, (y_1 - y_0)t + y_0, (z_1 - z_0)t + z_0 \rangle$;
   (b) $\mathbf{r}(t) = \langle (x_1 - x_0)t + x_0, (y_1 - y_0)t + y_0, (z_1 - z_0)t + z_0 \rangle$, $0 \le t \le 1$

37. $\mathbf{r}(t) = \langle -2t + 1, -9t + 7, 2t + 3 \rangle$, $0 \le t \le 1$

39. $\mathbf{r}(t) = \langle -4t + 3, 6t - 1, 5t + 4 \rangle$, $0 \le t \le 1$

41. The lines are parallel. They are $\sqrt{\frac{117}{7}}$ units apart.

43. The lines intersect at $(-8, 8, -9)$.

45. The lines are parallel. They are $5\sqrt{\frac{61}{21}}$ units apart.

47. $\sqrt{\frac{1084}{21}}$

49. $\frac{4\sqrt{5}}{5}$

51. (a) $\alpha = -3$, (b) $\alpha = 15$

53. To the nearest tenth of a kilometer, he is at an elevation of 4.7 km and is 0.7 km east of the summit.

55. See the proof preceding Theorem 10.38.

Section 10.6

1. F, F, F, F, T, T, F, T.

3. $\left(\frac{d}{a}, 0, 0\right)$, $\left(0, \frac{d}{b}, 0\right)$, $\left(0, 0, \frac{d}{c}\right)$. You can plot these three points and construct the triangle with these points as vertices to understand the plane that contains them.

5. If the vectors normal to the two planes are not parallel, the two planes intersect in a line.

7. It means that the three points lie on a single line. Let P, Q, and R be the points. If the two vectors $\overrightarrow{PQ}$ and $\overrightarrow{PR}$ are parallel, the points are collinear. If the three points don't lie on the same line, by definition they are noncollinear.

9. Three noncollinear points determine a unique plane. Any two distinct points on $\mathcal{L}$ along with P determine the plane. Choose any point Q on $\mathcal{L}$. Construct the vector $\overrightarrow{PQ}$. A normal vector to the plane is given by $\mathbf{N} = \overrightarrow{PQ} \times \mathbf{d}$, where $\mathbf{d}$ is the direction vector for $\mathcal{L}$. Use P and $\mathbf{N}$ to find the equation of the plane. In $\mathbb{R}^3$ there are infinitely many different planes containing every line.

11. Choose two distinct points on one of the lines and one point on the other. These three points determine a unique plane. (See the answer to Exercise 9.) Every point on both lines lies in the plane determined by the three chosen points.

13. Let $\mathbf{d}$ and $\mathbf{d}'$ be the direction vectors for the lines $\mathcal{L}$ and $\mathcal{L}'$, respectively. Since the lines are skew, $\mathbf{N} = \mathbf{d} \times \mathbf{d}' \ne \mathbf{0}$. Any point from $\mathcal{L}$ along with $\mathbf{N}$ and any point from $\mathcal{L}'$ along with $\mathbf{N}$ determine the unique pair of parallel planes.

15. The direction vector for $\mathcal{L}$ is $\mathbf{d} = \langle a, b, c \rangle$ and the normal vector for $\mathcal{P}$ is $\mathbf{N} = \langle \alpha, \beta, \gamma \rangle$. The line and plane are orthogonal if and only if $\mathbf{d}$ and $\mathbf{N}$ are scalar multiples.

17. Let Q be the point of intersection between $\mathcal{L}$ and $\mathcal{P}$. If $\mathcal{L}$ is orthogonal to $\mathcal{P}$, then every line in $\mathcal{P}$ passing through Q is orthogonal to $\mathcal{L}$, and the angle between $\mathcal{L}$ and $\mathcal{P}$ is $\frac{\pi}{2}$. Otherwise, the angle described is the minimum angle that $\mathcal{L}$ can form with any line in $\mathcal{P}$ through Q.

19. The formula builds upon the computation for the component of a vector in the direction of a given vector. We choose arbitrary points P_1 and P_2 on lines $\mathcal{L}_1$ and $\mathcal{L}_2$, respectively, and $|\text{comp}_{\mathbf{N}}\overrightarrow{P_1P_2}|$, where $\mathbf{N}$ is the cross product of the direction vectors of $\mathcal{L}_1$ and $\mathcal{L}_2$. The formula works because the skew lines lie on a unique pair of parallel planes.

21. $4x - y + 5z = 0$

23. $2x - y + 6z = 41$

25. $12x - 5y + 19z = 61$

27. $y_0z_0x + x_0z_0y + x_0y_0z = x_0y_0z_0$

29. $3x + 7y - 7z = -26$

31. The direction vectors for the lines $\langle -5, 2, 4\rangle$ and $\langle 15, -6, -12\rangle$ are scalar multiples of each other, so the lines are parallel; $14x + 55y - 10z = 137$.

33. The point $(7, -5, 14)$ is on both lines; $14x + 3y + 2z = 111$.

35. $\mathbf{r}(t) = \left\langle -\dfrac{8}{5} - 11t, \dfrac{14}{5} - 2t, 5t\right\rangle$

37. $\mathbf{r}(t) = \langle 4, 3 + 2t, 5t\rangle$

39. $\sqrt{2}$

41. The normal vectors are parallel, so the planes are parallel. They are $\dfrac{29}{3\sqrt{38}}$ units apart.

43. The direction vectors of the lines are $\langle 2, -4, 5\rangle$ and $\langle 3, 2, 3\rangle$. Since these vectors aren't scalar multiples of each other, the lines either intersect, or are skew. The planes containing the lines are $-22x + 9y + 16z = 81$ and $-22x + 9y + 16z = -97$, respectively, so the lines must be skew. The distance between the lines is $\dfrac{178}{\sqrt{821}}$.

45. $\cos^{-1}\left(\dfrac{4\sqrt{3}}{9}\right)$

47. $\dfrac{\pi}{2}$

49. The line and plane are parallel. They are $\dfrac{13}{15}\sqrt{30}$ units apart.

51. They intersect at $\left(\dfrac{143}{7}, \dfrac{3}{7}, 0\right)$ at an angle of $\dfrac{\pi}{2} - \cos^{-1}\left(\dfrac{7}{\sqrt{2738}}\right)$.

53. (a) $\dfrac{1}{2}x + \dfrac{1}{4}y + \dfrac{\sqrt{11}}{4}z = 1$, (b) $2x + 2y + z = 7$

55. The plane of the beach is $3x + y - 16z = 400$. The water is about 25 feet deep at the buoy.

57. The planes are perpendicular if and only if their normal vectors are orthogonal, that is if and only if $\langle a, b, c\rangle \cdot \langle \alpha, \beta, \gamma\rangle = a\alpha + b\beta + c\gamma = 0$.

59. (a) Let $\mathcal{P}$ be the plane. By Theorem 10.39 the distance from P to $\mathcal{P}$ is $\dfrac{|\mathbf{N}\cdot\overrightarrow{RP}|}{\|\mathbf{N}\|}$, where $\mathbf{N} = \langle a, b, c\rangle$ is the normal vector to $\mathcal{P}$ and R is a point on $\mathcal{P}$. At least one of the constants, a, b, and c is nonzero. Assume $c \neq 0$, the other cases are similar. Since R is on $\mathcal{P}$, it has coordinates $\left(x_1, y_1, -\dfrac{d+ax_1+by_1}{c}\right)$. Thus, $\overrightarrow{RP} = \left\langle x_0 - x_1, y_0 - y_1, z_0 + \dfrac{d+ax_1+by_1}{c}\right\rangle$. So, $\dfrac{|\mathbf{N}\cdot\overrightarrow{RP}|}{\|\mathbf{N}\|} = \dfrac{|ax_0 - ax_1 + by_0 - by_1 + cz_0 + d + ax_1 + by_1|}{\sqrt{a^2+b^2+c^2}} = \dfrac{|ax_0 + by_0 + cz_0 + d|}{\sqrt{a^2+b^2+c^2}}$.

(b) $\sqrt{2}$

61. If $\mathcal{L}_1$ and $\mathcal{L}_2$ lie in the same plane, then they intersect, are parallel, or are identical. If they are parallel or identical then $\mathbf{d}_1 \times \mathbf{d}_2 = \mathbf{0}$, so the result holds. If $\mathcal{L}_1$ and $\mathcal{L}_2$ intersect, then $\mathbf{d}_1 \times \mathbf{d}_2$ is orthogonal to every vector in their plane, so again the result holds. Conversely, if $(\mathbf{P}_0 - \mathbf{Q}_0) \cdot (\mathbf{d}_1 \times \mathbf{d}_2) = 0$, the vector $\mathbf{d}_1 \times \mathbf{d}_2$ is orthogonal to every vector connecting a point on line $\mathcal{L}_1$ and a point on $\mathcal{L}_2$, so $\mathbf{d}_1 \times \mathbf{d}_2$ is the normal vector to the plane containing the two lines.

For answers to odd-numbered Skill Certification exercises in the Chapter Review, please visit the Book Companion Web Site at www.whfreeman.com/tkcalculus.

Chapter 11

Section 11.1

1. F, F, T, T, T, F, F, F.

3. The limit $\lim\limits_{x\to c} f(x) = L$ means that for all $\epsilon > 0$, there exists $\delta > 0$ such that if $x \in (c - \delta, c) \cup (c, c + \delta)$, then $f(x) \in (L - \epsilon, L + \epsilon)$.

5. The limit of vector-valued function is defined in terms of the limits of the components of the functions. Each component is a function of a single variable. These limits are defined using an "epsilon-delta" definition.

7. The function $\mathbf{r}$ is said to be **continuous** at a point c in the domain of $\mathbf{r}$ if $\lim\limits_{t\to c} \mathbf{r}(t) = \mathbf{r}(c)$.

9. Since the functions $x(t)$ and $y(t)$ are both continuous functions, by the Extreme Value Theorem each of these functions has a minimum value and a maximum value on the interval $[a, b]$. Let M be the largest of the absolute values of these four numbers. The graph of $\mathbf{r}(t)$ on $[a, b]$ is contained within the circle $x^2 + y^2 = 2M^2$.

11. Since the functions $x(t)$, $y(t)$ and $z(t)$ are all continuous functions, by the Extreme Value Theorem each of these functions has a maximum value on the interval $[a, b]$. Let M be the maximum of these three maxima. The graph of $\mathbf{r}(t)$ on $[a, b]$ is contained within the sphere $x^2 + y^2 + z^2 = 3M^2$.

13. $\mathbf{r}(t) = \langle \cos 2t, \sin 2t, 2t\rangle, [0, \pi]$

15. We need $\lim_{t\to\infty} y(t) = L$, where L is the value of the horizontal asymptote and one of the limits; $\lim_{t\to\infty} x(t) = \infty$ or $\lim_{t\to\infty} x(t) = -\infty$. For example, the function $\mathbf{r}(t) = \left\langle t, 5 - \dfrac{1}{t} \right\rangle$.

17. $x_1(t)x_2(t) + y_1(t)y_2(t)$

19. 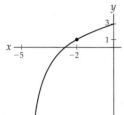   21.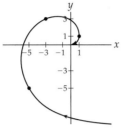

23. The graphs are the same. The rate at which the curve is drawn is faster when k is large.

25. Every point on the graph of $\mathbf{r}(t)$ satisfies the equations of both cylinders.

27. $x(t) = 2 - \sin t,\ y(t) = 4 + \cos t.$

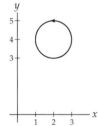

29. $x(t) = 1 + \sin t,\ y(t) = 3 - \cos 2t.$ The graph is the portion of the parabola shown below. The motion starts at $(1, 2)$ and the particle moves to the right for $t \in \left[0, \dfrac{\pi}{2}\right]$ and $t \in \left[\dfrac{3\pi}{2}, 2\pi\right]$ and to the left for $t \in \left[\dfrac{\pi}{2}, \dfrac{3\pi}{2}\right]$

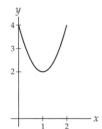

31. $x(t) = t,\ y(t) = t^2,\ z(t) = t^3$

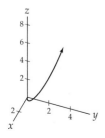

33. $x(t) = \cos^2 t,\ y(t) = \sin 2t$

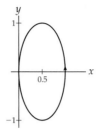

35. $\langle 1 + t, 3t + t^2, 2t^3 \rangle$ 37. $\langle 5\cos t, 5\sin t \rangle$

39. 0 41. $\mathbf{0}$

43. $\langle 0, -1, -1 \rangle$ 45. $\langle 1, 0, 1e \rangle$

47. $x(t) = \tan t,\ y(t) = \tan t - t + 1$

49. Every point on the graph of $\mathbf{r}(t)$ satisfies both the equation of the sphere with equation $x^2 + y^2 + z^2 = 25$ and the plane with equation $4x - 3z = 0$.

51. $A \approx 24.7$

53. These curves both lie on the cylinder with equation $x^2 + y^2 = 1$. One progresses clockwise around it, the other counterclockwise. Therefore, they must meet.

55. The parametric equations $x(t) = t\sin t,\ y(t) = t\cos t$ and $z(t) = t$ satisfy the equation $z = \sqrt{x^2 + y^2}$. Therefore, the graph of $\mathbf{r}(t)$ lies on the graph of the cone.

57. Let $\mathbf{r}_1(t) = \langle x_1(t), y_1(t) \rangle$ and $\mathbf{r}_2(t) = \langle x_2(t), y_2(t) \rangle$. The components $\mathbf{r}_1(t)$ and $\mathbf{r}_2(t)$ are continuous. We know that products and sums of continuous functions are continuous. Therefore, the function $\mathbf{r}_1(t) \cdot \mathbf{r}_2(t) = x_1(t)x_2(t) + y_1(t)y_2(t)$ is continuous. The result follows from this.

59. See the proof in Exercise 57.

61. We know that products and sums of differentiable functions are differentiable. Therefore, the function $\mathbf{r}_1(t) \cdot \mathbf{r}_2(t) = x_1(t)x_2(t) + y_1(t)y_2(t)$ is differentiable.

Section 11.2

1. F, F ,T, T, F, F, F, F.

3. The function is differentiable at c if $\lim_{h\to 0} \dfrac{f(c+h) - f(c)}{h}$ exists.

5. A vector function is differentiable at a point c if each of its component functions is differentiable at c. The differentiability of these scalar functions at a point has already been defined.

7. Let $\mathbf{r}(t) = \langle t, |t| \rangle$. The graph of this function is the same as the graph of $y = |x|$. The corner at the origin means that the function doesn't have a well-defined tangent vector at the origin.

9. Let $f(x)$ be a function defined on the interval $[a, b]$. The function is integrable if $\lim_{n\to\infty} \sum_{k=1}^{n} f(x_k^*)\Delta x$ exists, where $\Delta x = \dfrac{b-a}{n},\ x_k = a + k\Delta x,$ and $x_k^* \in [x_{k-1}, x_k]$.

11. A vector function is integrable on an interval $[a, b]$ if each of its component functions is integrable on $[a, b]$. The integrability of these scalar functions on $[a, b]$ has already been defined.

13. $\langle x'(t), y'(t), z'(t) \rangle$

15. $\frac{d}{d\tau}(\mathbf{r}(f(\tau))) = \mathbf{r}'(f(\tau))f'(\tau)$.

17. $\mathbf{r}'(t) = \langle 3t^2, 15t^2, -6t^2 \rangle$. If t_0 and t_1 are positive real numbers, then $\mathbf{r}'(t_0) = \langle 3t_0^2, 15t_0^2, -6t_0^2 \rangle = 3t_0^2 \langle 1, 5, -2 \rangle$ and $\mathbf{r}'(t_1) = \langle 3t_1^2, 15t_1^2, -6t_1^2 \rangle = 3t_1^2 \langle 1, 5, -2 \rangle$. Since $\mathbf{r}'(t_0)$ and $\mathbf{r}'(t_1)$ are both multiples of the same vector, they are scalar multiples of each other. The graph of $\mathbf{r}(t)$ is a portion of a straight line.

19. They differ by a constant vector.

21. $\langle \cos\tau, 2\sin\tau\cos\tau, 3\sin^2\tau\cos\tau \rangle$

23. $\left\langle \sin\tau + \tau\cos\tau, \frac{1}{2\sqrt{\tau}}, \cos\tau - \tau\sin\tau \right\rangle$

25. $x = e + 2et, y = t, z = 0$; $2ex + y = 1 + 2e^2$

27. $x = 1 + t, y = 0, z = -4t$; $x - 4z = 1$

29. $x = 3t, y = 4t, z = t + \pi/2$; $3x + 4y + z = \pi/2$

31. $\mathbf{v}(t) = \langle (t+1)e^t, 1 + \ln t \rangle$, $\mathbf{a}(t) = \langle (t+2)e^t, \frac{1}{t} \rangle$

33. $\mathbf{v}(t) = \langle \cos t, -\sin t, 4\cos 2t \rangle$, $\mathbf{a}(t) = \langle -\sin t. -\cos t, -8\sin 2t \rangle$

35. $\langle 3\tau^2, 6\tau^2(\tau^3 + 1), 9\tau^2(\tau^3 + 1)^2 \rangle$

37. $\left\langle -\frac{1}{\tau^2}\sec\frac{1}{\tau}\tan\frac{1}{\tau}, 1, \frac{1}{\tau^2}e^{1/\tau}(\ln\tau - \tau) \right\rangle$

39. $\langle -15\sin(15\tau - 6), 20\cos(20\tau - 8), 5 \rangle$

41. $\langle -\cos t + c_1, \sin t + c_2, -\ln|\cos t| + c_3 \rangle$

43. $\langle 0, 0, 2\pi^2 \rangle$

45. $\langle \frac{1}{2}t^2 + 3, \frac{1}{3}t^3 - 4 \rangle$

47. $\langle e^t + 1 - e, t\ln t - t - 5 \rangle$

49. $\left\langle \frac{1}{3}t^3 + t + 2, \frac{1}{4}t^4 - 2t - 3 \right\rangle$

51. $\langle 5t, -16t^2 + 5t + 26 \rangle$

53. $(1, 1, 1)$

55. The circles intersect at $\left(\frac{9}{4}, \pm\frac{\sqrt{63}}{4} \right)$. The angle of intersection at both points is $\cos^{-1}\frac{3}{4}$ radians.

57. The function is $\mathbf{r}(t) = \langle 0, e^t, 2e^t \rangle$. The graph is the ray with parametric equations $x = 0, y = s, z = 2s$ for $s > 0$.

59. (a) $\langle -0.5 dy(t), -9.8t - dy(t) \rangle$, (b) $-\frac{9.8t + y'(t)}{d}$, (c) $x'(t) = 4.9t + 0.5y'(t)$, (d) $\langle 2.45t^2 + 0.5y(t), y(t) \rangle$, (e) No, he should not jump.

61. If $\mathbf{a}(t) = \langle 0, 0, 0 \rangle$, then $\mathbf{v}(t) = \langle \alpha, \beta, \gamma \rangle$, where α, β, and γ are constants. We also have $\mathbf{r}(t) = \langle \alpha t + \delta, \beta t + \epsilon, \gamma t + \lambda \rangle$, where δ, ϵ, and λ are also constants. The graph of $\mathbf{r}(t)$ is a straight line.

63. We will prove the case where $\mathbf{r}(t) = \langle x(t), y(t) \rangle$. The case where $\mathbf{r}(t)$ has three components is similar. We have $k\mathbf{r}(t) = \langle kx(t), ky(t) \rangle$ and
$\frac{d}{dt}(k\mathbf{r}(t)) = \left\langle \frac{d}{dt}(kx(t)), \frac{d}{dt}(ky(t)) \right\rangle = \langle kx'(t), ky'(t) \rangle = k\langle x'(t), y'(t) \rangle = k\mathbf{r}'(t)$.

65. Let $\mathbf{r}_1(t) = \langle x_1(t), y_1(t), z_1(t) \rangle$ and $\mathbf{r}_2(t) = \langle x_2(t), y_2(t), z_2(t) \rangle$. Then
$$\frac{d}{dt}(\mathbf{r}_1(t) \cdot \mathbf{r}_2(t)) = \frac{d}{dt}(x_1x_2 + y_1y_2 + z_1 + z_2)$$
$$= x_1'x_2 + x_1x_2' + y_1'y_2 + y_1y_2' + z_1'z_2 + z_1z_2'$$
$$= \frac{d}{dt}(\langle x_1, y_1, z_1 \rangle) \cdot \langle x_2, y_2, z_2 \rangle + \langle x_1, y_1, z_2 \rangle \cdot \frac{d}{dt}(\langle x_2, y_2, z_2 \rangle)$$
$$= \mathbf{r}_1'(t) \cdot \mathbf{r}_2(t) + \mathbf{r}_1(t) \cdot \mathbf{r}_2'(t).$$

67. Let $\mathbf{r}(t) = \langle x(t), y(t), z(t) \rangle$. Then $\frac{d}{d\tau}(\mathbf{r}(t)) = \frac{d}{d\tau}(\langle x, y, z \rangle)$
$$= \left\langle \frac{dx}{dt}\frac{dt}{d\tau}, \frac{dy}{dt}\frac{dt}{d\tau}, \frac{dz}{dt}\frac{dt}{d\tau} \right\rangle = \left\langle \frac{dx}{dt}, \frac{dy}{dt}, \frac{dz}{dt} \right\rangle \frac{dt}{d\tau} = \frac{d\mathbf{r}}{dt}\frac{dt}{d\tau}.$$

69. We will show that $\|\mathbf{r}(t)\|^2$ is a constant. Since $\mathbf{r}$ is continuous, this is equivalent. Taking the derivative we have $\frac{d}{dt}(\|\mathbf{r}(t)\|^2) = \frac{d}{dt}(\mathbf{r}(t) \cdot \mathbf{r}(t)) = 2\mathbf{r}(t) \cdot \mathbf{r}'(t) = 0$. Since the derivative of $\|\mathbf{r}(t)\|^2$ is a zero, we must have $\|\mathbf{r}(t)\|$ is a constant.

71. Apply the result of Exercise 69.

Section 11.3

1. F, T, F, T, T, T, T, T.

3. All of these vectors have magnitude one. The unit tangent vector points straight ahead. The principal unit normal vector is orthogonal to the unit tangent vector and points "into" the curve. Together these two vectors determine the osculating plane that is the plane in which the curve fits "best" at the point of tangency. The binormal vector is normal to the osculating plane. The tangent vector, normal vector and binormal vector form a right-hand triple, so when the curve is bending to the left, it points up, and when the road bends to the right, it points down.

5. The unit tangent vector won't exist at any point t_0 at which $\mathbf{r}'(t_0) = \mathbf{0}$.

7. If $\mathbf{r}'(t_0) \neq \mathbf{0}$, then they point in the same direction, but $\mathbf{T}(t_0)$ is a unit vector, and $\mathbf{r}'(t_0)$ can have a different length.

9. $\mathbf{N}(t) = \frac{\mathbf{T}'(t)}{\|\mathbf{T}'(t)\|}$

11. All of these vectors have magnitude one. The unit tangent vector points straight in the direction of motion of the curve. The principal unit normal vector is orthogonal to the unit tangent vector and points "into" the curve. Together these two vectors determine the osculating plane that is the plane in which the curve fits "best" at the point of tangency. The derivative $\frac{d\mathbf{N}}{dt}\Big|_{t_0}$ is orthogonal to both the tangent vector and normal vector.

13. At each point t_0 at which $\mathbf{r}(t_0) \neq \mathbf{0}$, $\mathbf{B}(t_0) = \mathbf{T}(t_0) \times \mathbf{N}(t_0)$, where $\mathbf{T}$ and $\mathbf{N}$ are the unit tangent vector and principal unit normal vector, respectively.

15. It isn't defined when $\mathbf{r}'(t) = \mathbf{0}$.

17. Find $\mathbf{r}'(t_0)$. If $\mathbf{r}'(t_0) \neq \mathbf{0}$ continue. Find the unit tangent vector at t_0, $\mathbf{T}(t_0)$. If $\mathbf{T}(t_0) \neq \mathbf{0}$, continue. Find the principal unit normal vector at t_0, $\mathbf{N}(t_0)$. Find the binormal vector at t_0, $\mathbf{B}(t_0) = \mathbf{T}(t_0) \times \mathbf{N}(t_0)$. The osculating plane contains the point $\mathbf{r}(t_0)$ and has $\mathbf{B}(t_0)$ as its normal vector.

19. 3   21. $x = 1$

23. $\dfrac{1}{\sqrt{144t^4 + 4t^2 + 25}} \langle 2t, 5, 12t^2 \rangle$

25. $\dfrac{1}{|\sin(t)\cos(t)|} \langle -\cos^2 t \sin t, \sin^2 t \cos t \rangle$

27. $\dfrac{1}{5} \langle 3 \cos t, -5 \sin t, 4 \cos t \rangle$

29. $\mathbf{T}(1) = \left\langle \dfrac{\sqrt{5}}{5}, \dfrac{2\sqrt{5}}{5} \right\rangle$, $\mathbf{N}(1) = \left\langle -\dfrac{2\sqrt{5}}{5}, -\dfrac{\sqrt{5}}{5} \right\rangle$

31. $\mathbf{T}(\pi) = \langle -\sin \alpha\pi, \cos \alpha\pi \rangle$, $\mathbf{N}(\pi) = \langle -\cos \alpha\pi, -\sin \alpha\pi \rangle$

33. $\mathbf{T}(\pi) = \left\langle -\dfrac{3}{5}, 0, -\dfrac{4}{5} \right\rangle$, $\mathbf{N}(\pi) = \langle 0, 1, 0 \rangle$

35. $\mathbf{T}(1) = \left\langle \dfrac{\sqrt{14}}{14}, \dfrac{\sqrt{14}}{7}, \dfrac{3\sqrt{14}}{14} \right\rangle$, $\mathbf{N}(1) = \dfrac{1}{\sqrt{266}} \langle -11, -8, 9 \rangle$, $\mathbf{B}(1) = \dfrac{1}{\sqrt{123}} \langle 7, -7, 5 \rangle$, $7x - 7y + 5z = 5$

37. $\mathbf{T}(0) = \left\langle \dfrac{1}{2}, -\dfrac{1}{2}, \dfrac{\sqrt{2}}{2} \right\rangle$, $\mathbf{N}(0) = \left\langle \dfrac{\sqrt{2}}{2}, \dfrac{\sqrt{2}}{2}, 0 \right\rangle$, $\mathbf{B}(0) = \left\langle -\dfrac{1}{2}, \dfrac{1}{2}, \dfrac{\sqrt{2}}{2} \right\rangle$, $-x + y + \sqrt{2}z = 0$

39. $\mathbf{T}\left(\dfrac{\pi}{2} \right) = \left\langle -\dfrac{2\sqrt{5}}{5}, 0, \dfrac{\sqrt{5}}{5} \right\rangle$, $\mathbf{N}\left(\dfrac{\pi}{2} \right) = \left\langle \dfrac{\sqrt{2}}{2}, \dfrac{\sqrt{2}}{2}, 0 \right\rangle$, $\mathbf{B}\left(\dfrac{\pi}{2} \right) = \left\langle -\dfrac{\sqrt{5}}{5}, 0, -\dfrac{2\sqrt{5}}{5} \right\rangle$, $x + 2z = \pi$

41. Normal plane: $x - y + \sqrt{2}z = 0$. Rectifying plane: $x + y = 2$.

43. (a) Their tangent vector is $\langle 1.5t^{0.5}, 1.7 - 2t \rangle$, so that their speed is $4t^2 - 4.45t + 2.89$. This is concave up, so their maximum speed will occur when t is a maximum on the interval at $t = 1.7$. (b) Approximately $\langle 0.95, -0.30 \rangle$.

45. For any two vectors $\mathbf{a}$ and $\mathbf{b}$ in $\mathbb{R}^3$ $\|\mathbf{a} \times \mathbf{b}\| = \|\mathbf{a}\| \|\mathbf{b}\| \sin \theta$. If $\mathbf{a}$ and $\mathbf{b}$ are orthogonal unit vectors, then $\|\mathbf{a}\|$, $\|\mathbf{b}\|$ and $\sin \theta$ are all one. Thus, $\mathbf{a} \times \mathbf{b}$ is a unit vector.

47. The three properties follow from the definition of the three vectors and the geometric properties of the cross product. Since $\mathbf{T}(t_0)$ and $\mathbf{N}(t_0)$ are orthogonal unit vectors, their cross product $\mathbf{B}(t_0)$ is another unit vector and is orthogonal to them both. Furthermore, we know that for any two nonparallel vectors $\mathbf{a}$ and $\mathbf{b}$, $\mathbf{a}$, $\mathbf{b}$, and $\mathbf{a} \times \mathbf{b}$ form a right-hand triple.

Section 11.4

1. F, F, T, T, F, F, F, T.

3. Informally, it means that for every unit change in the parameter, a unit change occurs along the curve. For the formal definition see Defintion 11.19.

5. There are more ways for a curve to bend. Think of a spring. The concepts of "concave up" and "concave down" are insufficient to describe how it is bending.

7. So that curvature has a consistent definition, the rate of change of the tangent vector with respect to arc length provides a relatively simple definition. Unfortunately, it is often difficult to find an arc length parametrization for a curve.

9. When you have an arc length parametrization for the curve, use the definition. Use Theorem 11.23(a) when you have a space curve and $\mathbf{T}'(t)$ is easy to compute. Use Theorem 11.23(b) when you have a twice-differentiable parametrization for the space curve. Use Theorem 11.24(a) for a planar curve defined by a twice-differentiable function $y = f(x)$. Use Theorem 11.24(b) for a planar curve defined by a vector function.

11. (a) By Theorem 11.23, $\kappa = \dfrac{\|\mathbf{r}'(t) \times \mathbf{r}''(t)\|}{\|\mathbf{r}'(t)\|^3}$. So, if $\mathbf{r}(t) = \langle a + \alpha t, b + \beta t, c + \gamma t \rangle$, then $\mathbf{r}'(t) = \langle \alpha, \beta, \gamma \rangle$ and $\mathbf{r}''(t) = \langle 0, 0, 0 \rangle$, so $\kappa = 0$. Since $\kappa = 0$, the radius of curvature is undefined. (b) A straight line doesn't curve, so this is consistent with our concept of curvature.

13. Let (x_0, y_0) be a point on the curve. Find the curvature, κ, at (x_0, y_0) using the appropriate part of Theorem 11.24. If $\kappa \neq 0$, the radius of curvature $\rho = \dfrac{1}{\kappa}$. Find the normal vector, $\mathbf{N}$ at (x_0, y_0). The radius of the osculating circle is ρ. The center of the osculating circle is the terminal point of the position vector $\langle x_0, y_0 \rangle + \rho \mathbf{N}$.

15. $y'' = 2$ and by Theorem 11.24, $\kappa = \dfrac{2}{(1 + 4x^2)^{3/2}}$.

17.

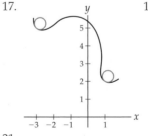

19.

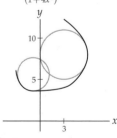

21.

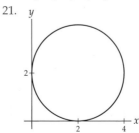

23. 6π

25. $\dfrac{488}{27}$

27. $\sqrt{3}(e^\pi - 1)$

29. The arc length of the curve on the interval $[0, 1]$ is $\sqrt{30}$;
$$\mathbf{r}(s) = \left\langle 3 + \frac{\sqrt{30}}{15}s, 4 - \frac{\sqrt{30}}{30}s, -1 + \frac{\sqrt{30}}{6}s \right\rangle$$

31. $\kappa = \dfrac{\sqrt{2}}{4}$

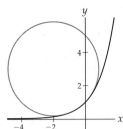

33. $\kappa = 1$

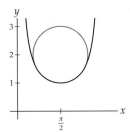

35. $\dfrac{\sqrt{3}}{6}$

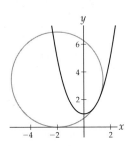

37. $\dfrac{ab}{(a^2\sin^2 t + b^2\cos^2 t)^{3/2}}$

39. $\dfrac{1}{2\sqrt{2(1-\cos t)}}$

41. $\sqrt{\dfrac{t^4 + 5t^2 + 8}{(t^2 + 2)^3}}$

43. $\sqrt{\dfrac{t^4 + (4+\alpha^2)t^2 + 4 + 4\alpha^2}{(t^2 + 1 + \alpha^2)^3}}$

45. $\pm\dfrac{1}{45^{1/4}}$

47. $\kappa = \dfrac{a}{a^2 + b^2}$

49. $\kappa = t^3\sqrt{\dfrac{t^6 + t^2 + 4}{(t^4 + 1)^3}}$ has the stated limits.

51. We differentiate:
$\dfrac{d}{ds}(\mathbf{B}) = \dfrac{d}{ds}(\mathbf{T} \times \mathbf{N}) = \dfrac{d\mathbf{T}}{ds} \times \mathbf{N} + \mathbf{T} \times \dfrac{d\mathbf{N}}{ds}$. Since the quantities on the right exist at every point on C, the derivative exists.

53. We differentiate: $\dfrac{d}{ds}(\mathbf{N}) = \dfrac{d}{ds}(\mathbf{B} \times \mathbf{T}) =$
$\dfrac{d\mathbf{B}}{ds} \times \mathbf{T} + \mathbf{B} \times \dfrac{d\mathbf{T}}{ds} = -\tau\mathbf{N} \times \mathbf{T} + \mathbf{B} \times (\kappa\mathbf{N}) = \tau\mathbf{B} - \kappa\mathbf{T}$

55. $\dfrac{6 + t^2}{4 + t^2 + t^4}$

57. By choosing the coordinate axes appropriately we may assume that the curve C lies in the xy-plane. Thus, the binormal vector $\mathbf{B}(t) = \pm\mathbf{k}$, and in either case is a constant. Therefore, $\dfrac{d\mathbf{B}}{ds} = 0$. Thus, the torsion is zero.

59.

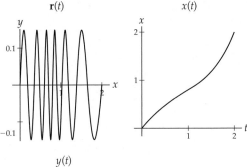

(b) When the x-coordinate changes slowly, that means that the y-coordinate is changing quickly relative to it. The greatest curvature should occur near the point where the slope of the x-coordinate is smallest, while the y-coordinate changes quickly. (c) The slope of the first component of the curve is a minimum at about $t = 0.91$, and the second component turns quickly at about $t = 0.92$. We expect the curvature to be greatest at about that point.

61. By Theorem 11.24 $\kappa = \dfrac{|f''(x)|}{\left(1 + \left(f'(x)\right)^2\right)^{3/2}}$. Let x be a point of inflection of a twice-differentiable function, f. Since $f''(x) = 0$ at each inflection point, we have $\kappa = 0$.

63. Let $\mathbf{r}(t)$ be a twice differentiable vector-valued function with graph C. By Theorem 11.23 the curvature at each point on C is $\kappa = \dfrac{\|\mathbf{r}'(t) \times \mathbf{r}''(t)\|}{\|\mathbf{r}'(t)\|^3}$. But recall that $\|\mathbf{r}'(t) \times \mathbf{r}''(t)\| = \|\mathbf{r}'(t)\| \, \|\mathbf{r}''(t)\| \sin\theta$ where θ is the angle between $\mathbf{r}'(t)$ and $\mathbf{r}''(t)$. Since the speed of the particle is constant, by Theorem 11.12 $\mathbf{r}'(t)$ and $\mathbf{r}''(t)$ are orthogonal. Therefore, $\kappa = \dfrac{\|\mathbf{r}''(t)\|}{\|\mathbf{r}'(t)\|^2}$. But the denominator of this expression is the square of the speed, which we are assuming is constant. Therefore, the curvature is proportional to the magnitude of the acceleration.

65. Let $\mathbf{r}(t) = \langle t, f(t), 0 \rangle$. So, $\mathbf{r}'(t) = \langle 1, f'(t), 0 \rangle$ and $\mathbf{r}''(t) = \langle 0, f''(t), 0 \rangle$. By Theorem 11.23
$\kappa = \dfrac{\|\mathbf{r}'(t) \times \mathbf{r}''(t)\|}{\|\mathbf{r}'(t)\|^3} = \dfrac{\|\langle 0,0,f''(t)\rangle\|}{\|\langle 1,f'(t),0\rangle\|} = \dfrac{|f''(t)|}{(1 + (f'(t))^2)^{3/2}}$.

Section 11.5

1. T, F, T, T, F, T, F, F.

3. When the downward acceleration due to the force of gravity is smaller, an object takes longer to fall to the ground. When the object is thrown horizontally, it still takes longer to hit the ground. Therefore, it travels farther.

5. You can include the wind velocity as a vector added to the velocity vector of the projectile.

7. $\mathbf{r}(b) - \mathbf{r}(a)$

9. These are the magnitudes of the acceleration vector, parallel to the tangent vector and parallel to the principal normal unit vector, respectively. Their values are given by $a_\mathbf{T} = \dfrac{\mathbf{v} \cdot \mathbf{a}}{\|\mathbf{v}\|}$ and $a_\mathbf{N} = \dfrac{\|\mathbf{v} \times \mathbf{a}\|}{\|\mathbf{v}\|}$.

11. The tangential component of acceleration always points in the instantaneous direction of your motion, tangent to the curve defined by your path. The normal component of acceleration is orthogonal to your direction of motion and points "into" the curve, so if you are curving to the left, it points to your left; if your are ascending, it points upward, etc.

13. (a) If you eliminate the parameter, the function may be expressed in the form $y = x^2$. (b) This will occur when $\|\mathbf{r}(t)\|$ is a constant and more generally when $f'(t)f''(t) + 4f(t)(f'(t))^3 + 4(f(t))^2 f'(t)f''(t) = 0$.

15. (a) $\langle 0, -\pi \rangle$, (b) π, (c) $\dfrac{1}{2}(\pi\sqrt{\pi^2 + 1} + \ln(\pi + \sqrt{\pi^2 + 1}))$

17. (a) $\langle \alpha\sin\beta, \alpha(\cos\beta - 1), \gamma \rangle$,
    (b) $\sqrt{\alpha^2 - 2\alpha\cos\beta + 1 + \gamma^2}$, (c) $\sqrt{\alpha^2\beta^2 + \gamma^2}$

19. $a_\mathbf{T} = \dfrac{-27\sin 3t\cos 3t + 64\sin 4t\cos 4t}{\sqrt{9\cos^2 3t + 16\sin^2 4t}}$,

    $a_\mathbf{N} = \dfrac{|48\cos 3t\cos 4t + 36\sin 3t\sin 4t|}{\sqrt{9\cos^2 3t + 16\sin^2 4t}}$

21. $a_\mathbf{T} = \sqrt{3}e^t$, $a_\mathbf{N} = \sqrt{2}e^t$

23. $a_\mathbf{T} = 0$, $a_\mathbf{N} = 1$

25. The values of t are the roots of $16t^4 - 129t^2 + 101 = 0$. These are approximately ± 0.94 seconds and ± 2.68 seconds. Only the positive values make sense in the context of the problem. When $t = 0.94$, $\theta \approx 14°$ and when $t = 2.68$, $\theta \approx 70°$.

27. (a) 10 meters per second, (b) 54 meters

29. (a) 0°, (b) 292 feet

31. 86,000 meters

33. The function $\mathbf{r}(t) = \langle x(t), y(t) \rangle =$ $\left\langle (\|\mathbf{v}_0\|\cos\theta)\,t, -\dfrac{1}{2}gt^2 + (\|\mathbf{v}_0\|\sin\theta)\,t \right\rangle$ models the motion of the ball. The ball will hit the ground when the y-component is zero. This will occur when $t = \dfrac{\|\mathbf{v}_0\|\sin\theta}{g}$. At this time it will have travelled $\dfrac{\|\mathbf{v}_0\|\sin\theta\cos\theta}{g}$ units horizontally. Thus, the x-component will be maximized when $\theta = 45°$.

35. If $\mathbf{a}(t) = \langle 0, 0, 0 \rangle$, then the velocity vector is $\mathbf{v}(t) = \langle \alpha, \beta, \gamma \rangle$ for constants α, β and γ. The position function will be $\mathbf{r}(t) = \langle a + \alpha t, b + \beta t, c + \gamma t \rangle$, where a, b, and c are three more constants. The graph of $\mathbf{r}(t)$ is a straight line.

37. Let $\mathbf{r}(t) = \langle t, f(t), 0 \rangle$. Then $\mathbf{v}(t) = \langle 1, f'(t), 0 \rangle$ and $\mathbf{a}(t) = \langle 0, f''(t), 0 \rangle$. We have $\mathbf{v}(t) \times \mathbf{a}(t) = \langle 0, 0, f''(t) \rangle$. At a point of inflection we have $a_\mathbf{T} = \dfrac{\|\mathbf{v}(t) \times \mathbf{a}(t)\|}{\|\mathbf{v}(t)\|} = 0$.

For answers to odd-numbered Skill Certification exercises in the Chapter Review, please visit the Book Companion Web Site at www.whfreeman.com/tkcalculus.

Chapter 12

Section 12.1

1. T, F, T, F, T, T, T, T.

3. The graph of f is the set $\{(x, f(x)) | x \in \mathbb{R}\}$.

5. The graph of f is the set $\{(x, y, z, f(x, y, z)) | (x, y, z) \in \mathbb{R}^3\}$.

7.

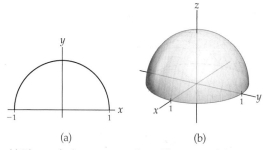

(a)   (b)

(c) The prefix *hyper* means *above*. The range of the square root function is nonnegative real numbers. $x^2 + y^2 + z^2 + w^2 = 1$. (d) Hypersphere, n, $\mathbb{R}^{n+1}$.

9. A point (x, y, z) is on the graph of f if and only if (x, y) is in the domain of the function f.

11. A point (x, y, z, w) is on the graph of f if and only if (x, y, z) is in the domain of the function f.

13.

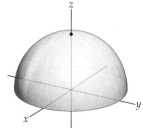

15.

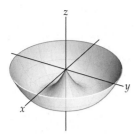

17. See the answer to Exercise 13

19. See the answer to Exercise 15

21.

23. $-24, 5$, Domain$(f) = \mathbb{R}^2$, Range$(f) = \mathbb{R}$

25. $(1 + e)^{-1/2}, 0$, Domain$(f) = \{(x, y) | 0 < x < 1$ and $y > 0\} \cup \{(x, y) | x < 0$ and $y < 0\}$, Range$(f) = \mathbb{R}$

27. $26, \dfrac{49}{36}$, Domain$(f) = \mathbb{R}^3$, Range$(f) = \{w \in \mathbb{R} | w \geq 0\}$

29. 1

31. $\sin(\cos t)$

33. $\dfrac{\cos t + \sin t}{\sin t + 1 - t}$

35. $\langle 1 - t, (1 - t)^2, (1 - t)^3 \rangle$

37. $F(x, y) = \sqrt{x^2 + y^2}$

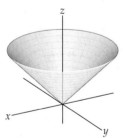

39. $F(x, y) = x^2 + y^2$

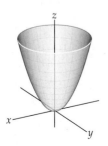

41. $F(x, y) = \sin \sqrt{x^2 + y^2}$

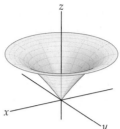

43.

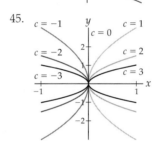

45.

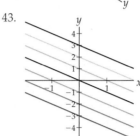

47.

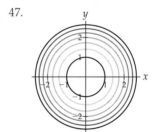

49. Only the level curves for $c = -1, 0$ and 1 are defined. The level curves for $c = -1$ are the parallel lines defined by $x + y = -\dfrac{\pi}{2} + 2k\pi$ where $k \in \mathbb{Z}$. The level curves for $c = 0$ are the parallel lines defined by $x + y = k\pi$ where $k \in \mathbb{Z}$. The level curves for $c = 1$ are the parallel lines defined by $x + y = \dfrac{\pi}{2} + 2k\pi$ where $k \in \mathbb{Z}$.

51.

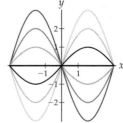

53. The level surfaces are the planes with equations
$x + 2y + 3z = c$ for $c = -3, -2, -1, 0, 1, 2, 3$.

55. The level surface $\frac{x}{y-z} = c$ consists of every point on the
plane $x - cy + cz = 0$ except those points where $y = z$.

57. The level surfaces for $c = -3, -2$ and -1 are undefined.
The level surface $x^2 + y^2 + z^2 = 0$ is the origin. The
level surfaces $x^2 + y^2 + z^2 = c$ for $c = 1, 2$, and 3 are the
spheres centered at the origin with radii $1, \sqrt{2}$, and $\sqrt{3}$,
respectively.

59. I A 61. III B

63. Domain$(V) = \{(r, h) | r > 0 \text{ and } h > 0\}$,
$S(r, h) = 2\pi r^2 + 2\pi rh$.

65. $V(x, y, z) = xyz$, $S(x, y, z) = 2xy + 2xz + 2yz$,
Domain(V)=Domain(S)= $\{(x, y, z) | x > 0, y > 0 \text{ and } z > 0\}$.

67. (a) From examining a plot, we can see that the caribou
are distributed up and down the valley along the line
$y = -0.5x + 2$. (b) From examining a plot, we can see
that the caribou are distributed up and down the valley
along the line $y = -0.5x + 2$. Thus the stream roughly
follows the bottom of the valley, $\langle 4, -2 \rangle$.

69. The graph of the equation $z = ax + by + c$ is a plane.
The normal vector to the plane $ax + by - z + c = 0$ is
$\langle a, b, -1 \rangle$. When $x = 0$ and $y = 0$ we have $z = c$.

71. If c_1 is not equal to c_2, then since f is a function, $f(x, y)$
cannot equal both c_1 and c_2 for the same input (x, y).
Thus, c_1 must equal c_2 in order for the level curves to
intersect. But this implies that the level curves are
identical.

73. You may model your proof on the solution of
Exercise 71.

Section 12.2

1. T, F, T, T, F, T, F, T.

3. Nothing, the limit may or may not exist.

5. $\lim_{(x,y)\to(3,-7)} f(x, y) = 5, f(3, -7)$ could be any real number,
or could even be undefined.

7. When $\lim_{x\to c} f(x) = L$, the function f must be defined on
$(c - \delta, c) \cup (c, c + \delta)$ for some $\delta > 0$.

9. When $\lim_{(x,y,z)\to(a,b,c)} f(x, y, z) = L$, the function f must be
defined on a punctured ball defined by the inequality
$0 < \sqrt{(x - a)^2 + (y - b)^2 + (z - c)^2} < \delta$ for some $\delta > 0$.

11. For f to be continuous at c, we must have
$\lim_{x\to c} f(x) = f(c)$. For the limit to exist, f must be defined
on $(c - \delta, c) \cup (c, c + \delta)$ for some $\delta > 0$.

13. Along the x-axis $\lim_{\substack{(x,y)\to(0,0) \\ C}} \frac{x^2 y}{x^4 + y^2} = \lim_{x\to 0} \frac{x^2 \cdot 0}{x^4 + 0} = 0$. Along

the y-axis $\lim_{\substack{(x,y)\to(0,0) \\ C}} \frac{x^2 y}{x^4 + y^2} = \lim_{y\to 0} \frac{0 \cdot y}{0 + y^2} = 0$.

15. We say that $\lim_{(x,y)\to(a,b)} f(x, y) = \infty$ if for every $M > 0$ there
is a $\delta > 0$ such that $f(x, y) > M$ whenever
$\sqrt{(x - a)^2 + (y - b)^2} < \delta$.

17. Let $f(x, y) = \frac{x^2}{x^2 + y^2}$ and $g(x, y) = \frac{y^2}{x^2 + y^2}$ then
$\lim_{(x,y)\to(0,0)} (f(x, y) + g(x, y)) = 1$; but neither of the
individual limits exists as $(x, y) \to (0, 0)$. This does not
contradict the sum rule for limits of a function of two
variables since neither of the individual limits exists.

19. (a) The quotient of two continuous functions is
continuous, as long as the denominator is not zero.
(b) For every real number a, every open disk containing
the point (a, a) contains infinitely many points of the
line $y = x$. Therefore, the quotient is not even defined
for infinitely many points in any open ball containing
(a, a).

21. (a) Open, (b) $\{(x, y) | x \leq 0 \text{ or } y \leq 0\}$,
(c) $\{(x, 0) | x \geq 0\} \cup \{(0, y) | y \geq 0\}$

23. (a) Closed, (b) $\{(x, y) | |x| + |y| > 1\}$,
(c) $\{(x, y) | |x| + |y| = 1\}$

25. (a) Both open and closed, (b) $\mathbb{R}^2$, (c) the empty set

27. (a) Open, (b) $\{(x, y, z) | x \leq 0, y \geq 0 \text{ or } z \geq 0\}$,
(c) $\{(x, y, 0) \mid x \geq 0, \ y \leq 0\} \cup \{(x, 0, z) \mid x \geq 0, z \leq 0\} \cup \{(0, y, z) \mid y \leq 0, \ z \leq 0\}$

29. (a) Closed, (b) $\{(x, y, z) | z \neq 0\}$ (c) the xy-plane

31. (a) Both open and closed, (b) $\mathbb{R}^3$, (c) the empty set

33. 0 35. $-\dfrac{5}{3}$

37. $\dfrac{9}{2}$ 39. does not exist

41. $\lim_{r\to 0} \dfrac{r^2 \cos^2 \theta}{r^2} = \lim_{r\to 0} \cos^2 \theta$. Since the limit is a
nonconstant function of θ, the limit does not exist.

43. $\lim_{r\to 0} \dfrac{(r^2 \cos^2 \theta)(r^2 \sin^2 \theta)}{r^2} = \lim_{r\to 0} r^2 \cos^2 \theta \sin^2 \theta = 0$.

45. $\lim_{r\to 0} \dfrac{(r\cos\theta)(r\sin\theta)}{r} = \lim_{r\to 0} r \cos\theta \sin\theta = 0$.

47. Domain$(f) = \{(x, y) | x^2 \neq y^2\}$. f is continuous on its
domain.

49. Domain$(f) = \{(x, y) | y \geq -x^2\}$. f is continuous on the
set $\{(x, y) | y > -x^2\}$.

51. Domain$(f) = \{(x, y, z) | x + y + z > 0\}$. f is continuous on
its domain.

53. Domain$(f) = \mathbb{R}^2$. f is continuous everywhere.

55. Domain$(f) = \mathbb{R}^2$. f is continuous everywhere.

57. The limit is zero. The pressure would decrease as the
gas cools off.

59. The limit is infinite. The pressure increases without
bound as the gas is compressed.

61. (a) $y = \frac{232}{119}x - \frac{4350}{17}$, (b) -40.38 feet, (c) There are many possibilities. The surface might not be as planar as her previous results led her to believe. She might have been unlucky and hit a pocket or hole in the basalt. Or again, there might be a joint in the basalt layer that would cause a discontinuity in its depth.

63. $x \in S$ if and only if $x \notin S^c$. $x \notin S^c$ if and only if $x \in (S^c)^c$. Therefore, $S = (S^c)^c$.

65. We prove the statement when $S \subseteq \mathbb{R}^2$. The proof for $\mathbb{R}^3$ is similar. First, assume S is open. Thus, for every $x \in S$ there is an open disk D such that $x \in D \subseteq S$. This implies that $x \notin \partial S$. Therefore, $\partial S \cap S = \emptyset$. Now, assume S is not open. There exists an $x \in S$ such that for every open disk D containing x, $D \cap S^c$ is nonempty. Therefore, $x \in \partial S \cap S$.

67. We prove the statement when $S \subseteq \mathbb{R}^2$. The proof for $\mathbb{R}^3$ is similar. $x \in \partial S$ if and only if for every open disk D containing x, D intersects both S and S^c. Since $S = (S^c)^c$, this occurs if and only if $x \in \partial(S^c)$.

69. By Exercise 68, ∂S is a closed set, and by Exercise 66 the boundary of a closed set is a subset of the set. Therefore, $\partial(\partial S) \subseteq \partial S$.

71. The complement of $\emptyset$ in $\mathbb{R}^2$ is $\mathbb{R}^2$. By Exercise 70, $\emptyset$ is both open and closed. Since the complement of an open set is closed and the complement of a closed set is open, $\mathbb{R}^2$ is both open and closed.

Section 12.3

1. T, T, F, F, T, F, F, F.

3. The slope remains zero.

5. $f_x(x_0, y_0)$ and $f_y(x_0, y_0)$ represent the slopes of the curves formed by the intersection of the surface and the planes $y = y_0$ and $x = x_0$, respectively. The lines tangent to these curves are also tangent to the surface.

7. (a) The graph of g is the "cylinder" created when the graph of f is translated in the x direction. (b) $\frac{\partial g}{\partial x}$ exists for every value of x and y if y is in the domain of f. $\frac{\partial g}{\partial y}$ exists for all values of x and y where $f'(y)$ exists. (c) $\frac{\partial g}{\partial x} = 0$ and $\frac{\partial g}{\partial y} = f'(y)$

9. $g_x(0, 0) = \lim_{h \to 0} \frac{h^2 - 0}{h} = 0 = \lim_{h \to 0} \frac{-h^2 - 0}{h} = g_y(0, 0)$. The tangent line in the x direction is the x-axis and the tangent line in the y direction is the y-axis. The plane containing these two lines is $z = 0$.

11. $f_x(0, 0) = \lim_{h \to 0} \frac{f(h, 0) - f(0, 0)}{h} = \lim_{h \to 0} \frac{0 - 0}{h} = 0$. The computation for $f_y(0, 0)$ is similar. The function is discontinuous at $(0, 0)$.

13. $f_x(0, 0, 0) = \lim_{h \to 0} \frac{f(h, 0, 0) - f(0, 0, 0)}{h} = \lim_{h \to 0} \frac{0 - 0}{h} = 0$. The computations for $f_y(0, 0, 0)$ and $f_z(0, 0, 0)$ are similar. The function is discontinuous at $(0, 0, 0)$.

15. (a) 2^2, (b) 2^3, (c) 2^n

17. (a) 3, (b) 4, (c) $n + 1$

19. $f(x, y) - g(x, y) = h(y)$, where $h(y)$ is a function of y

21. $f(x, y) - g(x, y) = C$, where C is a constant

23. -2

25. $\frac{\partial f}{\partial x} = \frac{y^2}{z}, \frac{\partial f}{\partial y} = \frac{2xy}{z}, \frac{\partial f}{\partial z} = -\frac{xy^2}{z^2}$

27. $\frac{\partial f}{\partial x} = e^x(\sin(xy) + y\cos(xy)), \frac{\partial f}{\partial y} = xe^x\cos(xy)$

29. $\frac{\partial f}{\partial x} = yx^{y-1}, \frac{\partial f}{\partial y} = (\ln x)x^y$

31. $\frac{\partial f}{\partial x} = \sin y, \frac{\partial f}{\partial y} = x\cos y$

33. $\frac{\partial f}{\partial r} = \sin\theta, \frac{\partial f}{\partial \theta} = r\cos\theta$

35. $\frac{\partial f}{\partial x} = \frac{y^2 z}{(x+z)^2}, \frac{\partial f}{\partial y} = \frac{2xy}{x+z}, \frac{\partial f}{\partial z} = -\frac{xy^2}{(x+z)^2}$

37. (a) $x = t, y = \frac{\pi}{2}, z = \frac{\pi}{2}t$, (b) $x = 0, y = \frac{\pi}{2} + t, z = 0$, (c) $\frac{\pi}{2}x = z$

39. (a) $x = e + t, y = 3, z = e^3 + 3e^2 t$, (b) $x = e, y = 3 + t, z = e^3 + e^3 t$, (c) $3e^2 x + e^3 y - z = 5e^3$

41. (a) $x = 2 + t, y = \frac{\pi}{3}, z = \sqrt{3} + \frac{\sqrt{3}}{2}t$, (b) $x = 2, y = \frac{\pi}{3} + t, z = \sqrt{3} + t$, (c) $\sqrt{3}x + 2y - 2z = \frac{2\pi}{3}$

43. $\frac{\partial^2 f}{\partial x^2} = e^x(\sin(xy) + 2y\cos(xy) - y^2\sin(xy))$, $\frac{\partial^2 f}{\partial x \partial y} = \frac{\partial^2 f}{\partial y \partial x} = e^x(\cos(xy) + x\cos(xy) - xy\sin(xy))$, $\frac{\partial^2 f}{\partial y^2} = -x^2 e^x \sin(xy)$

45. $\frac{\partial^2 f}{\partial x^2} = y(y-1)x^{y-2}, \frac{\partial^2 f}{\partial x \partial y} = \frac{\partial^2 f}{\partial y \partial x} = x^{y-1}(1 + y\ln x)$, $\frac{\partial^2 f}{\partial y^2} = (\ln x)^2 x^y$

47. $\frac{\partial^2 f}{\partial x^2} = 0, \frac{\partial^2 f}{\partial x \partial y} = \frac{\partial^2 f}{\partial y \partial x} = \cos y, \frac{\partial^2 f}{\partial y^2} = -x\sin y$

49. $\frac{\partial^2 f}{\partial r^2} = 0, \frac{\partial^2 f}{\partial r \partial \theta} = \frac{\partial^2 f}{\partial \theta \partial r} = \cos\theta, \frac{\partial^2 f}{\partial \theta^2} = -r\sin\theta$

51. $f(x, y) = h(y)$

53. $f(x, y) = xh_1(y) + h_2(y)$

55. $f(x, y, z) = h(y, z)$

57. $f(x, y, z) = xh_1(y, z) + h_2(y, z)$

59. $\frac{\partial g}{\partial y} = \frac{\partial h}{\partial x}, F(x, y) = e^x \cos y + y^2 + C$

61. $\frac{\partial g}{\partial y} = \frac{\partial h}{\partial x}, F(x, y) = \tan^{-1}(xy) + C$

63. $xe^y - 7y + C = 0$

65. $e^x \ln y + \frac{1}{4}x^4 + C = 0$

67. $\frac{\partial V}{\partial r} = 2\pi rh$. This tells us how fast the volume changes with a unit change in the radius. $\frac{\partial V}{\partial h} = \pi r^2$. This tells us how fast the volume changes with a unit change in the height.

69. (b) $0.04e^{\pi y/2}\sin\frac{\pi x}{2}$

71. Both results follow from the Chain Rule (Theorem 2.12), the power rule, and the definition of the partial derivative.

73. Both results follow from the Quotient Rule (Theorem 2.11) and the definition of the partial derivative.

75. $\frac{\partial h}{\partial y} = g'(y)$, so $\frac{\partial^2 h}{\partial x \partial y} = 0$. Similarly, $\frac{\partial h}{\partial x} = f'(x)$, so $\frac{\partial^2 h}{\partial y \partial x} = 0$.

Section 12.4

1. T, T, F, F, F, T, F, T.

3. There are two unit vectors in $\mathbb{R}^1$, $\mathbf{i}$ and $-\mathbf{i}$. There are infinitely many unit vectors in $\mathbb{R}^n$ when $n > 1$.

5. The directional derivative of $f(x, y)$ at (x_0, y_0) in the direction of the unit vector $\mathbf{u} = \langle a, b \rangle$ is the limit $\lim_{h \to 0} \frac{f(x_0 + a \cdot h, y_0 + b \cdot h) - f(x_0, y_0)}{h}$, provided that this limit exists.

7. $D_{\mathbf{u}}f(\mathbf{v}) = \lim_{h \to 0} \frac{f(\mathbf{v} + h\mathbf{u}) - f(\mathbf{v})}{h}$, provided that this limit exists.

9. (a) $D_{\mathbf{u}} = 3\alpha - \beta$, (b) $3k\alpha - k\beta$, (c) The limit from part (b) is different for each value of k. We want the value of the directional derivative to depend upon the direction of $\mathbf{u}$, not its magnitude.

11. $\lim_{h \to 0} \frac{f(c + \alpha h) - f(c)}{h}$, $\alpha = \pm 1$

13. See Definition 12.27.

15. If $f(x, y)$ is differentiable at the point (a, b), all of the lines tangent to the surface defined by f at (a, b) lie in the same plane, so any two distinct lines in that plane may be used to determine the equation of the plane.

17. If $f(x, y, z)$ is differentiable at the point (a, b, c), all of the lines tangent to the graph of f at (a, b, c) lie in the same hyperplane, so any three non-coplanar lines in that hyperplane may be used to determine the equation of the hyperplane.

19. Let $y = f(\mathbf{x})$ be a function of n variables defined on an open set containing the point $\mathbf{x_0}$ and let $\Delta y = f(\mathbf{x_0} + \Delta\mathbf{x}) - f(\mathbf{x_0})$. The function f is said to be **differentiable** at $\mathbf{x_0}$ if the partial derivatives $f_{x_i}(\mathbf{x_0})$ exist for each $1 \le i \le n$ and $\Delta y = \langle f_{x_1}(\mathbf{x_0}), f_{x_2}(\mathbf{x_0}), \ldots, f_{x_n}(\mathbf{x_0}) \rangle \cdot \Delta\mathbf{x} + \epsilon \cdot \Delta\mathbf{x}$, where $\epsilon \to \mathbf{0}$ as $\Delta\mathbf{x} \to \mathbf{0}$.

21. $-\sqrt{2}$

23. $-11\frac{\sqrt{10}}{10}$

25. $-7\frac{\sqrt{17}}{816}$

27. $-\frac{36}{5}$

29. $x = 2 + \frac{\sqrt{2}}{2}t, y = 3 + \frac{\sqrt{2}}{2}t, z = -5 - \sqrt{2}t$

31. $x = -2 + \frac{\sqrt{10}}{10}t, y = 1 - \frac{3\sqrt{10}}{10}t, z = -2 - 11\frac{\sqrt{10}}{10}t$

33. $x = 4 - \frac{\sqrt{17}}{17}t, y = 9 - \frac{4\sqrt{17}}{17}t, z = \frac{3}{2} - 7\frac{\sqrt{17}}{816}$

35. $-\frac{18}{13}\sqrt{26}$

37. $-\frac{33}{40}\sqrt{5}$

39. For every unit vector $\mathbf{u} = \langle \alpha, \beta \rangle$, $\lim_{h \to 0} \frac{(1+\alpha h)(-2+\beta h) + 2(1+\alpha h) - (-2+\beta h) - 2}{h} = 0$.

41. For every unit vector $\mathbf{u} = \langle \alpha, \beta \rangle$, $\lim_{h \to 0} \frac{\alpha h (\beta h)^2}{h} = 0$.

43. f has continuous first-order partial derivatives and is differentiable at every point in $\mathbb{R}^2$.

45. The function has continuous first-order partial derivatives and is differentiable at every point in $\mathbb{R}^2$ except when $x^2 + y^2 = 1$.

47. The function has continuous first-order partial derivatives and is differentiable at every point in $\mathbb{R}^2$.

49. The function has continuous first-order partial derivatives and is differentiable at every point in $\mathbb{R}^2$ at which the product xy is not an odd multiple of $\frac{\pi}{2}$

51. The function has continuous first-order partial derivatives and is differentiable at every point such that $x > 0$ and $y > 0$.

53. The function has continuous first-order partial derivatives and is differentiable at every point in $\mathbb{R}^3$.

55. $2(x - 1) + 6(y + 3) = z + 8$

57. $x + 9z = -6$

59. $\pi x + 4y + 2z = 4\pi$

61. $\pi x - 4y + 2z = 2\pi - 2$

63. $9x - y + 6z = 18$

65. $2(x - 1) - 10(y + 5) - 27(z - 3) = w + 1$

67. (a) At about a 10 degree slope. (b) He must move in the direction $(0, 1)$ or $(0, -1)$, due south or due north.

69. Let $\mathbf{u} = \langle \alpha, \beta \rangle$. By Definition 12.26, $D_{-\mathbf{u}}f(a, b) = \lim_{\eta \to 0} \frac{f(a - \alpha\eta, b - \beta\eta) - f(a, b)}{\eta}$. If we let $h = -\eta$, we have $D_{-\mathbf{u}}f(a, b) = \lim_{h \to 0} \frac{f(a + \alpha h, b + \beta h) - f(a, b)}{-h} = -D_{\mathbf{u}}f(a, b)$.

71. By Exercise 70, f cannot be differentiable at (a, b) if $D_{\mathbf{u}}f(a, b) = 1$ for every unit vector $\mathbf{u}$.

Section 12.5

1. F, F, F, F, F, T, T, T.

3. $e^{-\sin t}(2\sin t \cos t - 3\sin t \cos^2 t - \cos^3 t + 3\cos^2 t - 3\sin^2 t - 4\cos t)$.

5. When $n = m = 1$, $\frac{\partial z}{\partial t_j} = \frac{\partial z}{\partial x_1}\frac{\partial x_1}{\partial t_j} + \frac{\partial z}{\partial x_2}\frac{\partial x_2}{\partial t_j} + \cdots + \frac{\partial z}{\partial x_n}\frac{\partial x_n}{\partial t_j}$ simplifies to $\frac{dz}{dt_1} = \frac{dz}{dx_1}\frac{dx_1}{dt_1}$.

7. When $n = m = 2$,
$$\frac{\partial z}{\partial t_j} = \frac{\partial z}{\partial x_1}\frac{\partial x_1}{\partial t_j} + \frac{\partial z}{\partial x_2}\frac{\partial x_2}{\partial t_j} + \cdots + \frac{\partial z}{\partial x_n}\frac{\partial x_n}{\partial t_j} \text{ simplifies to}$$
$$\frac{\partial z}{\partial t_1} = \frac{\partial z}{\partial x_1}\frac{\partial x_1}{\partial t_1} + \frac{\partial z}{\partial x_2}\frac{\partial x_2}{\partial t_1} \text{ and } \frac{\partial z}{\partial t_2} = \frac{\partial z}{\partial x_1}\frac{\partial x_1}{\partial t_2} + \frac{\partial z}{\partial x_2}\frac{\partial x_2}{\partial t_2}.$$

9. (a) The level curves of f are the lines with equations $y = -\frac{2}{3}x + C$ where $C \in \mathbb{R}$. (b) The vector $\langle 3, -2 \rangle$ may be used as a direction vector for every level curve, $\nabla f(x, y) = \langle 2, 3 \rangle$, and $\langle 3, -2 \rangle \cdot \langle 2, 3 \rangle = 0$.

11. (a) The level curves of f are the lines with equations $y = -\frac{a}{b}x + C$ where $C \in \mathbb{R}$. (b) The vector $\langle b, -a \rangle$ may be used as a direction vector for every level curve, $\nabla f(x, y) = \langle a, b \rangle$, and $\langle b, -a \rangle \cdot \langle a, b \rangle = 0$.

13. $\nabla f(a, b, c) \cdot (\langle x, y, z \rangle - \langle a, b, c \rangle) = w - f(a, b, c)$.

15. The level curves are concentric circles centered at the origin. The gradient vectors are orthogonal to the level curves and point toward the origin. The gradient vectors increase in magnitude as you get further from the origin.

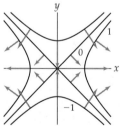

17. The level curves are hyperbolas. The gradient vectors are orthogonal to the level curves and point toward the x-axis and away from the y-axis. The gradient vectors increase in magnitude as you get further from the origin.

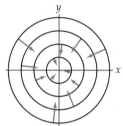

19. Staying on a contour line means that you are always staying at the same elevation, so the hike is relatively easy. Walking perpendicularly to the contour lines means you are always either ascending or descending, making the hike more rigorous.

21. $e^t \cos(e^t) \cos(t^3) - 3t^2 \sin(e^t) \sin(t^3)$

23. $2t \cos(t^3 + 1) + (15t^2 - 3t^4) \sin(t^3 + 1)$

25. $\dfrac{1 + 4t^3}{2\sqrt{t + t^4}}$

27. $2s^3 t^2 \cos s \cos^3 t \sin s + 3s^2 t^2 \cos^3 t \sin^2 s$

29. $e^{r \sin \theta}(2r \cos^2 \theta + 2r \sin \theta \cos \theta + r^2 \sin \theta \cos^2 \theta + r^2 \sin^2 \theta \cos \theta)$

31. $2t \sin(t^3) \cos(t^4) + 3t^4 \cos(t^3) \cos(t^4) - 4t^5 \sin(t^3) \sin(t^4)$

33. $e^{\rho \sin \phi \sin \theta}(2\rho \sin^2 \phi \cos^2 \theta + \rho^2 \sin^3 \phi \sin \theta \cos^2 \theta + \rho \sin \phi \cos \phi \sin \theta + \cos \phi)$

35. $\frac{1}{2}(1 + 4t^3 + 6t^5)(t + t^4 + t^6)^{-1/2}$

37. $\langle 2x \sin y + y \cos x, x^2 \cos y + \sin x \rangle$

39. $\dfrac{1}{\sqrt{x^2 + y^2}}\langle x, y \rangle$

41. $\dfrac{1}{\sqrt{x^2 + y^2 + z^2}}\langle x, y, z \rangle$

43. (a) $\left\langle \frac{3\pi}{2}, 0 \right\rangle$, (b) $\frac{3\pi}{2}$, (c) $-\left\langle \frac{3\pi}{2}, 0 \right\rangle$

45. (a) $\left\langle \frac{2}{\sqrt{13}}, -\frac{3}{\sqrt{13}} \right\rangle$, (b) 1, (c) $-\left\langle \frac{2}{\sqrt{13}}, -\frac{3}{\sqrt{13}} \right\rangle$

47. (a) $\left\langle \frac{2}{3}, -\frac{1}{3}, -\frac{2}{3} \right\rangle$, (b) 1, (c) $-\left\langle \frac{2}{3}, -\frac{1}{3}, -\frac{2}{3} \right\rangle$

49. $\dfrac{\sqrt{10}}{10}$

51. $\dfrac{55\sqrt{17}}{119}$

53. $\dfrac{9\sqrt{7}}{14}$

55. $f(x, y) = \ln\left|\dfrac{y}{x^3}\right| + C$

57. $f(x, y) = \tan^{-1}\left(\dfrac{y}{x}\right) + C$

59. $f(x, y) = \dfrac{x}{x + y} + C$

61. (a) $\langle 0.5, -0.35 \rangle$, (b) $\pm \langle 0.35, 0.5 \rangle$.

63. $\nabla f(x, y, z) = \left\langle \dfrac{x}{\sqrt{x^2 + y^2 + z^2}}, \dfrac{y}{\sqrt{x^2 + y^2 + z^2}}, \dfrac{z}{\sqrt{x^2 + y^2 + z^2}} \right\rangle$, so
$$\|\nabla f(x, y, z)\| = \left(\frac{x}{\sqrt{x^2 + y^2 + z^2}}\right)^2 + \left(\frac{y}{\sqrt{x^2 + y^2 + z^2}}\right)^2 + \left(\frac{z}{\sqrt{x^2 + y^2 + z^2}}\right)^2 = 1$$

65. We have $z = f(x, y)$, $x = u(s, t)$ and $y = v(s, t)$. Assume that u, v and $z = f(u(s, t), v(s, t))$ are differentiable at a point (s_0, t_0) and that $x_0 = u(s_0, t_0)$ and $y_0 = v(s_0, t_0)$. We will show that $\dfrac{\partial z}{\partial s} = \dfrac{\partial z}{\partial x}\dfrac{\partial x}{\partial s} + \dfrac{\partial z}{\partial y}\dfrac{\partial y}{\partial s}$. The remaining equality is quite similar. By the definition of the partial derivative at s_0:
$$\frac{\partial z}{\partial s} = \lim_{\Delta s \to 0} \frac{f(u(s_0 + \Delta s, t_0), v(s_0 + \Delta s, t_0)) - f(u(s_0, t_0), v(s_0, t_0))}{\Delta s} =$$
$$\lim_{\Delta s \to 0} \frac{f(u(s_0 + \Delta s, t_0), v(s_0 + \Delta s, t_0)) - f(x_0, y_0)}{\Delta s} = \lim_{\Delta s \to 0} \frac{\Delta z}{\Delta s}. \text{ Now,}$$
when Δs is sufficiently small, we may replace the numerator in the equation above by $f_x(x_0, y_0)\Delta x + f_y(x_0, y_0)\Delta y$. We now have
$$\frac{\partial z}{\partial s} = \lim_{\Delta s \to 0} \frac{f_x(x_0, y_0)\Delta x + f_y(x_0, y_0)\Delta y}{\Delta s} =$$
$$\lim_{\Delta s \to 0}\left(f_x(x_0, y_0)\frac{\Delta x}{\Delta s} + f_y(x_0, y_0)\frac{\Delta y}{\Delta s}\right) = \frac{\partial z}{\partial x}\frac{\partial x}{\partial s} + \frac{\partial z}{\partial y}\frac{\partial y}{\partial s}.$$

67. $\dfrac{\partial f}{\partial x} = 0 = \dfrac{\partial f}{\partial y}$, so $\nabla f = 0$.

69. $\nabla(f(x,y)+g(x,y))$
$= \langle f_x(x,y)+g_x(x,y), f_y(x,y)+g_y(x,y)\rangle$
$= \langle f_x(x,y), f_y(x,y)\rangle + \langle g_x(x,y), g_y(x,y)\rangle$
$= \nabla f(x,y) + \nabla g(x,y).$

71. $\nabla(f(x,y)g(x,y)) = \langle f_x(x,y)g(x,y)$
$+ f(x,y)g_x(x,y), f_y(x,y)g(x,y)+f(x,y)g_y(x,y)\rangle$
$= g(x,y)\langle f_x(x,y), f_y(x,y)\rangle + f(x,y)\langle g_x(x,y), g_y(x,y)\rangle$
$= f(x,y)\nabla g(x,y) + g(x,y)\nabla f(x,y).$

73. Use three component vectors rather than two component vectors.

Section 12.6

1. F, T, F, T, F, T, T, T.

3. A critical point of f is a point at which either $\nabla f = \mathbf{0}$ or ∇f does not exist. The only place a function can have an extreme value at an interior point of its domain is at a critical point.

5. A saddle point is a stationary point at which there is neither a maximum nor a minimum.

7. The first derivative test for a function of a single variable requires that you check the sign of the first derivative to the left and right of a critical point. To use an analogous test for a function of two variables, we would have to check the sign of the directional derivative at infinitely many points encircling each critical point.

9. If A and C have opposite signs then both terms AC and $-B^2$ are negative, so their sum is negative.

11. $g(x,y) = -f(x,y)$ from Exercise 10, so since $f(x,y)$ has a minimum at the origin, g has a maximum at the origin.

13. Plow through the algebra!

15. Plow through the algebra!

17. The distance from P to an arbitrary point on $\mathcal{L}$ is given by

$$d(t) = \sqrt{(x_0-x_1+at)^2 + (y_0-y_1+bt)^2 + (z_0-z_1+ct)^2}.$$

To find the distance from P to $\mathcal{L}$ we may minimize the function $d(t)$.

19. f will have an absolute minimum at $x=0$ if n is even and an inflection point if n is odd. There are no other possibilities.

21. $-4e^{4x}$

23. $-\frac{3}{4}s^4 e^t$

25. $\cos^2\theta\sin^2\phi - \sin^2\theta\cos^2\phi$

27. $\frac{14}{13}\sqrt{26}$

29. $\frac{8}{13}$

31. $f\left(-\frac{1}{2},0\right) = -\frac{31}{4}$ is the absolute minimum

33. $f(-2,-1) = 33$, local maximum; $f(-2,1) = 29$, saddle; $f(2,-1) = 1$, saddle; $f(2,1) = -3$, local minimum

35. $f(0,0) = 0$, saddle; $f(4,4) = -64$, local minimum

37. $f\left(-\frac{3}{2},-\frac{3}{2}\right) = -14$, absolute minimum

39. $f\left(\frac{1}{2},\frac{1}{2}\right) = 6$, local minimum

41. Every point of the form $(0,k\pi)$ where $k \in \mathbb{Z}$ is a stationary point. They are all saddle points.

43. There is a saddle point at every point of the form $(0,k\pi)$ where $k \in \mathbb{Z}$.

45. $f(0,0) = 1$, saddle

47. $f(0,0) = -1$, local maximum

49. f has an absolute minimum at every point of the form $(0,y)$ or $(x,0)$.

51. f has saddle point at every point of the form $(0,y)$, a local minimum at every point of the form $(x,0)$ for $x > 0$, and a local maximum at every point of the form $(x,0)$ for $x < 0$.

53. The box should have a square base measuring $\frac{10}{9}\sqrt[3]{\frac{81}{5}}$ feet on each side and a height of $\sqrt[3]{\frac{81}{5}}$ feet.

55. By Theorem 12.36, $D_{\mathbf{u}}f(x,y) = af_x(x,y) + bf_y(x,y)$. Taking the directional derivative, again, we obtain $D_{\mathbf{u}}(D_{\mathbf{u}}f(x,y)) = D_{\mathbf{u}}^2 f(x,y) = D_{\mathbf{u}}(af_x(x,y)+bf_y(x,y)) = a^2 f_{xx}(x,y) + abf_{yx}(x,y) + baf_{xy}(x,y) + b^2 f_{yy}(x,y)$. Since the function has continuous second-order partial derivatives, the mixed second-order partial derivatives are equal and we have the result.

57. Let (x,y) be a point on the line. The distance from (x_0,y_0) to (x,y) is given by $d = \sqrt{(x-x_0)^2+(y-y_0)^2}$ and the square of the distance from (x_0,y_0) to (x,y) is given by $D = (x-x_0)^2 + (y-y_0)^2$. The gradients of these two functions are
$\nabla d = \frac{1}{2\sqrt{(x-x_0)^2+(y-y_0)^2}}(2(x-x_0)\mathbf{i}+2(y-y_0)\mathbf{j})$ and $\nabla D = 2(x-x_0)\mathbf{i}+2(y-y_0)\mathbf{j}$. The critical points of these two functions are the same and would have the same minima.

Section 12.7

1. F, F, T, F, F, F, F, T.

3. The function f represents the quantity we wish to maximize or minimize. The constraint equation provides an interrelationship between the variables of f.

5. We can either eliminate one of the variables by solving the constraint equation for one of the variables, or use the method of Lagrange multipliers.

7. (a) $\{(x,y)|x > 0\}$, (b) $\{(x,y)|x \geq 0\}$, (c) $\{(x,y)|x^2+y^2 < 1\}$, (d) $\{(x,y)|x^2+y^2 \leq 1\}$

9. (a) $\{(x,y,z)|x > 0\}$, (b) $\{(x,y,z)|x \geq 0\}$, (c) $\{(x,y,z)|x^2+y^2+z^2 < 1\}$, (d) $\{(x,y,z)|x^2+y^2+z^2 \leq 1\}$

11. 4

13. $f_{yy}(x, y) = 12$ and the discriminant of f is $\det(H_f(x, y)) = 72(x + 3)$. (a) $\det(H_f(-6, 0)) < 0$ so there is a saddle point at $(-6, 0)$. (b) $\det(H_f(0, 0)) > 0$ and $f_{yy}(0, 0) > 0$ so there is a relative minimum at $(0, 0)$. (c) From Example 4 we already know (i) that $f(0, 0) = 0$ is the absolute minimum of f on the region defined by $x^2 + y^2 \leq 9$ and (ii) the point $(0, 0)$ is an interior point to that region, so there is, at least, a relative minimum of f at the origin. Since $f(-10, 0) = -100 < 0 = f(0, 0)$, the minimum is only relative, not absolute.

15. When $y = -x$ we are looking for the critical points of the function $g(x) = -x^3$. The only critical point of g is its inflection point at $x = 0$.

17. When $y = -\frac{a}{b}x$ we are looking for the critical points of the function $h(x) = -\frac{a}{b}x^3$. The only critical point of h is its inflection point at $x = 0$.

19. Find the extrema of the function f using the techniques of Section 12.6. Choose only the critical points in $\mathcal{R}$. Evaluate f at the critical points. Use the results of Exercise 18 to find the extrema on the boundary of $\mathcal{R}$. Choose the largest and smallest values of the function in the interior of $\mathcal{R}$ and on the boundary of $\mathcal{R}$.

21. Find the extrema of the function f using the techniques of Section 12.6. Choose only the critical points in $\mathcal{T}$. Evaluate f at the critical points. Use the results of Exercise 20 to find the extrema on the boundary of $\mathcal{T}$. Choose the largest and smallest values of the function in the interior of $\mathcal{T}$ and on the boundary of $\mathcal{T}$.

23. The Extreme Value Theorem guarantees that a maximum and minimum of f occurs on $\mathcal{R}$. If the gradient exists everywhere and is never zero, f does not have any critical points, so the maximum and minimum must occur on the boundary of $\mathcal{R}$.

25. $f\left(\frac{8\sqrt{5}}{5}, \frac{2\sqrt{5}}{5}\right) = 2\sqrt{5}$ is the maximum and $f\left(-\frac{8\sqrt{5}}{5}, -\frac{2\sqrt{5}}{5}\right) = -2\sqrt{5}$ is the minimum. Both exist because the ellipse is a closed and bounded set.

27. $f(2\sqrt{2}, \sqrt{2}) = f(-2\sqrt{2}, -\sqrt{2}) = 4$ is the maximum and $f(2\sqrt{2}, -\sqrt{2}) = f(-2\sqrt{2}, \sqrt{2}) = -4$ is the minimum. Both exist because the ellipse is a closed and bounded set.

29. $f\left(\frac{32}{\sqrt{21}}, \frac{8}{\sqrt{21}}, \frac{2}{\sqrt{21}}\right) = 2\sqrt{21}$ is the maximum and $f\left(-\frac{32}{\sqrt{21}}, -\frac{8}{\sqrt{21}}, -\frac{2}{\sqrt{21}}\right) = -2\sqrt{21}$ is the minimum. Both exist because the ellipsoid is a closed and bounded set.

31. The maxima and minima occur at the eight points $\left(\pm\frac{8}{\sqrt{3}}, \pm\frac{4}{\sqrt{3}}, \pm\frac{2}{\sqrt{3}}\right)$. When an even number of the signs are negative the function has a maximum of $\frac{64}{3\sqrt{3}}$. When an odd number of the signs are negative the function has a minimum of $-\frac{64}{3\sqrt{3}}$. Both exist because the ellipsoid is a closed and bounded set.

33. $(1, 0)$ and $(0, 1)$

35. $\left(\frac{1}{4}, \frac{1}{4}\right)$

37. $(1, 0)$ and $(0, 1)$

39. $(1, 1, 1), (-1, -1, 1), (-1, 1, -1), (1, -1, -1)$

41. $\left(\frac{ad}{a^2+b^2+c^2}, \frac{bd}{a^2+b^2+c^2}, \frac{cd}{a^2+b^2+c^2}\right)$

43. $\left(-\frac{5}{7}, \frac{39}{7}, \frac{15}{7}\right)$

45. The maximum of 8 occurs at each corner of the square and the minimum is $0 = f(0, 0)$.

47. The maximum is $1 = f(1, 0) = f(0, 1) = f(-1, 0)$. The minimum is $-1 = f(0, -1)$.

49. The maximum is $1 = f(1, 1)$. The minimum is $-1 = f(-1, 1)$.

51. $x = \frac{d}{3a}, y = \frac{d}{3b}, z = \frac{d}{3c}$

53. $\left(\frac{6}{5}, \frac{12}{5}, \frac{6\sqrt{5}}{5}\right)$

55. The base should be a square $5\sqrt[3]{2/5}$ feet on each side. The height should be $\sqrt[3]{2/5}$ feet.

57. Maximize $f(x, y) = xy$ subject to the constraint that $g(x, y) = 2x + 2y - P = 0$.

59. The distance is $d = \sqrt{x^2 + y^2 + z^2}$ and the square of the distance is $D = x^2 + y^2 + z^2$, $\nabla d = \frac{x}{\sqrt{x^2+y^2+z^2}}\mathbf{i} + \frac{y}{\sqrt{x^2+y^2+z^2}}\mathbf{j} + \frac{z}{\sqrt{x^2+y^2+z^2}}\mathbf{k}$ and $\nabla D = 2x\mathbf{i} + 2y\mathbf{j} + 2z\mathbf{k}$. The system of equations $\nabla d = \lambda \nabla g$ and $\nabla D = \lambda \nabla g$ have the same solutions.

61. Rotating, scaling or translating the point and the circle does not change the essential geometry of the system, if $P = (\alpha, \beta)$ is not the center of the circle (a, b), the point closest to P on the circle will be on the line containing both P and (a, b).

63. Rotating, scaling or translating the point and the sphere does not change the essential geometry of the system, if $P = (\alpha, \beta, \gamma)$ is not the center of the sphere (a, b, c), the point closest to P on the sphere will be on the line containing both P and (a, b, c).

65. Let $A > 0$, use the method of Lagrange multipliers to show that the function $f(x, y, z) = \sqrt[3]{xyz}$ has its maximum value when $x = y = z = A$, subject to the constraint equation $g(x, y, z) = \frac{1}{3}(x + y + z) - A = 0$. Thus, $\sqrt[3]{xyz} \leq \frac{1}{3}(x + y + z)$ when x, y and z are all positive.

For answers to odd-numbered Skill Certification exercises in the Chapter Review, please visit the Book Companion Web Site at www.whfreeman.com/tkcalculus.

Chapter 13

Section 13.1

1. T, T, F, T, F, T, T, T.

3. $\sum_{j=3}^{4}\sum_{k=2}^{4} je^{k^2}$.

5. 176

7. 3360

9. A definite integral is the limit of a Riemann sum of function of a single variable $f(x)$ over a closed interval $[a, b]$ as the number of subintervals goes to ∞. A double integral is the limit of a double Riemann sum of function of a two variables $f(x, y)$ over a rectangular region that has be subdivided into smaller subrectangles in both the x and y directions as the numbers of pieces in those two directions goes to ∞.

11. For each subrectangle $\mathcal{R}_{jk}$ choose
$(x_j^*, y_k^*) = \left(\frac{x_{j-1}+x_j}{2}, \frac{y_{k-1}+y_k}{2} \right)$.

13. See Theorem 13.7.

15. To evaluate the integral first find an antiderivative for f with respect to y. You use the FTC to evaluate the inner integral by evaluating this function at d and c and evaluating the difference. The result will be the definite integral of a function of the single variable x. Use the FTC to evaluate this integral, if possible.

17. 91

19. The first step in the integration will be fine, but the resulting integral will be one without a simple antiderivative. Say you integrate first with respect to y. You will obtain $\int_{\pi/4}^{\pi/2} \frac{1}{x} \left(\sin(\pi x) - \sin\left(\frac{\pi}{2}x\right) \right) dx$. The function $\frac{1}{x} \left(\sin(\pi x) - \sin\left(\frac{\pi}{2}x\right) \right)$ does *not* have a simple antiderivative.

21. $\int_0^1 \int_0^1 \frac{x-y}{(x+y)^3} dy\, dx = \frac{\pi}{2}$ and $\int_0^1 \int_0^1 \frac{x-y}{(x+y)^3} dx\, dy = -\frac{\pi}{2}$. This does not violate Fubini's Theorem because the function is not continuous everywhere on the rectangle $\mathcal{R} = \{(x, y)| 0 \le x \le 1 \text{ and } 0 \le y1\}$.

23. 20

25. -48

27. 1260

29. 15

31. 0

33. 15

35. 0

37. 26

39. $\frac{340}{3}$

41. 0

43. $\frac{4-\ln 5}{\ln 5}$

45. $\frac{9}{2} - 8\ln 2 + \frac{7}{2}\ln 7$

47. $\frac{1}{2}$

49. 18

51. $\frac{2}{\pi}$

53. $\frac{1}{2}$

55. $-\frac{1175650}{3}$

57. $\frac{e^{18}-e^8-10}{4}$

59. $\frac{169}{2}$

61. 4

63. $4\ln 5 + 12\ln 3$

65. 2.288

67. 1.386

69. Approximately 64533.3 cubic feet

71. The result follows from the commutative property of addition.

73. By Fubini's Theorem
$$\iint_{\mathcal{R}} g(x)h(y)dA = \int_a^b \int_c^d g(x)h(y)dy\, dx = \int_a^b g(x) \int_c^d h(y)dy\, dx = \left(\int_a^b g(x)dx \right)\left(\int_c^d h(y)dy \right).$$

75. Use the result of Exercise 73. The product $(b - a)(d - c)$ is the area of rectangle $\mathcal{R}$.

Section 13.2

1. T, F, F, T, T, T, F, T.

3. A type I region is bounded below by a function $g_1(x)$, above by a function $g_2(x)$, on the left by $x = a$ and on the right by $x = b$, for constants $a < b$. A type II region is bounded on the left by a function $h_1(y)$, on the right by a function $h_2(y)$, above by $y = d$ and below by $y = c$, for constants $c < d$.

5. Correct

7. Reverse dx and dy.

9. Correct

11. Correct

13. As a type I region, Ω is bounded below by the function $y = \frac{1}{2}x$ and above by the function $y = \sqrt{x}$ on the interval $[a, b] = [0, 4]$. As a type II region, Ω is bounded below on the left by the function $x = y^2$ and on the right by the function $x = 2y$ on the interval $[c, d] = [0, 2]$.

15. The area of $\Omega = \int_a^b \int_{g_1(x)}^{g_2(x)} dy\, dx = \int_a^b [y]_{g_1(x)}^{g_2(x)} dx = \int_a^b (g_2(x) - g_2(x))\, dx$.

17. $\frac{4}{3}$

19. $\int_{-3}^0 \int_{-x^3}^0 dy\, dx + \int_0^3 \int_0^{x^3} dy\, dx = \int_{-27}^0 \int_{-3}^{\sqrt[3]{y}} dx\, dy + \int_0^{27} \int_{\sqrt[3]{y}}^3 dx\, dy = 0$

21. (a) $\int_0^1 \int_0^{e^x} f(x, y)\, dy\, dx$,
(b) $\int_0^1 \int_0^1 f(x, y)\, dx\, dy + \int_1^e \int_{\ln y}^1 f(x, y)\, dx\, dy$

23. (a) $\int_{-2}^2 \int_0^{\sqrt{4-x^2}} f(x, y)\, dy\, dx + \int_{-2}^0 \int_{-\sqrt{4-x^2}}^0 f(x, y)\, dy\, dx$,
(b) $\int_{-2}^2 \int_{-\sqrt{4-y^2}}^0 f(x, y)\, dx\, dy + \int_0^2 \int_0^{\sqrt{4-y^2}} f(x, y)\, dx\, dy$

25. (a) $\int_0^{\pi/4} \int_{\sin x}^{\cos x} f(x, y)\, dy\, dx$,
(b) $\int_0^{\sqrt{2}/2} \int_0^{\sin^{-1} y} f(x, y)\, dx\, dy + \int_{\sqrt{2}/2}^1 \int_0^{\cos^{-1} y} f(x, y)\, dx\, dy$

27. (a) $\displaystyle\int_{-1}^{0}\int_{-x-1}^{x+1} f(x,y)\,dy\,dx + \int_{0}^{1}\int_{x-1}^{-x+1} f(x,y)\,dy\,dx,$

(b) $\displaystyle\int_{-1}^{0}\int_{-y-1}^{y+1} f(x,y)\,dx\,dy + \int_{0}^{1}\int_{y-1}^{-y+1} f(x,y)\,dx\,dy$

29. $\displaystyle\int_{0}^{1}\int_{1}^{e^{y}} f(x,y)\,dx\,dy + \int_{1}^{e}\int_{1}^{2} f(x,y)\,dx\,dy +$

$\displaystyle\int_{e}^{e^{2}}\int_{\ln y}^{2} f(x,y)\,dx\,dy$

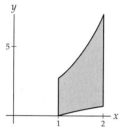

31. $\displaystyle\int_{0}^{1}\int_{\sin^{-1}x}^{\pi/2} f(x,y)\,dy\,dx$

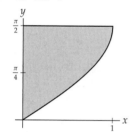

33. $\displaystyle\int_{0}^{1}\int_{0}^{\sin^{-1}x} f(x,y)\,dy\,dx$

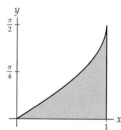

35. $\dfrac{e^{2}}{4} + 10e - \dfrac{49}{4}$ 37. 4π

39. 0   41. 4

43. $\dfrac{26}{3}$

45. $\dfrac{2}{9}(56\sqrt{7} - 1)$. The function $\sqrt{1+x^{3}}$ does not have a simple antiderivative.

47. 1. The function $\sec y$ has a simpler antiderivative when integrated with respect to x than it does when integrated with respect to y.

49. $-\dfrac{927}{8}$ 51. $\dfrac{101027}{90}$

53. $\dfrac{\pi}{8}$ 55. $\dfrac{1}{2}(e-1)$

57. 2   59. $\dfrac{1}{3}(e^{8} - 1)$

61. $\dfrac{3}{110}$ 63. 200 cubic meters

65. Using the notation of Definition 13.4, the equality holds since $\displaystyle\iint_{R} \alpha f(x,y)\,dA = \lim_{\Delta \to 0}\sum_{j=1}^{m}\sum_{k=1}^{n}\alpha f(x_{j}^{*}, y_{k}^{*})\Delta A =$

$\alpha \displaystyle\lim_{\Delta \to 0}\sum_{k=1}^{n}\sum_{j=1}^{m} f(x_{j}^{*}, y_{k}^{*})\Delta A = \alpha \iint_{R} f(x,y)\,dA.$

67. Let $\mathcal{R} = \{(x,y)\,|\,a \le x \le b \text{ and } c \le y \le d\}$ be a rectangle containing Ω. From Exercise 65, $\iint_{R} \alpha F(x,y)\,dA = \alpha \iint_{R} F(x,y)\,dA$, where $F(x,y) = f(x,y)$ on Ω and is zero everywhere else in $\mathcal{R}$. Thus, $\alpha F(x,y) = \alpha f(x,y)$ for every point in Ω and is zero everywhere else in $\mathcal{R}$, also, so the statement holds.

69. The volume is given by the integral

$$\int_{0}^{a}\int_{0}^{-b/ay+b}\left(c - \frac{c}{a}x - \frac{c}{b}y\right)dy\,dx = \frac{1}{6}abc.$$

Section 13.3

1. T, T, F, F, F, F, T, T.

3. The values $\theta = \alpha$, $\theta = \beta$, $r = a$ and $r = b$ provide constant boundaries for the region.

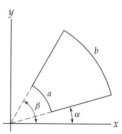

5.

7. In the rectangular coordinate system the integral represents the area between the graphs of $f_{1}(\theta)$ and $f_{2}(\theta)$ on the interval $[\alpha, \beta]$. In a polar system, the integral represents the volume of the solid bounded above by the function $f(r, \theta) = \dfrac{1}{r}$ on the polar region bounded by the two functions between the rays $\theta = \alpha$ and $\theta = \beta$.

9. Because the curve is traced twice on the interval $[0, 2\pi]$.

11. For any positive integer n, the integral represents the area of a sector of a circle whose central angle is $\dfrac{2\pi}{n}$. This area would be $\dfrac{\pi R^{2}}{n}$. The equation holds for every nonzero value of n.

13. $2\pi \displaystyle\int_0^b xf(x)\,dx$

15. $\displaystyle\int_{-b}^b \int_{-\sqrt{b^2-x^2}}^{\sqrt{b^2-x^2}} f\left(\sqrt{x^2+y^2}\right)\,dy\,dx$

17. From Exercise 16 we have the integral

$$\int_0^{2\pi}\int_0^b rf(r)\,dr\,d\theta = \left(\int_0^{2\pi} d\theta\right)\left(\int_0^b rf(r)\,dr\right) =$$

$$2\pi\int_0^b rf(r)\,dr,\text{ which equals the answer to Exercise 14.}$$

19. $\displaystyle\int_0^{2\pi}\int_a^b rf(r)\,dr\,d\theta$

21. $\dfrac{3\pi}{2}$

(1, 0)

23. $\dfrac{9\pi}{2}$

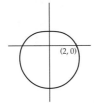

(2, 0)

25. $\dfrac{\pi}{4}$

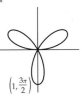

$\left(1, \frac{3\pi}{2}\right)$

27. The area between the inner and outer loops of the limaçon is $\dfrac{3+\pi}{2}$

$\left(1+\frac{\sqrt{2}}{2}, \frac{\pi}{2}\right)$

$\left(\frac{\sqrt{2}}{2}, 0\right)$

29. $\displaystyle\int_0^\pi \int_0^\theta r\,dr\,d\theta = \dfrac{\pi^3}{6}$

31. $2\displaystyle\int_0^{3\pi/4}\int_0^{1+\sqrt{2}\cos\theta} r\,dr\,d\theta - 2\int_\pi^{5\pi/4}\int_0^{1+\sqrt{2}\cos\theta} r\,dr\,d\theta =$ $3+\pi$

33. $2\displaystyle\int_{-\pi/2}^{\pi/6}\int_{1+\sin\theta}^{3-3\sin\theta} r\,dr\,d\theta = 8\pi + 9\sqrt{3}$

35. $2\displaystyle\int_0^{\pi/3}\int_{\sec\theta/2}^1 r\,dr\,d\theta = \dfrac{\pi}{3} - \dfrac{\sqrt{3}}{4}$

37. $\displaystyle\int_0^{2\pi}\int_0^{1+k\sin\theta} r\,dr\,d\theta = \pi\left(1+\dfrac{k^2}{2}\right)$. When $k=0$, "the limaçon" is a circle with radius one.

39. A sphere of radius 4 centered at the origin, $\dfrac{256}{3}\pi$.

41. The portion of the hyperboloid $z = 4 - r^2$ above the xy-plane, 8π.

43. The portion of the cone $z = 6 - 2r$ above the xy-plane, 18π.

45. The cylinder formed by the outer loop of the limaçon $r = \dfrac{\sqrt{2}}{2} + \sin\theta$ with height 1, $\dfrac{3}{4}(1+\pi)$.

47. $\displaystyle\int_0^{2\pi}\int_0^{2\sqrt{2}} (16 - 2r^2)r\,dr\,d\theta = 64\pi$

49. $\displaystyle\int_0^{2\pi}\int_0^{\sqrt{2}/2R} \left(\sqrt{R^2 - r^2} - r\right)r\,dr\,d\theta = \dfrac{\pi R^3}{3}(2 - \sqrt{2})$

51. $\displaystyle\int_0^{2\pi}\int_0^{4/5} \left(\sqrt{1 - r^2} - \dfrac{3}{5}\right)r\,dr\,d\theta = \dfrac{52}{375}\pi$

53. $\displaystyle\int_0^{2\pi}\int_0^{\sqrt{3}/2} (2\sqrt{1 - r^2} - 1)r\,dr\,d\theta = \dfrac{5}{12}\pi$

55. $\displaystyle\int_0^{2\pi}\int_0^h (h - r)\,r\,dr\,d\theta = \dfrac{1}{3}\pi h^3$

57. $\dfrac{9\pi}{8}$

59. $\dfrac{\pi}{4}e^{16}$

61. Approximately 14 wolves.

63. $\displaystyle\int_0^\pi \int_0^{2R\cos\theta} r\,dr\,d\theta = \pi R^2$

65. $2\displaystyle\int_0^{2\pi}\int_0^R \sqrt{R^2 - r^2}\,r\,dr\,d\theta = \dfrac{4}{3}\pi R^3$

67. $\displaystyle\int_0^\phi \int_0^R r\,dr\,d\theta = \dfrac{1}{2}\phi R^2$

69. $\displaystyle\int_0^{2\pi}\int_0^{\cos 2n\theta} r\,dr\,d\theta = \dfrac{\pi}{2}$ for every positive integer n

Section 13.4

1. F, F, F, T, F, F, T, T.

3. The area of Ω. The units are square centimeters.

5. The coordinates of the centroid of Ω. The units are centimeters.

7. The first moments M_y and M_x. The units are grams·centimeters.

9. The second moments I_y and I_x. The units are grams·square centimeters.

11. The radii of gyration R_y and R_x. The units are centimeters.

13. $\displaystyle\int_1^2 \int_{-x+2}^{2x-1} x\,dy\,dx = \int_1^2 \left[xy\right]_{-x+2}^{2x-1}\,dx =$ $\displaystyle\int_1^2 (3x^2 - 3x)\,dx = \left[x^3 - \dfrac{3}{2}x^2\right]_1^2 = (8-6) - \left(1 - \dfrac{3}{2}\right) = \dfrac{5}{2}.$

15. This is the same integral as in Exercise 13, with an extra factor of k.

17. $\displaystyle\int_1^2 \int_{-x+2}^{2x-1} kxy \, dy \, dx = \frac{k}{2}\int_1^2 x\left[y^2\right]_{-x+2}^{2x-1} dx$

$\displaystyle = \frac{k}{2}\int_1^2 (3x^3 - 3x)\, dx = \frac{k}{2}\left[\frac{3}{4}x^4 - \frac{3}{2}x^2\right]_1^2$

$\displaystyle = \frac{k}{2}\left((12-6) - \left(\frac{3}{4} - \frac{3}{2}\right)\right) = \frac{27}{8}k.$

19. $\displaystyle\int_1^2 \int_{-x+2}^{2x-1} kxy^2 \, dy \, dx = \frac{k}{3}\int_1^2 x\left[y^3\right]_{-x+2}^{2x-1} dx$

$\displaystyle = 3k\int_1^2 (x^4 - 2x^3 + 2x^2 - x)\, dx$

$\displaystyle = 3k\left[\frac{1}{5}x^5 - \frac{1}{2}x^4 + \frac{2}{3}x^3 - \frac{1}{2}x^2\right]_1^2$

$\displaystyle = 3k\left(\left(\frac{32}{5} - 8 + \frac{16}{3} - 2\right) - \left(\frac{1}{5} - \frac{1}{2} + \frac{2}{3} - \frac{1}{2}\right)\right) = \frac{28}{5}k.$

21. $I_y = \left(\frac{673}{175}\sqrt{2} - \frac{1}{5}\ln(\sqrt{2}+1)\right)k,$

$I_x = \left(\frac{821}{2100}\sqrt{2} + \frac{1}{20}\ln(\sqrt{2}+1)\right)k$

23. $\displaystyle\int_{\sec\theta}^{2\cos\theta} kr^3\cos\theta \, dr = \frac{k}{4}\left(16\cos^5\theta - \sec^3\theta\right),$

$\displaystyle\int_{-\pi/4}^{\pi/4} \frac{k}{4}\left(16\cos^5\theta - \sec^3\theta\right) d\theta$

$\displaystyle = \frac{k}{60}(157\sqrt{2} + 15\ln(\sqrt{2}-1)$

25. $\frac{2}{3}k$, where k is the constant of proportionality

27. $I_x = \frac{2}{15}k, I_y = \frac{2}{5}k, R_x = \frac{\sqrt{5}}{5}, R_y = \frac{\sqrt{15}}{5}$, where k is the constant of proportionality

29. $\left(\frac{3}{4}, 0\right)$

31. $\left(\frac{5}{3}, 0\right)$

33. $\left(\frac{17}{10}, 0\right)$

35. $\frac{17}{6}k$, where k is the constant of proportionality

37. $I_x = \frac{49}{90}k, I_y = \frac{43}{5}k, R_x = 7\frac{\sqrt{255}}{255}, R_y = \frac{\sqrt{21930}}{85}$, where k is the constant of proportionality

39. $\frac{1}{2}kb^2h$, where k is the constant of proportionality

41. $I_y = \frac{1}{4}kb^4h, I_x = \frac{1}{6}kb^2h^3, R_y = \frac{\sqrt{2}}{2}b, R_x = \frac{\sqrt{3}}{3}h$, where k is the constant of proportionality

43. $\left(\frac{3}{4}b, \frac{1}{2}h\right)$

45. $(0,0)$

47. $(0,0)$

49. $\frac{2}{3}\pi k$, where k is the constant of proportionality

51. $I_x = I_y = \frac{1}{5}k\pi, I_0 = \frac{2}{5}k\pi; R_x = R_y = \frac{\sqrt{30}}{10}; R_0 = \frac{\sqrt{15}}{5}.$

53. $\frac{2}{3}k$, where k is the constant of proportionality

55. $I_y = \frac{4}{15}k, I_x = \frac{2}{15}k, R_y = \frac{\sqrt{10}}{5}, R_x = \frac{\sqrt{5}}{5}$, where k is the constant of proportionality

57. $\left(\frac{3}{2\pi}, 0\right)$

59. The masses of Ω_1 and Ω_2 may be treated as point masses located at $(\bar{x}_1, \bar{y}_1)$ and $(\bar{x}_2, \bar{y}_2)$, respectively. The center of mass of this two-point system is given by $\bar{x} = \frac{m_1\bar{x}_1 + m_2\bar{x}_2}{m_1 + m_2}$ and $\bar{y} = \frac{m_1\bar{y}_1 + m_2\bar{y}_2}{m_1 + m_2}.$

61. $\left(\frac{5}{6}, \frac{5}{6}\right)$

63. $\left(\frac{b_1^2 h_2 + b_2^2 h_1 - b_1 b_2 h_1}{2(b_1 h_2 + b_2 h_1 - b_1 h_1)}, \frac{b_1 h_2^2 + b_2 h_1^2 - b_1 h_1^2}{2(b_1 h_2 + b_2 h_1 - b_1 h_1)}\right)$

65. $\left(0, \frac{260}{57}\right)$

67. $\left(0, \frac{15}{14}a\right)$

69. Approximately 1.79

71. We assume that $0 < c < a$ and leave the case where $0 < a \le c$ to you. Using the result of Exercise 70 we have

$$\bar{x} = \frac{\displaystyle\int_0^c \int_0^{d/cx} x\, dy\, dx + \int_0^c \int_0^{d/c - a(x-a)} x\, dy\, dx}{ad/2} = \frac{a+c}{3}$$

and

$$\bar{x} = \frac{\displaystyle\int_0^c \int_0^{d/cx} y\, dy\, dx + \int_0^c \int_0^{d/c - a(x-a)} y\, dy\, dx}{ad/2} = \frac{d}{3}.$$

73. Use polar coordinates and the circle $r = a$, where a is a positive constant. The first moments are

$$M_x = \int_0^{2\pi} \int_0^a r^2\cos\theta\, dr\, d\theta = 0 \text{ and}$$

$$M_y = \int_0^{2\pi} \int_0^a r^2\sin\theta\, dr\, d\theta = 0. \text{ Therefore, the origin is}$$

the centroid of the circle.

Section 13.5

1. F, T, F, T, T, F, T, F.

3. The two summations contain the same summands, although the summands are rearranged. By the commutative and associative properties of addition, they are equal.

5. For each subsolid $\mathcal{R}_{i,j,k}$ choose

$(x_i^*, y_j^*, z_k^*) = \left(\frac{x_{i-1}+x_i}{2}, \frac{y_{j-1}+y_j}{2}, \frac{z_{k-1}+z_k}{2}\right).$

7. A triple integral is the limit of a Riemann sum over a rectangular solid $\mathcal{R}$ as the limit of the mesh of subdivisions of $\mathcal{R}$ goes to zero. An iterated triple integral is the result of three consecutive integrations in the order specified by the placement of the variables of integration.

9. $\displaystyle\int_a^b \int_{h_1(y)}^{h_2(y)} \int_{g_1(y,z)}^{g_2(y,z)} f(x,y,z)\, dx\, dz\, dy,$

$\displaystyle\int_a^b \int_{h_1(z)}^{h_2(z)} \int_{g_1(y,z)}^{g_2(y,z)} f(x,y,z)\, dx\, dy\, dz$

11. $\displaystyle\int_{-1}^{3}\int_{0}^{2}\int_{2}^{7}\rho(x,y,z)\,dz\,dy\,dx,$

$\displaystyle\int_{0}^{2}\int_{-1}^{3}\int_{2}^{7}\rho(x,y,z)\,dz\,dx\,dy,$

$\displaystyle\int_{-1}^{3}\int_{2}^{7}\int_{0}^{2}\rho(x,y,z)\,dy\,dz\,dx,$

$\displaystyle\int_{2}^{7}\int_{-1}^{3}\int_{0}^{2}\rho(x,y,z)\,dy\,dx\,dz,$

$\displaystyle\int_{0}^{2}\int_{2}^{7}\int_{-1}^{3}\rho(x,y,z)\,dx\,dz\,dy,$

$\displaystyle\int_{2}^{7}\int_{0}^{2}\int_{-1}^{3}\rho(x,y,z)\,dx\,dy\,dz$

13. $\displaystyle\int_{0}^{2}\int_{0}^{4-2x}\int_{0}^{3-(3/2)x-(3/4)y}\rho(x,y,z)\,dz\,dy\,dx,$

$\displaystyle\int_{0}^{4}\int_{0}^{2-(1/2)y}\int_{0}^{3-(3/2)x-(3/4)y}\rho(x,y,z)\,dz\,dx\,dy,$

$\displaystyle\int_{0}^{2}\int_{0}^{3-(3/2)x}\int_{0}^{4-2x-(4/3)z}\rho(x,y,z)\,dy\,dz\,dx,$

$\displaystyle\int_{0}^{3}\int_{0}^{2-(2/3)z}\int_{0}^{4-2x-(4/3)z}\rho(x,y,z)\,dy\,dx\,dz,$

$\displaystyle\int_{0}^{4}\int_{0}^{4-(4/3)z}\int_{0}^{2-(1/2)y-(2/3)z}\rho(x,y,z)\,dx\,dz\,dy,$

$\displaystyle\int_{0}^{3}\int_{0}^{3-(3/4)y}\int_{0}^{2-(1/2)y-(2/3)z}\rho(x,y,z)\,dx\,dy\,dz$

15. $\displaystyle\int_{0}^{-x-y+1}k(y^2+z^2)\,dz$

$\displaystyle = k\left(y^2(-x-y+1)+\frac{1}{3}(-x-y+1)^3\right),$

$\displaystyle\int_{0}^{-x+1}k\left(y^2(-x-y+1)+\frac{1}{3}(-x-y+1)^3\right)dy$

$\displaystyle = \frac{k}{6}(x-1)^4,\ \int_{0}^{1}\frac{k}{6}(x-1)^4\,dx=\frac{k}{30}$

17. $\displaystyle\int_{0}^{-x-y+1}k(x^2+y^2)\,dz=k(-x-+1)(x^2+y^2),$

$\displaystyle\int_{0}^{-x+1}k(-x-+1)(x^2+y^2)\,dy=$

$\displaystyle\frac{k}{12}(7x^4-12x^3+12x^2-4x+1),$

$\displaystyle\int_{0}^{1}\frac{k}{12}(7x^4-12x^3+12x^2-4x+1)\,dx=\frac{k}{30}$

19. The volume of Ω. The units are cubic centimeters.

21. The first moment M_{yz}. The units are grams·centimeters.

23. The moment of inertia about the y-axis, I_y. The units are grams·square centimeters.

25. 36

27. $\dfrac{145}{48}$

29. $\dfrac{1}{3}+\dfrac{\pi}{8}$

31. 162

33. 8

35. The rectangular solid given by
$\mathcal{R}=\{(x,y,z)\mid 2\le x\le 6, 0\le y\le 5 \text{ and } -2\le z\le 4\}.$

37. The "half-cube":

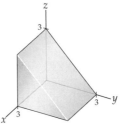

39. The tetrahedron with vertices $(0,0,0)$, $(2,0,0)$, $(0,3,0)$ and $(0,0,1)$.

41. A sphere with radius 3 centered at the origin.

43. The region bounded on the right by the paraboloid $x=9-y^2-z^2$ and bounded on the left by the yz-plane.

45. $\displaystyle\int_{-2}^{4}\int_{0}^{5}\int_{2}^{6}f(x,y,z)\,dy\,dx\,dz$

47. $\displaystyle\int_{0}^{3}\int_{0}^{3-z}\int_{0}^{3}f(x,y,z)\,dx\,dy\,dz$

49. $\displaystyle\int_{0}^{2}\int_{0}^{3-(3/2)x}\int_{0}^{1-(1/2)x-(1/3)y}f(x,y,z)\,dz\,dy\,dx$

51. $\displaystyle\int_{-3}^{3}\int_{-\sqrt{9-z^2}}^{\sqrt{9-z^2}}\int_{-\sqrt{9-x^2-z^2}}^{\sqrt{9-x^2-z^2}}f(x,y,z)\,dy\,dx\,dz$

53. $3k$, where k is the constant of proportionality

55. $\dfrac{31412}{315}k$, where k is the constant of proportionality

57. (a) The centroid of a rectangular box is its midpoint.
(b) For example, $\bar{x}=\dfrac{\int_{0}^{4}\int_{0}^{3}\int_{0}^{2}x\,dz\,dy\,dx}{\int_{0}^{4}\int_{0}^{3}\int_{0}^{2}dz\,dy\,dx}=2.$

59. (a) Since the density is proportional to the distance from the xy-plane, the location of the x- and y-coordinates will be the same as in Exercise 57. (b) $\bar{z}=\dfrac{4}{3}$

61. (a) The centroid of a rectangular box is its midpoint.
(b) For example, $\bar{x}=\dfrac{\int_{a_1}^{a_2}\int_{b_1}^{b_2}\int_{c_1}^{c_2}x\,dz\,dy\,dx}{\int_{a_1}^{a_2}\int_{b_1}^{b_2}\int_{c_1}^{c_2}dz\,dy\,dx}=\dfrac{a_1+a_2}{2}.$

63. (a) Since the density is proportional to the distance from the yz-plane, the location of the y- and z-coordinates will be the same as in Exercise 61. (b) $\bar{x}=\dfrac{2(a_2^3-a_1^3)}{3(a_2^2-a_1^2)}$

65. (a) $\bar{x}=\dfrac{a}{4}$. (b) The y-coordinate of the center of mass will also be one-fourth of the distance from the xz-plane to the point $(0,b,0)$, i.e., $\bar{y}=\dfrac{b}{4}$. Similarly, $\bar{z}=\dfrac{c}{4}$.

67. The mass of the tetrahedron is

$m = \int_0^a \int_0^{b-(b/a)x} \int_0^{c(1-(x/a)-(y/b))} ky\, dz\, dy\, dx$, where k is a constant of proportionality. The center of mass is given by $(\bar{x}, \bar{y}, \bar{z})$, where

$\bar{x} = \frac{1}{m} \int_0^a \int_0^{b-(b/a)x} \int_0^{c(1-(x/a)-(y/b))} kxy\, dz\, dy\, dx,$

$\bar{y} = \frac{1}{m} \int_0^a \int_0^{b-(b/a)x} \int_0^{c(1-(x/a)-(y/b))} ky^2\, dz\, dy\, dx,$ and

$\bar{z} = \frac{1}{m} \int_0^a \int_0^{b-(b/a)x} \int_0^{c(1-(x/a)-(y/b))} kyz\, dz\, dy\, dx.$

69. By Fubini's Theorem $\iiint_R \alpha(x)\beta(y)\gamma(z)\, dV$

$= \int_{a_1}^{a_2} \int_{b_1}^{b_2} \int_{c_1}^{c_2} \alpha(x)\beta(y)\gamma(z)dz\, dy\, dx$

$= \int_{a_1}^{a_2} \alpha(x) \int_{b_1}^{b_2} \beta(y) \int_{c_1}^{c_2} \gamma(z)dz\, dy\, dx$

$= \left(\int_{a_1}^{a_2} \alpha(x)\, dx \right) \left(\int_{b_1}^{b_2} \beta(y)\, dy \right) \left(\int_{c_1}^{c_2} \gamma(z)dz \right).$

71. Use the result of Exercise 69. The product $(a_2 - a_1)(b_2 - b_1)(c_2 - c_1)$ is the volume of rectangular solid R.

73. Using the notation of Definition 13.15, the equality holds since $\iiint_R (f(x, y, z) + g(x, y, z))\, dV$

$= \lim_{\Delta \to 0} \sum_{i=1}^{l} \sum_{j=1}^{m} \sum_{k=1}^{n} \left(f\left(x_i^*, y_j^*.z_k^*\right) + g\left(x_i^*, y_j^*.z_k^*\right) \right) \Delta V$

$= \lim_{\Delta \to 0} \sum_{i=1}^{l} \sum_{j=1}^{m} \sum_{k=1}^{n} f(x_i^*, y_j^*.z_k^*) \Delta V +$

$\lim_{\Delta \to 0} \sum_{i=1}^{l} \sum_{j=1}^{m} \sum_{k=1}^{n} g(x_i^*, y_j^*.z_k^*) \Delta V$

$= \iiint_R f(x, y, z)\, dV + \iiint_R g(x, y, z)\, dV.$

Section 13.6

1. T, T, F, T, T, T, F, F.

3. $x = x_0$ is a plane parallel to the yz-plane, $y = y_0$ is a plane parallel to the xz-plane and $z = z_0$ is a plane parallel to the xy-plane.

5. $\rho = \rho_0$ is a sphere centered at the origin, $\theta = \theta_0$ is a vertical half-plane starting at the z-axis, and $\phi = \phi_0$ is a cone whose axis of symmetry is the z-axis.

7. $r = \sqrt{x^2 + y^2}$, $\theta = \tan^{-1}\left(\frac{y}{x}\right)$, $z = z$.

9. $\rho = \sqrt{x^2 + y^2 + z^2}$, $\theta = \tan^{-1}\left(\frac{y}{x}\right)$,

$\phi = \cos^{-1}\left(\frac{z}{\sqrt{x^2+y^2+z^2}}\right).$

11. $\rho = \sqrt{r^2 + z^2}$, $\theta = \theta$, $\phi = \tan^{-1}\left(\frac{r}{z}\right)$.

13. $dx\, dy\, dz$, $dx\, dz\, dy$, $dy\, dx\, dz$, $dy\, dz\, dx$, $dz\, dx\, dy$, $dz\, dy\, dx$. You make your choice based on the shape of the region of integration.

15. $\rho^2 \sin\phi\, d\rho\, d\theta\, d\phi$. Since ρ is usually expressed as a function of θ and ϕ, this usually provides the simplest order of integration.

17. The region of integration Ω usually determines the coordinate system you choose. You usually start with the coordinate system that provides the simplest expression for Ω.

19. $\int_0^{\pi/2} \int_0^{\pi/2} \int_0^1 k\rho^2 \sin\phi\, d\rho\, d\theta\, d\phi$

$= \frac{k}{3} \int_0^{\pi/2} \int_0^{\pi/2} \sin\phi\, d\theta\, d\phi = \frac{\pi k}{6} \int_0^{\pi/2} \sin\phi\, d\phi = \frac{\pi k}{6}.$

21. $M_{xz} = \int_0^{\pi/2} \int_0^{\pi/2} \int_0^1 k\rho^3 \sin^2\phi \sin\theta\, d\rho\, d\theta\, d\phi$

23. Cylindrical $(1, 0, 0)$, spherical $\left(1, 0, \frac{\pi}{2}\right)$

25. Rectangular $(2\sqrt{3}, 6, 4)$, spherical $\left(8, \frac{\pi}{3}, \frac{\pi}{3}\right)$

27. Rectangular $(0, 0, -8)$, cylindrical $(0, 0, -8)$

29. A plane parallel to the yz-plane, cylindrical $r = 4\sec\theta$, spherical $\rho = 4\csc\phi\sec\theta$.

31. A right circular cylinder centered on the z-axis, rectangular $x^2 + y^2 = 4$, spherical $\rho = 2\csc\phi$.

33. The half of the yz-plane in which $y \geq 0$.

35. A sphere centered at the origin with radius 2, rectangular $x^2 + y^2 + z^2 = 4$, cylindrical $r^2 + z^2 = 4$.

37. The xy-plane, rectangular $z = 0$, cylindrical $z = 0$.

39. The region bounded below by the $r\theta$-plane and bounded above by the cone $z = r$ on the circle with radius 3 centered at the origin.

41. The region inside the right circular cylinder with equation $r = 2\sin\theta$, bounded below by the $r\theta$-plane and bounded above by the sphere with radius 4 centered at the origin.

43. The hemisphere below the xy-plane of the sphere with radius 2 centered at the origin.

45. The region inside the upward opening right circular cone that has the z-axis as its axis of symmetry, height 3, and base with radius 3.

47. $\frac{48\pi - 64}{9}$

49. $2\sqrt{3}\pi$

51. $\frac{\pi}{32}$

53. $\frac{16}{3}\pi$

55. $\frac{2}{3}\pi R^3 (1 - \cos\alpha)$. If $\alpha = 0$ the solid is "empty," so the volume should be zero. If $\alpha = \pi$, the solid is a sphere, so the volume should be $\frac{4}{3}\pi R^3$.

57. $\frac{1920\pi - 3328}{225}k$, where k is the constant of proportionality.

59. $(x, y, z) = \left(0, 0, \frac{94}{15\sqrt{3}+40\pi}\right)$.

61. $\frac{7\pi}{1024}k$, where k is the constant of proportionality.

63. $\frac{16\pi}{3}k$, where k is the constant of proportionality.

65. $x = y = 0, z = \frac{3}{8}R(\cos\alpha + 1)$.

67. Approximately 31800 cubic feet

69. Since $x = r\cos\theta = a$, when we solve for r we have
$r = a\sec\theta$.

71. Since $x = \rho\sin\phi\cos\theta = a$, when we solve for ρ we have $\rho = a\csc\phi\sec\theta$.

73. $\displaystyle\int_0^{2\pi}\int_0^R\int_{hr/R}^h r\,dz\,dr\,d\theta = \int_0^{2\pi}\int_0^R\left(hr - \frac{h}{R}r^2\right)dr\,d\theta$

$\displaystyle = \int_0^{2\pi}\frac{1}{6}R^2h\,d\theta = \frac{1}{3}\pi R^2 h.$

75. $\displaystyle\int_0^{\tan^{-1}(R/h)}\int_0^{2\pi}\int_0^{h\sec\phi}\rho^2\sin\phi\,d\rho\,d\theta\,d\phi$

$\displaystyle = \int_0^{\tan^{-1}(R/h)}\int_0^{2\pi}\frac{1}{3}h^3\frac{\sin\phi}{\cos^3\phi}\,d\theta\,d\phi$

$\displaystyle = \int_0^{\tan^{-1}(R/h)}\frac{2}{3}\pi h^3\frac{\sin\phi}{\cos^3\phi}\,d\phi = \frac{1}{3}\pi R^2 h.$

Section 13.7

1. T, T, T, T, T, F, T, F.

3. It is a scaling factor. For example, in the simplest case when the Jacobian is a constant, k, the area of the domain set Ω is k times the area of the target set Ω'.

5. $\displaystyle\int_0^1\int_{-y+1}^{y+1}x^2y\,dx\,dy + \int_1^2\int_{y-1}^{y+1}x^2y\,dx\,dy +$
$\displaystyle\int_2^3\int_{y-1}^{5-y}x^2y\,dx\,dy = \frac{4}{5} + \frac{17}{2} + \frac{97}{10} = 19.$

7. $x = \dfrac{u+v}{2}$, $y = \dfrac{u-v}{2}$, $\dfrac{\partial(x,y)}{\partial(u,v)} = -\dfrac{1}{2}$

9. $\displaystyle\frac{1}{16}\int_1^5\int_{-1}^1(u+v)^2(u-v)\,dv\,du = 19$

11. It translates the region a units horizontally and b units vertically. The rectangle is translated to the rectangle in the uv-plane where $3 \le u \le 4$ and $4 \le v \le 6$.

13. It would not be particularly useful because it only translates the region and would not provide a simpler expression for the limits of integration.

15. It scales the region by a factor of a units horizontally and b units vertically. The rectangle is scaled to the rectangle in the uv-plane where $0 \le u \le 3$ and $0 \le v \le \frac{1}{2}$.

17. If $a < 0$, in addition to a scaling factor, there is a reflection about the vertical axis, and if $b < 0$ there is a reflection about the horizontal axis. The sign of the Jacobian changes if either a or b is negative, but it does not change if both are negative.

19. $x = (\sin\theta)u + (\cos\theta)v$ and $y = -(\cos\theta)u + (\sin\theta)v$,
$\dfrac{\partial(x,y)}{\partial(u,v)} = 1$. The transformation rotates points about the origin. It does not change the area of the region, therefore the Jacobian should be, and is, 1.

21. $\displaystyle\iint_\Omega\frac{1}{(x+y)^2}\,dA =$
$\displaystyle\int_0^1\int_{1-x}^{4-x}\frac{1}{(x+y)^2}\,dy\,dx + \int_1^4\int_0^{4-x}\frac{1}{(x+y)^2}\,dy\,dx = \ln 4$

23. Area$(\Omega) = \dfrac{15}{2}$ square units and Area$(\Omega') = 15$ square units, Area$(\Omega) = \dfrac{\partial(x,y)}{\partial(u,v)}\cdot$Area$(\Omega')$

25. The quotient $\dfrac{u}{v} = xy\cdot\dfrac{x}{y} = x^2$. Therefore, $x = \sqrt{\dfrac{u}{v}}$. Similarly the product $uv = xy\cdot\dfrac{y}{x} = y^2$. Therefore,

$y = \sqrt{uv}$. So, $\dfrac{\partial(x,y)}{\partial(u,v)} = \begin{bmatrix} \dfrac{1}{2\sqrt{uv}} & \dfrac{1}{2}\sqrt{\dfrac{v}{u}} \\ -\dfrac{1}{2}\sqrt{\dfrac{u}{v^3}} & \dfrac{1}{2}\sqrt{\dfrac{u}{v}} \end{bmatrix} = \dfrac{1}{2v}.$

27. The rectangle with vertices $(0,0)$, $(2,2)$, $(1,3)$ and $(-1,1)$; $\dfrac{\partial(x,y)}{\partial(u,v)} = 2$.

29. The region in the first quadrant bounded by the hyperbolas with equations $y = \dfrac{1}{x}$ and $y = \dfrac{4}{x}$ and lines $y = x$ and $y = 9x$, $\dfrac{\partial(x,y)}{\partial(u,v)} = \dfrac{2u}{v}$.

31. The right half of the circle with radius two centered at the origin, $\dfrac{\partial(x,y)}{\partial(u,v)} = -u$.

33.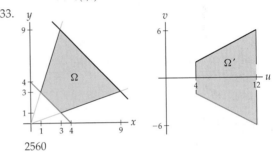

2560

35.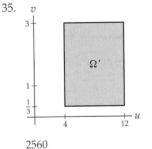

2560

37.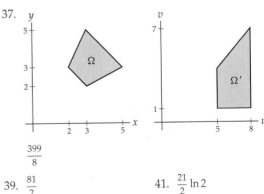

$\dfrac{399}{8}$

39. $\dfrac{81}{2}$

41. $\dfrac{21}{2}\ln 2$

43. $\dfrac{160}{3} + 6552\ln 3$

45. $\dfrac{15}{8}$

47. $\frac{5}{64}$

49. 1100 cubic feet.

51. $\dfrac{\partial(x,y,z)}{\partial(r,\theta,z)} = \det \begin{bmatrix} \cos\theta & \sin\theta & 0 \\ -r\sin\theta & r\cos\theta & 0 \\ 0 & 0 & 1 \end{bmatrix} = r.$

53. Using the transformation in the equation $ax + by = c$ we obtain $(a\alpha + b\gamma)u + (a\beta + b\delta)v = c$, which is a linear equation, if the coefficients of u and v are not both zero. This is the case when the Jacobian is nonzero.

55. $u = \dfrac{1}{\alpha\delta - \beta\gamma}(\delta x - \beta y)$ and $v = \dfrac{1}{\alpha\delta - \beta\gamma}(-\gamma x + \alpha y)$

57. Use the transformation given by $x = au$ and $y = bv$. You will transform the equation to the equation of the unit circle centered at the origin.

59. Show that $\det \begin{bmatrix} \frac{\partial x}{\partial u} & \frac{\partial y}{\partial u} \\ \frac{\partial x}{\partial v} & \frac{\partial y}{\partial v} \end{bmatrix} \det \begin{bmatrix} \frac{\partial u}{\partial s} & \frac{\partial v}{\partial s} \\ \frac{\partial u}{\partial t} & \frac{\partial v}{\partial t} \end{bmatrix} =$

$\det \begin{bmatrix} \frac{\partial x}{\partial u}\frac{\partial u}{\partial s} + \frac{\partial x}{\partial v}\frac{\partial v}{\partial s} & \frac{\partial y}{\partial u}\frac{\partial u}{\partial s} + \frac{\partial y}{\partial v}\frac{\partial v}{\partial s} \\ \frac{\partial x}{\partial u}\frac{\partial u}{\partial t} + \frac{\partial x}{\partial v}\frac{\partial v}{\partial t} & \frac{\partial y}{\partial u}\frac{\partial u}{\partial t} + \frac{\partial y}{\partial v}\frac{\partial v}{\partial t} \end{bmatrix}.$

For answers to odd-numbered Skill Certification exercises in the Chapter Review, please visit the Book Companion Web Site at www.whfreeman.com/tkcalculus.

Chapter 14

Section 14.1

1. F, T, F, T, T, F, F, T

3. A vector field in the Cartesian plane, or $\mathbb{R}^2$, has domain $D \subseteq \mathbb{R}^2$.

5. Vectors in the Cartesian plane.

7. A vector field $\mathbf{F}$ is conservative if it is the gradient of some function f.

9. For every point $(x, y) \in \mathbb{R}^2$, $\mathbf{F}(x, y)$ and $\mathbf{G}(x, y)$ are parallel; only for $(0, 0)$ are they the same vector.

11. For every point $(x, y) \in \mathbb{R}^2$, $\mathbf{F}(x, y) = -\mathbf{G}(x, y)$; at $(0, 0)$ they the same vector.

13. $\mathbf{G}(x, y, z) = \langle 2yz, 2xz, 2xy \rangle$

15. Let $\mathbf{F}(x, y) = \langle f_1(x, y), f_2(x, y) \rangle$. If $f_{1,y} \neq f_{2,x}$, then $\mathbf{F}$ is not conservative.

17. $f(x, y) = x^3 \cos y$

19. $g(x, y) = x^5 + xy - 3y^4$

21. $f(x, y, z) = xyz$

23. $g(x, y, z) = x \cos y + y \sin z$

25.

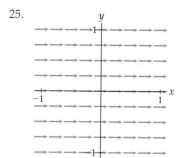

27.

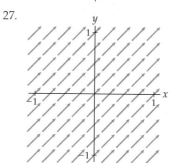

29.

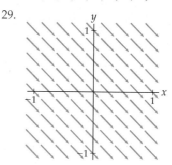

31.

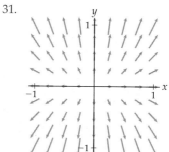

33. $\dfrac{\partial}{\partial y}(xy) = y \neq \dfrac{\partial}{\partial x}(-y) = 0$

35. $\dfrac{\partial}{\partial y}\left(\dfrac{1}{x^2 + y}\right) = -(x^2 + y)^{-2} \neq \dfrac{\partial}{\partial x}\left(\dfrac{y}{x}\right) = \dfrac{-y}{x^2}$

37. $\dfrac{\partial}{\partial z}(-z) = -1 \neq \dfrac{\partial}{\partial y}(e^{yz}) = ze^{yz}$

39. $\dfrac{\partial}{\partial z}(yz) = y \neq \dfrac{\partial}{\partial y}(z + 12)$

41. Not conservative

43. $g(x, y) = x^2 + \sin(xy) - y$

45. $f(x, y, z) = xye^{2z} + x$

47. Not conservative

49. (a) Since the fluid is moving at a constant rate, $\mathbf{F}(x, y)$ should have a constant magnitude. Since the motion is horizontal only, there is no $\mathbf{j}$ component to the field. Since the motion is in the negative $\mathbf{i}$ direction, a reasonable field is $\mathbf{F}(x, y) = -\alpha\mathbf{i}$, for any positive constant α. (b) $\mathbf{F}(x, y) = \langle -\alpha, -\alpha \rangle$, for any real number $\alpha > 0$. (c) $\mathbf{G}(x, y) = -y\mathbf{i} + x\mathbf{j}$

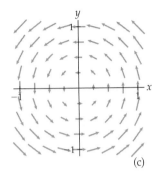

(c)

51. (b) It is conservative. Potential functions have the form $Q(x, y) = 0.9x - \dfrac{x^3}{3} + \dfrac{1}{2}xy^2 + k$. This function represents, in some sense, the amount of water that needs to get through a passage, or if you prefer, the pressure of the water that must pass through. (c) There are points of land due north and due south of this channel, forcing the water through a relatively narrower passage.

53. Suppose that $f(x, y)$ and $g(x, y)$ satisfy $\nabla f(x, y) = \nabla g(x, y)$. Then

$$f(x, y) = \int f_x(x, y)\, dx + B_f + \alpha_f$$

and $g(x, y) = \int g_x(x, y)\, dx + B_g + \alpha_g$, where B_f is the integral with respect to y of all the terms in $f_y(x, y)$ that do not involve an x term, B_g is the integral with respect to y of all the terms in $g_y(x, y)$ that do not involve an x term, and α_f, α_g are arbitrary constants. Since $g_y(x, y) = f_y(x, y)$, $B_f = B_g$, and $f(x, y) = g(x, y) + \alpha$.

Section 14.2

1. T, F, T, T, F, T, F, F

3. $\displaystyle\int_a^b e^{t^7}\sqrt{1 + 4t^2 + 16t^6}\,dt$

5. $\mathbf{F}(x, y) = (x + y)\mathbf{i} + xy\mathbf{j}$

7. $\mathbf{F}(x, y, z) = \langle xy^2, xy - z, \cos y \rangle$

9. $\displaystyle\int_C \mathbf{F}(x, y) \cdot d\mathbf{r} = \int_C (-y)\,dx + x\,dy$

11. $ds = \|\mathbf{r}'(t)\|\,dt$

13. $\displaystyle\int_a^b (4t^2 e^t + 2te^t - e^{2t})\,dt$

15. $\mathbf{r}(t) = \cos t\mathbf{i} + \sin t\mathbf{j}$, $0 \le t \le 2\pi$

17. $\mathbf{r}(t) = \langle 2\cos t, t, 2\sin t \rangle$, $0 \le t \le 4\pi$

19. $f_{avg}(C) = \dfrac{\int_C f(x, y, z)\,ds}{\int_C ds}$

21. $\dfrac{1}{12}(17^{3/2} - 1)$

23. $\mathbf{r}(t) = \langle \cos t, \sin t \rangle$, $0 \le t \le 2\pi$; $\|\mathbf{r}(t)\| = 1$, $\int_C f(x, y)\,ds = 2\pi$.

25. $\mathbf{r}(t) = \langle t, 2t, 3t \rangle$, $0 \le t \le 1$. Then
$$\int_C f\,ds = \sqrt{14}\int_0^1 e^{6t}\,dt = \frac{\sqrt{14}(e^6 - 1)}{6}.$$

27. $\sqrt{2}e\,(e^\pi - 1)$

29. $\mathbf{r}(t) = \langle t, t^2 + 1 \rangle$, $\mathbf{r}'(t) = \langle 1, 2t \rangle$,
$\displaystyle\int_C \mathbf{F}(x, y) \cdot d\mathbf{r} = \int_5^{10}(1 - 2t)\,dt = -70$

31. 70

33. 0 since $\mathbf{F}(x, y)$ is conservative and C is closed.

35. π 37. $\ln \pi - \pi - 1$

39. $\mathbf{F}$ is not conservative: $\dfrac{\partial f_1}{\partial z} = -1 \ne 1 = \dfrac{\partial f_3}{\partial x}$

41. $3^{32} - 1$, $\mathbf{F}(x, y, z) = \nabla(z^{xy})$; the integral is path independent.

43. $\mathbf{F}$ is not conservative: $\dfrac{\partial f_2}{\partial z} = 0 \ne 1 = \dfrac{\partial f_3}{\partial y}$

45. $\dfrac{7}{3}\pi^3$

47. $\|\mathbf{r}'(t)\| = \|\langle \sin t + t\cos t, \cos t - t\sin t \rangle\| = \sqrt{1 + t^2}$,
$\int_C f(x, y)\,ds = \dfrac{1}{3}((16\pi)^{3/2} - 1)$

49. $\mathbf{F}(x, y, z) = \nabla f(x, y, z)$, where $f(x, y, z) = e^{xyz} + 2x - y$. The intersection of the surfaces is an ellipse, in particular a closed curve. So the integral is zero.

51. 4π 53. 112

55. -2π 57. $\dfrac{1}{3}(\ln 2)^3 + \ln 2 - \dfrac{31}{24}e^{-3}$

59. $\dfrac{\pi^2}{12}$ 61. $2\pi e$

63. (a) No; (b) 0; (c) It is a measure of how much the current helps or hurts Annie's progress.

65. Suppose that $\mathbf{F}(x, y, z)$ is a conservative vector field. Then $\mathbf{F}(x, y, z) = \nabla f(x, y, z)$ for some $f(x, y, z)$. Let $\mathbf{r}(t)$, $a \le t \le b$ be a smooth parametrization of C. By the Fundamental Theorem of Line Integrals,

$$\int_C \mathbf{F}(x, y, z) \cdot d\mathbf{r}$$

$= f\big(x(b), y(b), z(b)\big) - f\big(x(a), y(a), z(a)\big)$, which depends only on the endpoints of C and not on C itself.

Section 14.3

1. F, T, T, T, F, T, F, T.

3. Surface integrals of multivariate functions only need to account for the distortion of the surface's area compared to a comparable region in the parameter plane (xy or uv); surface integrals of vector fields need to account for the action of $\mathbf{F}(x, y, z)$ through the surface, that is, in the normal direction.

5. $dS = \sqrt{\dfrac{a^2 + b^2 + c^2}{c^2}}$

7. $\mathbf{r}(x, y) = \langle x, y, \sqrt{1 - x^2 - y^2} \rangle$.

9. $\mathbf{r}(u, v) = \langle k \cos u \cos v, k \sin u \cos v, k \sin v \rangle$, $0 \le u \le 2\pi$, $0 \le v \le \pi$.

11. $\mathbf{n} = -f_x \mathbf{i} - f_y \mathbf{j} + \mathbf{k}$ 13. $\mathbf{n} = 2t\mathbf{i} - \mathbf{j}$

15. $f_{avg}(S) = \dfrac{\int_S f(x, y, z) dS}{\int_S 1 dS}$

17. $\|\mathbf{r}_u \times \mathbf{r}_v\|$ is the area of the parallelogram with sides $\mathbf{r}_u$ and $\mathbf{r}_v$; the area of this parallelogram approximates the area of the portion of the tangent plane used to approximate the area of $\mathcal{S}$; in the limit, the analogy is exact.

19. $\mathbf{n} = \mathbf{i} - f_y \mathbf{j} - f_z \mathbf{k}$ 21. $12\sqrt{2}$

23. $\dfrac{19\pi}{6}$ 25. $2\sqrt{6}$

27. $\pi - \dfrac{\pi}{e}$ 29. 0

31. $4e^2 + 14$ 33. $\dfrac{20}{21}\sqrt{3}$

35. $-\dfrac{12}{5}$ 37. 240π

39. $\mathbf{n} = \langle 2, -8, -10 \rangle$; $\int_S \mathbf{F} \cdot \mathbf{n}\, dS = -2 \int_\pi^\pi \int_0^1 4z \cos yz + 5z \sin yz\, dA = -20\pi$.

41. 0

43. $\int_S 1 dS = \int_0^3 \int_0^{2x} 2\sqrt{x^2 + 1}\, dy\, dx = \dfrac{4}{3}(10^{3/2} - 1)$.

45. $\dfrac{61}{324} + \dfrac{2^{-2/3}}{\ln 2} - \dfrac{1}{2\ln 2}$ 47. -832π

49. $\dfrac{k\pi}{6}(17^{3/2} - 1)$ 51. 3π

53. $\dfrac{596(37^{3/2} + 48)}{5(37^{3/2} - 5)}$ 55. 0.0643 square miles

57. Reversing the orientation of $\mathcal{S}$ means substituting $-\mathbf{n}$ for $\mathbf{n}$. Factoring -1 out of the integral gives the desired result.

59. In both cases, the surface area is $\int 1 dS$ with limits determined by R. Computing dS for each surface, we see that they are equal. $(dS = \sqrt{4x^2 + 4y^2 + 1})$

Section 14.4

1. F, F, T, T, T, T, F, F.

3. The terms in the integrand are the mixed partial derivatives of the component functions of the vector field.

5. Work done by $\mathbf{F}$ moving a particle along C, flux perpendicular to C.

7. Setting $G(x, y) = \dfrac{\partial f_2}{\partial x} - \dfrac{\partial f_1}{\partial y}$, allows the use of Green's Theorem to evaluate the integral.

9. At the point (x, y, z), div $\mathbf{F}$ will be negative.

11. At the point (x, y, z), div $\mathbf{F}$ will be positive.

13. $-\iint_R (\sin x \sin y + x^2 e^y) dA$, where R is the unit circle

15. $-\sin x \sin y - x^2 e^y$ 17. $2xy + x \sin y$

19. 0

21. $\dfrac{y}{\sqrt{1 - (xy)^2}} + \dfrac{1}{y + z} - \dfrac{5}{(2x + 3y + 5z + 1)^2}$

23. $-\left(\dfrac{3}{(2x + 3y + 5z + 1)^2} + \dfrac{1}{y + z}\right)\mathbf{i} + \left(\dfrac{2}{(2x + 3y + 5z + 1)^2}\right)\mathbf{j} - \dfrac{x}{\sqrt{1 - x^2 y^2}}\mathbf{k}$

25. $\langle xze^{xy} - xye^{xz}, xye^{yz} - yze^{xy}, yze^{xz} - xze^{yz} \rangle$

27. $(\cos(x - y) + \sin(x + y))\mathbf{k}$

29. Using Green's Theorem we have $\int_0^{2\pi} \int_0^1 4r^3\, dr\, d\theta = 2\pi$.

31. $-\dfrac{32}{3}$ 33. $\dfrac{1}{4}(e^2 - 2e + 3)$

35. 0. Note that $\mathbf{F}$ is not conservative, so this is not the result of applying the Fundamental Theorem of Line Integrals.

37. $\dfrac{1}{2}$ 39. 16

41. $4 - 2e + (e^2 - 1)(\ln 2 - 1)$ 43. $\dfrac{1}{\ln 2} - \dfrac{1}{e + 1}$

45. 0

47. Work is given by $\int_C \mathbf{F} \cdot d\mathbf{r}$. Using Green's Theorem to evaluate the integral, we have $W = -2\pi$.

49. $\mathbf{F}$ is conservative, so $W = 0$.

51. (a) 0; (c) They tell us that there is no creation of water inside the passage. The water flowing in and out is balanced.

53. (a) Since speed increases as we move left among the lanes, we can say that $\dfrac{\partial v_2}{\partial x} < 0$, while the fact that $v_1 \equiv 0$ means that $\dfrac{\partial v_1}{\partial y} = 0$. This means that the curl is always negative – the traffic tends to have a rotation to it. If the curl is low, it means all lanes are running nearly the same speed, but if it is high, then the variation in speeds among the lanes is high. Generally, low curl is safer. (b) The curl in this case is $-\alpha$, so the area integral on the right is approximately -0.00568α. The curve integral on the left divides into four segments. Since $d\mathbf{r}$ goes in opposite directions on the top and bottom and the speeds are the same there, those segments cancel. The integral on the left segment is $\int_{0.5}^0 75\, dy$, while that on the right is $\int_0^{0.5}(75 - 0.0113\alpha)\, dy$. Thus the integral on the left is $\int_0^{0.5} -0.0113\alpha$, which ends up the same as the other side. (c) These describe the average curl for the traffic flow on this stretch of road, i.e., the circulation of traffic on this stretch. The higher the number, the less safe the road. The curve integral is what engineers would be able to compute using road tubes they put down to measure the number and speed of passing cars. In other words, they could put road tubes along $y = 0$ and $y = 0.5$ to compute the curve integral.

55. Directly computing div curl $\mathbf{F}$, we find that div curl $\mathbf{F}$ is

$$\dfrac{\partial^2 F_3}{\partial x \partial y} - \dfrac{\partial^2 F_2}{\partial x \partial z} + \dfrac{\partial^2 F_1}{\partial y \partial z} - \dfrac{\partial^2 F_3}{\partial y \partial x} + \dfrac{\partial^2 F_2}{\partial z \partial x} - \dfrac{\partial^2 F_1}{\partial z \partial y} = 0.$$

57. If **F** is a conservative vector field, then by the Fundamental Theorem of Line Integrals, $\int_C \mathbf{F} \cdot d\mathbf{r} = 0$ for any closed simple smooth or piecewise-smooth curve. By Green's Theorem, we have

$$\iint_R \left(\frac{\partial F_2}{\partial x} - \frac{\partial F_1}{\partial y} \right) dA = \int_C \mathbf{F} \cdot d\mathbf{r} = 0$$

Section 14.5

1. F, T, F, F, T, T, T, T.

3. $\left\langle -\frac{\partial g}{\partial x}, -\frac{\partial g}{\partial y}, 1 \right\rangle$ and $\left\langle \frac{\partial g}{\partial x}, \frac{\partial g}{\partial y}, -1 \right\rangle$

5. By Stokes' Theorem, both integrals are equivalent to $\int_C \mathbf{F}(x, y, z) \cdot d\mathbf{r}$. So they are equal to each other.

7. $SA = \int_C \mathbf{F}(x, y, z) \cdot d\mathbf{r}$

9. The orientation of S determines the direction of travel along C; in Stokes' Theorem, C is traversed counterclockwise with respect to n. If the orientation is reversed, the integrals will change sign.

11. Any conservative vector field will have $\int_C \mathbf{F} \cdot d\mathbf{r} = 0$, so any conservative vector field is an example of this.

13. Both notations refer to area. In the case of Green's Theorem, dA refers to ordinary area. In the case of Stokes' Theorem, dS keeps track of any distortion to the area of the surface S caused by transforming a patch of the uv-plane to parametrize S.

15. Answers will vary. When the line integral has a piecewise-continuous boundary that requires the evaluation of several smooth pieces, but by using the right-hand side of Stokes' Theorem, we can obtain the same result with a single area integral.

17. Answers will vary. If a surface for an intended application of Stokes' Theorem is not smooth or piecewise smooth, there is no guarantee that the incremental application of Green's Theorem in the tangent planes is accurate, or even well defined.

19. $\frac{35}{12}$ 21. 0

23. $\frac{1}{12}$ 25. $\frac{16\pi}{81}$

27. $\frac{320}{3}$.   29. 8π

31. 36π

33. **F** is not defined on the z-axis.

35. S is not smooth; in particular, **n** is not well-defined at the cone point, $(0, 0, 0)$.

37. This vector field is conservative, so $\int_C \mathbf{F} \cdot d\mathbf{r} = 0$.

39. curl **F** = **0**

41. curl **F** = **0**   43. -60π

45. (a) 0; (b) $r = r(\theta) = \langle x(\theta), y(\theta) \rangle = \langle \cos\theta, \sin\theta \rangle$. Thus on ∂R we have $\mathbf{F} = \langle 0, 1.152 - 0.8\cos^2\theta \rangle$ and
$$d\mathbf{r} = \frac{d}{d\theta}\langle \cos\theta, \sin\theta \rangle d\theta = \langle -\sin\theta, \cos\theta \rangle d\theta,$$ hence
$\int_{\partial R} \mathbf{F} d\mathbf{r} = \int_0^{2\pi} (1.152 - 0.9\cos^2\theta)\cos\theta\, d\theta = 0$.; (c) They tell us that there is no net circulation of water in this zone.

47. Using Stokes' Theorem to compute the integral gives $\mathbf{n} = \pm\mathbf{k}$, depending on the direction in which C is traversed. The **k** component of curl **F** does not depend on F_3, so $\int_C \mathbf{F} \cdot d\mathbf{r}$ does not depend on F_3.

49. Since $\mathbf{n} = \pm\langle a, b, c \rangle$ on all of the surface, and the partial derivatives of F_1, F_2, F_3 are all constants, curl $\mathbf{F} \cdot \mathbf{n}\, dS$ will be some constant r. Then $\int_C \mathbf{F} \cdot d\mathbf{r} = r \iint_S dA$, which depends only on area.

51. This is a direct consequence of Stokes' Theorem. By the theorem, both sides of the equation are equal to $\int_C \mathbf{F}(x, y, z) \cdot d\mathbf{r}$.

Section 14.6

1. T, T, F, T, F, F, T, T.

3. No. Such surfaces have no interior (or infinite interior).

5. On the surface of the paraboloid, there is always a well-defined choice of normal vector; intuitively speaking, the graph has no sharp corners. The rectangular solid, however, is (piecewise) comprised of smooth planes, but the intersections of these planes form sharp corners on the surface, which do not have well-defined normal vectors.

7. (a) positive (b) zero   9. $f_{xx} + f_{yy} + f_{zz}$

11. $G(x, y, z) = 3\mathbf{i} - 4\mathbf{j} + 57\mathbf{k}$.   13. $\langle x + yz, y + 2xz, z + xy \rangle$

15. Not possible: S is neither smooth nor piecewise-smooth.

17. Not possible: $\mathbf{F}(x, y, z)$ is not defined throughout the region enclosed by S.

19. Possible: this integral satisfies the conditions of the Divergence Theorem.

21. div **F** = 3   23. div $\mathbf{F} = 3e^{xyz} + 3xyze^{xyz}$

25. $3\pi^7$ 27. 0

29. 12π 31. 0

33. $\frac{21\pi}{2}$ 35. 675π

37. $36\pi e$ 39. $\frac{8}{3}$

41. $18(e^5 + \ln 5 - e) + 4320$ 43. 16

45. (a) 0; (b) $\int_{\partial R} F \cdot n\, dS = \int_0^{2\pi} (1.152 - 0.9\cos^2\theta)\sin\theta\, d\theta = 0$; (c) They tell us that the same amount of water flows into the region as flows out. The fact that the divergence vanishes everywhere in the region shows that there are no sources or sinks of water in the region.

47. (a) No traffic can leave the roadway to either side, so this means that there is a net flux of cars into this stretch of road: traffic is slowing down. This condition could not persist indefinitely, since the road will fill up. (b) By the Divergence Theorem, it is the -0.0565 computed in part (a) divided by the area 0.0113 of the road; i.e., -5. The divergence here is $\dfrac{\partial v_2}{\partial y}$, which is the acceleration along the road. Since this is negative, it confirms that traffic is slowing down.

49. 12π

51. $a + b + c > 0$

53. Let $\mathcal{S}$ be the unit sphere and let $\mathbf{F}(x, y, z) = \langle -2xz, 2yz, xy \rangle$. Then $\mathbf{F}$ is not conservative, since $\dfrac{\partial f_1}{\partial z} = -2x \neq y = \dfrac{\partial f_3}{\partial x}$. However, div $\mathbf{F} = 0$, so by the Divergence Theorem, $\iint_{\mathcal{S}} \mathbf{F}(x, y, z) \cdot \mathbf{n} \, dS = 0$.

For answers to odd-numbered Skill Certification exercises in the Chapter Review, please visit the Book Companion Web Site at www.whfreeman.com/tkcalculus.

DIFFERENTIAL CALCULUS

Definition of the Limit

$\lim_{x \to c} f(x) = L$ means that:

For all $\epsilon > 0$, there exists a $\delta > 0$ such that if $x \in (c - \delta, c) \cup (c, c + \delta)$, then $f(x) \in (L - \epsilon, L + \epsilon)$.

Equivalently, in terms of absolute value inequalities, $\lim_{x \to c} f(x) = L$ means that:

For all $\epsilon > 0$, there exists a $\delta > 0$ such that if $0 < |x - c| < \delta$, then $|f(x) - L| < \epsilon$.

The Extreme Value Theorem

If f is continuous on a closed interval $[a, b]$, then there exist values M and m in the interval $[a, b]$ such that $f(M)$ is the maximum value of $f(x)$ on $[a, b]$ and $f(m)$ is the minimum value of $f(x)$ on $[a, b]$.

The Intermediate Value Theorem

If f is continuous on a closed interval $[a, b]$, then for any K strictly between $f(a)$ and $f(b)$, there exists at least one $c \in (a, b)$ such that $f(c) = K$.

Definition of the Derivative

The derivative of a function $f(x)$ is defined to be the function:

$$f'(x) = \lim_{h \to 0} \frac{f(x + h) - f(x)}{h}$$

or, equivalently:

$$f'(x) = \lim_{z \to x} \frac{f(z) - f(x)}{z - x}.$$

Derivative Rules

$$\frac{d}{dx}(kf(x)) = kf'(x)$$

$$\frac{d}{dx}(f(x) + g(x)) = f'(x) + g'(x)$$

$$\frac{d}{dx}(f(x)g(x)) = f'(x)g(x) + f(x)g'(x)$$

$$\frac{d}{dx}\left(\frac{f(x)}{g(x)}\right) = \frac{f'(x)g(x) - f(x)g'(x)}{(g(x))^2}$$

$$\frac{d}{dx}(f(g(x))) = f'(g(x))g'(x)$$

Rolle's Theorem

If f is continuous on $[a, b]$ and differentiable on (a, b), and if $f(a) = f(b) = 0$, then there exists at least one value $c \in (a, b)$ for which $f'(c) = 0$.

The Mean Value Theorem

If f is continuous on $[a, b]$ and differentiable on (a, b), then there exists at least one value $c \in (a, b)$ such that

$$f'(c) = \frac{f(b) - f(a)}{b - a}.$$

Derivatives of Basic Functions

$$\frac{d}{dx}\left(x^k\right) = kx^{k-1}$$

$$\frac{d}{dx}\left(e^{kx}\right) = ke^{kx}$$

If $b > 0$ and $b \neq 1$, then $\frac{d}{dx}(b^x) = (\ln b)b^x$

$$\frac{d}{dx}(\ln |x|) = \frac{1}{x}$$

If $b > 0$ and $b \neq 1$, then $\frac{d}{dx}(\log_b |x|) = \frac{1}{(\ln b)x}$

$$\frac{d}{dx}(\sin x) = \cos x$$

$$\frac{d}{dx}(\cos x) = -\sin x$$

$$\frac{d}{dx}(\tan x) = \sec^2 x$$

$$\frac{d}{dx}(\cot x) = -\csc^2 x$$

$$\frac{d}{dx}(\sec x) = \sec x \tan x$$

$$\frac{d}{dx}(\csc x) = -\csc x \cot x$$

$$\frac{d}{dx}(\sin^{-1} x) = \frac{1}{\sqrt{1 - x^2}}$$

$$\frac{d}{dx}(\tan^{-1} x) = \frac{1}{1 + x^2}$$

$$\frac{d}{dx}(\sec^{-1} x) = \frac{1}{|x|\sqrt{x^2 - 1}}$$

$$\frac{d}{dx}(\sinh x) = \cosh x$$

$$\frac{d}{dx}\cosh x = \sinh x$$

$$\frac{d}{dx}\tanh x = \operatorname{sech}^2 x$$